On the Stabilization of the Trace Formula

Stabilization of the trace formula,
Shimura varieties, and arithmetic applications

Volume 1

On the Stabilization of the Trace Formula

edited by

Laurent Clozel
Université Paris-Sud, Orsay

Michael Harris
Université Paris-Diderot Paris 7

Jean-Pierre Labesse
Université Aix-Marseille II

Bao-Châu Ngô
University of Chicago

International Press
www.intlpress.com

Stabilization of the trace formula, Shimura varieties, and arithmetic applications
Volume 1: On the Stabilization of the Trace Formula

Editors:
Laurent Clozel (Université Paris-Sud, Orsay)
Michael Harris (Université Paris-Diderot Paris 7)
Jean-Pierre Labesse (Université Aix-Marseille II)
Bao-Châu Ngô (University of Chicago)

2010 Mathematics Subject Classification. 11F72, 11F70, 22E50, 22E35, 11S37, 14D23.

ISBN: 978-1-57146-355-5

Paperback re-issue, 2017. Original hardcover published in 2011 under ISBN 978-1-57146-227-5.

Printed in the United States of America.

Contents

General Introduction: The stable trace formula, Shimura varieties, and arithmetic applications

The present volume is the first in a projected series of two or three collections of mainly expository articles on the arithmetic theory of automorphic forms. The books are primarily intended for two groups of readers. The first group is interested in the structure of automorphic forms on reductive groups over number fields, and specifically in qualitative information about the multiplicities of automorphic representations. The second group is interested in the problem of classifying ℓ-adic representations of Galois groups of number fields. Langlands' conjectures elaborate on the notion that these two problems overlap to a considerable degree. The goal of this series of books is to gather into one place much of the evidence that this is the case, and to present it clearly and succinctly enough so that both groups of readers are not only convinced by the evidence but can pass with minimal effort between the two points of view.

More than a decade's worth of progress toward the stabilization of the Arthur-Selberg trace formula, culminating in the recent proof by Laumon and Ngô[1] of the Langlands-Shelstad conjecture for unitary groups over function fields, better known as the *fundamental lemma*, has made this series timely. The 1980s saw the formulation of increasingly explicit conjectures, primarily by Langlands, Shelstad, Arthur, and Kottwitz, concerning both the ultimate form of the stable trace formula and its consequences for multiplicities of automorphic representations and the structure of Galois representations occurring in the ℓ-adic cohomology of Shimura varieties. The test case of the group $U(3)$ was treated comprehensively in Rogawski's book on the subject, and the applications to Galois representations were derived in the volume edited by Langlands and Ramakrishnan. Although progress on these conjectures subsequently stalled, their statements became widely known. New applications to the arithmetic of Galois representations were derived, assuming the truth of these conjectures. Simultaneously, work of Waldspurger and Arthur succeeded in reducing these conjectures to variants of the fundamental lemma. The latter thus became the bottleneck limiting progress on a host of arithmetic questions.

[1] References are to be found in the bibliographies of individual chapters.

The breakthrough article of Laumon and Ngô, building on earlier work of Goresky, Kottwitz, and MacPherson, and of Laumon, removed that obstacle in the crucial case of unitary groups over local fields of positive characteristic. Subsequent work of Ngô extended the geometric techniques of his work with Laumon and succeeded in proving the fundamental lemma for all groups, again over local fields. At roughly the same time, Waldspurger proved that the fundamental lemma depends only on the residue field, thus deriving the fundamental lemma for p-adic fields from the results of Laumon and Ngô. More recently, Cluckers, Hales, and Loeser obtained similar results on independence of characteristic using methods of motivic integration and model theory. All these results are only valid when the residue characteristic is sufficiently large, but earlier work of Hales showed that this implies the general case.

In the broadest possible terms, the Arthur-Selberg trace formula, and its stable variant, is an identity between expressions of two sorts. We fix a reductive group G over a number field F. The *spectral side* is an expression for the traces of a family of operators that determine the representation of $G(\mathbf{A}_F)$ on the discrete spectrum of $L_2(G(F)\backslash G(\mathbf{A}_F))$.[2] The spectral side, in other words, contains the information of ultimate interest about spaces of automorphic forms. It can be viewed as a sum (or more generally an integral) of characters (more generally weighted characters) of irreducible representations of $G(\mathbf{A}_F)$. The *geometric side* is a sum of orbital integrals and their generalizations that can be understood in terms of harmonic analysis on $G(F_v)$, as v runs over the completions of F. It can be viewed as a sum indexed by conjugacy classes in $G(F)$. The principle is that the geometric side lends itself to comparisons between different groups, but only after the sums over conjugacy classes are replaced by sums over *stable* conjugacy classes, a notion discussed in detail in the first section of this book. Such comparisons are then used to compare the spectral sides for different groups, which could not otherwise be compared directly.

It may be helpful to think of the stable trace formula not as a single equality but rather as a collection of techniques for generating formulas that can be used for a variety of purposes: to compare the automorphic representations of different groups, in the spirit of Langlands' functoriality conjectures; to obtain qualitative information about representations of local (generally p-adic) groups from global constructions; and in certain cases to relate automorphic representations to Galois representations. The purpose of these books, then, is to present the stable trace formula in a manner that should be accessible not only to specialists but to anyone who may benefit from application of these techniques.

Before going on, we should list some of the things these books are not. They are certainly *not* a detailed introduction to the general trace formula.

[2]This is not strictly accurate if the center of G contains a torus split over F, but there are several standard substitutes for $L_2(G(F)\backslash G(\mathbf{A}_F))$ that will be considered in the text and we ignore this issue here.

The stable trace formula presents two quite distinct sorts of difficulties. Analytic problems, some of them very deep, arise as soon as one attempts to understand the discrete automorphic spectrum of a non-compact (isotropic) reductive group. The resolution of these problems was carried out in a long series of articles, mainly by Jim Arthur. We cannot hope to improve on Arthur's recent expository presentation of this material, to which we refer the reader.

On the other hand, the difficulties connected with stabilization of the trace formula already appear in their full complexity for anistropic groups. The simple version of the trace formula, derived by Arthur from his general trace formula for appropriate choices of data, can also be stabilized, and the result is practically identical to that obtained for anisotropic groups. For the applications we have in mind, these cases will suffice. In particular, we will only be considering orbital integrals (not weighted orbital integrals) and characters (not weighted characters). Thus this book also does *not* contain a complete treatment of the stable trace formula. The stabilization of the trace formula was treated by Arthur in a recent series of articles. For the most important applications (including in this book, see in particular Labesse's chapter) one needs to stabilize a twisted trace formula. We refer to Arthur's forthcoming book on functorial transfer between classical groups and $GL(n)$ for the comparison in this case; however, Arthur assumes, in this work, the stabilization of the twisted trace formula, which remains to be achieved in general.

To summarize the two preceding paragraphs, it will be enough for us to present the stabilization of the *elliptic terms* of the geometric side of the trace formula. After the pioneering article of Labesse-Langlands on the stable trace formula for $SL(2)$, the stabilization of the elliptic terms was initiated by Langlands and pursued by Kottwitz in a series of articles written during the 1980s. A different point of view, making systematic use of non-abelian cohomology, was developed by Labesse and is used either implicitly or explicitly in a number of chapters. The two introductory chapters by Harris and Labesse present this material from somewhat different perspectives, described below. Here it should only be mentioned that these chapters do *not* contain complete proofs of the intermediate steps in stabilization of even the elliptic terms of the trace formula; but it does provide detailed references for those who want to read the proofs.

Most importantly, it should be stressed this book does *not* pretend to provide a general treatment of Langlands' functoriality conjectures, even for the special groups considered, even in the special situation of endoscopy that has motivated much of the work on the trace formula over the past three decades. For the most part we restrict our attention to the stabilization of the trace formula for unitary groups of vector spaces over CM fields. In the first place, the initial breakthrough of Laumon-Ngô was the proof of the fundamental lemma for unitary groups, and even now that the fundamental lemma is known in general, unitary groups are somewhat easier to understand because of their intimate relation to general linear groups. In particular, the geometric constructions of Laumon and Ngô, involving the Hitchin fibration, can be expressed in the intuitively satisfying terms of matrices and their characteristic polynomials. On the other hand, the Shimura varieties attached to unitary groups are the most fruitful source for construction of compatible systems of ℓ-adic Galois representations, again because of the close connection to general linear groups via stable base change. If one is merely interested in the construction of Galois representations attached to certain classes of automorphic representations of $GL(n)$, it is unnecessary to concern oneself with the fine points of representation theory of general reductive groups over p-adic fields; in particular, we can and do restrict our attention to automorphic representations of unitary groups for which the problem of local L-packets does not arise. Moreover, we can arrange matters so that only *tempered* automorphic representations appear, thus avoiding the subtle analytic problems connected with the general case. This suffices in the first instance for the construction of Galois representations, but non-tempered automorphic representations are needed for certain applications, and we expect to return to this topic in later books. On the other hand, as Moeglin realized, the simple stable trace formula considered in the first book is sufficiently flexible to allow her to construct and classify local L-packets for classical groups; in particular her construction for local unitary groups requires nothing more than the theory developed here. Thus, while this book does *not* work out the general theory of endoscopic transfer even for unitary groups, it does establish the theory in sufficient generality for several applications of importance in arithmetic as well as representation theory.

It deserves to be mentioned separately that the only automorphic representations considered in this book are those of cohomological type at archimedean places. This is in part because the test functions used to identify cohomological representations can be used to simplify the trace formula, and even to simplify the stabilization of the twisted trace formula for $GL(n)$ for the base change from unitary groups. This theory is worked out in Labesse's chapter [L.IV.A]. The techniques now exist to extend base change relating general automorphic representations of $U(n)$ to automorphic representations of $GL(n)$, as in Rogawski's book, which develops the case $n = 3$. The present series of books, intended as an introduction to the stable trace formula, are not the appropriate place for such an ambitious undertaking. Again, only cohomological representations can be associated to cohomology of the Shimura varieties attached to $U(n)$, the only systematic source of the Galois representations attached to automorphic representations. Thus this book does *not* shed any light whatsoever on questions of functoriality connected with Maass forms and their generalizations in higher dimension.

Finally, as Langlands has stressed increasingly in recent years, endoscopy, including the twisted endoscopy of base change and transfer between classical groups and $GL(n)$, only accounts for very special cases of functoriality. This book does *not* look beyond endoscopy; for that, readers are strongly encouraged to read Langlands' recent articles on the subject.

The remainder of this introduction will use the tables of contents of the first book in the series as a pretext for reviewing in detail the developments described in the preceding paragraphs. Most of the material in the first book should be familiar to the first group of intended readers, less so to the second group. For the material in the second book, whose table of contents should be available in the near future, the proportions should be reversed.

Book 1. The stable trace formula

Section I. Introduction to the stable trace formula. As mentioned above, this portion of the book consists of two chapters by Harris and Labesse. The bulk of Harris's chapter [H.I.A] is a review of Kottwitz' theory. The results are stated for the most part without proof, since Kottwitz' proofs are extremely clear. The main purpose of this chapter is to assemble all this material in a single place. In this Harris has followed the unpublished notes of the IHES seminar of 2003-2004 on the stable trace formula, especially the lectures of Ngô. Labesse's chapter [L.I.B] contains a more sustained treatment of the first non-trivial case, that of $SL(2)$.

Section II.A, B. Endoscopy: the local theory. In Kottwitz' articles of the 1980s, and in Arthur's later work on the full stabilization of the trace formula, the fundamental lemma is assumed as a working hypothesis. Thanks especially to the work of Laumon and Ngô, the fundamental lemma is now a theorem. The second portion of the first book provides a detailed introduction to the fundamental lemma and to the problem of endoscopic transfer in general. This breaks up logically into three chapters, treating separate aspects of the theory of endoscopic transfer. Sections II.A and B. present the local theory of endoscopy. The theory for real groups was developed in a series of fundamental papers by Shelstad in the 1970s. Renard's chapter [R.II.A] reviews this material in the light of subsequent work on representation theory, and on the systematic definitions of transfer factors due to Langlands, Shelstad, and Kottwitz.

The basic technical problem of endoscopy is to show how to relate the test functions that enter into the trace formula for different groups (a group and its endoscopic groups). We have already seen that this *transfer problem* was solved for real groups by Shelstad. For p-adic groups the fundamental lemma, which gives an explicit formula for transfer in the simplest case, plays a special role. In the 1980s, Clozel and Labesse independently showed how to use the global trace formula to reduce the general transfer problem for stable base change of Hecke functions to the special case of the fundamental lemma, proved in this case by Kottwitz. This method was adapted to endoscopy by Hales and especially by Waldspurger, who reduced the general endoscopic transfer problem to the fundamental lemma for endoscopy. These results are reviewed in Chaudouard's chapter [C.II.B.1]. More recently,

Waldspurger has shown how to reduce transfer for twisted endoscopy, as required in Arthur's work on classical groups, to what he calls the "nonstandard fundamental lemma." Waldspurger's chapter [W.II.B.2] provides an introduction to this work.

II.C. The fundamental lemma, geometric techniques. The work of Laumon and Ngô has introduced an unfamiliar array of concepts into the heart of the theory of automorphic forms. As Ngô's most recent article makes clear, the central concept is that of the Hitchin fibration on the moduli stack of vector bundles with additional structure on a fixed algebraic curve. As Langlands wrote recently, a key insight is that the formalism of stacks allows provides a geometric meaning to global orbital integrals. These orbital integrals can be interpreted in particular in terms of certain perverse sheaves on the Hitchin fibration. For local fields of sufficiently large positive characteristic, endoscopy is obtained in the papers of Laumon and Ngô by analyzing the functorial behavior of these perverse sheaves, using the basic properties of perversity to reduce to the situation where explicit calculation is possible.

Section II.C. presents this material in two chapters. Chapter [DT.II.C.1], due to Dat and Tuan, develops the general framework; Ngô's chapter [N.II.C.2] explains in more detail one of the crucial steps in Ngô's argument.

II.D, E. The fundamental lemma, independence of characteristic. The geometric techniques of Laumon and Ngô only apply to local fields of positive characteristic, whereas the global applications of this series of books are primarily to number fields. The bridge is provided by a theorem of Waldspurger that shows that the fundamental lemma in (sufficiently large) mixed characteristic can be derived from the corresponding statement for local fields of positive characteristic. Earlier work of Hales shows that this suffices for the general case.

Waldspurger's theorem, presented in Lemaire's chapter [L.II.D], is proved using a detailed analysis of the fine structure of the lattice of open compact subgroups of a reductive group over a local field, based in turn on earlier results of DeBacker and Kim-Murnaghan. A very different proof of independence of characteristic has been subsequently obtained by Cluckers, Hales, and Loeser. This proof is based on a *transfer principle* in model theory that uses formal properties of integrals on groups over local fields to transfer statements from one family of local fields to another. The basic framework is the theory of motivic integration, as developed in the work of Denef-Loeser and extended by Cluckers-Loeser. This perspective makes use of notions of mathematical logic that are undoubtedly unfamiliar to most specialists in arithmetic and automorphic forms, including the editors of this volume. Once these notions have been introduced, however, the technique proves to be highly adaptable. The chapter [CHL.II.E] of Cluckers, Hales, and Loeser reviews the basic constructions and then explains how they can be applied to three different sorts of fundamental lemmas.

III. Local representation theory of unitary groups. In the earlier chapters, special attention has been given to unitary groups, and their relations to $GL(n)$, as an example to illustrate the general principles of endoscopy. In the remainder of the book we restrict our attention to unitary groups. Section III is a survey of basic results in the theory of representations of unitary groups over local fields, as it intervenes in the spectral and geometric sides of the simple trace formula. Most of this material is standard but some is new, and in any case it has not previously been assembled in one place.

Clozel's chapter [C.III.A] is a review of the local theory of representations of real unitary groups and base change to $GL(n, \mathbb{C})$, and works out Shelstad's theory of endoscopy explicitly in this setting. It can also serve as an introduction to the local Langlands parametrization in a concrete situation where much of the subtlety of the general case is already apparent. Adams' chapter [A.III.B] treats the same material from a different perspective, namely that of the author's book with Barbasch and Vogan on stable distributions on real reductive groups. While Adams only considers discrete series representations of unitary groups, whose theory had already been developed by Shelstad in the 1970s, the formalism he presents is in some ways more intuitive. In particular, the signs in endoscopic transfer of discrete series representations can be given a uniform treatment when all (strong) inner forms of unitary groups are considered simultaneously.

Chapter [M.III.C], written by Minguez, develops the analogous theory for unitary groups over p-adic fields. For the purposes of the applications to construction of Galois representations, only the most elementary aspects of the theory, such as the explicit base change of spherical representations of quasi-split unitary groups, need to be treated. However, the literature on representations of p-adic unitary groups is scattered, and it seemed useful to present more of the theory than is strictly necessary for immediate applications. Finally, Labesse's chapter [L.III.D] works out the local transfer factors explicitly, up to a sign that proves to be irrelevant for the global theory, in the case studied in detail in the final section of the book.

IV. Base change and endoscopic transfer for unitary groups. The three chapters of this section contain the only genuinely new results of this book. Labesse's chapter [L.IV.A] develops the theory of base change for cohomological automorphic representations of inner forms of the quasi-split group $U(n)$, attached to a quadratic extension E/F, to the corresponding $GL(n)$, extending his earlier work with Clozel and Harris. In Labesse's article, it is assumed that F is totally real and of degree at least 2, and that E is totally imaginary and unramified over F at all finite primes. It is moreover assumed that the automorphic representations Π to which one applies base change are spherical at all finite places that remain inert in E/F. For some applications, it is also assumed that the infinitesimal parameter of Π at any archimedean place is sufficiently regular. These hypotheses are restrictive and certainly unnecessary for stable base change, and indeed no such hypotheses are made in Rogawski's book on $U(3)$. However, the comparison of stable trace formulas simplifies considerably under these hypotheses, which suffice for applications to construction of Galois representations as well as for all purely local applications.

Chapter [CHL.IV.B], by Clozel, Harris, and Labesse, analyzes the endoscopic transfer in the simplest possible case, from the endoscopic group $U(n) \times U(1)$ to certain inner forms of $U(n+1)$. Again, simplifying assumptions are made on the local behavior at finite primes (to avoid complications from L-packets) and on infinitesimal characters at archimedean primes. The results in this section, inspired by an article of Blasius and Rogawski, are used in the final chapter [CHL.IV.C], by the same three authors, to construct n-dimensional Galois representations in the cohomology of Shimura varieties obtained as quotients of the unit ball in $\mathbb{C}^n$, following the strategy outlined by Kottwitz at the 1988 Ann Arbor conference on automorphic representations. S. W. Shin has obtained a stronger result by a somewhat different method, to be described in Book 2; in particular, Shin obtains almost complete results at primes of bad reduction.

The editors—Laurent Clozel, Michael Harris, Jean-Pierre Labesse, and Ngô Bau-Châu—are pleased to express our gratitude to the many colleagues whose contributions helped bring this volume into being. We thank the speakers at the seminar "Groupes réductifs et formes automorphes" at the Institut Mathématique de Jussieu, on whose talks many of the chapters of this book and the next are based. We thank the authors of all the chapters—in many cases the same as the speakers—for participating actively in the project and for making an effort to make their texts accessible to the broadest possible audience. We thank the referees—every chapter was sent to a referee—for taking the project seriously and for holding the authors to a high standard. We thank the Banff International Research Station for hosting a two-week program in 2008, devoted to the subject of this book project as well as Jim Arthur's forthcoming book on functoriality for classical groups, thus providing us with an opportunity to test the material before a demanding audience. This project was conceived in response to an invitation by the editors at International Press to submit a book on the Langlands program; we thank them for their willingness to publish a series of books aimed primarily at non-specialists, and for their consistent support.

Finally, we thank all the participants in the seminar and in the Banff meeting, and especially Jan Nekovář. He has brought to this project the perspective of a number theorist who insists on the relevance, for the field at large, of the results issuing from the theory of automorphic forms; and, as much as possible, on their intelligibility for a (relative) outsider. We hope we have have produced a book that, in part at least, meets his high standards.

Section I

Introduction to the stabilization of the trace formula

An introduction to the stable trace formula
[H.I.A]

Michael Harris

The goal of this introduction is to motivate the problem of stabilization of the trace formula and to present the approach to stabilization of the elliptic terms carried out by Langlands and Kottwitz in the 1980s. The first two sections are devoted to describing the trace formula and presenting two typical applications. In this series of books we studiously avoid all reference to the analytic problems resolved by Arthur in incorporating parabolic terms into the invariant trace formula. The reader who wishes to learn more about this material should consult Arthur's article [A], which we have also used as a reference for introductory material. In this introduction, we restrict our attention to automorphic representations of groups that are anisotropic (or at least anisotropic modulo center); then all terms on the geometric side of the trace formula are necessarily elliptic. Other chapters in this volume will make use of Arthur's simple trace formula, in which only elliptic terms occur even when the group is not anisotropic; the description of stabilization works identically in this situation.

The third section explains how the problem of stabilization arises naturally when one wants to compare trace formulas for different groups. The remaining sections present the various steps involved in stabilization, mainly following Kottwitz' articles on the subject, as reinterpreted by Ngô in his seminar talks in 2003. At various points in his articles, Kottwitz' invokes conjectures in order to proceed with stabilization of the two sides of the trace formula. The stabilization of the elliptic part of the geometric side, involving orbital integrals, depends crucially on the Fundamental Lemma, as we explain in §6. Twenty years later, this is now a theorem, so Kottwitz' stabilization of the geometric side of the trace formula for anisotropic groups is now complete. The stabilization of the spectral side depends on the partition of the irreducible admissible representations of a p-adic group into L-packets (more generally, Arthur packets) of a specific form. There

---

Institut des Mathématiques de Jussieu, U.M.R. 7586 du CNRS. Membre, Institut Universitaire de France.

has been considerable recent progress on this question but much remains to be done. In this introduction we concentrate on the geometric side. Some of the subsequent articles will consider the spectral side of the trace formula for unitary groups, with applications in Book 2 to the cohomology of related Shimura varieties.

Conventions for harmonic analysis on reductive groups over local fields and adèle groups, and for automorphic representations, are as in Arthur's article [A].

1. Automorphic representations and the trace formula

Here and in what follows, E denotes a number field and G denotes a connected reductive group over E, with center Z_G. We choose a maximal compact subgroup $K_\infty \subset G_\infty := G(E \otimes_{\mathbb{Q}} \mathbb{R})$. This choice is not always innocent, but little or nothing in the present volume will depend on that choice. We also let $\mathfrak{g}$ denote the complexified Lie algebra of $G(\mathbb{R})$, $\mathbf{A}_E$ the adèle ring of E, and $\mathbf{A}_E^f$ the ring of finite adèles of E. Thus $G(\mathbf{A}_E) = G_\infty \times G(\mathbf{A}_E^f)$, and any Haar measure dg on $G(\mathbf{A}_E)$ decomposes correspondingly as a product $dg_\infty dg^f$. Let $C_c^\infty(G(\mathbf{A}_E))$ (resp. $C_c^\infty(G_\infty)$, resp. $C_c^\infty(G(\mathbf{A}_E^f)))$ denote the space of compactly supported functions on $G(\mathbf{A}_E)$ that are C^∞ in the archimedean variables and locally constant in the non-archimedean variables. Likewise, for any place v of E, we define $C_c^\infty(G(E_v))$. The map

$$(\phi_\infty, \phi_f) \mapsto [(g_\infty, g_f) \mapsto \phi_\infty(g_\infty) \cdot \phi_f(g_f)]$$

defines an isomorphism

$$C_c^\infty(G_\infty) \otimes C_c^\infty(G(\mathbf{A}_E^f)) \xrightarrow{\sim} C_c^\infty(G(\mathbf{A}_E)).$$

These spaces of *test functions* are algebras under convolution (cf. [K7, 1.2]).

Our main class of examples will be unitary groups. Let $\mathcal{K}/E$ be a quadratic extension, and let V be an n-dimensional $\mathcal{K}$-vector space with a non-degenerate hermitian form $<,>_V$ relative to the extension $\mathcal{K}/E$; the group $G = U(V)$ of symmetries of this hermitian form is naturally a group scheme over E, whose A-valued points for any E-algebra A is given by

$$G(A) = \{g \in GL(V \otimes_E A) \mid < gv, gv' >_V = < v, v' >_V, \forall v, v' \in V \otimes_E A\}$$

If v is a place of E that splits as a product $w \cdot w'$ in $\mathcal{K}$, then

$$V_v = V \otimes_E E_v = V_w \oplus V_{w'}$$

where $V_w = V \otimes_{\mathcal{K}} \mathcal{K}_w$. Then V_w and $V_{w'}$ are maximal isotropic subspaces of V_v, and the hermitian form defines an isomorphism $V_{w'} \xrightarrow{\sim} V_w^\vee$. Then $G(E_v) \subset GL(V_w \oplus V_{w'})$ fixes both V_w and $V_{w'}$ and its action on the former defines a canonical isomorphism

$$G(E_v) \xrightarrow{\sim} GL(V_w).$$

The center Z_G is the diagonal subgroup $U(1)_{\mathcal{K}/E}$, whose group of A-valued points is the subgroup of $(\mathcal{K} \otimes_E A)^\times$ of elements of norm 1 down to $A^\times$;

in particular, Z_G is anisotropic. We can also consider $G = GL(n)_E$, with $Z_G = GL(1)_E$, embedded diagonally in G.

Suppose G is anisotropic; for example $G = U(V)$, where E has a real place v that becomes complex in $\mathcal{K}$, such that $G(E_v)$ is isomorphic to the compact unitary group $U(n)$. Then the adèlic quotient $G(E)\backslash G(\mathbf{A}_E)$ is *compact.* Right-translation defines a unitary action of $G(\mathbf{A}_E)$ on the Hilbert space $L_2(G(E)\backslash G(\mathbf{A}_E))$. This action gives rise to an action of $C_c^\infty(G(\mathbf{A}_E))$ by integration: for $f \in L_2(G(E)\backslash G(\mathbf{A}_E))$, $\phi \in C_c^\infty(G(\mathbf{A}_E))$, define

$$R(\phi)f(g) = \int_{G(\mathbf{A}_E)} \phi(gh)f(h)dh.$$

The map $\phi \mapsto R(\phi)$ defines an algebra homomorphism from the algebra $C_c^\infty(G(\mathbf{A}_E))$ of *test functions* to the C^* algebra of bounded operators on $L_2(G(E)\backslash G(\mathbf{A}_E))$. In fact, $R(\phi)$ is a *compact* operator for all ϕ (cf. [A], p.8). It follows formally [A, p. 9] that

PROPOSITION 1.1. When G is anisotropic, $L_2(G(E)\backslash G(\mathbf{A}_E))$ decomposes as a countable Hilbert space direct sum

$$L^2(G(E)\backslash G(\mathbf{A}_E)) = \bigoplus_\pi m_\pi \, \pi,$$

where π runs over the set of equivalence classes of irreducible unitary representations of $G(\mathbf{A}_E)$—or equivalently, of $C_c^\infty(G(\mathbf{A}_E))$—and the multiplicities m_π are non-negative integers that take positive values on a countable set of π.

The operators $R(\phi)$ are of trace class, [A, pp. 14–15]. The multiplicities m_π can be determined, in principle, by calculating the distribution $\phi \mapsto tr R(\phi)$, as follows. The irreducible representation π of $G(\mathbf{A}_E)$ can be written, non-canonically, as a restricted tensor product [cf. A, (2.2)]

$$\pi \xrightarrow{\sim} \bar{\pi}_\infty \otimes \bar{\pi}_f = \bigotimes_v{}' \bar{\pi}_v$$

where $\bar{\pi}_\infty$, resp. $\bar{\pi}_f$, resp. $\bar{\pi}_v$, is an irreducible unitary representation of G_∞, resp. $G(\mathbf{A}_E^f)$, resp. $G(E_v)$. Let $K^f \subset G(\mathbf{A}_E^f)$ be a compact open subgroup, which we assume to be of the form $K^f = \prod_v K_v$, where v runs over non-archimedean places of E and each K_v is a compact open subgroup of $G(E_v)$, maximal compact for almost all v.

FACT 1.2. For almost all v, π_v is unramified: the space $\bar{\pi}_v^{K_v}$ of K_v-invariant vectors in $\bar{\pi}_v$ is one-dimensional.

We let $\pi_v \subset \bar{\pi}_v$ be the subspace of K_v-finite vectors, and we also write π for the restricted tensor product of the π_v. For v-non-archimedean, π_v is a *smooth* representation—every vector is fixed by an open compact subgroup of K_v—and is moreover always an *admissible* representation—for every open subgroup $U \subset K_v$, the subspace π_v^U is finite-dimensional.

The algebra $C_c^\infty(G(\mathbf{A}_E))$ can be written as a restricted tensor product of local algebras $C_c^\infty(G(E_v))$; every ϕ can be written as a finite sum of *factorizable* test functions: infinite tensor products $\otimes\phi_v$ where, for almost all non-archimedean v, ϕ_v is the characteristic function 1_{K_v} of the chosen K_v. We decompose $dg^f = \prod dg_v$, where dg_v is a Haar measure on $G(E_v)$. We can arrange that, for almost all non-archimedean v,

$$\int_{G(E_v)} 1_{K_v} dg_v = 1.$$

For non-archimedean v, the action of $C_c^\infty(G(E_v))$ on π_v, defined by

$$\pi_v(\phi_v)(w) = \int_{G(E_v)} \pi_v(g)(w)\phi_v(g)dg_v, w \in \pi_v, \phi_v \in C_c^\infty(G(E_v))$$

is a finite sum with finite-dimensional image. In particular, $\pi_v(\phi_v)$ is a trace class operator. The same is true for archimedean v, provided on takes ϕ_v to be left and right K_v-finite. We assume this to be the case henceforward. The *character* of π_v is the distribution $\phi_v \mapsto tr\ \pi_v(\phi_v)$. It is invariant under conjugation by $G(E_v)$, or simply an *invariant distribution.*

When π_v is unramified $tr\ \pi_v(1_{K_v}) = \int_{G(E_v)} 1_{K_v} dg_v = 1$ for almost all v. Thus when $\phi = \otimes\phi_v$, the infinite product

$$tr\ \pi(\phi) = \prod_v tr\ \pi_v(\phi_v)$$

is actually finite and well-defined, and extends linearly to all K_∞-finite elements in $C_c^\infty(G(\mathbf{A}_E))$.

FACT 1.3. (a) The distribution $\phi \mapsto tr\ \pi$ determines π up to isomorphism.

(b) The sum $\sum m_\pi\ tr\ \pi$ of distributions converges to the distribution $tr\ R$.

(c) (linear independence of characters) The distribution $tr\ R$ determines the m_π uniquely.

We emphasize that the distributions $tr\ \pi$ and $tr\ R$ depend on the choice of Haar measure dg.

What we have described so far is a straightforward generalization of the theory of characters of finite-dimensional representations of finite groups, though the proofs are in general not so straightforward. In the case of finite groups, the class function analogous to $tr\ R$ is the character of the regular representation on a coset space, and thus can be written as the sum of characteristic functions of conjugacy classes with explicit coefficients. The corresponding assertion for the automorphic trace trR is the *Selberg trace formula.* The characteristic functions of conjugacy classes are replaced by *orbital integrals.* Suppose $\gamma \in G(\mathbf{A})$ and $\phi \in C_c^\infty(G(E_v))$ is a K_∞-finite test function. Let $H_\gamma \subset G(\mathbf{A})$ denote the centralizer of γ. Suppose for the moment that H_γ has a Haar measure dg_γ, and let $d\dot{g} = dg/dg_\gamma$ denote the quotient measure on $G_\gamma\backslash G(\mathbf{A})$. We define the orbital integral

$$O_\gamma(\phi) = \Phi_{G(\mathbf{A})}(\gamma,\phi) = \int_{H_\gamma\backslash G(\mathbf{A})} \phi(x^{-1}\gamma x)d\dot{x}. \tag{1.4}$$

The two expressions on the left correspond to two systems of notation used in the literature.

Convergence of the integral (1.4) is a hypothesis that is automatically satisfied in favorable circumstances, for example when $\gamma \in G(E)$ for anisotropic G. In that case H_γ is the group of $\mathbf{A}_E$-points of the centralizer G_γ of γ, an E-rational algebraic subgroup of G. Since G is anisotropic over E, $\gamma \in G(E)$ is necessarily semi-simple, and G_γ is thus reductive.[1] It is often convenient to assume the derived subgroup $G^{der} \subset G$ is *simply connected*; then G_γ is even a connected reductive group. With or without this hypothesis, the map $\phi \mapsto O_\gamma(\phi)$ defines a $G(\mathbf{A}_E)$-invariant distribution on $C_c^\infty(G(\mathbf{A}_E))$. It depends both on the choice of measure dg_γ on $G_\gamma(\mathbf{A}_E)$ as well as on dg. On the other hand, dg_γ defines a volume

$$v_\gamma = \int_{G_\gamma(E)\backslash G_\gamma(\mathbf{A}_E)} dg_\gamma.$$

THEOREM 1.5 (SELBERG TRACE FORMULA, ANISOTROPIC CASE). *There is an identity of distributions:*

$$tr R(\phi) = \sum_\gamma v_\gamma O_\gamma(\phi)$$

where γ runs over $G(E)$-conjugacy classes in $G(E)$ and, for any fixed choice of ϕ, the sum of orbital integrals is finite. In other words

$$\sum_\gamma v_\gamma O_\gamma(\phi) = \sum_\pi m_\pi \; tr\; \pi(\phi).$$

The left-hand side is called the *geometric side* of the trace formula, the right-hand side is the *spectral side.*

Remark. We write $\phi = \phi_\infty \otimes \phi_f$. The function ϕ_f is biinvariant under some open compact subgroup $K_f \subset G(\mathbf{A}^f)$, and thus the sum on the right hand side only involves representations with K_f-fixed vectors. In practice, we will choose ϕ_∞ such that $Tr(\pi(\phi_\infty))$ is non-zero only when π is a cohomological representation with fixed infinitesimal character. It then (almost) follows that the spectral side is finite for the given ϕ. This is not quite true if G is anisotropic modulo its center but if Z_G has a non-trivial split component. There are various ways to modify the trace formula to avoid this problem, one of which are illustrated in Labesse's article [L.IV.A.]. We will ignore the question in this introduction, which is mainly concerned with local properties of the geometric side.

[1] More generally, the integral (1.4) converges when γ is semi-simple.

The finiteness of the geometric side in the anisotropic case is a consequence of reduction theory.

The coefficients v_γ and m_π are global and cannot be simplified further, although when dg_γ is chosen to be Tamagawa measure v_γ can be expressed in terms of cohomological data. When dg_γ and dg are Tamagawa measures, we write $\tau(G_\gamma)$ in place of v_γ; thus

$$\sum_\gamma \tau(G_\gamma) O_\gamma(\phi) = \sum_\pi m_\pi \; tr \; \pi(\phi). \tag{1.6}$$

The terms $tr\ \pi(\phi)$ and $O_\gamma(\phi)$, on the other hand, are local: if ϕ is factorizable, then so are the distributions. We have already seen this for $tr\ \pi$, and for the orbital integrals it is even easier:

$$O_\gamma(\phi) = \prod_v O_\gamma(\phi_v)$$

where γ is viewed as an element of $G(E_v)$ and

$$O_\gamma(\phi_v) = \int_{G_\gamma(E_v)\backslash G(E_v)} \phi_v(x^{-1}\gamma x) d\dot{x}_v,$$

with $d\dot{x} = \prod d\dot{x}_v$ a factorization of the quotient measure.

When G is itself not anisotropic, Arthur, extending Selberg's constructions for $GL(2)$ and some other rank one groups, generalized Theorem 1.5 by generalizing both sides of the equality. Although $L_2(G(E)\backslash G(\mathbf{A}_E))$ is no longer a direct sum of irreducible automorphic representations, it does decompose as a direct integral of unitary representations. In particular, the subspace $L_2^d(G(E)\backslash G(\mathbf{A}_E))$ defined as the direct sum of all irreducible direct summands of $L_2(G(E)\backslash G(\mathbf{A}_E))$—the *discrete spectrum* of $G(E)\backslash G(\mathbf{A}_E)$—has a character that can be placed on the spectral side of the Arthur-Selberg trace formula. It is unfortunately not alone, and needs to be accompanied by quite complicated integral terms. On the geometric side, the problem is that it is no longer true that every $\gamma \in G(E)$ is elliptic semisimple; the failure of either of these conditions leads to the appearance of complicated terms on this side as well. Under favorable choices of test functions, all these bad terms disappear and we are left with a formula very similar to Theorem 1.5, called the *simple trace formula*. For the heuristic purposes of this introduction, it will be enough to restrict our attention to the anisotropic case.

2. Simple applications of the trace formula

It should be taken as axiomatic that the spectral side of the trace formula is intrinsically interesting, whereas the geometric side lends itself to calculation. Thus it is in principle possible to obtain information about the spectral side by calculating the geometric side for carefully chosen test functions ϕ. In what follows we write G_v in place of $G(E_v)$.

Example 2.1: Limit multiplicities and pseudocoefficients. The first example develops a theme that goes back to Langlands' early work on the dimensions of spaces of automorphic forms. Here we follow Clozel's approach in [C1], though only in the simplest cases. As before, we assume G is anisotropic, and to simplify the discussion we assume G semisimple as well. Let S be a finite set of places of E, including all archimedean places, and for each $v \in S$ choose an irreducible admissible representation π_v^0 of G_v. Suppose for each v we can choose a test function ϕ_v such that

$$tr\ \pi_v^0(\phi_v) = 1,\ tr\ \pi_v(\phi_v) = 0, \pi_v \neq \pi_v^0, \tag{2.1.1}$$

where π_v runs over all irreducible admissible representations of G_v. For nonarchimedean places $w \notin S$ we let $\phi_w = 1_{K_w}$, so that

$$tr\ \pi_w(\phi_w) = vol(K_w) \dim \pi_w^{K_w}, \forall \pi_w. \tag{2.1.2}$$

Let $K^S = \prod_{w \notin S} K_w$, $\pi_S^0 = \otimes_{v \in S} \pi_v^0$. With this choice of test functions the spectral side of the trace formula would then be

$$\sum_{\pi = \pi_S^0 \otimes \pi^S} m_\pi vol(K^S) \dim(\pi^S)^{K^S}. \tag{2.1.3}$$

In other words, this choice of ϕ counts the automorphic forms of type π_S at S and of level K^S away from S. For example, suppose S equals the set of archimedean places, G_v is the automorphism group of a hermitian symmetric domain D_v for every archimedean v, and π_v^0 is the representation corresponding to holomorphic modular forms of some fixed weight on D_v. Then $tr \sim R(\phi)$ counts the holomorphic modular forms of given weight and level K^S.

Functions with property (2.1.1) can be found when v is nonarchimedean, π_v^0 is supercuspidal, and the center $Z_G(E_v)$ is compact. Indeed, supercuspidal representations π_v^0 are characterized by the condition that their *matrix coefficients*

$$\phi_{e,e^\vee}(g) = e^\vee(\pi_v^0(g)e), e \in \pi_v^0, e^\vee \in (\pi_v^0)^\vee$$

are compactly supported modulo $Z_G(E_v)$, which we are assuming finite. Thus under these hypotheses, such a $\phi_{e,e^\vee} \in C_c^\infty(G_v)$. On the other hand, just as for finite groups, the matrix coefficient of a supercuspidal representation π_v^0 acts as zero, hence *a fortiori* has trace zero, on any representation other than π_v^0, but for appropriate choices of $e, e^\vee$ has non-zero trace on π_v^0. Thus, up to multiplication by a scalar, matrix coefficients of supercuspidal representations satisfy (2.1.1). The same is true when v is archimedean and G_v is compact.

Assume for the moment, then, that G_v is compact for all archimedean v, and that π_v^0 is supercuspidal for finite v in S. When $K^S = K_f$ is sufficiently small an argument based on the discreteness of $G(E)$ in $G(\mathbf{A})$ implies that the only conjugacy class that meets the support of ϕ is the class of the identity element e [C1, Lemma 5]. Thus the geometric side of the trace formula is reduced to the single term $v_e O_e(\phi) = v_e \phi(e)$. We thus obtain

$$(2.1.4) \qquad \sum_{\pi} vol(K^S)\dim(\pi^S)^{K^S} \text{ is constant for sufficiently small } K^S.$$

When the compactness and supercuspidality hypotheses are not satisfied, there are in general no test functions satisfying (2.1.1). Matrix coefficients of discrete series representations can be used, provided they are integrable, but they are not compactly supported, and a different version of the trace formula is needed. However, one can make do with a weaker hypothesis

$$(2.1.1(\text{bis})) \qquad tr\ \pi_v^0(\phi_v) = 1,\ tr\ \pi_v(\phi_v) = 0, \pi_v \neq \pi_v^0, \pi_v \textbf{ tempered.}$$

Test functions satisfying (2.1.1)(bis) are called *pseudocoefficients* and are known to exist for all square integrable representations; this is due to Clozel and Delorme for real groups and to Bernstein, Deligne, and Kazhdan for p-adic groups. On the other hand, it can be shown that the non-tempered representations that have non-zero trace for a given pseudocoefficient have strictly smaller limit multiplicities than discrete series representations. (The relevant arguments are in [C1].) This can be made precise, but the upshot is that (2.1.4) remains true when π_S is merely discrete series, with the word "constant" replaced by "asymptotically constant." See [C1] for details.

Example 2.2: Jacquet-Langlands transfer. We present the Jacquet-Langlands correspondence in the simplest possible setting. Let E be a totally real field of even degree. Then up to isomorphism there exists a unique quaternion algebra D over E such that, for all archimedean places v, D_v is ramified, i.e., isomorphic to the Hamiltonian quaternions, and such that for all non-archimedean v $D_v \simeq M(2, E_v)$. Let D' be a second quaternion algebra over E that is ramified at all archimedean places but not isomorphic to D. Then there exists a non-empty finite set S of non-archimedean places of E such that D'_v is a division algebra if and only if $v \in S$ or v is archimedean; moreover, S has even cardinality. We let $G = D^\times$, $G' = D'^{,\times}$, viewed as group schemes over $Spec(E)$. Thus $G_v \xrightarrow{\sim} GL(2, E_v)$ for all finite v, and $G'_v \xrightarrow{\sim} G_v$ for $v \notin S$, but G'_v is an inner form of $GL(2, E_v)$, anistropic modulo its center $E_v^\times$, for $v \in S$.

For $v \notin S$ we may identify

$$(2.2.1) \qquad C_c^\infty(G_v) \xrightarrow{\sim} C_c^\infty(G'_v),$$

The isomorphism (2.2.1) is based on an isomorphism of groups and is unique up to inner automorphism. In particular, (2.2.1) determines a canonical isomorphism of spaces of invariant distributions (such as orbital integrals).

We let $N : G \to GL(1)$ denote the reduced norm, and use the same notation for G'. For v archimedean all irreducible representations of $G_v \xrightarrow{\sim} G'_v$ are finite-dimensional. The space of C^∞ K_v-finite functions is spanned, up to the action of the center, by the matrix coefficients of these finite-dimensional representations, and we would like to use these as test functions. However, these matrix coefficients are not compactly supported. There are three ways to get around this problem: (a) work with the class of test functions that transform under a fixed character of the center, and with the version of the trace formula adapted to this situation; (b) truncate the matrix coefficient ϕ_v^1 by multiplying it by $c_V(N(g))$, where $V \subset \mathbb{R}^\times$ is a compact subset and c_V is the characteristic function of V; (c) ignore it. In other situations we will be tempted to follow the strategy (c), but here we can use (b), it being understood that as V varies among compact sets and ϕ_v^1 varies among matrix coefficients of finite-dimensional representations, the functions $\phi_{V,v}^1 = c_V \circ N \cdot \phi_v^1$ separate representations. To simplify further, we will choose as test functions for G_v as well as for G'_v the (truncations of the) function $\phi_v^1 \equiv 1$, the matrix coefficient of the trivial representation, for all archimedean v.

For $v \in S$ the groups G_v and G'_v are not isomorphic. However, the conjugacy classes in G'_v can be identified with a subset of the conjugacy classes of G_v. Every element $\gamma' \in G'_v$ generates the subalgebra $E_v[\gamma'] \subset D'_v$. Since D'_v is a division algebra, $E_v[\gamma']$ is a field, which equals E_v if $\gamma' \in Z_{G'_v}$; otherwise, $[E_v[\gamma'] : E_v] = 2$. The conjugacy class of γ' in D'_v is completely determined by the minimal polynomial of γ' over E_v, which is irreducible of degree 1 or 2. Given such a minimal polynomial $P_{\gamma'}$, we can associate to it a conjugacy class $[\gamma] \subset G_v = GL(2, E_v)$. If $\gamma \in [\gamma]$, then its minimal polynomial equals $P_{\gamma'}$, and indeed $E_v[\gamma] \xrightarrow{\sim} E_v[\gamma']$. Such conjugacy classes in G_v are called *elliptic*; they are characterized by the fact that their minimal polynomials are irreducible, or equivalently by the fact that $E_v[\gamma] \subset M(2, E_v)$ is a field. We say that the conjugacy classes $[\gamma'] \subset G'_v$ and $[\gamma] \subset G_v$ are *associated*, and write $\gamma \leftrightarrow \gamma'$. A conjugacy class in G_v has an associated conjugacy class in G'_v if and only if it is elliptic.

The keys to comparing automorphic representations of G and of G' are contained in the following list of facts:

Fact 2.2.2 (Existence of transfer). To any function $\phi'_v \in C_c^\infty(G'_v)$ we can associate a function $\phi_v \in C_c^\infty(G_v)$ with the properties

(2.2.2.1) $$O_\gamma(\phi_v) = O_{\gamma'}(\phi'_v), \gamma \leftrightarrow \gamma' \text{ regular}$$

(2.2.2.2) $$O_\gamma(\phi_v) = 0 \text{ if } \gamma \in G_v \text{ is semisimple non-elliptic.}$$

The function ϕ_v is called a transfer of ϕ'_v, and we write $\phi_v \leftrightarrow \phi'_v$.

Fact 2.2.3 (Density of orbital integrals). The function ϕ_v of (2.2.2) is not unique but the properties (2.2.2.1) and (2.2.2.2) suffice to determine $tr\ \pi(\phi_v)$ for any irreducible admissible representation π of G_v.

Implicit in (2.2.2) is the possibility to associate Haar measures on G'_v and G_v. Since the centralizers of associated regular elements are isomorphic—$(G'_v)_\gamma \xrightarrow{\sim} (G_v)_\gamma$—this suffices to pin down the quotient measures except when γ is central.[2]

FACT 2.2.4 (LOCAL JACQUET-LANGLANDS CORRESPONDENCE). To any irreducible admissible representation π'_v of G'_v, there exists an irreducible admissible representation π_v of G_v, defined up to isomorphism, with the following properties.

(a) If $\dim \pi'_v > 1$, then π_v is supercuspidal; if $\pi'_v = \chi \circ N$ for some character $\chi : E_v^\times \to \mathbb{C}^\times$, then $\pi_v \xrightarrow{\sim} St \otimes \chi \circ \det$, where St is the Steinberg representation of $GL(2, E_v)$; we write $\pi_v = JL(\pi'_v)$.

(b) Let $\phi'_v \in C_c^\infty(G'_v)$ with transfer $\phi_v \in C_c^\infty(G_v)$. Then
$$tr\ JL(\pi'_v)(\phi_v) = -tr\ \pi'_v(\phi'_v);$$
$tr\ \pi_v(\phi_v) = 0$ if π_v is neither one-dimensional nor of the form $JL(\pi'_v)$.

(c) The space of distributions $tr\ JL(\pi'_v)$, as π'_v varies over all irreducible admissible representations of G'_v, separates the transfers to G_v of functions in $C_c^\infty(G'_v)$. More precisely, suppose $\{\pi'_{i,v}, i \in I\}$ is a finite set and $a_i \in \mathbb{C}$ for all $i \in I$. Suppose
$$\sum a_i tr\ JL(\pi'_{i,v})(\phi_v) = 0$$
for all transfers ϕ_v from $C_c^\infty(G'_v)$. Then $a_i = 0$ for all i.

This fact is assumed for simplicity of exposition. In fact, it was obtained by Jacquet and Langlands as a consequence of the constructions we are about to present. Note again that it is assumed implicitly that measures are associated. Condition (c) is a simple consequence of the facts already presented; one can even let ϕ_v run over the space of all test functions on G_v whose orbital integrals vanish off the elliptic set, since these spaces are the same.

We need one more fact:

FACT 2.2.5 (GLOBAL TRANSFER OF CONJUGACY CLASSES). Let $(\gamma_v) \in G(\mathbf{A}_E)$ be a conjugacy class. Suppose there exists $\gamma' \in G'(E)$ such that, for all $v \notin S$, γ' is conjugate to γ_v in $G_v \simeq G'_v$ and such that, for $v \in S$, $\gamma_v \leftrightarrow \gamma'$. Then there exists $\gamma \in G(E)$ such that γ is conjugate to γ_v for all v.

Conversely, suppose $\gamma \in G(E)$, and suppose γ is elliptic at all $v \in S$. Then there exists $\gamma' \in G'(E)$ such that γ' is conjugate to γ at all $v \notin S$, and $\gamma \leftrightarrow \gamma'$ at all $v \in S$.

We write $\gamma \leftrightarrow \gamma'$. Then $\tau(G_\gamma) = \tau(G_{\gamma'})$ (equality of Tamagawa measure).

In the present case most of this is easy: γ' generates a quadratic extension $E[\gamma']$ of E that embeds locally in D' everywhere. Since there is no additional

---

[2]Centralizers of central elements are inner twists of one another, and one compares Haar measures is as in [K4,§1].

local obstruction to embedding in D, $E[\gamma']$ also embeds globally in D, and we let γ denote the image of γ' under any such embedding; it is well-defined up to conjugacy. Conversely, if γ is as in the second paragraph, the local hypotheses at v guarantee that $E[\gamma]$ embeds locally in D' everywhere, hence there is a global embedding, which defines γ' up to conjugacy. For groups other than multiplicative groups of division algebras, this is a much more serious issue.

The equality of Tamagawa measures is a special case of a theorem due in complete generality to Kottwitz. At this point it suffices to remark that when Tamagawa measures are taken on both sides, then it can be assumed that local measures are associated; this is implicit in (2.2.2) and (2.2.4).

Now let $\phi' \in C_c^\infty(G'(\mathbf{A}_E))$ be a factorizable test function, with $\phi'_v = \phi^1_{V,v} = c_V \circ N \cdot 1$ for every archimedean v. Define a factorizable function $\phi \in C_c^\infty(G(\mathbf{A}_E))$ by

$$\phi_v = \phi'_v, v \notin S \text{ (via (2.2.1))}; \quad \phi_v \leftrightarrow \phi'_v, v \in S.$$

We write $\phi \leftrightarrow \phi'$.

Theorem 2.2.6 (global Jacquet-Langlands correspondence). *Let π' be an automorphic representation of G', and suppose $\dim \pi' > 1$, π'_v trivial for all archimedean v. Then there exists an automorphic representation $\pi = JL(\pi')$ of G, with*

$$\pi_v \xrightarrow{\sim} \pi'_v, v \notin S; \quad \pi_v = JL(\pi'_v), v \in S.$$

Moreover, $m(JL(\pi')) = m_{\pi'}$.

Proof. We first apply the Selberg trace formula to G', with test functions ϕ' as above. Write $\phi' = \phi'_S \otimes \phi'^{,S}$. Automorphic representations of G and G' can be factorized analogously. The spectral side of the trace formula is then

(2.2.6.1)
$$\sum_{\dim \pi' > 1, \pi'_\infty = 1} m_{\pi'}\, tr\, \pi'_S(\phi'_S) tr\, \pi'^{,S}(\phi'^{,S}) + \sum_{\dim \pi' = 1, \pi'_\infty = 1} m_{\pi'}\, tr\, \pi'_S(\phi'_S) tr\, \pi'^{,S}(\phi'^{,S}).$$

The restriction to π'_∞ is imposed by our choice of ϕ'_v at archimedean v, and is in fact completely unnecessary. Similarly, for $\phi \leftrightarrow \phi'$, the spectral side of the trace formula for G is

(2.2.6.2)
$$\sum_{\dim \pi > 1, \pi_\infty = 1} m_\pi\, tr\, \pi_S(\phi_S) tr\, \pi^S(\phi^S) + \sum_{\dim \pi = 1, \pi_\infty = 1} m_\pi\, tr\, \pi_S(\phi_S) tr\, \pi^S(\phi^S).$$

The geometric side of the trace formula for G' is

$$\sum_{\gamma'} v_{\gamma'} O_{\gamma'}(\phi'_S) \cdot O_{\gamma'}(\phi'^{,S}). \tag{2.2.6.3}$$

Likewise, the geometric side of the trace formula for G is

$$\sum_{\gamma} v_\gamma O_\gamma(\phi_S) \cdot O_\gamma(\phi^S), \tag{2.2.6.4}$$

where the sum is taken over all conjugacy classes γ in $G(E)$. Now G is anisotropic over E, so every $\gamma \in G(E)$ is semisimple, hence is still semisimple in G_v for every v. But (2.2.2.2) implies that $O_\gamma(\phi_S) = 0$ unless γ is elliptic at all $v \in S$. Thus (2.2.5) implies that (2.2.6.4) equals

$$\sum_{\gamma',\gamma\leftrightarrow\gamma'} v_\gamma O_\gamma(\phi_S) \cdot O_\gamma(\phi^S)$$
$$= \sum_{\gamma'} v_\gamma O_{\gamma'}(\phi'_S) \cdot O_{\gamma'}(\phi'^{,S})$$
$$(2.2.6.5) \qquad = \sum_{\gamma'} v_{\gamma'} O_{\gamma'}(\phi'_S) \cdot O_{\gamma'}(\phi'^{,S})$$

The second equality is a consequence of the last assertion in (2.2.5).

Now the last line is equal to (2.2.6.3). Thus the Selberg trace formula implies equality of (2.2.6.1) and (2.2.6.2). The one dimensional automorphic representations of G' and G correspond in an obvious way, and we thus obtain

(2.2.6.6)
$$\sum_{\dim \pi'>1,\pi'_\infty=1} m_{\pi'}\, tr\, \pi'_S(\phi'_S) tr\, \pi'^{,S}(\phi'^{,S}) = \sum_{\dim \pi>1,\pi_\infty=1} m_\pi\, tr\, \pi_S(\phi_S) tr\, \pi^S(\phi^S).$$

Apart from the (unnecessary) restriction on π'_∞, the representations as well as the test functions on the left-hand side are completely general. By (2.2.4) (b) the right-hand side is the sum over π with π_v in the image of the local Jacquet-Langlands correspondence for all $v \in S$. The theorem is then a simple consequence of (2.2.4)(c) and general results on linear independence of characters of $G(\mathbf{A}_E^S)$. More precisely, fix an irreducible representation $\pi'_{S,0}$ of $G'(\mathbf{A}_{E,S})$ with trivial archimedean components, and let $\pi_{S,0}$ be the representation of $G(\mathbf{A}_{E,S})$ associated to π'_S by Fact 2.2.4. It follows from (2.2.4)(c) that the relation (2.2.6.6) specializes to an identity

$$(2.2.6.7) \qquad \sum_{\dim \pi'>1,\pi'_S=\pi'_{S,0}} m_{\pi'}\, tr\, \pi'^{,S}(\phi'^{,S}) = \sum_{\dim \pi>1,\pi'_S=\pi'_{S,0}} m_\pi\, tr\, \pi^S(\phi^S).$$

But this is an identity of distributions on the group $G'(\mathbf{A}_E^S) \simeq G(\mathbf{A}_E^S)$. The theorem then follows from linear independence of characters (Fact 1.3(c)). □

3. Stable conjugacy

Example 2.2 is perhaps the most elementary use of the trace formula to transfer automorphic representations from one group to another. In this case G' and G are inner forms of each other, but similar arguments are used when $G' = GL(n, E)$ and $G = GL(n, E')$, with E'/E a cyclic extension (cyclic base change, due to Langlands for $n = 2$ and Arthur-Clozel in general) or when $G' = GL(n, E)$, E is a cyclic extension of F of degree d, and $G = GL(nd, F)$ (automorphic induction, due to Arthur-Clozel, and revisited by Henniart-Herb). Each such case follows the general pattern of Example 2.2, especially the transfer of local and global conjugacy classes, and the transfer of orbital integrals. There are local difficulties, especially in obtaining the correct form of base change to substitute for (2.2.2.1) and (2.2.2.2), but the main difficulties are global and analytic and arise from the presence of non-elliptic and non-cuspidal terms in the general trace formula for isotropic groups.

A quite different sort of difficulty arises when G and G' are not inner forms of $GL(n)$. The main relevant property of $GL(n)$ is concealed in the discussion of transfer of associated conjugacy classes preceding (2.2.2). Semi-simple conjugacy classes in inner forms of $GL(n)$ are completely characterized by their characteristic polynomials. Similar characterizations are possible for other groups, but only over an algebraically closed field. In other words, two semi-simple elements of $G(F)$, where F is either a global or a local field, can belong to the same $G(\bar{F})$-conjugacy class but to distinct $G(F)$-conjugacy classes. It is often possible to associate the former for different groups, but there is no natural way to associate the latter. However, the geometric expansion of the trace formula is a sum of integrals over $G(F)$-conjugacy classes. The discrepancy between $G(F)$-conjugacy and $G(\bar{F})$-conjugacy, first studied when $G = SL(2)$ by Labesse and Langlands, is the origin of the stable trace formula.

Strictly speaking, the above remarks only apply when the derived subgroup $G^{der} \subset G$ is assumed simply connected, and we will make this assumption henceforward. We then say that two semi-simple elements $\gamma_1, \gamma_2 \in G(F)$ are *stably conjugate* if they belong to the same $G(\bar{F})$-conjugacy class; the intersections with $G(F)$ of semi-simple $G(\bar{F})$-conjugacy classes are called *stable conjugacy classes*. The *stable orbital integral* of a test function f attached to the semi-simple element $\gamma \in G(F)$ is the sum, weighted by sign factors to be described below, of the orbital integrals of f over the conjugacy classes in the stable conjugacy class associated to γ. For F local as well as global, we will soon see that this sum is finite, and its terms can be parametrized in terms of Galois cohomology. A *stable distribution* on $G(F)$, or $G(\mathbf{A}_F)$, is a distribution that is invariant over $G(\bar{F})$ (resp. $G(\bar{\mathbf{A}}_F)$) and not only over $G(F)$ (resp. $G(\mathbf{A}_F)$). The trace formula would be *stable* if it were a sum of stable distributions.

For some very special groups G, stable conjugacy and conjugacy coincide. We have already seen this for inner forms of $GL(n)$. It is also effectively the case globally for certain inner forms of unitary groups. This is the basis of the results of Kottwitz on the zeta functions of "simple Shimura varieties" and of Clozel and Labesse on stable base change from unitary groups to $GL(n)$. The work of Harris-Taylor on Galois representations at places of bad reduction of Shimura varieties was carried out in this setting, as was Harris' proof of a version of the Jacquet-Langlands correspondence between two such unitary groups. Certain test functions are also insensitive to the difference between stable conjugacy and $G(F)$-conjugacy. This was first observed by Kottwitz in his proof of Weil's conjecture on Tamagawa numbers, and was developed elsewhere, most recently in Labesse's book on stable base change [Lab] and in an article of Harris-Labesse [HL].

For general G and for general test functions these simplifications are not possible. Hence the problem arises of *stabilizing* the trace formula. The trace formula is not itself a sum of stable distributions. The error, however, can conjecturally be expressed in terms of stable distributions on other groups: these are the *endoscopic groups.* The stable trace formula is this conjectural expression. Several decades of work by a number of people have reduced the stabilization of the trace formula to a collection of conjectures, known collectively and informally as the "fundamental lemma," concerning a specific class orbital integrals on p-adic groups. The breakthrough in the proof of the fundamental lemma was finally made by Laumon and Ngô in the spring of 2004, and this is the motivation for the present collection of books.

The subsequent sections of the introduction roughly follow the notes of the lectures of Ngô at IHES in the spring of 2003. Ngô's notes sketch proofs of a number of the theorems stated here, and they are available for consultation at various locations on the internet, notably at http://www.ihes.fr/IHES/Scientifique/Seminaires/seminaire-LL.html. Complete proofs are of course in the articles quoted below.

The first step, due to Langlands and Kottwitz, is to determine the set of conjugacy classes in a stable conjugacy class. When F is a non-archimedean local field this is not bad at all. Let $\gamma \in G(F)$ be a semi-simple element, and let $I = I_\gamma \subset G$ denote its centralizer (also denoted G_γ). Since γ is semi-simple, I is a reductive group, and since G^{der} is simply connected, I is *connected.* This is the main reason we assume G^{der} simply connected; the general case can also be handled by reduction to this case.

Suppose γ' is stably conjugate to γ, i.e. there exists $g \in G(\bar{F})$ such that $g^{-1}\gamma g = \gamma'$. Let $\Gamma = Gal(\bar{F}/F)$. For any Γ-module M, we write $H^\bullet(F, M)$ for $H^\bullet(\Gamma, M)$. For any $\sigma \in \Gamma$,

$$g^{-1}\gamma g = \gamma' = \sigma(\gamma') = \sigma(g^{-1})\gamma\sigma(g),$$

hence $\sigma \mapsto g\sigma(g)^{-1}$ defines a 1-cocycle

$$c_\sigma \in H^1(F, I) := H^1(F, I(\bar{F})).$$

It is easy to see that the cohomology class of c_σ is independent of the choice of g, and that the conjugacy class of γ' depends only on this cohomology class. If γ is a regular element of G then I is a torus, and thus $H^1(F, I)$ is obviously an abelian group. On the other hand, the image of c_σ in $H^1(F, G)$ is obviously a coboundary; thus

$$c_\sigma \in \ker[H^1(F, I) \to H^1(F, G)] \tag{3.1}$$

In fact, because G^{der} is simply connected, it is a result of Kneser that $H^1(F, G^{der}) = 1$, and after some work it follows that $H^1(F, G)$ can be interpreted in terms of $H^1(F,D)$, where $D = G/G^{der}$ is a torus; in particular, the kernel in (3.1) is an abelian subgroup of $H^1(F,I)$.

We write $inv(\gamma,\gamma')$ for the class in $\ker[H^1(F,I) \to H^1(F,G)]$ corresponding to the conjugacy class of γ'.

After a bit more work, Kottwitz showed in [K3] that (3.1) is an abelian group even if γ is not regular, and indeed even if G^{der} is not simply connected. This has been reformulated systematically by Labesse in terms of a generalization of what Borovoi called the *abelianized Galois cohomology* of I and G. Labesse defined in a functorial way a set $H^0(F, I\backslash G)$ that fits into a short exact sequence:

$$(3.2) \quad 0 \to I(F)\backslash G(F) \to H^0(F, I\backslash G) \to \ker[H^1(F,I) \to H^1(F,G)] \to 0.$$

The term $I(F)\backslash G(F)$ is naturally the $G(F)$-conjugacy class of γ, whereas the kernel on the right is a finite abelian group that parametrizes the conjugacy classes in the stable conjugacy class. Thus the term in the middle is the stable conjugacy class. We write $\gamma' \overset{st}{\sim} \gamma$ if γ' and γ are stably conjugate.

FACT 3.3. For any $\gamma' \overset{st}{\sim} \gamma$, the centralizer $I_{\gamma'}$ is the inner form of I_γ determined by the image of the cocycle c_σ defined by γ' under the natural map $H^1(F, I_\gamma) \to H^1(F, I_{\gamma,ad})$. There is a map (the *Kottwitz sign*):

$$e : \{\text{ Inner forms of } I_\gamma\} = H^1(F, I_{\gamma,ad}) \to \{\pm 1\}$$

The stable orbital integral $SO_\gamma(f)$ of $f \in C_c^\infty(G(F))$ is the sum

$$\sum_{\gamma' \overset{st}{\sim} \gamma} e(I_{\gamma'}) O_{\gamma'}(f) = \int_{H^0(F, I\backslash G)} e(x) f(x^{-1}\gamma x)\, d\dot{x}.$$

The right-hand side is here treated as a suggestive expression whose precise definition can be found on p. 67 of [Lab].

Stable conjugacy over global fields has an analogous but more complicated treatment, to which we turn after introducing the L-group.

4. L-groups and stable conjugacy

We let F be either a local field or a number field, and let Γ denote either $Gal(\bar{F}/F)$ or the Weil group $W_{\bar{F}/F}$. Let G be a connected reductive group over F. Langlands' theory of the L-group begins with the observation that a connected reductive algebraic group $\mathcal{G}$ over an algebraically closed field (of characteristic zero, for simplicity) can be reconstructed up to canonical isomorphism in terms of the *based root datum* $\Psi_0(\mathcal{G}) = (X^*, \Delta^*, X_*, \Delta_*)$. Here T is a maximal torus of $\mathcal{G}$, $X^* = X^*(T)$, resp. X_* is its group of characters (resp. cocharacters), $B \supset T$ is a Borel subgroup, Δ^* is the set of positive simple roots of T in B, Δ_* the set of positive simple coroots. These data depend on a number of choices, but for any two choices the corresponding data are *canonically* isomorphic, hence the notation $\Psi_0(\mathcal{G})$ is justified.

The quadruple $(X^*, \Delta^*, X_*, \Delta_*)$ satisfies a collection of axioms including as a subset the axioms for a root datum satisfied by the first two items. It suffices to mention that any quadruple Ψ of the same type, with X^* and X_* finitely generated free abelian groups, and $\Delta^* \subset X^*$ and $\Delta_* \subset X_*$, that satisfy these axioms comes from a connected reductive group. The dual based root datum $\hat{\Psi} = (X_*, \Delta_*, X^*, \Delta^*)$, obtained by switching the first two items with the last two, also satisfies these axioms, and therefore defines up to canonical isomorphism a connected reductive algebraic group $\hat{\mathcal{G}}$ over $\mathbb{C}$ with $\Psi_0(\hat{\mathcal{G}}) \xrightarrow{\sim} \hat{\Psi}$.

We write $\hat{G}$ for $\hat{G}_{\bar{F}}$. The F-rational structure on $G_{\bar{F}}$ translates to an action of Γ on $\Psi_0(G_{\bar{F}})$, preserving the natural pairing between characters and cocharacters, and thus, almost, to an action of Γ on $\hat{G}$. To define such an action unambiguously one needs to choose a *splitting* ("épinglage") of $\hat{G}$, i.e., a triple $(\hat{T}, \hat{B}, \{X_\alpha, \alpha \in \hat{\Delta}\})$, where $\hat{T}$ is a maximal torus of $\hat{G}$, $\hat{B} \supset \hat{T}$ a Borel subgroup, $\hat{\Delta}$ the set of positive simple roots of $\hat{T}$ in $\hat{B}$, and X_α is a non-zero element of the root space $\hat{\mathfrak{g}}_\alpha \subset Lie(\hat{B})$ for every $\alpha \in \hat{\Delta}$. An action of Γ on $\hat{G}$ that fixes some splitting is called an L-action, and an L-group ${}^L G$ of G is the semi-direct product of $\hat{G}$ with Γ with respect to a fixed L-action.

If $\mathcal{G}_1 \to \mathcal{G}_2$ is an F-homomorphism whose image is a normal subgroup, then the dual construction yields a homomorphism $\hat{\mathcal{G}}_2 \to \hat{\mathcal{G}}_1$ that is canonical up to $(\hat{\mathcal{G}}_1)^\Gamma$-conjugacy.

In the following section, we will begin with an object on the L-group side and use it to reconstruct endoscopic groups of H.

This construction has a number of subtle and unexpected properties that ought ideally to be mentioned here but are probably too difficult to motivate at this stage. The skeptical reader is advised that motivation will come through experience working with these notions (as I keep telling myself). It should not be hard to see that the construction guarantees that L-groups of inner forms are isomorphic. For classical groups the L-groups are easy to calculate explicitly.

Example 4.1. Unitary groups. Let Φ_n be the matrix whose ij entry is $(-1)^{i+1}\delta_{i,n-j+1}$:

$$\Phi_n = \begin{pmatrix} 0 & 0 & \dots & 0 & 1 \\ 0 & 0 & \dots & -1 & 0 \\ \dots & \dots & \dots & \dots & \dots \\ (-1)^{n-1} & 0 & \dots & 0 & 0 \end{pmatrix}$$

Let $\mathcal{K}/E$ be a quadratic extension as above, and define an action of $Gal(\mathcal{K}/E)$ on $\hat{G} = GL(n, \mathbb{C})$ by letting the non-trivial element c act as

$$c(g) = \Phi_n {}^t g^{-1} \Phi_n^{-1}.$$

Conjugation by Φ_n is necessary in order to preserve the standard splitting of $\hat{G}$ defined by the standard upper triangular Borel subgroup with its standard diagonal torus, and the standard basis of the simple root subspaces in $Lie(\hat{G})$ given by the matrices $X_{i,i+1}$, $i = 1, \ldots, n-1$, with entry 1 in the $(i, i+1)$ place and zero elsewhere: one checks that $c(X_{i,i+1}) = X_{n-i,n-i+1}$ for all i. Without this conjugation by Φ_n, the Langlands parametrization of representations (local or automorphic) of G is just wrong. For example, in [M.III.C, 4.2], the introduction of auxiliary unramified characters μ_a, μ_b in formulas for endoscopic transfer is necessary in order to make the map ξ_{μ_a,μ_b} an L-homomorphism. One can define a homomorphism more simply by eliminating the conjugations by Φ_n, Φ_a, Φ_b, but the resulting map would not be compatible with Langlands functoriality.

An action of $Gal(\bar{E}/E)$, or of the Weil group W_E, on $\hat{G}$ is defined by projection onto $Gal(\mathcal{K}/E)$. The L-group in the Weil (resp. Galois) normalization is the semi-direct product of $\hat{G}$ with W_E (resp. $Gal(\bar{E}/E)$) with respect to this action.

The endoscopic groups that contribute to the stable trace formula are naturally defined in terms of the L-group. The first indication of this relation is given by the following determination, due to Langlands and Kottwitz [L,K2], of the cohomology group parametrizing the stable conjugacy class of a semi-simple element. We let $Z(\hat{G})$ denote the center of $\hat{G}$ and ^{D} denote Pontryagin dual.

Proposition 4.2 [K2]. For any local field F, there is a canonical map

$$H^1(F, G) \to \pi_0(Z(\hat{G})^\Gamma)^D,$$

functorial in G, that is an isomorphism if F is p-adic. In particular, if $\gamma \in G(F)$ is semi-simple, $I = I_\gamma$, then the stable conjugacy class of γ pairs canonically with

$$\pi_0(Z(\hat{I})^\Gamma / Z(\hat{G})^\Gamma) = \pi_0(Z(\hat{I})^\Gamma Z(\hat{G})/Z(\hat{G}))$$

and this pairing is a duality when F is p-adic.

Let F be any local field. If $\gamma \in G(F)$ is semi-simple, $\gamma' \overset{st}{\sim} \gamma$, we define

$$s \mapsto < inv(\gamma, \gamma'), s >$$

for the character $\pi_0(Z(\hat{I})^\Gamma / Z(\hat{G})^\Gamma) \to \mathbb{C}^\times$ defined by the pairing in (4.2). The determination of the set of conjugacy classes in a stable conjugacy class (in the p-adic case) in terms of the Langlands dual proceeds by replacing the cohomology of the reductive group G by the cohomology of a complex of tori. Proposition 4.2 is then derived as a consequence of, and generalization of, local Tate-Nakayama duality for cohomology of tori.

Global Tate-Nakayama duality for tori associates all the local duality maps by means of a reciprocity law. The generalization to arbitrary reductive groups is due to Langlands and Kottwitz. We begin by generalizing the notion of elliptic elements to arbitrary reductive groups.

Definition 4.3. Let F be either a p-adic field or a number field. Let $\gamma_0 \in G(F)$ be a semi-simple element, with centralizer I. The element γ_0 is

elliptic if $Z(I)^0/Z(G)^0$ is anisotropic. Equivalently, γ_0 is elliptic if it belongs to a maximal torus of G contained in no proper parabolic subgroup.

Following the method pioneered by Harish-Chandra, harmonic analysis on p-adic and adelic groups is reduced by a procedure known as descent to the study of elliptic elements. When F is a number field we will only be considering anisotropic groups, all of whose elements are elliptic, and descent will not be an issue.

DEFINITION 4.4. Let F be a number field. Let $\gamma_0 \in G(F)$ be a semi-simple elliptic element, with centralizer I. Define

$$(4.4.1) \quad \mathfrak{K}(\gamma_0) = \{c \in \pi_0((Z(\hat{I})/Z(\hat{G}))^\Gamma) \mid \forall v \; c \in \pi_0(Z(\hat{I})^{\Gamma_v} Z(\hat{G})/Z(\hat{G}))\}$$

where v runs over places of F, and Γ_v is a decomposition group at v.

THEOREM 4.5 [K3, THEOREM 6.6]. Let $\gamma_0 \in G(F)$ be a semi-simple elliptic element, Let $\gamma = (\gamma_v) \in G(\mathbf{A}_F)$ and suppose γ_v is stably conjugate to γ_0 for all v. Then there exists $\gamma' \in G(F)$ in the $G(\mathbf{A}_F)$-conjugacy class of γ if and only if

$$(4.5.1) \qquad \sum_v inv(\gamma_0, \gamma_v) \mid_{\mathfrak{K}(\gamma_0)} = 0.$$

Kottwitz' statement has a slightly different form but is equivalent to this one. The proof of this theorem makes use of the Hasse principle for simply connected semi-simple groups. (When Kottwitz' article was written this had not yet been established for E_8, so he includes the hypothesis, no longer necessary, that G contain no E_8 factors.) The sum of local invariants is finite because at almost all places, γ_v and γ_0 belong to the same hyperspecial maximal compact subgroup. Excluding a finite set of places defined in terms of the divisors of $1 - \alpha(\gamma_0)$, where α runs through the roots of G not orthogonal to γ_0, it then follows that γ_v and γ_0 are in the same conjugacy class [K3, Prop. 7.1].

It is now relatively easy to complete the first stage of the stabilization of the trace formula. Details are contained in Ngô's notes as well as in the original sources. Theorem 4.5 provides a necessary and sufficient condition for the intersection of a $G(\mathbf{A}_F)$-conjugacy class with the $G(\mathbf{A}_F)$-stable conjugacy class of γ_0 to contain a $G(F)$-conjugacy class. The next proposition completes this condition by determining the number of $G(F)$-conjugacy classes in the $G(\mathbf{A}_F)$-conjugacy class of γ.

PROPOSITION 4.6 [K3, §9] . Let γ_0 and γ be as in Theorem 4.5, and suppose γ satisfies the reciprocity condition (4.5.1). Then the $G(\mathbf{A}_F)$-conjugacy class of γ contains

$$|\mathfrak{K}(\gamma_0)|\tau(G)\tau(G_{\gamma_0})^{-1}$$

$G(F)$-conjugacy classes.

The geometric side of the trace formula can thus be simplified. First, let E_0 denote a set of representatives in $G(F)$ of the set of stable conjugacy classes in $G(F)$. Write $T_e(\phi)$ for the geometric side of the trace formula, i.e, for the sum of terms on the left of (1.6). These can be grouped according to stable conjugacy:

$$T_e(\phi) = \sum_{\gamma} \tau(G_\gamma) O_\gamma(\phi) = \sum_{\gamma_0 \in E_0} \tau(G_{\gamma_0}) \sum_{\gamma \overset{st}{\sim} \gamma_0} O_\gamma(\phi).$$

Here we have used the important theorem of Kottwitz that

$$\tau(G_\gamma) = \tau(G_{\gamma'}) \text{ if } \gamma \overset{st}{\sim} \gamma'.$$

Now $O_\gamma(\phi)$ is purely local: it only depends on the $G(\mathbf{A}_F)$-conjugacy class of γ. Proposition 4.6 allows us to group together the terms in the inner sum in terms of adelic conjugacy:

$$T_e(\phi) = \tau(G) \sum_{\gamma_0 \in E_0} |\mathfrak{K}(\gamma_0)| \sum_{\gamma \overset{st\mathbf{A}}{\sim} \gamma_0} O_\gamma(\phi) \tag{4.7}$$

where $\gamma \overset{st\mathbf{A}}{\sim} \gamma_0$ means $\gamma \in G(F)$ is stably conjugate to γ_0 but is taken only up to $G(\mathbf{A}_F)$-conjugacy.

The inner sum in (4.7) is a sum of local terms, indexed by a condition that is part local and part global. Theorem 4.5 allows us to separate these conditions by Fourier analysis on the group $\mathfrak{K}(\gamma_0)$:

$$T_e(\phi) = \tau(G) \sum_{\gamma_0 \in E_0} \sum_{\kappa \in \mathfrak{K}(\gamma_0)} \prod_v \sum_{\gamma_v \overset{st}{\sim} \gamma_0} < \kappa, inv(\gamma_0, \gamma_v) > O_{\gamma_v}(\phi_v) \tag{1}$$

$$= \tau(G) \sum_{\gamma_0 \in E_0} \sum_{\kappa \in \mathfrak{K}(\gamma_0)} \prod_v O^\kappa_{\gamma_0}(\phi_v). \tag{4.8}$$

The expression $O^\kappa_{\gamma_0}(\phi_v)$, the κ-orbital integral, is defined locally by

$$O^\kappa_{\gamma_v}(\phi_v) = \sum_{\gamma_v \overset{st}{\sim} \gamma_0} e(I_{\gamma_v}) < \kappa, inv(\gamma_0, \gamma_v) > O_{\gamma_v}(\phi_v). \tag{4.9}$$

Thus when $\kappa = 1$ is the trivial character, we recover the stable orbital integral:

$$O^1_{\gamma_0}(\phi_v) = SO_{\gamma_0}(\phi_v).$$

The Kottwitz signs in (4.9) disappear in the product because all terms come from global groups and the product over all places of Kottwitz signs for a global group is $+1$.

In certain rare situations all the O^κ vanish except for $\kappa = 1$, and then (4.8) is a sum of stable distributions. In the case studied by Kottwitz in [K5], and taken up again in [HT], $\mathfrak{K}(\gamma_0)$ is always trivial, thus there are no non-trivial κ-orbital integrals. This is also possible for more general groups with appropriate choices of test functions.

Remark 4.10. Note that the abelian group $\mathfrak{K}(\gamma_0)$ is attached to the centralizer I rather than to the element γ_0. In the notation of [Lab], recalled in the appendix to this section,

$$\mathfrak{K}(\gamma_0) = \mathfrak{K}(I, G; F) = H^0_{ab}(\mathbf{A}_F/F, I\backslash G)^D.$$

Here and in the remainder of this section one could have followed Labesse and expressed all calculations in terms of abelianized Galois cohomology, without reference to the L-group. For example, Labesse's formalism expresses the global κ-orbital integral concisely as

$$\int_{H^0(\mathbf{A}_F, I\backslash G)} < \kappa, \dot{x} > \phi(x^{-1}\gamma_0 x) d\dot{x}.$$

The expression (4.4.1) will be necessary in subsequent sections in order to express the stabilization in terms of endoscopic groups.

Example 4.11: Stable conjugacy and κ in the case of unitary groups. We let $G = U(V)$ as in §1, and we suppose G to be anisotropic. Then any maximal torus of G is anisotropic. Let L_i be an extension of E of degree n_i, $\mathcal{K}_i = L_i \otimes_E \mathcal{K}$, and assume $\mathcal{K}_i$ is a field for all i. Let $T_i \subset R_{\mathcal{K}_i/E}\mathbb{G}_m$ be the kernel of the norm map $R_{\mathcal{K}_i/E}\mathbb{G}_m \to R_{L_i/E}\mathbb{G}_m$. Any maximal torus of G is isomorphic to one of the form $\prod_i T_i$, with T_i as above and $\sum_i n_i = n$. Conversely, any such torus embeds in the quasi-split inner form of G, and embeds in G provided certain local and global obstructions vanish.

Suppose $\gamma \in G(F)$ is a regular (semi-simple) element, with centralizer I isomorphic to $\prod_{i=1}^r T_i$ as above. Then $Z(\hat{I}) = \hat{I} = \prod \hat{T}_i$. The non-trivial element of the Galois group $Gal(\mathcal{K}/E) = Gal(\mathcal{K}/L_i)$ acts on each $\hat{T}_i$ by $x \mapsto x^{-1}$. Thus

$$\pi_0(Z(\hat{I})^\Gamma) = Z(\hat{I})^\Gamma = \prod_{i=1}^r \mathbb{Z}/2\mathbb{Z} = (\mathbb{Z}/2\mathbb{Z})^r.$$

It follows that

$$\mathfrak{K}_0(I) = \pi_0(Z(\hat{I})^\Gamma / Z(\hat{G})^\Gamma) \xrightarrow{\sim} (\mathbb{Z}/2\mathbb{Z})^{r-1}.$$

Whether or not γ is regular, $\mathfrak{K}_0(G_\gamma)$ is of the form $(\mathbb{Z}/2\mathbb{Z})^{r-1}$ for some $r \leq n$.

Remark 4.12. In the literature there are frequent references to "strongly regular" semisimple elements. In general groups, a semisimple element is **regular** if its centralizer is a diagonalizable subgroup, and **strongly regular** if it is a torus. In other words, an element is strongly regular if and only if it is regular and its centralizer is connected. In this introduction we are exclusively concerned with G for which G^{der} is simply connected, thus every regular element is strongly regular.

Appendix to §4: Labesse's cohomological formalism for endoscopy

In this introduction we have mainly adopted the notation of Langlands and Kottwitz for the Galois cohomology groups relevant to stable conjugacy, because most of the literature on the subject is written using this notation. Kottwitz' constructions have been given a more functorial treatment by Borovoi in [B]; this work was extended by Milne in [M]. Building on this approach, Labesse [Lab] found a unified cohomological formalism that applies to twisted as well as standard endoscopy; it is flexible as well as intuitively satisfying. Moreover, Labesse's cohomological objects can be defined without reference to Tate-Nakayama duality, though the latter remains the best tool for calculation. In this appendix, we state some of the main results of [Lab]. As an application, we use Labesse's formalism to classify unitary groups over local and global fields, a calculation that will be needed in the body of this text.

Inner forms of unitary groups. The abelianized Galois cohomology in the sense of Borovoi [B] and [Lab] is easy to compute for U_n^* and $U_{n,ad}^*$. The abelianized cohomology of U_n^* is the cohomology of its cocenter U_1^*. The abelianized cohomology of $U_{n,ad}^*$ is, by definition, the hypercohomology of the crossed module $SU_n^* \to U_{n,ad}^*$. This complex is quasi-isomorphic to the complex of tori $U_1^* \xrightarrow{n} U_1^*$ and hence, up to a shift by 1, to the diagonalisable group $D_n = \ker[U_1^* \xrightarrow{n} U_1^*]$. In particular there are isomorphisms

$$\mathbf{H}^i_{ab}(F, U_n^*) \xrightarrow{\sim} \mathbf{H}^i(F, U_1^*) \quad \text{and} \quad \mathbf{H}^i_{ab}(F, U_{n,ad}^*) \xrightarrow{\sim} \mathbf{H}^{i+1}(F, D_n) \ .$$

LEMMA 4.A.1.

(i) We have

$$\mathbf{H}^1_{ab}(\bullet, U_n^*) = \mathbb{Z}/2\mathbb{Z} \qquad \text{and} \qquad \mathbf{H}^2_{ab}(\bullet, U_n^*) = 1$$

where $\bullet = F$ (resp. $\bullet = \mathbb{A}_F/F$) when E/F is a quadratic extension of local (resp. global) fields. Moreover, if F is global, we have $\ker^i(F, U_1^*) = 1$ for $i \geq 0$.

(ii) We have

$$\mathbf{H}^1_{ab}(\bullet, U_{n,ad}^*) = \mathbf{H}^2(\bullet, D_n) = \begin{matrix} 1 & \text{if } n \text{ is odd} \\ \mathbb{Z}/2\mathbb{Z} & \text{if } n \text{ is even} \end{matrix}$$

where $\bullet = F$ (resp. $\bullet = \mathbb{A}_F/F$) when E/F is a quadratic extension of local (resp. global) fields. This is also the case if $F = \mathbb{R}$ and $E = \mathbb{R} \oplus \mathbb{R}$. When F is a non archimedean local field and $E = F \oplus F$ then

$$\mathbf{H}^1_{ab}(F, U_{n,ad}^*) = \mathbf{H}^2(F, D_n) = \mathbb{Z}/n\mathbb{Z} \ .$$

When $F = \mathbb{C}$ then $\mathbf{H}^2(F, D_n) = 1$.

(iii) When E/F is a quadratic extension of global fields the map

$$\mathbf{H}^2(F_v, D_n) \to \mathbf{H}^2(\mathbb{A}_F/F, D_n)$$

is surjective, unless maybe when $F_v = \mathbb{C}$. Moreover $\ker^i(F, D_n) = 1$ for $i \geq 2$.

Proof. Assertions (i) and (ii) follow easily from the Tate-Nakayama isomorphism and from the above remarks. We still have to prove (iii). The co-localization map can be computed using Poitou-Tate duality [Ltam, Corollaire 2.2]; it is known that $\ker^i(F, D_n) = 1$ for $i \geq 3$ [Ltam, Corollaire 2.4]. We are left to prove that $\ker^2(F, D_n) = 1$. Using that

$$D_n = \ker[U_1^* \xrightarrow{n} U_1^*]$$

we obtain, when n is odd, a commutative and exact diagram

$$\begin{array}{ccccccc} & & & & 1 & & 1 \\ & & & & \downarrow & & \downarrow \\ 1 & \to & \ker^2(F, D_n) & \to & \mathbf{H}^2(F, D_n) & \to & \mathbf{H}^2(\mathbb{A}_F, D_n) \\ & & & & \downarrow & & \downarrow \\ & & 1 & \to & \mathbf{H}^2(F, U_1^*) & \to & \mathbf{H}^2(\mathbb{A}_F, U_1^*) \end{array}$$

and the conclusion is clear in this case. When n is even we get a commutative and exact diagram

$$\begin{array}{ccccccc} & & & & 1 & & 1 \\ & & & & \downarrow & & \downarrow \\ & & 1 & \to & \mathbf{H}^1(F, U_1^*) & \to & \mathbf{H}^1(\mathbb{A}_F, U_1^*) \\ & & & & \downarrow & & \downarrow \\ 1 & \to & \ker^2(F, D_n) & \to & \mathbf{H}^2(F, D_n) & \to & \mathbf{H}^2(\mathbb{A}_F, D_n) \\ & & & & \downarrow & & \downarrow \\ & & 1 & \to & \mathbf{H}^2(F, U_1^*) & \to & \mathbf{H}^2(\mathbb{A}_F, U_1^*) \end{array}$$

and again the conclusion is easy. □

Inner forms of U_n^* (resp. GU_n^*) will be denoted U_n (resp. GU_n) or simply U (resp. GU). The set of their isomorphism classes is in bijection with $\mathbf{H}^1(F, U_{n,ad}^*)$. An inner form of U_n^* (resp. GU_n^*) defines a class, called its invariant, in $\mathbf{H}^1_{ab}(F, U_{n,ad}^*)$ via the abelianization map

$$\mathbf{H}^1(F, U_{n,ad}^*) \to \mathbf{H}^1_{ab}(F, U_{n,ad}^*) \ .$$

There is also a natural map

$$\mathbf{H}^1(F, U_{n,ad}^*) \to \mathbf{H}^1(F, G_{n,ad}^*)$$

and hence an inner form U_n of U_n^* defines an inner form $Res_{E/F}B^\times$ of G_n^* where B is a simple algebra over E and we may regard U_n as the unitary group of the simple E-algebra B with an involution of the second kind, denoted $b \mapsto b^\dagger$. Then $U_n = \{x | xx^\dagger = 1\}$ and the F_0-similitude group GU_n is the subgroup of $Res_{E/F_0} B^\times$ such that, for any F_0-algebra R,

$$GU_n(R) = \{x \in (B \otimes_{F_0} R)^\times \mid x \cdot x^\dagger = \nu(x) \in R^\times\} .$$

We shall now recall the classification of inner forms of unitary groups over local and global fields (see also [C2]).

If F is local non-archimedean, the invariant suffices to determine the inner form, since the abelianization map

$$\mathbf{H}^1(F, U^*_{n,ad}) \to \mathbf{H}^1_{ab}(F, U^*_{n,ad})$$

is an isomorphism in this case. When $F = \mathbb{R}$ and $E = \mathbb{C}$ the set $\mathbf{H}^1(F, U^*_{n,ad})$ is in natural bijection with signatures (p, q) with $p + q = n$ and $p \geq q$. The invariant is $[\frac{p-q}{2}]$ modulo 2 if n is even. The description in other cases is left to the reader.

Now let E/F be a quadratic extension of global fields. We want to describe the local-global obstructions i.e. the description of the image of the map $\mathbf{H}^1(F, U^*_{n,ad}) \to \mathbf{H}^1(\mathbb{A}_F, U^*_{n,ad})$. This will be an immediate consequence of the next lemma.

LEMMA 4.A.2. The adjoint unitary group satisfies the Hasse principle i.e.

$$\ker^1(F, U^*_{n,ad}) = 1 .$$

We have

$$\mathbf{H}^1_{ab}(\mathbb{A}_F/F, U^*_{n,ad}) = \begin{matrix} 1 & \text{if } n \text{ is odd} \\ \mathbb{Z}/2\mathbb{Z} & \text{if } n \text{ is even} \end{matrix}$$

and the natural map

$$\mathbf{H}^1_{ab}(F_v, U^*_{n,ad}) \to \mathbf{H}^1_{ab}(\mathbb{A}_F/F, U^*_{n,ad})$$

is always surjective.

Proof. We first observe that, according to [Lab, Corollaire 1.6.11],

$$\ker^1(F, U^*_{n,ad}) = \ker^1_{ab}(F, U^*_{n,ad}) .$$

The assertions then follow from 4.A.1. □

PROPOSITION 4.A.3. Let F be a global field. There is at most one global inner form of U^*_n when the local ones are prescribed. There is no local-global obstruction for inner forms when n is odd. There is a parity condition when n is even: the sum of images of local invariants must equal zero in $\mathbb{Z}/2\mathbb{Z}$.

Proof. The exact sequence

$$\ker^1(F, G) \to \mathbf{H}^1(F, G) \to \mathbf{H}^1(\mathbb{A}_F, G) \to \mathbf{H}^1_{ab}(\mathbb{A}_F/F, G) \to \ker^2_{ab}(F, G)$$

valid for any reductive group (see [Lab, Proposition 1.6.12]) reads in our case, in view of 4.A.1 and 4.A.2,

$$1 \to \mathbf{H}^1(F, U^*_{n,ad}) \to \mathbf{H}^1(\mathbb{A}_F, U^*_{n,ad}) \to \mathbf{H}^1_{ab}(\mathbb{A}_F/F, U^*_{n,ad}) \to 1 .$$

The Hasse principle implies that the local inner forms determine uniquely the global one. The local-global obstruction can be represented by a cohomology class in

$$\mathbf{H}^1_{ab}(\mathbb{A}_F/F, U^*_{n,ad})$$

and the proposition follows from 4.A.2. □

5. Endoscopic groups

The goal of stabilization is to rewrite the expression (4.8) as a sum of stable trace formulas for groups over F, without reference to the finite groups $\mathfrak{K}(\gamma_0)$. The new sets of groups over F, called *endoscopic groups*, are derived directly from the elements $\kappa \in \mathfrak{K}(\gamma_0)$. There are two obvious obstacles to this program. In the first place, the conjugacy classes γ_0 belong to $G(F)$. In the second place, the test functions ϕ are defined on $G(\mathbf{A}_F)$. We have already seen in example 2.2 how to overcome these obstacles in a special case. Transfer of conjugacy classes to endoscopic groups was solved in the 1980s by Kottwitz, following the first steps of Langlands. It is fair to say that the endoscopic groups were devised in order to make this possible, given the expression (4.8). The difficulty has thus been shifted to the problem of transferring test functions to endoscopic groups. This is the main topic of the present volume and will only be discussed briefly in this introduction.

Version (4.8) of the trace formula writes it as a sum over pairs (γ_0, κ), where γ_0 is a stable elliptic conjugacy class in $G(F)$ and κ is an element of the group (4.4.1) defined in terms of the dual group $\hat{G}$. The goal is to rewrite it as a sum over pairs (γ^H, H) where H runs through a family of connected reductive groups over F and γ^H is a stable elliptic conjugacy class in $H(F)$ whose contribution to the sum is its stable orbital integral. This turns out to be too naïve, and the object H needs to be endowed with additional structure in order to qualify as an endoscopic group.

Recall that κ is an element of the center of the dual group of the centralizer I of the elliptic element γ_0. With a great deal of hindsight, this might suggest that one look for H containing γ^H among groups whose L-groups contain $Z(\hat{I})$ as central subgroups. Here is a precise definition:

Definition 5.1. An **endoscopic triple** for G is a datum of the form (H, s, ξ) where

(i) $s \in \hat{G}$ is a semi-simple element;
(ii) H is a quasi-split connected reductive group over F;
(iii) $\xi : \hat{H} \to \hat{G}$ is an injective homomorphism identifying $\xi(\hat{H}) = Z_{\hat{G}}(s)^0$.

These data satisfy the following conditions:

(a) ξ is Γ-equivariant up to $\hat{G}$-conjugation; i.e., for all $\sigma \in \Gamma$, there exists $x_\sigma \in \hat{G}$ such that

$$\sigma_{\hat{G}} \circ \xi \sigma_{\hat{H}}^{-1} = ad(x_\sigma) \circ \xi$$

where $\sigma_{\hat{G}}$ and $\sigma_{\hat{H}}$ are the respective L-actions of Γ on the dual groups;

(b) The image of s in $Z(\hat{H})/Z(\hat{G})$ is Γ-invariant and its image in $H^1(F, Z(\hat{G}))$ under the connecting homomorphism associated to the short exact sequence of Γ-modules:
$$1 \to Z(\hat{G}) \to Z(\hat{H}) \to Z(\hat{H})/Z(\hat{G}) \to 1$$
is everywhere locally trivial (i.e., belongs to the Tate-Shafarevich group $ker^1(F, Z(\hat{G}))$).

Condition (b) should be recognized as a variant of the condition satisfied by κ in (4.4.1), see below. Denote by $\mathfrak{A}(H/F)$ the subgroup of $\pi_0(Z(\hat{H})/Z(\hat{G})^\Gamma)$ satisfying this condition. Condition (a) implies that ξ extends to an L-homomorphism, but this extension is not unique. The group H will be called an **endoscopic group**, though it should always be kept in mind that the full triple is needed for the sake of classification.

DEFINITION 5.2. An isomorphism between two endoscopic triples (H, s, ξ) and (H', s', ξ') for G is an F-isomorphism $\alpha : H \to H'$ such that

(i) $\xi \circ \hat{\alpha} : \hat{H}' \to \hat{G}$ and ξ' are $\hat{G}$-conjugate;

(ii) The isomorphism $\mathfrak{A}(H'/F) \xrightarrow{\sim} \mathfrak{A}(H/F)$ determined by (i) takes s' to s.

In particular, the image $H^{ad}(F)$ of $H(F)$ in $Aut(H)$ is contained in the group $Aut((H, s, \xi))$ of automorphisms of the endoscopic triple. We let

$$\Lambda_H = Aut((H, s, \xi))/H^{ad}(F). \tag{5.3}$$

This is obviously a finite group, and we let λ_H denote its order.

DEFINITION 5.4. The endoscopic triple (H, s, ξ) is **elliptic** if
$$\xi(Z(\hat{H})^\Gamma)\, Z(\hat{G})/Z(\hat{G})$$
is a finite group.

We note that Definitions 5.1, 5.2, and 5.4 all make sense when F is a local field as well as a global field; we just have to replace "locally trivial" by "trivial" in 5.1(b).

5.5. Example: elliptic endoscopic triples for unitary groups. Let $G = U(V)$, as in the previous examples. The following proposition is proved in Rogawski's book [R]:

PROPOSITION 5.5.1 [R, 4.6.1]. Let (H, s, ξ) be an elliptic endoscopic triple for G. Then H is isomorphic to $U(a) \times U(b)$ where a and b are non-negative integers such that $a + b = n$. The triple is determined by $\{a, b\}$ up to isomorphism. Furthermore, Λ_H is a group of order 2 if $a = b$ and is trivial otherwise.

Sketch of proof. Since $\hat{G}^{der}$ is simply connected, ξ identifies $\hat{H}$ with the centralizer of the semisimple element $s \in \hat{G}$. Thus $\hat{H} = \prod_{i=1}^{t} GL(n_i)$ for some partition $n = \sum n_i$. A simple reduction using property (b) of (5.1) shows we can assume $s \in Z(\hat{H})^{\Gamma}$, and that the action of Γ on H factors through $Gal(\mathcal{K}/E)$. Now Definition 5.4 implies that the identity component $Z(\hat{H})^{\Gamma,0}$ is contained in $Z(\hat{G})^{\Gamma} = \{\pm 1\}$, hence that $Z(\hat{H})^{\Gamma}$ is itself a finite group. This in turn implies that the non-trivial element $c \in Gal(\mathcal{K}/E)$ fixes the individual blocks $GL(n_i) \subset \hat{H}$ and acts non-trivially on the center $GL(1) \subset GL(n_i)$ for each i. In other words, $c(z_i) = z_i^{-1}$ for $z_i \in Z(GL(n_i))$. Thus $s = (z_1, \ldots, z_t)$ with each $z_i = \pm 1$. Since $\hat{H}$ is the centralizer of s, it follows that $t \leq 2$. Similar arguments prove the remaining assertions.

Application of the stable trace formula to unitary groups is complicated by the fact that the isomorphism between H, whose L-group is canonically the connected centralizer of s in $\hat{G}$, and a fixed product $U(a) \times U(b)$ of unitary groups, depends on an additional arbitrary choice. In the following discussion we let H denote the quasi-split E-group $U(a)^* \times U(b)^*$ (here the superscript * is standard notation for the quasi-split inner form, unique up to isomorphism). The transfer from H to G of parameters, and of automorphic representations, depends on the choice of an L-homomorphism

$$\xi : {}^{L}H \to {}^{L}G \tag{5.5.2}$$

which we normalize following Rogawski, p. 68 of [LR]. Let $\eta_{\mathcal{K}/E}$ denote the quadratic character of $\mathbf{A}_E^{\times}$ corresponding to the quadratic extension $\mathcal{K}$. Fix Hecke characters μ_a and μ_b of $\mathcal{K}$ which extend the characters $\eta_{\mathcal{K}/E}^{a}$ and $\eta_{\mathcal{K}/E}^{b}$ of the idèles of E. On $\hat{H} = {}^{L}H^0 = GL(a) \times GL(b)$, we have

$$\xi(h_a, h_b) = \begin{pmatrix} h_a & 0 \\ 0 & h_b \end{pmatrix} \in \hat{G} \subset {}^{L}G, \quad h_a \in GL(a), h_b \in GL(b) \tag{5.5.3}$$

The Hecke character μ_i can be viewed as a character of $W_{\mathcal{K}}$ by class field theory, and for $w \in W_{\mathcal{K}}$ we let

$$\xi(w) = \begin{pmatrix} \mu_b(w) I_a & 0 \\ 0 & \mu_a(w) I_b \end{pmatrix} \times w \in \hat{G} \ltimes W_{\mathcal{K}} \tag{5.5.4}$$

Finally, if $w_\sigma \in W_E$ is a representative of the non-trivial coset of $W_{\mathcal{K}}$, we set

$$\xi(w_\sigma) = \begin{pmatrix} \Phi_a & 0 \\ 0 & \Phi_b \end{pmatrix} \cdot \Phi_n^{-1} \times w_\sigma. \tag{5.5.5}$$

Here Φ_m, $m = a, b, n$, is defined as in (4.1).

One verifies that the formulas (5.5.3-5.5.5) define an L-homomorphism ξ, which we can denote ξ_{μ_a,μ_b} to emphasize the choices. The insertion of these characters is required in order to make the homomorphisms well-defined with respect to the Φ_m, and we have seen in (4.1) that these latter are required by the need to preserve the standard splittings.

Remark 5.6. The classification of elliptic endoscopic triples for other classical groups is worked out in various places, for example in Waldspurger's Astérisque volume [W2].

Let (H, s, ξ) be an elliptic endoscopic triple for G. and let $\gamma_H \in H$ be an elliptic semi-simple element, T_H a maximal elliptic torus of the connected centralizer $I_H = I^0_{\gamma_H}$ of γ_H in H. There is a canonical G-conjugacy class of embeddings $j : T_H \to G$, defined over $\bar{F}$. Choose such a j and let $\gamma = j(\gamma_H)$. The stable conjugacy class of γ is independent of the choice of j and depends only on the stable conjugacy class of γ_H. We say $\gamma_0 \in G(F)$ is associated to γ_H, or "comes from" γ_H, or (more commonly) γ_H is **an image** of γ_0, if γ_0 is in the stable conjugacy class of $j(\gamma_H)$. If G is quasi-split, then a theorem of Steinberg and Kottwitz [K1] guarantees there is always such a γ_0. In general, there are both local and global obstructions. An example of local obstruction: at a real place v of F, γ_H has eigenvalues of absolute value > 1, but G_v is compact. The global obstruction belongs to a certain H^2.

The image $j(T_H) \subset G_{\bar{F}}$ is a maximal torus $T \subset G$. Thus roots of T in G define characters of T_H. We say γ_H is **(G, H)-regular**. if $\alpha(\gamma_H) \neq 1$ for any root of T in G that does not restrict to a root of H. This condition is independent of the choice of T_H and j.

We write $I_0 = G_{\gamma_0}$ if γ_0 comes from γ_H. If γ_H is (G, H)-regular, then j extends to an isomorphism $I_H \xrightarrow{\sim} I$ of centralizers, unique up to conjugacy by T. The element $s \in Z(\hat{H})$ can be viewed as an element of $Z(\hat{I}_H)$ and thus as an element of $Z(\hat{I}_0)$. Condition 5.1(b) then implies that the corresponding element of $\pi_0((Z(\hat{I})/Z(\hat{G}))^\Gamma)$ belongs to $|\mathfrak{K}(I_0)|$. We denote this element κ.

To the quadruple (H, s, ξ, γ_H), with γ_H (G, H)-regular, we have thus associated a pair (γ_0, κ), with γ_0 a stable conjugacy class in $G(F)$ and $\kappa \in \mathfrak{K}(I_0)$, provided G is quasi-split. If not, the pair (γ_0, κ) may or may not exist. We let

$$\pi : \{ \text{ quadruples } (H, s, \xi, \gamma_H)\} \to \{ \text{ pairs } (\gamma_0, \kappa)\}$$

be the partially defined map. The double sum over pairs (γ_0, κ) is transformed into a double sum over stable conjugacy classes in elliptic endoscopic groups by means of the following proposition:

LEMMA 5.7. The (partially defined) map π is surjective, and the fiber above any pair (γ_0, κ) in the image is either empty or contains λ_H quadruples.

Following Langlands, define

$$i(G, H) = \tau(G)\tau(H)^{-1}\lambda_H.$$

Let $\mathfrak{E}$ denote a set of representatives of equivalence classes of of elliptic endoscopic triples. For any $(H, s, \xi) \in \mathfrak{E}$, let $E_{H,0}$ denote a set of representatives of (G, H)-regular stable conjugacy classes in $H(F)$. The above constructions allow us to rewrite the double sum over $E_0 \times \mathfrak{K}(I_0)$ in (4.8) as a double sum over $\mathfrak{E} \times E_{H,0}$:

$$T_e(\phi) = \sum_{(H,s,\xi)\in\mathfrak{E}} i(G,H)\tau(H) \sum_{\gamma_H \in E_{H,0}} O^{\kappa}_{\gamma_0}(\phi), \tag{5.8}$$

where $(\gamma_0, \kappa) = \pi((H,s,\xi),\gamma_H)$. This is obviously useless, because the term $O^{\kappa}_{\gamma_0}(\phi)$ is still defined in terms of G rather than H. The next section explains how to rewrite $O^{\kappa}_{\gamma_0}(\phi)$ as a stable orbital integral on H.

6. Transfer and the fundamental lemma

The expression of $SO^{\kappa}_{\gamma_0}(\phi)$ in terms of stable orbital integrals on endoscopic groups is given explicitly by the following conjecture when γ_0 is a regular elliptic element.

CONJECTURE 6.1 (LANGLANDS-SHELSTAD). Let F be a local field. Let (H,s,ξ) be an elliptic endoscopic triple for G, Let $\phi \in C^{\infty}_c(G(F))$. Then there is a function $\phi^H \in C^{\infty}_c(H(F))$, well-defined as a functional on stably invariant distributions on H, such that, for γ_H a stable semi-simple conjugacy class in H, $(\gamma_0,\kappa) = \pi((H,s,\xi),\gamma_H)$, with γ_0 regular in G, we have

$$SO_{\gamma_H}(\phi^H) = \sum_{\gamma \overset{st}{\sim} \gamma_0} \Delta(\gamma_H,\gamma) O_{\gamma}(\phi) \tag{6.1.1}$$

Here $\Delta(\gamma_H,\gamma)$ is the **transfer factor** to be defined in §7.

This conjecture—which is now a theorem, as we will see below—has been stated in various versions in the literature, and one of the objectives in Chapter II of this book is to explain the relations between these versions. The analogous result for real groups is due to Shelstad, and is explained in Chapter II.A. We immediately derive the expected conclusion. Let

$$ST^*_e(\phi^H) = \tau(H) \sum_{\gamma_H \in E_{H,0}} SO_{\gamma_H}(\phi^H)$$

where the sum runs over (G,H)-regular elliptic γ_H.

THEOREM 6.2 (KOTTWITZ). Assume Conjecture 6.1 (in Kottwitz' version, see Remark 6.3 (b)). Then (5.8) can be written

$$T_e(\phi) = \sum_{(H,s,\xi)\in\mathfrak{E}} i(G,H) ST^*_e(\phi^H)$$

for a certain function $\phi^H \in C^{\infty}_c(H(\mathbf{A}_F))$, defined as the tensor product of local transfers of ϕ.

Remark 6.3. (a) Conjecture 6.1 is purely local, whereas Theorem 6.2 is global. The global test function ϕ can be considered factorizable as $\otimes_v \phi_v$, and then for almost all v ϕ_v is necessarily the characteristic function of a hyperspecial maximal compact subgroup. In order for the right-hand side of the expression in (6.2) to make sense, each ϕ^H has to have the same property. Thus Conjecture 6.1 needs to be supplemented by the Fundamental Lemma (Conjecture 6.4), to be stated below.

(b) Kottwitz' version (Conjecture 5.5 of [K3]) is an extension of Conjecture 6.1 to (G, H)-regular γ_H. Conjecture 6.1 as stated only suffices for an identity between the regular terms on both sides. Kottwitz' version is derived by Langlands and Shelstad from Conjecture 6.1 [LS2,Lemma 2.4.A].

(c) Theorem 6.2 is derived from the transfer conjectures by identifying the right-hand side of (6.1.1) with the κ-orbital integral of ϕ over γ. This depends on formal properties of the transfer factors, which will be explained in §7, at which point it will be explained in greater detail how one passes from (6.1.1) to (6.2).

The main theorem of [W1] is that the Langlands-Shelstad transfer conjecture is a consequence of the following special case, now a theorem thanks primarily to Ngô:

Theorem 6.4 (Fundamental Lemma). *Let G be an unramified group over the local field F, and let (H, s, ξ) be as in conjecture 6.1, with H also unramified. Let $K \subset G(F)$ (resp. $K_H \subset H(F)$ be a hyperspecial maximal compact subgroup of G (resp. H), $\phi = 1_K$ the characteristic function of K. Then there is a constant c, depending only on the measures of K and K_H, such that (6.1.1) holds with $\phi^H = c \cdot 1_{K_H}$ with 1_{K_H} the characteristic function of K_H.*

As mentioned in (6.3)(a), something of this sort is necessary even in order to formulate the stabilization globally. Waldspurger's method actually does not require the full strength of the Fundamental Lemma; in order to deduce the transfer conjecture, it suffices to know Theorem 6.4 locally in almost all residue characteristics. (See Chapter [C.II.B.1] for an account of this and related work.) Moreover, Hales showed in [H2] that it suffices to know the fundamental lemma for unit elements in the Hecke algebra, for sufficiently large residue characteristics; cf. §9 for a complete statement. Thanks to Laumon-Ngô [LN] and Waldspurger [W3], this has now been established for unitary groups. The results of Laumon and Ngô are presented in Chapters [DT.II.C.1] and [N.II.C.2], following the more recent work of Ngô [N1, N2] which proves the fundamental lemma for general reductive groups over local fields in positive characteristic. The results of [W3], that show that the Fundamental Lemma depends only on the residue field, are treated in Chapter [L.II.D]. An alternative approach to the independence of the characteristic is presented in Chapter [CHL.II.E].

More fundamentally, perhaps, Conjecture 6.4 is necessary in order to characterize the image of endoscopic transfer at (almost all) unramified places. This topic is treated in §9 and in more detail in Chapter [M.III.C].

7. Formal properties of transfer factors

Transfer factors are combinatorially extremely involved, but the most important terms are roots of unity and the main complication is in the correct definition of signs. We begin by assuming G to be quasi-split; the endoscopic groups H are by definition quasi-split. Transfer factors can be derived for general G from those for quasi-split G. In the present section F is a non-archimedean local field and (H, s, ξ) is an elliptic endoscopic triple for G.

Let $\mathfrak{h}$ and $\mathfrak{g}$ denote the F-Lie algebras of H and G, respectively. If G is quasi-split, it admits a Γ-stable splitting (épinglage) $(T, B, \{X_\alpha, \alpha \in \Delta\})$. The data are as in §4; the Borel subgroup B and its maximal torus T are defined over F, and the root data X_α are Γ-stable in the sense that $\sigma(X_\alpha) = X_{\sigma(\alpha)}$ if $\sigma \in \Gamma$. Let $\mathfrak{b}$ be the Lie algebra of B. Let $X_+ = \sum_{\alpha\in\Delta} X_\alpha$. Then $X_+ \in \mathfrak{g}(F)$. For each $\alpha \in \Delta$ one can define $sl(2)$ triples $(X_\alpha, H_\alpha, X_{-\alpha})$ so that

$$[H_\alpha, X_{\pm\alpha}] = \pm 2X_{\pm\alpha}, [X_\alpha, X_{-\alpha}] = H_\alpha.$$

Then let $X_- = \sum_{\alpha\in\Delta} X_{-\alpha}$; X_- is also in $\mathfrak{g}(F)$.

We begin by observing that stable conjugacy can be defined for semisimple elements of $\mathfrak{g}$ just as for G, and are classified in the same way in terms of the cohomology of centralizers in G. Moreover, if $X, X' \in \mathfrak{g}$ are stably conjugate, then one can define an invariant

$$inv(X, X') \in \ker[H^1(F, I) \to H^1(F, G)]$$

if $I \subset G$ is the stabilizer of X.

Transfer factors for Lie algebras:

$$\Delta_{G,H}(\bullet, \bullet) : \mathfrak{h}^{G-reg}(F) \times \mathfrak{g}^{reg}(F) \to \mathbb{C}$$

were defined by Waldspurger. These factors are simpler than the corresponding factors for groups, and satisfy the relation

$$\Delta_{G,H}(X_H, X) = \Delta'(exp(X_H), exp(X)) \tag{7.1}$$

when X_H and X are sufficiently close to 0 in the corresponding Lie algebras. Here *exp* is the exponential function in the Lie algebra and Δ' is the Langlands-Shelstad transfer factor with the volume term removed (this will be explained below).

THEOREM 7.2 (KOTTWITZ, [K6]). *Assume G is quasi-split.*

(a) *(Kostant) Any stable semisimple conjugacy class in $\mathfrak{g}$ contains an element in $X_- + \mathfrak{b}$; i.e. of the form $X_- + Y$ where $Y \in \mathfrak{b}$.*
(b) *Let $X_H \in H$ be a semisimple element that transfers to a regular stable conjugacy class $[X_G]$ in $\mathfrak{g}$. Let $X_G^{Kostant} \in \mathfrak{g}(F) \cap (X_- + \mathfrak{b})$ be stably conjugate to X_G. Then $\Delta_{G,H}(X_H, X_G^{Kostant}) = 1$.*

(c) More generally

$$\Delta_{G,H}(X_H, X_G) =< inv(X_G, X_G^{Kostant}), s >$$

(See Ngô's notes on transfer factors, §5.)

Remark 7.3. Let $\mathfrak{g}$ be a reductive Lie algebra over $\bar{F}$ with maximal torus $\mathfrak{t}$ and Weyl group W. The scheme $Spec(\bar{F}[\mathfrak{t}]^W)$ is isomorphic to the scheme-theoretic quotient of $\mathfrak{g}$ by the adjoint action of G and thus gives a coarse parametrization of conjugacy classes in $\mathfrak{g}$. The Kostant map defined above on semisimple conjugacy classes extends to a map from $Spec(\bar{F}[\mathfrak{t}]^W)$ to $\mathfrak{g}$ with image contained in the open subscheme of *regular* elements (including regular unipotent elements as well as regular semisimple elements). For example, when $\mathfrak{g} = \mathfrak{gl}_n$, $Spec(\bar{F}[\mathfrak{t}]^W)$ is the affine space of (coefficients of) characteristic polynomials, so the Kostant section gives a way of picking out an element with given characteristic polynomial, once a splitting has been fixed. The Kostant section is central in the geometric approach to endoscopy and should be recalled when reading the chapters on the fundamental lemma.

The general Langlands-Shelstad transfer factor $\Delta(\gamma_H, \gamma)$ is given as a product of five terms, denoted Δ_I, Δ_{II}, $\Delta_1 = \Delta_{III_1}$, $\Delta_2 = \Delta_{III_2}$, and Δ_{IV}. Various of these factors are only defined in terms of additional arbitrary choices, most of which disappear in the product. The term Δ_I also depends on the choice of splitting. When G is not quasi-split, the term Δ_1 is only defined up to an arbitrary constant. The choice of constants at places of a fixed number field is normalized globally by the condition that almost all of the constants can be taken to equal 1 and the product of the local constants is forced to equal 1. This is the "global hypothesis", cf. (7.3.7) below. For this reason the literature sometimes describes the local transfer factors as being defined only up to a constant multiple, the quotient

$$\Delta(\gamma_H, \gamma; \gamma'_H, \gamma') = \Delta(\gamma_H, \gamma)/\Delta(\gamma'_H, \gamma')$$

being independent of all choices. Note that the simplified factor in (7.2) also depends on the choice of splitting.

All terms in the product are roots of unity, with the exception of Δ_{IV}, which is a power of the order of the residue field. Waldspurger's transfer factor for Lie algebras omits the volume factor Δ_{IV}, which reappears elsewhere, and on elements close to the origin Δ_2 is trivial. The main terms are in any case the remaining terms. These are described briefly below. Here it suffices to recall some of the main formal properties satisfied by the transfer factors.

7.3. Formal properties of transfer factors. The field F is assumed to be local until (7.3.7), where it is a global field. The group G is assumed to be a non-archimedean connected reductive group, except in (7.3.4) and (7.3.5) where it can be a real algebraic group.

Property 7.3.1 (change of inner form). Up to now we have been supposing G quasi-split. We drop that hypothesis. This means, in particular, that a given γ_H may not correspond to a stable class in G containing a point in $G(F)$. If it does, the transfer factor $\Delta(\gamma_H, \gamma)$ is defined for any $\gamma \in G(F)$ in the stable class. Let G^* be the quasi-split inner form of G, $\gamma^* \in G^*(F)$ an element of the stable class. If γ_H is sufficiently close to the identity then one can use the Kostant section of the Lie algebra to define a base point; one can also find a base point when the groups are unramified (see (7.3.3)). In general there is no natural choice. On the other hand, G^* is also an endoscopic group of G, and γ is an image in $G(F)$ of γ^*, so there is also a transfer factor $\Delta(\gamma^*, \gamma)$. The relation is described in [LS1]. Let

$$\Delta_{G|G^*}(\gamma_H, \gamma, \gamma^*) = \Delta(\gamma_H, \gamma)/\Delta(\gamma_H, \gamma^*)$$

and let $\gamma'_H \in H(F), \gamma' \in G(F), \gamma'^{,*} \in G^*(F)$ be another triple with the same property. Then

$$\Delta_{G|G^*}(\gamma_H, \gamma, \gamma^*)/\Delta_{G|G^*}(\gamma'_H, \gamma', \gamma'^{,*}) = \lambda_H(\gamma, \gamma^*; \gamma', \gamma'^{,*})$$

where the term on the right hand side is given by a cohomological invariant whose expression can be found in [LS1, 4.2].

Property 7.3.2 (cocycle property). Suppose γ and γ' are stably conjugate in $G(F)$ with image γ_H in $H(F)$ The relation (7.2)(c) generalizes as follows:

$$\Delta(\gamma_H, \gamma) =< inv(\gamma, \gamma'), s > \Delta(\gamma_H, \gamma').$$

There seem to be different sign conventions in the literature for the definition of the invariant, or maybe I have just got the order wrong. This property allows us to rewrite (6.1.1) as

$$SO_{\gamma_H}(\phi^H) = \sum_{\gamma \overset{st}{\sim} \gamma_0} < inv(\gamma, \gamma'), s > \Delta(\gamma_H, \gamma_0) O_\gamma(\phi) \tag{2}$$

$$= c \sum < inv(\gamma, \gamma'), \kappa > O_\gamma(\phi) \tag{3}$$

with $c = \Delta(\gamma_H, \gamma_0)$. This is the same as $cO^\kappa_{\gamma_0}(\phi)$ (when γ is regular its centralizer is a torus, so the Kottwitz signs are trivial) and the product over the different places of the constants c turns out to equal 1.

Property 7.3.3 (transfer factors for unramified groups). In [H], Hales provides explicit formulas for transfer factors when G and H are both unramified. This is possible because the a-data and χ-data of Langlands-Shelstad can be chosen more or less canonically in this case. The formulas are given by a series of reductions, including (7.3.6) below, so in fact there is no simple way to write them down. The reader is referred to [H] for more details.

Property 7.3.4 (transfer factors for real groups). Conjugacy classes can be described more explicitly in real groups than in p-adic groups, and the transfer factors are also more explicit. Details can be found in §5 of Renard's chapter [R.II.A].

Property 7.3.5 (descent to centralizers of semisimple elements). This is the topic of the article [LS2]. Let F be a local field, $\epsilon \in G(F)$ a semisimple element, $\epsilon_H \in H(F)$ an image of ϵ. Let G_ϵ, H_{ϵ_H} denote their respective centralizers. We assume ϵ_H is chosen so that H_{ϵ_H} is quasi-split, so that it is an endoscopic group for G_ϵ. Let $\gamma, \gamma' \in G_\epsilon(F)$ be semisimple elements that are regular in $G(F)$, and let γ_H, γ'_H be their images in $H(F)$, which can be chosen in $H_{\epsilon_H}(F)$. Thus there are two transfer factors: $\Delta(\gamma_H, \gamma; \gamma'_H, \gamma')$, taken relative to G and H, and $\Delta_\epsilon(\gamma_H, \gamma; \gamma'_H, \gamma')$, taken relative to G_ϵ and H_{ϵ_H}. Let $\Theta = \Delta/\Delta_\epsilon$.

THEOREM [LS2,1.6.A]. As γ_H and γ'_H tend to ϵ_H and γ, γ' tend to ϵ,

$$\lim \Theta(\gamma_H, \gamma; \gamma'_H, \gamma') = 1.$$

This is the property that allows the reduction of the transfer conjecture to the analogous conjecture for Lie algebras.

Property 7.3.6 (descent to Levi subgroups). Here G is quasi-split and unramified and M is a Levi subgroup of a parabolic subgroup $P \subset G$. Then M is also quasi-split and unramified. There is an endoscopic group H_M associated to M.

LEMMA [H,9.2]. Let $\gamma \in M(F)$, $\gamma_H \in H_M(F)$. Then

$$\Delta^G(\gamma_H, \gamma) = \Delta^M(\gamma_H, \gamma) \cdot \prod_\alpha |(\alpha(\gamma) - 1)|^{\frac{1}{2}}$$

where α ranges over roots of G outside M and H.

Hales verifies that all factors of both sides agree except for the volume factor Δ_{IV}, where the difference is obviously compensated by the product indicated.

Property 7.3.7 (global reciprocity). Suppose F is a number field, $\{v\}$ the set of places of F. Let $\gamma_H \in H(F)$, up to stable conjugacy, $\gamma = (\gamma_v) \in G(\mathbf{A}_F)$, and suppose γ_H, viewed as an element of $H(F_v)$, is an image of γ_v for all v. The normalizations of the local transfer factors, denoted $\Delta^v(\bullet, \bullet)$, can be chosen so that $\Delta^v(\gamma_H, \gamma_v) = 1$ for almost all v and

$$\prod_v \Delta^v(\gamma_H, \gamma_v) = \prod_v < inv(\gamma_0, \gamma_v), \kappa >$$

if $(\gamma_0, \kappa) = \pi(\gamma_H, s, \xi) \in G(F) \times \mathfrak{K}_0(I_{\gamma_0})$.

Property 7.3.8 (extension to (G,H)-regular classes). As mentioned in Remark (6.3)(b), Kottwitz's stabilization of the elliptic part of the trace formula requires matching of orbital integrals for all (G,H)-regular classes, and not only for G-regular classes. This in turn requires extending the definition of transfer factors to pairs (γ_H, γ) where γ_H is (G,H)-regular. This is carried out in [LS2,2.4], using the descent Property 7.3.5. Formula (6.1.1) is replaced by

$$SO_{\gamma_H}(\phi^H) = \sum_{\gamma \overset{st}{\sim} \gamma_0} e(I_\gamma)\Delta(\gamma_H, \gamma)O_\gamma(\phi)$$

([LS2], (2.4.1)) and the cocycle property (7.3.2) remains valid in this more general setting, as required for comparison with (4.9).

8. Explicit expressions for transfer factors for unitary Lie algebras and groups

The regular semisimple conjugacy classes in classical Lie algebras over local fields can be described in an elementary way in terms of elements of field extensions and linear algebra. In §10 of [W2], Waldspurger provides explicit formulas for transfer factors for unitary, orthogonal, and symplectic Lie algebras. Here we present his formulas in the unitary case.

Let $G = U(V, q_V)$, where q_V is the hermitian form on the n-dimensional vector space V over $\mathcal{K}$. In Waldspurger's normalization q_V is anti-linear in the first variable and linear in the second. The parametrization of conjugacy classes in $\mathfrak{g} = Lie(G)$ parallels the discussion of (4.11). Let I be a finite set. For each $i \in I$ let E_i be an extension of E of degree n_i, $\mathcal{K}_i = E_i \otimes_E \mathcal{K}$; choose $a_i, b_i \in \mathcal{K}_i^\times$. Let c denote the non-trivial element of $Gal(\mathcal{K}/E)$, $c_i = 1 \otimes c$ acting on $\mathcal{K}_i$. In order to follow Waldspurger we temporarily allow $\mathcal{K}_i$ to be a direct sum of two fields, in which case c_i exchanges the two factors. We assume

1. For all i, $\mathcal{K}_i = \mathcal{K}[a_i]$;
2. For all $i \neq j \in I$, there is no $\mathcal{K}$-linear morphism $\mathcal{K}_i \to \mathcal{K}_j$ taking a_i to a_j (i.e., the minimal polynomials of a_i and a_j have no common factors);
3. For all i, $c_i(a_i) = -a_i$, $c_i(b_i) = b_i$;
4. $\sum_i [\mathcal{K}_i : \mathcal{K}] = \sum_i n_i = n$

Set $W = \oplus_i \mathcal{K}_i$ and define a hermitian form q_W on the $\mathcal{K}$-vector space W by

$$q_W(\sum_i w_i, \sum_i w_i') = \sum_i n_i^{-1} trace_{\mathcal{K}_i/\mathcal{K}}(c_i(w_i)w_i' b_i). \tag{8.1}$$

Define an element $X_{(a_i),W} \in End(W)$ by

$$X_{(a_i),W}(\sum_i w_i) = \sum a_i w_i.$$

Then $X_{(a_i),W}$ is a regular semisimple element of the Lie algebra of $U(W, q_W)$. Now suppose there is an isomorphism $(W, q_W) \xrightarrow{\sim} (V, q_V)$. Fixing such an isomorphism, $X_{(a_i),W}$ defines an element $X_{(a_i)} \in \mathfrak{g}^{reg}$ whose conjugacy class is denoted $\mathcal{O}(I, (a_i), (b_i))$. Two such conjugacy classes $\mathcal{O}(I, (a_i), (b_i))$ and $\mathcal{O}(I', (a_i'), (b_i'))$ coincide if and only if there are a bijection $I \xrightarrow{\sim} I'$ denoted $i \mapsto i'$ and isomorphisms $\sigma_i : \mathcal{K}_{i'} \xrightarrow{\sim} \mathcal{K}_i$ taking a_i' to a_i and such that

$$\varepsilon_{\mathcal{K}_i/E_i}(\sigma_i(b_i')) = \varepsilon_{\mathcal{K}_i/E_i}(b_i) \text{ for all } i \tag{8.2}$$

where ε_* is the quadratic character associated to the extension (so there is no condition if $\mathcal{K}_i$ is not a field). On the other hand, $\mathcal{O}(I,(a_i),(b_i))$ and $\mathcal{O}(I',(a'_i),(b'_i))$ are in the same stable conjugacy class if $I \xrightarrow{\sim} I'$ and the σ_i exist as indicated, but (8.1.2) is dropped. The isomorphism $(W,q_W)\xrightarrow{\sim}(V,q_V)$ imposes one constraint on the signs $\varepsilon(b_i)$. Thus the set of conjugacy classes in the stable conjugacy class is in bijection with $(\mathbb{Z}/2\mathbb{Z})^{|I^*|-1}$ where $I^* \subset I$ is the subset for which $\mathcal{K}_i$ is a field, the term -1 corresponding to the constraint given by the isometry class of (V,q_V).

As in (4.11), $I^* = I$ if and only if $\mathcal{O}(I,(a_i),(b_i))$ is elliptic. An elliptic endoscopic datum (H,s,ξ) is defined by a partition $I = I_1 \coprod I_2$. Let $I^* = I_1^* \coprod I_2^*$ be the corresponding partition. Let $\daleth \in \mathcal{K}$ be an element such that $c(\daleth) = (-1)^{n+1}\daleth$, and assume the matrix of q_V is $2\daleth\Phi_n$ with Φ_n as in (4.1).

Let P denote the characteristic polynomial of $X = X_{(a_i)}$, viewed as an element of $End_{\mathcal{K}}(V)$, P' its derivative. For $i \in I$, set

$$C_i = \daleth[\mathcal{K}_i : E]b_i^{-1}P'(a_i). \tag{8.3}$$

One checks that $C_i \in E_i$ for all i. Now let $Y \in \mathfrak{h}^{reg}$ and suppose the stable classes of X and Y correspond. Let

$$D_G(X) = |\prod_{\alpha} d\alpha(X)|_E^{\frac{1}{2}}$$

where α runs over the roots of G and define $D_H(Y)$ likewise. The following proposition is a special case of [W2,X.8]:

PROPOSITION 8.4. Under these hypotheses

$$\Delta_{G,H}(Y,X) = D_G(X)D_H(Y)^{-1}\prod_{i\in I_2^*}\varepsilon_{\mathcal{K}_i/E_i}(C_i).$$

The transfer factor in the above proposition evidently includes the volume term Δ_{IV}. However, the term $\Delta_{III,2}$ is trivial for Lie algebras and Waldspurger indicates that he has made implicit choices that trivialize the term Δ_{II} as well as the significant term $\Delta_{III,1}$. So only Δ_I (a sign) and Δ_{IV} (a power of the residue characteristic) are present.

8.5. On global reciprocity. Labesse's Chapter [L.III.D] includes an explicit determination of the local and global transfer factors, up to sign, in the case of interest in the final part of the book. More precisely, by restricting to the values of the transfer factor on the simplest elliptic torus common to a unitary group and a specific endoscopic group, he determines the normalization of the local transfer factors compatible with our choice of endoscopic data (in particular, with the embedding ξ) and with the global property. This provides a link between the determination of local transfer at unramified places in [M.III.C], which correspond to Waldspurger's calculations recalled above, and at archimedean places in [C.III.A].

9. Endoscopic transfer of representations

9.1. Formalism for transfer of stable distributions. We start from the stable trace formula in the form of Theorem 6.2:

$$T_e(\phi) = \sum_{(H,s,\xi)\in\mathfrak{E}} i(G,H)ST_e^*(\phi^H)$$

and assume the subscript $_e$ and asterisks can be removed:

$$T(\phi) = \sum_{(H,s,\xi)\in\mathfrak{E}} i(G,H)ST_H(\phi^H) \tag{9.1.1}$$

The subscript H has been added for clarity. If G is anisotropic, as we assume in the applications, there is no difference between $T_e(\phi)$ and $T(\phi)$. In the applications we will also choose the local test functions ϕ_∞ so that a similar equality holds on the right-hand side; see Labesse's chapter [L.IV.A], where this choice is made and justified.

For each triple (H, s, ξ), the map

$$\phi \mapsto ST_H(\phi^H)$$

defines an invariant distribution on $C_c^\infty(G(\mathbf{A}_E))$. Although ϕ^H is not uniquely determined, the distribution is well defined by the properties of the transfer map. More generally, let t be a *stable* (stably invariant) distribution on H. Then the map

$$\phi \mapsto \xi_*(t)(\phi) := t(\phi^H) \tag{9.1.2}$$

is a well-defined invariant distribution on $C_c^\infty(G(\mathbf{A}_E))$. The notation ξ_* may not be optimal, but I have not seen alternative notation for this map of distributions and the present choice, in emphasizing the choice of ξ_*, is useful for our applications to unitary groups, where the exact normalization of ξ, as in (5.5), plays an important role in the explicit formulas.

Formula (9.1.1) then becomes an equality of distributions.

$$T(\phi) = \sum_{(H,s,\xi)\in\mathfrak{E}} i(G,H)\xi_*(ST_H)(\phi) \tag{9.1.3}$$

Endoscopic transfer then comes down to the explicit calculation of the distributions $\xi_*(ST_H)(\phi)$. Since G is anisotropic, we can write

$$T(\phi) = T_d(\phi) = \sum_{\pi\subset\mathcal{A}_d(G)} m_\pi Tr\ \pi(\phi) \tag{9.1.4}$$

as in §1. Here $\mathcal{A}_d(G)$ is the sum of automorphic representations occurring in the discrete spectrum L_2^d, defined as at the end of §1. Since G is anisotropic, $L_2^d = L_2$, but we also assume we are in the situation of the *simple trace formula*, as in [L.IV.A], so that the terms on the right-hand side of (9.1.1) can also be written

$$ST_H(\phi^H) = \sum_{[\tau]\subset\mathcal{A}_d(H)} m_{[\tau]}Tr[\tau](\phi^H) \tag{9.1.5}$$

Here $[\tau]$ is ad hoc notation for a stable character of H, a finite linear combination $[\tau] = \sum a_i\tau_i$ of characters of irreducible representations which is stably invariant, and then

$$Tr[\tau] = \sum a_i Tr\ \tau_i.$$

The existence of such a decomposition is still conjectural, but it is known in the applications to be considered in [CHL.IV.B], and we can admit it as a hypothesis. More refined decompositions were considered in [LL] and a general expression was conjectured in [K2], but this is not what we choose to emphasize here. The upshot is that

$$(9.1.6)\qquad \sum_{\pi\subset\mathcal{A}_d(G)} m_\pi Tr\ \pi(\phi) = \sum_{(H,s,\xi)\in\mathfrak{E}} i(G,H) \sum_{[\tau]\subset\mathcal{A}_d(H)} m_{[\tau]}\xi_*(Tr\ [\tau])(\phi).$$

By linear independence of characters (1.3(c)), in order to determine the m_π it suffices to know the $m_{[\tau]}$ and to calculate the maps ξ_* explicitly. The stable character $[\tau]$ is a (restricted) tensor product $\otimes'_v[\tau]_v$ over places v of E, so the latter reduces to the local problem of calculating the $\xi_{*,v}[\tau]_v$ for all v. For archimedean v this problem was posed and solved by Shelstad for tempered representations and, more generally, by Adams, Barbasch, and Vogan; these results are reviewed in Renard's chapter [R.II.A] and Adams' chapter [A.III.B]. When $[\tau]_v$ is an unramified (spherical) representation, there is a natural conjectural expression for $\xi_{*,v}[\tau]_v$ in terms of Satake parameters. In [H2] Hales showed how to reduce this conjecture to the special case treated by the fundamental lemma; thus this is now also a theorem thanks to [LN, N1, N2]. In §9.2 we describe these formulas when G is a unitary group, with the elliptic endoscopic groups parametrized as in (5.5) in terms of auxiliary choices of Hecke characters.

Remark 9.1.7. The terms on the right hand side of (9.1.6) are not linearly independent. Cancellation occurs between the terms occurring in the expansions of different (H, s, ξ) and these cancellations often have interesting arithmetic applications. Moreover, it should be noted that the coefficients $i(G, H)$ are in general fractions and not integers. This forces some of the distributions $m_{[\tau]}\xi_*(Tr\ [\tau])$ to be *fractional* linear combinations of irreducible characters, frequently with negative coefficients. Of course, the final coefficients m_π are necessarily non-negative integers.

9.2. Functoriality and Arthur parameters. Here and below we write $\xi_*([\tau]) = \xi_*(Tr\ [\tau])$ for brevity. The basic stable discrete automorphic distributions on H are supposed to be associated to admissible maps (Arthur parameters)

$$\Phi : \mathcal{L}_E \times SU(2) \to {}^LH$$

Here $\mathcal{L}_E$ is the hypothetical Langlands group and the factor $SU(2)$ measures the failure of the packet $\Pi(\Phi)$ of discrete automorphic representations of H indexed by Φ to be tempered. We only consider Φ trivial on the factor $SU(2)$ and replace the map of the Langlands group by the collection Φ_v of restrictions of Φ, as v varies over places of E, to the local Weil-Deligne group $WD_v = WD(E_v)$. Composition of Φ with the L-homomorphism $\xi : {}^LH \to {}^LG$ defines an Arthur parameter $\tilde{\xi}_*(\Phi)$ for G. If Φ corresponds to the stable character $[\tau]$, then the corresponding packet $\Pi(\tilde{\xi}_*(\Phi))$ of admissible irreducible representations of $G(\mathbf{A})$ is conjecturally related to the transfer $\xi_*([\tau])$, as discussed above. However, $\Pi(\tilde{\xi}_*(\Phi))$ is no longer a stable character, and the characters of the various members of the collection $\Pi(\tilde{\xi}_*(\Phi))$ are expected to occur in $\xi_*([\tau])$ with multiplicities determined by an explicit formula depending on the interaction of the datum s with the parameter $\tilde{\xi}_*(\Phi)$; this is the content of Arthur's multiplicity conjectures. Since these conjectures can only be stated in terms of local L-packets, since the complete analysis of local L-packets for unitary groups is not yet available as of this writing, and since in any case the structure of these L-packets, constructed by Moeglin, is too intricate for the purposes of this introduction, I will limit my explicit description of the endoscopic transfer to the simplest possible cases.

9.3. Explicit formulas for endoscopic transfer. We work in the setting of unitary groups, with the elliptic endoscopic groups described in (5.5). Let $n = a + b$ be a partition of n, with $a \geq b$, and let $H_{a,b} = U(a)^* \times U(b)^*$ (quasi-split inner forms); thus $H_{n,0} = G^*$ is the quasi-split inner form of G. Choose characters μ_a, μ_b as in (5.5) and define the L-homomorphism

$$\xi = \xi_{\mu_a,\mu_b} : {}^LH_{a,b} \to {}^LG$$

by the formulas (5.5.3-5.5.5). Let v be a place of E and consider a stable character $[\tau]_v$ of $H_{a,b}(E_v)$. Our goal is to describe the character $\xi_{\mu_a,\mu_b,v,*}([\tau]_v)$ in a few simple cases. Note that

LEMMA 9.3.1. For all places v, the characters $\mu_{a,v}$ and $\mu_{b,v}$ of $(\mathcal{K}\otimes_E E_v)^\times$ are unitary.

It suffices to show that μ_a and μ_b are unitary characters; but this is clear because they restrict trivially to the kernel in $\mathbf{A}_E^\times$ of $\eta_{\mathcal{K}/F}$, and this kernel is cocompact in the idèles of $\mathcal{K}$.

9.3.2 Split places. If v splits in $\mathcal{K}/E$ then $H_{a,b}(E_v) \simeq GL(a, E_v) \times GL(b, E_v)$ and $G(E_v) \simeq GL(n, E_v)$. Consider the standard maximal parabolic $P_{a,b} \subset G(E_v)$ corresponding to the partition $n = a + b$, and let $M_{a,b} \subset P_{a,b}$ denote its standard Levi subgroup. Then $H_{a,b}(E_v)$ is clearly isomorphic to $M_{a,b}$, so we can identify their corresponding sets of irreducible admissible representations

$$i : \mathcal{A}(H_{a,b}(E_v)) \xrightarrow{\sim} \mathcal{A}(M_{a,b}).$$

The set of such identifications is a homogeneous space under the action of the group $X(M_{a,b})$ of continuous characters of $M_{a,b}$, which can in turn be identified with the group $X(E_v^\times)^2$ of ordered pairs of characters of E_v, via composition with the determinant $M_{a,b} \to E_v^\times \times E_v^\times$. Every irreducible representation $\tau_v = (\tau_{a,v}, \tau_{b,v})$ of $H_{a,b}(E_v)$ is stable and therefore defines a stable character $[\tau_v]$. We assume τ_v is a *unitary* representation (in practice it will in fact be tempered). If we identify $M_{a,b}$ with $GL(a, E_v) \times GL(b, E_v)$, then let

(9.3.2.1) $$i(\xi_{\mu_a,\mu_b})_*([\tau_v]) = (\tau_{a,v} \otimes \mu_b \circ \det, \tau_{b,v} \otimes \mu_a \circ \det)$$

and define

(9.3.2.2) $$\xi_{\mu_a,\mu_b,v,*}([\tau_v]) = I_{P_{a,b}}^{GL(n,E_v)} i(\xi_{\mu_a,\mu_b})_*([\tau_v])$$

where the representation of $M_{a,b}$ in (9.3.2.1) has been inflated to a representation of $P_{a,b}$. In particular $\xi_{\mu_a,\mu_b,v,*}([\tau_v])$ is an irreducible admissible representation of $G(E_v)$, and we have the following identity of standard L-functions

(9.3.2.3) $$L(s, \xi_{\mu_a,\mu_b,v,*}([\tau_v])) = L(s, \tau_{a,v} \otimes \mu_b \circ \det)L(s, \tau_{b,v} \otimes \mu_a \circ \det)$$

and the analogous identity for standard ε-factors.

9.3.3 Unramified representations. Suppose v is inert (and unramified) in $\mathcal{K}/E$, and suppose the unitary group G is unramified at v. Thus $G(E_v)$ as well as $H_{a,b}(E_v)$ are quasi split and have hyperspecial maximal compact subgroups K_v and $K_{a,b}$, respectively. Irreducible admissible representations τ_v of $H_{a,b}(E_v)$ with $K_{a,b}$-fixed vectors are in one-to-one correspondence with characters of the unramified Hecke algebra $\mathcal{H}(H_{a,b}) = C_c^\infty(K_{a,b}\backslash H_{a,b}(E_v)/K_{a,b})$ of $K_{a,b}$-bi-invariant test functions on $H_{a,b}(E_v)$. These latter, in turn, are in one-to-one correspondence with unramified admissible homomorphisms (Satake parameters)

(9.3.3.1) $$\Phi : W_{E_v} \to {}^L H_{a,b}.$$

Here *unramified* means that the composition of Φ with projection on $\hat{H}_{a,b}$ is trivial on the inertia subgroup, hence factors through the quotient $E_v^\times/\mathcal{O}_{E_v}^\times$ of W_{E_v}; *admissible* means that the composition of Φ with the natural map ${}^L H_{a,b} \to W_{E_v}$ is the identity map. Let

(9.3.3.2) $$\lambda_{a,b}(\Phi) : \mathcal{H}(H_{a,b}) \to \mathbb{C}$$

denote the corresponding character of the unramified Hecke algebra. Of course the Satake parameter Φ is also an Arthur parameter, in the sense of (9.2), and the associated spherical representation τ_v is denoted $\Pi(\Phi)$, following the notation of (9.2). This is consistent:

PROPOSITION 9.3.3.3. *Any such spherical $\Pi(\Phi)$ has a stable character, hence defines a stable distribution $[\Pi(\Phi)]$.*

For details, see Minguez' chapter [M.III.C].

The same Satake parametrization is valid for the quasi-split unitary group $G(E_v)$. Define $\tilde{\xi}_{\mu_a,\mu_b,*}(\Phi) : W_{E_v} \to {}^L G$ as in (9.2). This is now a Satake parameter for G, hence defines a spherical representation $\Pi(\xi_{\mu_a,\mu_b,*}(\Phi))$.

Define the unramified Hecke algebra $\mathcal{H}(G)$ with respect to K_v, as above, and let

$$\lambda_n(\tilde{\xi}_{\mu_a,\mu_b,*}(\Phi)) : \mathcal{H}(G) \to \mathbb{C}$$

be the character of $\mathcal{H}(G)$ corresponding to the indicated Satake parameter. The map $\tilde{\xi}_{\mu_a,\mu_b,*}$ on Satake parameters is dual to a homomorphism

$$\tilde{\xi}^*_{\mu_a,\mu_b} : \mathcal{H}(G) \to \mathcal{H}(H_{a,b}). \tag{9.3.3.4}$$

The following theorem is a reformulation in abstract terms of Theorem 4.3 of [M.III.C].

THEOREM 9.3.3.5. We have the equality

$$\xi_{\mu_a,\mu_b,*}([\Pi(\Phi)]) = \Pi(\tilde{\xi}_{\mu_a,\mu_b,*}(\Phi)),$$

where $\xi_{\mu_a,\mu_b,*}$ is the transfer map on distributions defined by (9.1.2).

This theorem is the basic fact about endoscopic transfer, specialized to the case of unitary groups. The analogous theorem is valid for elliptic endoscopic transfer of unramified representations in general.

Sketch of proof. It follows from the definitions that the theorem is equivalent to the following assertion:

$$\forall \phi \in \mathcal{H}(G), \phi^{H_{a,b}} = \tilde{\xi}^*_{\mu_a,\mu_b}(\phi). \tag{9.3.3.6}$$

Allowing for differences of notation, this is what is called in [H2] (and elsewhere) the *fundamental lemma for standard endoscopy*, in the case under consideration. The article [H2] of Hales reduces this equality to the special case in which ϕ is the unit element 1_{K_v} and therefore the right-hand side of (9.3.3.6) is just $1_{K_{a,b}}$. In other words, Hales reduces (9.3.3.6) to Conjecture 6.4; indeed he obtains the equality assuming Conjecture 6.4 has been established for sufficiently large residue characteristic. The theorem thus follows by combining [H2] with the results of [LN] and [W3].

The analogue of the relation (9.3.2.3) holds for the Langlands L-functions attached to the standard representations of ${}^LH_{a,b}$ and ${}^LG$, more or less tautologically.

9.3.4 Global endoscopic transfer. Now suppose that $\mathcal{K}/E$ and the group G are unramified at all finite places and consider the restriction of the transfer maps $\phi \mapsto \phi^{H_{a,b}}$ to ϕ which are biinvariant under an appropriately chosen compact open subgroup $K_f \subset G(\mathbf{A}^f)$ containing K_v for all inert places v. The transfers are then defined on automorphic representations τ with fixed vectors under the corresponding compact open subgroups of $H_{a,b}$; in particular, each such τ is spherical at every inert place v. Under these hypotheses (corresponding to the "simplifying hypotheses" to be considered in [L.IV.A,CHL.IV.B, CHL.IV.C] and Book 2), any stable L-packet of $H_{a,b}$ is of the form

$$[\tau] = [\tau]_\infty \otimes \tau_f$$

where τ_f is a (single) irreducible representation of $H_{a,b}(\mathbf{A}^f)$. Then

$$\xi_{\mu_a,\mu_b,*}([\tau]) = \xi_{\mu_a,\mu_b,\infty,*}([\tau]_\infty) \otimes \bigotimes_{v\nmid\infty}{}' \xi_{\mu_a,\mu_b,v,*}(\tau_v),$$

where the restricted tensor product is taken over finite places. The terms $\xi_{\mu_a,\mu_b,v,*}(\tau_v)$ are explicitly computed by the formulas in (9.3.2) and (9.3.3). The calculation of $\xi_{\mu_a,\mu_b,\infty,*}([\tau]_\infty)$ is the topic of the chapters of Renard and Adams.

In the applications, especially in Part IV, $[\tau]_\infty$ is a discrete series L-packet, and $\xi_{\mu_a,\mu_b,\infty,*}([\tau]_\infty)$ is a sum of discrete series characters of $G(E_\infty)$, weighted by signs determined long ago by Shelstad. More details are provided in [C.III.A, CHL.IV.B].

9.4 Twisted endoscopy and base change for unitary groups. Notation in this section differs from that of the previous section. Here the unitary group is denoted $\mathbf{U}^*$, following [L.IV.A], and the letter G is reserved for the group appearing in the twisted trace formula. From the third paragraph G is assumed to be a general linear group.

In the twisted trace formula, the group G is replaced by a disconnected group $G\rtimes < \theta >$ whose component group is cyclic and generated by an element θ (usually) of prime order, or alternately by the component $\tilde{G} = G \rtimes \theta$ of this group whose image generates the component group. The spectral side of the twisted trace formula roughly counts automorphic representations of the identity component G invariant under the action of the component group; the geometric side takes the twisting into account in a natural way by summing over integrals over *twisted conjugacy classes* (for example the θ-conjugacy classes discussed below). Twisted endoscopy [KS,Lab2] is the theory that leads to an expression analogous to Theorem 6.2, or (9.1.1), for the geometric side of the twisted trace formula. The endoscopic groups of $\tilde{G}$ are defined by a procedure analogous to that discussed above.

A complete treatment of twisted endoscopy would require (at least) another chapter, and is beyond the scope of these books. However, the topic inevitably arises in the global applications in Part IV, especially in [L.IV.A], and an important aspect of the local theory is addressed in [W.II.A.2]. In the case of concern to us here, $G = GL(n)_E$, where E/F is a quadratic extension of number fields, with F totally real and E totally imaginary, $< \theta >= Gal(E/F)$, and the action of the non-trivial element $c \in Gal(E/F)$ on $GL(n)_E$ is given by

(9.4.1) $$\theta(g) = J_n{}^t(\bar{g})^{-1}J_n^{-1},$$

where J_n is the anti-diagonal matrix with alternating -1s and 1s, cf. [L.IV.A]; here $\bar{g}$ is the result of applying c to the coefficients of g. Conjugation of the element $(\delta, \theta) \in \tilde{G}$ by en element $g \in G$ takes the following form:

$$g(\delta, \theta)g^{-1} = (g\delta\theta(g)^{-1}, \theta) \tag{9.4.2}$$

where the first component on the right is the θ-conjugate of δ.

The elliptic part of the geometric side of the twisted trace formula is an expression in θ-conjugacy classes in G, which are identified via (9.4.2) with conjugacy classes in $\tilde{G}(\mathbf{A}) = G(\mathbf{A}) \rtimes \theta$. Let ϕ be a test function on $\tilde{G}(\mathbf{A})$. The twisted trace formula is thus the equality of spectral and geometric expressions for

$$T^{\tilde{G}}(\phi) = tr\, R_{disc}(\phi) = J(\tilde{G}) \sum_{\delta} \tau(I_\delta) O_\delta(\phi) = \sum_{\tilde{\pi}} tr \tilde{\pi}(\phi) \tag{9.4.3}$$

where we are adopting the notation of [L.IV.A, (3.2)]; here δ runs over conjugacy classes in $\tilde{G}(E)$, or θ-conjugacy classes in $G(E)$, and I_δ is what was denoted G_γ above. The $\tilde{\pi}$ are indexed by automorphic representations of G that are isomorphic to their θ-conjugates, but the action of $\tilde{G}$ needs to be made more precise, and that is done in [L.IV.A]. The multiplicities m_π are absent because of strong multiplicity one for $GL(n)$.

On the other hand, it is important to stress that $G = GL(n)$ is not anisotropic, and (9.4.3) is unrealistically simple. The choice of ϕ_∞ made in [L.IV.A], where it is moreover assumed that $[F : \mathbb{Q}] \geq 2$, eliminates the difficult terms of the geometric side—so the sum really runs over elliptic δ—as well as most of the continuous spectrum. However, there remain isolated terms in the continuous spectrum of G that contribute non-trivially to the spectral side. These questions are treated carefully in [L.IV.A] and [CHL.IV.B], and we need not dwell on them here.

The stabilization of the twisted trace formula rewrites the geometric side of (9.4.3) as a sum over (twisted) endoscopic groups for $\tilde{G}$. These twisted endoscopic groups are listed in [L.IV.A, §4.2]. Here the choice of archimedean test functions in [L.IV.A] leads to a real simplification; only one twisted endoscopic group contributes to the result, namely the quasi-split unitary group of size n, denoted $\mathbf{U}^*$ in [L.IV.A]. The key formula is then Theorem 4.13 of [L.IV.A], which is here stated in simplified form:

$$T^{\tilde{G}}(\phi) = ST^{\mathbf{U}^*}(f) \tag{9.4.4}$$

when f and ϕ are associated functions in the sense of having matching orbital integrals for a pair $(\delta, \gamma) \in G(E) \times \mathbf{U}^*(F)$ where γ is a norm of δ; we make this more explicit in (9.4.5), below. The orbital integral is strictly speaking defined not for δ but rather for (δ, θ), but we ignore this distinction. The condition that f and ϕ have matching orbital integrals is exactly as in standard endoscopy, for example as in (6.1.1).

The fundamental lemma in this situation was proved for unit elements by Kottwitz and for general Hecke functions by Clozel and Labesse. At unramified places there is a well-defined homomorphism $\phi \mapsto f$ from the spherical Hecke algebra of G to that of $\mathbf{U}^*$, dual to the map of Satake parameters denoted $\widetilde{BC}$ in [M.III.C, Theorem 4.1]. As in (9.3.3), let

$$\Phi : W_{F_v} \to {}^L\mathbf{U}^*$$

be a local Langlands parameter. Then the restriction of Φ to W_{E_v} is mapped to the L-group of G, and the base change corresponds to this restriction. A general Φ is associated to an L-packet, but if Φ is unramified then the restriction is precisely the map of Satake parameters. Details are provided in [M.III.C].

9.4.5. Norms. The norm is defined carefully in [L.IV.A, 4.4]; roughly it means that γ is conjugate to $\delta\theta(\delta)$ in $GL(n)(\bar{E})$. The illuminating case $n = 1$ is discussed in [W.II.B.2], for example at the end of his section 5.2. There $\delta \in G(E_v) = E_v^\times$ and $\gamma = \delta \cdot \theta(\delta) = \delta \cdot c(\delta)^{-1}$. For general n a semisimple conjugacy class δ is characterized by its collection of eigenvalues λ_i, in some algebraic extension of E_v. Then the typical eigenvalue of the corresponding γ is of the form $\lambda_i \cdot c(\lambda_j)^{-1}$ for an appropriate extension of c. See §5 of [W.II.B.2] for a more detailed account of the relation taking stable conjugacy into account.

References

References of the form [X.Y.Z.n], where X denotes one or more initials, Y is a Roman numeral, Z a letter, and the optional n a small integer, are to chapters of this book, with X the initials of the authors, Y the section heading, Z the chapter heading, and n the possible subchapter (only present when Y = II). Thus [W.II.B.2] refers to Waldspurger's chapter.

[A] J. Arthur, An introduction to the trace formula, in J. Arthur, D. Ellwood, R. Kottwitz, eds, *Harmonic Analysis, the trace formula, and Shimura varieties, Clay Mathematics Proceedings*, **Vol. 4** (2005).

[BR] D. Blasius & J. Rogawski, Motives for Hilbert modular forms, *Invent. Math.*, **114** (1993), 55–87.

[C1] L. Clozel, On the limit multiplicities of discrete series representations in spaces of automorphic forms, *Invent. Math.*, **83** (1986), 265–284.

[C2] L. Clozel, Représentations galoisiennes associées aux représentations automorphes autoduales de GL(n), *Publ. Math. I.H.E.S.*, **73** (1991) 97–145.

[H1] T. Hales, A simple definition of transfer factors for unramified groups, in *Representation theory of groups and algebras, Contemp. Math.*, **145**, Amer. Math. Soc., Providence, RI, (1993) 109–134.

[H2] T. Hales, On the fundamental lemma for standard endoscopy: reduction to unit elements, *Canad. J. Math.* , **47** (1995) 974–994.

[HL] M. Harris, & J.-P. Labesse, Conditional base change for unitary groups, *Asian J. Mathematics*, **8** (2004) 653–683.

[HT] M. Harris & R. Taylor, The geometry and cohomology of some simple Shimura varieties, *Annals of Mathematics Studies*, **151** (2001).

[K1] R. Kottwitz, Rational conjugacy classes in reductive groups, *Duke Math. J.*, **49** (1982) 785–806.

[K2] R. Kottwitz, Stable trace formula: cuspidal tempered terms, *Duke Math. J.* 51 (1984), 611–650.

[K3] R. Kottwitz, Stable trace formula: elliptic singular terms, *Math. Ann.* 275 (1986), 365–399.

[K4] R. Kottwitz, Tamagawa Numbers, *Annals of Math.*, **127** (1988) 629–646.

[K5] R. Kottwitz, On the λ-adic representations associated to some simple Shimura varieties, *Invent. Math.*, **108** (1992) 653–665.

[K6] R. Kottwitz, Transfer factors for Lie algebras, *Representation Theory*, **3** (1999) 127–138.

[K7] R. Kottwitz, Harmonic analysis on reductive p-adic groups and Lie algebras, in J. Arthur, D. Ellwood, R. Kottwitz, eds, *Harmonic Analysis, the trace formula, and Shimura varieties*, *Clay Mathematics Proceedings*, **Vol. 4** (2005), 393–522.

[Lab] J.-P. Labesse, Cohomologie, stabilisation et changement de base, *Astérisque*, **257** (1999) 1–116.

[Lab 2] J.-P. Labesse, Stable twisted trace formula: elliptic terms. *J. Inst. Math. Jussieu*, **3** (2004), 473–530.
[Ltam] J.-P. Labesse, Nombres de Tamagawa des groupes réductifs quasi-connexes *Manuscripta Math.*, **104** (2001) 407–430.

[LL] J.-P. Labesse & R.P. Langlands L-indistinguishability for $SL(2)$, *Can. J. Math.*, **31** (1979) 726–785.

[L] R.P. Langlands, *Les débuts d'une formule des traces stables, Publications de l'Université Paris 7*, **13** (1983).

[LR] R.P. Langlands & D. Ramakrishnan, eds., *The zeta functions of Picard modular surfaces*, Université de Montreal, Centre de Recherches Mathématiques, Montreal (1992).

[LS1] R.P. Langlands & D. Shelstad, On the definition of transfer factors, *Math. Ann.*, **278** (1987), 219–271.

[LS2] R.P. Langlands & D. Shelstad, Descent for transfer factors, *The Grothendieck Festschrift, Vol. II, Progr. Math.*, **87**, Birkhäuser Boston, Boston, MA, (1990) 485–563.

[LN] G. Laumon & B.C. Ngô, Le lemme fondamental pour les groupes unitaires, *Annals of Math.* (in press).

[N1] B.C. Ngô, Fibration de Hitchin et endoscopie. *Invent. Math.* **164** (2006), 399–453.

[N2] B.C. Ngô, Le lemme fondamental pour les algèbres de Lie. *Publ. Math. IHES,* **111** (2010) 1–169.

[R] J. Rogawski, Automorphic Representations of Unitary Groups in Three Variables, *Annals of Math. Studies*, **123** (1990).

[W1] J-L. Waldspurger, Le lemme fondamental implique le transfert, *Compositio Math.*, **105** (1997) 153–236.

[W2] J-L. Waldspurger, Intégrales orbitales nilpotentes et endoscopie pour les groupes classiques non ramifiés, *Astérisque*, **269** (2001).

[W3] J.-L. Waldspurger, Endoscopie et changement de caractéristique, *J. Inst. Math. Jussieu*, **5** (2006).

UFR de Mathématiques, Université Paris 7, 2 Pl. Jussieu, 75251 Paris cedex 05, FRANCE

Introduction to Endoscopy
Snowbird Lectures
Revised version, May 2010
[L.I.B]

Jean-Pierre Labesse

CONTENTS

1. Introduction

1.1. What is endoscopy about? Many questions about noncommutative Lie groups boil down to questions in invariant harmonic analysis: the study of distributions on the group that are invariant by conjugacy. The fundamental objects of invariant harmonic analysis are orbital integrals and characters. For various reasons (among which is the trace formula), one calls "geometric" the objects related to conjugacy classes and orbital integrals and "spectral" the objects related to representations and characters.

In the Langlands program a cruder form of conjugacy called **stable conjugacy** plays a role. The study of Langlands functoriality often leads to correspondences that are defined only up to stable conjugacy. Endoscopy is the name given to a series of techniques aimed to investigate the difference between ordinary and stable conjugacy. The word "endoscopy" has been coined to express that we want to see ordinary conjugacy **inside** stable conjugacy. The aim is to recover orbital integrals and characters from their stable avatars on a family of auxiliary groups called endoscopic groups.

One also expects a parametrization of irreducible admissible representations in term of the so-called dual picture where one considers Langlands parameters that are classes of admissible homorphisms of the Weil group W_F (see 2.6), or more generally the Langlands group $\mathcal{L}_F$, into the L-group. Moreover endoscopy should be encoded in this dual picture. Such a dual parametrization is conjectural in general but has been completely established for real groups by Langlands and Shelstad.

1.2. The contents of these lectures. In this series of lectures we shall introduce the basic notions of local endoscopy: stable conjugacy, κ-orbital integrals, endoscopic groups, endoscopic transfer of orbital integrals and its dual for characters with an emphasis on the case of real groups, following the work of Diana Shelstad.

We start in section 2 with a brief review of some basic objects: we recall the definition of pseudo-coefficients, then we discuss stable conjugacy, L-packets, Weil groups, L-groups and Langlands parameters. In section 3 we review the classification and the Langlands parameters for admissible representations of $GL(2,\mathbb{R})$ in order to fix the notation to be used for this basic example. In section 4 we describe the endoscopy for $SL(2,\mathbb{R})$ which is the simplest non-trivial example. We describe in this case the stable conjugacy, the L-packets as well as the endoscopic transfer of functions and of characters. In section 5 we briefly describe the endoscopy for the real unitary group $U(2,1)$. In section 6 we introduce some simple objects from Galois cohomology in order to define stable conjugacy, κ-orbital integrals and endoscopic groups in general. In section 7 we discuss the endoscopic transfer of pseudo-coefficients of discrete series. We give a simple direct proof of it. In the last section we discuss some extensions of the preceding material.

2. Some basic definitions

The Langlands theory and in particular endoscopy deals with connected reductive algebraic groups defined over local or global fields.

One should be warned that even if G is defined over $\mathbb{R}$ and is connected as an algebraic group it may happen that $G(\mathbb{R})$ is disconnected as a topological group. For example the algebraic group $GL(2)$ is connected while $GL(2,\mathbb{R})$ has two connected components.

2.1. Orbital integrals. Let G be a connected reductive algebraic group defined over a local field F. Let $\gamma \in G(F)$ be a strongly regular semisimple element: this means that its centralizer G_γ is a (maximal) torus. Let $f \in \mathcal{C}_c^\infty(G(F)$, the orbital integral of f on the orbit of γ is given by

$$\mathcal{O}_\gamma(f) = \int_{G_\gamma(F)\backslash G(F)} f(x^{-1}\gamma x)\, d\dot{x}$$

where $d\dot{x}$ is a $G(F)$-invariant measure on the quotient $G_\gamma(F)\backslash G(F)$.

2.2. Pseudo-coefficients for discrete series. Let F be a local field. Assume that the center of $G(F)$ is compact. Let π be a discrete series representation of $G(F)$ we say that a function $f \in \mathcal{C}_c^\infty(G(\mathbb{R}))$ is a (normalized) pseudo-coefficient for π if for any **tempered** irreducible representation π' we have

$$\text{trace } \pi'(f) = \begin{cases} 1 & \text{if } \pi' \simeq \pi \\ 0 & \text{otherwise} \end{cases}$$

Observe that it would be more canonical to consider $f\, d\mu$ where $d\mu$ is a Haar measure in $G(\mathbb{R})$ since the choice of the Haar measure enters in the definition of trace $\pi(f)$.

A suitably normalized diagonal matrix coefficient of π (or rather its complex conjugate) would satisfy these requirements but for one condition: matrix coefficients are not compactly supported unless the group is compact. The interest of using pseudo-coefficients instead of matrix coefficients is that since they are compactly supported trace $\pi(f)$ makes sense even if π is not tempered and they can be used in the trace formula. This is important to observe since non-tempered representations show up in the spectral decomposition of the space of square integrable automorphic forms.

When $F = \mathbb{R}$ the existence of pseudo-coefficients is an easy consequence of the existence of multipliers due to Arthur [A] and Delorme [D]. We refer the reader to [CD] or [Lab2] for the construction. In [Lab2] more general functions, called index functions are constructed. They allow in particular to deal with certain linear combinations of limits of discrete series although individual ones do not have pseudo-coefficients. For p-adic fields the existence of pseudo-coefficients is due to Kazhdan [Ka].

We shall denote by f_π a pseudo-coefficient for π, although it is highly non-unique. But as regards invariant harmonic analysis this plays no role. In particular the orbital integrals are independent of the choice of the pseudo-coefficient; they are also independent of the choice of the Haar measure on $G(F)$ but one has to use the measure on the compact torus $G_\gamma(F)$ for which it has volume 1. The orbital integrals of f_π are easily computed for γ regular semisimple:

$$\mathcal{O}_\gamma(f_\pi) = \begin{cases} \Theta_\pi(\gamma^{-1}) & \text{if } \gamma \text{ is elliptic} \\ 0 & \text{otherwise} \end{cases}$$

where Θ_π is the character of π.

2.3. Stable conjugacy. Let G be a connected reductive group over a local field F. Recall that one says that γ and γ' in $G(F)$ are conjugate if there exists $x \in G(F)$ such that $\gamma' = x\gamma x^{-1}$. Roughly speaking, stable conjugacy amounts to conjugacy over the algebraic closure $\overline{F}$. At least for strongly regular semisimple elements, one says that γ and γ' in $G(F)$ are stably conjugate if there is an $x \in G(\overline{F})$ such that $\gamma' = x\gamma x^{-1}$. This definition is also valid when $\gamma \in G(F)$ is any semisimple element if G has a simply connected derived group. In general the definition is slightly more technical and will be given in 6.3.

Let $f \in \mathcal{C}_c^\infty(G(F))$, the stable orbital integral for $\gamma \in G(F)$ strongly regular is by definition

$$S\mathcal{O}_\gamma(f) = \sum_{\gamma' \in S(\gamma)} \mathcal{O}_{\gamma'}(f)$$

where $S(\gamma)$ is a set of representatives for conjugacy classes in the stable conjugacy class of γ. Implicit here is the choice of a natural measure on the various conjugacy classes: one uses that the various centralizers $G_{\gamma'}$ are canonically isomorphic.

Stably invariant harmonic analysis is the study of stable distributions: the distributions that are, roughly speaking, constant on stable conjugacy classes. This is the closure of the space spanned by stable orbital integrals.

The miracle is that one can recover ordinary orbital integrals as a linear combination of stable orbital integrals on endoscopic groups (see 6.5). This supposes one is able to transfer κ-orbital integrals on the group G itself to stable orbital integrals endoscopic groups. The transfer involves the so-called "transfer factors" of Langlands and Shelstad. The existence of the transfer is due to Shelstad [Sh4] when $F = \mathbb{R}$ (see 6.7 and the paper of Shelstad in this volume). For arbitrary local fields it has been obtained only recently thanks to Ngô and Waldspurger (see section 8.3).

2.4. *L*-packets. To define L-packets for representations of $G(F)$ we consider first the case of discrete series. Consider a discrete series representation π and let us denote by f_π a pseudo-coefficient. Two discrete series representations π and π' for $G(F)$ will be said to belong to the same L-packet if for any strongly regular semisimple γ one has

$$S\mathcal{O}_\gamma(f_\pi) = c(\pi, \pi') S\mathcal{O}_\gamma(f_{\pi'})$$

where $c(\pi, \pi')$ is some non-zero constant.

Over the reals, an L-packet of discrete series of a reductive Lie group is the set of all (equivalence classes of) discrete series with the same infinitesimal character and then $c(\pi, \pi') \equiv 1$. The discrete series L-packets are in bijection with the (equivalence classes of) finite dimensional representations of the compact inner form.

One can construct all tempered irreducible representations in term of the discrete series representations of Levi subgroups using unitary parabolic induction and by taking subrepresentations. Using this it is now easy to define

L-packets for tempered representations. Two tempered irreducible representations π and π' are said to be in the same L-packet if, up to equivalence, π and π' are subrepresentations of parabolically induced representations from discrete series σ and σ' in the same L-packet for some Levi subgroup. One should get a stable distributions by taking a suitable linear combination of characters in an L-packet (namely the sum when $F = \mathbb{R}$).

The notion of L-packets of representations is the stable avatar for irreducible tempered representations. The second expected miracle is that one should be able to express tempered characters in terms of stable characters on endoscopic groups.

2.5. Arthur packets. Using parabolic induction from quasi-tempered[1] irreducible representations of Levi subgroups and by taking appropriate subquotients (the Langlands quotients) one can construct all admissible irreducible representations of $G(F)$ and thus one could extend the notion of L-packets to this more general case. But, in general, no non-trivial linear combination of character in the L-packet will yield a stable distribution. For the non-tempered case one needs the notion of Arthur packets; they are bigger packets than L-packets and the linear combination to be taken is more subtle even when $F = \mathbb{R}$. This will be examined in Vogan's lectures.

2.6. The Weil and the Langlands groups. Let F be a local field. One denotes by W_F the Weil group of F defined as follows. The Weil group of $\mathbb{C}$ is simply $\mathbb{C}^\times$:

$$W_{\mathbb{C}} = \mathbb{C}^\times .$$

The Weil group for $\mathbb{R}$ can be defined as the subgroup of matrices in $SU(2)$ generated by

$$\left\{ \mathbf{z} = \begin{pmatrix} z & 0 \\ 0 & \overline{z} \end{pmatrix} , \ z \in \mathbb{C}^\times \right\} \quad \text{and} \quad w_\sigma = \begin{pmatrix} 0 & -1 \\ 1 & 0 \end{pmatrix}$$

The element w_σ acts by conjugacy as the non-trivial element in the Galois group $\mathrm{Gal}(\mathbb{C}/\mathbb{R})$ on $\mathbb{C}^\times$. In other words there is an exact sequence

$$1 \to W_{\mathbb{C}} \to \mathbb{W}_{\mathbb{R}} \to \mathrm{Gal}(\mathbb{C}/\mathbb{R}) \to 1$$

and a section of the map $\mathbb{W}_{\mathbb{R}} \to \mathrm{Gal}(\mathbb{C}/\mathbb{R})$ is defined by

$$\sigma \mapsto w_\sigma$$

Observe that $w_\sigma^2 = -\mathbf{1}$ and hence the above extension of $W_{\mathbb{C}} = \mathbb{C}^\times$ by $\mathrm{Gal}(\mathbb{C}/\mathbb{R})$ is the non-trivial one. There is a surjective map $W_{\mathbb{C}/\mathbb{R}} \to \mathbb{R}^\times$ given by $z \mapsto z\overline{z}$ and $w_\sigma \mapsto -1$.

When F is a p-adic field the Weil group W_F can be viewed as a subgroup of $\mathrm{Gal}(\overline{F}/F)$. It sits in the following diagram with exact lines

$$\begin{array}{ccccccccc} 1 & \to & I_F & \to & W_F & \to & \mathbb{Z} & \to & 1 \\ & & \downarrow & & \downarrow & & \downarrow & & \\ 1 & \to & I_F & \to & \mathrm{Gal}(\overline{F}/F) & \to & \overline{\mathbb{Z}} & \to & 1 \end{array}$$

[1] i.e. tempered up to a twist by a quasi-character (a non-necessarily unitary character).

where I_F is the inertia group and $\overline{\mathbb{Z}}$ is the completion of $\mathbb{Z}$ for the topology of finite index subgroups. The group $\mathbb{Z}$ in the first line is endowed with the discrete topology. The vertical arrows are continuous inclusions with dense image, the first one being a topological isomorphism.

If K/F is a finite Galois extension one has an exact sequence

$$1 \to W_K^{ab} \to W_F \to \mathrm{Gal}(K/F) \to 1$$

where W_K^{ab} is the abelianized quotient of W_K. By class field theory one has an isomorphism $W_K^{ab} \to K^\times$.

The Langlands group $\mathcal{L}_F$ is defined to be

$$\mathcal{L}_F = W_F$$

when $F = \mathbb{C}$ or $\mathbb{R}$. When F is p-adic one must take

$$\mathcal{L}_F = W_F \times SL(2, \mathbb{C}) \ .$$

2.7. L-groups and Langlands parameters. Let F be a local field and G a connected reductive group over F. Denote by G^* the quasisplit inner form. Choose a Borel pair (B, T) in G^*. We get a based root datum:

$$(X, \Delta, \check{X}, \check{\Delta})$$

Denote by $\check{G}$ the **complex** reductive Lie group defined by the dual based root datum

$$(\check{X}, \check{\Delta}, X, \Delta)$$

For example

$$\text{if} \quad G = GL(n) \quad \text{then} \quad \check{G} = GL(n, \mathbb{C})$$

$$\text{if} \quad G = SL(n) \quad \text{then} \quad \check{G} = PGL(n, \mathbb{C})$$

$$\text{if} \quad G = Sp(2n) \quad \text{then} \quad \check{G} = SO(2n+1, \mathbb{C})$$

There is a canonical bijection between conjugacy classes of parabolic subgroups of G^* and of $\check{G}$. Those coming from G form a subset.

The Galois group $\mathrm{Gal}(\overline{F}/F)$ acts on the based root datum and hence it acts on $\check{G}$ via **holomorphic automorphisms** assumed to preserve a splitting. If G is split this action is trivial. Now W_F acts via its natural map to $\mathrm{Gal}(\overline{F}/F)$ and one denotes by ${}^L G$ the semidirect product $\check{G} \rtimes W_F$.

Let us give a non-trivial example: if $F = \mathbb{R}$ and $G(\mathbb{R}) = U(2,1)$ then $\check{G} = GL(3, \mathbb{C})$ and the holomorphic action of the non-trivial element σ in the Galois group $\mathrm{Gal}(\mathbb{C}/\mathbb{R})$ on $\check{G}$ is given by

$$\check{\sigma} : g \mapsto J {}^t g^{-1} J^{-1} \quad \text{with} \quad J = \begin{pmatrix} 0 & 0 & 1 \\ 0 & -1 & 0 \\ 1 & 0 & 0 \end{pmatrix} .$$

A Langlands parameter is a $\check{G}$-conjugacy class of homomorphisms

$$\varphi \ : \ \mathcal{L}_F \ \to \ {}^L G$$

holomorphic on the $SL(2)$ factor in the p-adic case, such that the compositum with the natural projection of $\mathcal{L}_F \to W_F$

$$\mathcal{L}_F \quad \to \quad {}^LG \quad \to \quad W_F$$

is natural projection of $\mathcal{L}_F$ onto W_F. One asks moreover that the images of elements in W_F are semisimple. A parameter is said to be relevant (for G) if the image of φ in $\check{G}$ does not lie in a parabolic subgroup unless it comes from G.

One of the goals of the theory is to show that one has a natural parametrization of irreducible admissible representations of $G(F)$ when F is a local field (and of automorphic representations when F is a global field) using the Langlands parameters: to each Langlands parameter there should be associated an L-packet $\Pi(\varphi)$ of representations of $G(F)$. More precisely, relevant Langlands parameters should parametrize L-packets and the constituents of an L-packet should be attached to representations of some finite group defined by the parameter.

When $F = \mathbb{R}$ Langlands has established such a correspondence for an arbitrary connected reductive group G in a paper that remained unpublished for a long time but is now available in print [Lan2]. The correspondence between constituents of L-packets and representations of a finite group defined by the parameter is due to Shelstad (see 6.13). To obtain a nice formulation it is better to use K-groups (see 8.1). We explain these correspondences for $GL(2,\mathbb{R})$ and $SL(2,\mathbb{R})$ in 3.2 and 4.3.

For non-archimedean local fields little is known except for $GL(n)$, but rapid progress is to be expected at least for classical groups.

3. GL(2)

3.1. Representations of $GL(2,\mathbb{R})$. By representations we understand admissible representations. We shall use Jacquet-Langlands notation. The classification is as follows: all admissible irreducible representations of $GL(2,\mathbb{R})$ are subquotients of principal series

$$\rho(\mu_1,\mu_2)$$

where μ_i are characters of $\mathbb{R}^\times$. The principal series representations are induced by characters from the Borel subgroup: $\rho(\mu_1,\mu_2)$ is the right regular representation in the space of smooth functions such that

$$f(\begin{pmatrix}\alpha & x\\ 0 & \beta\end{pmatrix} g) = \mu_1(\alpha)\mu_2(\beta)\left|\frac{\alpha}{\beta}\right|^{1/2} f(g)\ .$$

We shall restrict ourselves to representations that induce unitary characters on the center and hence we assume that the product $\mu_1\mu_2$ is unitary. We have three types of subquotients according to the value of

$$\mu = \mu_1\mu_2^{-1}$$

(1) Irreducible principal series

$$\pi(\mu_1,\mu_2)$$

where

$$\mu(x)\neq x^n\,.\,\mathrm{sign}\,(x)$$

for some $n\in\mathbb{Z}-\{0\}$. These representations are unitarizable if μ is unitary or if $\mu(x)=|x|^s$ with s real and $-1<s<1$.

(2) Finite dimensional subquotients

$$\pi(\mu_1,\mu_2)$$

when $\mu(x)=x^n\,.\,\mathrm{sign}\,(x)$ with $n\in\mathbb{Z}-\{0\}$. It is unitarizable if $n=\pm1$.

(3) Discrete series subquotients

$$\sigma(\mu_1,\mu_2)$$

when $\mu(x)=x^n\,.\,\mathrm{sign}\,(x)$ with $n\in\mathbb{Z}-\{0\}$. These representations are unitarizable.

The various representations are equivalent under permutation of the μ_i: for instance $\pi(\mu_1,\mu_2)$ is equivalent to $\pi(\mu_2,\mu_1)$ and so on. Moreover $\sigma(\mu_1,\mu_2)$ is equivalent to $\sigma(\mu_1\varepsilon,\mu_2\varepsilon)$ where $\varepsilon(x)=\mathrm{sign}\,(x)$.

3.2. Langlands parameters for $GL(2,\mathbb{R})$. For $z=\rho e^{i\theta}\in\mathbb{C}^\times$ let

$$\chi_{s,n}(z)=\rho^{2s}e^{in\theta}\ .$$

The Langlands correspondence for $GL(1,\mathbb{R})$ is given as follows: to a character μ of $GL(1,\mathbb{R})=\mathbb{R}^\times$ of the form

$$\mu(x)=|x|^s\,\mathrm{sign}\,(x)^m$$

corresponds the character of the Weil group given by

$$\mathbf{z}\mapsto\chi_{s,0}(z)=|z\overline{z}|^s\quad\text{and}\quad w_\sigma\mapsto(-1)^m\ .$$

A Langlands parameter for $GL(2,\mathbb{R})$ is a conjugacy classes of homomorphisms of $\mathbb{W}_\mathbb{R}$ in $GL(2,\mathbb{C})$ with semisimple images. Up to conjugacy, the admissible maps φ are of the following form:

(1) for some $s_i\in\mathbb{C}$ and $m_i\in\mathbb{Z}_2$

$$\varphi_{s_1,m_1,s_2,m_2}(\mathbf{z})=\begin{pmatrix}\chi_{s_1,0}(z)&0\\0&\chi_{s_2,0}(z)\end{pmatrix}$$

with

$$\varphi_{s_1,m_1,s_2,m_2}(w_\sigma)=\begin{pmatrix}(-1)^{m_1}&0\\0&(-1)^{m_2}\end{pmatrix}$$

The map $\varphi_{s_1,m_1,s_2,m_2}$ is conjugate to $\varphi_{s_2,m_2,s_1,m_1}$.

(2) for some $s \in \mathbb{C}$ and $n \in \mathbb{Z}$

$$\varphi_{s,n}(\mathbf{z}) = \begin{pmatrix} \chi_{s,n}(z) & 0 \\ 0 & \chi_{s,-n}(z) \end{pmatrix}$$

with

$$\varphi_{s,n}(w_\sigma) = \begin{pmatrix} 0 & (-1)^n \\ 1 & 0 \end{pmatrix}$$

The map $\varphi_{s,n}$ is conjugate to $\varphi_{s,-n}$.

The intersection of the two sets of conjugacy classes of maps is the class of parameters of the form

$$\varphi_{s,0} \simeq \varphi_{s,1,s,0} \simeq \varphi_{s,0,s,1}$$

Let us denote by ε the homomorphism from $\mathbb{W}_{\mathbb{R}}$ to $\mathbb{C}^\times$ defined by

$$\varepsilon(\mathbf{z}) = 1 \quad \text{and} \quad \varepsilon(w_\sigma) = -1$$

LEMMA 3.1. *If φ is a Langlands parameter, then*

$$\varphi \otimes \varepsilon \simeq \varphi$$

if and only if φ is in the class of $\varphi_{s,n}$ for some s and some n.

Proof. First it is clear that

$$\varphi_{s,n} \otimes \varepsilon = \alpha\, \varphi_{s,n}\, \alpha^{-1} \quad \text{where} \quad \alpha = \begin{pmatrix} -1 & 0 \\ 0 & 1 \end{pmatrix}$$

Conversely we have to observe that

$$\varphi_{s_1,m_1,s_2,m_2} \not\simeq \varphi_{s_1,m_1+1,s_2,m_2+1}$$

unless $s_1 = s_2$ and $m_1 + m_2 \equiv 1$ □

The correspondence between irreducible representations and Langlands parameters for $GL(2,\mathbb{R})$ is obtained as follows. We have a natural bijection between equivalence classes of admissible irreducible representations of $GL(2,\mathbb{R})$ and conjugacy classes of admissible homomorphisms of $\mathbb{W}_{\mathbb{R}}$ in $GL(2,\mathbb{C})$ as follows:

$$\pi(\mu_1,\mu_2) \mapsto \varphi_{s_1,m_1,s_2,m_2} \quad \text{with} \quad \mu_i(x) = |x|^{s_i}\, \text{sign}\,(x)^{m_i}$$

and

$$\sigma(\mu_1,\mu_2) \mapsto \varphi_{s,n} \quad \text{with} \quad \mu_1\mu_2(x) = |x|^{2s}\, \text{sign}\,(x)^{n+1}$$

where we assume moreover that

$$\mu(x) = \mu_1\mu_2^{-1}(x) = x^n\, \text{sign}\,(x) = |x|^n \text{sign}\,(x)^{n+1} \ .$$

A Langlands parameter corresponds to a tempered representation if the image of the map is bounded i.e. if the s_i are imaginary.

3.3. Endoscopy for $GL(2,F)$. Inside $GL(2,F)$ conjugacy over F and over $\overline{F}$ coincide. Hence there is no non-trivial endoscopy for $GL(2)$ in the sense that conjugacy and stable conjugacy coincide and L-packets are singletons. Nevertheless, if we consider an inner form G of $G^* = GL(2)$, then there are correspondences for conjugacy classes and for representations between G and G^*.

Let G be the multiplicative group of a quaternion algebra over a local field F. There is a bijection between elliptic conjugacy classes in $G(F)$ and $G^*(F)$: this correspondence matches elements with the same eigenvalues in $\overline{F}$. The bijection $\pi \mapsto \pi^*$ between discrete series representations of $G(F)$ and of $G^*(F)$ is characterized by the equality of character distributions up to a sign:

$$\Theta_\pi(\gamma) = -\Theta_{\pi^*}(\gamma^*)$$

whenever $\gamma \in G(F)$ and $\gamma^* \in G^*(F)$ are regular elliptic elements with the same eigenvalues in $\overline{F}$.

These correspondences, often called Jacquet-Langlands correspondences, are nowadays considered as an instance of endoscopy.

4. $SL(2)$

Besides Jacquet-Langlands correspondence for $GL(2)$, endoscopy for $SL(2)$ is the simplest instance of it and in fact it is in this case that it was first studied. As far as I know endoscopy was discovered by Langlands while he was trying to understand the Zeta function of some simple Shimura varieties: those attached to inner forms of groups between $GL(2)$ and $SL(2)$ over a totally real field.

4.1. Endoscopy for $SL(2,\mathbb{R})$. Endoscopy arises because conjugacy over F and over $\overline{F}$ (the algebraic closure of F) may be different. For example, for $\theta \notin \mathbb{Z}\pi$, the two rotations

$$r(\theta) = \begin{pmatrix} \cos\theta & \sin\theta \\ -\sin\theta & \cos\theta \end{pmatrix}$$

and

$$r(-\theta) = \begin{pmatrix} \cos\theta & -\sin\theta \\ \sin\theta & \cos\theta \end{pmatrix}$$

are not conjugate in $SL(2,\mathbb{R})$ although they are conjugate inside $SL(2,\mathbb{C})$ and also inside $GL(2,\mathbb{R})$ by

$$w = \begin{pmatrix} -i & 0 \\ 0 & i \end{pmatrix} \quad \text{and} \quad \alpha = \begin{pmatrix} -1 & 0 \\ 0 & 1 \end{pmatrix}$$

respectively. The union of the sets of conjugates of elements in the pair $(r(\theta), r(-\theta))$ is said to be a **stable conjugacy class**. Similarly the unipotent elements

$$u_0 = \begin{pmatrix} 1 & 1 \\ 0 & 1 \end{pmatrix} \quad \text{and} \quad u_0^{-1} = \begin{pmatrix} 1 & -1 \\ 0 & 1 \end{pmatrix}$$

are conjugate in $SL(2,\mathbb{C})$ but not in $SL(2,\mathbb{R})$. There are two conjugacy classes but only one stable conjugacy class of non trivial unipotent elements.

Endoscopy on the spectral side is also easily described for $SL(2,\mathbb{R})$: discrete series and limits of discrete series representations come by pairs called L-packets.

In all above examples the objects inside pairs are exchanged by conjugacy under $w = i\alpha$ an element in the normalizer of $SO(2)$ in $SL(2,\mathbb{C})$. We observe that if σ is the non trivial element of the Galois group, then

$$a_\sigma = w\sigma(w)^{-1} = \begin{pmatrix} -1 & 0 \\ 0 & -1 \end{pmatrix}$$

generates a subgroup of order 2 that can be identified with $H^1(\mathbb{C}/\mathbb{R}, SO(2))$. The characters of this 2-group are called endoscopic character. The endoscopic groups for $SL(2,\mathbb{R})$ correspond to these two characters. The endoscopic group corresponding to the trivial endoscopic character is $SL(2,\mathbb{R})$ himself while the endoscopic group corresponding to the non-trivial one is

$$T(\mathbb{R}) = SO(2,\mathbb{R})$$

the compact torus.

4.2. Representations of $SL(2,\mathbb{R})$. The representations of $SL(2,\mathbb{R})$ and their L-packets are easily understood in terms of those for $GL(2,\mathbb{R})$ recalled in 3.1. In fact it is an easy exercise to show that any irreducible representation of $SL(2,\mathbb{R})$ occurs in the restriction of an irreducible representation of $GL(2,\mathbb{R})$. Those restrictions either remain irreducible (which is the case for principal series for generic values of the parameter) or split into two irreducible components whose union is an L-packet for $SL(2,\mathbb{R})$.

Two representations π and π' are in the same L-packet if and only if, up to equivalence, they are conjugated by α:

$$\pi' \simeq \pi \circ \mathrm{Ad}(\alpha) \quad \text{where} \quad \alpha = \begin{pmatrix} -1 & 0 \\ 0 & 1 \end{pmatrix}$$

We have the following classification:

(1) Irreducible principal series: they are the $\pi(\mu)$ obtained by restriction of $\pi(\mu_1,\mu_2)$ with $\mu(x) \neq x^n \,.\, \mathrm{sign}\,(x)$ with $n \in \mathbb{Z}$.

(2) Finite dimensional representations of dimension n: they are the $\pi(\mu)$ obtained by restriction of $\pi(\mu_1,\mu_2)$ with $\mu(x) = x^n \,.\, \mathrm{sign}\,(x)$ and $n \neq 0$.

(3) Discrete series L-packets

$$\sigma(\mu) = \{D^+_{|n|}, D^-_{|n|}\}$$

are obtained by restriction of $\sigma(\mu_1,\mu_2)$ with

$$\mu(x) = x^n \,.\, \mathrm{sign}\,(x)$$

and $n \in \mathbb{Z} - 0$.

(4) Limits of discrete series L-packet

$$\sigma(\mu) = \{D_0^+, D_0^-\}$$

is obtained by restriction of $\pi(\mu_1, \mu_2)$ with

$$\mu(x) = \mu_1(x)\mu_2(x)^{-1} = \operatorname{sign}(x)$$

The L-packets of representations that are indexed by characters μ and μ^{-1} are equivalent.

The minimal K-type for $D_n^{\pm}$ with $n \in \mathbb{N}$ is $\pm(n+1)$ i.e.

$$r(\theta) \mapsto \exp(\pm i\,(n+1)\,\theta)$$

if

$$r(\theta) = \begin{pmatrix} \cos\theta & \sin\theta \\ -\sin\theta & \cos\theta \end{pmatrix}$$

The character of D_n^+ on $K = SO(2,\mathbb{R})$ is given by

$$\Theta_n^+(r(\theta)) = \frac{e^{i(n+1)\theta}}{1-e^{2i\theta}} = \frac{-e^{in\theta}}{e^{i\theta}-e^{-i\theta}}$$

while the character of D_n^- is the complex conjugate:

$$\Theta_n^-(r(\theta)) = \frac{e^{-i(n+1)\theta}}{1-e^{-2i\theta}} = \frac{e^{-in\theta}}{e^{i\theta}-e^{-i\theta}}$$

4.3. Langlands parameters for $SL(2,\mathbb{R})$. From the bijection between equivalence classes of representations and conjugacy classes of Langlands parameters for $GL(2,\mathbb{R})$ recalled in 3.2 one derives a bijection between equivalence classes of L-packets of admissible irreducible representations of $SL(2,\mathbb{R})$ and conjugacy classes of admissible homomorphisms of $\mathbb{W}_\mathbb{R}$ in $PGL(2,\mathbb{C})$. We need the

LEMMA 4.1. *Any projective representation of $\mathbb{W}_\mathbb{R}$ lifts to a representation.*

Proof. This is a particular case of a result of [Lab1]. This particular case can also be found in [Lan2]. We shall prove it only for two dimensional representations. Consider a two-dimensional projective parameter. The image of $\mathbb{C}^\times$ is inside a torus. Now this image is either trivial or is a maximal torus in $PGL(2,\mathbb{C})$ and hence of dimension 1. If this image is trivial the lemma is easy to prove. If this image is a torus the map restricted to $\mathbb{C}^\times$ is given by a non-trivial character $\chi_{s,n}$. If $n = 0$ the existence of a lift is easy. If $n \neq 0$ then the image of w_σ must lie in the normalizer of the torus and act non trivially and hence its square is central i.e. projectively trivial and $s = 0$. Since $w_\sigma^2 = -1$ this shows that n is even and hence $\varphi_{0,n/2}$ is a lift. □

The correspondence is a follows:

(1) the parameter for $\pi(\mu)$ is the conjugacy class of the projective parameter $\phi_{s,m}$ defined by $\varphi_{s,m,0,0}$ with

$$\mu(x) = |x|^s \operatorname{sign}(x)^m$$

(2) the parameter for $D_n^{\pm}$ is the conjugacy class of the projective parameter ϕ_n defined by $\varphi_{0,n}$ with $n \in \mathbb{N}$.

We have seen in lemma 3.1 that

$$\varphi_{0,n} \otimes \varepsilon = \alpha \varphi_{0,n} \alpha^{-1} \quad \text{where} \quad \alpha = \begin{pmatrix} -1 & 0 \\ 0 & 1 \end{pmatrix}$$

But ε has a central image and hence the projective parameters defined by $\varphi_{0,n}$ and $\varphi_{0,n} \otimes \varepsilon$ are equal. This shows that the projective image of α belongs to the centralizer of the projective image of $\varphi_{0,n}$.

Let ϕ_n be the projective parameter defined by $\varphi_{0,n}$ and let S_{ϕ_n} denote the centralizer of the image of ϕ_n and $\mathfrak{S}_{\phi_n}$ the quotient of S_{ϕ_n} by its connected component $S^0_{\phi_n}$ times the center $Z_{\check{G}}$ of $\check{G}$ (trivial here). When $n \neq 0$ we have

$$\mathfrak{S}_{\phi_n} = S_{\phi_n} \simeq \{1, \alpha\}$$

When $n = 0$ the group $S^0_{\phi_0}$ is a torus but again $\mathfrak{S}_{\phi_0}$ is generated by the image of α.

4.4. Character identities. We observe that the sum

$$S\Theta_n = \Theta_n^+ + \Theta_n^-$$

verifies

$$S\Theta_n(r(\theta)) = -\frac{e^{in\theta} - e^{-in\theta}}{e^{i\theta} - e^{-i\theta}}$$

and hence is invariant under stable conjugacy; we say that $S\Theta_n$ is a stable character, while

$$\Delta(r(\theta))(\Theta_n^+ - \Theta_n^-)(r(\theta)) = e^{in\theta} + e^{-in\theta}$$

where

$$\Delta(r(\theta)) = -2i \sin\theta$$

is the sum of two characters of $T(\mathbb{R})$. This is the spectral endoscopic transfer for $SL(2, \mathbb{R})$. For further generalization we observe that

$$\Delta(r(\theta)) = -e^{i\theta}(1 - e^{-2i\theta}) = -i \text{ sign}(\sin(\theta))|e^{i\theta} - e^{-i\theta}|$$

The group $G^* = SL(2)$ over $\mathbb{R}$ has an inner form G such that

$$G(\mathbb{R}) = SU(2)$$

Observe that $SL(2, \mathbb{R})$ and $SU(2)$ have in common the compact Cartan subgroup

$$T(\mathbb{R}) = SO(2)$$

The conjugacy classes in $SU(2)$ are classified by the pairs of eigenvalues $\{e^{i\theta}, e^{-i\theta}\}$. Hence they are in bijection with elliptic stable conjugacy classes in $SL(2, \mathbb{R})$. This correspondence is dual to the correspondence between representation F_n of dimension n of $SU(2)$ and L-packets of discrete series $D_n^{\pm}$ and there is the character identity

$$\text{trace } F_n(r(\theta)) = -S\Theta_n(r(\theta))$$

4.5. Asymptotic behaviour of orbital integrals and geometric transfer. Let f be a smooth and compactly supported function on $G = SL(2,\mathbb{R})$. We are to study its orbital integrals. For $\gamma \in G$ let us denote by G_γ its centralizer in G. The orbital integral is

$$\mathcal{O}_\gamma(f) = \int_{G_\gamma \backslash G} f(x^{-1}\gamma x)\, d\dot{x}$$

Observe it depends on the choice of Haar measures on G and G_γ. We shall study the asymptotic behaviour of $\mathcal{O}_\gamma(f)$ when γ is diagonal and $\gamma \to 1$ and when $\gamma = r(\theta)$ and $\theta \to 0$. Up to integration over the compact group K we may and will assume that f is K-central i.e. $f(kx) = f(xk)$.

Consider the first case where γ is diagonal:

$$\gamma = \begin{pmatrix} a & 0 \\ 0 & b \end{pmatrix}$$

with $ab = 1$. Then

$$\mathcal{O}_\gamma(f) = \int_U f(u^{-1}\gamma u)\, du$$

for a standard choice of Haar measures, where U is the group of unipotent matrices and hence

$$\mathcal{O}_\gamma(f) = \int_{\mathbb{R}} f\begin{pmatrix} a & (a-b)x \\ 0 & b \end{pmatrix} dx$$

which yields the

LEMMA 4.2. *Let $\Delta(\gamma) = |a-b|$ then*

$$h(\gamma) = \Delta(\gamma)\mathcal{O}_\gamma(f)$$

extends to a smooth function on the group $A(\mathbb{R})$ of diagonal real matrices.

This is the simplest example of endoscopic geometric transfer for a non-elliptic endoscopic group.

When $\gamma = r(\theta)$ the orbital integral can be computed using the Cartan decomposition $G = KAK$. Here we may take $K = T(\mathbb{R})$ the group of rotations. We have

$$\mathcal{O}_{r(\theta)}(f) = c\, F(\sin\theta)$$

with c a constant that depends on the choice of Haar measures and

$$F(\lambda) = \int_1^\infty f\begin{pmatrix} a(\lambda) & t\lambda \\ -t^{-1}\lambda & a(\lambda) \end{pmatrix} |t - t^{-1}|\, \frac{dt}{t}$$

where

$$a(\lambda) = \sqrt{1-\lambda^2}$$

when θ is close to 0. Observe that since f is K-central we have for any a, b and c

$$f\begin{pmatrix} a & b \\ c & a \end{pmatrix} = f\begin{pmatrix} a & -c \\ -b & a \end{pmatrix}$$

and hence

$$F(\lambda) = \int_0^\infty \varepsilon(t-1) f \begin{pmatrix} a(\lambda) & t\lambda \\ -t^{-1}\lambda & a(\lambda) \end{pmatrix} dt$$

with $\varepsilon(x) = \operatorname{sign}(x)$. To study its asymptotic behaviour when $\lambda \to 0$ we consider

$$A(\lambda) = \int_0^\infty \varepsilon(t-1) f \begin{pmatrix} a(\lambda) & \lambda t \\ 0 & a(\lambda) \end{pmatrix} dt$$

By Taylor-Lagrange formula

$$F(\lambda) = A(\lambda) + \lambda B(\lambda)$$

where

$$B(\lambda) = \int_0^\infty \varepsilon(t-1) g \begin{pmatrix} a(\lambda) & t\lambda \\ -t^{-1}\lambda & a(\lambda) \end{pmatrix} \frac{dt}{t}$$

for some smooth function g compactly supported in the upper right variable and that has a $O(u)^{-1}$ decay in the lower left variable u so that the integral is absolutely convergent.

Observe that

$$A(\lambda) = |\lambda|^{-1} \int_0^\infty f \begin{pmatrix} 1 & \varepsilon(\lambda) u \\ 0 & 1 \end{pmatrix} du - 2f(\mathbf{1}) + o(\lambda)$$

Since $B(\lambda)$ is at most of logarithmic growth we see that the even functions

$$G(\lambda) = |\lambda|(F(\lambda) + F(-\lambda))$$

and

$$H(\lambda) = \lambda(F(\lambda) - F(-\lambda))$$

extend to continuous functions at the $\lambda = 0$.

We need more precise informations on the asymptotic behaviour. Hence we have to look more carefully to the error term $B(\lambda)$ which, we recall, equals:

$$\int_0^\infty \varepsilon(t-1) g \begin{pmatrix} a(\lambda) & t\lambda \\ -t^{-1}\lambda & a(\lambda) \end{pmatrix} \frac{dt}{t}$$

But this is the difference of two terms whose leading terms are equivalent to $\ln(|\lambda|^{-1}) g(\mathbf{1})$ up to continuous terms. Hence B is continuous. Generalizing this process one gets asymptotic expansions of the following form:

$$G(\lambda) = \sum_{n=0}^{N} (a_n + b_n\,|\lambda|)\,\lambda^{2n} + o(\lambda^{2N})$$

and

$$H(\lambda) = \sum_{n=0}^{N} h_n\,\lambda^{2n} + o(\lambda^{2N})$$

and hence $H(\lambda)$ is smooth. We have proved the following

LEMMA 4.3. *There is a smooth fonction h on $T(\mathbb{R})$ such that:*

$$h(\gamma) = \Delta(\gamma)\left(O_\gamma(f) - O_{w(\gamma)}(f)\right)$$

for $\gamma = r(\theta) \in T(\mathbb{R})$ *and*

$$\Delta(r(\theta)) = -2i\sin\theta$$

This lemma establishes the simplest case of a non-trivial geometric endoscopic transfer for an elliptic endoscopic group. A variant of this lemma establishes the transfer between orbital integrals weighted by the character of order two

$$\omega = \text{sign} \circ \det$$

for functions on $GL(2,\mathbb{R})$:

$$\mathcal{O}^\omega_\gamma(f) = \int_{G_\gamma\backslash G} \omega(x) f(x^{-1}\gamma x)\, d\dot{x}$$

and functions on the elliptic maximal torus

$$\widetilde{T}(\mathbb{R}) \simeq \mathbb{C}^\times = \left\{\begin{pmatrix} a & b \\ -b & a\end{pmatrix} \in GL(2,\mathbb{R})\right\}$$

LEMMA 4.4. *There is a smooth fonction h on $\widetilde{T}(\mathbb{R})$ such that:*

$$h(\gamma) = \Delta(\gamma) O^\omega_\gamma(f)$$

for $\gamma = \rho\, r(\theta) \in \widetilde{T}(\mathbb{R})$ *and*

$$\Delta(r(\theta)) = -2i\sin\theta$$

An other simple example is given by the transfer between inner forms. Consider the compact group $G(\mathbb{R}) = SU(2)$. The group G is an inner form of $G^* = SL(2)$. In fact one has $g \in SU(2)$ iff

$$g \in SL(2,\mathbb{C}) \quad \text{and} \quad g = J\sigma(g)J^{-1}$$

with

$$J = \begin{pmatrix} 0 & -1 \\ 1 & 0 \end{pmatrix}$$

and $\sigma(g)$ is the complex conjugate of g.

There is an injection of the set of conjugacy classes inside $G(\mathbb{R})$ into the set of stable conjugacy classes in $G^*(\mathbb{R})$ induced by the following correspondence: given $\gamma \in SU(2)$ we denote by $\tilde{\gamma}$ any semisimple element $\tilde{\gamma} \in SL(2,\mathbb{R})$ with the same eigenvalues.

LEMMA 4.5. *Given $f \in \mathcal{C}^\infty_c(SU(2))$ there is a function $\tilde{f} \in \mathcal{C}^\infty_c(SL(2,\mathbb{R}))$ such that for $\gamma \in SU(2)$ non-central*

$$SO_{\tilde{\gamma}}(\tilde{f}) = -O_\gamma(f)$$

where

$$SO_\gamma(\tilde{f}) = \sum SO_{w(\tilde{\gamma})}(\tilde{f})$$

and the sum is over the complex Weyl group for $SO(2) \subset SL(2,\mathbb{R})$ *and*

$$SO_{\tilde{\gamma}}(\tilde{f}) = O_{\tilde{\gamma}}(\tilde{f}) = 0$$

if $\tilde{\gamma} \in SL(2,\mathbb{R})$ *has distinct real eigenvalues.*

An easy proof is obtained using pseudo-coefficients. This will be explained in a fairly general case in 7.2.

In the case of $GL(2)$ a similar result is used for all local fields in Jacquet-Langlands. This lemma is a particular case of the general result due to Shelstad [She1] on the transfer between inner forms of real groups.

5. $U(2,1)$

5.1. Endoscopy for $U(2,1)$. Consider the quasisplit unitary group in three variables $G(\mathbb{R}) = U(2,1)$. A maximal compact torus $T(\mathbb{R})$ in $G(\mathbb{R})$ is such that

$$T(\mathbb{R}) \simeq U(1)^3$$

and can be represented by matrices

$$g(u,v,w) = \begin{pmatrix} u\cos\theta & 0 & iu\sin\theta \\ 0 & v & 0 \\ iu\sin\theta & 0 & u\cos\theta \end{pmatrix}$$

where u and v are complex numbers with $|u| = |v| = 1$ and $w = r(\theta)$. More precisely

$$(u,v,w) \mapsto g(u,v,w)$$

defines a twofold cover T_1 of T. The root system can be represented by the $\alpha_{i,j} = \phi_i - \phi_j$ with

$$e^{i\phi_1} = ue^{i\theta} \quad e^{i\phi_2} = v \quad e^{i\phi_3} = ue^{-i\theta}$$

We take as positive roots those with $i < j$. We observe that the half sum of the roots verifies $\rho = \alpha_{1,3}$.

The complex Weyl group is isomorphic to $\mathfrak{S}_3$ while the real Weyl group is isomorphic to $\mathfrak{S}_2$. The set of conjugacy classes inside a strongly regular stable elliptic conjugacy class is in bijection with the pointed set

$$\mathfrak{D}(\mathbb{R},T,G) = \mathfrak{S}_3/\mathfrak{S}_2$$

(see 6.4) that can be viewed as a sub-pointed-set of the group

$$\mathfrak{E}(\mathbb{R},T,G) = (\mathbb{Z}_2)^2$$

We shall denote by $\mathfrak{K}(\mathbb{R},T,G)$ its Pontryagin dual.

Choose as a maximal compact subgroup K in $G(\mathbb{R})$

$$K = \left\{ k = \frac{1}{2}\begin{pmatrix} \lambda+\nu & -\sqrt{2}\sigma & \lambda-\nu \\ \sqrt{2}\tau & 2\mu & -\sqrt{2}\tau \\ \lambda-\nu & \sqrt{2}\sigma & \lambda+\nu \end{pmatrix} \right\}$$

Observe that k is a conjugate of

$$\begin{pmatrix} \lambda & 0 & 0 \\ 0 & \mu & \tau \\ 0 & \sigma & \nu \end{pmatrix} \in U(1) \times U(2) \subset U(3)$$

The compact group K has been so chosen that it contains $T(\mathbb{R})$. The Weyl group of K is generated by w_0 the symmetry with respect to ρ.

5.2. Discrete series and transfer for $U(2,1)$. Now, consider a dominant weight λ. One attaches to it a discrete series representation π_μ with

$$\mu = \lambda + \rho$$

The L-packet containing π_μ contains also the various $\pi_{w\mu}$. Moreover $\pi_{w\mu}$ is equivalent to $\pi_{w_0 w\mu}$ and hence we may parametrize the elements in the L-packet by the elements in the Weyl group such that sign $(w) = 1$. For such a w the character of $\pi_{w\mu}$ is given by

$$\Theta_{w\mu}(\gamma) = \frac{\gamma^{w(\mu)} - \gamma^{w_0 w(\mu)}}{\gamma^\rho \Delta_B(\gamma)}$$

with

$$\Delta_B(\gamma) = \prod_{\alpha>0} (1 - \gamma^{-\alpha})$$

Consider $\kappa \neq 1$ in $\mathfrak{K}(\mathbb{R}, T, G)$ such that $\overline{\kappa}(\check{\alpha}_{1,3}) = 1$. Such a κ is unique: in fact one has necessarily

$$\overline{\kappa}(\check{\alpha}_{1,2}) = \overline{\kappa}(\check{\alpha}_{2,3}) = -1$$

The endoscopic group H one associates to κ is isomorphic to

$$U(1,1) \times U(1)$$

the positive root of T in H (for a compatible order) being

$$\alpha_{1,3} = \rho$$

The group H can be embedded in G as the subgroup of matrices in G of the form

$$g(u,v,w) = \begin{pmatrix} ua & 0 & iub \\ 0 & v & 0 \\ -iuc & 0 & ud \end{pmatrix}$$

with

$$w = \begin{pmatrix} a & b \\ c & d \end{pmatrix} \quad \text{and} \quad ad - bc = 1$$

It will be useful to consider also the twofold cover

$$H_1 = U(1) \times U(1) \times SL(2)$$

(with maximal torus T_1) defined by

$$(u,v,w) \mapsto g(u,v,w)$$

Let f_μ be a pseudo-coefficient for π_μ then the κ-orbital integral of a γ regular in $T(\mathbb{R})$ is given by

$$\mathcal{O}_\gamma^\kappa(f_\mu) = \sum_{\mathrm{sign}\,(w)=1} \kappa(w)\Theta_\mu^G(\gamma_w^{-1})$$

but using the bijection

$$\Omega_G/\Omega_K \simeq \Omega_K\backslash\Omega_G$$

this can be rewritten

$$\mathcal{O}_\gamma^\kappa(f_\mu) = \sum_{\mathrm{sign}\,(w)=1} \kappa(w)\Theta_{w\mu}(\gamma^{-1})$$

The transfer factor $\Delta(\gamma, \gamma_H)$ is given by

$$(-1)^{q(G)+q(H)}\chi_{G,H}(\gamma)\Delta_B(\gamma^{-1}) \,.\, \Delta_{B_H}(\gamma_H^{-1})^{-1}$$

for some character $\chi_{G,H}$ defined as follows. It would be canonical to take

$$\chi_{G,H}(\gamma^{-1}) = \gamma^{\rho-\rho_H}$$

but this is does not make sense since ρ_H defines a character of the twofold cover $T_1(\mathbb{R})$ but not of $T(\mathbb{R})$. A way out is to consider a weight ξ that defines a character of the cocenter of the twofold cover $H_1(\mathbb{R})$ of $H(\mathbb{R})$ so that

$$\chi_{G,H}^{-1} = e^{\rho-\rho_H+\xi}$$

descends to a character of $T(\mathbb{R})$. With such a choice we get when $\mathrm{sign}\,(w) = 1$

$$\Delta(\gamma^{-1}, \gamma_H^{-1})\Theta_{w\mu}^G(\gamma) = -\frac{\gamma_H^{w(\mu)+\xi} - \gamma_H^{w_0 w(\mu)+\xi}}{\gamma^{\rho_H}\Delta_{B_H}(\gamma_H)}$$

We observe that $\kappa(w) = -1$ if $\mathrm{sign}\,(w) = 1$ and $w \neq 1$ and that $w\mu$ is positive or negative with respect to B_H according to the sign of $\kappa(w)$. Hence

$$\Delta(\gamma, \gamma_H)\Theta_{w\mu}^G(\gamma^{-1}) = \kappa(w)^{-1} S\Theta_\nu^H(\gamma_H^{-1})$$

where $S\Theta_\nu^H$ is the character of the L-packet Σ_ν of discrete series for H defined by the parameter

$$\nu = w(\mu) + \xi$$

Altogether we get

$$\Delta(\gamma, \gamma_H)\mathcal{O}_\gamma^\kappa(f_\mu) = \sum_\nu S\Theta_\nu^H(\gamma_H^{-1})$$

the sum being over the $\nu = w(\mu) + \xi$ with $\mathrm{sign}\,(w) = 1$ and this can be rewritten

$$\Delta(\gamma, \gamma_H)\mathcal{O}_\gamma^\kappa(f_\mu) = \sum_\nu S\mathcal{O}_{\gamma_H}(g_\nu)$$

where g_ν is a pseudo-coefficient for any one of the discrete series of $H(\mathbb{R})$ in the L-packet Σ_ν attached to ν. In other words we have

$$f_\mu^H = \sum_\nu g_\nu$$

Notice that the set of parameters ν depends on the choice of ξ. Alternatively we could use the canonical choice

$$\chi_{G,H}(\gamma^{-1}) = \gamma^{\rho-\rho_H}$$

if we replace H by H_1 and then the set of parameters would be canonical. This second solution is not the classical one but it seems after all more natural and at any rate considering central extensions H_1 instead of H cannot be avoided in general.

Now more generally we get

$$f^H_{w\mu} = \sum_\nu a(w,\nu) g_\nu$$

with

$$a(w_1, w_2\mu) = \kappa(w_2)\kappa(w_2 w_1)^{-1} .$$

Since pseudo-coefficients give a dual basis of the set of characters, in the space they generate, we get

$$\text{trace } \Sigma_\nu(f^H) = \sum_w a(w,\nu) \text{ trace } \pi_{w\mu}(f)$$

at least when f is a linear combination of pseudo-coefficients. Now, observe that

$$a(w_1, w_2\mu) = \kappa(w_2)\kappa(w_2 w_1)^{-1} = \kappa^{w_2}(w_1)^{-1}$$

Let us denote by κ_ν the character

$$x \mapsto \kappa^w(x)^{-1}$$

when $\nu = w(\mu) + \xi$. Now, using the bijection

$$\Pi_\mu \simeq \mathfrak{D}(\mathbb{R}, T, G)$$

we get a pairing denoted $<,>$ between Π_μ and $\mathfrak{K}(\mathbb{R}, T, G)$ and the transfer equation can be written

$$\text{trace } \Sigma_\nu(f^H) = \sum_{\pi\in\Pi_\mu} <\kappa_\nu, \pi> \text{ trace } \pi(f)$$

The above formula for the dual transfer can be rewritten

$$\text{trace } \Sigma_s(f^H) = \sum_{\pi\in\Pi_\Sigma} <s, \pi> \text{ trace } \pi(f)$$

where the pairing $<s,\pi>$ is the Shelstad pairing between

$$\mathbf{S}_\varphi \simeq \mathfrak{K}(\mathbb{R}, T, G)$$

and Π_μ. Observe that Π_μ is of order 3 and is identified by this pairing to a subset of the group of characters of $\mathfrak{K}(\mathbb{R}, T, G)$; the missing character corresponds to a representation of the compact inner form that has to be added to get a K-group (see 8.1).

5.3. The dual picture for $U(2,1)$. Let us now describe the dual picture. Observe that $\overline{\kappa}$ is the image, via Tate-Nakayama duality (see 6.5), of

$$s = \begin{pmatrix} 1 & 0 & 0 \\ 0 & -1 & 0 \\ 0 & 0 & 1 \end{pmatrix} \in \check{T} \subset \check{G}$$

and hence

$$\check{H} = \begin{pmatrix} * & 0 & * \\ 0 & * & 0 \\ * & 0 & * \end{pmatrix}$$

The holomorphic Galois action in ${}^LG$ is given by

$$g \mapsto J^t g^{-1} J^{-1}$$

with

$$J = \begin{pmatrix} 0 & 0 & 1 \\ 0 & -1 & 0 \\ 1 & 0 & 0 \end{pmatrix}$$

while in ${}^LH$ it is $h \mapsto J_H {}^t h^{-1} J_H^{-1}$ with

$$J_H = \begin{pmatrix} 0 & 0 & 1 \\ 0 & 1 & 0 \\ -1 & 0 & 0 \end{pmatrix}$$

As a consequence of the difference between J and J_H there is no canonical admissible embedding of ${}^LH$ in ${}^LG$.

6. Galois cohomology and endoscopy

6.1. Non abelian hypercohomology. Consider first a set X with an action of a group Γ. By definition $H^0(\Gamma, X)$ is the fixed point set X^Γ: the set of $x \in X$ such that

$$x = \sigma(x) \quad \text{for all} \quad \sigma \in \Gamma$$

Now let G be an algebraic group with an action of Γ compatible with the group structure. Then $H^0(\Gamma, G)$ is a group. Moreover we may define a set $H^1(\Gamma, G)$ as the quotient of the set $Z^1(\Gamma, G)$ of 1-cocycles i.e. maps

$$\sigma \in \Gamma \mapsto a_\sigma \in G$$

such that

$$a_\sigma \sigma(a_\tau) a_{\sigma\tau}^{-1} = 1$$

by the equivalence relation

$$a'_\sigma \simeq a_\sigma \Longleftrightarrow a'_\sigma = b\, a_\sigma\, \sigma(b)^{-1}$$

for some $b \in G$. The set $H^1(\Gamma, G)$ has no natural group structure, but the trivial class makes it a pointed set. Finally, if A is an abelian group, we have the familiar cohomology groups in all degrees

$$H^i(\Gamma, A)$$

This can be slightly generalized as follows. One can define a hyper-cohomological theory in degrees ≤ 0 for complexes of the form

$$[G \Rightarrow X]$$

where G is a group acting (on the left) on a set X with a group Γ acting on the complex. The cohomology set

$$H^0(\Gamma, G \Rightarrow X)$$

it the quotient of the set of pairs (x, a) with $x \in X$ and a a map from Γ to X with

$$x = a_\sigma \,.\, \sigma(x) \quad \text{and} \quad a_\sigma \sigma(a_\tau) a_{\sigma\tau}^{-1} = 1$$

by the equivalence relation

$$(x', a') \simeq (x, a) \Longleftrightarrow x' = bx \quad \text{and} \quad a'_\sigma = b\, a_\sigma\, \sigma(b)^{-1}$$

There is also an hyper-cohomological theory in degrees ≤ 1 for complexes of groups of the form

$$[G' \to G]$$

that are "left crossed modules". This means that G and G' are groups and we are given not only a homomorphism $G' \to G$, which yields an action of G' on G by left translations, but also an action of G on G' and that those two actions are compatible with adjoint actions. This implies in particular that

$$\ker[G' \to G]$$

is an abelian group.

A morphism of crossed modules

$$[G' \to G] \to [H' \to H]$$

is a pair of maps $G' \to H'$ and $G \to H$ that intertwines the various actions, in particular we have a commutative diagram

$$\begin{array}{ccc} G' & \to & G \\ \downarrow & & \downarrow \\ H' & \to & H \end{array}$$

Such a morphism is a quasi-isomorphism if it induces isomorphisms for kernels and cokernels.

If we have an action of a group Γ, compatible with the structure of crossed module, one has a cohomology theory

$$H^i(\Gamma, G' \to G)$$

in degrees $i \leq 1$. The H^1 is defined as a quotient of the set of 1-hyper-cocycles (a, b) with

$$a_\sigma \sigma(a_\tau) a_{\sigma\tau}^{-1} = \rho(b_{\sigma,\tau})$$

where ρ is the homomorphism $G' \to G$ and

$$a_\sigma * \sigma(b_{\tau,\mu}) \,.\, b_{\sigma,\tau\mu} = b_{\sigma,\tau} \,.\, b_{\sigma\tau,\mu}$$

where $*$ denotes the action of G on G'. The equivalence relation is

$$a'_\sigma = \alpha\,\rho(\beta_\sigma)a_\sigma\,\sigma(\alpha)^{-1}$$

and

$$b'_{\sigma,\tau} = \alpha * (\beta_\sigma(a_\sigma * \sigma(\beta_\tau))b_{\sigma,\tau}\beta_{\sigma\tau}^{-1})$$

There is a natural group structure on H^0 but again H^1 is simply a pointed set while

$$H^{-1}(\Gamma, G' \to G) := H^0(\Gamma, \ker[G' \to G])$$

is an abelian group. These cohomological objects are functorial for the category of crossed modules and, as for usual hypercohomology theories, they are invariant under quasi-isomorphisms. Cohomology in higher degrees cannot be defined (in a functorial way) without further structures.

We observe that G can be seen as the crossed module

$$[1 \to G]$$

and we have

$$H^i(\Gamma, G) = H^i(\Gamma, 1 \to G)$$

There is an exact sequence

$$1 \to H^{-1}(\Gamma, G' \to G) \to H^0(\Gamma, G') \to H^0(\Gamma, G) \to$$

$$H^0(\Gamma, G' \to G) \to H^1(\Gamma, G') \to H^1(\Gamma, G) \to H^1(\Gamma, G' \to G)$$

with

$$H^{-1}(\Gamma, G' \to G) := H^0(\Gamma, \ker \rho)$$

The exactness has to be understood for pointed sets. In general the exact sequence cannot be continued since the H^2 does not make sense for a non commutative group.

If we denote by K the kernel of $[G' \to G]$ and by L the cokernel we have another exact sequence:

$$1 \to H^0(\Gamma, K) \to H^{-1}(\Gamma, G' \to G) \to H^{-1}(\Gamma, L) \to$$

$$H^1(\Gamma, K) \to H^0(\Gamma, G' \to G) \to H^0(\Gamma, L) \to H^2(\Gamma, K)$$

$$\to H^1(\Gamma, G' \to G) \to H^1(\Gamma, L) \to H^3(\Gamma, K)$$

where $H^{-1}(\Gamma, H) = 1$. Observe that since K is abelian the $H^i(\Gamma, K)$ are defined for all i.

If moreover the complex $G' \to G$ is quasi-isomorphic to a complex of abelian groups $B \to A$ (in a way compatible with Γ-actions) then one can define hyper-cohomology groups

$$H^i(\Gamma, G' \to G) := H^i(\Gamma, B \to A)$$

in all degrees and the above exact sequence extends indefinitely.

6.2. Galois cohomology and abelianized cohomology. Let F be a field and let $\overline{F}$ be an separable closure. Consider an algebraic variety X defined over F then the set $X(F)$ of points with value in F is the fixed point set of $\mathrm{Gal}(\overline{F}/F)$ on $X(\overline{F})$. This is the 0-th cohomology set of the Galois group with values in $X(\overline{F})$:

$$X(F) = H^0(\mathrm{Gal}(\overline{F}/F), X(\overline{F}))$$

We shall also use the standard notation for Galois cohomology

$$H^*(F, X) := H^*(\mathrm{Gal}(\overline{F}/F), X(\overline{F}))$$

whenever defined. In particular if G is an algebraic group over F we have the pointed set

$$H^1(F, G) := H^1(\mathrm{Gal}(\overline{F}/F), G(\overline{F}))$$

and if A is an abelian algebraic group we have cohomology groups

$$H^i(F, A) := H^i(\mathrm{Gal}(\overline{F}/F), A(\overline{F}))$$

for all i.

Let G be a connected reductive group over a field F and let G_{SC} be the simply connected cover of its derived subgoup. We observe that complexes

$$[G_{SC} \to G]$$

are crossed modules for the obvious adjoint actions and hence we have at hand

$$H^*(F, G_{SC} \to G)$$

in degrees ≤ 1. But in fact such a complex is quasi-isomorphic to a complex of abelian groups:

$$[Z_{sc} \to Z]$$

where Z is the center of G and Z_{sc} is the center of G_{SC}. Since hypercohomology is invariant by quasi-isomorphisms this allows to define abelian groups

$$H^*(F, G_{SC} \to G) = H^*(F, Z_{sc} \to Z)$$

in all degrees. Another quasi-isomophic complex of abelian groups is useful: let T be a torus in G and let T_{sc} its preimage in G_{SC} then

$$H^*(F, G_{SC} \to G) = H^*(F, T_{sc} \to T)$$

We shall use the following compact notation

$$H^*_{ab}(F, G) := H^*(F, G_{SC} \to G)$$

THEOREM 6.1. *Let G be a connected reductive group over a field F and let G_{SC} be the simply connected cover of its derived subgoup. There is a family of abelian group $H^*_{ab}(F, G)$ with natural maps*

$$H^i(F, G) \to H^i_{ab}(F, G) \quad \textit{for} \quad i \leq 1$$

giving rise to an exact sequence

$$\cdots \to H^0_{ab}(F, G) \to H^1(F, G_{SC}) \to H^1(F, G) \to H^1_{ab}(F, G)$$

Moreover, when F is a local field

$$H^1(F,G) \to H^1_{ab}(F,G)$$

is surjective and even bijective if F is non-archimedean.

Proof. The surjectivity follows from the existence of fundamental tori and that such tori have a vanishing H^2. The injectivity follows from a theorem due to Kneser quoted below. □

THEOREM 6.2. *When F is a non-archimedean local field then*

$$H^1(F,G_{SC}) = 1 .$$

We observe that using the universal property of simply connected spaces one can show that the abelianization map

$$H^1(F,G) \to H^1_{ab}(F,G)$$

is functorial in G (cf.[Lab3] lemme I.6.3).

Before the introduction of abelianized cohomology by Borovoi [Bo] (see also [Lab3]), Kottwitz had found a substitute for the abelianization map but that was more subtle to define, not obviously functorial and was restricted to reductive groups over local fields. Given G reductive over some field F consider the abelian group

$$\pi_0(Z(\check{G})^\Gamma)$$

where $\check{G}$ is the complex dual group, $Z(\check{G})$ its center, Γ the Galois group $\mathrm{Gal}(\overline{F}/F)$ and $\pi_0(\bullet)$ the group of connected components of $\bullet$. Kottwitz has shown that when F is a local field there exists a canonical map

$$H^1(F,G) \to \pi_0(Z(\check{G})^\Gamma)^D$$

where the exponent D denotes the Pontryagin duality. The map can be constructed as follows. Observe that

$$[\check{G} \to (G_{SC})\check{}\,]$$

is quasi-isomorphic to

$$[Z(\check{G}) \to 1]$$

then, by Tate-Nakayama duality one gets an injective homomorphism

$$H^1_{ab}(F,G) \to \pi_0(Z(\check{G})^\Gamma)^D$$

which is bijective when F is non-archimedean ([Lab3] Proposition 1.7.3). This homomorphism composed with the abelianization map $H^1(F,G) \to H^1_{ab}(F,G)$ gives Kottwitz' map.

We still have to introduce the Kottwitz signs [Ko1]. Given a reductive group G consider its quasisplit inner form G^* and G^*_{ad} the adjoint group. Let $a(G)$ be the cohomology class in $H^1(F,G^*_{ad})$ defining G as an inner form of G^*. There is an isomorphism from

$$H^*_{ab}(F,G^*_{ad}) \to H^2(F,Z(G^*_{SC}))$$

and by composition we get a class

$$\overline{a}(G) \in H^2(F, Z(G^*_{SC}))$$

Now the half sum of positive roots (for some order) defines a map from $Z(G^*_{SC})$ to the group of elements of order 2 in the multiplicative group and hence if F is local we get a number $e(G) \in \{\pm 1\}$ called the Kottwitz sign. It can be shown that when $F = \mathbb{R}$ one has

$$e(G) = (-1)^{q(G^*)-q(G)}$$

where $q(G)$ is half the dimension of the symmetric space attached to G.

6.3. Stable conjugacy and κ-orbital integrals. We shall now define stable conjugacy in general. Let G be a connected reductive group over some field F. For simplicity we assume F is a field of characteristic zero.

We denote by G_{SC} the simply connected cover of its derived subgroup. Given $\gamma \in G(F)$ we denote by I_γ the subgroup of G generated by $Z(G)$ (the center of G) and the image in G of the centralizer of γ in G_{SC}. We call I_γ the stable centralizer of γ.

When γ is strongly regular semisimple i.e. if the centralizer is a torus, then I_γ is this torus. If γ is regular semisimple i.e. if the centralizer has a torus as connected component, then again I_γ is this torus. But in general I_γ is not a torus and can even be disconnected as seen in the example of $n_0 \in SL(2, \mathbb{R})$ whose centralizer is $\pm N$ with N the group of upper-triangular unipotent matrices. Nevertheless we have a

LEMMA 6.3. *Let γ be semisimple in $G(F)$. Then I_γ is a connected reductive subgroup, namely the connected component of the centralizer.*

Proof. The centralizer in G_{SC} of a semisimple element γ is a connected reductive group; this is due to Steinberg. Hence, its image in G is contained in the connected centralizer which, in turn, contains a maximal torus T, but any T contains $Z(G)$. □

Consider γ and γ' in $G(F)$. We shall say that γ and γ' are stably conjugate if there is an $x \in G(\overline{F})$ such that

(*i*) $$x^{-1}\gamma\, x = \gamma'$$

and

(*ii*) $$a_\sigma = x\sigma(x)^{-1} \in I_\gamma \quad \text{for all} \quad \sigma \in \mathrm{Gal}(\overline{F}/F)$$

where $\mathrm{Gal}(\overline{F}/F)$ is the Galois group of $\overline{F}/F$. In other words we assume that a_σ is a 1-cochain with values in I_γ.

The same relations hold if we replace x by $y = tx$ with $t \in I_\gamma$. The second condition is automatically satisfied if the first is, whenever $G = G_{SC}$: in fact a_σ always belongs to the centralizer of γ.

The pair (x, a_σ) defines a 0-cocycle with values in the quotient set

$$I_\gamma \backslash G$$

and hence a class in the Galois cohomology set associated to this set

$$H^0(\mathrm{Gal}(\overline{F}/F), I_\gamma \backslash G)$$

also denoted

$$H^0(F, I_\gamma \backslash G) \quad \text{or} \quad (I_\gamma \backslash G)(F)$$

This is the set of rational points of the quotient $I_\gamma \backslash G$. This cohomology set is the quotient of the set of 0-hypercocycles as above by the equivalence relation:

$$(x, a_\sigma) \simeq (y, b_\sigma)$$

whenever there exists $t \in I_\gamma$ such that

$$y = tx \quad \text{and} \quad b_\sigma = t a_\sigma \sigma(t)^{-1} .$$

We observe that there is an exact sequence of pointed sets

$$H^0(F, I_\gamma \backslash G) \to H^1(F, I_\gamma) \to H^1(F, G)$$

induced by the map $(x, a_\sigma) \mapsto a_\sigma$ and the inclusion $I_\gamma \to G$. One denotes by $\mathfrak{D}(F, I_\gamma \backslash G)$ the image of $H^0(F, I_\gamma \backslash G)$ into $H^1(F, I_\gamma)$ or equivalently the kernel of the next map

$$\mathfrak{D}(F, I_\gamma \backslash G) = \ker \left[H^1(F, I_\gamma) \to H^1(F, G)\right]$$

With this we have a small exact sequence of pointed sets

$$1 \to I_\gamma(F) \backslash G(F) \to (I_\gamma \backslash G)(F) \to \mathfrak{D}(F, I_\gamma \backslash G) \to 1$$

To continue, let us assume first that γ is regular semisimple and hence I_γ is a maximal torus T in G defined over F. We shall now introduce the abelian groups $\mathfrak{E}(F, T \backslash G)$. We first recall that

$$\mathfrak{D}(F, T \backslash G) = \ker \left[H^1(F, T) \to H^1(F, G)\right]$$

the group $\mathfrak{E}(F, T \backslash G)$ is an abelianized version of it namely

$$\mathfrak{E}(F, T \backslash G) = \ker \left[H^1(F, T) \to H^1_{ab}(F, G)\right]$$

There is a natural injective map

$$\mathfrak{D}(F, T \backslash G) \to \mathfrak{E}(F, T \backslash G)$$

Observe that by composition we get a map

$$(T \backslash G)(F) \to \mathfrak{E}(F, T \backslash G)$$

This can also be seen as follows: we have

$$\mathfrak{E}(F, T \backslash G) = \ \mathrm{Im}\ \left[H^1(F, T_{sc}) \to H^1(F, T)\right]$$

and that there is a natural map

$$(T \backslash G)(F) \to H^1(F, T_{sc})$$

since $H^1(F, T_{sc})$ can be identified with an abelianized avatar of

$$H^0(F, T \backslash G) = (T \backslash G)(F)$$

Assume from now on that F is a local field. When F is non-archimedean the map

$$\mathfrak{D}(F, T\backslash G) \to \mathfrak{E}(F, T\backslash G)$$

is bijective since

$$H^1(F, G) \to H^1_{ab}(F, G)$$

is bijective in this case. Over the reals, as we shall see in the next section, the set $\mathfrak{D}(\mathbb{R}, T\backslash G)$ is a subset but not usually a subgroup of $\mathfrak{E}(\mathbb{R}, T\backslash G)$.

We shall denote by $\mathfrak{K}(F, T\backslash G)$ the Pontryagin dual of the finite abelian group $\mathfrak{E}(F, T\backslash G)$. We call the elements of $\mathfrak{K}(F, T\backslash G)$ endoscopic characters. Let $\kappa \in \mathfrak{K}(F, T\backslash G)$; it defines by composition with the map

$$(T\backslash G)(F) \to \mathfrak{E}(F, T\backslash G)$$

a function, again denoted κ on $(T\backslash G)(F)$. The function κ has the following multiplicative property: assume that x and $z = xy$ in G define elements in $(T\backslash G)(F)$. This means that $x\sigma(x)^{-1} \in T$ and $z\sigma(z)^{-1} \in T$ for all $\sigma \in \mathrm{Gal}(\overline{F}/F)$ then

$$y\sigma(y)^{-1} \in T_x := x^{-1}Tx \quad \text{and} \quad \kappa(xy) = \kappa^x(y)\kappa(x)$$

where κ^x is the character of $\mathfrak{E}(F, T_x\backslash G)$ obtained from κ via the F-isomorphism

$$T \to x^{-1}Tx$$

A κ-orbital integral for γ regular in $T(\mathbb{R})$ is defined by

$$\mathcal{O}^\kappa_\gamma(f) = \int_{(T\backslash G)(F)} \kappa(x) f(x^{-1}\gamma\, x)\, d\dot{x}$$

Implicit in the definition is a compatible choice of Haar measures on the various stable conjugates $x^{-1}Tx$ of T for $x \in (T\backslash G)(F)$ obtained by transporting invariant differential forms of maximal degree.

The stable orbital integrals are the $\mathcal{O}^1_\gamma(f)$ i.e. κ-orbital integrals when κ is trivial. They are often denoted $S\mathcal{O}_\gamma(f)$.

Now consider the case of an arbitrary semisimple element γ and let $I = I_\gamma$ its stable centralizer. We put again

$$\mathfrak{D}(F, I\backslash G) = \ker\left[H^1(F, I) \to H^1(F, G)\right]$$

it has an abelianized avatar

$$\mathfrak{E}(F, I\backslash G) = \ker\left[H^1_{ab}(F, I) \to H^1_{ab}(F, G)\right]$$

with a map

$$\mathfrak{D}(F, I\backslash G) \to \mathfrak{E}(F, I\backslash G)$$

which need not be injective nor surjective in general. Now given κ a character of $\mathfrak{E}(F, I\backslash G)$ the κ-orbital integral is defined as follows

$$\mathcal{O}^\kappa_\gamma(f) = \int_{(I\backslash G)(F)} e(I_x)\kappa(x) f(x^{-1}\gamma\, x)\, d\dot{x}$$

where $e(I_x)$ is the Kottwitz sign of the group

$$I_x := x^{-1} I\, x$$

the stable centralizer of $x^{-1}\gamma\, x$.

6.4. Stable conjugacy and compact Cartan subgroups over $\mathbb{R}$. We shall now consider the special case where $F = \mathbb{R}$ and γ is elliptic and regular in G which means that

$$T = I_\gamma$$

is an elliptic torus i.e. such that T_{sc} is an $\mathbb{R}$-anisotropic torus.

LEMMA 6.4. *Let T be an elliptic torus in a real reductive group G. For any n in $N_G(T)$ the automorphism $w = \mathrm{Ad}(n)$ restricted to T is defined over $\mathbb{R}$.*

Proof. Any automorphism w of T is defined by a $\mathbb{Z}$-linear automorphism w^* of $X^*(T_{ad})$; it suffices to prove that w^* is real. But σ the non-trivial element in $\mathrm{Gal}(\mathbb{C}/\mathbb{R})$ acts as -1 on $X^*(T_{ad})$ and hence commutes with w^*. □

We denote by $\Omega_{\mathbb{C}}(G,T)$, or simply Ω_G, the complex Weyl group, which is the group generated by the automorphisms of T induced by $\mathrm{Ad}(n)$ with $n \in N_G(T)$ and $\Omega_{\mathbb{R}}(G,T)$, or Ω_K, the subgroup generated by the automorphisms of T induced by $\mathrm{Ad}(n)$ with

$$n \in G(\mathbb{R}) \cap N_G(T) = K \cap N_G(T)$$

where K is the maximal compact subgroup of $G(\mathbb{R})$ containing $T(\mathbb{R})$.

PROPOSITION 6.5. *Let T be an elliptic torus in G over $\mathbb{R}$. There is a natural bijection*

$$\mathfrak{D}(\mathbb{R}, T\backslash G) \to \Omega_{\mathbb{C}}(G,T)/\Omega_{\mathbb{R}}(G,T)$$

Proof. Consider γ strongly regular in T and in particular $T = I_\gamma$. Then, we observe that if γ' is stably conjugate to γ then, $I_{\gamma'}$ is again an elliptic torus; but all elliptic tori are conjugate in $G(\mathbb{R})$. Hence, up to ordinary conjugacy we may replace γ' by $\gamma'' \in T$ and now there is n in the normalizer $N_G(T)$ of T in G such that

$$n^{-1}\gamma\, n = \gamma''$$

The image of n in the Weyl group is uniquely defined by the pair (γ, γ'') since γ is strongly regular but the element γ'' is defined by γ' only up to the action of the real Weyl group. This yields an injective map

$$\mathfrak{D}(\mathbb{R}, T\backslash G) \to \Omega_{\mathbb{C}}(G,T)/\Omega_{\mathbb{R}}(G,T)$$

Now the map is surjective thanks to lemma 6.4. □

More generally we observe that all conjugacy classes inside the stable conjugacy class of an elliptic element in T, intersect T.

REMARK. The set $\mathfrak{D}(\mathbb{R}, T\backslash G)$ is not usually a group. In fact $\Omega_{\mathbb{R}}(G,T)$ is not an invariant subgroup of $\Omega_{\mathbb{C}}(G,T)$ in general although $\mathfrak{D}(\mathbb{R}, T\backslash G)$ can be embedded naturally in the abelian group $\mathfrak{E}(\mathbb{R}, T\backslash G)$. For example consider $G = U(2,1)$, then

$$\Omega_{\mathbb{C}}(G,T) = \mathfrak{S}_3 \quad \text{while} \quad \Omega_{\mathbb{R}}(G,T) = \mathfrak{S}_2$$

In this case $\mathfrak{D}(\mathbb{R}, T\backslash G)$ has three elements and the group $\mathfrak{E}(\mathbb{R}, T\backslash G)$ is isomorphic to $\mathbb{Z}_2 \times \mathbb{Z}_2$.

6.5. Endoscopic groups. We have to recall the Tate-Nakayama isomophism for tori. Let S be a torus defined over F and split over a Galois extension K; we denote by $\check{X}(S)$ the $\mathbb{Z}$-free module of finite rank of its cocharacters.

THEOREM 6.6. *Assume that F is a local field. There is a canonical isomorphism between Tate cohomology groups*

$$\hat{H}^i(K/F, \check{X}(S)) \to \hat{H}^{i+2}(K/F, S(K))$$

In particular there is a canonical isomorphism

$$\hat{H}^{-1}(K/F, \check{X}(S)) \to H^1(K/F, S(K)) = H^1(F, S)$$

where, by definition

$$\hat{H}^{-1}(\Gamma, X) := \hat{H}_0(\Gamma, X)) = X^{N_\Gamma}/I_\Gamma X$$

where X^{N_Γ} is the kernel of the norm endomorphism of X:

$$N_\Gamma : x \mapsto \sum_{\sigma\in\Gamma} \sigma(x)$$

and I_Γ is the augmentation ideal in the group algebra $\mathbb{Z}[\Gamma]$

$$I_\Gamma = \left\{ \tau = \sum_{\sigma\in\Gamma} n_\sigma \, \sigma \;\middle|\; \sum_{\sigma\in\Gamma} n_\sigma = 0 \right\}$$

Altogether we have the

PROPOSITION 6.7. *There is a canonical isomorphism*

$$\check{X}(S)^{N_\Gamma}/I_\Gamma \check{X}(S) \to H^1(F, S)$$

with $\Gamma = \mathrm{Gal}(K/F)$.

Now there is a surjective map

$$H^1(F, T_{sc}) \to \mathfrak{E}(F, T\backslash G)$$

and if moreover we assume T elliptic then T_{sc} is anisotropic and hence we have a bijection

$$\check{X}(T_{sc})^{N_\Gamma} \to \check{X}(T_{sc}) \,.$$

The above proposition yields a surjective map

$$\check{X}(T_{sc}) \to \mathfrak{E}(F, T\backslash G)$$

and hence if T is elliptic any character κ of $\mathfrak{E}(F, T\backslash G)$ defines a character $\overline{\kappa}$ of $\check{X}(T_{sc})$.

Let R_G and $\check{R}_G$ denote the set of roots and coroots of T in G. We observe that the coroots are homomorphisms of the multiplicative group $\mathbb{G}_m$ into T that factor through T_{sc} and hence

$$\check{R}_G \subset \check{X}(T_{sc}) .$$

Let us denote by $\check{R}_{\overline{\kappa}}$ the subset of coroots $\check{\alpha} \in \check{R}$ such that

$$\overline{\kappa}(\check{\alpha}) = 1 .$$

LEMMA 6.8. *This is a (co)root system*

Proof. In fact $\check{R}_{\overline{\kappa}}$ is the root system for the connected centralizer $\check{H}$ of $s \in \check{T} \subset \check{G}$ whose image in $\check{X}(T_{sc})$ by duality is $\overline{\kappa}$. □

Let $X = X(T)$ and $\check{X} = \check{X}(T)$, there is a root datum

$$(X, R_{\overline{\kappa}}, \check{X}, \check{R}_{\overline{\kappa}})$$

which inherits, from the Galois action on T, a natural Galois action and we get a quasi-split reductive group denoted $H_{\overline{\kappa}}$ (or simply H) with maximal torus $T_H \simeq T$ and coroot system

$$\check{R}_H \simeq \check{R}_{\overline{\kappa}}$$

This is the elliptic endoscopic group attached to κ.

More generally, if we do not assume T elliptic, we can associate, as above, an endoscopic group H to any characters $\overline{\kappa}$ of $\check{X}(T_{sc})/I_\Gamma\check{X}(T_{sc})$. Such $\overline{\kappa}$'s naturally occur through local-global constructions for the stabilization of the trace formula. Now $\overline{\kappa}$ defines, by restriction to $\check{X}(T_{sc})^{N_\Gamma}$, a character κ of

$$H^1(F, T_{sc})$$

which descends to a character of $\mathfrak{E}(F, T\backslash G)$ if κ is trivial on the kernel of the map

$$H^1(F, T_{sc}) \to H^1(F, T)$$

and in all cases κ defines a function on $(T\backslash G)(F)$ via the natural map

$$(T\backslash G)(F) \to H^1(F, T_{sc})$$

It gives rise, when κ is non-trivial on the above kernel, to the variant of ordinary endoscopy that deals with orbital integrals weighted by a character ω as in 4.4 above.

We observe that H and G share a torus and that the Weyl group $\Omega(T_H, H)$ is canonically isomorphic to a subgroup of $\Omega(T, G)$. Nevertheless there may not exist any homomorphism from H to G that extends the isomorphism $T_H \simeq T$.

6.6. The dual picture. Consider T, $\overline{\kappa}$ and H as above. Recall that the dual groups $\check{G}$ and $\check{H}$ are equipped with a "splitting" (un épinglage) and a Galois action that preserves the splitting. The Galois action on the characters of T defines by duality an holomorphic action of Galois on $\check{T}$ (chosen as the torus in the splitting of $\check{G}$). For σ in the Galois group let us denote by $\check{\sigma}_T$ the corresponding automorphism of $\check{T}$ and by $\check{\sigma}_G$ the corresponding automorphism of $\check{G}$; then for $t \in \check{T}$ one has

$$\check{\sigma}_T(t) = n_\sigma \check{\sigma}_G(t)\, n_\sigma^{-1}$$

with n_σ in the normalizer of $\check{T}$ in $\check{G}$. Thanks to a vanishing property of inflation from Galois groups to Weil groups this Galois action that can be lifted to an homomorphism ξ of W_F into ${}^LG$

$$w \mapsto \xi(w) = n_w \rtimes w$$

with

$$\check{\sigma}_T(t) = \xi(w_\sigma) t\, \xi(w_\sigma)^{-1}$$

whenever w_σ projects to σ, allowing to embedd ${}^LT$ in ${}^LG$. This means that n_σ and $\xi(w_\sigma)$ have the same image in the Weyl group for $\check{T}$ in $\check{G}$.

Consider an element $s \in \check{T}$ giving $\overline{\kappa}$ by duality; s is defined only up to an element in the center of $\check{G}$ and its class is Galois-invariant. The connected centralizer of s in $\check{G}$ can be identified with $\check{H}$. The action of the Weil group W_F on $\check{T}$ induces an action on $\check{H}$ which coincides, up to inner automorphisms, with the action of W_F on $\check{H}$ in ${}^LH$. Now consider the subgroup $\mathcal{H}$ of ${}^LG$ generated by $\check{H}$ and the image of the Weil group by the above homomorphism ξ. The group $\mathcal{H}$ is independent of the choice of s and ξ and is a split extension:

$$1 \to \check{H} \to \mathcal{H} \to W_F \to 1$$

(as a split extension it depends on the choice of ξ).

An admissible embedding of ${}^LH$ in ${}^LG$ is an L-homomorphism

$$\eta : {}^LH \to {}^LG$$

that extends the natural inclusion

$$\check{H} \to \check{G}$$

(by L-homomorphism we understand an homomorphism whose restriction to $\check{H}$ is holomorphic and induces the identity on $\mathbb{W}_{\mathbb{R}}$). The existence of η is tantamount to the existence of an L-isomorphism

$$\mathcal{H} \to {}^LH$$

which is the identity on $\check{H}$.

But in general $\mathcal{H}$ and ${}^LH$ are not isomorphic and hence there is no admissible embedding (see [Sh3]) and even if they are isomorphic there may not exist a canonical one (an example is given in 5.3). Nevertheless one can always construct a central extension H_1 of H, such that

$$1 \to Z_1(F) \to H_1(F) \to H(F) \to 1$$

is an exact sequence, and an injective L-homomorphism

$$\mathcal{H} \to {}^L H_1$$

instead of an isomorphism. We refer the reader to [KS] for further details.

Consider now a Langlands parameter

$$\varphi : \mathcal{L}_F \to {}^L G$$

Let S_φ denote the centralizer in $\check{G}$ of $\varphi(\mathbb{W}_\mathbb{R})$. We observe that any $s \in S_\varphi$ defines an endoscopic group H. As above, the group $\mathcal{H}$ generated in ${}^L G$ by the connected centralizer $\check{H}$ of s in $\check{G}$ and the image of φ, is a split extension:

$$1 \to \check{H} \to \mathcal{H} \to W_F \to 1$$

and by construction φ factors through $\mathcal{H}$. If there exists an admissible embedding $\eta : {}^L H \to {}^L G$ then one may factor φ through it and one thus defines a Langlands parameter for H. In general one only obtains a parameter for H_1.

Now, denote by $\mathbf{S}_\varphi$ the quotient of S_φ by its connected component S_φ^0 times $Z(\check{G})^\Gamma$ the center of ${}^L G$. In [She4] Diana Shelstad has established the

PROPOSITION 6.9. *For discrete series parameters there is an isomorphism between* $\mathbf{S}_\varphi$ *and* $\mathfrak{K}(\mathbb{R}, T\backslash G)$.

We give the proof when G is semisimple and simply connected. Consider an elliptic torus T in G. Let $\check{T}$ be the goup of complex characters of the lattice $\check{X}(T)$. Now consider a discrete series parameter φ. We may and will assume that $\check{T}$ contains $\varphi(\mathbb{C}^\times)$ the image of the subgoup $\mathbb{C}^\times \subset \mathbb{W}_\mathbb{R}$ but then S_φ is the subgroup of invariant elements in $\check{T}$ under the Galois action from ${}^L T$ and hence S_φ is the set of elements of order 2 in $\check{T}$. We observe that an element in $\mathfrak{K}(\mathbb{R}, T\backslash G)$ defines a complex character of $\check{X}(T)$. Hence to $\kappa \in \mathfrak{K}(\mathbb{R}, T\backslash G)$ corresponds an element $s \in \check{T}$ that moreover commutes with the Galois action. We thus get a bijective homomorphism

$$\mathfrak{K}(\mathbb{R}, T\backslash G) \to \mathbf{S}_\varphi$$

For $\mathfrak{p}$-adic fields the Langlands parametrization and the structure of L-packets of discrete series are not known in general but it can be checked in examples that the group $\mathbf{S}_\varphi$ is often bigger than $\mathfrak{K}$ and can even be non-abelian.

6.7. Endoscopic transfer. Consider T an elliptic torus and κ an endoscopic character. Let H be the endoscopic group defined by (T, κ). We need some notation. We let B_G be a Borel subgroup of G containing T.

$$\Delta_B(\gamma) = \prod_{\alpha>0} (1 - \gamma^{-\alpha})$$

where the product is over the positive roots for for the order defined by B. There is only one choice of a Borel subgroup B_H in H, containing T_H compatible with the isomorphism $j : T_H \simeq T$.

PROPOSITION 6.10. *Assume there is an admissible embedding*

$$\eta : {}^L H \to {}^L G$$

One can attach to the triple (G, H, η) *a character* $\chi_{G,H}$ *of* $T(\mathbb{R})$ *with the following property. Given* f *a pseudo-coefficient for a discrete series on* G*, there is a function* f^H *which is a linear combination of pseudo-coefficients for discrete series on* H *such that for a* $\gamma = j(\gamma_H)$ *regular in* $T(\mathbb{R})$

$$S\mathcal{O}_{\gamma_H}(f^H) = \Delta_H^G(\gamma_H, \gamma)\mathcal{O}_\gamma^\kappa(f)$$

where $\Delta_H^G(\gamma_H, \gamma)$ *the "transfer factor" has the following expression:*

$$(-1)^{q(G)+q(H)}\chi_{G,H}(\gamma)\Delta_B(\gamma^{-1}) \,.\, \Delta_{B_H}(\gamma_H^{-1})^{-1}$$

When $G(\mathbb{R}) = U(2,1)$ a proof has been given in 5.2, and a proof is given in 7.2 in general. The expression for the transfer factor is borrowed from Kottwitz [Ko2].

The character $\chi_{G,H}$ depends on the choice of the admissible embedding η. To obtain an unconditional and more canonical transfer we may change a little the requirements: using the "canonical" transfer factor one gets, instead of a transfer to $H(\mathbb{R})$, a transfer to $H_1(\mathbb{R})$ some covering group of $H(\mathbb{R})$ and thus one gets functions f^{H_1} that transform according to some character of the kernel of

$$H_1(\mathbb{R}) \to H(\mathbb{R})$$

If an admissible embedding exists the character on the kernel is the restriction of some character of $H_1(\mathbb{R})$ which allows to twist the transfer so that it descends to $H(\mathbb{R})$. This has been explained in the case $G = U(2,1)$.

The transfer $f \mapsto f^H$ of pseudo-coefficients can be extended to all functions in $\mathcal{C}_c^\infty(G(\mathbb{R}))$; to define it one has to extend the correspondence $\gamma \mapsto \gamma_H$, called the norm, to all semisimple regular elements and the definition of the transfer factors to all tori. This is established by Shelstad in the series of four papers [She1], [She2], [She3] and [She4].

THEOREM 6.11. *Assume there is an admissible embedding*

$$\eta : {}^L H \to {}^L G \,.$$

One can define transfer factors $\Delta_H^G(\gamma_H, \gamma)$ *such that for any* $f \in \mathcal{C}_c^\infty(G(\mathbb{R})$ *there exists a function* $f^H \in \mathcal{C}_c^\infty(H(\mathbb{R})$ *with*

$$S\mathcal{O}_{\gamma_H}(f^H) = \Delta_H^G(\gamma_H, \gamma)\mathcal{O}_\gamma^\kappa(f)$$

whenever γ_H *is a norm of* γ *semisimple regular and*

$$S\mathcal{O}_{\gamma_H}(f^H) = 0$$

if γ_H *is not a norm.*

It should be observed that the correspondence

$$f \mapsto f^H$$

is not a map since f^H is not uniquely defined. In fact f^H is only prescribed through its orbital integrals. But as regards invariant harmonic analysis this is a well defined object.

The geometric transfer

$$f \mapsto f^H$$

is dual of a transfer for representations. To any admissible irreducible representation σ of $H(\mathbb{R})$ it corresponds an element σ_G in the Grothendieck group of virtual representations of $G(\mathbb{R})$ as follows. Given φ a Langlands parameter for H then $\eta \circ \varphi$ is a Langlands parameter for G^* where η is the admissible embedding

$$\eta : {}^L H \to {}^L G$$

as above. Now consider Σ the L-packet of admissible irreducible representation of $H(\mathbb{R})$ corresponding to φ and Π the L-packet of representations of $G(\mathbb{R})$, corresponding to $\eta \circ \varphi$ (that can be the empty set if this parameter is not relevant for G).

THEOREM 6.12. *There is a function*

$$\varepsilon : \Pi \to \pm 1$$

such that, if we consider σ_G in the Grothendieck group defined by

$$\sigma_G = \sum_{\pi \in \Pi} \varepsilon(\pi)\, \pi$$

then $\sigma \mapsto \sigma_G$ is the dual of the geometric transfer:

$$\text{trace } \sigma_G(f) = \text{ trace } \sigma(f^H)$$

This is the Theorem 4.1.1 of [She4]. We proved it for discrete series of $U(2,1)$ in section 5.2.

We shall now give an expression for $\varepsilon(\pi)$ following section 5 of [She4]. Suppose we are given a complete set of inequivalent endoscopic groups H and for each H an admissible embedding

$$\eta : {}^L H \to {}^L G$$

Now consider a parameter

$$\varphi : \mathbb{W}_{\mathbb{R}} \to {}^L G$$

The connected centralizer of $s \in \mathbf{S}_\varphi$ is a group $\check{H}_s$ conjugate to $\check{H}$ for some H and hence a conjugate of φ factors through $\eta({}^L H)$ and defines an L-paquet Σ_s of representations of $H(\mathbb{R})$. (The L-packet is not unique in general: it depends on the choice of the conjugate which may not be unique). On the other hand, using proposition 6.9 above Shelstad defines a pairing $< s, \pi >$ between $\mathbf{S}_\varphi$ and $\Pi(\varphi)$ and shows that

$$\varepsilon(\pi) = c(s) < s, \pi >$$

and hence the above identity

$$\sum_{\sigma\in\Sigma} \text{trace } \sigma(f^H) = \sum_{\pi\in\Pi} \varepsilon(\pi) \text{ trace } \pi(f)$$

reads

$$\widetilde{\Sigma}_s(f^H) = \sum_{\pi\in\Pi} < s, \pi > \text{ trace } \pi(f)$$

where

$$\widetilde{\Sigma}_s(f^H) = c(s)^{-1} \sum_{\sigma\in\Sigma_s} \text{trace } \sigma(f^H)$$

This can be inverted to give the theorem 5.2.9 of [She4]:

THEOREM 6.13.

$$\text{trace } \pi(f) = \frac{1}{\#\mathbf{S}_\varphi} \sum_{s\in\mathbf{S}_\varphi} < s, \pi > \widetilde{\Sigma}_s(f^H) \, .$$

7. Discrete series and endoscopy

7.1. *L*-packets of discrete series over $\mathbb{R}$. We recall that, according to Harish-Chandra, the group $G(\mathbb{R})$ has a discrete series representation if and only if there exist an $\mathbb{R}$-elliptic torus T in G. Assume from now on this is the case. They are parametrized as follows.

Given T, an $\mathbb{R}$-elliptic torus T, we consider K a maximal compact subgroup of $G(\mathbb{R})$ containing $T(\mathbb{R})$. Consider a Borel subgroup B_K in $K_\mathbb{C}$ containing T and a Borel subgroup B in G containing B_K. Now consider a character of $T(\mathbb{R})$ defined by a

$$\lambda \in X^*(T) \otimes \mathbb{C}$$

which is B-dominant. There is a discrete series representation $\pi_{\lambda+\rho}$, where ρ is half the sum of positive roots of T in B, characterized by its character given on $T(\mathbb{R})$ by

$$\Theta_{\lambda+\rho} = (-1)^q \, \frac{\sum_{w\in\Omega_K} \varepsilon(w) e^{w(\lambda+\rho)}}{\sum_{w\in\Omega_G} \varepsilon(w) e^{w\rho}}$$

where

$$q = \frac{1}{2}(\dim G(\mathbb{R}) - \dim K)$$

An L-packet of discrete series is a set of representations π_μ where

$$\mu = w(\lambda + \rho) \quad \text{with} \quad w \in \Omega_\mathbb{C}$$

such that μ belongs to the Weyl chamber defined by B_K. We shall denote by $\Omega(B_K)$ the subset of such w. Clearly, $\Omega(B_K)$ is a set of representatives in $\Omega_\mathbb{C}(G,T)$ of the quotient $\Omega_\mathbb{R}(G,T)\backslash\Omega_\mathbb{C}(G,T)$.

PROPOSITION 7.1. *The set of representations in an L-packet of discrete series is in bijection with $\mathfrak{D}(\mathbb{R}, T\backslash G)$; the bijection depends on the choice of B.*

Proof. This follows from 6.5 and the above remarks. □

Recall that the set $\mathfrak{D}(\mathbb{R}, T\backslash G)$ embeds in the abelian group $\mathfrak{E}(\mathbb{R}, T\backslash G)$. Thus we get a pairing between the L-packet and $\mathfrak{K}(\mathbb{R}, T\backslash G)$ the Pontryagin dual of $\mathfrak{E}(\mathbb{R}, T\backslash G)$. This pairing depends on the choice of B. We shall see that this pairing has another formulation using L-groups.

Some choices of B seem to be better than others at least when G is a quasi-split group: they are those that correspond to generic representations i.e. with a Whittaker model (with respect to some further choice). For example, for $G = U(2,1)$ and $K = U(2) \times U(1)$ once B_K is chosen there are three Weyl chambers for G that are contained in the Weyl chamber defined by B_K in the Lie algebra of T. A Weyl chamber for G that contains ρ_K is a "good" choice. There is only one for $G = U(2,1)$. But for $G = SL(2)$, $\rho_K = 0$ and any choice is good.

7.2. General discrete transfer. Consider a reductive Lie group $G(\mathbb{R})$ with compact maximal torus $T(\mathbb{R})$. In particular, $G(\mathbb{R})$ has discrete series. We choose an endoscopic character κ, this defines an endoscopic group H. We choose a Borel subgroup B in G containing T and a compatible Borel subgroup B_H in H. To simplify the discussion we assume that $\rho - \rho_H$ the difference of half sums of positive roots for G and H respectively defines a character of $T(\mathbb{R})$ and hence the canonical transfer factor:

$$\Delta(\gamma^{-1}) = (-1)^{q(G)-q(H)} \frac{\sum_{w \in \Omega_G} \varepsilon(w)\gamma^{w\rho}}{\sum_{w \in \Omega_H} \varepsilon(w)\gamma^{w\rho_H}}$$

is a well defined function.

Let μ be parameter for a discrete series and let f_μ be a pseudo-coefficient for π_μ. Let $w_0\rho$ be the half sum of positive roots for the Weyl chamber defined by μ i.e. $\mu = w_0\mu_0$ where μ_0 is B-dominant and regular. We have

$$(-1)^{q(G)} \mathcal{O}_\gamma^\kappa(\check{f}_\mu) = (-1)^{q(G)} \sum_{w \in \Omega_G/\Omega_K} \kappa(w) \mathcal{O}_{\gamma_w}(\check{f}_\mu)$$

$$= \sum_{w \in \Omega_G/\Omega_K} \kappa(w) \frac{\sum_{w' \in \Omega_K} \varepsilon(w')\gamma^{ww'\mu}}{\sum_{w_1 \in \Omega_G} \varepsilon(w_1)\gamma^{ww_1w_0\rho}}$$

so that

$$(-1)^{q(G)} \mathcal{O}_\gamma^\kappa(\check{f}_\mu) = \varepsilon(w_0) \frac{\sum_{w \in \Omega_G} \kappa(w)\varepsilon(w)\gamma^{w\mu}}{\sum_{w \in \Omega_G} \varepsilon(w)\gamma^{w\rho}}$$

Hence, we have

$$(-1)^{q(H)} \Delta(\gamma^{-1}) \mathcal{O}_\gamma^\kappa(\check{f}_\mu) = \varepsilon(w_0) \frac{\sum_{w \in \Omega_G} \kappa(w)\varepsilon(w)\gamma^{w\mu}}{\sum_{w \in \Omega_H} \varepsilon(w)\gamma^{w\rho_H}}$$

LEMMA 7.2. *The function κ is left invariant under Ω_H:*

Proof. Let $w \in \Omega_G$ and $w_0 \in \Omega_H$ then the multiplicativity property of κ shows that

$$\kappa(w_0 w) = \kappa^{w_0}(w)\kappa(w_0)$$

But, by definition of H, for any reflexion $s_\alpha \in \Omega_H$ one has

$$\kappa(s_\alpha) = \overline{\kappa}(\check{\alpha}) = 1 \quad \text{and} \quad \kappa^{s_\alpha} = \kappa$$

and hence, by induction on the length of w_0, one has $\kappa(w_0) = 1$ and $\kappa^{w_0}(w) = \kappa(w)$ so that

$$\kappa(w_0 w) = \kappa(w)$$

□

Now introduce $\Omega_*(\mu)$ the set of representatives of the quotient $\Omega_H \backslash \Omega_G$ such that $w_*\mu$ is B_H dominant. The above lemma shows that

$$(-1)^{q(H)}\Delta(\gamma^{-1})\mathcal{O}^\kappa_\gamma(\check{f}_\mu)$$

$$= \sum_{w_* \in \Omega_*(\mu)} \kappa(w_*)\varepsilon(w_* w_0)\frac{\sum_{w\in\Omega_H}\varepsilon(w)\gamma^{ww_*\mu}}{\sum_{w\in\Omega_H}\varepsilon(w)\gamma^{w\rho_H}}$$

Altogether, if we denote by g_ν a pseudo-coefficient for the discrete series of $H(\mathbb{R})$ with parameter ν we have

$$\Delta(\gamma)\mathcal{O}^\kappa_\gamma(f_\mu) = \sum_{w_* \in \Omega_*(\mu)} \kappa(w_*)\varepsilon(w_* w_0)S\mathcal{O}_\gamma(g_{w_*\mu})$$

and hence

$$f^H_\mu = \sum_{w_* \in \Omega_*(\mu)} a(w,\mu)\, g_{w\mu}$$

with

$$a(w,\mu) = \kappa(w_*)\varepsilon(w_* w_0)$$

and where w is any element in the Ω_H-class defined by $w_* \in \Omega_*(\mu)$.

Consider $\nu = w_*\mu = w_* w_0 \mu_0$ and let Σ be the L-packet of representations of $H(\mathbb{R})$ with parameter ν. We have $w_\Sigma = w_* w_0 \in \Omega_*(\mu_0)$ and hence

$$a(w_\Sigma,\mu_0) = \kappa(w_\Sigma)\varepsilon(w_\Sigma) = \kappa^{w_\Sigma}(w_0^{-1})^{-1}\kappa(w_*)\varepsilon(w_\Sigma)$$

so that

$$a(w,\mu) = \kappa^{w_\Sigma}(w_0^{-1})\, a(w_\Sigma,\mu_0)$$

Let

$$< \kappa_\Sigma, \pi >= \kappa^{w_\Sigma}(w_0^{-1})$$

where π is the discrete series for G with parameter $\mu = w_0\mu_0$; then

$$a(w,\mu) =< \kappa_\Sigma, \pi > c(\Sigma,\pi_0)$$

with

$$c(\Sigma,\pi_0) = a(w_\Sigma,\mu_0)$$

and π_0 defined by μ_0; altogether we get

$$f^H_\pi = \sum_\Sigma < \kappa_\Sigma, \pi > c(\Sigma,\pi_0) g_\sigma$$

the sum being over L-packets Σ of discrete series for $H(\mathbb{R})$ with parameters in the orbit of μ_0 and we have chosen some $\sigma \in \Sigma$. This is equivalent to

$$c(\Sigma, \pi_0)^{-1} \text{ trace } \Sigma(f^H) = \sum_{\pi} < \kappa_\Sigma, \pi > \text{ trace } \pi(f)$$

the sum being over discrete series for $G(\mathbb{R})$ with parameters in the orbit of μ_0.

8. Further developments

8.1. K-groups. To get a more uniform treatment of endoscopy when F is any local field it is better to introduce, following Adams-Barbasch-Vogan, Kottwitz, and Arthur, the K-group $\widetilde{G}$ associated to G. A K-group is a disjoint union of groups

$$\widetilde{G} = \coprod_{\alpha \in A} G_\alpha \, .$$

indexed by

$$A = \text{ Im } [H^1(F, G_{SC}) \to H^1(F, G)]$$

Each group G_α is the inner form of G defined by α (or rather its image in the adjoint group) and we are given maps between G_β and G_α:

$$\psi_{\alpha\beta} : G_\beta \to G_\alpha$$

satisfying the obvious composition rules and such that $\psi_{\alpha\beta}$ defines G_β as an inner form of G_α: we are also given 1-cocycles $u_{\alpha\beta;\sigma} \in G_\alpha$ with

$$\psi_{\alpha\beta}\sigma(\psi_{\alpha\beta})^{-1} = \text{ Int } u_{\alpha\beta;\sigma}$$

and whose cohomology class belong to the image of the H^1 of the simply connected group $G_{\alpha,SC}$. When F is non-archimedean Kneser's theorem 6.2 shows that A is trivial and hence one has always $\widetilde{G} = G$; but for $F = \mathbb{R}$ this is not so in general.

A stable conjugacy class of $\gamma \in G_\alpha(F)$ in the K-group $\widetilde{G}$ associated to G is the set of $\gamma' \in G_\beta(F)$ for some β and such that there is $x \in G_\alpha$ with

$$\psi_{\alpha\beta}(\gamma') = x^{-1}\gamma x \quad \text{and} \quad a_\sigma = x u_{\alpha\beta;\sigma}\sigma(x)^{-1} \in I_\gamma$$

The set of classes of such 1-cocycles a build a set we denote by

$$\mathfrak{D}(F, I_\gamma \backslash \widetilde{G})$$

Assume that γ is regular semisimple, then I_γ is a torus T.

PROPOSITION 8.1. *Let F be a local field and let T be a torus. There is a bijective map*

$$\mathfrak{D}(F, T\backslash\widetilde{G}) \to \mathfrak{E}(F, T\backslash G)$$

Proof. Recall that

$$\mathfrak{E}(F, T\backslash G) = \ker[H^1(F,T) \to H^1_{ab}(F,G)]$$

By definition of the abelianized cohomology the class of a 1-cocycle a with values in T has a trivial image in $H^1_{ab}(F,G)$ if and only if

$$a_\sigma = x u_\sigma \sigma(x)^{-1} \in T$$

for some $x \in G$ and some 1-cocycle u image of a 1-cocycle with values in G_{SC}. Hence the map is well defined and is obviously bijective. □

8.2. The twisted case. Another genralization is important for applications: the twisted case. We consider a group G acting on the left on a space $\widetilde{G}$

$$(x,\delta) \mapsto x\,\delta \quad , \quad x \in G \qquad \delta \in \widetilde{G}$$

which is a principal homogeneous space under this action. We say that $\widetilde{G}$ is a twisted G-space if we are given a G-equivariant map

$$\mathrm{Ad}_{\widetilde{G}} : \widetilde{G} \to \mathrm{Aut}\; G$$

This allows to define a right action of G on $\widetilde{G}$ such that

$$x\,\delta\,y = x\left(\mathrm{Ad}_{\widetilde{G}}(\delta)\,y\right)\delta$$

By choosing a point $\delta_0 \in \widetilde{G}$ we get an isomorphism

$$\widetilde{G} \to G \rtimes \theta$$

where

$$\theta = \mathrm{Ad}_{\widetilde{G}}(\delta_0)$$

but this isomorphism is not canonical when G has a non trivial center. This is why it is better to work with $\widetilde{G}$ rather than with $G \rtimes \theta$. One can develop as above the notion of stable conjugacy inside $\widetilde{G}(F)$ once stable centralizers are defined: given $\delta \in \widetilde{G}(F)$ we say that $\delta' \in \widetilde{G}(F)$ is stably conjugate if there is $x \in G(\overline{F})$ such that

$$x^{-1}\delta x = \delta' \quad \text{and} \quad x\sigma(x)^{-1} \in I_\delta$$

for all $\sigma \in \mathrm{Gal}(\overline{F}/F)$ and where I_δ is the group generated by $Z_{\widetilde{G}} := Z(G)^\theta$ and the image of G^δ_{SC} the fixed points in G_{SC} under the adjoint action of δ. The stable centralizer I_δ may not be connected even if δ is regular semisimple i.e. when G^δ_{SC} is a torus.

We refer to [KS], [Lab3] and [Lab4] for a study of the twisted case in the trace formula context. As regards the real case Shelstad's results have been extended to the twisted case by Renard [Ren].

Remark. We warn the reader that the definition given for the stable centralizer in [Lab3] differs slightly from the one given above and used in [Lab4].

8.3. Trace formula stabilization. Last but not least, the main motivation for studying endoscopy is the adelic trace formula. In the old paper [LL] the trace formula for $SL(2)$ was expressed as a sum of stable trace formulas for its endoscopic groups. The stabilization gives a better understanding of the space of automorphic forms and allows one to establish cases of Langlands functoriality. This was the beginning of all this game.

Later on, various other instances of endoscopy and of twisted endoscopy were treated like the Base Change for $GL(2)$ by Saito-Shintani and Langlands [Lan1] and for unitary groups in three variables by Rogawski [Rog]. Endoscopy also played a role at the place where it was first discovered: for computing the Zeta function of Shimura varieties.

At the present time[2] the problem of stabilizing the trace formula which is to express the trace formula for G (or more generally the trace formula for a twisted space) as a linear combination of stable trace formulas for its endoscopic groups is essentially solved.

Wonderful progress have been made by Jim Arthur in first establishing the trace formula for arbitrary reductive groups and then by developing techniques that have allowed him to get a conditional stabilization of the (non-twisted) trace formula. The condition is about the endoscopic transfer for non-archimedean fields. Observe that transfer factors have been defined for all local fields by Langland and Shelstad [LS]. One needs the existence of the transfer: given f there exists f^H satisfying the transfer identity recalled below and one needs also that the so-called fundamental lemma holds.

The fundamental lemma: *Assume that f is the characteristic function of an hyperspecial maximal compact subgroup in $G(F)$ (when this makes sense) and let f^H be the similar function for H. Then the transfer identity holds:*

$$SO_{\gamma_H}(f^H) = \Delta_H^G(\gamma_H, \gamma) O_\gamma^\kappa(f)$$

whenever γ_H is a norm of γ and it vanishes if γ_H is not a norm.

These questions have resisted the attacks until recently although many important advances have been made by various people.The situation has now changed completely.

Waldspurger has proved that the fundamantal lemma for large residual characteristics implies the existence of the transfer for p-adic groups for all residual characteristics [W1]. He has also proved that fundamental lemma for p-adic groups holds if the fundamental lemma holds for Lie algebras over function fields [W2]. More recently Waldspurger again has announced results that essentially reduce the twisted case to the ordinary one [W3].

A proof of the fundamental lemma for the Lie algebra (in large positive characteristic) for unitary groups has been obtained in 2005 by Laumon and Ngô [LN]. More recently Ngô has announced (during the winter of 2007) results [N] that imply the fundamental lemma and the transfer in full generality, even in the twisted case, thanks to Waldspurger's reductions.

[2]May 2010

One also needs a weighted variant of these results for the stabilization of the trace formula. The weighted case is now a theorem due to Chaudouard and Laumon [CL].

A complete proof of the stabilization in the twisted case has not been written up. It also remains to extract the spectral informations from the stabilisation of the (twisted) trace formula.

As a last word I would like to say that endoscopy is not the final target, fand some are looking "beyond endoscopy"—but this is another story (see [Lan3]).

References

[Bo] M.V. Borovoi, Abelian Galois cohomology of Reductive groups, *Mem. Amer. Math. Soc.* **132** (1998), No. 626.

[A] J. Arthur, A Paley-Wiener theorem for real reductive groups, *Acta Math.* **150** (1983), 1–89.

[CL] P.-H. Chaudouard & G. Laumon, Le lemme fondamental pondéré. II. Énoncés cohomologiques, *Preprint* **arXiv:0912.4512v1** [math.AG] (2009).

[CD] L. Clozel & P. Delorme, Pseudo-coefficients et cohomologie des groupes réductifs, *C.R. Acad. Sci. Paris* **300** No. 11 (1985), 385–387.

[D] P. Delorme, Multipliers for the convolution algebra of left and right K-finite compactly supported smooth functions on a semi-simple Lie group, *Invent. Math.* **75** (1984), 9–23.

[Ka] D. Kazhdan, Cuspidal geometry of p-adic groups, *J. Analyse Math* **47** (1986), 1–36.

[Ko1] R. Kottwitz, Sign changes in harmonic analysis on reductive groups, *Trans. Amer. Math. Soc.* **278** (1983), p. 289–297.

[Ko2] R. Kottwitz, Shimura Varieties and λ-adic Representations, *Perspectives in Math.* **10** Academic Press (1988), pp. 161–210.

[KS] R. Kottwitz & D. Shelstad, Foundations of Twisted Endoscopy, *Astérisque* **255** (1999).

[Lab1] J.-P. Labesse, Cohomologie, L-groupes et fonctorialité, *Compositio* **55** (1985), p. 163–184.

[Lab2] J.-P. Labesse, Pseudo-coefficients très cuspidaux et K-théorie, *Math. Ann.* **291** (1991), p. 607–616.

[Lab3] J.-P. Labesse, Cohomologie, Stabilisation et changement de base , *Astérisque* **257** (1999).

[Lab4] J.-P. Labesse, Stable twisted trace formula: elliptic terms, *Journ. Inst. Math. Jussieu* **3** (2004), p. 473–530.

[LL] J.-P. Labesse & R.P. Langlands, L-indistinguishability for SL_2, *Can. J. Math.* **31** (1979), p. 726–785.

[Lan1] R.P. Langlands, Base Change for $GL(2)$, *Annals of Math. Studies* **96** Princeton Univ. Press (1980).

[Lan2] R.P. Langlands, On the classification of irreducible representations of real algebraic groups, *Math. Surveys and Monographs* **31** AMS (1989). p. 101–170

[Lan3] R.P. Langlands, Beyond endoscopy, *Contributions to automorphic forms, geometry, and number theory, Johns Hopkins Univ. Press, Baltimore, MD* (2004) p. 611–697.

[LS] R.P. Langlands & D. Shelstad, On the Definition of Transfer Factors, *Math. Annalen* **278** (1987). p. 219–271.

[LN] G. Laumon & B.C. Ngô, Le lemme fondamental pour les groupes unitaires, *Ann. of Math.* **168** (2008) no. 2, 477–573.

[N] B.C. Ngô, Le lemme fondamental pour les algèbres de Lie, *preprint* **arXiv:0801.0446v3** [math.AG](2008)

[Ren] D. Renard, Twisted endoscopy for real groups, *Journ. Inst. Math. Jussieu* **2** (2003), p. 529–566.

[Rog] J. Rogawski, Automorphic Representations of Unitary Groups in Three Variables, *Annals of Math. Studies* **123** (1990).

[She1] D. Shelstad, Characters and inner forms of a quasi-split group over $\mathbb{R}$, *Compositio Math.* **39** (1979), p. 11–45.

[She2] D. Shelstad, Orbital integrals and a family of groups attached to a real reductive group, *Ann. Sci. E.N.S.* **12** (1979), p. 1–31.

[She3] D. Shelstad, Embeddings of L-groups, *Can. J. Math.* **33** (1981), p. 513–558.

[She4] D. Shelstad, L-Indistinguishability for Real Groups, *Math. Annalen* **259** (1982), p. 385–430.

[W1] J.-L. Waldspurger, Le lemme fondamental implique le transfert, *Compositio Math.* **105** (1997), p. 153–236.

[W2] J.-L. Waldspurger, Endoscopie et changement de caractéristique, *J. Inst. Math. Jussieu* **5** (2006), p. 423–525.

[W3] J.-L. Waldspurger, L'endoscopie tordue n'est pas si tordue, *Mem. Amer. Math. Soc.* **194** (2008) no. 908.

Institut Mathématique de Luminy, UMR 6206, Marseille, France

Section II

Endoscopy and transfer

Endoscopy for real reductive groups

[R.II.A]

David Renard

0. Introduction

The purpose of this paper is to review Diana Shelstad's results on endoscopy for real reductive groups ([**Sh1**], [**Sh2**], [**Sh3**], [**Sh4**]). This impressive body of work is a part of the general Langlands program in automorphic representation theory, where it gives the first examples of Langlands' (local) functoriality principle. It establishes a transfer of orbital integrals between a real reductive group $G(\mathbb{R})$ and one of its endoscopic groups $H(\mathbb{R})$. With hindsight, and somewhat artificially, it is easiest for us to break down the difficulties Shelstad encountered into two parts. Her first problem was to find the correct definition of the transfer factors; the second problem was to show that they indeed give the transfer of orbital integrals. In her series of papers on the subject, these two problems are intertwined, since the second one obviously gives necessary conditions for the first. Moreover, the definitions of the transfer factors grew more elaborate and conceptual as increasingly general situations were considered. Finally, Shelstad's work in the real case served as a guide to the (perhaps definitive) treament of transfer factors for all local fields in [**LS**], and [**KS**] for the twisted case. In [**LS2**], a somewhat indirect argument shows that, up to a global sign, the transfer factors of [**LS**], when specialized to the real case, are the same as those in [**Sh4**].

For anybody who wants to study the subject today in some depth, the natural path would be first to absorb the definitions in [**LS**], specialize them to the real case, and then use [**Sh1**], [**Sh2**], [**Sh3**] and [**Sh4**] to establish the transfer of orbital integrals. This requires substantial effort, and it is this effort we want to reduce as much as possible.[1] Our point of view is to take

[1]The first version of the present chapter was written in 2006. The situation has changed dramatically since then, with the appearance of Shelstad's new series of articles [**Sh5**] that reformulate the theory of endoscopy for real groups on the basis of the canonical transfer factors of [**LS**]. Shelstad's new articles are indispensable for anyone who wants to make serious use of endoscopy for real groups, but they make few concessions to the reader

[**LS**], [**LS2**] and [**KS**] as the foundations of the theory of endoscopy, and this will be our starting point. We explain in some detail how the definitions there specialize to the real case. Then we reduce the proof of the transfer of orbital integrals to a set of properties of the transfer factors (Proposition 4.4 and Lemma 4.6). Some of these properties are already established in [**LS**], [**LS2**] or [**KS**], and we don't repeat the proofs here. Other properties are extracted from Shelstad's articles. We hope that the treatment given here is simpler. It follows the lines of [**R**], where the transfer of orbital integrals is established for real groups in the context of *twisted* endoscopy, assuming some (very subtle) properties of the transfer factors.

In the general setting of twisted endoscopy, as formulated in [**KS**] and specialized to real groups, one starts with a datum (G,θ,ω) where G is an algebraic reductive group defined over $\mathbb{R}$, θ is an algebraic $\mathbb{R}$-automorphism of G, and ω a quasicharacter of $G(\mathbb{R})$, the group of real points of G. Shelstad's series of papers, culminating in [**Sh4**], deals with the case $\theta = 1$, $\omega = 1$. In [**R**], we deal with the case $\omega = 1$ and θ is of finite order. In this paper, we take $\theta = 1$, but we allow ω to be non-trivial. The technical complications it induces are minor (unlike the case where θ is non-trivial). It could be argued that the results of [**R**] and those here could be combined to give the general case. But our motivation here is to simplify the existing literature, so we won't go in this direction. Another difference with Shelstad's approach is that we work with the space of compactly supported smooth functions rather than the Schwartz space. For this, we use Bouaziz' results on orbital integrals of such functions ([**B1**]).

For a general discussion of endoscopy and the Langlands functoriality principle, we refer to [**LTF**], [**Kn**], [**G**].

Let us now describe in more detail the contents of the paper. The first section introduces the basic notions: reductive algebraic groups defined over $\mathbb{R}$, their L-groups, the Langlands classification of irreducible representations, stable conjugacy and finally the definition of endoscopic data. The second section describes the correspondences of points (or rather conjugacy classes) between an algebraic group G defined over $\mathbb{R}$ and one of its endoscopic groups H. The main notion here is that of Cartan subgroups of $H(\mathbb{R})$ originating in $G(\mathbb{R})$. When a Cartan subgroup $T_H(\mathbb{R})$ of $H(\mathbb{R})$ originates in $G(\mathbb{R})$, there is an isomorphism $T_H(\mathbb{R}) \simeq T_G(\mathbb{R})$ with a Cartan subgroup of $G(\mathbb{R})$, compatible with the correspondence of regular conjugacy classes. Some properties are established: if $T_H(\mathbb{R})$ originates in $G(\mathbb{R})$, all Cartan subgroups in $H(\mathbb{R})$ smaller than $T_H(\mathbb{R})$ in the Hirai order also originate in $G(\mathbb{R})$ (Lemma 2.10), and for a Cartan subgroup immediately bigger than $T_H(\mathbb{R})$ in the Hirai order, a necessary and sufficient condition is given in terms of Cayley transforms (Lemma 2.12).

---

who is easily distracted. We hope the present article will provide a useful introduction to this important subject. —*The Editors*

In the third section, we recall properties of orbital integrals. The space of orbital integrals on $G(\mathbb{R})$ is defined. It is a locally convex topological vector space, and its dual is isomorphic to the space of (conjugation-) invariant distributions on $G(\mathbb{R})$. It can be viewed as the space of smooth compactly supported functions on the "variety" of conjugacy classes in $G(\mathbb{R})$. Cartan subgroups are transversal to regular orbits and regular elements are dense in $G(\mathbb{R})$. This explain why orbital integrals are characterized by the properties of their restrictions to Cartan subgroups ([**B1**]). Stable orbital integrals on $H(\mathbb{R})$ are also introduced. The main theorem of the paper is then stated (Theorem 3.6): starting from an orbital integral on $G(\mathbb{R})$, one can define a function on a dense open set of regular elements in $H(\mathbb{R})$, by a formula involving the so-called transfer factors. This formula can be extended smoothly to all regular elements, and it gives a stable orbital integral on $H(\mathbb{R})$. In the fourth section, we reduce the proof of the main theorem to a set of properties of the transfer factors. This is purely formal, as the transfer factors are not even defined at this point. This is done in the next section, where the definitions of [**LS**] are specialized to the real case. In the last section, we prove the desired properties of the transfer factors. In an appendix, we recall the Langlands correspondence and the Tate-Nakayama duality for real tori, since these are ingredients of the definition of transfer factors.

1. Notation and basic definitions

1.1. Notation for group actions. Let A be a group, and X a set on which A acts. For all subset B in X, set:

$$Z(A,B) = \{a \in A \,|\, \forall b \in B,\ a \cdot b = b\}$$
$$N(A,B) = \{a \in A \,|\, \forall b \in B,\ a \cdot b \in B\}.$$

Different actions of a group on itself will be considered, so unless otherwise stated, the above notations will refer to the usual action by conjugation. The inner automorphism of A given by an element $a \in A$ is denoted by $\operatorname{Int} a$ and $A^a := \{b \in A \,|\, \operatorname{Int} a(b) = aba^{-1} = b\}$. If A is a topological group, the connected component of the trivial element in A is denoted by A_0. The center of A is denoted by $Z(A)$.

1.2. Weil group. Let $\Gamma = \{1, \sigma\} = \mathrm{Gal}(\mathbb{C}/\mathbb{R})$ be the Galois group of $\mathbb{C}$ over $\mathbb{R}$. The Weil group $W_{\mathbb{R}}$ is a non-split extension

$$1 \to \mathbb{C}^{\times} \to W_{\mathbb{R}} \to \Gamma \to 1.$$

More precisely, $W_{\mathbb{R}}$ is the group generated by $\mathbb{C}^*$ and an element j which projects to $\sigma \in \Gamma$ with relations

$$j^2 = -1, \quad jz = \bar{z}j.$$

1.3. Reductive algebraic groups over $\mathbb{C}$. For the proof of the results in this paragraph, we refer to [**ABV**], Chapter 2. Let G be a connected reductive algebraic group defined over $\mathbb{C}$. The group G is identified with the group of its complex points.

Following [**LS**], by a pair in G, we mean a couple (B,T) where B is a Borel subgroup of G and T a maximal torus in B, and by a splitting of G, we mean a triple $\mathbf{spl}_G = (B,T,\{X_\alpha\})$ where (B,T) is a pair in G and $\{X_\alpha\}$ a collection of non-zero root vectors, one for each simple root α of T in B. Two spittings $\mathbf{spl}_G = (B,T,\{X_\alpha\})$ and $\mathbf{spl}'_G = (B',T',\{X'_\alpha\})$ are conjugate under G by an element uniquely determined modulo $Z(G)$, two pairs (B,T) and (B',T') are conjugate under G by an element uniquely determined modulo T, and two Borel subgroups B and B' are conjugate under G by an element uniquely determined modulo B.

Associated to a pair (B,T), there is a based root datum

$$\Psi(G,B,T) = (X^*(T), \Delta(B,T), X_*(T), \Delta\check{}(B,T))$$

where $X^*(T)$ is the group of algebraic characters of T, $X_*(T)$ the group of 1-parameter subgroups of T, $\Delta(B,T)$ are the simple roots of the positive root sytem $R(B,T)$ of T in B, and $\Delta\check{}(B,T)$ are the simple coroots. The structure on this datum is the lattice structure of $X^*(T)$ and $X_*(T)$, together with the perfect pairing

$$X^*(T) \times X_*(T) \to \mathbb{Z},$$

the containment $\Delta(B,T) \subset X^*(T)$ and $\Delta\check{}(B,T) \subset X_*(T)$ and the bijection $\alpha \leftrightarrow \alpha\check{}$ between roots and coroots. By an automorphism of $\Psi(G,B,T)$ we mean a bijective map respecting this structure.

Since two pairs (B,T) and (B',T') are conjugate under G by an element uniquely determined modulo T, there is a canonical isomorphism between $\Psi(G,B,T)$ and $\Psi(G,B',T')$. Let us define

$$\Psi_G = (X^*, \Delta, X_*, \Delta\check{})$$

as the projective limit of the $\Psi(G,B,T)$. Thus, for each pair (B,T), there is a canonical isomorphism between $\Psi(G,B,T)$ and Ψ_G. There is an exact sequence:

$$1 \to \operatorname{Int} G \to \operatorname{Aut} G \to \operatorname{Aut} \Psi_G \to 1$$

where $\operatorname{Aut} G$ denotes the group of complex algebraic automorphisms of G.

1.4. ***L*-group data.** The dual of Ψ_G is by definition $\Psi_G\check{} = (X_*, \Delta\check{}, X^*, \Delta)$. A dual group for G is a connected reductive algebraic complex group $\hat{G}$ such that $\Psi_{\hat{G}} \simeq \Psi_G\check{}$. It is unique up to isomorphism.

Let us consider extensions (of topological groups) of the form

$$1 \to \hat{G} \to \mathcal{G} \xrightarrow{p} W_{\mathbb{R}} \to 1.$$

A splitting of this extension is a continuous homomorphism

$$s : W_{\mathbb{R}} \to \mathcal{G}$$

such that $p \circ s$ is the identity of $W_{\mathbb{R}}$. Each splitting defines a homomorphism η_s of $W_{\mathbb{R}}$ into the group of automorphism of $\hat{G}$. A splitting will be called admissible if for all $w \in W_{\mathbb{R}}$, $\eta_s(w)$ is complex analytic and the associated linear transformation of the Lie algebra of $\hat{G}$ is semisimple. It will be called distinguished if there is a splitting $\mathbf{spl}_{\hat{G}} = (\mathcal{B}, \mathcal{T}, \{\mathcal{X}_\alpha\})$ of $\hat{G}$ such that elements of $\mathbb{C}^\times \subset W_{\mathbb{R}}$ acts trivially on $\hat{G}$, and $\eta_s(j)$ is an automorphism of $\hat{G}$ preserving $\mathbf{spl}_{\hat{G}}$ (recall that j is an element of $W_{\mathbb{R}}$ such that $W_{\mathbb{R}} = \mathbb{C}^\times \coprod j\mathbb{C}^\times$). Two distinguished splitting will be called equivalent if they are conjugate under $\hat{G}$.

Following Langlands we consider the category $\mathfrak{G}\hat{}$ whose objects are extensions of the above type, with $\hat{G}$ a connected reductive algebraic group defined over $\mathbb{C}$, together with an equivalence class of distinguished splittings (called special). A morphism in this category between two extensions:

$$1 \to \hat{G} \to \mathcal{G} \xrightarrow{p} W_{\mathbb{R}} \to 1.$$

$$1 \to \hat{G}' \to \mathcal{G}' \xrightarrow{p'} W_{\mathbb{R}} \to 1.$$

is an equivalence class of L-homomorphisms, *i.e.* continuous group morphisms $\xi : \mathcal{G} \to \mathcal{G}'$ such that $p' \circ \xi = p$, such that the restriction of ξ to $\hat{G}$ (with value in $\hat{G}'$) is a morphism of complex algebraic groups and such that ξ preserves admissible splittings. Two L-homomorphisms ξ_1 and ξ_2 as above are in the same class if there exists $g \in \hat{G}$ such that

$$\xi_2 = \operatorname{Int} g \circ \xi_1.$$

Suppose now that G is a connected reductive algebraic group defined over $\mathbb{R}$. The group G is identified with the group of its complex points. Let σ be the non trivial element of $\Gamma = \mathrm{Gal}(\mathbb{C}/\mathbb{R})$. We will denote by σ_G the action of σ on G, by $G(\mathbb{R})$ the group of real points of G. The automorphism σ_G on G induces an automorphism a_σ of Ψ_G via the isomorhisms $\Psi_G \simeq \Psi(G, B, T)$, $\Psi_G \simeq \Psi(G, \sigma_G(B), \sigma_G(T))$ and $\sigma_G : \Psi(G, B, T) \simeq \Psi(G, \sigma_G(B), \sigma_G(T))$. The automorphism a_σ of Ψ_G induces an automorphism of $\Psi_G\check{}$ that we still denote by a_σ.

If we fix a splitting $\mathbf{spl}_{\hat{G}} = (\mathcal{B}, \mathcal{T}, \{\mathcal{X}_\alpha\})$ of $\hat{G}$, we can transfer the automorphism a_σ from $\Psi_{\hat{G}}$ to $\Psi(\hat{G}, \mathcal{B}, \mathcal{T})$ via the canonical isomorphism $\Psi_{\hat{G}} \simeq \Psi(\hat{G}, \mathcal{B}, \mathcal{T})$. Then there is a unique algebraic automorphism $\sigma_{\hat{G}}$ of $\hat{G}$ preserving $\mathbf{spl}_{\hat{G}}$ and inducing a_σ on $\Psi(\hat{G}, \mathcal{B}, \mathcal{T})$. (Notice that $\sigma_{\hat{G}}$ is an algebraic automorphism, while σ_G is not).

An L-group datum for G is an object

$$1 \to \hat{G} \to {}^L G \xrightarrow{p} W_{\mathbb{R}} \to 1.$$

in $\mathfrak{G}\hat{}$, such that the action of $\eta_s(j)$ on $\hat{H}$ coincide with $\sigma_{\hat{G}}$ constructed as above from a special splitting $\mathbf{spl}_{\hat{G}}$.

A realization of the L-group ${}^L G$ of G is then a semi-direct product $\hat{G} \rtimes_{\rho_G} W_{\mathbb{R}}$, where the action ρ_G of $W_{\mathbb{R}}$ on $\hat{G}$ factors through the projection on $\mathrm{Gal}(\mathbb{C}/\mathbb{R})$, with σ acting by $\sigma_{\hat{G}}$, where $\sigma_{\hat{G}}$ is constructed as above from a special splitting. Two realizations of ${}^LG$ given by two different choices of a splitting of $\hat{G}$ are isomorphic as objects in $\mathfrak{G}^{\hat{}}$..

REMARK. Suppose that $\mathcal{G}$ is a split extension of $W_{\mathbb{R}}$ by $\hat{G}$, i.e. that we have a split exact sequence

$$1 \to \hat{G} \to \mathcal{G} \to W_{\mathbb{R}} \to 1. \tag{1.1}$$

Then $\mathcal{G}$ is not always an L-group for some real form of G. Nethertheless, we can attach to $\mathcal{G}$ an L-action $\rho_{\mathcal{G}}$ of $W_{\mathbb{R}}$ on $\hat{G}$ as follows. If $c : W_{\mathbb{R}} \to G$ splits (1.1), then for $w \in W_{\mathbb{R}}$, $\mathrm{Int}\, c(w)$ acts by conjugacy on $\hat{G}$. Composing with an inner automorphism of $\hat{G}$, we obtain an automorphism $\rho_{\mathcal{G}}(w)$ preserving $\mathbf{spl}_{\hat{G}}$. It is straightforward to check that $\rho_{\mathcal{G}}$ doesn't depend on the choices (for instance, if we fix a splitting $\mathbf{spl}_{\hat{G}}$ of $\hat{G}$ from which we construct ${}^LG$ with L-action ρ_G, and if $\mathcal{G}$ is another realization of the L-group obtained from another choice of splitting, we get $\rho_{\mathcal{G}} = \rho_G$). This will be used in the definition of endoscopic data.

1.5. Inner forms. The group G is quasi-split if and only if it has an $\mathbb{R}$-splitting, i.e. one preserved by σ_G. Two connected reductive algebraic groups G and G^* defined over $\mathbb{R}$ are inner forms if there exists an isomorphism $\psi : G \to G^*$ defined over $\mathbb{C}$, and an element $u_\sigma \in G^*$ such that

$$\psi \circ \sigma(\psi)^{-1} = \psi \circ \sigma_G \circ \psi^{-1} \circ \sigma_{G^*} = \mathrm{Int}\, u_\sigma \tag{1.2}$$

This define an equivalence relation between real forms of connected algebraic reductive groups defined over $\mathbb{R}$. Every equivalence class contains a quasi-split form ([**ABV**], Proposition 2.7). It is easy to check that up to isomorphism, ${}^LG$ depends only on the inner class of real forms of G.

1.6. Langlands classification. Let us denote by $\mathcal{HC}_G$ the category of finite length Harish-Chandra modules of $G(\mathbb{R})$ (technically this requires the choice of a maximal compact subgroup $K(\mathbb{R})$ of $G(\mathbb{R})$, but we will ignore this and quite often we will loosely refer to this category as the category of representations of $G(\mathbb{R})$). We denote by $\Pi(G(\mathbb{R}))$ the set of irreducible representations of $G(\mathbb{R})$.

A Langlands parameter is an L-morphism $\phi : W_{\mathbb{R}} \to {}^L G$. A $\hat{G}$-conjugacy class of Langlands parameters is then a morphism in the category $\mathfrak{G}^{\hat{}}$. Let us denote by $\Phi(G)$ the set conjugacy classes of Langlands parameters, and by $[\phi]$ the conjugacy class of ϕ. . To each $[\phi] \in \Phi(G)$ is attached a packet Π_ϕ (sometimes empty) of representations, with the following properties:

- the Π_ϕ, with $[\phi]$ running over $\Phi(G)$ form a partition of $\Pi(G(\mathbb{R}))$,
- each packet Π_ϕ is finite,
- all representations in a packet have the same central and infinitesimal character,

- if a representation in some Π_ϕ is in the discrete series (resp. is tempered), then all representations in Π_ϕ are in the discrete series (resp. are tempered).

For further details about the Langlands classification, we refer to [**L1**], [**Bo**] or [**ABV**].

1.7. Quasicharacter. The action of $W_{\mathbb{R}}$ on $Z(\hat{G})$ is well defined, and so is $H^1(W_{\mathbb{R}}, Z(\hat{G}))$, where it is understood that only continuous cocycles are to be considered. There is an action of $H^1(W_{\mathbb{R}}, Z(\hat{G}))$, defined as folows. If ϕ is a Langlands parameter, and if $\mathbf{a} \in H^1(W_{\mathbb{R}}, Z(\hat{G}))$, one defines $\mathbf{a}[\phi]$ to be the class of

$$a\phi(w) = a(w)\phi(w), \quad (w \in W_{\mathbb{R}}),$$

where a is any choice of a cocycle representing $\mathbf{a}$ (it is easily checked that the class of $a\phi$ doesn't depend on the choice of a). Thus it also acts on the set of L-packets. To describe this, we recall a construction made in [**L1**], before Lemma 2.12: to each $\mathbf{a} \in H^1(W_{\mathbb{R}}, Z(\hat{G}))$ is attached a quasicharacter $\omega_{\mathbf{a}}$ of $G(\mathbb{R})$, satisfying the following property:

$$\Pi_\phi \otimes \omega_{\mathbf{a}} := \{\pi \otimes \omega_{\mathbf{a}} \mid \pi \in \Pi_\phi\} = \Pi_{\mathbf{a}\phi}.$$

Our interest is in the study of L-packets Π_ϕ of irreducible admissible representations of $G(\mathbb{R})$ such that

$$\Pi_\phi = \Pi_\phi \otimes \omega_{\mathbf{a}} = \Pi_{\mathbf{a}\phi}. \tag{1.3}$$

Since all representations in a L-packet have same infinitesimal character, we see that this forces $\omega_{\mathbf{a}}$ to have trivial infinitesimal character. Thus we suppose that $\omega_{\mathbf{a}}$ is trivial on $G(\mathbb{R})_0$. Set $G(\mathbb{R})^+ = \ker \omega_{\mathbf{a}}$. This is an normal open subgroup of $G(\mathbb{R})$ and one can view $\omega_{\mathbf{a}}$ as a character of the finite group $G(\mathbb{R})/G(\mathbb{R})^+$.

1.8. Endoscopic data. Let G be a connected reductive algebraic group defined over $\mathbb{R}$. We fix a quasi-split group G^* in the inner class of real forms of G with an inner inner twist $\psi : G \to G^*$ and an element u_σ in G^* as in (1.2).

We denote by G_{sc} the universal covering of the derived group G_{der} of G^*, and if T is a maximal torus in G, T_{sc} denote the inverse image of $T \cap G_{der}$ in G_{sc} under the natural projection. We will often identify elements in G_{sc} or T_{sc} with their image in G or T without comment.

The group G^* being quasi-split, let us fix once for all a $\mathbb{R}$-splitting $\mathbf{spl}_{G^*} = (\mathbf{B}, \mathbf{T}, \{\mathbf{X}_\alpha\})$. Let us also fix a splitting $\mathbf{spl}_{\hat{G}} = (\mathcal{B}, \mathcal{T}, \{\mathcal{X}_\alpha\})$ of $\hat{G}$, from which we construct a realization of the L-group of G. Finally, let $\mathbf{a} \in H^1(W_{\mathbb{R}}, Z(\hat{G}))$ and let $\omega = \omega_{\mathbf{a}}$ be the quasicharacter of $G(\mathbb{R})$, attached to it as in 1.7.

Definition 1.1. Following [**KS**], we call the quadruple $(H, \mathcal{H}, s, \xi)$ an endoscopic datum for $(G, \mathbf{a})$ if :

(i) H is a quasi-split connected reductive algebraic group defined over $\mathbb{R}$.
(ii) s is a semi-simple element in $\hat{G}$.
(iii) $\mathcal{H}$ is a split extension of $W_{\mathbb{R}}$ by $\hat{H}$ such that for a choice of a splitting $\mathbf{spl}_{\hat{H}}$ of $\hat{H}$, ρ_H and $\rho_{\mathcal{H}}$ coincide (see remark 1.4). (Both ρ_H and $\rho_{\mathcal{H}}$ depend on the choice of $\mathbf{spl}_{\hat{G}}$, but not the property $\rho_H = \rho_{\mathcal{H}}$.)
(iv) $\xi : \mathcal{H} \to {}^L G$ is an L-homomorphism.
(v) $\operatorname{Int} s \circ \xi = a \cdot \xi$ where a is a 1-cocycle of $W_{\mathbb{R}}$ in $Z(\hat{G})$ representing $\mathbf{a}$.
(vi) ξ maps isomorphically $\hat{H}$ into $(\hat{G}^s)_0$.

The purpose of endoscopic data for $(G, \mathbf{a})$ is to study representations of $G(\mathbb{R})$ in L-packets satisfying (1.3). But $\Pi_\phi = \Pi_{a\phi}$ if and only if there exists $g \in \hat{G}$ such that

$$\operatorname{Int} g \circ \phi = a\phi,$$

i.e. the set $S_\phi = \{g \in \hat{G} \,|\, \operatorname{Int} g \circ \phi = a\phi\}$ is not empty. Assume $g \in S_\phi$ has Jordan decomposition $g = su$. Then $s \in S_\phi$ and $u \in \operatorname{Cent}_{\hat{G}}(\phi)_0$, a connected group which acts by translations on S_ϕ. Therefore, every connected component of S_ϕ contains a semisimple element.

Assume that $s \in S_\phi$ is semisimple, so that $\hat{H} := \hat{G}^s_0$ is reductive. Let $\mathcal{H}$ be the subgroup of ${}^L G$ generated by $\hat{H}$ and the image of ϕ, endowed with the induced topology, and let ξ be the inclusion of $\mathcal{H}$ in ${}^L G$. Then there is a split exact sequence

$$1 \to \hat{H} \to \mathcal{H} \to W_{\mathbb{R}} \to 1.$$

Define $\rho_{\mathcal{H}}$ as in remark 1.4, and construct an L-group ${}^L H$ from $\hat{H}$, $\rho_{\mathcal{H}}$ and a choice of a splitting $\mathbf{spl}_{\hat{H}}$. Let H be a quasi-split reductive group defined over $\mathbb{R}$ with L-group ${}^L H$. Then $(H, \mathcal{H}, s, \xi)$ is an endoscopic datum for $(G, \mathbf{a})$. If $\mathcal{H}$ is isomorphic to ${}^L H$, then $\xi : W_{\mathbb{R}} \to {}^L G$, which factorizes through $\mathcal{H}$, gives a Langlands parameter

$$\xi_H : W_{\mathbb{R}} \to \mathcal{H}.$$

In line with Langlands functoriality principle, there should a transfer from Π_{ϕ_H} to Π_ϕ, *i.e.* a way to relate characters of representations in Π_{ϕ_H} to characters of of representations in Π_ϕ. This relationship between characters of the two groups will be obtained from a dual transfer of orbital integrals from $G(\mathbb{R})$ to $H(\mathbb{R})$. The goal of the paper is to explain how this transfer of orbital integrals is obtained. When $\mathcal{H}$ is not an L-group for H, the situation is more complicated. We need the notion of z-pair introduced below, and we get a tranfer of orbital integrals from $G(\mathbb{R})$ to some extension $H_1(\mathbb{R})$ of $H(\mathbb{R})$.

DEFINITION 1.2. The endoscopic data $(H, s, \mathcal{H}, \xi)$ and $(H', s', \mathcal{H}', \xi')$ are isomorphic if there exists $g \in \hat{G}$, an $\mathbb{R}$-isomorphism $\alpha : H \to H'$ and an L-isomorphism $\beta : \mathcal{H}' \to \mathcal{H}$ such that:

- The maps induced by α and β on the based root data, $\Psi_H \xrightarrow{\alpha} \Psi_{H'}$ and $\Psi_{\hat{H}'} \xrightarrow{\beta} \Psi_{\hat{H}}$ are dual to each other.
- $\operatorname{Int} g \circ \xi \circ \beta = \xi'$.
- $gsg^{-1} = s'$ modulo $Z(\hat{G})Z(\xi')_0$, where $Z(\xi')$ is the centralizer in $\hat{G}$ of the image of $\mathcal{H}'$ under ξ'.

We may replace (s, ξ) by $(gsg^{-1}, \operatorname{Int} g \circ \xi)$ for any $g \in \hat{G}$ and get another equivalent endoscopic datum. Thus, up to equivalence, we can assume that $s \in \mathcal{T}$. Then $\operatorname{Int} s \cdot (\mathcal{B}, \mathcal{T}) = (\mathcal{B}, \mathcal{T})$. Making another such replacement with a $g \in (\hat{G}^s)_0$, we may also assume that:

$$\xi(\mathcal{B}_H) = \mathcal{B} \cap (\hat{G}^s)_0 \text{ and } \xi(\mathcal{T}_H) = \mathcal{T}. \tag{1.4}$$

For the rest of the paper, we fix G, $\hat{G}$, a splitting $\mathbf{spl}_{\hat{G}} = (\mathcal{B}, \mathcal{T}, \{\mathcal{X}\})$, an endoscopic datum $(H, \mathcal{H}, s, \xi)$, and a splitting $\mathbf{spl}_{\hat{H}} = (\mathcal{B}_H, \mathcal{T}_H, \{\mathcal{X}_H\})$ of $\hat{H}$. We assume that $s \in \mathcal{T}$ and that $(H, \mathcal{H}, s, \xi)$ satisfies (1.4).

1.9. *z*-pair. Since $\mathcal{H}$ is not necessarily an L-group for H (see [**KS**], §2.1) we need to introduce a z-extension H_1 of H. We recall the definition ([**K**]):

DEFINITION 1.3. A z-extension of a connected reductive algebraic quasi-split real group H is a central extension H_1 of H:

$$1 \to Z_1 \to H_1 \to H \to 1$$

where H_1 is a connected reductive algebraic quasi-split real group whose derived group is simply connected and Z_1 is a central torus in H_1, isomorphic to a product of $\operatorname{Res}_{\mathbb{C}/\mathbb{R}} \mathbb{R}^\times$.

Note that since $H^1(\Gamma, Z_1) = \{1\}$, we have also:

$$1 \to Z_1(\mathbb{R}) \to H_1(\mathbb{R}) \to H(\mathbb{R}) \to 1.$$

Dual to the exact sequence in the above definition we have

$$1 \to \hat{H} \to \hat{H}_1 \to \hat{Z}_1 \to 1,$$

so we regard $\hat{H}$ as a subgroup of $\hat{H}_1$. This inclusion can be extended to a L-homomorphism $\xi_{H_1} : \mathcal{H} \to {}^L H_1$ (see [**KS**] lemma 2.2.A).

DEFINITION 1.4. By a z-pair for $\mathcal{H}$, we mean a pair (H_1, ξ_{H_1}) where H_1 is a z-extension of H and $\xi_{H_1} : \mathcal{H} \to {}^L H_1$ a L-homomorphism that extends $\hat{H} \hookrightarrow \hat{H}_1$.

Observe that ξ_{H_1} determines a character λ_{H_1} of $Z_1(\mathbb{R})$. This character has Langlands parameter:

$$W_{\mathbb{R}} \xrightarrow{c} \mathcal{H} \xrightarrow{\xi_{H_1}} {}^L H_1 \to {}^L Z_1, \tag{1.5}$$

where c is any section of $\mathcal{H} \to W_{\mathbb{R}}$ and ${}^L H_1 \to {}^L Z_1$ is the natural extension of $\hat{H}_1 \to \hat{Z}_1$. For a discussion of the significance of λ_{H_1} in terms of Langlands functoriality principle, see [**KS**], end of section 2.2. Let us just say here that if $\mathcal{H}$ is not an L-group, then there is no matching between compactly supported smooth functions on $G(\mathbb{R})$ and compactly supported smooth functions on $H(\mathbb{R})$. Instead, we will establish a matching between compactly supported smooth functions on $G(\mathbb{R})$ and smooth functions on $H_1(\mathbb{R})$, compactly supported modulo $Z_1(\mathbb{R})$ and transforming under translations by elements of $Z_1(\mathbb{R})$ according to λ_{H_1}.

2. Correspondences of points

2.1. Correspondences of semi-simple conjugacy classes. Recall the inner twist $\psi : G \to G^*$ and the element $u_\sigma \in G^*$ such that $\psi\sigma(\psi)^{-1} = \operatorname{Int} u_\sigma$. Then ψ induces a bijective map $\mathcal{A}_{G,G^*}$ from conjugacy classes in G to conjugacy classes in G^* which is defined over $\mathbb{R}$. Let us denote by $\mathcal{A}_{G^*,G}$ its inverse.

We recall now the points correspondences between H and G. We have fixed splitings $\mathbf{spl}_{\hat{G}^*} = (\mathcal{B}, \mathcal{T}, \{\mathcal{X}\})$ and $\mathbf{spl}_{\hat{H}} = (\mathcal{B}_H, \mathcal{T}_H, \{\mathcal{X}_H\})$.

THEOREM 2.1. *([**LS**] 1.3.A) There is a canonical map defined over* $\mathbb{R}$*:*

$$\mathcal{A}_{H/G} : Cl_{ss}(H) \to Cl_{ss}(G) \tag{2.1}$$

between semi-simple conjugacy classes in H and semi-simple conjugacy classes in G.

This map is obtained in the following way: suppose (B_H, T_H) is a pair in H and that (B,T) is a pair in G^*. Attached to (B_H, T_H) and $(\mathcal{B}_H, \mathcal{T}_H)$ is an isomorphism $\hat{T}_H \simeq \mathcal{T}_H$ and attached to (B,T) and $(\mathcal{B}, \mathcal{T})$ is an isomorphism $\hat{T} \simeq \mathcal{T}$. We have therefore a chain of isomorphisms:

$$\hat{T}_H \simeq \mathcal{T}_H \xrightarrow{\xi} \mathcal{T} \simeq \hat{T}$$

which yields $T_H \simeq T$. This isomorphism transports the coroots of T_H in H into a subsystem of the coroots of T in G^* and the Weyl group $W_H := W(H, T_H)$ into a subgroup of the Weyl group $W := W(G^*, T)$, and so induce a map:

$$T_H/W_H \to T/W.$$

Therefore we have:

$$\mathrm{Cl}_{ss}(H) \simeq T_H/W_H \to T/W \simeq \mathrm{Cl}_{ss}(G^*) \xrightarrow{\mathcal{A}_{G^*,G}} \mathrm{Cl}_{ss}(G)$$

yielding the map (2.1).

If T_H is defined over $\mathbb{R}$, we may choose (B,T) and B_H such that both T and $T_H \simeq T$ are defined over $\mathbb{R}$ (see [**LS**], 1.3.A). An $\mathbb{R}$-isomorphism $\eta : T_H \to T$ as above will be called an admissible embedding of T_H in G^*. It is uniquely determined up to $\mathcal{A}(T)$-conjugacy, that is up to composition with $\operatorname{Int} g$ where g lies in

$$\mathcal{A}(T) := \{g \in G^* : \sigma_{G^*}(g)^{-1} g \in T\}.$$

2.2. Norms. Let $\delta \in G(\mathbb{R})$ be a semi-simple element. Recall that δ is regular (resp. strongly regular) if $(G^\delta)_0$ (resp. G^δ) is a maximal torus of G.

Let us denote by $\mathcal{O}_\gamma$ the conjugacy class of an element $\gamma \in H$, and by $\mathcal{O}_\delta$ the conjugacy class of an element $\delta \in G$.

DEFINITION 2.2. An element $\gamma \in H$ is G-regular (resp. strongly G-regular if $\mathcal{A}_{H/G}(\mathcal{O}_\gamma)$ is a regular (resp. strongly regular) conjugacy class in G.

LEMMA 2.3. *(**[KS]**,3.3.C)*
(i) G-regular implies regular,
(ii) strongly G-regular implies strongly regular.

DEFINITION 2.4. The stable conjugacy class of a strongly regular element $\delta \in G(\mathbb{R})$ is $\mathcal{O}_\delta \cap G(\mathbb{R})$.

We extend this definition to regular elements. Let $\delta \in G(\mathbb{R})$ be such a element. Suppose that $\delta' = g\delta g^{-1} \in G(\mathbb{R})$. Then we have $\sigma_G(g)^{-1}g \in G^\delta$. Let us denote $T_\delta = (G^\delta)_0$. Since δ is regular, T_δ is a torus.

DEFINITION 2.5. In the above setting, we say that δ' is in the stable conjugacy class of δ if and only if $\sigma_G(g)^{-1}g \in T_\delta$.

We will see below the reason for this requirement. Note that if δ is strongly regular, then $G^\delta = T_\delta$ and the two definitions agree.

DEFINITION 2.6. Let $\gamma \in H(\mathbb{R})$ be G-regular, and let T_H be the maximal torus of H containing γ. Fix an admissible embedding $T_H \xrightarrow{\eta} T$ of T_H in G^*. We say that γ is a norm of $\delta \in G(\mathbb{R})$ if:
(i) δ lies in the image of $\mathcal{O}_\gamma$ under $\mathcal{A}_{H/G}$.
Then, by definition, there exist $x \in G^*$ and $\delta^* \in T$ such that $\delta^* = x\psi(\delta)x^{-1}$ and $\delta^* = \eta(\gamma)$.
(ii) $\operatorname{Int} x \circ \psi : T_\delta \to T$ is defined over $\mathbb{R}$.

We will see in the proof of the next theorem that if γ is strongly G-regular, then the condition (ii) is automatically fulfilled. When it is not possible to find such an element δ, we say that γ is not a norm.

THEOREM 2.7. *Let $\gamma \in H(\mathbb{R})$ be G-regular. Then γ is a norm of exactly one stable conjugacy class in $G(\mathbb{R})$ or is not a norm.*

Proof. Recall that

$$\psi = \operatorname{Int} u_\sigma \circ \sigma_{G^*} \circ \psi \circ \sigma_G = \operatorname{Int} u_\sigma \circ \sigma(\psi),$$

$$\text{thus } \sigma(\operatorname{Int} x \circ \psi) = \operatorname{Int}(\sigma_{G^*}(x)u_\sigma^{-1}) \circ \psi.$$

Therefore, $\operatorname{Int} x \circ \psi : T_\delta \to T$ is defined over $\mathbb{R}$ if and only if

$$\sigma(\operatorname{Int} x \circ \psi) = \operatorname{Int}(\sigma_{G^*}(x)u_\sigma^{-1}) \circ \psi = \operatorname{Int} x \circ \psi.$$

Set $v_\sigma = xu_\sigma\sigma_{G^*}(x)^{-1}$. Thus $\operatorname{Int} x \circ \psi$ is defined over $\mathbb{R}$ if and only if $\operatorname{Int} v_\sigma^{-1}$ is trivial on T, i.e. $v_\sigma \in T$.

Furthermore, since

$$v_\sigma^{-1}\delta^* v_\sigma = \sigma_{G^*}(x)u_\sigma^{-1}\psi(\delta)u_\sigma\sigma_{G^*}(x)^{-1} = \sigma_{G^*}(x)\sigma(\psi)(\delta)\sigma_{G^*}(x)^{-1}$$
$$= \sigma_{G^*}(\delta^*) = \delta^*$$

we get $v_\sigma \in (G^*)^{\delta^*}$.

From this discussion, we see that if γ is stronly G-regular, then δ^* is strongly regular, $(G^*)^{\delta^*} = T$ and $\operatorname{Int} x \circ \psi : T_\delta \to T$ is defined over $\mathbb{R}$.

Suppose now that δ_1 is stably conjugate to δ, i.e. there exists $g \in G$, with $\delta_1 = g\delta g^{-1} \in G$ and $\sigma_G(g)^{-1}g \in T_\delta$. Suppose also that γ is a norm for δ. Then we compute

$$\delta^* = x\psi(g)^{-1}\psi(\delta_1)(x\psi(g)^{-1})^{-1}.$$

We want to prove that γ is a norm for δ_1, so we have to check that $\operatorname{Int}(x\psi(g)^{-1}) \circ \psi : T_{\delta_1} \to T$ is defined over $\mathbb{R}$. From the discussion above, this is the case if and only if $\operatorname{Int}(x\psi(g)^{-1}u_\sigma\sigma_{G^*}(x\psi(g)^{-1})^{-1})$ is trivial on T. We compute:

$$x\psi(g)^{-1}u_\sigma\sigma_{G^*}(x\psi(g)^{-1})^{-1} = (x\psi(g)^{-1}u_\sigma)(u_\sigma^{-1}\psi(\sigma_G(g))u_\sigma)\sigma_{G^*}(x)^{-1}$$
$$= x\psi(g^{-1}\sigma_G(g))u_\sigma\sigma_{G^*}(x)^{-1}$$
$$= x\psi(g^{-1}\sigma_G(g))x^{-1}v_\sigma$$

Since γ is a norm for δ, $\operatorname{Int} v_\sigma$ is trivial on T, hence we only have to check that $\operatorname{Int}(x\psi(g^{-1}\sigma_G(g))x^{-1})$ is trivial on T, i.e. $\operatorname{Int}(g^{-1}\sigma_G(g))$ is trivial on T_δ. This is now obvious by the definition of stable conjugacy and the requirement that $g^{-1}\sigma_G(g) \in T_\delta$.

Let us now prove the other inclusion, and so suppose that γ is a norm for δ, δ_1, i.e. there exist δ^* in T and x, $x_1 \in G^*_{sc}$ such that

$$\delta^* = x\psi(\delta)x^{-1} = x_1\psi(\delta_1)x_1^{-1}$$

We are led to :

$$\delta_1 = \psi^{-1}(x_1^{-1}x)\delta(\psi^{-1}(x_1^{-1}x))^{-1}.$$

We have then to show that $w^{-1}\sigma_G(w) \in T_\delta$, where $w = \psi^{-1}(x_1^{-1}x)$. This is equivalent to : $\operatorname{Int} x \circ \psi(w^{-1}\sigma_G(w)) \in T$, and:

$$\operatorname{Int} x \circ \psi(w^{-1}\sigma_G(w)) = \sigma_{G^*}(v_1)^{-1}\sigma_{G^*}(v),$$

where $v = v_\sigma$ as above and $v_1 = x_1u_\sigma\sigma_{G^*}(x_1)^{-1}$. This is a product of elements in T, hence $\operatorname{Int} x \circ \psi(w^{-1}\sigma_G(w)) \in T$. The proof of the theorem is now complete. □

REMARK. Suppose that γ is a norm of δ, and that $\eta : T_H \to T$, $\delta^* = \eta(\gamma)$ and x are fixed as in the definition. We may replace x by an element $x_1 = tx$ with $t \in (G^*)^{\delta^*}$, provided that $\operatorname{Int} x_1 \circ \psi$ is defined over $\mathbb{R}$. As we have seen in the proof of the theorem, this is equivalent to $\operatorname{Int} v_\sigma^1$ trivial on T, with $v_\sigma^1 = txu_\sigma\sigma_{G^*}(x)^{-1}\sigma_{G^*}(t)^{-1}$. Since $t \in (G^*)^{\delta^*}$, t normalizes T and since v_σ is trivial on T, we get that $t\sigma_{G^*}(t)^{-1}$ must act trivially on T, i.e. $t\sigma_{G^*}(t)^{-1} \in T$.

2.3. Admissible embeddings in G. Let T_H be a maximal torus of H defined over $\mathbb{R}$. If there exists a G-regular element $\gamma \in T_H(\mathbb{R})$ which is the norm of an element $\delta \in G(\mathbb{R})$, we say that T_H originates in G. Recall this means that given an admissible embedding $T_H \xrightarrow{\eta} T$ of T_H in G^* defined over $\mathbb{R}$, there exist $x \in G^*_{sc}$ and $\delta^* \in T$ such that $\delta^* = x\psi(\delta)x^{-1}$, $\delta^* = \eta(\gamma)$ and

$$\operatorname{Int} x \circ \psi : T_\delta \to T$$

is defined over $\mathbb{R}$. Let us denote T_δ by T_G.

Let $\gamma_1 \in T_H$ be another G-regular element, $\delta_1{}^* = \eta(\gamma_1) \in T$ and $\delta_1 \in T_G$ such that $\operatorname{Int} x \circ \psi(\delta_1) = \delta_1{}^*$. Since $\operatorname{Int} x \circ \psi : T_G \to T$ is defined over $\mathbb{R}$, γ_1 is a norm of δ_1. Conversely, for every regular element δ_1 in $T_G(\mathbb{R})$, there is a G-regular element γ_1 of $T_H(\mathbb{R})$ which is a norm of δ_1.

We summarize the discussion.

PROPOSITION 2.8. *Let T_H as above be a maximal torus of H defined over $\mathbb{R}$ originating in G and let $\gamma \in T_H(\mathbb{R})$ be a G-regular element which is a norm of a regular $\delta \in G(\mathbb{R})$. Let T_G the maximal torus G containing δ, and fix T, η and x as above. Then there is an isomorphism defined over $\mathbb{R}$:*

$$\eta_x : T_H \to T_G, \qquad \eta_x = (\operatorname{Int} x \circ \psi)^{-1} \circ \eta,$$

such that $\eta_x^{-1}(\delta_1)$ is a norm of δ_1 for all regular element $\delta_1 \in T_G$. As suggested by the notation, this isomorphism does not depend on the initial choice of δ and γ but only on the admissible embedding η of T_H in G^* and the element x. We call such an isomorphism an admissible embedding of T_H in G.

Graphically,

$$T_H \xrightarrow{\eta} T \xleftarrow{\operatorname{Int} x \circ \psi} T_G.$$

Suppose that we are in the setting above, i.e. $\gamma \in T_H(\mathbb{R})$ is a norm of a regular element in $G(\mathbb{R})$. Then γ is a norm of exactly one stable conjugacy class $\mathcal{O}^{st}$ in $G(\mathbb{R})$. Let us denote by Σ_γ a system of representatives for the conjugacy classes in $G(\mathbb{R})$ contained in $\mathcal{O}^{st}$. Let us first fix an admissible embedding η of T_H in G^*. Then, for all $\delta_i \in \Sigma_\gamma$, let us choose $x_i \in G^*$ such that x_i and η define an admissible embedding $\eta_{x_i} : T_H \to T_{\delta_i}$ as above. Put $T_{G,i} = T_{\delta_i}$. Suppose that γ_1 is another G-regular element in $T_H(\mathbb{R})$. Then, we have seen that γ_1 is a norm of the $\delta_{1,i} := \eta_{x_i}(\gamma_1)$.

LEMMA 2.9. *The set $\{\delta_{1,i}\}$ forms a system of representatives for the conjugacy classes in $G(\mathbb{R})$ contained in the stable conjugacy class of elements in $G(\mathbb{R})$ for which γ_1 is a norm.*

Proof. This is straightforward and more or less contained in the proof of Theorem 2.7.

If T_H originates in G, a set of isomorphisms η_{x_i} as in the lemma will be called a complete system of admissible embeddings of T_H in G.

2.4. Admissible embeddings in $G(\mathbb{R})^+$. Suppose that the maximal torus T_H of H originates in G. Let us fix a complete system of admissible embeddings $\eta_{x_j} : T_H \to T_j$ as in the previous paragraph. We say that T_H (we should say $T_H(\mathbb{R})$) originates in $G(\mathbb{R})^+$ if one of the Cartan subgroups $T_j(\mathbb{R})$ of $G(\mathbb{R})$ is contained in $G(\mathbb{R})^+$. We first remark that if one of the $T_j(\mathbb{R})$ is contained in $G(\mathbb{R})^+$, then all of them are. Indeed, the $T_j(\mathbb{R})$ are conjugate under G, so a well known result asserts they are conjugate under $G(\mathbb{R})$. Since $G(\mathbb{R})^+$ is normal in $G(\mathbb{R})$, this proves the assertion. The same argument also proves that the notion "originates in $G(\mathbb{R})^+$" does not depend on the choices of the η_{x_i}.

2.5. Roots originating in H. Suppose that we have a maximal torus T_H of H defined over $\mathbb{R}$, and an admissible embedding

$$\eta : T_H \to T.$$

We say that $\alpha \in R(G^*, T)$ originates in H when there exists a root $\alpha_H \in R(H, T_H)$ such that $\eta(\alpha_H\check{}) = \alpha\check{}$.

Another way to say this is to notice that

$$\alpha_H\check{} \in R\check{}(H, T_H) \subset X_*(T_H) \simeq X^*(T_H) \simeq X^*(T) \simeq X_*(T),$$

but, as it is obvious from the definition of an endoscopic datum, we get $R\check{}(H, T_H) \subset R\check{}(G, T)$. Thus $\alpha\check{}$ originates in H if it coincide with a root $\alpha_H\check{}$ through these identifications.

Let $\gamma \in T_H$ be G-regular, and $\delta^* \in G^*$ such that $\eta(\gamma) = \delta$. We have then:

$$\alpha_H(\gamma) = \alpha(\delta^*) \tag{2.2}$$

Let $\eta_x : A := T_H \to T_G$ be an admissible embedding of T_H in G. We can define roots of $R(G, T_G)$ originating in H in a similar way. It is clear from the construction that the set of coroots in $R\check{}(G, T)$ (or $R\check{}(G, T_G)$) originating in H form a subsystem of coroots.

2.6. Cayley tranforms. Let $\mathfrak{g}$ be a real reductive Lie algebra and $\mathfrak{g}_\mathbb{C}$ its complexification. We denote by σ the conjugation of $\mathfrak{g}_\mathbb{C}$ with respect to $\mathfrak{g}$. Let $\mathfrak{b} \subset \mathfrak{g}$ be a Cartan subalgebra. Let $\alpha \in R(\mathfrak{g}_\mathbb{C}, \mathfrak{b}_\mathbb{C})$ be an imaginary root, choose a root vector X_α for α and fix a root vector $X_{-\alpha}$ of $-\alpha$ such that $[X_\alpha, X_{-\alpha}] = H_\alpha$, where H_α is another notation for the coroot $\alpha\check{} \in \mathfrak{b}_\mathbb{C}$. Then $\mathfrak{s}_\mathbb{C} = \mathbb{C} \cdot X_\alpha + \mathbb{C} \cdot X_{-\alpha} + \mathbb{C} \cdot H_\alpha$ is a simple complex Lie algebra invariant under conjugation, $\sigma(H_\alpha) = -H_\alpha = H_{-\alpha}$ and $\sigma(X_\alpha) = cX_{-\alpha}$ for some $c \in \mathbb{R}^*$. If $c < 0$, we can renormalize to get $\sigma(X_\alpha) = -X_{-\alpha}$ or if $c > 0$, to get $\sigma(X_\alpha) = X_{-\alpha}$. In the former case, α is compact, $\mathfrak{s} = \mathfrak{s}_\mathbb{C}^\sigma \simeq \mathfrak{su}(2)$. In the latter case, α is non-compact and $\mathfrak{s} \simeq \mathfrak{sl}(2, \mathbb{R})$. Suppose that α is non-compact. We define a standard Cayley transform with respect to α to be an element of the adjoint group of $\mathfrak{s}_\mathbb{C}$ of the form $c_\alpha = \exp(-i\pi(X_\alpha + X_{-\alpha})/4)$, where X_α, $X_{-\alpha}$ are normalized as explained above (they are unique up to a scalar factor of absolute value 1, and all the standard Cayley transforms for α are conjugate in the adjoint group of $\mathfrak{s}_{\mathbb{C}}$). We have:

$$\mathfrak{b}_{\mathbb{C}} = \ker\alpha \oplus \mathbb{C}\cdot H_\alpha$$
$$\mathfrak{b} = \ker\alpha_{|\mathfrak{b}} \oplus i\mathbb{R}\cdot H_\alpha.$$

Let $\mathfrak{a}_{\mathbb{C}} := c_\alpha \cdot \mathfrak{b}_{\mathbb{C}} = \ker\alpha \oplus \mathbb{C}\cdot(X_\alpha - X_{-\alpha})$. This is a Cartan subalgebra defined over $\mathbb{R}$ and

$$\mathfrak{a} = \ker\alpha_{|\mathfrak{b}} \oplus i\mathbb{R}\cdot(X_\alpha - X_{-\alpha}).$$

The root $\beta := c_\alpha\cdot\alpha$ of $R(\mathfrak{g}_{\mathbb{C}}, \mathfrak{a}_{\mathbb{C}})$ is real and $c_\alpha\cdot H_\alpha = H_\beta = i(X_\alpha - X_{-\alpha})$. Furthermore:

$$\sigma(c_\alpha) = \exp(i\pi(X_\alpha + X_{-\alpha})/4) = c_\alpha^{-1}.$$

A standard Cayley transform is a particular case of a generalized Cayley transform defined by Shelstad (see [**Sh1**], §2), in particular, as it is easy to check, $\sigma(c_\alpha)^{-1}c_\alpha = c_\alpha^2$ realizes the Weyl reflection s_α with respect to the root α.

It will be useful to reverse the process, and define Cayley transforms with respect to real roots. If $\mathfrak{a}_{\mathbb{C}}$ is a Cartan subalgebra of $\mathfrak{g}_{\mathbb{C}}$ defined over $\mathbb{R}$, and β a real root, we take root vectors X_β and $X_{-\beta}$ in $\mathfrak{g}$ such that $[X_\beta, X_{-\beta}] = H_\beta$ and define $c_\beta := \exp(i\pi(X_\beta + X_{-\beta})/4)$. The root $c_\beta\cdot\beta$ of $\mathfrak{b}_{\mathbb{C}} := c_\beta\cdot\mathfrak{a}_{\mathbb{C}}$ is imaginary non-compact and we can make the choices such that $c_\beta = c_\alpha^{-1}$.

2.7. Jump data. Let us denote by $\mathfrak{g}_{\mathbb{C}}$ (resp. $\mathfrak{g}$) the Lie algebra of G (resp. $G(\mathbb{R})$). We say that $x \in G(\mathbb{R})$ is semi-regular when the derived algebra of $\mathfrak{g}^x$ is isomorphic to $\mathfrak{sl}(2,\mathbb{R})$ or $\mathfrak{su}(2)$. Suppose it is $\mathfrak{sl}(2,\mathbb{R})$. Let $\mathfrak{b}$ be a fundamental Cartan subalgebra of $\mathfrak{g}^x$, and $\pm\alpha$ the roots of $\mathfrak{b}_{\mathbb{C}}$ in $\mathfrak{g}_{\mathbb{C}}^x$: they are non-compact imaginary and satisfy

$$\det(\mathrm{Id} - \mathrm{Ad}x^{-1})_{|\mathfrak{g}_{\mathbb{C}}^\alpha} = 0.$$

Let c_α be a Cayley transform with respect to α as in the previous section and let us also denote by $\mathfrak{a}$ the maximally split Cartan subalgebra of $\mathfrak{g}^x$ obtained from the Cayley transform (ie. $\mathfrak{a}_{\mathbb{C}} = c_\alpha\cdot\mathfrak{b}_{\mathbb{C}}$). Let $B(\mathbb{R})$ and $A(\mathbb{R})$ be the Cartan subgroups corresponding respectively to $\mathfrak{b}$ and $\mathfrak{a}$. We refer to these notations by saying that $(x, B(\mathbb{R}), A(\mathbb{R}), c_\alpha)$ is a jump datum for $G(\mathbb{R})$. Jump data for the group $H(\mathbb{R})$ are defined in the same way.

2.8. Jump data and admissible embedding. Let $(\gamma_0, T_H, T'_H, c_{\alpha_H})$ be a jump datum on H.

LEMMA 2.10. *(see* [**Sh1**], *§2) If T_H does not originate in G, neither does T'_H.*

<u>*Proof.*</u> Suppose that T'_H does originates in G. We want to prove that T_H also originates in G.

Our hypothesis is that there exist a maximal torus T' of G^* defined over $\mathbb{R}$, with an admissible embedding $\eta' : T'_H \to T'$, elements $\gamma' \in T'_H$, $\delta^{*\prime} \in T'$

such that $\eta'(\gamma') = \delta^{*\prime}$, an element $x \in G^*_{sc}$, a maximal torus T'_G of G defined over $\mathbb{R}$ and an element $\delta' \in T'_G(\mathbb{R})$ such that $\delta^{*\prime} = x\psi(\delta')x^{-1}$ and

$$\mathrm{Int}\, x \circ \psi : T'_G = G_0^{\delta} \to T'$$

is defined over $\mathbb{R}$ (see section 4). From these data, we get an admissible embedding:

$$\eta_x : T'_H \to T'_G.$$

Let us denote by β_H the real root of $R(G, T'_H)$ obtained by taking the Cayley transform $c_{\alpha_H} \cdot \alpha_H$, and by β the corresponding root of $R(G, T'_G)$. Since η' is defined over $\mathbb{R}$, it is clear that β is real.

Set $\delta_0 = \eta_x(\gamma_0) \in T'_G(\mathbb{R})$ and $\delta_0^* = \eta(\gamma_0)$. We take a standard Cayley transform c_β with respect to β and and define $\mathfrak{t}_{G,\mathbb{C}} = c_\beta \cdot \mathfrak{t}'_{G,\mathbb{C}}$. Let T_G be the corresponding torus of G. We need also to introduce the corresponding roots and Cayley transforms in G^*. Let us denote these by β^*, c_{β^*}...

The map $\mathrm{Int}\, x \circ \psi$ realizes isomorphisms from G^{δ_0} onto $(G^*)^{\delta_0^*}$ and from T_G onto T. The following diagram is easily seen to commute:

$$\begin{array}{ccc} T'_G & \xrightarrow{\mathrm{Int}\, x\circ\psi} & T' \\ \Big\downarrow \mathrm{Int}\, c_\beta & & \Big\downarrow Intc_{\beta^*} \\ T_G & \xrightarrow{\mathrm{Int}\, x\circ\psi} & T \end{array}$$

From this, we conclude that $\mathrm{Int}\, x \circ \psi : T_G \to T$ is defined over $\mathbb{R}$. Indeed, since $\mathrm{Int}\, x \circ \psi : T'_G \to T'$ is defined over $\mathbb{R}$ and since $\sigma(c_\beta) = c_\beta^{-1}$, $\sigma(c_\beta) = c_\beta^{-1}$, this amounts to

$$\mathrm{Int}\, c_{\beta^*}^{-1} (\mathrm{Int}\, x \circ \psi)_{|T'_G} \mathrm{Int}\, c_\beta = \mathrm{Int}\, c_{\beta^*} (\mathrm{Int}\, x \circ \psi)_{|T'_G} \mathrm{Int}\, c_\beta^{-1},$$

or equivalently

$$\mathrm{Int}\, c_{\beta^*}^{-2} (\mathrm{Int}\, x \circ \psi)_{|T'_G} \mathrm{Int}\, c_\beta^{2} = (\mathrm{Int}\, x \circ \psi)_{|T'_G} \tag{2.3}$$

But the action of $c_{\beta^*}^{-2}$ on T is given by the action of the Weyl group reflection $s_{\beta^*} \in W(G^*, T)$ and the action of c_β^2 on T_G is given by the action of the Weyl group reflection $s_\beta \in W(G, T_G)$. It is thus obvious that (2.3) holds.

We define η by the following commutative diagram:

$$\begin{array}{ccc} T'_H & \xrightarrow{\eta'} & T' \\ \Big\downarrow \mathrm{Int}\, c_{\beta_H} & & \Big\downarrow \mathrm{Int}\, c_{\beta^*} \\ T_H & \xrightarrow{\eta} & T \end{array}$$

It remains to check that η is defined over $\mathbb{R}$ to complete the proof of the lemma. The argument is similar to the one just given.

$$\begin{aligned}\sigma_{G^*}(\eta) &= \sigma_{G^*}(\operatorname{Int} c_{\beta^*} \circ \eta' \circ \operatorname{Int} c_{\beta_H}^{-1}) \\ &= \operatorname{Int}(\sigma_{G^*}(c_{\beta^*})) \circ \sigma_{G^*}(\eta') \circ (\operatorname{Int} \sigma_H(c_{\beta_H}^{-1}) \\ &= \operatorname{Int} c_{\beta^*}^{-1} \circ \eta' \circ (\operatorname{Int} c_{\beta_H}) \\ &= \operatorname{Int} c_{\beta^*} \circ s_{\beta^*} \circ \eta' \circ s_{\beta_H} \circ \operatorname{Int} c_{\beta_H}^{-1} \\ &= \operatorname{Int} c_{\beta^*} \circ \eta' \circ \operatorname{Int} c_{\beta_H}^{-1} = \eta\end{aligned}$$

□

COROLLARY 2.11. *If T_H does not originates in $G(\mathbb{R})^+$, neither does T'_H.*

Proof. If T_H does not originate in G, this is clear from the previous lemma. Suppose that T'_H does originate in $G(\mathbb{R})^+$ and that T_H originates in G, but not in $G(\mathbb{R})^+$. Fix an admissible embedding $\eta_x : T_H \to T_G$. This means that some connected components of $T_G(\mathbb{R})$ are not in $G(\mathbb{R})^+$. Suppose T'_H originates in $G(\mathbb{R})^+$ and fix an admissible embedding $\eta_y : T'_H \to T'_G$, with $T'_G(\mathbb{R}) \subset G(\mathbb{R})^+$. We can arrange the choices such that T'_G is obtained from T_G by a standard Cayley transform. But then it is well known that all connected components of $T_G(\mathbb{R})$ intersect $T'_G(\mathbb{R})$, and we would get $T_G(\mathbb{R}) \subset G(\mathbb{R})^+$, and obtain a contradiction. Thus T'_H does not originate in $G(\mathbb{R})^+$. □

2.9. Let $(\gamma_0, T_H, T'_H, c_{\alpha_H})$ be a jump datum on H and suppose now that T_H does originate in G. Fix a complete system of admissible embeddings $\{\eta_{x_i} : T_H \to T_{G_i}\}$. Set $\delta_{0,i} = \eta_{x_i}(\gamma_0)$ and $\alpha_i = \eta_{x_i}(\alpha_H)$, and set

$$I_{\gamma_0} := \{i \,|\, \alpha_i \text{ is imaginary non-compact }\}$$

LEMMA 2.12. *T'_H originates in G if and only if $I_{\gamma_0} \neq \emptyset$.*

Proof. Suppose T'_H originates in G, and let $\eta'_y : T'_H \to T'_G$ be an admissible embedding. Let us denote by β_H the real root in $R(T'_H, H)$ which is obtained from α_H by the Cayley transform c_{α_H}, and denote by β the corresponding real root in $R(T'_G, G)$. By the construction made in the previous paragraph, we get an admissible embedding $\eta_y : T_H \to T_G$, and a non compact imaginary root β in $R(T_G, G)$. Since our system of admissible embedding η_{x_i} was complete, there is one i such that η_y is conjugate in $G(\mathbb{R})$ to η_{x_i}, and so α_i is imaginary non-compact. Conversely, suppose that I_{γ_0} is not empty, and for some $i \in I_{\gamma_0}$, consider the data attached to the admissible embedding η_{x_i}. We want to construct from it an admissible embedding $T'_H \to T'_G$. It is straightforward to check that the construction in the proof of lemma 2.8 can be adapted to the present situation, by exchanging the role of T_H and T'_H, the Cayley transforms being with respect to non-compact imaginary roots rather than real roots. This will be refined below in lemma 2.14. □

In the setting above, suppose I_{γ_0} is not empty.

LEMMA 2.13. *If for some $i \in I_{\gamma_0}$, the reflection s_{α_i} is realized in the group $G(\mathbb{R})^{\delta_{0,i}}$, then the same is true for all j in I_{γ_0}*

Proof. Suppose $n_i \in G(\mathbb{R})^{\delta_{0,i}}$ realizes s_{α_i}. Take an arbitrary $j \in I_{\gamma_0}$, and let us take $g \in G$ such that $g \cdot T_{G_i} = T_{G_j}$ with $\sigma_G(g)^{-1}g = t \in T_{G,i}$. Then,

$$\sigma_G(gn_ig^{-1}) = \sigma_G(g)n_i\sigma_G(g)^{-1} = gt^{-1}n_itg^{-1} = gn_in_i^{-1}t^{-1}n_itg^{-1}$$
$$= g(s_{\alpha_i} \cdot t^{-1})tg^{-1}.$$

Set $t = \exp X$ for some $X \in \mathfrak{t}_{G,i}$. We get

$$(s_{\alpha_i} \cdot t^{-1})t = \exp(X - s_{\alpha_i} \cdot X) = \exp(\alpha_i(X)H_{\alpha_i}).$$

Since $(\sigma \cdot \alpha_i)(t) = (-\alpha_i)(t) = \alpha_i(t^{-1})$ because α_i is imaginary, and that on the other hand

$$(\sigma \cdot \alpha_i)(t) = \overline{\alpha_i(\sigma_G(t))} = \overline{\alpha_i(t^{-1})},$$

we see that $\alpha_i(t) = e^{\alpha_i(X)} \in \mathbb{R}$. Thus $\alpha_i(X) = 0$ or $\pi \mod (2\pi)$. Let us now use the fact that α_j is imaginary non-compact to rule out the case $\alpha(X) = \pi \mod (2\pi)$. Let us choose a $\mathfrak{sl}_2$-triple $(X_{\alpha_i}, X_{-\alpha_i}, H_{\alpha_i})$ with $\sigma_G(X_{\alpha_i}) = X_{-\alpha_i}$ as we may, since α_i is imaginary non-compact. Set

$$(X_{\alpha_j}, X_{-\alpha_j}, H_{\alpha_j}) = (g \cdot X_{\alpha_i}, g \cdot X_{-\alpha_i}, g \cdot H_{\alpha_i}).$$

We have

$$\sigma_G(X_{\alpha_j}) = \sigma_G(g)X_{-\alpha_i} = \sigma_G(g)g^{-1} \cdot X_{-\alpha_j}$$
$$= gt^{-1}g^{-1} \cdot X_{-\alpha_j} = g(t^{-1} \cdot X_{-\alpha_i})g^{-1} = ((-\alpha)(t^{-1}))X_{-\alpha_j} = \alpha(t)X_{-\alpha_j}.$$

Since α_j is imaginary non-compact, $\alpha_i(t)$ must be positive, and thus $\alpha(X) = 0 \mod (2\pi)$. We can now conclude that $\exp(\alpha(X)H_{\alpha_i}) = 0$ since $2\pi\mathbb{Z}H_{\alpha_i}$ is in the kernel of the exponential map, and finally, $\sigma_G(gn_ig^{-1}) = gn_ig$. Then s_{α_j} is realized by $n_j = gn_ig^{-1} \in G(\mathbb{R})^{\delta_{0,j}}$. □

2.10. Suppose that $\delta_0 \in T_G(\mathbb{R})$ is semi-regular, such that the roots $\pm\alpha$ of $\mathfrak{t}_{G,\mathbb{C}}$ in $\mathfrak{g}_{\mathbb{C}}^{\delta_0}$ are imaginary. Let $H_\alpha \in i\mathfrak{t}_G$ be the coroot of α, and $\delta_\nu = \delta_0 \exp(i\nu H_\alpha) \in T_G(\mathbb{R})$. Then for ν sufficiently small and non-zero, δ_ν is a regular element in T_G. Let $w_\alpha \in (G^{\delta_0})_0$ be an element realizing the Weyl reflection s_α with respect to α. We have then :

$$w_\alpha\delta_0 \exp(i\nu H_\alpha)w_\alpha^{-1} = \exp \delta_0(-i\nu H_\alpha).$$

Hence, δ_ν and $\delta_{-\nu}$ are stably conjugate, and there are two possibilities: either they are in the same conjugacy class in $G(\mathbb{R})$, or they are not. Let us consider the second case.

Let ν be small enough and non-zero, so that δ_ν is regular. Let $\gamma_\nu = \eta_x^{-1}(\delta_\nu)$, and Σ_{γ_ν} be a set of representatives of conjugacy classes under $G(\mathbb{R})$ in the stable conjugacy class of elements for which γ_ν is a norm. Then we may assume that δ_ν and $\delta_{-\nu}$ are in Σ_{γ_ν}.

We can construct another admissible embedding:

$$\eta_{x\psi(w_\alpha^{-1})} : T_H \to T_G$$
$$\gamma^\nu \mapsto \delta^{-\nu}$$

using the same admissible embedding $T_H \xrightarrow{\eta} T$ of T_H in G^* and the element $x\psi(w_\alpha)$. It is indeed easy to show that $(\mathrm{Int}\,(x\psi(w_\alpha)) \circ \psi)_{|T_G}$ is defined over $\mathbb{R}$ from the fact that $(\mathrm{Int}\, x \circ \psi)_{|T_G}$ and the reflection s_α on T_G are defined over $\mathbb{R}$. We will denote $\eta_{x\psi(w_\alpha)}$ by $\bar{\eta}_x$ for short.

In the situation of Section 2.9, suppose that for all $i \in I_{\gamma_0}$, the reflection s_{α_i} is not realized in $G(\mathbb{R})^{\delta_{0,i}}$ (see Lemma 2.13). Replacing some admissible embeddings in the complete system $\{\eta_{x_i}\}$ by conjugate ones, we can choose a subset J_{γ_0} consisting of half the indices $i \in I_{\gamma_0}$, such that:

$$\{\eta_{x_i}\}_{i\in I_{\gamma_0}} = \{\eta_{x_j}, \bar{\eta}_{x_j}\}_{j\in J_{\gamma_0}}.$$

If the reflections s_{α_i}, $i \in I_{\gamma_0}$, are realized in $G(\mathbb{R})^{\delta_{0,i}}$, we set $J_{\gamma_0} = I_{\gamma_0}$.

LEMMA 2.14. *In the setting of Section 2.9, let us suppose that I_{γ_0} is not empty. For all $j \in J_{\gamma_0}$, construct an admissible embedding $\eta'_{x_j} : T'_H \to T'_G$ as in the end of the proof of lemma 2.12. Then the set $\{\eta'_{x_j}\}_{j\in J_{\gamma_0}}$ is a complete system of admissible embeddings of T'_H in G.*

Proof. First, let us notice that that constructions of lemmas 2.8 and 2.12 are inverse of each other in an obvious sense. Notice also that if α is real or non-compact imaginary, and if w_α is an element in G^{δ_0} realizing the reflection s_α we can always define $\bar{\eta}_x$ as above, replacing x by $x\psi(w_\alpha)$. But it is clear that when w_α can be realized in $G(\mathbb{R})^{\delta_0}$, $\bar{\eta}_x$ is conjugate to η_x. This is always the case if α is real. The result follows easily. □

3. Transfer of orbital integrals

3.1. Normalization of measures. In order to define the transfer of orbital integrals we have to normalize invariant measures on the various groups in a consistent way. We chose the normalization of Duflo-Vergne, defined as follows: let A be a reductive group (complex or real), and pick an A-invariant symmetric, non-degenerate bilinear form κ on $\mathfrak{a}$. Then $\mathfrak{a}$ will be endowed with the Lebesgue measure dX such that the volume of a parallelotope supported by a basis $\{X_1, \ldots, X_n\}$ of $\mathfrak{a}$ is equal to $|\det(\kappa(X_i, X_j)|^{\frac{1}{2}}$ and A will be endowed with the Haar measure tangent to dX. If M is a closed unimodular subgroup of A, such that κ is non-degenerate on $\mathfrak{m}$, we endow M with the Haar measure determined by κ as above. If $M' \subset M$ are two closed subgroups of A such that κ is non-degenerate on their respective Lie algebras, we endow M/M' with the M-invariant measure, which is the quotient of the Haar measures on M and M' defined as above. We will denote it by $d\dot{m}$.

3.2. Orbital integrals on $G(\mathbb{R})$. Let $f \in \mathcal{C}_c^\infty(G(\mathbb{R}))$. Its orbital integral is the function defined on $G(\mathbb{R})_{reg}$ by:

$$J_{G(\mathbb{R})}(f)(x) = |\det(\mathrm{Id} - \mathrm{Ad}x^{-1})_{\mathfrak{g}_\mathbb{C}/\mathfrak{t}_\mathbb{C}}|^{\frac{1}{2}} \int_{G(\mathbb{R})/T(\mathbb{R})} f(gxg^{-1})\, d\dot{g}$$

where $T(\mathbb{R})$ is the Cartan subgroup of $G(\mathbb{R})$ containing x and $d\dot{g}$ is the invariant measure on $G(\mathbb{R})/T(\mathbb{R})$ normalized with our conventions. Note that if x is strongly regular, then $G^x(\mathbb{R}) = T(\mathbb{R})$. These objects have been studied in [**HC1**] and later in [**B1**]. We recall their properties and for this we need some notation.

Recall that if $T(\mathbb{R})$ is a Cartan subgroup of $G(\mathbb{R})$ we have a decomposition:

$$\mathfrak{g}_\mathbb{C} = \mathfrak{t}_\mathbb{C} \oplus \sum_{\beta \in R(\mathfrak{g}_\mathbb{C}, \mathfrak{t}_\mathbb{C})} \mathfrak{g}_\mathbb{C}^\beta$$

where $R(\mathfrak{g}_\mathbb{C}, \mathfrak{t}_\mathbb{C})$ is the root system of $\mathfrak{t}_\mathbb{C}$ in $\mathfrak{g}_\mathbb{C}$ and $\mathfrak{g}_\mathbb{C}^\beta$ is the root space for the root β.

Let P be a system of positive imaginary roots in $R(\mathfrak{g}_\mathbb{C}, \mathfrak{t}_\mathbb{C})$. We introduce Harish-Chandra's normalizing factor b_P on $T(\mathbb{R})_{reg}$:

$$b_P(a) = \frac{\prod_{\alpha \in P} \det(\mathrm{Id} - \mathrm{Ad}a^{-1})_{|\mathfrak{g}_\mathbb{C}^\alpha}}{|\prod_{\alpha \in P} \det(\mathrm{Id} - \mathrm{Ad}a^{-1})_{|\mathfrak{g}_\mathbb{C}^\alpha}|}$$

DEFINITION 3.1. Let $T(\mathbb{R})$ be a Cartan subgroup of $G(\mathbb{R})$. We will denote by $T(\mathbb{R})_{I-reg}$ (resp. $T(\mathbb{R})_{In-reg}$) the set of $a \in T(\mathbb{R})$ such that the root system of $\mathfrak{t}_\mathbb{C}$ in $\mathfrak{g}_\mathbb{C}^a$ has no imaginary (resp. non-compact imaginary) roots. This implies (in both cases) that $\mathfrak{t}$ is a maximally split Cartan subalgebra of $\mathfrak{g}^a$.

With notations as above, we denote by $S(\mathfrak{t}_\mathbb{C})$ the symmetric algebra of $\mathfrak{t}_\mathbb{C}$, and we identify it with the algebra of differential operators on $T(\mathbb{R})$ which are invariant under left translations by elements of $T(\mathbb{R})$. We denote by $\partial(u)$ the differential operator corresponding to $u \in S(\mathfrak{t}_\mathbb{C})$.

Let $A(\mathbb{R})$ be a Cartan subgroup of $G(\mathbb{R})$, $y \in A(\mathbb{R})$ and ϕ a function on $A(\mathbb{R})_{reg}$. Let β be a imaginary root of $\mathfrak{a}_\mathbb{C}$ in $\mathfrak{g}_\mathbb{C}$, and $H_\beta \in i\mathfrak{a}$ its coroot. Then, when the limits in the following formula exist we set:

$$[\phi]_\beta^+(y) = \lim_{t \to 0^+} \phi(y \exp tiH_\beta) + \lim_{t \to 0^-} \phi(y \exp tiH_\beta).$$

DEFINITION 3.2. Let $\mathcal{I}(G(\mathbb{R}))$ be the subspace of $\mathcal{C}^\infty(G(\mathbb{R})_{reg})$ of functions ψ which are constant on the conjugacy classes and have the properties I_1, I_2, I_3, I_4 which we now define:

I_1: if $A(\mathbb{R})$ is a Cartan subspace of $G(\mathbb{R})$, for all compact subset K of $A(\mathbb{R})$ and for all $u \in S(\mathfrak{a}_\mathbb{C})$ we have:

$$\sup_{a \in K_{reg}} |\partial(u) \cdot \psi_{|A(\mathbb{R})}(a)| < \infty.$$

I_2: if $A(\mathbb{R})$ is a Cartan subspace of $G(\mathbb{R})$, for all system P of positive imaginary roots in $R(\mathfrak{a}_\mathbb{C}, \mathfrak{g}_\mathbb{C})$, $b_P \psi_{|A(\mathbb{R})}$ has a smooth extension on $A(\mathbb{R})_{In-reg}$. This is equivalent to:

I_2': $\psi_{|A(\mathbb{R})}$ has a smooth extension on $A(\mathbb{R})_{I-reg}$, and for all semi-regular element $x \in A(\mathbb{R})$ such that the roots $\pm\alpha$ of $\mathfrak{a}_\mathbb{C}$ in $\mathfrak{g}_\mathbb{C}^x$ are compact imaginary, for all $u \in S(\mathfrak{a}_\mathbb{C})$,

$$[\partial(u) \cdot \psi_{|A(\mathbb{R})}]_\alpha^+(x) = 0$$

I_3: for all jump data $(x, A(\mathbb{R}), A_1(\mathbb{R}), c_\alpha)$ and for all $u \in S(\mathfrak{a}_\mathbb{C})$,

$$[\partial(u) \cdot \psi_{|A(\mathbb{R})}]_\alpha^+(x) = \mathrm{d}(x)\, \partial(c_\alpha \cdot u) \cdot \psi_{|A_1(\mathbb{R})}(x),$$

where $\mathrm{d}(x)$ is equal to 2 if the reflection $s_\alpha \in W(G, A)$ is realized in $G(\mathbb{R})^x$ and 1 otherwise.

I_4: if $A(\mathbb{R})$ is a Cartan subspace of $G(\mathbb{R})$, $\mathrm{Supp}\,(\psi_{|A(\mathbb{R})})$ is a compact subspace of $A(\mathbb{R})$.

The space $\mathcal{I}(G(\mathbb{R}))$ is endowed with a topology of an inductive limit of Fréchet spaces, and we denote by $\mathcal{I}(G(\mathbb{R}))'$ its dual.

THEOREM 3.3. *(Bouaziz* [**B1**]*) The map $J_{G(\mathbb{R})}$ is linear, continuous and surjective from $\mathcal{C}_c^\infty(G(\mathbb{R}))$ onto $\mathcal{I}(G(\mathbb{R}))$, and its transpose ${}^tJ_{G(\mathbb{R})}$ realizes a bijection from $\mathcal{I}(G(\mathbb{R}))'$ onto the space of invariant distributions on $G(\mathbb{R})$.*

3.3. Twisted orbital integrals on $G(\mathbb{R})$. Recall the character ω of $G(\mathbb{R})$ with Langlands parameter $\mathbf{a} \in H^1(W_\mathbb{R}, Z(\hat{G}))$.

DEFINITION 3.4. Let $x \in G(\mathbb{R})$ be a regular element and let $T(\mathbb{R})$ be the Cartan subgroup of $G(\mathbb{R})$ containing x. For all $f \in \mathcal{C}_c^\infty(G(\mathbb{R}))$, define $J_{G(\mathbb{R})}^\omega(f)(x)$ as follows. If $T(\mathbb{R})$ is not included in $G(\mathbb{R})^+$, put

$$J_{G(\mathbb{R})}^\omega(f)(x) = 0.$$

If $T(\mathbb{R})$ is included in $G(\mathbb{R})^+$, put

$$J_{G(\mathbb{R})}^\omega(f)(x) = |\det(\mathrm{Id} - \mathrm{Ad}x^{-1})_{\mathfrak{g}_\mathbb{C}/\mathfrak{t}_\mathbb{C}}|^{\frac{1}{2}} \int_{G(\mathbb{R})/T(\mathbb{R})} \omega(g)\, f(gxg^{-1})\, d\dot{g}.$$

REMARKS. 1 - Suppose that $J_{G(\mathbb{R})}^\omega$ is not always 0. Thus there is a Cartan subgroup of $G(\mathbb{R})$ in the kernel of ω. Let us decompose $\mathfrak{g}_\mathbb{C}$ as the sum of its center and its semi-simple part $\mathfrak{g}_\mathbb{C} = \mathfrak{c}_\mathbb{C} \oplus \mathfrak{g}_\mathbb{C}^{der}$. Since ω is a character, its differential $d\omega$ is 0 on $\mathfrak{g}_\mathbb{C}^{der}$ and since $\mathfrak{c}_\mathbb{C} \subset \mathfrak{t}_\mathbb{C}$, $d\omega$ is also 0 on $\mathfrak{c}_\mathbb{C}$. Thus $d\omega$ is 0 on $\mathfrak{g}_\mathbb{C}$, and ω is 0 on $G(\mathbb{R})_0$. Thus $G(\mathbb{R})^+$ is a normal open subgroup of $G(\mathbb{R})$. This condition was already noticed in Section 1.7 from other considerations.

2 - Let us fix a system of representatives $\{g_1, \ldots, g_r\}$ for the cosets of $G(\mathbb{R})^+$ in $G(\mathbb{R})$. If $T(\mathbb{R})$ is included in $G(\mathbb{R})^+$ and $x \in T(\mathbb{R})$ is regular,

$$J_{G(\mathbb{R})}^\omega(f)(x) = \sum_{i=1}^r \omega(g_i)\, J_{G(\mathbb{R})^+}(f^{g_i^{-1}})(x)$$

where $f^{g_i^{-1}}$ is the function $x \mapsto f(g_i x g_i^{-1})$.

3.4. Stable orbital integrals on $H_1(\mathbb{R})$. Recall the z-pair (H_1, ξ_{H_1}) from Section 1.9, the resulting central extension of $H(\mathbb{R})$:

$$1 \to Z_1(\mathbb{R}) \to H_1(\mathbb{R}) \to H(\mathbb{R}) \to 1$$

and the character λ_{H_1} of $Z_1(\mathbb{R})$ (see Eq. 1.5).

Let $\mathcal{C}^\infty_{c,Z_1}(H_1(\mathbb{R}), \lambda_{H_1})$ be the space of smooth functions f^{H_1} on $H_1(\mathbb{R})$ with compact support modulo $Z_1(\mathbb{R})$ and such that:

$$f^{H_1}(zh) = \lambda_{H_1}(z)^{-1} f^{H_1}(h) \quad (h \in H_1(\mathbb{R})_{reg},\ z \in Z_1(\mathbb{R}))$$

The orbital integral of such a function is given by:

$$J_{H_1(\mathbb{R})}(f^{H_1})(\gamma_1) = |\det(\mathrm{Id} - \mathrm{Ad}\gamma_1^{-1})_{\mathfrak{h}_\mathbb{R}/\mathfrak{t}_\mathbb{R}}|^{\frac{1}{2}} \int_{H_1(\mathbb{R})/T_1(\mathbb{R})} f(h\gamma_1 h^{-1})\, d\dot{g}$$

where $\gamma_1 \in H_1(\mathbb{R})$ is regular and $d\dot{h}$ is the invariant measure on $H_1(\mathbb{R})/T_1(\mathbb{R})$ normalized with our conventions. This is a well-defined converging integral since $\mathrm{Supp}\, f \cap \mathcal{O}_{\gamma_1}$ is compact.

The stable orbital integral of the function $f^{H_1} \in \mathcal{C}^\infty_{c,Z_1}(H_1(\mathbb{R}), \lambda_{H_1})$ is defined by:

$$J^{st}_{H_1}(f^{H_1})(\gamma_1) = \sum_{\gamma_i} J_{H_1(\mathbb{R})}(f^{H_1})(\gamma_i)$$

where the sum is taken over a system of representatives of conjugacy class in $H_1(\mathbb{R})$ in the stable conjugacy class of γ_1.

In [**B1**], A. Bouaziz gave a characterization of stable orbital integrals of compactly supported functions on a real algebraic reductive connected group. We rephrase his results for functions in $\mathcal{C}^\infty_{c,Z_1}(H_1(\mathbb{R}), \lambda_{H_1})$, indicating briefly how the proof can be adapted.

The map $\gamma_1 \mapsto J^{st}_{H_1(\mathbb{R})}(f^{H_1})(\gamma_1)$ is smooth on $H_1(\mathbb{R})_{reg}$, stably invariant (i.e. constant on stable conjugacy classes) and satisfies for all $h \in H_1(\mathbb{R})_{reg}$, $z \in Z_1(\mathbb{R})$

$$J^{st}_{H_1(\mathbb{R})}(f^{H_1})(zh) = \lambda_{H_1}(z)^{-1} J^{st}_{H_1}(f^{H_1})(h). \tag{3.1}$$

Let $\mathcal{I}^{st}(H_1(\mathbb{R}), \lambda_{H_1})$ be the subspace of $\mathcal{C}^\infty(H_1(\mathbb{R})_{reg})$ of functions ψ which are constant on the stable conjugacy classes satisfing the properties $I_1^{st} = I_1$, I_2^{st}, I_3^{st}, I_4^{st}, I_5^{st}.

I_2^{st}: if $A(\mathbb{R})$ is a maximal torus of H_1 defined over $\mathbb{R}$, for any system P of positive imaginary roots in $R(\mathfrak{a}_\mathbb{C}, \mathfrak{h}_{1,\mathbb{C}})$, $b_P \psi_{|A(\mathbb{R})}$ has a smooth extension on $A(\mathbb{R})_{st-In-reg}$, where $A(\mathbb{R})_{st-In-reg}$ is the set of $a \in A(\mathbb{R})$ such that all elements in $A(\mathbb{R})$ stably conjugate to a are in $A(\mathbb{R})_{In-reg}$. Note that this rather subtle definition is not really necessary here since for H_1 quasi-split, we have $A(\mathbb{R})_{st-In-reg} = A(\mathbb{R})_{I-reg}$ ([**Sh1**], prop. 4.11). So in fact I_2^{st} reduces to :

$I_2^{st'}$: $\psi_{|A(\mathbb{R})}$ has a smooth extension on $A(\mathbb{R})_{I-reg}$.

I_3^{st}: for all jump data $(x, A(\mathbb{R}), A_1(\mathbb{R}), c_\alpha)$, and for all $u \in S(\mathfrak{a}_\mathbb{C})$,

$$[\partial(u) \cdot \psi_{|A(\mathbb{R})}]_\alpha^+(x) = 2\partial(c_\alpha \cdot u) \cdot \psi_{|A_1(\mathbb{R})}(x)$$

Note that the right-hand side is well defined because $x \in A(\mathbb{R})_{1,I-reg}$.

I_4^{st}: if A is a maximal torus of H_1 defined over $\mathbb{R}$, Supp $(\psi_{|A(\mathbb{R})})$ is a compact subspace of $A(\mathbb{R})$ modulo $Z_1(\mathbb{R})$.

I_5^{st}: for all $h \in H_1(\mathbb{R})$, $z \in Z_1(\mathbb{R})$:

$$\psi(zh) = \lambda_{H_1}(z)^{-1}\psi(h) \tag{3.2}$$

The space $\mathcal{I}^{st}(H_1(\mathbb{R}), \lambda_{H_1})$ is endowed with a topology of inductive limit of Fréchet spaces and we denote by $\mathcal{I}^{st}(H_1(\mathbb{R}), \lambda_{H_1})'$ its dual. For all functions $f^{H_1} \in \mathcal{C}^\infty_{c,Z_1}(H_1(\mathbb{R}), \lambda_{H_1})$, $J^{st}_{H_1(\mathbb{R})}(f^{H_1}) \in \mathcal{I}^{st}(H_1(\mathbb{R}), \lambda_{H_1})$ (see [**Sh1**] and [**B1**] §6 for the case $H_1 = H$, i.e. orbital integrals of smooth functions with compact support, and see below for an argument of how this can be adapted to the general case). The last property is a easy consequence of (3.1).

We can now define stable distributions on $H_1(\mathbb{R})$ as the closure of the subspace of $\mathcal{C}^\infty_{c,Z_1}(H_1(\mathbb{R}), \lambda_{H_1})'$ generated by distributions of the form

$$f \mapsto J^{st}_{H_1}(f)(x)$$

for regular x in $H_1(\mathbb{R})$. We will denote this space by $\mathrm{Dist}(H_1(\mathbb{R}), \lambda_{H_1})^{st}$.

THEOREM 3.5. *(see [**B1**], théorème 6.1) The map:*

$$J^{st}_{H_1}; \mathcal{C}^\infty_{c,Z_1}(H_1(\mathbb{R}), \lambda_{H_1}) \to \mathcal{I}^{st}(H_1(\mathbb{R}), \lambda_{H_1})$$

is linear, continuous, surjective and its transpose ${}^tJ^{st}_{H_1}$ realizes an isomorphism between $\mathcal{I}^{st}(H_1(\mathbb{R}), \lambda_{H_1})'$ and the space of stable invariant distributions $\mathrm{Dist}(H_1(\mathbb{R}), \lambda_{H_1})^{st}$.

Proof. Suppose that the extension

$$1 \to Z_1(\mathbb{R}) \to H_1(\mathbb{R}) \to H(\mathbb{R}) \to 1 \tag{3.3}$$

is split, and let $c : H(\mathbb{R}) \to H_1(\mathbb{R})$ be a section. This section provides an isomorphism:

$$\mathrm{Restr} : \mathcal{C}^\infty_{c,Z_1}(H_1(\mathbb{R}), \lambda_{H_1}) \to \mathcal{C}^\infty_c(H(\mathbb{R}))$$

by restricting a function to $c(H(\mathbb{R})) \simeq H(\mathbb{R})$.

Thus $\mathcal{C}^\infty_{c,Z_1}(H_1(\mathbb{R}), \lambda_{H_1})' \simeq \mathrm{Distr}(H(\mathbb{R}))$ and we have the following commuting diagram:

$$\begin{array}{ccc} \mathcal{C}^\infty_{c,Z_1}(H_1(\mathbb{R}), \lambda_{H_1}) & \xrightarrow{\mathrm{Restr}} & \mathcal{C}^\infty_c(H(\mathbb{R})) \\ \big\downarrow J^{st}_{H_1(\mathbb{R})} & & \big\downarrow J^{st}_{H(\mathbb{R})} \\ \mathcal{I}^{st}(H_1(\mathbb{R}), \lambda_{H_1}) & \xrightarrow{\mathrm{Restr}} & \mathcal{I}^{st}(H(\mathbb{R})) \end{array}$$

The results follow easily from Bouaziz' results in this case.

When (3.3) is not split, we use the fact that $H_1(\mathbb{R}) \simeq H_{1,der}(\mathbb{R}) \times Z(H_1)(\mathbb{R})/F$ where F is a finite subgroup and $Z(H_1)(\mathbb{R})$ is the center of $H_1(\mathbb{R})$. The theorem is established for functions in $\mathcal{C}^\infty_{c,Z_1}(H_{1,der}(\mathbb{R}) \times Z(H_1)(\mathbb{R}), \lambda_{H_1})$. We deduce the statement for $\mathcal{I}^{st}(H_1(\mathbb{R}), \lambda_{H_1})$ from the following commutative diagram:

$$\begin{array}{ccc} \mathcal{C}^\infty_{c,Z_1}(H_{1,der}(\mathbb{R}) \times Z(H_1)(\mathbb{R}), \lambda_{H_1}) & \xrightarrow{\mathrm{M}} & \mathcal{C}^\infty_c(H_1(\mathbb{R}), \lambda_{H_1}) \\ \Big\downarrow J^{st}_{H_{1,der}(\mathbb{R}) \times Z(H_1)(\mathbb{R})} & & \Big\downarrow J^{st}_{H_1(\mathbb{R})} \\ \mathcal{I}^{st}(H_{1,der}(\mathbb{R}) \times Z(H_1)(\mathbb{R}), \lambda_{H_1}) & \xrightarrow{\mathrm{M}} & \mathcal{I}^{st}(H_1(\mathbb{R}), \lambda_{H_1}) \end{array}$$

where $M(\phi)(\gamma) = \sum_{z \in F} \phi(\gamma z)$ for any function ϕ on $H_{1,der}(\mathbb{R}) \times Z(H_1)(\mathbb{R})$. $\square$

Let $T_{H_1}(\mathbb{R})$ be a Cartan subgroup of $H_1(\mathbb{R})$ with projection $T_H(\mathbb{R})$ on $H(\mathbb{R})$. There is an exact sequence:

$$\{0\} \to \mathfrak{z}_1 \to \mathfrak{h}_1 \to \mathfrak{h} \to \{0\} \tag{3.4}$$

of Lie algebras, with $\mathfrak{z}_1$ central in $\mathfrak{h}_1$. Since such a sequence always splits, we may, by fixing a section of 3.4 identify $\mathfrak{h}$ with a subalgebra of $\mathfrak{h}_1$, and:

$$\mathfrak{h}_1 = \mathfrak{z}_1 \oplus \mathfrak{h} \tag{3.5}$$

This decomposition (3.5) induces:

$$\mathfrak{t}_{H_1} = \mathfrak{z}_1 \oplus \mathfrak{t}_H \tag{3.6}$$

Furthermore, the decompositions:

$$\mathfrak{h}_{1,\mathbb{C}} = \mathfrak{t}_{H_1,\mathbb{C}} \oplus \sum_{\alpha \in R(T_{H_1}, H_1)} \mathfrak{h}^\alpha_{1,\mathbb{C}}$$

$$\mathfrak{h}_{\mathbb{C}} = \mathfrak{t}_{H,\mathbb{C}} \oplus \sum_{\alpha \in R(T_H, H)} \mathfrak{h}^\alpha_{\mathbb{C}}$$

provide identification between $R(T_{H_1}, H_1) \simeq R(T_H, H)$ and $\mathfrak{h}^\alpha_{1,\mathbb{C}} \simeq \mathfrak{h}^\alpha_{\mathbb{C}}$.

Let $\gamma_1 \in T_{H_1}(\mathbb{R})$ and γ its projection on $T_H(\mathbb{R})$. Let $\alpha \in R(T_{H_1}, H_1) \simeq R(T_H, H)$; then $\alpha(\gamma_1) = \alpha(\gamma)$. Thus, if P is a system of positive imaginary roots in $R(T_{H_1}, H_1) \simeq R(T_H, H)$, we get $b_P(\gamma_1) = b_P(\gamma)$.

We end this section by the following remark concerning differential operators from $S(\mathfrak{t}_{H_1,\mathbb{C}})$. It is clear that for all smooth function ψ on $H_1(\mathbb{R})_{reg}$ satisfying (3.2) and for all $u \in S(\mathfrak{z}_{1,\mathbb{C}})$,

$$\partial(u) \cdot \psi = \mathrm{d}\lambda^{-1}_{H_1}(u)\psi$$

i.e. ψ is a eigenfunction for all $u \in S(\mathfrak{z}_{1,\mathbb{C}})$. Thus, to check properties $I^{st}_1, I^{st}_2, I^{st}_3$ we have only to consider differential operators coming from $S(\mathfrak{t}_{H,\mathbb{C}})$ (since (3.6) yields $S(\mathfrak{t}_{H_1,\mathbb{C}}) = S(\mathfrak{t}_{H,\mathbb{C}}) \otimes S(\mathfrak{z}_{1,\mathbb{C}})$).

3.5. Transfer of orbital integrals. If γ_1 is an element in $H_1(\mathbb{R})$, γ will denote its projection on $H(\mathbb{R})$. Notation and terminology relative to $H(\mathbb{R})$ will be transferred to $H_1(\mathbb{R})$ using the projection. (For instance, the notion of G-regular element.) Langlands and Shelstad have defined absolute transfer factors $\Delta(\gamma_1, \delta)$ where $\gamma_1 \in H_1(\mathbb{R})$ is G-regular and $\Delta(\gamma_1, \delta) = 0$ if γ is not a norm of $\delta \in G(\mathbb{R})$. This transfer factor is a product of four terms Δ_I, Δ_{II}, Δ_{III_1} and Δ_{III_2} (we omit their term Δ_{IV} since it is already included in our definition of orbital integrals). We will recall the properties of these transfer factors when we need them.

We say that the function $f \in \mathcal{C}_c^\infty(G(\mathbb{R}))$ and the function $f^{H_1} \in \mathcal{C}_c^\infty(H(\mathbb{R}), \lambda_{H_1})$ have matching orbital integrals if

$$J^{st}_{H_1(\mathbb{R})}(f^{H_1})(\gamma_1) = \sum_{\delta \in \Sigma_\gamma} \Delta(\gamma_1, \delta) J^\omega_G(f)(\delta) \tag{3.7}$$

for every G-regular $\gamma_1 \in H_1(\mathbb{R})$. The sum (which might be empty, in which case the right-hand side is 0) is taken over a set of representatives of conjugacy classes under $G(\mathbb{R})$ of elements $\delta \in G(\mathbb{R})$ for which γ is a norm and such that the unique Cartan subgroup in $G(\mathbb{R})$ containing δ is contained in $G(\mathbb{R})^+$.

In this case, we can rewrite (3.7) as

$$J^{st}_{H_1(\mathbb{R})}(f^{H_1})(\gamma_1) = \sum_{\delta \in \Sigma_\gamma} \Delta(\gamma_1, \delta) \sum_{i=1}^{r} \omega(g_i) J_{G(\mathbb{R})^+}(f^{g_i^{-1}})(\delta). \tag{3.8}$$

The principal result of the whole theory is that for every function $f \in \mathcal{C}_c^\infty(G(\mathbb{R}))$, there is a function $f^{H_1} \in \mathcal{C}_c^\infty(H_1(\mathbb{R}), \lambda_{H_1})$ having matching orbital integrals with f. Using the terminology of the previous paragraphs, we can rephrase it in the following form:

THEOREM 3.6. *For all $f \in \mathcal{C}_c^\infty(G(\mathbb{R}))$, there is a function $f^{H_1} \in \mathcal{C}_c^\infty(H_1(\mathbb{R}), \lambda_{H_1})$ having matching orbital integrals with f, ie. there exists an element* Trans $(f) \in \mathcal{I}^{st}(H_1(\mathbb{R}), \lambda_{H_1})$ *such that:*

$$\mathrm{Trans}\,(f)(\gamma_1) = \sum_{\delta \in \Sigma_\gamma} \Delta(\gamma_1, \delta) \sum_{i=1}^{r} \omega(g_i) \psi^i(\delta) \tag{3.9}$$

for any G-regular element γ_1 of $H_1(\mathbb{R})$ originating in $G(\mathbb{R})^+$, with $\psi^i = J_{G(\mathbb{R})^+}(f^{g_i^{-1}})$, and

$$\mathrm{Trans}\,(f)(\gamma_1) = 0 \tag{3.10}$$

if γ_1 does not originate in $G(\mathbb{R})^+$. Futhermore, Trans (f) *is defined on regular, non-G-regular elements of $H_1(\mathbb{R})$ by smooth extension.*

The next section will be devoted to the proof of theorem 3.6.

4. Proof of theorem 3.6

4.1. We sketch briefly the proof of the theorem before going into details. Let us first remark that the right-hand side of (3.9) is well-defined, i.e. does not depend on the choices of representatives in Σ_γ. This is a consequence of the following lemma :

LEMMA 4.1. *([**LS**], Lemma 4.1.C).*

$$\Delta(\gamma_1, \delta) = \Delta(\gamma_1, \delta')$$

when δ and δ' are conjugate in G.

The fact that Trans (ψ) is constant on stable conjugacy classes is proved in the same lemma of [**LS**]:

LEMMA 4.2. *$\Delta(\gamma_1, \delta)$ is unchanged when γ_1 is replaced with a stably conjugate element in $H_1(\mathbb{R})$.*

We will show how Trans (f) is defined on regular element of $H_1(\mathbb{R})$ (but not G-regular) by a smooth extension of (3.9). This is lemma 4.3 of [**Sh2**].

To prove the theorem, we have to establish that properties $I_1^{st}, \ldots, I_5^{st}$ hold for Trans (f). As notations suggest, properties $I_1^{st}, \ldots, I_4^{st}$ for Trans (ψ) are consequences of properties $I_1, \ldots, I_4$ for the ψ^i. Some are immediate $I_1^{st}, I_2^{st}, I_4^{st}$, the other one I_3^{st} requiring extra work. The last property I_5^{st}, is established by the following lemma:

LEMMA 4.3. *([**KS**], lemma 5.1.C)*
$\Delta(z\gamma_1, \delta) = \lambda_{H_1}(z)^{-1}\Delta(\gamma_1, \delta)$ where $\gamma_1 \in H_1(\mathbb{R})$ is a regular element and $z \in Z_1(\mathbb{R})$.

4.2. Transfer of differential operators. If $T_H(\mathbb{R})$ originates in $G(\mathbb{R})^+$, then there exists a Cartan subgroup $T_G(\mathbb{R})$ in $G(\mathbb{R})$, and an isomorphism:

$$\eta_x : T_H \rightarrow T_G$$

defined over $\mathbb{R}$ such that $\eta_x^{-1}(\delta)$ is a norm of δ for all regular $\delta \in T_G(\mathbb{R})$. Recall that this map depends on the choice of a admissible embedding $T_H \xrightarrow{\eta} T$ of T_H in G^* and an element $x \in G^*$. Differentiating and complexifying, we get isomorphisms over $\mathbb{R}$:

$$\mathfrak{t}_{H,\mathbb{C}} \xrightarrow{\eta} \mathfrak{t}_{\mathbb{C}} \xleftarrow{(\mathrm{Int}\, x \circ \psi)} \mathfrak{t}_{G,\mathbb{C}} \tag{4.1}$$

Let us denote by η_x again the isomorphism between the extreme terms of (4.1) and also the induced isomorphism between $S(\mathfrak{t}_{G,\mathbb{C}})$ and $S(\mathfrak{t}_{H,\mathbb{C}})$. This will enable us to transfer differential operators.

Let ϕ be a smooth function defined on an open subset $\mathcal{V}$ of $T_G(\mathbb{R})$ and let $\tilde{\phi}$ be its pull-back on by η_x. Take $u \in S(\mathfrak{t}_{H,\mathbb{C}})$ and $u' = \eta_x(u) \in S(\mathfrak{t}_{G,\mathbb{C}})$. It is clear that we have for all $\delta \in \mathcal{V}$:

$$\partial(u') \cdot \phi(\delta) = \partial(u) \cdot \tilde{\phi}(\eta_x^{-1}(\delta)). \tag{4.2}$$

4.3. Local behaviour of transfer factors. We continue in the setting of the previous paragraphs, ie. $T_{H_1}(\mathbb{R})$ is a Cartan subgroup of $H_1(\mathbb{R})$ with projection $T_H(\mathbb{R})$ on $H(\mathbb{R})$ originating in $G(\mathbb{R})^+$. Let γ_{01} be an element in $T_{H_1}(\mathbb{R})$. We need to study $\varphi_{|T_{H_1}(\mathbb{R})}$ in a neighbourhood of γ_{01}. Let γ_0 be the projection of γ_1 in $T_H(\mathbb{R})$. Fix an admissible embedding η_x of T_H in G and let $\delta_0 = \eta_x(\gamma_0) \in T_G(\mathbb{R})$.

Because of the property I_5^{st} of φ, it is sufficient to study its restriction to a neighbourhood $\mathcal{U}_{\gamma_{01}}$ of γ_{01} in

$$\gamma_{01} \exp \mathfrak{t}_{H,\mathbb{R}} \subset T_{H_1}(\mathbb{R}).$$

It is clear that the projection map induces a local topological isomorphim

$$\gamma_{01} \exp \mathfrak{t}_H \to \gamma_0 \exp \mathfrak{t}_H. \tag{4.3}$$

Suppose that $\mathcal{U}_{\gamma_{01}}$ is sufficiently small, so that the above local isomorphism restrict to a topological isomorphism from $\mathcal{U}_{\gamma_{01}}$ onto its image that we call $\mathcal{U}_{\gamma_0}$. Let us still denote by η_x the embedding of $\mathcal{U}_{\gamma_{01}}$ in $T_G(\mathbb{R})$ obtained from the composition (4.3) and $\eta_x : T_H \to T_G$ and call $\mathcal{V}_{\delta_0}$ its image. Then, for any regular element $\delta \in \mathcal{V}_{\delta_0}$, we define

$$\begin{aligned} \Delta_{\eta_x} : \mathcal{V}_{\delta_0} &\to \mathbb{C} \\ \delta &\mapsto \Delta(\eta_x^{-1}(\delta), \delta) \end{aligned}$$

We will need the following properties of this function:

PROPOSITION 4.4. *In the setting above:*

(i) Δ_{η_x} *is smooth on* $(\mathcal{V}_{\delta_0})_{reg}$. *For any compact subset* K *of* $\mathcal{V}_{\delta_0}$, Δ_{η_x} *is bounded on* K_{reg}. *There exist* $\lambda \in \mathfrak{t}^*_{G,\mathbb{C}}$ *and* $c \in \mathbb{C}$ *such that*

$$\Delta_{\eta_x}(\delta_0 \exp X) = c\, e^{\lambda(X)}$$

for all $X \in \mathfrak{t}_G$ *such that* $\delta_0 \exp X \in (\mathcal{V}_{\delta_0})_{reg}$. *Furthermore,* Δ_{η_x} *has a smooth continuation to* $(\mathcal{V}_{\delta_0})_{I-reg}$. *Let* τ_λ *be the algebra automorphism of* $S(\mathfrak{t}_{G,\mathbb{C}})$ *mapping* $X \in \mathfrak{t}_{G,\mathbb{C}}$ *to* $X + \lambda(X)$. *We have the following identity of differential operators on* $T_G(\mathbb{R})$: *for all* $u \in S(\mathfrak{t}_{G,\mathbb{C}})$

$$\partial(u) \circ \Delta_{\eta_x} = \Delta_{\eta_x} \circ \partial(\tau_\lambda(u)).$$

Suppose δ_0 *is a semi-regular element, such that the roots* $\pm\alpha$ *of* $\mathfrak{t}_{G,\mathbb{C}}$ *in* $\mathfrak{g}_{\mathbb{C}}^{\delta_0}$ *are imaginary. Let* $H_\alpha \in i\mathfrak{t}_G$ *be the coroot of* α, *and* $\delta_\nu = \delta_0 \exp(i\nu H_\alpha)$. *For* ν *small enough and non-zero,* δ_ν *is regular. Set* $\gamma_{1,\nu} = \eta_x^{-1}(\delta_\nu) \in H_1(\mathbb{R})$.

(ii) if α *is compact and* $\gamma_0 \in T_H(\mathbb{R})_{reg}$, *then*

$$\lim_{\nu \to 0^+} \Delta(\gamma_{1,\nu}, \delta_\nu) = - \lim_{\nu \to 0^-} \Delta(\gamma_{1,\nu}, \delta_\nu),$$

(iii) if α *is compact and* γ_0 *is semi-regular in* $H(\mathbb{R})$, *then*

$$\lim_{\nu \to 0^+} \Delta(\gamma_{1,\nu}, \delta_\nu) = \lim_{\nu \to 0^-} \Delta(\gamma_{1,\nu}, \delta_\nu),$$

(iv) if α is non-compact and $\gamma_0 \in T_H(\mathbb{R})_{reg}$, then the reflection s_α with respect to the root α is not realized in $G(\mathbb{R})^{\delta_0}$, i.e. δ_ν and $\delta_{-\nu}$ are not conjugate in $G(\mathbb{R})$. Furthermore

$$\lim_{\nu \to 0^+} \Delta(\gamma_{1,\nu}, \delta_\nu) = - \lim_{\nu \to 0^-} \Delta(\gamma_{1,\nu}, \delta_\nu),$$

$$\Delta(\gamma_{1,\nu}, \delta_\nu) = -\Delta(\gamma_{1,\nu}, \delta_{-\nu}),$$

(v) if α is non-compact imaginary and γ_0 is semi-regular, then

$$\lim_{\nu \to 0^+} \Delta(\gamma_{1,\nu}, \delta_\nu) = \lim_{\nu \to 0^-} \Delta(\gamma_{1,\nu}, \delta_\nu),$$

$$\Delta(\gamma_{1,\nu}, \delta_\nu) = \Delta(\gamma_{1,\nu}, \delta_{-\nu}),$$

The proof relies on the fine analysis of the local properties of transfer factors. It is postponed to Section 6.

4.4. Smooth extension to $H_1(\mathbb{R})_{reg}$. In the setting of the previous sections, suppose that $\gamma_0 \in H(\mathbb{R})$ is regular but not G-regular. If $T_H(\mathbb{R})$ does not originate in $G(\mathbb{R})^+$ then by definition, the restriction of φ to G-regular elements in $T_{H_1}(\mathbb{R})$ is zero, and there is a smooth extention of $\varphi_{|T_{H_1}(\mathbb{R})}$ around γ_{10}. Since γ_{10} is regular, there is a neighborhood $\mathcal{U}$ of γ_{10} in $T_{H_1}(\mathbb{R})$ such that $H_1(\mathbb{R}) \cdot \mathcal{U}$ is a neighborhood of γ_{10} in $H_1(\mathbb{R})$. Thus, by invariance, φ is zero in a neighborhood of γ_{10}.

Suppose that $T_H(\mathbb{R})$ originates in $G(\mathbb{R})^+$. Fix a complete system of admissible embeddings $\eta_{x_j} : T_H \to T_j$. Put $\delta_{0,j} = \eta_{x_j}(\gamma_0)$.

Suppose that one of the $\delta_{0,j}$ is in $T_j(\mathbb{R})_{I-reg}$. Then it will be the case for all of them. To see this, take two of them, say $\delta_{0,j}$ and $\delta_{0,k}$. Then, they are conjugate by an element $g \in G$, such that $\operatorname{Int} g$ sends $G^{\delta_{0,j}}$ isomorphically onto $G^{\delta_{0,k}}$ and $\mathfrak{t}_{j,\mathbb{C}}$ isomorphically onto $\mathfrak{t}_{k,\mathbb{C}}$, the latter being defined over $\mathbb{R}$. Thus $\operatorname{Int} g$ sends the roots of $\mathfrak{t}_{j,\mathbb{C}}$ in $\mathfrak{g}_{\mathbb{C}}^{\delta_{0,j}}$ bijectively onto the roots of $\mathfrak{t}_{k,\mathbb{C}}$ in $\mathfrak{g}_{\mathbb{C}}^{\delta_{0,k}}$, respecting their types (real, complex or imaginary). It follows from proposition 4.4 (i) and property I_2 of the ψ^i for all $i = 1, \ldots, r$ that there is a smooth extension of $\varphi_{|T_{H_1}(\mathbb{R})}$ around γ_{10}.

Now suppose that $\delta_{0,j} \notin T_j(\mathbb{R})_{I-reg}$. We want to study the behaviour of $\delta_j \mapsto (\Delta_{\eta_{x_j}} \psi^i_{|T_j(\mathbb{R})})(\delta_j)$ around $\delta_{0,j}$. We drop locally the indices 'i' and 'j', since we are working with only one of them at a time. Assume that δ_0 is semi-regular. Then the roots $\pm\alpha$ of $\mathfrak{t}_{\mathbb{C}}$ in $\mathfrak{g}_{\mathbb{C}}^{\delta_0}$ are imaginary. Suppose they are of compact type. Then for all $u \in S(\mathfrak{t}_{H,\mathbb{C}})$, setting $u' = \eta_x(u)$,

$$\begin{aligned}
&[\partial(u') \cdot (\Delta_{\eta_x} \psi_{|T(\mathbb{R})})]^-_\alpha(\delta_0) \\
&= \lim_{\nu\to 0^+} \partial(u') \cdot (\Delta_{\eta_x} \psi_{|T(\mathbb{R})})(\delta_\nu) - \lim_{\nu\to 0^-} \partial(u') \cdot (\Delta_{\eta_x} \psi_{|T(\mathbb{R})})(\delta_\nu) \\
&= \lim_{\nu\to 0^+} \Delta_{\eta_x}(\delta_\nu)\partial(\tau_\lambda(u)) \cdot \psi_{|T(\mathbb{R})}(\delta_\nu) - \lim_{\nu\to 0^-} \Delta_{\eta_x}(\delta_\nu)\partial(\tau_\lambda(u')) \cdot \psi_{|T(\mathbb{R})}(\delta_\nu) \\
&= \left(\lim_{\nu\to 0^+} \Delta_{\eta_x}(\delta_\nu)\right) [\partial(\tau_\lambda(u')) \cdot \psi_{|T(\mathbb{R})}]^+_\alpha = 0
\end{aligned}$$

We have used Proposition 4.4, (i), (ii) and property I_2 of the ψ. Thus, the contribution of η_x to the right-hand side of (3.9) is smooth around δ_0.

Let us see what happens in the other case, i.e. when the roots $\pm\alpha$ are non-compact. Lemma 4.4 (iv) asserts that in that case δ_ν and $\delta_{-\nu}$ are not conjugate in $G(\mathbb{R})$, i.e. the reflection s_α with respect to the root α is not realized in $G(\mathbb{R})^{\delta_0}$.

Then, we are in the situation considered at the end of paragraph 2.10. To check that φ has a smooth extension around γ_0, we have to look at the contributions to the right-hand side of (3.9) of η_x and $\overline{\eta}_x$. We need the following lemma:

LEMMA 4.5. *Let $u' = \eta_x(u)$ and $\overline{u}' = \overline{\eta}_x(u)$. Then $\overline{u}' = (u')^{s_\alpha}$*

The proof is obvious. Let us resume our computation:

$$\begin{aligned}
&\lim_{\nu\to 0^+} \partial(u') \cdot (\Delta_{\eta_x} \psi_{|T(\mathbb{R})})(\delta_\nu) - \lim_{\nu\to 0^-} \partial(u') \cdot (\Delta_{\eta_x} \psi_{|T(\mathbb{R})})(\delta_\nu) \\
&+ \lim_{\nu\to 0^+} \partial(\overline{u}') \cdot (\Delta_{\overline{\eta}_x} \psi_{|T(\mathbb{R})})(\delta_{-\nu}) - \lim_{\nu\to 0^-} \partial(\overline{u}') \cdot (\Delta_{\overline{\eta}_x} \psi_{|T(\mathbb{R})})(\delta_{-\nu}) = \\
&\lim_{\nu\to 0^+} \Delta_{\eta_x}(\delta_\nu)\, \partial(\tau_\lambda(u')) \cdot \psi_{|T(\mathbb{R})}(\delta_\nu) - \lim_{\nu\to 0^-} \Delta_{\eta_x}(\delta_\nu)\, \partial(\tau_\lambda(u')) \cdot \psi_{|T(\mathbb{R})}(\delta_\nu) \\
&+ \lim_{\nu\to 0^-} \Delta_{\overline{\eta}_x}(\delta_\nu)\partial(\tau_\lambda(\overline{u}')) \cdot \psi_{|T(\mathbb{R})}(\delta_\nu) - \lim_{\nu\to 0^+} \Delta_{\overline{\eta}_x}(\delta_\nu)\partial(\tau_\lambda(\overline{u}')) \cdot \psi_{|T(\mathbb{R})}(\delta_\nu).
\end{aligned}$$

We have $\lim_{\nu\to 0^+} \Delta_{\overline{\eta}_x}(\delta_\nu) = \lim_{\nu\to 0^+}(\Delta_{\eta_x})(\delta_\nu)$ (Proposition 4.4, (iv)). Hence, if $\partial(\tau_\lambda(u')) = \partial(\tau_\lambda(\overline{u}')) = \partial(\tau_\lambda(u'))^{s_\alpha}$, the whole expression cancels. To complete the proof, it remains to check what happens when $\partial(\tau_\lambda(u')) = -\partial(\tau_\lambda(\overline{u}')) = -\partial(\tau_\lambda(u'))^{s_\alpha}$, the general case being deduced by linearity. Under this latter assumption we use Proposition 4.4(iv) to obtain:

$$[\partial(u') \cdot (\Delta_{\eta_x} \psi_{|T_G(\mathbb{R})})]^-_\alpha(\delta_0) = (\lim_{\nu\to 0^+} \Delta_{\eta_x}(\delta_\nu))[\partial(\tau_\lambda(u')) \cdot \psi_{|T_G(\mathbb{R})})]^+_\alpha(\delta_0)$$

and a well-know principle of Harish-Chandra ([**HC1**]), asserts that this is zero. For the same reason,

$$[\partial(\overline{u}') \cdot (\Delta_{\overline{\eta}_x} \psi_{|T_G(\mathbb{R})})]^-_\alpha(\delta_0) = 0.$$

To conclude, we have proved that if $\delta_{0,j} \notin T_j(\mathbb{R})_{I-reg}$ is semi-regular for one j, then the same is true for all of them, and that the various contributions to the right-hand side of (3.9), when suitably grouped, extends to smooth functions around the $\delta_{0,j}$'s. Another principle of Harish-Chandra ([**HC2**]) asserts that these results still hold if the $\delta_{0,j}$'s are not semi-regular. Thus, Trans $(\psi) = \varphi$ has a smooth extension to $H_1(\mathbb{R})_{reg}$.

4.5. Properties $I_1^{st}, I_2^{st}, I_4^{st}$. Let T_{H_1} be a maximal torus of H_1 defined over $\mathbb{R}$. If $T_{H_1}(\mathbb{R})$ does not originate in $G(\mathbb{R})^+$, then the support of $\varphi_{|T_{H_1}(\mathbb{R})}$ is empty. If $T_{H_1}(\mathbb{R})$ originates in $G(\mathbb{R})^+$, we choose a complete system of admissible embeddings of T_H. For each Cartan subgroup $T_G(\mathbb{R})$ in $G(\mathbb{R})$ (and contained in $G(\mathbb{R})^+$) in this complete system, the restriction of the ψ^i to $T_G(\mathbb{R})$ has compact support (I_4), for all $i = 1, \dots, r$. It follows easily from the definition that the support of $\varphi_{|T_H}$ is compact.

The two other properties are local, so it is sufficient to check them in the setting of Proposition 4.4. Then, they are immediate consequences of I_1, I_2 for the ψ^i and Proposition 4.4 (i).

4.6. Jump relations. In this section, we will prove the jump relations for φ = Trans (f). Let $(\gamma_{10}, T_{H_1}(\mathbb{R}), T'_{H_1}(\mathbb{R}), c_{\alpha_H})$ be a jump datum on $H_1(\mathbb{R})$.

Corollary 2.11 asserts that if T_H does not originate in $G(\mathbb{R})^+$, neither does T'_H. In that case, $\varphi_{|T_{H_1}} \equiv 0$ and $\varphi_{|T'_{H_1}} \equiv 0$, and I_3^{st} is satisfied at γ_{10}.

We suppose now that T_{H_1} does originate in $G(\mathbb{R})^+$. We fix a complete system of admissible embeddings $\{\eta_{x_j}\} : T_H \to T_j$ as before. We suppose first that $\delta_{0,j} = \eta_{x_j}^{-1}(\gamma_0)$ is semi-regular (recall that if one is, all of them are) and we denote by $\pm\alpha_j$ the roots of $(\mathfrak{t}_{i,\mathbb{C}})$ in $\mathfrak{g}_{\mathbb{C}}^{\delta_{j,0}}$. It means that α_H and α_j have corresponding coroots through η_{x_j}. Recall that $\delta_{j,\nu} = \delta_{j,0}\exp(i\nu H_{\alpha_j})$. Let $\gamma_{j,\nu} = \eta_{x_i}^{-1}(\delta_{j,\nu}) = \gamma_{01}\exp(i\nu H_{\alpha_H})$.

Lemma 2.12 asserts that if for all j, $\delta_{j,0} \in T_{j,In-reg}$, then T'_{H_1} does not originate in G. It is then easily checked that the two sides of the jump relation are 0, and so that I_3^{st} is satisfied.

We are now dealing with the case where both T_{H_1} and T'_{H_1} originate in $G(\mathbb{R})^+$. Let $u \in S(\mathfrak{t}_{H,\mathbb{C}})$ and $v_j = \eta_{x_j}(u)$. We have then:

$$[\partial(u)\cdot\varphi_{|T_{H_1}(\mathbb{R})}]_{\alpha_H}^+(\gamma_{10}) = \sum_i \omega(g_i)\sum_j [\partial(v_j)\cdot(\Delta_{\eta_{x_j}}\psi^i_{|T_j(\mathbb{R})})]_{\alpha_j}^+(\delta_{j,0})$$

As in the previous paragraph, thanks to Proposition 4.4 (ii), if $\delta_{j,0} \in T_j(\mathbb{R})_{In-reg}$ the contribution of this term to the right-hand side is 0, so we are left only with the ones such that the root α_j is imaginary and non-compact in $\Delta(\mathfrak{t}_{j,\mathbb{C}}, \mathfrak{g}_{\mathbb{C}}^{\delta_{j,0}})$, *i.e.* with the set I_{γ_0}.

A general principle of Harish-Chandra asserts that if $\partial(u)^{s_{\alpha_H}} = -\partial(u)$, then the jump relations are satisfied, the jump being 0. In the following computations we assume that $\partial(u)^{s_{\alpha_H}} = \partial(u)$, the general case being deduced by linearity. In particular, in the second case considered above, we have $v_j = \overline{v}_j$. Assuming that J_{γ_0} is 'half of' I_{γ_0}, *ie.* that the reflections with respect to the relevant roots are not realized in the real groups, we compute

$$[\partial(u)\cdot\varphi_{|T_{H_1}(\mathbb{R})}]^+_{\alpha_H}(\gamma_{01}) = \sum_i \omega(g_i)\sum_j [\partial(v_j)\cdot(\Delta_{\eta_{x_j}}\psi^i_{|T_j(\mathbb{R})})]^+_{\alpha_j}(\delta_{j,0}) \tag{4.4}$$

$$= \sum_i \omega(g_i)\sum_{j\in I_{\gamma_0}} [\partial(v_j)\cdot(\Delta_{\eta_{x_j}}\psi^i_{|T_j(\mathbb{R})})]^+_{\alpha_j}(\delta_{j,0}) \tag{4.5}$$

$$= \sum_i \omega(g_i)\sum_{j\in J_{\gamma_0}} (\lim_{\nu\to 0}\Delta_{\eta_{x_j}}(\delta_{j,\nu}))\,\partial(c_{\alpha_j}\cdot\tau_\lambda(v_j))\cdot\psi^i_{|T'_j(\mathbb{R})}(\delta_{j,0}) \tag{4.6}$$

$$+(\lim_{\nu\to 0}\Delta_{\overline{\eta}_{x_j}}(\delta_{j,\nu}))\,\partial(c_{\alpha_j}\cdot\tau_\lambda(\overline{v}_j))\cdot\psi^i_{|T'_j(\mathbb{R})}(\delta_{j,0})$$

$$= \sum_i \omega(g_i)\sum_{j\in J_{\gamma_0}} 2(\lim_{\nu\to 0}\Delta_{\eta_{x_j}}(\delta_{j,\nu}))\,\partial(c_{\alpha_j}\cdot\tau_\lambda(v_j))\cdot\psi^i_{|T'_j(\mathbb{R})}(\delta_{j,0}) \tag{4.7}$$

$$= \sum_i \omega(g_i)\sum_{j\in J_{\gamma_0}} 2\partial(c_{\alpha_j}\cdot v_j)\cdot(\Delta_{\eta'_{x_j}}\psi^i_{|T'_j(\mathbb{R})})(\delta_{j,0}) \tag{4.8}$$

$$= 2\,\partial(c_{\alpha_H}\cdot u)\cdot\varphi_{|T'_{H_1}(\mathbb{R})}(\gamma_{01}) \tag{4.9}$$

Let us make some comments on these computations. The first line (4.4) is obtained by using the local expression of φ around γ_{01}. The contribution of the indices not in I_{γ_0} being 0, we get (4.5). To get (4.6) and then (4.7), we use Proposition 4.4 (*i*) and (*iv*) and the jump relations I_3 for the ψ^i's. Lemma 2.14 and the following Lemma yield (4.8). A similar computation proves the jump relations when $J_{\gamma_0} = I_{\gamma_0}$.

LEMMA 4.6. *We have*

$$\lim_{\nu\to 0}\Delta_{\eta_{x_j}}(\delta_{j,\nu}) = \Delta_{\eta'_{x_j}}(\delta_{j,0})$$

and $(\eta'_{x_j})^{-1}(c_{\alpha_j}\cdot v_j) = c_{\alpha_H}\cdot u$.

The proof is also postponed to Section 6.

So far, we have supposed that the $\delta_{j,0}$'s are semi-regular. A well-known result of Harish-Chandra enable us to relax this asumption. Thus the φ satifies the jump relations, and complete the proof of theorem 3.6. □

5. Transfer factors

We will give in this section some details on the definition of transfer factors in [**LS**], in order to be able to establish some of their properties that we need to complete the proof of the transfer. This also makes this paper more self-contained. On the other hand, we will need also to use some other properties of transfer factors that are established in [**LS**], [**LS2**] or [**KS**]. For these, we simply recall the results and we refer the reader to these papers for the proofs. We also take advantage of the fact that our base field is $\mathbb{R}$ to give somehow simpler or more explicit expressions for the various factors.

In particular, the general definitions of **[LS]**, **[LS2]** or **[KS]** use the notions of a-data and χ-data which are redundant in our case. Indeed let (B,T) be a pair for G^*, with T defined over $\mathbb{R}$. Let us now explain how we fix the a-data and χ-data on $R(T,G^*)$ or $R(\mathbf{T},G^*)$. If $\alpha \in R(T,G^*)$ is imaginary positive (*ie.* if $\alpha \in R(T,B)$ and $\sigma_T(\alpha)=-\alpha$), set $a_\alpha = i$, $a_{-\alpha}=-i$, and for all $z\in\mathbb{C}^*$, $\chi_\alpha(z)=z/|z|$, $\chi_{-\alpha}(z)=\bar{z}/|z|$. If α is not imaginary, set $a_\alpha=1$. and $\chi_\alpha(z)=1$. Take $h\in G^*$ such that $(B,T)=\operatorname{Int} h\cdot(\mathbf{B},\mathbf{T})$ and transport χ and a-data to $R(\mathbf{T},G^*)$.

It will be convenient to use the following notation: if $z=\exp w\in\mathbb{C}^*$, and if $\alpha\in R(T,G^*)$, we sometimes write $z^{\alpha^\vee}$ for $\exp(wH_\alpha)$.

5.1. An overview of the transfer factors. In this section, we briefly discuss the role of the various factors in the transfer of orbital integrals, as a motivation for the definitions to come, but also to shed some light on the path Shelstad took to find them.

Recall from the statement of theorem 3.6 that we want to match certain orbital integrals on $G(\mathbb{R})$ (or rather some linear combination of orbital integrals in a stable orbit) to stable orbital integrals on $H_1(\mathbb{R})$, that orbital integrals are determined by their restrictions to Cartan subgroups, and that we have a correspondence between Cartan subgroups of $G(\mathbb{R})$ and of $H_1(\mathbb{R})$. From definition 3.2, we see that restriction of orbital integrals to Cartan subgroups become more regular when multiplied by a certain factor (denoted b_P there). It is therefore natural to multiply orbital integrals on $G(\mathbb{R})$ and $H_1(\mathbb{R})$ by the corresponding factors, to make them as smooth as possible. The quotient of this factor for $G(\mathbb{R})$ and the one for $H_1(\mathbb{R})$ is essentially what is called Δ_{II} below, and we see that it appears as a requirement from invariant harmonic analysis.

The basic idea of endoscopy is that we should put some weights on different orbital integrals on $G(\mathbb{R})$ belonging to a single stable orbit, to form a linear combination which should match some stable orbital integrals on $H_1(\mathbb{R})$. Conjugacy classes in a stable conjugacy class are parametrized by some first cohomology group and the weights will be given by a pairing (Tate-Nakayama) between this cohomology group and a group obtained from the endoscopic data. The corresponding transfer factor is denoted Δ_{III_1} below.

As mentioned in section 1.8, the transpose of the transfer of orbital integrals, which is a map from stable invariant distributions on $H_1(\mathbb{R})$ to invariant distributions on $G(\mathbb{R})$, should induce a correspondence of characters of representations of $H_1(\mathbb{R})$ in some L-packet to characters of representations of $G(\mathbb{R})$ in the corresponding L-packet for G. Consideration of infinitesimal characters of representations of these packets shows that one needs to incorporate a correction character (in Shelstad's terminology) to match orbital integrals. This character (of a Cartan subgroup) is the Δ_{III_2} factor.

So far, all the factors are there to deal with the behaviour of orbital integrals on a given Cartan subgroup. The most subtle factor is the remaining one, Δ_I, which deals with the compatibility between pairs of adjacent Cartan subgroups, *i.e.* jump relations. This factor is just a sign, again given by a Tate-Nakayama pairing as for Δ_{III}, but it accomplishes a miracle : not only jump relations will be satisfied, but also the global transfer factor Δ will become canonical, independent of all the choices made to define each factor.

5.2. Some general constructions. G^* is quasi-split and $\mathbf{spl}_{G^*} = \{\mathbf{B}, \mathbf{T}, \{\mathbf{X}_\alpha\}\}$ is a splitting over $\mathbb{R}$. Then Γ acts on the Weyl group $W(G^*, \mathbf{T})$ by automorphisms and one can form the semi-direct product

$$W(G^*, \mathbf{T}) \rtimes \Gamma.$$

For all $\theta \in W(G^*, \mathbf{T}) \rtimes \Gamma$ one can lift θ to an element $n(\theta)$ in $G^* \rtimes \Gamma$ as follows:

• For any simple root α in $R(\mathbf{T}, G^*)$ notice that its coroot H_α and the element X_α which are part of the splitting $\mathbf{spl}_{G^*}$ determines an $\mathfrak{sl}_2$-triple $\{X_\alpha, X_{-\alpha}, H_\alpha\}$ and an embedding of $\mathfrak{sl}_2$ in the Lie algebra of G^*. This embedding exponentiates to a homomorphism $\phi_\alpha : SL_2 \to G^*$. Set

$$n(s_\alpha) = \phi_\alpha \begin{pmatrix} 0 & 1 \\ -1 & 0 \end{pmatrix}.$$

It is clear that $\operatorname{Int} n(s_\alpha)$ realizes the reflection s_α on $\mathbf{T}$. Another expression for $n(s_\alpha)$ is

$$n(s_\alpha) = \exp X_\alpha \exp X_{-\alpha} \exp X_\alpha.$$

• If $w = s_{\alpha_1} \dots s_{\alpha_r}$ is a reduced decomposition in $W(G^*, \mathbf{T})$, set

$$n(w) = n(s_{\alpha_1}) \dots n(s_{\alpha_r}).$$

A well-known result of Chevalley-Steinberg implies that $n(w)$ is independent of the chosen reduced decomposition of w.

• If $\theta = w \times \epsilon$, set $n(\theta) = n(w) \times \epsilon$.

There is an exact sequence

$$1 \to \mathbf{T} \to N(G^*, \mathbf{T}) \rtimes \Gamma \to W(G^*, \mathbf{T}) \rtimes \Gamma \to 1$$

and $n : W(G^*, \mathbf{T}) \rtimes \Gamma \to\to N(G^*, \mathbf{T}) \rtimes \Gamma$ is a section. This gives a 2-cocycle t from $W(G^*, \mathbf{T}) \ltimes \Gamma$ with values in $\mathbf{T}$, defined by

$$n(\theta_1)n(\theta_2) = t(\theta_1, \theta_2)n(\theta_1\theta_2), \quad \theta_1, \theta_2 \in W(G^*, \mathbf{T}) \rtimes \Gamma$$

LEMMA 5.1. [**LS**] *Lemma 2.1.A. The cocycle t is given by*

$$t(\theta_1, \theta_2) = \prod_{\alpha>0,\, \theta_2\alpha<0,\, \theta_1\theta_2\alpha>0} \exp(i\pi H_\alpha), \quad \theta_1, \theta_2 \in W(G^*, \mathbf{T}) \ltimes \Gamma$$

Notice that since $\exp(2i\pi H_\alpha) = 1$, $t(\theta_1, \theta_2)^{-1} = t(\theta_1, \theta_2)$.

5.3. Definition of an invariant. Let (B,T) be a pair for G^*, with T defined over $\mathbb{R}$. Choose $h \in G^*$ such that $(B,T) = \operatorname{Int} h \cdot (\mathbf{B},\mathbf{T})$. Denote by σ_T both the action of Γ on T and its transport to $\mathbf{T}$ by $\operatorname{Int} h$. Let $w(\sigma_T) \in W(G^*,\mathbf{T})$ be the class of $h^{-1}\sigma_{G^*}(h)$. Then on $\mathbf{T}$, $\sigma_T = w(\sigma_T) \times \sigma$. By Lemma 5.1 $n(\sigma_T) := n(w(\sigma_T)\times\sigma) = n(w(\sigma_T)) \times \sigma$ satisfies

$$n(\sigma_T)\sigma(n(\sigma_T)) = \prod_{\alpha>0,\sigma_T\alpha<0} \exp(i\pi H_\alpha) = \partial x_\sigma$$

where

$$x_\sigma = \prod_{\alpha>0,\sigma_T\alpha<0} \exp\left(i\frac{\pi}{2}H_\alpha\right),$$

the product being over positive roots α in $R(\mathbf{T},G^*)$ such that $\sigma_T\alpha = -\alpha$. Then

$$m_\sigma = x_\sigma n(\sigma_T)$$

is a 1-cocycle of Γ in $N(G^*,\mathbf{T})$. Because x_σ lies in T, the image of m_σ in $W(G^*,T)$ coincide with $w(\sigma_T)$, therefore

$$hm_\sigma\sigma_{G^*}(h)^{-1} = h(m_\sigma\sigma_{G^*}(h)^{-1}h)h^{-1}$$

lies in T and is evidently a 1-cocycle. Let us denote by $\lambda(T)$ its class in $H^1(\Gamma,T)$. It is readily verified that $\lambda(T)$ does not depend on the choice of h, which is defined modulo left translation by an element in T. Thus $\lambda(T)$ depends on the choice of B and the choice of $\mathbf{spl}_{G^*}$ only.

LEMMA 5.2. *The cohomology class $\lambda(T)$ in $H^1(\Gamma,T)$ does not depend on the choice of B. If $\mathbf{spl}_{G^*}$ is replaced by $\mathbf{spl}_{G^*} = g\cdot\mathbf{spl}_{G^*}$ with $g \in G^*$ such that $g\sigma_{G^*}(g)^{-1} \in Z(G^*)$, then $\lambda(T)$ is replaced by $\lambda'(T) = \lambda(T)\mathbf{g}$ where $\mathbf{g}$ is the class of $g\sigma_{G^*}(g)^{-1}$ in $H^1(\Gamma,T)$. Suppose that $g \in \mathcal{A}(T)$ and $T' = g\cdot T$. Let $\mathbf{g}$ be the class of the cocycle $\sigma_{G^*}(g)^{-1}g$ in $H^1(\Gamma,T)$. Then*

$$\lambda(T') = \mathbf{g}\operatorname{Int} g\cdot(\lambda(T)) \in H^1(\Gamma,T').$$

Proof. See [**LS**] (2.3).

5.4. Embeddings of L-groups. We continue with a pair (B,T) in G^*, with T defined over $\mathbb{R}$.

If $w \in W_{\mathbb{R}}$, let us write $w = w_0 \times \epsilon_w$, with $\epsilon_w = 1$ or σ and $w_0 \in \mathbb{C}^*$. Define

$$r(w) = \prod_{\alpha\in R(\mathbf{T},\mathbf{B}),\sigma_T\alpha<0} (w_0/|w_0|)^{\alpha^\vee}.$$

LEMMA 5.3. *The coboundary of r is the inflation of t to $W_{\mathbb{R}}$.*

Proof. $\partial r(w,y) = r(w)w\cdot r(y)r(wy)^{-1}$

$$= \prod_{\alpha\in R(\mathbf{T},\mathbf{B}),\sigma_T\alpha<0} (w_0/|w_0|)^{\alpha^\vee}\epsilon_w\cdot\left((y_0/|y_0|)^{\alpha^\vee}\right)((wy)_0/|(wy)_0|)^{-\alpha^\vee}$$

This equals 1 if $w = w_0\times 1$ or $y = y_0 \times 1$ and $(-1)^{-\alpha^\vee}$ if $w = w_0\times\sigma$ and $y = y_0\times\sigma$. □

We want to construct a L- homomorphism $\xi_T : {}^LT \to {}^LG$ such that ξ_T maps $\hat{T}$ isomorphically onto $\mathcal{T}$, this isomorphism being the one attached to the pairs (B, T) and $(\mathcal{B}, \mathcal{T})$.

To specify ξ_T, we have only to give a homomorphim

$$w \mapsto \xi_T(w) = \xi_T^0(w) \times w,$$

where $\xi_T^0(w) \in N(\hat{G}, \mathcal{T})$, and where, in addition, if $w = w_0 \times \sigma$, then $\operatorname{Int} \xi_T(w)$ acts on $\mathcal{T}$ as the transport by ξ_T of the action σ_T on T and if $w = w_0 \times 1$, then $\xi_T(w)$ acts trivially on $\mathcal{T}$. As above, we write $\sigma_T = w(\sigma_T) \times \sigma$ in $W(\hat{G}, \mathcal{T}) \ltimes \Gamma$.

If $w = w_0 \times \epsilon_w \in W_{\mathbb{R}}$, let us define $n(w)$ to be $n(w(\sigma_T)) \times w$ if $\epsilon_w = \sigma$ and $1 \times w$ if $\epsilon = 1$.

From Lemma 5.1, if $w_i = w_{i0} \times \epsilon_i \in \Gamma$, $i = 1, 2$,

$$n(w_1)n(w_2)n(w_1w_2)^{-1} = t(\epsilon_1, \epsilon_2).$$

Since t splits over $W_{\mathbb{R}}$ (Lemma 5.3)

$$w \mapsto r(w)n(w)$$

defines an admissible embedding ξ_T of ${}^LT$ in ${}^LG$, depending on B and $\mathbf{spl}_{\hat{G}}$.

LEMMA 5.4. *The $\hat{G}$-conjugacy class of ξ_T is independent of the choices of B and $\mathbf{spl}_{\hat{G}}$. Suppose that $g \in \mathcal{A}(T)$ and $T' = g \cdot T$. It gives an L-isomorphism ${}^LT \to {}^LT'$. The $\hat{G}$-conjugacy classes of ξ_T and $\xi_{T'}$ are canonically identified via this isomorphism.*

Proof. See [**LS**] (2.4)

5.5. Suppose that $\gamma_1 \in H_1(\mathbb{R})_{reg}$ projects on a G-regular element $\gamma \in H(\mathbb{R})$, which is a norm of $\delta \in G(\mathbb{R})^+$ (otherwise we set $\Delta(\gamma_1, \delta) = 0$). Let us denote by T_{H_1} (resp. T_H) the maximal torus of H_1 where γ_1 (resp. γ) lies. Suppose that (B_H, T_H) is a pair in H and that (B, T) is a pair in G^* from which we get an admissible embedding $\eta : T_H \to T$ of T_H in G^*. Set $\delta^* = \eta(\gamma)$. Consider $x \in G_{sc}$ giving an admissible embedding $\eta_x : T_H \to T_G$ of T_H in G and let us assume that $T_G(\mathbb{R}) \subset G(\mathbb{R})^+$. The transfer factor $\Delta(\gamma_1, \delta)$ of [**LS**] is a product of four terms:

$$\Delta(\gamma_1, \delta) = \Delta_I(\gamma_1, \delta)\Delta_{II}(\gamma_1, \delta)\Delta_{III_1}(\gamma_1, \delta)\Delta_{III_2}(\gamma_1, \delta).$$

The last term Δ_{IV} of [**LS**] is already included in our definition of orbital integrals, so it does not appear here. This absolute transfer factor is not canonical. What is canonical is the relative transfer factor $\Delta(\gamma_1, \delta; \gamma_1', \delta_1')$ defined for two pairs of elements (γ_1, δ) and (γ_1', δ_1') as above. For all terms except Δ_{III_1}, the corresponding factor of $\Delta(\gamma_1, \delta; \gamma_1', \delta_1')$ is just the quotient of the corresponding factors of $\Delta(\gamma_1, \delta)$ and $\Delta(\gamma_1', \delta')$ respectively. Only Δ_{III_1} is a genuine relative factor. The absolute and non-canonical Δ is obtained by fixing a pair (γ_1', δ_1') satisfying the conditions of the begining of the paragraph, and a non-zero value $\Delta(\gamma_1', \delta_1')$. Then for all pair (γ_1, δ),

$$\Delta(\gamma_1, \delta_1) = \Delta(\gamma_1, \delta; \gamma_1', \delta_1')\Delta(\gamma_1', \delta_1').$$

5.6. Δ_I. Let us denote by $\sigma_{\mathcal{T}_H}$ the action of σ on $\mathcal{T}_H$, given by the transport of the action of σ on $\hat{T}_H$ through the isomorphism $\hat{T}_H \simeq \mathcal{T}_H$ given by the pairs (B_H, T_H) and $(\mathcal{B}_H, \mathcal{T}_H)$. There is another action of σ on $\mathcal{T}_H$, namely $\rho_{\mathcal{H}}(1 \times \sigma) = \rho_H(1 \times \sigma) = \sigma_{\hat{H}}$. These two actions differ by the action of an element of the Weyl group of $\mathcal{T}_H$ in $\hat{H}$, thus the action $\sigma_{\mathcal{T}_H}$ is induced by the adjoint action of an element $h \in \mathcal{H}$ normalizing $\mathcal{T}_H$. Let us write

$$\xi(h) = g \times w \in {}^L G = \hat{G} \rtimes_{\rho_G} W_{\mathbb{R}}$$

Since $\xi(h)$ and w normalize $\mathcal{T}$, g also normalizes $\mathcal{T}$. From the property $\operatorname{Int} s \circ \xi = a \otimes \xi$ we get

$$\xi(h) s \xi(h^{-1}) = s a(w).$$

Writing $s_H = \xi^{-1}(s)$, we get:

$$\sigma_{\mathcal{T}_H}(s_H) = h s_H h^{-1} = s_H \xi^{-1}(a(w)). \tag{5.1}$$

Since s is central in $\xi(\hat{H})$, $s \in \mathcal{T}$, and the preimage of s in $\hat{T}_H$ through the isomorphisms $\hat{T}_H \simeq \mathcal{T}_H \overset{\xi}{\simeq} \mathcal{T}$ is independent of the choice of B_H. Thus its image s_T in $\hat{T}$ depends only on the admissible embedding $T_H \to T$. Equation (5.1) is an equality in $\mathcal{T}_H$. We can push it to an equality in $\hat{T}$ and get

$$s_T^{-1} \sigma_T(s_T) = a(w) \in Z(\hat{G}).$$

Notice that the embedding of $Z(\hat{G})$ in $\hat{T}$ is canonical. Let $\hat{T}_{ad} := T/Z(\hat{G})$. It is easy to check that $\hat{T}_{ad} = \widehat{T_{sc}}$. Furthermore, from the previous equation, we see that the image of s_T in $\hat{T}_{ad}$ is Γ-invariant, and thus define a class $\mathbf{s}_T$ in

$$\pi_0 = \pi_0(\hat{T}_{ad}^{\Gamma}) = \hat{T}_{ad}^{\Gamma} / (\hat{T}_{ad}^{\Gamma})_0.$$

Recall the Tate-Nakayama pairing (see Appendix 6.4)

$$\langle ., . \rangle : H^1(\Gamma, T_{sc}) \times \pi_0(\hat{T}_{ad}^{\Gamma}) \to \{\pm 1\}.$$

Then $\Delta_I(\gamma, \delta) = \langle \lambda(T_{sc}), \mathbf{s}_T \rangle$, where $\lambda(T_{sc})$ was defined in Section 5.3.

5.7. Δ_{II}. Let us denote by $R_I(T, G^*)$ the set of imaginary roots in $R(T, G^*)$, and by $R_I(T, B)$ the set of positive imaginary roots. We use similar notation $R_I(T_H, H)$ and $R_I(T_H, B_H)$ for the roots of T_H in H. Set

$$\begin{aligned}
\Delta_{II}(\gamma_1, \delta) &= \prod_{\alpha \in R_I(T,B)} \frac{\alpha(\delta^*) - 1}{|\alpha(\delta^*) - 1|} \times \prod_{\alpha_H \in R_I(T_H, B_H)} \frac{|\alpha_H(\gamma) - 1|}{\alpha_H(\gamma) - 1} \\
&= \prod_{\alpha \in R_I(T,B),\ \text{not from } H} \frac{\alpha(\delta^*) - 1}{|\alpha(\delta^*) - 1|}.
\end{aligned}$$

The equality results from the fact that $\alpha_H(\gamma_1) = \alpha(\delta^*)$ if $\eta(\alpha_H) = \alpha$.

5.8. $\Delta_{III_1} = \Delta_1$**.** This is the only relative term in the transfer factors. Let γ_1, γ, δ^*, δ, x, T_H, B_H, T, B, T_G, η, η_x as above and suppose we have also data γ_1', γ', $\delta^{*\prime}$, δ', x', T_H', B_H', T', B', T_G', η', η_x' satisfying the same properties. The relative transfer factor $\Delta_1(\gamma_1', \delta'; \gamma_1, \delta)$ is defined as follows. Set $v_\sigma = x u_\sigma \sigma_{G^*}(x)^{-1}$ and $v_\sigma' = x' u_\sigma \sigma_{G^*}(x')^{-1}$. Then v_σ (resp. v_σ') is a cochain of Γ in T_{sc} (resp. T_{sc}'), which is well-defined up to a coboundary since x (resp. x') is defined up to left multiplication by an element in T (resp. T'). Further, $\partial v_\sigma = \partial v_\sigma' = \partial u_\sigma$ takes values in Z_{sc}, the center of G_{sc}^*. Let U be the torus

$$U = T_{sc} \times T_{sc}' / \{(z^{-1}, z),\ z \in Z_{sc}\}.$$

Then $(v_\sigma^{-1}, v_\sigma')$ defines an element of $H^1(\Gamma, U)$ which is independent of the choices of u_σ, x and x'. Let us write this class as

$$\mathbf{inv}(\gamma_1, \delta; \gamma_1', \delta').$$

Recall that $\hat{T}_{ad}$ is $\hat{T}/Z(\hat{G})$ and that $\hat{T}_{ad} = \widehat{T_{sc}}$. Dually, $T_{ad} = T/Z(G)$ has dual group $\hat{T}_{sc}$. Then the center $\hat{Z}_{sc}$ of the simply-connected covering of the derived group of $\hat{G}$ (a finite group isomorphic to Z_{sc}) is canonically embedded in $\hat{T}_{sc}$ and $\hat{T}_{sc}'$. Set

$$\hat{U} = \hat{T}_{sc} \times \hat{T}_{sc}' / \{(z, z),\ z \in \hat{Z}_{sc}\}.$$

One checks readily that $\hat{U}$ is indeed the dual of U.

To the endoscopic datum s, we attach $\mathbf{s}_U \in \pi_0(\hat{U}^\Gamma)$. Suppose that $\tilde{s}$ lies in the preimage of s under $\mathcal{T}_{sc} \to \mathcal{T}$. From the isomorphisms $\mathcal{T} \simeq \hat{T}$ and and $\hat{T} \simeq \hat{T}'$, we obtain $\mathcal{T}_{sc} \simeq \hat{T}_{sc}$ and $\mathcal{T}_{sc} \simeq \hat{T}_{sc}'$. We denote by $\tilde{s}_T$ and $\tilde{s}_{T'}$ the images of $\tilde{s}$ under these isomorphisms. Then s_U is the image of $(\tilde{s}_T, \tilde{s}_{T'})$ in $\hat{U}$. It is independent of the choice of $\tilde{s}$ and it is Γ-invariant. We denote by $\mathbf{s}_U$ the image of s_U in $\pi_0(\hat{U}^\Gamma)$. We can now define Δ_1:

$$\Delta_1(\gamma_1, \delta; \gamma_1', \delta') = \langle \mathrm{inv}(\gamma_1, \delta; \gamma_1', \delta'), \mathbf{s}_U \rangle.$$

5.9. $\Delta_{III_2} = \Delta_2$**.** To get a better idea of this factor, it is useful to start with the case where $\mathcal{H} \simeq {}^L H$, which makes the definition much simpler. In this case, we take obviously $(H_1, \xi_{H_1}) = (H, \mathrm{Id})$. Consider data γ, δ^*, δ, x, T_H, B_H, T, B, T_G, η, η_x as above. Then, we have the following diagram

$$\begin{array}{ccc} {}^L T_H & \xrightarrow{\ \xi_{T_H}\ } & {}^L H \\ \eta \big\downarrow & & \big\downarrow \xi \\ {}^L T & \xrightarrow{\ \xi_T\ } & {}^L G \end{array}$$

Notice that we have denoted again by η the L-isomorphim induced from the $\mathbb{R}$-isomorphism $\eta : T_H \simeq T$. The L-embeddings ξ_T and ξ_{T_H} are constructed in Section 5.4

This diagram is not commutative. Identifying t in ${}^LT$ with its preimage in ${}^LT_H$, we can write

$$\xi \circ \xi_{T_H}(t \times w) = a_T(w)\xi_T(t \times w)$$

for all $t \times w$ in ${}^LT$. Then a_T is a cocycle of $W_{\mathbb{R}}$ in $\mathcal{T}$ (for the $W_{\mathbb{R}}$-action through σ_T), and it defines a class $\mathbf{a}_T$ in $H^1(W_{\mathbb{R}}, \mathcal{T})$. Let us transport this through the isomorphism $\mathcal{T} \simeq \hat{T}$ to a class in $H^1(W_{\mathbb{R}}, \hat{T})$ that we still denote by $\mathbf{a}_T$. Then

$$\Delta_2(\gamma, \delta) = \langle \mathbf{a}_T, \delta^* \rangle$$

where $\langle ., . \rangle$ is the Langlands pairing between T and $H^1(W_{\mathbb{R}}, \hat{T})$ (see Appendix 6.4).

Let us now consider the general case. We cannot compare directly the L-embeddings $\xi_T : {}^LT \to {}^LG$ and $\xi_{T_H} : {}^LT_H \to {}^LH$ as above. We should rather compare ξ_T and $\xi_{T_{H_1}} : {}^LT_{H_1} \to {}^LH_1$, but this will be indirect. Let us first remark that $\mathbf{spl}_{\hat{H}} = \{\mathcal{B}_H, \mathcal{T}_H, \{\mathcal{X}\}\}$ determines a splitting $\mathbf{spl}_{\hat{H}_1} = \{\mathcal{B}_{H_1}, \mathcal{T}_{H_1}, \{\mathcal{X}\}\}$ of $\hat{H}_1$, and that any pair (B_H, T_H) in H determines a pair (B_{H_1}, T_{H_1}) in H_1. Thus the isomorphism $\hat{T}_H \simeq \mathcal{T}_H$ given by these choices of pairs extends uniquely to an isomorphism $\hat{T}_{H_1} \simeq \mathcal{T}_{H_1}$. The L-embedding $\xi_{T_{H_1}} : {}^LT_{H_1} \to {}^LH_1$ is the extension of this isomorphism which was constructed in Section 5.4.

Let us consider the subgroup $\mathcal{U}$ of $\mathcal{H}$ consisting of elements u normalizing $\mathcal{T}_H$ and acting by conjugacy on $\mathcal{T}_H$ as σ_{T_H} if u projects on $\sigma \in \Gamma$, and trivially if u projects trivially to Γ. It is easy to see that $\mathcal{U}$ projects surjectively onto $W_{\mathbb{R}}$ and that the kernel of this projection is $\mathcal{T}_H$, *ie.* we have an exact sequence

$$1 \to \mathcal{T}_H \to \mathcal{U} \to W_{\mathbb{R}} \to 1 \tag{5.2}$$

The restriction of $\xi_{H_1} : \mathcal{H} \to {}^LH_1$ to $\mathcal{U}$ is easily seen to have its image included in the image of $\xi_{T_{H_1}}$, thus there exists an unique L-homomorphism $\alpha_0 : \mathcal{U} \to {}^LT_{H_1}$ such that $\xi_{H_1} = \xi_{T_{H_1}} \circ \alpha_0$ on $\mathcal{U}$. Let us also set α to be the composition of α_0 and $t \mapsto t^{-1}$ on ${}^LT_{H_1}$.

For $u \in \mathcal{U}$, write $\xi(u) \in {}^LG$ as $\xi(u) = g \times w \in \hat{G} \times W_{\mathbb{R}}$. Since T and T_H are isomorphic other $\mathbb{R}$, $\xi(u)$ acts on $\mathcal{T}$ by conjugacy as σ_T if $w = w_0 \times \sigma$ and trivially if $w = w_0 \times 1$. Thus it is also clear that $\xi(\mathcal{U})$ is included in the image of ξ_T and therefore there exists a unique L-embedding $\beta : \mathcal{U} \to {}^LT$ such that $\xi = \xi_T \circ \beta$.

So far, we have defined

$$(\alpha, \beta) : \mathcal{U} \to {}^LT_{H_1} \times {}^LT \simeq {}^L(T_{H_1} \times T). \tag{5.3}$$

Let us consider the fiber product $T_1 := T_{H_1} \times_{T_H} T$. This is an algebraic torus defined over $\mathbb{R}$. Since $T_H \simeq T$ over $\mathbb{R}$, $T_1 \simeq T_{H_1}$ and

$$\hat{T}_1 = \hat{T}_{H_1} \times \hat{T}/\{(t^{-1}, \eta^{-1}(t)), t \in T_H\}.$$

The natural projection $\hat{T}_{H_1} \times \hat{T} \to \hat{T}_1$ extends to an L-homomorphism

$$ {}^L(T_{H_1} \times T_H) \to {}^LT_1. \tag{5.4}$$

The composition of (5.3) and (5.4) gives a L-homomorphism

$$\mathcal{U} \to {}^L T_1$$

which has $\mathcal{T}_H$ in its kernel. Since $\mathcal{U}/\mathcal{T}_H \simeq W_{\mathbb{R}}$ (Eq (5.2)), we get a L-homomorphism

$$W_{\mathbb{R}} \to {}^L T_1$$

or equivalently, a 1-cocycle of $W_{\mathbb{R}}$ in $\hat{T}_1$. As we have observed, $T_1 \simeq T_{H_1}$, so we can transport this cocycle to a cocycle with value in $\hat{T}_{H_1}$. Let $\mathbf{a}_T$ be its class in $H^1(W_{\mathbb{R}}, \hat{T}_{H_1})$. We can now give the definition of Δ_2:

$$\Delta_2(\gamma_1, \delta) = \langle \mathbf{a}_T, \gamma_1 \rangle$$

where the pairing is the Langlands pairing for tori.

6. Proof of Proposition 4.4 and Lemma 4.6

6.1. Proof of (*i*). The setting is that of Proposition 4.4. Let us fix $\delta \in (\mathcal{V}_{\delta_0})_{reg}$, $\gamma_1 = \eta_x^{-1}(\delta)$ and $X \in \mathfrak{t}_G$ sufficiently close to 0. The rest of the notation is recalled in the following diagram:

$$\begin{array}{ccccc} \gamma_1 \exp X_H \in T_{H_1}(\mathbb{R}) & & & & \\ \downarrow & & & & \\ \gamma \exp X_H \in T_H(\mathbb{R}) & \xrightarrow{\ \eta\ } & \delta^* \exp X^* \in T(\mathbb{R}) & \xleftarrow{\operatorname{Int} x \circ \psi} & \delta \exp X \in T_G(\mathbb{R}) \end{array}$$

The transfer factor we want to examine is:

$$\begin{aligned} \Delta(\exp X) &:= \Delta(\gamma_1 \exp X_H, \delta \exp X) \\ &= \Delta(\gamma_1 \exp X_H, \delta \exp X; \gamma_1, \delta)\Delta(\gamma_1, \delta) \end{aligned}$$

where $\Delta(.,.;.,.)$ is the canonically defined relative transfer factor of [**LS**].

As before, we write $\Delta(X)$ as a product of four terms (recall that the last term of [**LS**] has been included in our definition of orbital integrals).

$$\Delta(X) = \Delta_I(X)\Delta_{II}(X)\Delta_1(X)\Delta_2(X)$$

The term Δ_1 is a quotient ([**LS**], section 3.2). Because the two elements $\gamma \exp X_H$ and γ lie in the same maximal torus T_H and the numerator and denominator of Δ_I each depend on the torus rather than the individual elements, we find that $\Delta_I(X) = 1$.

Let us now examine $\Delta_{II}(X)$. It is again a quotient

(6.1)
$$\begin{aligned} \Delta_{II}(X) &= \frac{\Delta_{II}(\gamma_1 \exp X_H, \delta \exp X)}{\Delta_{II}(\gamma_1, \delta)} \\ &= \prod \frac{\alpha(\delta^* \exp X^*) - 1}{|\alpha(\delta^* \exp X^*) - 1|} \frac{|\alpha(\delta^*) - 1|}{\alpha(\delta^*) - 1} \end{aligned}$$

where the product is over the imaginary roots in $R(T, B)$ not from H.

Let us write $\alpha(\delta^*) = z^2$ and $\alpha(\exp X^*) = w^2$ with z and w on the unit circle. Then

$$\frac{\alpha(\delta^* \exp X^*) - 1}{|\alpha(\delta^* \exp X^*) - 1|} \frac{|\alpha(\delta^*) - 1|}{\alpha(\delta^*) - 1} = \frac{zw(zw - \bar{z}\bar{w})}{|zw(zw - \bar{z}\bar{w})|} \frac{|z(z - \bar{z})|}{z(z - \bar{z})} = \frac{w}{|w|}$$

because X is close to 0. Thus

$$\Delta_{II}(X) = \prod_{\alpha \in R_I(B,T), \text{ not from } H} \exp \alpha(X^*/2). \tag{6.2}$$

We now take a closer look at $\Delta_1(X) = \Delta_1(\gamma_1 \exp X_H, \delta \exp X; \gamma_1, \delta)$. Here, the general construction of 5.8 simplifies because $T' = T$, $v'_\sigma = v_\sigma$ and thus the class $\mathbf{inv}(\gamma_1 \exp X_H, \delta \exp X; \gamma_1, \delta)$ in $H^1(\Gamma, U)$ is the class of the cocycle $(v_\sigma^{-1}, v_\sigma)$. Since in this case $\mathbf{s}_U$ is the class of (s_T, s_T) in $\pi_0(\hat{U}^\Gamma)$, we see that

$$\Delta_1(\gamma_1 \exp X_H, \delta \exp X; \gamma_1, \delta) = \langle \text{inv}(\gamma_1 \exp X_H, \delta \exp X; \gamma_1, \delta), \mathbf{s}_U \rangle = 1.$$

Finally, $\Delta_2(X)$ is obtained by evaluating the character given by the class $\mathbf{a}_T$ on $\exp X_H$.

The obstruction to extending Δ_{η_x} to a smooth function on $(\mathcal{V}_{\delta_0})$ comes from the Δ_{II} term. More precisely, with the notation of section 5.7, there might be an obstruction at the point $\delta_0 \exp X$ if $\alpha(\delta_0^* \exp X^*) = 1$ with α imaginary, not from H. Thus Δ_{η_x} extends to a smooth function on $(\mathcal{V}_{\delta_0})_{I-reg}$. Since Δ_{II} is locally bounded, Δ is also locally bounded. We see also that Δ has the form $\Delta(X) = c \exp \lambda(X)$ for some constant c and some $\lambda \in \mathfrak{t}_\mathbb{C}$, from direct consideration of the explicit form of the non-trivial factors Δ_{II} and Δ_2. This finishes the proof of (i).

6.2. The notation is now as in the rest of Proposition 4.4. If γ_0 is semi-regular, then α is not from H and Δ_{η_x} has a smooth continuation at δ_0. This can be rewritten

$$\lim_{\nu \to 0^+} \Delta(\gamma_1^\nu, \delta^\nu) = \lim_{\nu \to 0^-} \Delta(\gamma_1^\nu, \delta^\nu)$$

proving the corresponding assertions in (iii) and (v). If γ_0 is regular, we are in the opposite situation, *i.e.* α is from H. The obstruction to the smoothness of Δ_{η_x} at δ_0 comes from the factor

$$\frac{\alpha(\delta^{\nu,*}) - 1}{\alpha(\delta^{\nu,*}) - 1}$$

and it is easy to see that

$$\lim_{\nu \to 0^+} \frac{\alpha(\delta^{\nu,*}) - 1}{\alpha(\delta^{\nu,*}) - 1} = - \lim_{\nu \to 0^-} \frac{\alpha(\delta^{\nu,*}) - 1}{\alpha(\delta^{\nu,*}) - 1},$$

proving the assertions in (ii) and (iv).

6.3. For the remaining assertions of Proposition 4.4, we need to compare

$$\Delta(\gamma_{1,\nu},\delta_\nu) \quad \text{and} \quad \Delta(\gamma_{1,\nu},\delta_{-\nu}).$$

By [**KS**], Theorem 5.1.D, the quotient of these two terms is given by the quantity

$$\langle \mathbf{inv}(\delta^\nu,\delta^{-\nu}), \mathbf{s}_T\rangle,$$

where $\mathbf{inv}(\delta^\nu,\delta^{-\nu}) \in H^1(\Gamma, T)$ is defined as follows. Let w_α be an element in $G_{sc}^{\delta_0}$ realizing the Weyl group reflection s_α. We have

$$w_\alpha \delta^\nu w_\alpha^{-1} = \delta^{-\nu}.$$

Thus $w_\alpha^{-1}\sigma_G(w_\alpha)$ is a 1-cocycle of Γ in $T_{G,sc}$. Its class in $H^1(\Gamma, T_{G,sc})$ and identifying $H^1(\Gamma, T_{sc}) \simeq H^1(\Gamma, T_{G,sc})$ we get $\mathbf{inv}(\delta^\nu,\delta^{-\nu})$. The element $\mathbf{s}_T \in \pi_0(\hat{T}^\Gamma)$ was defined in Section 5.6. Let us identify the cohomology class of $w_\alpha^{-1}\sigma_G(w_\alpha)$ in $H^1(\Gamma, T_{G,sc})$ using the isomorphism

$$H^1(\Gamma, T_{G,sc}) \simeq X_*(T_{G,sc})^{-\sigma_*}/(1-\sigma_*)X_*(T_{G,sc})$$

of Proposition B.2. For this, choose a morphism

$$\phi_\alpha : SL(2,\mathbb{C}) \to G^*_{sc}$$

defined over $\mathbb{R}$ and sending the imaginary non compact root of the circle torus in $SL(2,\mathbb{R})$ to α. An easy computation in $SL(2,\mathbb{C})$ shows that we can choose w_α such that $w_\alpha^{-1}\sigma_G(w_\alpha)$ is the image of $\begin{pmatrix} -1 & 0 \\ 0 & -1 \end{pmatrix}$, and thus one has $w_\alpha^{-1}\sigma_G(w_\alpha) = \exp(i\pi H_\alpha)$ which gets identified with the class of $\alpha\check{}$ in $X_*(T_{G,sc})^{-\sigma_*}/(1-\sigma_*)X_*(T_{G,sc})$. Using the explicit form of the Tate-Nakayama pairing in Proposition B.2, we see that $\langle \mathbf{inv}(\delta^\nu,\delta^{-\nu}), \mathbf{s}_T\rangle = 1$ if and only if $\tilde{\alpha}\check{}(s) = 1$, where $\tilde{\alpha}$ is the transport of α in $R(\mathbf{T}, G^*)$, i.e. if and only if $\alpha \in R(T_{G,sc}, G^*_{sc}) \simeq R(T_G, G^*)$ comes from H. The values of the Tate-Nakayama pairing are in $\{\pm 1\}$. Thus $\Delta(\gamma_{1,\nu},\delta_\nu) = \Delta(\gamma_{1,\nu},\delta_{-\nu})$ when γ_0 is singular in $H(\mathbb{R})$, and $\Delta(\gamma_{1,\nu},\delta_\nu) = -\Delta(\gamma_{1,\nu},\delta_{-\nu})$ when γ_0 is regular in $H(\mathbb{R})$. In this case, by Lemma 4.1 δ_ν and $\delta_{-\nu}$ cannot be conjugate in $G(\mathbb{R})$. This finishes the proof of Proposition 4.4.

6.4. Proof of Lemma 4.6. The second assertion of the lemma follows from an easy computation. We concentrate now on the first. The statement and the proof of this result is the main technical improvment from Shelstad treatment in her series of papers. The reason is that at the time, she was missing the conceptual definition of the Δ_I factor, in contrast with the other factors which are already implicit in her work. The virtue of the definition of Δ_I is that it makes Lemma 4.6 true, and it makes the global transfer factor Δ independent of all the choices. Concerning the proof of the Lemma, the approach is also sensibly different. It becomes now a simple consequence of the descent technique in [**LS2**]. Indeed, the main theorem there (Theorem 1.6.A) allows us to reduce the proof of Lemma 4.6 to a group of rank one, namely the centralizer of the semi-regular elements $\delta_{0,j}$ in the lemma. Explicit computations for $SL(2,\mathbb{R})$ are in [**LS**]. This finishes the proof.

Appendix A. Langlands correspondence for real tori

A.1. Characters. Let $\mathbb{T}$ be a torus defined over $\mathbb{R}$, with character module $X^*(\mathbb{T})$ and set $X_*(\mathbb{T}) = \mathrm{Hom}(X^*(\mathbb{T}), \mathbb{Z})$ for the lattice of one-parameter subgroups. Identify $\mathbb{T}$ with the group of its complex points. We have identifications $\mathfrak{t}_{\mathbb{C}} = X_*(\mathbb{T}) \otimes_{\mathbb{Z}} \mathbb{C}$ and $\mathfrak{t}_{\mathbb{C}}^* = X^*(\mathbb{T}) \otimes_{\mathbb{Z}} \mathbb{C}$ such that the canonical pairing between $X^*(\mathbb{T})$ and $X_*(\mathbb{T})$ gives the pairing between $\mathfrak{t}_{\mathbb{C}}$ and $\mathfrak{t}_{\mathbb{C}}^*$. Recall that $\ker[\exp : \mathfrak{t}_{\mathbb{C}} \to \mathbb{T}] = 2i\pi X_*(\mathbb{T})$. Let us denote by σ the Galois involution of $\mathbb{T}$, by $d\sigma$ its differential, which is a Galois involution of $\mathfrak{t}_{\mathbb{C}} = \mathrm{Lie}(\mathbb{T}) = X_*(\mathbb{T}) \otimes_{\mathbb{Z}} \mathbb{C}$ and respectively by σ^* and σ_* the induced actions on $X^*(\mathbb{T})$ and $X_*(\mathbb{T})$. With these notations we have

$$d\sigma(\check{\lambda} \otimes z) = \sigma_*(\check{\lambda}) \otimes \overline{z}. \tag{A.1}$$

Set $T := \mathbb{T}(\mathbb{R})$, the group of real points of $\mathbb{T}$. An element $h = \exp X$, $X \in \mathfrak{t}_{\mathbb{C}}$ belongs to T if and only if

$$d\sigma(X) - X \in 2i\pi X_*(\mathbb{T}).$$

We write $X = X_I + X_R$ where

$$X_I = \frac{1}{2}(X - d\sigma(X)) \text{ and } X_R = \frac{1}{2}(X + d\sigma(X)).$$

Then $X_R \in \mathfrak{t}$ and since $d\sigma(X) - X = 2X_I$, we find that $X_I \in i\pi X_*(\mathbb{T})$. Furthermore, writing $X_I = \check{\lambda} \otimes i\pi$, we see that

$$d\sigma(X_I) = -X_I = \sigma_*(\check{\lambda}) \otimes -i\pi,$$

and thus $\sigma_*(\check{\lambda}) = \check{\lambda}$.

Thus we decompose h as a product $h_1 h_2$ where $h_1 = \exp(X_R) \in T_0 = \exp \mathfrak{t}$ and

$$h_2 = \exp X_I \in F = \{\exp i\pi\check{\lambda} \,|\, \check{\lambda} \in X_*(\mathbb{T}),\ \sigma_*(\check{\lambda}) = \check{\lambda}\}. \tag{A.2}$$

Therefore $T = T_0 F$, with

$$T_0 \cap F = \{\exp i\pi\check{\lambda} \,|\, \check{\lambda} = \check{\mu} + \sigma_*(\check{\mu}),\ \check{\mu} \in X_*(\mathbb{T})\}\} \tag{A.3}$$

Let us prove this assertion. Suppose that $\exp i\pi\check{\lambda} = \exp X$ with $X = \check{\nu} \otimes z \in \mathfrak{t}$, i.e. $d\sigma(X) = \sigma_*(\check{\nu}) \otimes \overline{z} = \check{\nu} \otimes z$. For some $\check{\mu} \in X_*(\mathbb{T})$, we have

$$\check{\lambda} \otimes i\pi - \check{\nu} \otimes z = \check{\mu} \otimes 2i\pi. \tag{A.4}$$

Applying σ_* and complex conjugacy to (A.4) successively on the first and second factor of $X_*(\mathbb{T}) \otimes \mathbb{C}$ we get:

$$\check{\lambda} \otimes i\pi - \sigma_*(\check{\nu}) \otimes z = \sigma_*(\check{\mu}) \otimes 2i\pi \tag{A.5}$$

$$\check{\lambda} \otimes -i\pi - \check{\nu} \otimes \overline{z} = \check{\mu} \otimes -2i\pi. \tag{A.6}$$

The sum of (A.5) and (A.6) gives:

$$-(\sigma_*(\check{\nu}) \otimes z + \check{\nu} \otimes \overline{z}) = (\sigma_*(\check{\mu}) - \check{\mu}) \otimes 2i\pi \tag{A.7}$$

But since $\sigma_*(\check{\nu})\otimes \bar{z} = \check{\nu}\otimes z$, we have $\sigma_*(\check{\nu})\otimes z = \check{\nu}\otimes \bar{z}$, and by (A.5) and (A.7) we get

$$\check{\lambda}\otimes i\pi + \frac{1}{2}(\sigma_*(\check{\mu}) - \check{\mu})\otimes 2i\pi = \sigma_*(\check{\mu})\otimes 2i\pi. \tag{A.8}$$

We conclude that $\check{\lambda} = \sigma_*(\check{\mu}) + \check{\mu})$. The converse is obvious. □

Given a pair of elements (μ,λ) in $\mathfrak{t}^*_{\mathbb{C}}$ we set, for all $X\in\mathfrak{t}_{\mathbb{C}}$ such that $\exp X\in T$,

$$\xi_{(\mu,\lambda)}(\exp X) = e^{\mu(X_R)+2\lambda(X_I)}.$$

Then $\xi_{(\mu,\lambda)}$ is a well-defined character on T if and only if $\xi_{(\mu,\lambda)}(\exp X) = 1$ when $X\in 2i\pi X_*(\mathbb{T})$. This condition is satisfied if and only if $\mu(X_R) + 2\lambda(X_I)\in 2i\pi\mathbb{Z}$. Set $X = 2i\pi\check{\nu}$, with $\check{\nu}\in X_*(\mathbb{T})$. From the definitions, we get that $X_R = i\pi(\check{\nu} - \sigma_*(\check{\nu}))$ whereas $X_I = i\pi(\check{\nu} + \sigma_*(\check{\nu}))$. So a necessary and sufficient condition for (μ,λ) to define a character of T is that for all $\check{\nu}\in X_*(\mathbb{T})$,

$$\mu(i\pi(\check{\nu} - \sigma_*(\check{\nu}))) + 2\lambda(i\pi(\check{\nu} + \sigma_*(\check{\nu})))\in 2i\pi\mathbb{Z}. \tag{A.9}$$

An easy computation shows that this amounts to

$$\frac{1}{2}(\mu - \sigma^*(\mu)) + (\lambda + \sigma^*(\lambda))\in X^*(\mathbb{T}). \tag{A.10}$$

Furthermore, $\xi_{(\mu,\lambda)} = \xi_{(\mu',\lambda')}$ if and only if

$$\mu = \mu' \quad \text{and } \lambda - \lambda'\in X^*(\mathbb{T}) + \{\nu - \sigma^*(\nu)\,|\,\nu\in\mathfrak{t}^*_{\mathbb{C}}\} \tag{A.11}$$

Let us summarize what we have obtained:

PROPOSITION A.1. *The characters of T are all of the form $\xi_{(\mu,\lambda)}$ with μ and λ satisfying*

$$\frac{1}{2}(\mu - \sigma^*(\mu)) + (\lambda + \sigma^*(\lambda))\in X^*(\mathbb{T}),$$

and $\xi_{(\mu,\lambda)} = \xi_{(\mu',\lambda')}$ if and only if

$$\mu = \mu' \quad \textit{and } \lambda - \lambda'\in X^*(\mathbb{T}) + \{\nu - \sigma^*(\nu)\,|\,\nu\in\mathfrak{t}^*_{\mathbb{C}}\}$$

A.2. Langlands parameters. We recall some definitions. The Weil group $W_{\mathbb{R}}$ of $\mathbb{R}$ is an extension of $\Gamma := \mathrm{Gal}(\mathbb{C}/\mathbb{R}) = \{1,\sigma\}$ by $\mathbb{C}^*$, i.e. we have an exact sequence:

$$1\to\mathbb{C}^*\to W_{\mathbb{R}}\to\Gamma\to 1. \tag{A.12}$$

As a set $W_{\mathbb{R}}\simeq\mathbb{C}^*\times\Gamma$ and multiplication is defined by

$$(1\times\sigma)(z\times 1) = (\bar{z}\times\sigma);\ (z\times 1)(1\times\sigma) = (z,\sigma);\ (1\times\sigma)^2 = (-1\times 1). \tag{A.13}$$

Let us denote by $\hat{\mathbb{T}}$ the dual torus of $\mathbb{T}$, i.e. $\hat{\mathbb{T}}$ is a complex algebraic torus with identifications $X^*(\hat{\mathbb{T}}) = X_*(\mathbb{T})$ and $X_*(\hat{\mathbb{T}}) = X^*(\mathbb{T})$. It particular, we have $\mathrm{Lie}(\hat{\mathbb{T}}) = \mathfrak{t}^*_{\mathbb{C}}$ and $\mathrm{Lie}(\hat{\mathbb{T}})^* = \mathfrak{t}_{\mathbb{C}}$. Let $\sigma_{\hat{\mathbb{T}}}$ denote the *algebraic* action on $\hat{\mathbb{T}}$ inducing respectively σ^* and σ_* on $X^*(\mathbb{T})$ and $X_*(\mathbb{T})$. The L-group of $\mathbb{T}$,

${}^LT$ is an the semi-direct product of $\hat{\mathbb{T}}$ and $W_{\mathbb{R}}$, the action ρ_T of $W_{\mathbb{R}}$ being given by

$$\rho_T(z\times 1)=1\,;\quad \rho_T(z\times\sigma)=\sigma_{\hat{\mathbb{T}}} \tag{A.14}$$

DEFINITION A.2. An homomorphism $\phi : W_R \to {}^LT$ is called an L-homomorphism if it satisfies:

(i) ϕ is continous

(ii) Let π be the projection from ${}^LT$ on $W_{\mathbb{R}}$. Then $\pi\circ\phi=\mathrm{Id}_{W_{\mathbb{R}}}$.

The group $\hat{\mathbb{T}}$ acts on L-homomorphisms by conjugation on the image. A Langlands parameter is a conjugacy class of L-homomorphisms.

To specify an L-homomorphism, we need to specify two things: the restiction ϕ_0 of ϕ to $\mathbb{C}^*$, and the element $\phi(1\times\sigma)$. These are subject to various constraints coming from the defining relations in $W_{\mathbb{R}}$. First of all, let us write

$$\phi(1\times\sigma)=\exp 2i\pi\lambda\times(1\times\sigma)$$

for some λ in $\mathfrak{t}_{\mathbb{C}}^*$ (λ is defined up to an element of $X^*(\mathbb{T})$). The restriction ϕ_0 is given by a pair of (μ,ν) of elements of $\mathfrak{t}_{\mathbb{C}}^*$, with $\mu-\nu\in X^*(\mathbb{T})$, such that

$$\phi_0(\exp z)=\exp(z\mu+\overline{z}\nu). \tag{A.15}$$

The first two relations in $W_{\mathbb{R}}$ force

$$\phi(1\times\sigma)\phi_0(z)\phi(1\times\sigma)^{-1}=\phi(\overline{z}). \tag{A.16}$$

Using (A.15), we find that $\nu=\sigma^*(\mu)$. The third relation in $W_{\mathbb{R}}$ gives:

$$\phi(1\times\sigma)^2=\phi_0(-1). \tag{A.17}$$

Using $\phi(1\times\sigma)=\exp 2i\pi\lambda\times(1\times\sigma)$, we find that the left hand side of (A.17) is $\exp(2i\pi(\lambda+\sigma^*(\lambda)))$. The right hand side may be computed by setting $z=i\pi$ in (A.15). What we obtain is:

$$\exp(2i\pi(\lambda+\sigma^*(\lambda)))=\exp(i\pi(\mu-\sigma^*(\mu))). \tag{A.18}$$

Thus we find that

$$(\lambda+\sigma^*(\lambda))-\frac{1}{2}(\mu-\sigma^*(\mu))\in X^*(\mathbb{T}). \tag{A.19}$$

We will denote by $\phi_{\mu,\lambda}$ the L-homomorphism given by such a pair (μ,λ).

LEMMA A.3. *(i) Suppose we have pairs (μ,λ) and (μ',λ') satisfying the congruence (A.19). Then $\phi_{\mu,\lambda}=\phi_{\mu',\lambda'}$ if and only if $\mu=\mu'$ and $\lambda-\lambda'\in X^*(\mathbb{T})$.*

(ii) The L-homomorphism $\phi_{\mu,\lambda}$ and $\phi_{\mu',\lambda'}$ define the same Langlands parameter if and only if $\mu=\mu'$ and $\lambda-\lambda'\in X^*(\mathbb{T})+\{\nu-\sigma^*(\nu)|\nu\in\mathfrak{t}_{\mathbb{C}}^*\}$

The proof is straightforward. Notice that (A.19) is the same as (A.10) and that (ii) is the same as A.11 and thus that we have a bijection between Langlands parameters and characters of T: this is the Langlands correspondance for (real) tori.

THEOREM A.4. *(Langlands correspondence for real tori). The characters of T are in one-to-one correspondence with the $\hat{\mathbb{T}}$-conjugacy classes of L-morphisms $\phi : W_{\mathbb{R}} \to {}^L T$, the correspondence is given by*

$$\xi_{(\lambda,\mu)} \longleftrightarrow \phi_{\lambda,\mu}$$

where λ, μ are elements in $\mathfrak{t}_{\mathbb{C}}$ satisfying (A.10).

Let us remark the following equivalent definition of the set of Langlands parameters.

PROPOSITION A.5. *The set of Langlands parameters is in bijection with $H^1(W_{\mathbb{R}}, \hat{\mathbb{T}})$.*

This is immediate from the definitions.

Appendix B. Tate-Nakayama duality

We state some results as one can find them in chapter 9 of [**ABV**]. Let $\mathbb{T}$ be an algebraic torus over $\mathbb{C}$. There is a natural inclusion-reversing bijection between subgroups of $\mathbb{T}$ and sublattices of $X^*(\mathbb{T})$. To an algebraic subgroup $\mathbb{S}$ corresponds the sublattice L of characters which are trivial on $\mathbb{S}$. Dually, to a sublattice L of $X^*(\mathbb{T})$ corresponds the algebraic subgroup of elements annihilating the characters in L.

Suppose that $\mathbb{S}_1 \subset \mathbb{S}_2$ correspond to $L_2 \subset L_1$. Then the restriction of characters defines a natural isomorphism

$$\text{(B.1)} \qquad \mathrm{Hom}_{alg}(\mathbb{S}_1/\mathbb{S}_2, \mathbb{C}^*) \simeq L_2/L_1.$$

Suppose that $\mathbb{S}$ correspond to a sublattice L. Then $\mathbb{S}_0$, the identity connected component of $\mathbb{S}$, corresponds to the lattice

$$\text{(B.2)} \qquad L^0 = \{\tau \in X^*(\mathbb{T}) \text{ such that } n\tau \in L \text{ for some positive integer } n\}.$$

Suppose that θ is an algebraic automorphism of $\mathbb{T}$ and θ^* is the transpose automorphism of $X^*(\mathbb{T})$. Then $\mathbb{T}^\theta$ correspond to

$$\text{(B.3)} \qquad (1-\theta^*)X^*(\mathbb{T}) = \{\tau - \theta^*(\tau),\, \tau \in X^*(\mathbb{T})\}$$

Dually the sublattice of fixed points $X^*(\mathbb{T})^{\theta^*}$ corresponds to the (connected) subgroup

$$\text{(B.4)} \qquad (1-\theta)\mathbb{T} = \{t\theta(t)^{-1},\, t \in \mathbb{T}\} = (\mathbb{T}^{-\theta})_0$$

Suppose $\mathbb{T}$ is defined over $\mathbb{R}$, that is, that we are given an Galois action σ on $\mathbb{T}$. The notations are as in the previous section, and $\sigma_{\hat{\mathbb{T}}}$ will denote the *algebraic* action on $\hat{\mathbb{T}}$ inducing respectively σ^* and σ_* on $X^*(\mathbb{T})$ and $X_*(\mathbb{T})$. We consider the following component group:

$$\text{(B.5)} \qquad \pi_0(\hat{\mathbb{T}}^{\sigma_{\hat{\mathbb{T}}}}) = \hat{\mathbb{T}}^{\sigma_{\hat{\mathbb{T}}}}/(\hat{\mathbb{T}}^{\sigma_{\hat{\mathbb{T}}}})_0 = \hat{\mathbb{T}}^{\sigma_{\hat{\mathbb{T}}}}/(1+\sigma_{\hat{\mathbb{T}}})\hat{\mathbb{T}}.$$

From (B.3) and (B.4) we find that

$$\text{(B.6)} \qquad \mathrm{Hom}_{alg}(\pi_0(\hat{\mathbb{T}}^{\sigma_{\hat{\mathbb{T}}}}), \mathbb{C}^*) \simeq X_*(\mathbb{T})^{-\sigma_*}/(1-\sigma_*)X_*(\mathbb{T}).$$

Let us write $\mathbb{T}^{fin}$ for the subgroup of $\mathbb{T}$ consisting of elements of finite order. Define

$$\mathfrak{t}_{\mathbb{Q}} := X_*(\mathbb{T}) \otimes_{\mathbb{Z}} \mathbb{Q} \subset \mathfrak{t}_{\mathbb{C}}. \tag{B.7}$$

Then we have:

LEMMA B.1. *(*[**ABV**]*, lemma 9.9)*
(i) The normalized exponential mapping e, given by $e(\tau) = \exp(2i\pi\tau)$ defines an isomorphism

$$\mathfrak{t}_{\mathbb{C}}/X_*(\mathbb{T}) \simeq \mathbb{T}.$$

(ii) The preimage of $\mathbb{T}^{fin}$ under this isomorphism is precisely $\mathfrak{t}_{\mathbb{Q}}$, so we have a natural isomorphism:

$$\mathfrak{t}_{\mathbb{Q}}/X_*(\mathbb{T}) \simeq \mathbb{T}^{fin}.$$

Consider now the following subgroups:

$$\mathbb{T}^{-\sigma_T,fin} = \{t \in \mathbb{T} \text{ such that } t\sigma_T(t) \text{ has finite order }\}$$

$$\mathbb{T}^{-\sigma_T} = \{t \in \mathbb{T} \text{ such that } t\sigma_T(t) = 1\}$$

$$\mathbb{T}_0^{-\sigma_T} = \{s\sigma_T(s)^{-1},\ s \in \mathbb{T}\}$$

PROPOSITION B.2. *The mapping $\tau \mapsto e(\tau/2)$ maps the -1 eigenspace $\mathfrak{t}_{\mathbb{Q}}^{-\sigma_*}$ into $\mathbb{T}^{-\sigma_T,fin}$. The preimages of the subgroups $\mathbb{T}_0^{-\sigma_T} \subset \mathbb{T}^{-\sigma_T}$ are the lattices*

$$(1-\sigma_*)X_*(\mathbb{T}) \subset X_*(\mathbb{T})^{-\sigma_*}.$$

There is a natural isomorphism:

$$X_*(\mathbb{T})^{-\sigma_*}/(1-\sigma_*)X_*(\mathbb{T}) \simeq \mathbb{T}^{-\sigma_T}/\mathbb{T}_0^{-\sigma_T}$$

Notice that $\mathbb{T}^{-\sigma_T}/\mathbb{T}_0^{-\sigma_T}$ is nothing but $H^1(\Gamma,\mathbb{T})$. From (B.6) and (B.2) we get a perfect pairing

$$\pi_0(\hat{\mathbb{T}}^{\hat{\sigma}_{\mathbb{T}}}) \times H^1(\Gamma,\mathbb{T}) \to \mathbb{C}^* \tag{B.8}$$

which is the Tate-Nakayama pairing.

References

[ABV] J. Adams, D. Barbasch & D.Vogan, The Langlands classification and irreducible characters for real reductive groups, *Progress in Math.* **104**, Birkhauser, (1992).

[Art] J. Arthur, Unipotent automorphic representations: conjectures, p. 13-71 in *Orbites unipotentes et représentations II. Groupes p-adique et réels*, Asterisque 171–172 (1989).

[B1] A. Bouaziz, Intégrales orbitales sur les groupes de Lie reductifs réels *Ann. Scient. Ec. Norm. Sup.* t **27** (1994), p. 573–609.

[Bo] A. Borel, Automorphic L-functions, *Proc. Sympos. Pure Math.* **33**, part 2, Am. Math. Soc (1979), p. 27–61.

[G] S. Gelbart, An elementary introduction to the Langlands program, *Bull. Amer. Math. Soc. (N.S.)* **10** (1984), no. 2, p. 177–219.

[HC1] Harish-Chandra, Some results on invariant integral on a semi-simple Lie group, *Ann. of Math.* **80** (1964), p. 551–593.

[HC2] Harish-Chandra, Harmonic analysis on real reductive groups I, *Jour. of Func. An.* **19** (1975), p. 104–204.

[Kn] A.W. Knapp, Introduction to the Langlands program, Representation theory and automorphic forms (Edinburgh, 1996), p. 245–302, *Proc. Sympos. Pure Math.* **61**, Amer. Math. Soc., Providence, RI, 1997.

[K] R. Kottwitz, Rational conjugacy classes in reductive groups, *Duke Math. J.* **49** (1982), p. 785–806.

[KS] R. Kottwitz & D. Shelstad, Twisted Endoscopy I: Definitions, Norm Mappings and Transfer Factors, *Astérisque* **255** (1999).

[L1] R.P. Langlands, On the classification of representations of real algebraic groups, in Representation theory and Harmonic analysis on semi-simple Lie groups, *Math. Surveys and Monographs* **31**, p. 101–170. A.M.S, Providence, Rhode-Island (1989).

[LTF] R.P. Langlands, Les débuts d'une formule des traces stable, *Publications Mathématiques de l'Université Paris VII*, **13**, Université de Paris VII, U.E.R. de Mathématiques, Paris, 1983.

[LS] R.P Langlands & D. Shelstad, On the definition of transfer factors, *Math. Ann.* **278** (1987), p. 219–271.

[LS2] R.P Langlands & D. Shelstad, Descent for transfer factors, *The Grothendieck Festschrift*, Vol. II, p. 485–563, *Progr. Math.* **87**, Birkhauser Boston, Boston, MA, (1990).

[R] D. Renard, Twisted Endoscopy for Real Groups, *Journal de l'institut mathématique de Jussieu 2,* vol. 4 (2003), p. 529–566.

[Sh1] D. Shelstad, Characters and inner forms of a quasi-split group over $\mathbb{R}$, *Comp. Math*, vol 39, fasc 1 (1979), p. 11–45.

[Sh2] D. Shelstad, Orbitals integrals and a familly of groups attached to a real reductive group, *Ann. Scient. Ec. Norm. Sup. t* **12** (1994), p. 1–31.

[Sh3] D. Shelstad, Embeddings of L-groups, *Canadian J. of Math.* vol 33. (1981), p. 513–558.

[Sh4] D. Shelstad, L-indistinguishability for real groups, *Math. Ann.* **259** (1982), p.385–430.

[Sh5] D. Shelstad, Tempered endoscopy for real groups, I: geometric transfer with canonical factors, *Contemp. Math.* vol. 472 (2008), 215–248; II. Spectral transfer factors, to appear in Automorphic Forms and the Langlands Program, 243–282, Int. Press, Boston, MA; III. Inversion of transfer and L-packet structure, *Represent. Theory* **12** (2008), 369–402.

Subsection II.B

The fundamental lemma and transfer

Le transfert lisse des intégrales orbitales d'après Waldspurger
[C.II.B.1]

Pierre-Henri Chaudouard

Table des matières

1. Introduction

Cet article se veut une introduction au travail [**Wif**] de Waldspurger qui démontre l'existence du transfert modulo la validité du lemme fondamental sur les algèbres de Lie qui, aujourd'hui, est un théorème dû à Ngô (cf. [**Ngo2**], cas de la caractéristique positive) et à Waldspurger (cf. [**W-car**] pour le passage à la caractéristique nulle). En fait, l'existence du transfert est un problème local au voisinage d'éléments semi-simples. Des arguments de descente montre qu'il suffit de résoudre le même problème pour l'algèbre de Lie. L'approche de Waldspurger exploite un analogue sur les algèbres de Lie de la partie géométrique elliptique de la formule des traces d'Arthur-Selberg. On dispose d'un côté spectral donné par la transformée de Fourier du côté géométrique, les deux étant reliés par la formule sommatoire de Poisson. On peut alors appliquer les techniques de stabilisation, dues à Kottwitz et Langlands, pour comparer ces formules des traces sur les algèbres de Lie de différents groupes.

Décrivons plus en détail le contenu de cet article. Dans la section 2, le lecteur trouvera les notations utilisées. On y reprend les notions de cohomologie abélianisée pour lesquelles on renvoie à [**LabBreen**]. Les données endoscopiques sont introduites dans la section 3. On s'est livré à l'exercice oulipien cher à Labesse qui consiste à ne pas parler de dual de Langlands, d'où l'aspect inhabituel des définitions. Sauf erreur de ma part, leur formulation devrait être équivalente à la définition habituelle. Dans la section 4, on explique comment se transfèrent les classes de conjugaison entre un groupe et son groupe endoscopique. L'accent est mis sur la section de Kostant. Dans les sections 5 et 6, on procède à des rappels sur l'analyse harmonique p-adique sur les algèbres de Lie. Dans la section 7, on donne une définition des facteurs de transfert, qui s'inspire du résultat de Kottwitz dans [**KoTF**]. Les énoncés principaux de la théorie endoscopique sur les algèbres de Lie p-adiques se trouvent dans la section 8. On y trouve pêle-même l'énoncé de transfert lisse des intégrales orbitales, le fait que le transfert commute à la transformation de Fourier et le lemme fondamental. Le point des travaux de Waldspurger est de montrer que chaque énoncé est impliqué par le suivant. Dans la section 9, on revoit la théorie analogue sur les algèbres de Lie complexes, dont les résultats reposent sur ceux de Bouaziz. La préstabilisation dans notre contexte est revue à la section 10. Dans la section finale 11, on explique quelques points de la preuve de Waldspurger du fait suivant : la validité du lemme fondamental implique que le transfert commute à la transformation de Fourier. Par conséquent, *dixit* l'article éponyme, «le lemme fondamental implique le transfert».

Remerciements. — Je remercie le rapporteur de cet article pour ses commentaires et sa relecture qui ont épargné à ce texte quelques inexactitudes.

2. Notations

2.1. Soit F un corps de caractéristique 0. On suppose que F est soit un corps p-adique c'est-à-dire une extension finie de $\mathbb{Q}_p$ soit un corps de nombres. Soit $\bar{F}$ une clôture algébrique de F et $\Gamma = \mathrm{Gal}(\bar{F}/F)$ le groupe de Galois associé.

2.2. Soit G un groupe réductif connexe sur F. Soit $\mathfrak{g}$ son algèbre de Lie. Soit $G_{\mathrm{der}} \subset G$ son sous-groupe dérivé et G_{sc} le revêtement simplement connexe de G_{der}. Soit Z_G le centre de G et $G_{\mathrm{ad}} = G/Z_G$ le groupe adjoint. Pour tout sous-tore maximal T de G, soit $\mathrm{Norm}_G(T)$ le normalisateur de T dans G et $W^G(T) = \mathrm{Norm}_G(T)/T$ le groupe de Weyl. Soit $T_{G_{\mathrm{sc}}}$ l'image réciproque de T dans G_{sc} et $T_{G_{\mathrm{ad}}} = T/Z_G$. Lorsque le contexte est clair, on pose $T_{\mathrm{sc}} = T_{G_{\mathrm{sc}}}$

et $T_{\mathrm{ad}} = T_{G_{\mathrm{ad}}}$. On commettra parfois l'abus suivant : on notera simplement G au lieu de $G(\bar{F})$ le groupe des points de G à valeurs dans $\bar{F}$.

2.3. Le groupe G opère sur son algèbre de Lie par adjonction. Soit $X \in \mathfrak{g}(F)$. Par définition, X est régulier si son centralisateur G_X dans G est de dimension minimale, égale au rang de G. Si X est de plus semi-simple et régulier, son centralisateur G_X est un tore. On pose alors $T_X = G_X$: c'est un sous-tore maximal de G défini sur F.

Soit $\mathfrak{g}_{\mathrm{reg}}(F)$ l'ensemble des éléments semi-simple et réguliers de $\mathfrak{g}(F)$. On introduit l'ensemble des classes de conjugaison semi-simples régulières

$$\Gamma(\mathfrak{g}) = \mathfrak{g}_{\mathrm{reg}}(F)/G(F).$$

On introduit son analogue stable

$$\Sigma(\mathfrak{g}) = \mathfrak{g}_{\mathrm{reg}}(F)/G(\bar{F})$$

qui est le quotient de $\mathfrak{g}_{\mathrm{reg}}(F)$ pour la relation d'équivalence suivante : deux éléments X et Y de $\mathfrak{g}_{\mathrm{reg}}(F)$ définissent la même classe dans $\Sigma(\mathfrak{g})$ si et seulement s'il existe $g \in G$ tel que $X = \mathrm{Ad}(g)Y$. On dit encore que X et Y sont stablement conjugués.

2.4. Soit A_G le sous-tore déployé et central maximal de G. Pour tout tore T, le tore A_T est le sous-tore déployé maximal de T.

On note $\mathfrak{g}_{\mathrm{ell}}(F) \subset \mathfrak{g}_{\mathrm{reg}}(F)$ le sous-ensemble des éléments *elliptiques* : ce sont les éléments $X \in \mathfrak{g}_{\mathrm{reg}}(F)$ pour lesquels le centralisateur T est *elliptique* c'est-à-dire vérifie l'une des deux conditions équivalentes suivantes

– $A_T = A_G$;
– $X_*(T_{G_{\mathrm{sc}}})^\Gamma = 1$.

Cette notion d'ellipticité est stable par conjugaison et par conjugaison stable. On définit de manière évidente alors les ensembles $\Gamma_{\mathrm{ell}}(\mathfrak{g})$ et $\Sigma_{\mathrm{ell}}(\mathfrak{g})$.

2.5. Dans la suite, par abus, on ne distingue pas dans les notations une variété définie sur F, les $\overline{F}$-points de cette même variété ou encore la variété qui s'en déduit par extension des scalaires à $\overline{F}$. Cette dernière est munie d'une action de Γ.

Soit (G^*, φ) une forme intérieure quasi-déployée de G ce par quoi on entend un couple (G^*, φ) formé d'un groupe G^* réductif connexe et quasi-déployé sur F et d'un torseur intérieur φ c'est-à-dire d'un isomorphisme sur $\bar{F}$

$$\varphi : G \to G^* \tag{2.1}$$

tel que pour tout $\sigma \in \Gamma$ l'automorphisme de G^* défini par $\varphi\sigma(\varphi)^{-1}$ est intérieur. En particulier, il existe un cocycle u à valeurs dans G^*_{ad} tel que pour tout $\sigma \in \Gamma$ on ait

$$\varphi\sigma(\varphi)^{-1} = \mathrm{Ad}(u_\sigma).$$

La classe de u dans $H^1(F, G^*_{ad})$ ne dépend que de la classe de G^*-conjugaison de φ. Réciproquement, les éléments de $H^1(F, G^*_{ad})$ paramètrent, à F-isomorphisme près, les couples (G, φ) où G est un groupe réductif connexe sur F et φ est une classe de G^*-conjugaison de torseurs intérieurs $\varphi : G \to G^*$. Le cocycle trivial correspond au couple (G^*, Id_{G^*}).

2.6. Soit T un sous-tore maximal de G défini sur F. La suite exacte courte

$$0 \to T \to G \to G/T \to 0$$

induit une suite exacte longue en cohomologie

$$0 \to H^0(F,T) \to H^0(F,G) \to H^0(F,G/T) \to H^1(F,T) \to H^1(F,G)$$

Suivant Langlands, Kottwitz et Labesse, on pose

$$\begin{aligned}\mathcal{D}(T,G,F) &= \ker(H^1(F,T) \to H^1(F,G))\\ &= \mathrm{coker}(H^0(F,G) \to H^0(F,G/T)),\end{aligned}$$

On prendra garde, dans ce contexte non abélien, à la signification des notations ker et coker : l'ensemble $H^1(F,G)$ n'est pas un groupe mais il est pointé par la classe du cocycle trivial et le noyau ker désigne l'image inverse de cette classe triviale. De même, $H^0(F, G/T)$ est simplement un ensemble muni d'une action du groupe $G(F) = H^0(F,G)$ et coker désigne l'ensemble des orbites.

Soit $X \in \mathfrak{g}(F)$ semi-simple et régulier. Alors l'ensemble $\mathcal{D}(T_X, G, F)$ est en bijection avec l'ensemble des classes de conjugaison dans la classe de conjugaison stable de X. Rappelons que la bijection s'obtient de la manière suivante : soit $Y \in \mathfrak{g}(F)$ stablement conjugué à X. Il existe $g \in G(\bar{F})$ tel que $Y = \mathrm{Ad}(g)X$. Pour tout $\sigma \in \Gamma$, on vérifie que $g^{-1}\sigma(g)$ centralise X et donc appartient à T_X. On en déduit un élément de $\mathcal{D}(T_X, G, F)$.

La suite exacte longue ci-dessous est une version «abélianisée» de la suite exacte longue précédente

$$H^0(F,T) \to H^0(F, T_{\mathrm{sc}} \to T) \to H^1(F, T_{\mathrm{sc}}) \to H^1(F,T) \to H^1(F, T_{\mathrm{sc}} \to T).$$

D'où la définition

$$\begin{aligned}\mathcal{E}(T,G,F) &= \ker(H^1(F,T) \to H^1(F, T_{\mathrm{sc}} \to T))\\ &= \mathrm{coker}(H^0(F, T_{\mathrm{sc}} \to T) \to H^1(F, T_{\mathrm{sc}})),\end{aligned}$$

Rappelons que Labesse a défini pour le module croisé $[G_{\mathrm{sc}} \to G]$ des ensembles de cohomologie $H^i(F, G_{\mathrm{sc}} \to G)$ qui sont isomorphes à $H^i(F, T_{\mathrm{sc}} \to T)$. On dispose en degrés $i = 0, 1$ d'applications d'abélianisation

$$H^i(F,G) \to H^i(F, G_{\mathrm{sc}} \to G)$$

et donc d'une application

$$\mathcal{D}(T,G,F) \hookrightarrow \mathcal{E}(T,G,F)$$

qui est évidemment une injection. C'est de plus une bijection si F est local non-archimédien. Dans ce cas, en effet, l'application d'abélianisation

$$H^1(F,G) \to H^1(F, G_{\mathrm{sc}} \to G)$$

est injective : en général son noyau est l'image de $H^1(F, G_{\mathrm{sc}}) \to H^1(F,G)$ mais, lorsque F est local non-archimédien, il est bien connu qu'on a

$$H^1(F, G_{\mathrm{sc}}) = 1.$$

Si F est un corps de nombres, on note $\mathbb{A}_F$ l'anneau des adèles de F et on pose

$$\mathcal{E}(T, G, \mathbb{A}_F/F) = \mathrm{coker}(H^0(\mathbb{A}_F, T_{\mathrm{sc}} \to T) \to H^1(\mathbb{A}_F/F, T_{\mathrm{sc}})).$$

Soit C_F le corps F si F est local ou la F-algèbre des classes d'adèles $\mathbb{A}_F/F$ si F est global. Dans les deux cas on définit *le groupe des caractères endoscopiques*

$$\mathcal{K}(T, G, F) = \mathcal{E}(T, G, C_F)^D$$

où l'exposant D désigne la dualité de Pontryagin.

3. Données endoscopiques

3.1. La définition. — Soit G comme dans la section précédente. Une *donnée endoscopique*[1] de G est la donnée d'un triplet (H, κ, ξ) où

- H est un groupe réductif connexe quasi-déployé sur F ;
- ξ est la donnée d'un F-isomorphisme

$$\xi : T_H \to T_G \tag{3.1}$$

  entre des sous-tores maximaux définis sur F notés T_H et T_G respectivement de H et G. On suppose de plus que l'isomorphisme qui s'en déduit entre groupes de cocaractères

$$X_*(T_H) \to X_*(T_G)$$

  envoie le réseau engendré par les coracines de (H, T_H) dans celui de (G, T_G) autrement dit ξ induit une injection

$$X_*(T_{H_{\mathrm{sc}}}) \to X_*(T_{G_{\mathrm{sc}}}) \ ;$$

- κ est un caractère du groupe $H_0(F, X_*(T_{H_{\mathrm{sc}}})\backslash X_*(T_{G_{\mathrm{sc}}}))$ (on identifie $X_*(T_{H_{\mathrm{sc}}})$ à son image dans $X_*(T_{G_{\mathrm{sc}}})$).

Ces objets doivent satisfaire en outre *deux conditions* que nous allons expliciter.

Soit R_G et $R_G^\vee$ respectivement les ensembles des racines et des coracines de T dans G. On pose

$$R_\kappa^\vee = \{\alpha^\vee \in R_G^\vee \mid \langle \kappa, \alpha^\vee \rangle = 1\}$$

[1]Comme le souligne le rapporteur de l'article, dans la définition classique, toutes les formes intérieures de G se partagent les mêmes données endoscopiques. On se limite en fait ici aux données endoscopiques de G qui possèdent un tore maximal en commun avec G.

et

$$R_\kappa = \{\alpha \in R_G \mid \alpha^\vee \in R_\kappa^\vee\},$$

où $\alpha^\vee$ est la coracine de α.

La première condition s'exprime alors de la manière suivante. L'isomorphisme ξ de (3.1) induit un F-isomorphisme de données radicielles

$$(X^*(T_H), R_H, X_*(T_H), R_H^\vee) \simeq (X^*(T_G), R_\kappa, X_*(T_G), R_\kappa^\vee). \tag{3.2}$$

Pour exprimer la seconde, fixons E une extension galoisienne de F qui déploie les tores $T_{G_{\mathrm{sc}}}$ et $T_{H_{\mathrm{sc}}}$. On peut alors voir κ comme un caractère du groupe $H_0(E/F, X_*(T_{H_{\mathrm{sc}}})\backslash X_*(T_{G_{\mathrm{sc}}}))$. En utilisant le composé des morphismes évidents

$$\begin{aligned}\hat{H}^{-1}(E/F, X_*(T_{G_{\mathrm{sc}}})) \to \hat{H}^{-1}(E/F, X_*(T_{H_{\mathrm{sc}}})\backslash X_*(T_{G_{\mathrm{sc}}}))\\ \to H_0(E/F, X_*(T_{H_{\mathrm{sc}}})\backslash X_*(T_{G_{\mathrm{sc}}})),\end{aligned}$$

on voit que κ induit un caractère de $\hat{H}^{-1}(E/F, X_*(T_{G_{\mathrm{sc}}}))$ et donc, via l'isomorphisme de Tate-Nakayama

$$H^1(C_F, T_{G_{\mathrm{sc}}}) \simeq \hat{H}^{-1}(E/F, X_*(T_{G_{\mathrm{sc}}})),$$

un caractère κ_0 de $H^1(C_F, T_{G_{\mathrm{sc}}})$. La seconde condition porte sur ce caractère κ_0. On demande que κ_0 appartienne à $\mathcal{K}(T_G, G, F)$ c'est-à-dire que κ_0 soit trivial sur l'image de $H^0(\mathbb{A}_F, T_{G_{\mathrm{sc}}} \to T_G)$ si F est global, resp. de $H^0(F, T_{G_{\mathrm{sc}}} \to T_G)$ si F est local.

Remarque 3.1. — On peut utiliser la dualité de Tate-Nakayama pour un complexe de tores de longueur 1

$$H^1(C_F, T_{H_{\mathrm{sc}}} \to T_{G_{\mathrm{sc}}}) \simeq \hat{H}^{-1}(E/F, X_*(T_{H_{\mathrm{sc}}})\backslash X_*(T_{G_{\mathrm{sc}}})),$$

pour déduire de κ un caractère de $H^1(C_F, T_{H_{\mathrm{sc}}} \to T_{G_{\mathrm{sc}}})$. Ce complexe de tores apparaît dans le triangle distingué

$$\ldots \to [T_{H_{\mathrm{sc}}} \to T] \to [T_{G_{\mathrm{sc}}} \to T] \to [T_{H_{\mathrm{sc}}} \to T_{G_{\mathrm{sc}}}][1] \to \ldots.$$

On a donc une application

$$H^0(C_F, T_{G_{\mathrm{sc}}} \to T) \to H^1(C_F, T_{H_{\mathrm{sc}}} \to T_{G_{\mathrm{sc}}})$$

et la précédente condition sur κ_0 est que le caractère de $H^1(C_F, T_{H_{\mathrm{sc}}} \to T_{G_{\mathrm{sc}}})$ ainsi obtenu soit trivial sur l'image de $H^0(\mathbb{A}_F, T_{G_{\mathrm{sc}}} \to T)$ dans le cas global, resp. de $H^0(F, T_{G_{\mathrm{sc}}} \to T)$ dans le cas local.

3.2. Équivalence. — Soit (H, κ, ξ) et (H', κ', ξ') deux données endoscopiques de G.

Par définition, ces données sont équivalentes s'il existe un F-isomorphisme

$$\alpha \, : H' \to H$$

et des éléments $h \in H_{\rm sc}$ et $g \in G_{\rm sc}$ qui vérifient les deux propriétés suivantes.

(1) On a

$$(\mathrm{Ad}(h) \circ \alpha)(T_{H'}) = T_H$$

et

$$\mathrm{Ad}(g)(T'_G) = T_G$$

ainsi qu'un diagramme commutatif

$$\begin{array}{ccc} T_{H'} & \xrightarrow{\ \xi'\ } & T'_G \\ {\scriptstyle \mathrm{Ad}(h)\circ\alpha}\big\downarrow & & \big\downarrow{\scriptstyle \mathrm{Ad}(g)} \\ T_H & \xrightarrow{\ \xi\ } & T_G \end{array}$$

Avant d'énoncer la seconde propriété, observons que la commutativité du diagramme ci-dessus implique que $\mathrm{Ad}(g)$ induit un isomorphisme

$$X_*(T_{H'_{\rm sc}})\backslash X_*(T'_{G_{\rm sc}}) \to X_*(T_{H_{\rm sc}})\backslash X_*(T_{G_{\rm sc}}). \tag{3.3}$$

qui est en fait Γ-équivariant. En effet, comme ξ et ξ' sont définis sur F, il revient au même de montrer que l'isomorphisme

$$X_*(T_{H'_{\rm sc}})\backslash X_*(T_{H'}) \to X_*(T_{H_{\rm sc}})\backslash X_*(T_H)$$

induit par $\mathrm{Ad}(h) \circ \alpha$ est Γ-équivariant. Or ce dernier vérifie

$$\mathrm{Ad}(h) \circ \alpha \circ \sigma(\mathrm{Ad}(h) \circ \alpha)^{-1} = \mathrm{Ad}(h\sigma(h)^{-1})$$

et $h\sigma(h)^{-1}$ appartient au normalisateur de T_H dans H. Il suffit donc de vérifier que tout élément du groupe de Weyl $W^H(T_H)$ agit trivialement sur le quotient $X_*(T_{H_{\rm sc}})\backslash X_*(T_H)$. Il suffit encore de le vérifier sur la partie génératrice des réflexions simples s_α où α parcourt l'ensemble des racines de T_H dans H. Or cela est évident sur la formule suivante valable pour tout $\chi \in X_*(T_H)$

$$s_\alpha(\chi) - \chi = \alpha(\chi)\alpha^\vee$$

puisque $\alpha^\vee \in X_*(T_{H_{\rm sc}})$.

(2) Identifions κ' à un caractère de $H_0(F, X_*(T_{H_{\rm sc}})\backslash X_*(T_{G_{\rm sc}}))$ via l'isomorphisme

$$H_0(F, X_*(T_{H'_{\rm sc}})\backslash X_*(T'_{G_{\rm sc}})) \to H_0(F, X_*(T_{H_{\rm sc}})\backslash X_*(T_{G_{\rm sc}}))$$

induit par (3.3). La seconde propriété stipule que les caractères κ et κ' coïncident sur le sous-groupe de torsion de $H_0(F, X_*(T_{H_{\rm sc}})\backslash X_*(T_{G_{\rm sc}}))$.

Remarque 3.2. — D'après la remarque 3.1, κ et κ' définissent chacun un caractère de $H^1(C_F, T_{H_{\rm sc}} \to T_{G_{\rm sc}})$. La seconde condition ci-dessous n'est autre que l'égalité de ces caractères.

3.3. Ellipticité. — Une donnée endoscopique (H, κ, ξ) de G est *elliptique* si $\xi(A_H) = A_G$ (l'inclusion $\supset$ est toujours vraie).

4. Transfert de classes de conjugaison.

4.1. Soit $\bar{F}[\mathfrak{g}]$ la $\bar{F}$-algèbre des fonctions polynomiales régulières sur $\mathfrak{g}$. Elle est munie d'une action évidente de Γ. Soit $\bar{F}[\mathfrak{g}]^G$ la sous-algèbre des fonctions G-invariantes : elle est stable par Γ et c'est une algèbre de polynômes sur $\bar{F}$ en rang(G) variables. Le schéma Spec$(\bar{F}[\mathfrak{g}]^G)$ est un espace affine de dimension rang(G) muni d'une action de Γ. Il se descend donc en un F-schéma affine $\mathfrak{car}_G$. Soit

$$\chi = \chi_G \ : \ \mathfrak{g} \to \mathfrak{car} = \mathfrak{car}_G$$

le F-morphisme dual de l'inclusion $\bar{F}[\mathfrak{g}]^G \subset \bar{F}[\mathfrak{g}]$. Lorsque $X \in \mathfrak{g}$ est semi-simple et régulier, la fibre $\chi^{-1}(\chi(X))$ est la classe de G-conjugaison de X. Lorsque $X \in \mathfrak{g}_{\text{reg}}(F)$, les points sur F de la fibre $\chi^{-1}(\chi(X))$ forment la classe de conjugaison stable de X.

Notons que χ induit des applications encore notées χ de $\Gamma(\mathfrak{g})$ et $\Sigma(\mathfrak{g})$ dans $\mathfrak{car}_G$.

4.2. Soit T un sous-tore maximal de G et $W = W^G(T)$. Alors le morphisme de restriction $\bar{F}[\mathfrak{g}] \to \bar{F}[\mathfrak{t}]$ induit un isomorphisme (dû à Chevalley)

$$\bar{F}[\mathfrak{g}]^G \to \bar{F}[\mathfrak{t}]^W$$

sur la $\bar{F}$-algèbre des polynômes sur $\mathfrak{t}$ invariants par W. Notons que si T est défini sur F, cet isomorphisme est Γ-équivariant.

4.3. Transfert de classes de conjugaison. — Soit (H, κ, ξ) une donnée endoscopique de G. Soit T_H et T_G les sous-F-tores maximaux respectivement de H et G et

$$\xi \ : \ T_H \to T_G$$

le F-isomorphisme de tores maximaux sous-jacent à cette donnée. Notons qu'il résulte de la définition d'une donnée endoscopique que $W^H = W^H(T_H)$ est naturellement un sous-groupe de $W^G = W^G(T_G)$ et que ξ est équivariant sous W^H. En particulier, ξ induit un morphisme Γ-équivariant

$$\bar{F}[\mathfrak{t}_G]^{W^G} \to \bar{F}[\mathfrak{t}_H]^{W^H}.$$

En combinant ce dernier avec l'isomorphisme de Chevalley, on obtient un F-morphisme de schémas affines :

$$\mathfrak{car}_H \to \mathfrak{car}_G \,. \tag{4.1}$$

Soit $Y \in \mathfrak{h}$ et $X \in \mathfrak{g}$. On dit que Y est une image de X et on note $Y \to X$ si $\chi_H(Y)$ a pour image $\chi_G(X)$ par le morphisme ci-dessus. On dit que $Y \in \mathfrak{h}$ est G-régulier s'il existe $X \in \mathfrak{g}$ semi-simple G-régulier tel que $Y \to X$.

Définition 4.1. — Soit $Y \in \mathfrak{h}(F)$ semi-simple G-régulier . On dit que Y se transfère à $\mathfrak{g}$ s'il existe $X \in \mathfrak{g}_{\text{reg}}(F)$ tel que $Y \to X$.

Si $Y \to X$, il en est de même pour tous $Y' \in \mathfrak{h}(F)$ et $X' \in \mathfrak{g}(F)$ tels que Y' est H-stablement conjugué à Y et X' est G-stablement conjugué à X ; réciproquement tout $X' \in \mathfrak{g}(F)$ tel que $Y \to X'$ est G-stablement conjugué à X.

Soit $\Sigma_G(\mathfrak{h}) \subset \Sigma(\mathfrak{h})$ le sous-ensemble des classes G-régulières. On a donc défini une fonction (de domaine les classes des images de $\mathfrak{g}_{\mathrm{reg}}(F)$)

$$\Sigma_G(\mathfrak{h}) \to \Sigma(\mathfrak{g}).$$

On vérifie qu'elle est à fibres finies.

4.4. Soit $Y \in \mathfrak{h}(F)$ G-régulier et $X \in \mathfrak{g}_{\mathrm{reg}}(F)$. On peut montrer qu'on a $Y \to X$ si et seulement s'il existe $h \in H_{\mathrm{sc}}$ et $g \in G_{\mathrm{sc}}$ tels que

- $\mathrm{Ad}(h^{-1})T_Y = T_H$;
- $\mathrm{Ad}(g)T_G = T_X$
- le morphisme

$$\xi' : T_Y \to T_X,$$

défini comme le composé

$$T_Y \overset{\mathrm{Ad}(h^{-1})}{\longrightarrow} T_H \overset{\xi}{\longrightarrow} T_G \overset{\mathrm{Ad}(g)}{\longrightarrow} T_X \tag{4.2}$$

est un F-isomorphisme qui envoie Y sur X.

On a alors un diagramme commutatif

$$\begin{array}{ccc} T_H & \xrightarrow{\ \xi\ } & T_G \\ {\scriptstyle \mathrm{Ad}(h)}\big\downarrow & & \big\downarrow{\scriptstyle \mathrm{Ad}(g)} \\ T_Y & \xrightarrow{\ \xi'\ } & T_X \end{array}$$

On a vu (cf. ce qui suit la ligne (3.3)) que cela implique que $\mathrm{Ad}(g^{-1})$ induit un isomorphisme Γ-équivariant

$$X_*(T_{Y,H_{\mathrm{sc}}})\backslash X_*(T_{X,G_{\mathrm{sc}}}) \to X_*(T_{H_{\mathrm{sc}}})\backslash X_*(T_{G_{\mathrm{sc}}}) \tag{4.3}$$

Via cet isomorphisme, κ définit un caractère noté κ' de $H_0(F, X_*(T_{Y,H_{\mathrm{sc}}})\backslash X_*(T_{X,G_{\mathrm{sc}}}))$. On vérifie que κ' est endoscopique i.e. définit sur $H^1(C_F, T_{X,G_{\mathrm{sc}}})$ un caractère noté κ_Y qui appartient à $\mathcal{K}(T_X, G, F)$. On dispose donc d'une donnée endoscopique (H, κ', ξ'). L'identité de H se complète en une équivalence entre les données (H, κ, ξ) et (H, κ', ξ').

4.5. Pour deux données endoscopiques (H, κ, ξ) et (H, κ', ξ') équivalentes, on obtient le même morphisme $\mathfrak{car}_H \to \mathfrak{car}_G$, cf. I.(4.1) et la même fonction $\Sigma_G(\mathfrak{h}) \to \Sigma(\mathfrak{g})$.

4.6. Dans ce paragraphe, on suppose que $G = G^*$ c'est-à-dire que G est quasi-déployé. Soit B un sous-groupe de Borel de G défini sur F. Soit T un sous-tore de B défini sur F. Pour toute racine α de T dans G, le sous-espace propre

$$\mathfrak{g}_\alpha = \{X \in \mathfrak{g} \mid \forall t \in T \ \ \mathrm{Ad}(t)X = \alpha(t)X\}$$

est de dimension 1 sur $\overline{F}$. Soit Δ l'ensemble des racines simples de T dans B. Pour tout $\alpha \in \Delta$, soit $X_\alpha \in \mathfrak{g}_\alpha$ un élément non nul tel que $\sigma(X_\alpha) = X_{\sigma(\alpha)}$ pour tout $\sigma \in \Gamma$. Le triplet $(B, T, \{X_\alpha\})$ est un épinglage Γ-stable.

Pour tout $\alpha \in \Delta$, soit $X_{-\alpha}$ l'unique élément de $\mathfrak{g}_{-\alpha}$ défini par la condition $[X_\alpha, X_{-\alpha}]$ est la coracine $\alpha^\vee$ vue comme élément de $\mathfrak{t}$. Posons

$$X_+ = \sum_{\alpha \in \Delta} X_\alpha$$

et

$$X_- = \sum_{\alpha \in \Delta} X_{-\alpha}.$$

Soit $\mathfrak{g}_{X_+}$ le centralisateur de X_+ dans $\mathfrak{g}$. Énonçons le théorème suivant dû à Kostant.

Théorème 4.2. — *La restriction à $X_- + \mathfrak{g}_{X_+}$ du morphisme χ induit un F-isomorphisme de $X_- + \mathfrak{g}_{X_+}$ sur $\mathfrak{car}$. En particulier, on a une bijection de $X_- + \mathfrak{g}_{X_+}(F)$ sur $\mathfrak{car}(F)$.*

On en déduit les corollaires suivants.

Corollaire 4.3. — *Toute classe stable dans $\Sigma(\mathfrak{g})$ contient un et un seul élément dans $X_- + \mathfrak{g}_{X_+}(F)$.*

Remarque 4.4. — La classe de $G(F)$-conjugaison de cet élément dépend du choix de l'épinglage. Cependant comme les différents épinglages sont conjugués sous $G_{\mathrm{ad}}(F)$, la classe de $G_{\mathrm{ad}}(F)$-conjugaison de cet élément est uniquement déterminée.

Corollaire 4.5. — *Soit (H, κ, ξ) une donnée endoscopique d'un groupe quasi-déployé G. Tout élément G-régulier de $\mathfrak{h}(F)$ se transfère à $\mathfrak{g}$.*

4.7. Nous ne faisons plus l'hypothèse que G est quasi-déployé. Le $\overline{F}$-isomorphisme φ (cf. l. (2.1)) induit un F-isomorphisme

$$\mathfrak{car}_G \simeq \mathfrak{car}_{G^*}.$$

Le corollaire 4.5 nous conduit à étudier les conditions pour qu'un élément $X^* \in \mathfrak{g}^*_{\mathrm{reg}}(F)$ se transfère à $\mathfrak{g}$ c'est-à-dire pour que la fibre $\chi_G^{-1}(\chi_{G^*}(X^*))$ ait des éléments rationnels.

Lemme 4.6. — *Soit $X^* \in \mathfrak{g}^*_{\mathrm{reg}}(F)$. Les conditions suivantes sont équivalentes.*

(1) *X^* se transfère à $\mathfrak{g}$;*

(2) *Il existe $g \in G^*_{\mathrm{sc}}$ et $X \in \mathfrak{g}_{\mathrm{reg}}(F)$ tel que*

$$\mathrm{Ad}(g) \circ \varphi(X) = X^* ;$$

(3) *la classe de φ dans $H^1(F, G^*_{\mathrm{ad}})$ appartient à l'image de*

$$H^1(F, T_{X^*,\mathrm{ad}}) \longrightarrow H^1(F, G^*_{\mathrm{ad}}).$$

Si F est local non-archimédien, tous les éléments elliptiques de $\mathfrak{g}^*(F)$ se transfèrent à $\mathfrak{g}$.

Démonstration. — Clairement les assertions 1 et 2 sont équivalentes et 2 implique 3. On se contente donc de prouver que 3 implique 2. Pour alléger les notations, on pose $T = T_{X^*}$. Soit u_σ le cocycle à valeurs dans G^*_{ad} défini par $\varphi\sigma(\varphi)^{-1}$. L'assertion 3 affirme que la classe de u appartient à l'image de

$$H^1(F, T_{\mathrm{ad}}) \longrightarrow H^1(F, G^*_{\mathrm{ad}})$$

c'est-à-dire qu'il existe $g \in G^*_{\mathrm{sc}}$ tel que $gu_\sigma\sigma(g)^{-1}$ appartient à T/Z_{G^*}. Soit $X \in \mathfrak{g}$ tel que $\mathrm{Ad}(g)\circ\varphi(X) = X^*$. On vérifie que $X \in \mathfrak{g}_{\mathrm{reg}}(F)$ d'où l'assertion 2.

Prouvons enfin la dernière affirmation. On suppose F local non-archimédien, X^* elliptique et on note encore $T = T_{X^*}$. Par l'assertion 3, il s'agit de prouver que la flèche

$$H^1(F, T_{\mathrm{ad}}) \longrightarrow H^1(F, G^*_{\mathrm{ad}})$$

est surjective. La suite exacte

$$H^1(F, T_{\mathrm{ad}}) \to H^1(F, T_{\mathrm{sc}} \to T_{\mathrm{ad}}) \to H^2(F, T_{\mathrm{sc}})$$

et l'annulation $H^2(F, T_{\mathrm{sc}}) = 1$ (puisque X^* est elliptique, T est elliptique et l'annulation résulte de l'isomorphisme de Tate-Nakayama) impliquent que la flèche $H^1(F, T_{\mathrm{ad}}) \to H^1(F, T_{\mathrm{sc}} \to T_{\mathrm{ad}})$ est surjective. La surjectivité cherchée résulte alors du diagramme commutatif ci-dessus

$$\begin{array}{ccc} H^1(F, T_{\mathrm{ad}}) & \longrightarrow & H^1(F, G_{\mathrm{ad}}) \\ \downarrow & & \downarrow \\ H^1(F, T_{\mathrm{sc}} \to T_{\mathrm{ad}}) & \longrightarrow & H^1(F, G_{\mathrm{sc}} \to G_{\mathrm{ad}}) \end{array}$$

dont la flèche horizontale du bas est un isomorphisme, la flèche verticale gauche est surjective (comme on vient de le voir) et la flèche verticale droite est injective (c'est une flèche d'abélianisation de H^1 qui, pour un corps local non-archimédien, est injective puisque, pour un tel corps, $H^1(F, G_{\mathrm{sc}}) = 1$). □

4.8. Le lemme suivant est une conséquence évidente du corollaire 4.5 et du lemme 4.6 ci-dessus.

Lemme 4.7. — *Soit F un corps local non-archimédien. Supposons que la donnée endoscopique (H, κ, ξ) soit elliptique. Alors tous les éléments elliptiques et G-réguliers de $\mathfrak{h}(F)$ se transfèrent à G.*

5. Analyse harmonique sur les algèbres de Lie p-adique

5.1. Dans toute cette section, F désigne un corps p-adique. Dans cette section, on rappelle quelques résultats d'analyse harmonique sur $\mathfrak{g}(F)$ qui nous seront utiles pour la suite. Une excellente référence est [**HC-Debacker**].

5.2. Soit $C_c^\infty(\mathfrak{g}(F))$ l'espace vectoriel des fonctions complexes sur $\mathfrak{g}(F)$ lisses (c'est-à-dire localement constantes) et à support compact. Soit $\mathcal{O}_F$ l'anneau des entiers de F. Par réseau de $\mathfrak{g}(F)$, on entend un sous-$\mathcal{O}_F$-module de $\mathfrak{g}(F)$ qui est à la fois ouvert et compact. Soit R un réseau de $\mathfrak{g}(F)$. On note $C_c^\infty(\mathfrak{g}(F)/R)$ le sous-espace de $C_c^\infty(\mathfrak{g}(F))$ formé des fonctions invariantes par R.

5.3. Par distribution (sur $\mathfrak{g}(F)$), on entend une forme linéaire sur $C_c^\infty(\mathfrak{g}(F))$. Si D est une distribution, on note $\mathrm{supp}(D)$ son support. L'action de $G(F)$ sur $\mathfrak{g}(F)$ définit par dualité une action sur $C_c^\infty(\mathfrak{g}(F))$ et sur l'espace des distributions. Une distribution est invariante (par $G(F)$) si elle est fixe sous cette action. Soit J l'espace des distributions invariantes.

Pour tous sous-ensembles, $\omega \subset \mathfrak{g}(F)$ et $H \subset G(F)$, on pose

$$\omega^H = \{\mathrm{Ad}(x)X \mid x \in H \text{ et } X \in \omega\}.$$

Soit $\omega \subset \mathfrak{g}(F)$ un sous-ensemble. On note $J(\omega)$ l'ensemble des distributions invariantes T dont le support est inclus dans l'adhérence de $\omega^{G(F)}$.

Voici un premier résultat fondamental de finitude dû à Howe et Harish-Chandra. (c'est le théorème 12.1 de [**HC-Debacker**]).

Théorème 5.1. — *Soit $\omega \subset \mathfrak{g}(F)$ un sous-ensemble* compact *et R un réseau de $\mathfrak{g}(F)$. Soit $J_R(\omega)$ l'espace engendré par les restrictions à $C_c^\infty(\mathfrak{g}(F)/R)$ des éléments de $J(\omega)$. Alors*

$$\dim_{\mathbb{C}}(J_R(\omega)) < \infty.$$

5.4. Soit

$$\langle \cdot, \cdot \rangle \;:\; \mathfrak{g}(F) \times \mathfrak{g}(F) \to F$$

une forme bilinéaire symétrique non dégénérée et invariante par l'action de $G(F)$. Soit

$$\psi \;:\; F \to \mathbb{C}^\times$$

un caractère continu et non trivial. Tout sous-espace de $\mathfrak{g}(F)$ pour lequel la restriction de cette forme est non dégénérée est muni de la mesure de Haar auto-duale pour le bicaractère $\psi(\langle \cdot, \cdot \rangle)$.

On définit une transformation de Fourier qui est une bijection linéaire de $C_c^\infty(\mathfrak{g}(F))$ dans lui-même par

$$\hat{f}(X) = \int_{\mathfrak{g}(F)} f(Y)\psi(\langle X, Y \rangle)\, dY.$$

Par dualité, on définit une transformée de Fourier sur l'espace des distributions. L'action de $G(F)$ commute à la transformation de Fourier. En particulier, cette dernière préserve J.

5.5. On peut alors énoncer le théorème de régularité de Harish-Chandra (cf. théorème 4.4 de [**HC-Debacker**]).

Théorème 5.2. — *Soit ω un compact de $\mathfrak{g}(F)$ et $D \in J(\omega)$. Il existe une fonction $\hat{d}$ localement constante sur $\mathfrak{g}_{\mathrm{reg}}(F)$ et localement sommable sur $\mathfrak{g}(F)$ telle que*

$$\hat{D}(f) = \int_{\mathfrak{g}(F)} \hat{d}(X) f(X)\, dX.$$

6. Intégrales orbitales : cas d'un corps p-adique

6.1. On continue avec les notations de la section précédente.

6.2. On fixe une mesure de Haar sur $G(F)$. Soit $\mathcal{T}^G$ l'ensemble des classes de $G(F)$-conjugaison des sous-F-tores maximaux de G. On identifie dans la suite cet ensemble avec un système de représentants. On munit les tores dans $\mathcal{T}^G$ de mesures de Haar. Tout sous-F-tore maximal de G est conjugué à un tore dans $\mathcal{T}^G$: il est alors muni de la mesure de Haar obtenue par transport.

6.3. Soit $X \in \mathfrak{g}_{\mathrm{reg}}(F)$. L'intégrale orbitale de X est la distribution invariante I_X définie par

$$I_X(f) = |D^{\mathfrak{g}}(X)|_F^{\frac{1}{2}} \int_{T_X(F)\backslash G(F)} f(\mathrm{Ad}(x^{-1})X)\, dx$$

où

- dx est la mesure de Haar quotient ;
- $D^{\mathfrak{g}}(X) = \det(\mathrm{ad}(x)_{|\mathfrak{g}/\mathfrak{t}_X})$ est le discriminant de Weyl.

Notons que l'intégrale est convergente. Cela résulte du fait que l'application

$$x \in T_X(F)\backslash G(F) \mapsto f(\mathrm{Ad}(x^{-1})X)$$

est à support compact : en effet, $x \mapsto \mathrm{Ad}(x^{-1})X$ induit un homéomorphisme de $T_X(F)\backslash G(F)$ sur l'orbite de X et celle-ci est fermée dans $\mathfrak{g}(F)$ puisque X est semi-simple.

L'importance des intégrales orbitales (semi-simples régulières) dans l'analyse harmonique provient du théorème suivant de Harish-Chandra : l'ensemble des intégrales orbitales (semi-simples régulières) est faiblement dense dans J c'est-à-dire si $I_X(f) = 0$ pour tout $X \in \mathfrak{g}_{\mathrm{reg}}(F)$ alors $T(f) = 0$ pour tout $T \in J$ (cf. théorème 3.1 de [**HC-Debacker**]).

6.4. Un problème fondamental est de caractériser les fonctions

$$X \in \mathfrak{g}_{\mathrm{reg}}(F) \mapsto I_X(f)$$

obtenues avec $f \in C_c^\infty(\mathfrak{g}(F))$. C'est en général un problème très difficile. Lorsqu'on se restreint aux $f \in C_c^\infty(\mathfrak{g}_{\mathrm{reg}}(F))$ c'est-à-dire aux fonctions lisses qui ont un support compact inclus dans $\mathfrak{g}_{\mathrm{reg}}(F)$, ce problème est facile.

Lemme 6.1. — *Soit I une application de $\mathfrak{g}_{\mathrm{reg}}(F)$ dans $\mathbb{C}$. On a les équivalences suivantes :*

(1) *I est localement constante, invariante sous l'action de $G(F)$ et à support inclus dans un ensemble $\omega^{G(F)}$ tel que ω soit compact ;*

(2) *il existe $f \in C_c^\infty(\mathfrak{g}_{\mathrm{reg}}(F))$ tel que*

$$\forall X \in \mathfrak{g}_{\mathrm{reg}}(F) \quad I(X) = I_X(f).$$

Démonstration. — Soit $T \in \mathcal{T}^G$ et

$$\begin{array}{rcccc} \Phi_T & : & T(F)\backslash G(F) \times \mathfrak{t}_{\mathrm{reg}}(F) & \to & \mathfrak{g}_{\mathrm{reg}}(F) \\ & & (g,X) & \mapsto & \mathrm{Ad}(g^{-1})X \end{array}$$

Un peu de calcul différentiel montre que Φ_T induit un revêtement d'espace analytique de $T(F)\backslash G(F) \times \mathfrak{t}_{\mathrm{reg}}(F)$ sur un ouvert noté $\mathfrak{g}_T(F)$ de $\mathfrak{g}_{\mathrm{reg}}(F)$: ce revêtement est galoisien de groupe $W^G(T)$. Les ouverts $\mathfrak{g}_T(F)$ pour $T \in \mathcal{T}^G$ sont deux à deux disjoints et leur union est $\mathfrak{g}_{\mathrm{reg}}(F)$.

Soit $f \in C_c^\infty(\mathfrak{g}_{\mathrm{reg}}(F))$. Montrons que l'application $X \in \mathfrak{g}_{\mathrm{reg}}(F) \mapsto I_X(f)$ vérifie l'assertion 1. Seul le fait qu'elle soit localement constante n'est pas évident. Pour cela il suffit de montrer que pour tout $T \in \mathcal{T}^G$ l'application $I_\bullet(f) \circ \Phi_T$ est localement constante. Or $f \circ \Phi_T$ appartient à $C_c^\infty(T(F)\backslash G(F) \times \mathfrak{t}_{\mathrm{reg}}(F))$. Il existe donc une famille finie de fonctions $\alpha_i \in C_c^\infty(T(F)\backslash G(F))$ et $\theta_i \in C_c^\infty(\mathfrak{t}_{\mathrm{reg}}(F))$ telle que

$$f \circ \Phi_T = \sum_i \alpha_i \otimes \theta_i.$$

On calcule alors pour $(g,X) \in T(F)\backslash G(F) \times \mathfrak{t}_{\mathrm{reg}}(F)$

$$\begin{aligned} I_{\Phi_T(g,X)}(f) &= |D^{\mathfrak{g}}(X)|_F^{\frac{1}{2}} \int_{T_{\mathrm{Ad}(g^{-1})X}(F)\backslash G(F)} f(\mathrm{Ad}(x^{-1}g^{-1})X)\, dx \\ &= |D^{\mathfrak{g}}(X)|_F^{\frac{1}{2}} \int_{T_X(F)\backslash G(F)} f(\mathrm{Ad}(x^{-1})X)\, dx \\ &= |D^{\mathfrak{g}}(X)|_F^{\frac{1}{2}} \sum_i \int_{T_X(F)\backslash G(F)} \alpha_i(x)\, dx \times \theta_i(X) \end{aligned}$$

et le résultat est alors évident.

Réciproquement, soit I qui vérifie la conditions 1. Soit $T \in \mathcal{T}^G$. La fonction $I \circ \Phi_T$ est constante sur la première variable et définit une fonction $\theta \in C_c^\infty(\mathfrak{t}_{\mathrm{reg}}(F))$ en la seconde variable. Le fait que θ soit à support régulier et compact vient de l'inclusion évidente $\mathrm{supp}(\theta) \subset \mathfrak{t}_{\mathrm{reg}}(F) \cap \omega^{G(F)}$ et ce dernier ensemble est compact. Soit $\alpha \in C_c^\infty(T(F)\backslash G(F))$ une fonction W-invariante qui vérifie

$$\int_{T_X(F)\backslash G(F)} \alpha_i(x)\, dx = 1.$$

Alors la fonction $\alpha \otimes |D^{\mathfrak{g}}(\cdot)|_F^{-\frac{1}{2}}\theta$ est W-invariante et appartient à $C_c^\infty(T(F)\backslash G(F) \times \mathfrak{t}_{\mathrm{reg}}(F))$. Elle définit donc une fonction $f_T \in C_c^\infty(\mathfrak{g}_T(F))$. En faisant varier T, on obtient une fonction f qui satisfait 2.

□

6.5. Voici une formulation précisée et légèrement différente du théorème de régularité 5.2 de Harish-Chandra pour les intégrales orbitales.

Théorème 6.2. — *Il existe une unique fonction*

$$\hat{\imath}^G : \mathfrak{g}_{\mathrm{reg}}(F) \times \mathfrak{g}_{\mathrm{reg}}(F) \to \mathbb{C}$$

localement constante et invariante par $G(F) \times G(F)$ telle que pour tout $X \in \mathfrak{g}_{\mathrm{reg}}(F)$ et tout $f \in C_c^\infty(\mathfrak{g}(F))$ on ait

$$\hat{I}_X(f) = \sum_{T \in \mathcal{T}^G} |W^G(T)|^{-1} \int_{\mathfrak{t}(F)} I_Y(f)\hat{\imath}^G(X,Y)\, dY$$

Le lien entre les deux formulations se fait à l'aide de la formule d'intégration de Weyl : il existe des constantes c_T pour $T \in \mathcal{T}^G$ qui vérifient pour tout $f \in C_c^\infty(\mathfrak{g}(F))$

$$\int_{\mathfrak{g}(F)} f(X)\, dX$$
$$= \sum_{T \in \mathcal{T}^G} c_T \cdot |W^G(T)|^{-1} \int_{\mathfrak{t}(F)} |D^{\mathfrak{g}}(X)|_F \int_{T(F)\backslash G(F)} f(\mathrm{Ad}(x^{-1})X)\, dx\, dX.$$

Le fait que $\hat{\imath}^G$ est localement constante en la première variable résulte du lemme 6.1.

Remarque 6.3. —Il est clair que la fonction $\hat{\imath}^G$ ne dépend pas du choix des mesures de Haar sur $G(F)$ et les tores $T(F)$.

7. Facteurs de transfert

7.1. Dans cette section, F désigne un corps local.

7.2. Soit (H, κ, ξ) une donnée endoscopique de H. L'objet de cette section est de définir le facteur de transfert de Langlands-Shelstad. C'est une application

$$\Delta : \Sigma_G(\mathfrak{h}) \times \Gamma(\mathfrak{g}) \to \mathbb{C}$$

qui vérifie : $\Delta(Y, X) \neq 0$ si et seulement si $Y \to X$. On va préciser les valeurs non nulles de Δ. Pour cela, on définit dans la suite un facteur de transfert relatif canonique $\Delta(Y_1, X_1, Y_2, X_2)$ pour des éléments $Y_i \in \Sigma_G(\mathfrak{h})$

et $X_i \in \Gamma(\mathfrak{g})$ qui vérifient $Y_i \to X_i$ pour $i = 1, 2$. On fixe un couple base (Y_0, X_0) tel que $Y_0 \to X_0$ (s'il existe un tel couple). On fixe arbitrairement $\Delta(Y_0, X_0)$ dans $\mathbb{C}^\times$. Pour tout couple (Y, X) tel que $Y \to X$ on pose

$$\Delta(Y, X) = \Delta(Y, X, Y_0, X_0)\Delta(Y_0, X_0).$$

Cette définition fait sens car comme on le verra pour un tel couple (Y, X) on a

$$\Delta(Y, X, Y, X) = 1. \tag{7.1}$$

Pour définir le facteur de transfert relatif, on a besoin d'une forme intérieure quasi-déployée G^* de G (cf. §2.5 et l'isomorphisme φ de l.(2.1)) ainsi que d'un épinglage de G^* (cf. §4.6). On a alors une section de χ_{G^*} (cf. théorème 4.2) qui composée avec $\mathfrak{car}_H \to \mathfrak{car}_{G^*}$ (cf. l.4.1) et χ_H donne une application

$$\Sigma_G(\mathfrak{h}) \to X_- + \mathfrak{g}_{X_+}(F)$$

qu'on note $Y \mapsto Y^*$. On laisse au lecteur le soin de vérifier que le facteur de transfert relatif, que nous allons maintenant définir, ne dépend pas de ces choix auxiliaires.

7.3. Sous-jacents à la donnée (H, κ, ξ), on a des sous-tores maximaux $T_G \subset G$ et $T_H \subset H$. D'après la remarque 3.1, le caractère κ induit un caractère du groupe $H^1(F, T_{H_{\mathrm{sc}}} \to T_{G_{\mathrm{sc}}})$. Posons

$$U = T_{G_{\mathrm{sc}}} \times T_{G_{\mathrm{sc}}} / \{(z, z^{-1}) | z \in Z_{G_{\mathrm{sc}}}\}.$$

On a un morphisme $U \to T_{G_{\mathrm{sc}}}$ induit par le produit et plus généralement un morphisme de complexes

$$[T_{H_{\mathrm{sc}}} \times T_{H_{\mathrm{sc}}} \to U] \to [T_{H_{\mathrm{sc}}} \to T_{G_{\mathrm{sc}}}].$$

On a donc une flèche en cohomologie

$$H^1(F, T_{H_{\mathrm{sc}}} \times T_{H_{\mathrm{sc}}} \to U) \to H^1(F, T_{H_{\mathrm{sc}}} \to T_{G_{\mathrm{sc}}}).$$

Soit κ_U le caractère que κ induit sur le premier groupe.

On veut définir $\Delta(Y_1, X_1, Y_2, X_2)$ pour un quadruplet (Y_1, X_1, Y_2, X_2) comme ci-dessus. Par définition de Y_i^*, il existe $g_i \in G_{\mathrm{sc}}^*$ tel que $\mathrm{Ad}(g_i)\varphi(T_G)$ $= T_{Y_i^*}$ et qui vérifie les conditions du paragraphe §4.4. Posons

$$U_{1,2} = T_{Y_1^*, G_{\mathrm{sc}}} \times T_{Y_2^*, G_{\mathrm{sc}}} / \{(z, z^{-1}) | z \in Z_{G_{\mathrm{sc}}}\}.$$

Par une variante de ce qui suit la ligne (3.3), on vérifie que le couple $(\mathrm{Ad}(g_1^{-1}), \mathrm{Ad}(g_2^{-1}))$ induit un isomorphisme Γ-équivariant

$$X_*(T_{Y,H_{\mathrm{sc}}} \times T_{Y,H_{\mathrm{sc}}}) \backslash X_*(U_{1,2}) \simeq X_*(T_{H_{\mathrm{sc}}} \times T_{H_{\mathrm{sc}}}) \backslash X_*(U)$$

d'où un isomorphisme

$$H^1(F, T_{Y,H_{\mathrm{sc}}} \times T_{Y,H_{\mathrm{sc}}} \to U_{1,2}) \simeq H^1(F, T_{H_{\mathrm{sc}}} \times T_{H_{\mathrm{sc}}} \to U).$$

Donc κ_U induit un caractère sur $H^1(F, T_{Y,H_{\mathrm{sc}}} \times T_{Y,H_{\mathrm{sc}}} \to U_{1,2})$ donc un caractère noté $\kappa_{1,2}$ sur $H^1(F, U_{1,2})$.

Puisque $Y_i \to X_i$ par hypothèse, on a $\chi_G(X_i) = \chi_{G^*}(Y_i^*)$ et donc il existe $x_i \in G^*_{\mathrm{sc}}$ tel que

$$\mathrm{Ad}(x_i) \circ \varphi(X_i) = Y_i^*.$$

Pour tout $\sigma \in \Gamma$, soit u_σ un élément de G^*_{sc} qui vérifie

$$\varphi\sigma(\varphi)^{-1} = \mathrm{Ad}(u_\sigma).$$

Alors le couple $(x_1 u_\sigma \sigma(x_1)^{-1}, \sigma(x_2) u_\sigma^{-1} x_2^{-1})$ définit une 1-cochaîne à valeurs dans $U_{1,2}$ dont on note $\mathrm{inv}(Y_1, X_1, Y_2, X_2)$ la classe dans $H^1(F, U_{1,2})$. On pose[2]

$$\Delta(Y_1, X_1, Y_2, X_2) = \langle \kappa_{1,2}, \mathrm{inv}(Y_1, X_1, Y_2, X_2) \rangle.$$

7.4. Soit $X^* \in \mathfrak{g}^*_{\mathrm{reg}}(F)$. Pour $i = 1, 2$ soit $X_i \in \mathfrak{g}(F)$ tel qu'il existe $x_i \in G^*_{\mathrm{sc}}$ tel que

$$\mathrm{Ad}(x_i) \circ \varphi(X_i) = X^*.$$

Alors le produit

$$x_1 u_\sigma \sigma(x_1)^{-1} \sigma(x_2) u_\sigma^{-1} x_2^{-1}$$

définit une 1-cochaîne galoisienne à valeurs dans $T_{X^*,\mathrm{sc}}$. Sa classe dans $H^1(F, T_{X^*,\mathrm{sc}})$ est notée $\mathrm{inv}(X^*; X_1, X_2)$. On vérifie la formule suivante : pour tout $Y \in \mathfrak{h}_{G-\mathrm{reg}}(F)$ tel que $Y \to X^*$, on a

$$\Delta(Y, X_1, Y, X_2) = \langle \kappa_Y, \mathrm{inv}(X^*; X_1, X_2) \rangle.$$

où κ_Y est défini à la fin du paragraphe 4.4 et donc en particulier on a bien (7.1) qui justifie la définition du facteur de transfert relatif. Pour tout quadruplet (Y_1, X_1, Y_2, X_2) du paragraphe précédent, on a

$$\Delta(Y_1, X_1) = \Delta(Y_1, X_1, Y_2, X_2) \Delta(Y_2, X_2).$$

On a donc aussi

$$\Delta(Y, X_1) = \Delta(Y, X_2) \langle \kappa_Y, \mathrm{inv}(X^*; X_1, X_2) \rangle. \tag{7.2}$$

8. Transfert lisse et transformée de Fourier : cas p-adique.

8.1. Dans toute cette section, F désigne un corps p-adique.

8.2. Deux éléments de $\mathfrak{g}_{\mathrm{reg}}(F)$ stablement conjugués ont leurs tores centralisateurs «stablement conjugués» et donc isomorphes. On impose que les mesures de Haar (cf. §6.2) sur ces tores se correspondent par ces isomorphismes.

8.3. Soit $X \in \Sigma(\mathfrak{g})$. On définit S_X l'intégrale orbitale stable comme la distribution invariante définie par

$$S_X(f) = \sum_{\{Z \in \Gamma(\mathfrak{g}) \ | \ \chi(Z) = \chi(X)\}} I_Z(f).$$

[2]Comme le remarque le rapporteur de cet article, ce qu'on définit ci-dessous est en fait *l'inverse* du facteur de Langlands-Shelstad (cf. [**LS**] §(3.4).

Une distribution D est stable si elle appartient à l'adhérence faible de l'espace des intégrales orbitales stables c'est-à-dire si $S_X(f) = 0$ pour tout $X \in \Sigma(\mathfrak{g})$ implique $D(f) = 0$.

Pour rappeler que ces objets sont relatifs à G, on écrira I^G, S^G etc.

8.4. Soit (H, κ, ξ) une donnée endoscopique de G. Pour tout $Y \in \Sigma_G(\mathfrak{h})$, on définit $I_Y^{G,H}$ l'intégrale orbitale endoscopique comme la distribution invariante définie par

$$I_Y^{G,H}(f) = \sum_{X \in \Gamma(\mathfrak{g})} \Delta(Y, X) I_X(f).$$

Remarque 8.1. — Cette expression dépend de la normalisation du facteur de transfert (cf. section 7).

8.5. Voici le théorème de transfert lisse dû à Waldspurger.

Théorème 8.2. — *Pour tout $f \in C_c^\infty(\mathfrak{g}(F))$, il existe $f^H \in C_c^\infty(\mathfrak{h}(F))$ tel que pour tout $Y \in \Sigma_G(\mathfrak{h})$, on ait*

$$I_Y^{G,H}(f) = S_Y^H(f^H).$$

On dit que f^H est un transfert de f.

Remarque 8.3. — Bien que la notation ne le rappelle pas, les fonctions f^H qui conviennent dépendent aussi de la normalisation du facteur de transfert (cf. section 7).

On peut montrer que si D une distribution stable sur $\mathfrak{h}(F)$ et si f^H est un transfert de f alors l'expression $D(f^H)$ ne dépend en fait que de f. On obtient ainsi une distribution $f \mapsto D(f^H)$ sur $\mathfrak{g}(F)$: c'est le transfert endoscopique dual.

8.6. Supposons que le groupe G est non ramifié; par définition, cela signifie qu'il est quasi-déployé et déployé sur une extension non ramifiée. La théorie de Bruhat-Tits montre qu'il existe un schéma en groupes $\mathcal{G}$ lisse sur $\mathcal{O}_F$ l'anneau des entiers de F, à fibres réductives connexes, dont la fibre générique est G. Soit 1_G la fonction caractéristique du réseau $\mathrm{Lie}(\mathcal{G})(\mathcal{O}_F)$ dans $\mathfrak{g}(F)$. On dit qu'un tel réseau est *hyperspécial*. Supposons que H est non ramifiée, on a alors une fonction 1_H. Le théorème suivant est vrai pour tout choix de modèle entier comme ci-dessus : c'est le lemme fondamental de Langlands-Shelstad. Il est dû à Ngô (cf. [**Ngo2**] pour la caractéristique positive) et à Waldspurger (cf. [**W-car**] pour le passage en caractéristique nulle). Ces travaux imposent une condition sur la caractéristique résiduelle. Cette restriction est levée pour des corps p-adiques par un résultat de Hales (cf. [**Hales**]).

Théorème 8.4. — *Il existe une constante $c \neq 0$ tel que pour tout $Y \in \Sigma_G(\mathfrak{h})$, on ait*

$$c\, I_Y^{G,H}(1_G) = S_Y^H(1_H).$$

8.7. Harish-Chandra a mis en évidence l'analogie entre l'analyse harmonique sur le groupe G et sur son algèbre de Lie $\mathfrak{g}$. Évidemment, de part et d'autre on dispose des intégrales orbitales associées aux éléments semi-simples réguliers. Il y a cependant d'autres distributions sur le groupe à savoir les traces des représentations (tempérées). Leurs analogues sur l'algèbre de Lie sont les transformées de Fourier des intégrales orbitales (des éléments semi-simples réguliers).

Waldspurger a poussé plus en avant cette analogie. Sur le groupe, les représentations tempérées se regroupent en L-paquets dont les traces associées devraient être des distributions stables. Selon Waldspurger, l'analogue d'un L-paquet est une classe de conjugaison stable. En particulier, on s'attend à ce que la transformée de Fourier $\hat{S}_X^G$ pour $X \in \Sigma(\mathfrak{g})$ soit stable. Waldspurger a montré le théorème suivant.

Théorème 8.5. — *La transformée de Fourier d'une distribution stable est stable.*

8.8. On fixe comme forme bilinéaire sur $\mathfrak{h}(F)$ (cf. §5.4) la forme qui se restreint à $\mathfrak{t}_H$ en la forme $\langle \iota(\cdot), \iota(\cdot) \rangle$ (cf. §3.1). On prend le même caractère ψ d'où une transformation de Fourier. On peut alors regarder le transfert réciproque de $\hat{S}_Y^H$. Quel est-il ? Voici la réponse due à Waldspurger.

Théorème 8.6. — *Il existe une constante $c \neq 0$ telle que pour tout $Y \in \Sigma_G(\mathfrak{h})$ et tout $f \in C_c^\infty(\mathfrak{g}(F))$, on ait*

$$c\, \hat{I}_Y^{G,H}(f) = \hat{S}_Y^H(f^H)$$

où f^H est un transfert de f. Autrement dit, le transfert commute à la transformée de Fourier à une constante près.

8.9. Le théorème 8.6 est équivalent au théorème suivant (dû aussi à Waldspurger). L'affirmation «implique» est une conséquence du fait que les fonctions $\hat{\imath}^G$ sont localement constantes et du lemme 6.1 qui permet d'isoler les orbites.

Théorème 8.7. — *Il existe une constante $c \neq 0$ telle que pour tout $Y \in \Sigma_G(\mathfrak{h})$ et tout $Z \in \Gamma(X)$ on ait l'égalité entre*

$$\sum_{X \in \Gamma(\mathfrak{g})} \Delta(Y, X) \hat{\imath}^G(X, Z)$$

et

$$c \sum_{Z' \in \Gamma(\mathfrak{h})} \sum_{Y'} \hat{\imath}^H(Y', Z') \Delta(Z', Z) |\ker(H^1(F, T_{Z'}) \to H^1(F, H))|^{-1}. \tag{8.1}$$

où la somme sur Y' porte sur les $Y' \in \Gamma(\mathfrak{h})$ stablement conjugués à Y.

8.10. Le théorème 8.7 implique l'existence du transfert du théorème 8.2. Esquissons une explication. Fixons ω un compact de $\mathfrak{h}$. Pour tout compact K de $\mathfrak{g}(F)$, il résulte du théorème 5.1 de finitude de Howe que l'espace noté $V_{\omega,K}$ engendré par les fonctions indexées par $Z \in K \cap \mathfrak{g}_{\mathrm{reg}}(F)$

$$Y \in \omega \cap \mathfrak{h}_{G-\mathrm{reg}}(F) \mapsto \sum_{X \in \Gamma(\mathfrak{g})} \Delta(Y, X)\hat{i}^G(X, Z)$$

est de dimension finie. Par le théorème 8.7, cet espace est inclus dans celui engendré par les fonctions de la ligne (8.1) et ce dernier est inclus dans

$$Y \in \omega \cap \mathfrak{h}_{G-\mathrm{reg}}(F) \mapsto S_Y^H(f')$$

lorsque f' parcourt $C_c^\infty(\mathfrak{h}(F))$ (pour le voir, on utilise le lemme 6.1.

Prenons $f \in C_c^\infty(\mathfrak{g}(F))$. Il résulte de la formule d'intégration du théorème 6.2 que la fonction

$$Y \in \omega \cap \mathfrak{h}_{G-\mathrm{reg}}(F) \mapsto I_Y^{G,H}(f)$$

appartient à l'espace $V_{\omega,K}$ défini ci-dessus pour un K convenable. Mais comme on l'a dit tout élément de $V_{\omega,K}$ est de la forme $S_Y^H(f')$ pour $f' \in C_c^\infty(\mathfrak{h}(F))$.

Cela donne le transfert local au voisinage de tout point. À l'aide de partition de l'unité, on peut recoller les transferts obtenus.

8.11. L'usage d'une exponentielle et d'arguments de descente au centralisateur montre que le transfert au niveau des groupes se déduit du transfert au niveau des algèbres de Lie des centralisateurs des éléments semi-simples du groupe. Cela nécessite de vérifier différentes formules de descente pour les facteurs de transfert.

9. Intégrales orbitales sur les algèbres de Lie complexes

9.1. Dans cette section, le corps considéré est $\mathbb{C}$. Sur un tel corps la conjugaison stable se confond avec la conjugaison. Soit (H, κ, ξ) une donnée endoscopique de G. Le facteur de transfert vaut soit une constante non nul soit est nul.

9.2. Soit $B \subset G$ un sous-groupe de Borel, $T \subset B$ un sous-tore maximal et $K \subset G(\mathbb{C})$ un sous-groupe compact maximal «en bonne position» relativement à T. Soit U_B le radical unipotent de B. Soit $\mathcal{S}(\mathfrak{g}(\mathbb{C}))$ l'espace de Schwartz de $\mathfrak{g}(\mathbb{C})$. Pour $X \in \Gamma(\mathfrak{g})$ et $f \in \mathcal{S}(\mathfrak{g}(\mathbb{C}))$, on définit une intégrale orbitale $I_X(f)$ comme en §6.3. L'intégrale converge. Dans le cas complexe, une intégrale orbitale se calcule via la décomposition d'Iwasawa comme un terme constant. Plus précisément pour des choix de mesure de Haar que nous ne précisons pas ici : pour tout $X \in \mathfrak{t}_{reg}(\mathbb{C})$ on a

$$I_X(f) = f_B(X)$$

où l'on a défini pour *tout* $X \in \mathfrak{t}(\mathbb{C})$ le «terme constant»

$$f_B(X) = \int_{K \times U_B(\mathbb{C})} f(\mathrm{Ad}(k)(X+N))\, dN\, dk.$$

On a alors le théorème suivant (cas particulier d'un théorème de Bouaziz, cf. [**Bouaziz1**]).

Théorème 9.1. — *Soit $W = W^G(T)$ le groupe de Weyl. L'application*

$$f \in \mathcal{S}(\mathfrak{g}(\mathbb{C})) \mapsto f_B \in \mathcal{S}(\mathfrak{t}(\mathbb{C}))^W$$

est surjective.

9.3. On fixe un caractère ψ et une forme comme en §5.4 d'où une transformation de Fourier sur $\mathcal{S}(\mathfrak{g}(\mathbb{C}))$. On demande que les formes sur $\mathfrak{g}$ et $\mathfrak{h}$ soient compatibles (cf. §8.7) c'est-à-dire qu'elles coïncident sur l'algèbre de Lie «du» tore commun («du» puisque tous les tores maximaux sont conjugués). Il est alors facile de déduire du théorème ci-dessus le résultat suivant.

Théorème 9.2. — *Soit $f \in \mathcal{S}(\mathfrak{g}(\mathbb{C}))$. Alors*

- *Il existe un transfert $f^H \in \mathcal{S}(\mathfrak{h}(\mathbb{C}))$ tel que pour tout $Y \in \Gamma_G(\mathfrak{h})$*

$$I_Y(f^H) = I_Y^{G,H}(f)\ ;$$

- *le transfert commute à la transformation de Fourier.*

10. Préstablisation de la formule des traces pour les algèbres de Lie.

10.1. Dans cette section et sauf mention contraire, F est un corps de nombres.

10.2. Analogue de la partie elliptique de la formule des traces d'Arthur-Selberg. Soit $\mathcal{S} = \mathcal{S}(\mathfrak{g}(\mathbb{A}))$ l'espace de Schwartz-Bruhat de $g(\mathbb{A})$: c'est l'espace engendré par les fonctions $f_\infty \otimes f^\infty$ où f_∞ est une fonction de Schwartz sur $\mathfrak{g}(F\otimes\mathbb{R})$ et f^∞ est la fonction caractéristique d'un sous-groupe compact ouvert de $\mathfrak{g}(\mathbb{A}_f)$ où $\mathbb{A}_f$ désigne l'ensemble des adèles finis. On note $G(\mathbb{A})^1$ le sous-groupe de $G(\mathbb{A})$ qui s'insère dans la suite exacte

$$1 \to G(\mathbb{A})^1 \to G(\mathbb{A}) \to \mathrm{Hom}(X^*(G), \mathbb{R}_+^\times) \to 1$$

où $X^*(G)$ est le groupe des caractères rationnels de G définis sur F et la flèche non évidente est celle qui envoie $g \in G(\mathbb{A})$ sur le morphisme $\chi \mapsto |\chi(g)|_{\mathbb{A}}$.

On munit $G(\mathbb{A})$ de la mesure de Tamagawa et $\mathrm{Hom}(X^*(G), \mathbb{R}_+^\times)$ de la mesure déduite de la mesure usuelle de $\mathbb{R}_+^*$. On en déduit une mesure sur $G(\mathbb{A})^1$. Le quotient $G(F)\backslash G(\mathbb{A})^1$ est de volume fini et on définit le nombre de Tamagawa de G par

$$\tau(G) = \mathrm{vol}(G(F)\backslash G(\mathbb{A})^1).$$

Le calcul des nombres de Tamagawa a été achevé avec la preuve par Kottwitz de la conjecture de Weil. Donnons le résultat sous la forme suivante (cf. [**LabBreen**] corollaire 1.7.4 p.39)

$$\tau(G) = \frac{|H^1_{ab}(\mathbb{A}/F, G)|}{|\operatorname{Ker}^1_{ab}(F, G)|}.$$

Cette expression vaut aussi pour les sous-tores de G.

Théorème 10.1. — *Pour tout $f \in \mathcal{S}$, l'intégrale*

$$\int_{G(F)\backslash G(\mathbb{A})^1} \sum_{X \in \mathfrak{g}_{\mathrm{ell}}(F)} |f(x^{-1}Xx)|\, dx < \infty$$

est convergente.

Pour la preuve de ce théorème, nous renvoyons à [**Wif**] paragraphe 10.8. Elle repose sur la théorie de la réduction qui fournit un ersatz de domaine fondamental au quotient $G(F)\backslash G(\mathbb{A})^1$.

On peut donc poser

$$J^G(f) = \int_{G(F)\backslash G(\mathbb{A})^1} \sum_{X \in \mathfrak{g}_{\mathrm{ell}}(F)} f(\mathrm{Ad}(x^{-1})X)\, dx.$$

On écrit encore

$$J^G(f) = \sum_{X \in \Gamma_{\mathrm{ell}}(\mathfrak{g})} \tau(T_X)\, I^G_X(f)$$

où l'on a introduit l'intégrale orbitale globale

$$I^G_X(f) = \int_{T_X(\mathbb{A})\backslash G(\mathbb{A})} f(\mathrm{Ad}(x^{-1})X)\, dx. \tag{10.1}$$

10.3. Au paragraphe précédent, on a exprimé $J^G(f)$ à l'aide de classes de conjugaison rationnelles. On aimerait trouver une expression en termes de classes de conjugaison stable locales. Ce sera possible mais à un terme d'erreur près. Il nous faut pour cela d'abord introduire l'obstruction de Langlands.

10.4. L'obstruction de Langlands. — On cherche à trouver une condition cohomologique pour qu'un élément de $\mathfrak{g}^*(F)$ qui localement partout se transfère à $\mathfrak{g}$ se transfère aussi globalement.

Plus précisément, soit $X^* \in \mathfrak{g}^*(F)$ un élément semi-simple régulier tel qu'il existe $x \in G^*(\bar{\mathbb{A}})$ et $X \in \mathfrak{g}(\mathbb{A})$ tel que

$$\mathrm{Ad}(x) \circ \varphi(X) = X^*.$$

On peut et on va supposer que $x \in G^*_{\mathrm{sc}}(\bar{\mathbb{A}})$ (cf. [**Wif**] 10.2). Soit u_σ une cochaîne à valeurs dans G^*_{sc} qui vérifie

$$\varphi\sigma(\varphi)^{-1} = \mathrm{Ad}(u_\sigma)$$

pour tout $\sigma \in \Gamma$. Soit T le centralisateur de X^* dans G^*. Alors la cochaîne $xu_\sigma\sigma(x)^{-1} \in T_{\mathrm{sc}}(\bar{\mathbb{A}})$ définit une classe de 1-cocycle dans $H^1(\mathbb{A}_F/F, T_{\mathrm{sc}})$

notée $\widetilde{\mathrm{inv}}(X^*, X)$. Comme x est bien défini à translation à gauche près par un élément de $T_{\mathrm{sc}}(\bar{\mathbb{A}})$, la classe $\widetilde{\mathrm{inv}}(X^*, X)$ ne dépend que de X^* et de X. On note $\mathrm{inv}(X^*, X)$ l'image de $\widetilde{\mathrm{inv}}(X^*, X)$ dans $\mathcal{E}(T, G, \mathbb{A}/F)$

Théorème 10.2. — *(Langlands) Il existe $g \in G(\mathbb{A})$ tel que $\mathrm{Ad}(g)(X) \in \mathfrak{g}(F)$ si et seulement si $\mathrm{inv}(X^*, X) = 1$.*

Démonstration. — La condition est nécessaire. En effet si $g \in G(\mathbb{A})$ vérifie $\mathrm{Ad}(g)(X) \in \mathfrak{g}(F)$, on est immédiatement ramené au cas où $X \in \mathfrak{g}(F)$. Mais alors l'ensemble des $y \in G_{\mathrm{sc}}$ tels que

$$\mathrm{Ad}(y) \circ \varphi(X) = X^*$$

est une sous-variété de G_{sc} non vide puisqu'elle possède un point dans $\bar{\mathbb{A}}$. On peut donc prendre $x \in G_{\mathrm{sc}}(\bar{F})$. On voit donc que $\widetilde{\mathrm{inv}}(X^*, X) = 1$ d'où la première implication.

La réciproque est plus élaborée. Puisque $\mathrm{inv}(X^*, X) = 1$, il existe un 1-cocycle β_σ à valeurs dans $T_{\mathrm{sc}}(\bar{\mathbb{A}})$, $y \in T(\bar{\mathbb{A}})$ et une 1-cochaîne t_σ à valeurs dans $T_{\mathrm{sc}}(\bar{F})$ tels que

$$\beta_\sigma = y^{-1}\sigma(y)$$

(par abus, on ne distingue pas dans l'écriture β_σ et son image dans $T(\bar{\mathbb{A}})$) et

$$xu_\sigma\sigma(x)^{-1} = \beta_\sigma t_\sigma.$$

pour tout $\sigma \in \Gamma$. Puisque le cobord $\partial\beta$ est trivial, on a $\partial u = \partial t$. On en déduit que la cochaîne $t_\sigma u_\sigma^{-1}$ est un cocycle «à valeurs dans G_{sc}» expression un peu abusive qui signifie que $\varphi^{-1}(t_\sigma u_\sigma^{-1})$ est un cocycle à valeurs dans G_{sc}. La cochaîne $yxu_\sigma\sigma(yx)^{-1}u_\sigma^{-1}$ définit un 1-cocycle à valeurs dans G localement trivial : sa classe est donc un élément de $\mathrm{Ker}^1(F, G)$. Son image dans l'abélianisé $\mathrm{Ker}^1_{ab}(F, G)$ est triviale : en effet c'est la classe de l'hypercochaîne $(1, t_\sigma u_\sigma^{-1})$ qui est cohomologue à la classe triviale. Mais la flèche d'abélianisation $\mathrm{Ker}^1(F, G) \to \mathrm{Ker}^1_{ab}(F, G)$ est un isomorphisme. On en déduit que la cochaîne $yxu_\sigma\sigma(yx)^{-1}u_\sigma^{-1}$ est triviale : il existe donc $a \in G^*$ tel que

$$yxu_\sigma\sigma(yx)^{-1}u_\sigma^{-1} = au_\sigma\sigma(a)^{-1}u_\sigma^{-1}.$$

Soit $Y \in \mathfrak{g}(\overline{F})$ tel que $\mathrm{Ad}(a) \circ \varphi(Y) = X^*$. Alors $Y = \mathrm{Ad}(b)X$ avec $b = \varphi^{-1}(a^{-1}yx)$ et on vérifie que $b \in G(\mathbb{A})$. Il s'ensuit que $Y \in \mathfrak{g}(\overline{F}) \cap \mathfrak{g}(\mathbb{A}) = \mathfrak{g}(F)$ ce qu'il fallait vérifier.□

10.5. Pré-stabilisation. Pour exprimer $J^G(f)$ à l'aide de classes de conjugaison stable locales, on va utiliser une inversion de Fourier sur le groupe $\mathcal{E}(T_{X^*}, G^*, \mathbb{A}/F)$.

Par abus, on confond $\Gamma_{\mathrm{ell}}(\mathfrak{g})$ et $\Sigma_{\mathrm{ell}}(\mathfrak{g}^*)$ avec des systèmes de représentants. Pour tout $X^* \in \Sigma_{\mathrm{ell}}(\mathfrak{g}^*)$, soit

$$\Gamma_{X^*}(\mathfrak{g}) = \{X \in \Gamma_{\mathrm{ell}}(\mathfrak{g}) \mid \exists g \in G^*_{\mathrm{sc}} \quad \mathrm{Ad}(g) \circ \varphi(X) = X^*\}.$$

Il se peut que cet ensemble soit vide. S'il ne l'est pas, ses éléments sont des représentants des classes de conjugaison ordinaire dans une même classe stable de $\mathfrak{g}(F)$ déterminée par X^*. On a

$$\Gamma_{\text{ell}}(\mathfrak{g}) = \bigcup_{X^* \in \Sigma_{\text{ell}}(\mathfrak{g}^*)} \Gamma_{X^*}(\mathfrak{g})$$

et la réunion est évidemment disjointe. Introduisons alors

$$\Gamma_{X^*}(\mathfrak{g}(\mathbb{A})) = \{X \in \mathfrak{g}(\mathbb{A}) \mid \exists g \in G^*_{\text{sc}}(\bar{\mathbb{A}}) \quad \text{Ad}(g) \circ \varphi(X) = X^*\}/G(\mathbb{A}).$$

Le théorème 10.2 implique que l'application évidente

$$\Gamma_{X^*}(\mathfrak{g}) \to \{X \in \Gamma_{X^*}(\mathfrak{g}(\mathbb{A})) \text{ tel que } \text{inv}(X^*, X) = 1\} \tag{10.2}$$

est surjective. Elle n'est pas injective : soit $X \in \Gamma_{X^*}(\mathfrak{g})$. Alors l'ensemble $\Gamma_{X^*}(\mathfrak{g})$ est en bijection naturelle avec $\mathcal{D}(T_X, G, F)$. Un élément de $\Gamma_{X^*}(\mathfrak{g})$ s'envoie sur la classe de X par (10.2) si et seulement si il correspond à un élément de

$$\text{Ker}(\mathcal{D}(T_X, G, F) \to \mathcal{D}(T_X, G, \mathbb{A})) = \text{Ker}(\text{Ker}^1(F, T_X) \to \text{Ker}^1(F, G)).$$

En utilisant le fait que $\text{Ker}^1(F, G)$ est en bijection naturelle avec $\text{Ker}^1_{ab}(F, G)$ et que la cohomologie abélianisée reste insensible aux torsions intérieures, on voit que les fibres de (10.2) ont toutes même cardinal à savoir

$$d(X^*) = |\,\text{Ker}(\text{Ker}^1(F, T_{X^*}) \to \text{Ker}^1(F, G^*))|.$$

L'expression (10.1) qui définit $I^G_X(f)$ vaut aussi pour tout $X \in \mathfrak{g}(\mathbb{A})$. Comme fonction de X, l'intégrale orbitale est invariante sous l'action de $G(\mathbb{A})$. Enfin si $X \in \Gamma_{X^*}(\mathfrak{g})$, on vérifie que $\tau(T_X) = \tau(T_{X^*})$.

On peut donc écrire

$$\begin{aligned} J^G(f) &= \sum_{X^* \in \Sigma_{\text{ell}}(\mathfrak{g}^*)} \tau(T_{X^*}) d(X^*) \sum_{\{X \in \Gamma_{X^*}(\mathfrak{g}(\mathbb{A})) \mid \text{inv}(X^*, X) = 1\}} I^G_X(f) \\ &= \sum_{X^* \in \Sigma_{\text{ell}}(\mathfrak{g}^*)} \frac{\tau(T_{X^*}) d(X^*)}{|\mathcal{E}(T_{X^*}, G^*, \mathbb{A}/F)|} \sum_{X \in \Gamma_{X^*}(\mathfrak{g}(\mathbb{A}))} \sum_{\kappa \in \mathcal{K}(T_{X^*}, G^*, F)} \langle \kappa, \text{inv}(X^*, X) \rangle I^G_X(\end{aligned}$$

On utilise alors la suite exacte suivante ([**LabBreen**] proposition 1.8.4 p.42)

$$\begin{aligned} 1 &\to \text{Ker}(\text{Ker}^1(F, T_{X^*}) \to \text{Ker}^1_{ab}(F, G^*)) \to \text{Ker}^1(F, T_{X^*}) \\ &\to \text{Ker}^1_{ab}(F, G^*) \to \mathcal{E}(T_{X^*}, G^*, \mathbb{A}/F) \to H^1(\mathbb{A}/F, T_{X^*}) \\ &\to H^1_{ab}(\mathbb{A}/F, G^*) \to H^2(\mathbb{A}/F, T_{X^*,\text{sc}}) \to 1. \end{aligned}$$

Comme $T_{X^*,\text{sc}}$ est elliptique, on voit en utilisant l'isomorphisme de Tate-Nakayama que $H^2(\mathbb{A}/F, T_{X^*,\text{sc}}) = 1$. En utilisant la bijection naturelle entre $\text{Ker}^1(F, G^*)$ et son abélisanisé, on en déduit que

$$\frac{\tau(T_{X^*}) d(X^*)}{|\mathcal{E}(T_{X^*}, G^*, \mathbb{A}/F)|} = \tau(G^*).$$

L'ensemble $\Gamma_{X^*}(\mathfrak{g}(\mathbb{A}))$ est infini. Il existe un ensemble fini S de places tel que G soit défini sur $\mathfrak{o}^S$ (l'anneau des éléments de F entiers hors de S) et que pour tous X et X' dans $\Gamma_{X^*}(\mathfrak{g}(\mathbb{A}))$ tels que les deux intégrales $I_X^G(f)$ et $I_{X'}^G(f)$ soient non nulles, X_v et X'_v soient conjugués sous un élément de $G(\mathfrak{o}_v)$ pour $v \notin S$ (cf. [**Wif**] lemme 7.2). Il en résulte que l'ensemble

$$\{X \in \Gamma_{X^*}(\mathfrak{g}(\mathbb{A})) \mid I_X^G(f) \neq 0\}$$

est fini et on peut intervertir l'ordre de sommation entre X et κ dans la somme précédente. On trouve alors

$$J^G(f) = \tau(G^*) \sum_{X^* \in \Sigma_{\text{ell}}(\mathfrak{g}^*)} \sum_{\kappa \in \mathcal{K}(T_{X^*}, G^*, F)} I_{X^*}^{G,\kappa}(f) \tag{10.3}$$

où l'on a posé

$$I_{X^*}^{G,\kappa}(f) = \sum_{X \in \Gamma_{X^*}(\mathfrak{g}(\mathbb{A}))} \langle \kappa, \text{inv}(X^*, X) \rangle I_X^G(f)$$

10.6. Dans ce paragraphe F est local ou global.

Partons d'un triplet endoscopique (H, κ, ξ) et de $Y \in \mathfrak{h}_{G-\text{reg}}(F)$. Alors il existe $X^* \in \Sigma(\mathfrak{g}^*)$ tel que $Y \to X^*$. On a expliqué au paragraphe §4.4 comment on définissait un élément $\kappa_Y \in \mathcal{K}(T_{X^*}, G^*, F)$. Énonçons le lemme suivant.

Lemme 10.3. — *Pour tous $X^* \in \Sigma_{\text{ell}}(\mathfrak{g}^*)$ et $\kappa_0 \in \mathcal{K}(T_{X^*}, G^*, F)$, il existe un triplet endoscopique elliptique (H, κ, ξ) et $Y \in \Sigma(\mathfrak{h})$ tel que $Y \to X^*$ et $\kappa_Y = \kappa_0$.*
Deux telles données (H, κ, ξ, Y) et (H', κ', ξ', Y') sont équivalentes comme données endoscopiques de G par un isomorphisme qui envoie Y sur Y'. Un tel isomophisme est unique à un automorphisme intérieur près.

Démonstration. — On ne traite que la partie existence (on trouve une preuve complète dans [**Ko2**] lemme 9.7). Soit $X^* \in \mathfrak{g}^*_{\text{ell}}(F)$. Posons $T = T_{X^*}$. Soit $\kappa_0 \in \mathcal{K}(T_{X^*}, G^*, F)$. En particulier, κ_0 définit un caractère de $H^1(C_F, T_{\text{sc}})$. Fixons E une extension galoisienne finie de F qui déploie T. La dualité de Tate-Nakayama identifie $H^1(C_F, T_{\text{sc}})$ à $\hat{H}^{-1}(E/F, X_*(T_{\text{sc}}))$ mais comme $X_*(T_{\text{sc}})^\Gamma = 1$ par ellipticité de T on a aussi

$$H^1(C_F, T_{\text{sc}}) \simeq H_0(E/F, X_*(T_{\text{sc}})).$$

On peut donc voir κ_0 comme un caractère de $H_0(E/F, X_*(T_{\text{sc}}))$. La donnée radicielle

$$(X^*(T), R_\kappa, X_*(T), R_\kappa^\vee)$$

est stable par Γ. C'est donc la donnée radicielle d'un groupe H défini sur F. On complète aisément ces données en un triplet (H, κ, ξ) endoscopique. Le tore T s'identifie à un sous-tore de H et cette identification envoie X^* sur un élément $Y \in \mathfrak{h}(F)$ qui est G-régulier. Cela construit le quadruplet (H, κ, ξ, Y).

□

10.7. Normalisation des facteurs de transfert. — Soit F un corps global. Soit (H,κ,ξ) une donnée endoscopique de G. Soit V l'ensemble des places de F. De manière évidente, on en déduit une donnée endoscopique locale (H_v,κ,ξ) de G_v pour toute place $v\in V$: on a noté G_v le groupe sur F_v, le complété de F en v, obtenu par extension des scalaires.

Soit $Y_i\in\mathfrak{h}_{G-\text{reg}}(F)$ et $X_i^*\in\mathfrak{g}^*_{\text{reg}}(F)$ tels que $Y_i\to X_i^*$ pour $i=1,2$. Supposons que $\Gamma_{X_i^*}(\mathfrak{g}(\mathbb{A}))\neq\emptyset$. Soit $X_i\in\Gamma_{X_i^*}(\mathfrak{g}(\mathbb{A}))$.

En chaque place v, on a un facteur de transfert relatif noté Δ_{G_v,H_v} ou simplement Δ_v (cf. section 7). On vérifie alors la formule suivante (cf. [**LS**] section 6)

$$\langle\kappa_{Y_1},\text{inv}(X_1^*,X_1)\rangle\langle\kappa_{Y_2},\text{inv}(X_2^*,X_2)\rangle^{-1}=\prod_{v\in V}\Delta_v(Y_1,X_1,Y_2,X_2)$$

où pour presque tout v

$$\Delta_v(Y_1,X_1,Y_2,X_2)=1.$$

On normalise alors le facteur de transfert en prenant $(Y,X_{2,v})$ pour point base. On impose que

– pour presque tout v, $\Delta_v(Y,X_{2,v})=1$;
– $\prod_{v\in V}\Delta_v(Y_2,X_2)=\langle\kappa_{Y_2},\text{inv}(X_2^*,X_2)\rangle$.

Une fois le facteur de transfert ainsi normalisé, les deux formules ci-dessus valent pour *tous* $Y\in\mathfrak{h}_{G-\text{reg}}(F)$, $X^*\in\mathfrak{g}^*_{\text{reg}}(F)$ et $X\in\Gamma_{X^*}(\mathfrak{g}(\mathbb{A}))$ tels que $Y\to X^*$.

10.8. Reprenons les notations de la fin du paragraphe 10.5. Soit $X^*\in\Sigma_{\text{ell}}(\mathfrak{g}^*)$ et $\kappa\in\mathcal{K}(T_{X^*},G^*,F)$. Soit (H,κ,ξ) une donnée endoscopique et $Y\in\mathfrak{h}_{G-\text{reg}}(F)$ tels que $Y\to X^*$ et $\kappa_Y=\kappa$ (cf. lemme 10.3). Supposons $f=\otimes_{v\in V}f_v$. En utilisant la normalisation des facteurs de transfert ci-dessus on voit que

$$I^{G,\kappa}_{X^*}(f)=\prod_{v\in V}I^{G_v,H_v}_Y(f_v)$$

où pour presque tout $v\in V$

$$I^{G_v,H_v}_Y(f_v)=1$$

(cf. [**Wif**] section 7 et §10.2). Ici, nous passons sous silence les relations de compatibilité entre les mesures de Haar locales et globales.

11. Le résultat de Waldspurger

11.1. Soit F un corps global et u une place finie de F. Soit (H,κ,ξ) une donnée endoscopique. On suppose que les hypothèses suivantes sont vérifiées :

(1) F est totalement imaginaire ;

(2) pour tout $v\in V$, il existe $Y\in\mathfrak{h}_{G-\text{reg}}(F_v)$ et $X\in\mathfrak{g}_{\text{reg}}(F_v)$ tel que $Y\to X$;

(3) u est inerte dans $F[H]$, la plus petite extension sur laquelle H se déploie ;

(4) la donnée (H_u, κ, ξ) est elliptique.

Voici le théorème que démontre Waldspurger dans [**Wif**].

Théorème 11.1. — *Soit w une place finie de V distincte de u. Alors le théorème 8.6 vaut pour la paire (G_w, H_w).*

De là, par des arguments de globalisation ou de descente, on montre que le théorème 8.6 vaut en toute généralité (cf. sections 4,5,6 et 11 de [**Wif**]). Notons que pour démontrer le théorème 11.1, on a besoin de connaître le lemme fondamental (théorème 8.4) pour presque toute paire (G_v, H_v).

11.2. On fixe $w \neq u$ une place finie de F. Soit

$$\langle \cdot, \cdot \rangle \, : \, \mathfrak{g}(F) \times \mathfrak{g}(F) \to F$$

une forme bilinéaire symétrique non dégénérée et invariante par l'action de $G(F)$. Soit

$$\psi \, : \, \mathbb{A}_F/F \to \mathbb{C}^\times$$

un caractère continu et non trivial. Comme au paragraphe 8.8, on en déduit une forme analogue sur $\mathfrak{h}(F)$.

On fixe deux $\mathcal{O}_F$-réseaux $a \subset \mathfrak{g}(F)$ et $b \subset \mathfrak{h}(F)$. On en déduit des $\mathcal{O}_{F_v}$ réseaux $a_v \subset \mathfrak{g}(F)$ et $b_v \subset \mathfrak{h}(F)$. Il existe un ensemble fini de places S_1 qui contient les places archimédiennes, les places u et w et tels que pour tout $v \notin S_1$ les réseaux a_v et b_v sont hyperspéciaux et autoduaux pour les bicaractères $\psi(\langle \cdot, \cdot \rangle)$.

11.3. L'ensemble des classes d'équivalence de données endoscopiques de G^* (forme intérieure quasi-déployée de G) qui sont non ramifiées en dehors de S_1 est fini (cf. [**Langlands-formule-stable**] lemme 8.12). Fixons un ensemble $\mathcal{H}$ de représentants de ces classes. Comme H est non ramifiée hors S_1, on peut et on va supposer que $(H, \kappa, \xi) \in \mathcal{H}$. Soit $\mathcal{H}' \subset \mathcal{H}$ le sous-ensemble formé des données (H', κ', ξ') telles que $F[H'] \not\subset F[H]$ (c'est-à-dire telles que H' n'est pas déployé sur $F[H]$). Soit $v = v_{(H', \kappa', \xi')} \in V - S_1$ tel que le Frobenius associé à v dans $\mathrm{Gal}(F[H]F[H']/F)$ fixe $F[H]$ mais pas $F[H']$. Dans ce cas H_v est déployé alors que H'_v ne l'est pas. Soit

$$S_2 = \{v_{(H', \kappa', \xi')} \text{ pour } (H', \kappa', \xi') \in \mathcal{H}'\}.$$

On pose

$$S = S_1 \cup S_2.$$

11.4. On veut prouver qu'il existe une constante $c \neq 0$ qui vérifie l'assertions suivante. Pour tout $f_w \in C_c^\infty(\mathfrak{g}(F_w))$ et tout $f_w^H \in C_c^\infty(\mathfrak{h}(F_w))$ qui est un transfert de f_w (c'est-à-dire f_w^H vérifie la conclusion du théorème 8.2), pour tout $Y_w \in \Sigma_G(\mathfrak{h})$ on a l'égalité

$$c \, \hat{I}_{Y_w}^{G_w, H_w}(f) = \hat{S}_{Y_w}^{H_w}(f_w^H). \tag{11.1}$$

On verra d'ailleurs que $c = \prod_{v \in S-V_\infty, v\neq w} c_v^{-1}$ où la constante c_v est introduite à l'assertion 7 de la proposition 11.2.

Soit $Y_w \in \mathfrak{h}_{G-\mathrm{reg}}(F_w)$. On fixe un sous-ensemble ouvert compact Ω_w inclus dans $\mathfrak{h}_{G-\mathrm{reg}}(F_w)$ tel que $Y_w \in \Omega_w$ et que les deux membres de l'égalité à prouver soient constants sur Ω_w.

11.5. Soit V_∞ l'ensemble des places archimédiennes. Pour tout $v \in S - V_\infty$ et $v \neq w$, on fixe un sous-ensemble ouvert compact Ω_v inclus dans $\mathfrak{h}_{G-\mathrm{reg}}(F_v)$ tel que

(1) tout point de Ω_v est l'image d'un élément de $\mathfrak{g}_{\mathrm{reg}}(F_v)$ (c'est possible par la condition 2 du paragraphe 11.1) ;

(2) si $v \in S_2$ tout point de Ω_v a pour centralisateur un tore déployé (c'est possible puisque pour $v \in S_2$ on a vu ci-dessus que H_v était déployé) ;

(3) tout point de Ω_u est elliptique (c'est compatible avec la condition 1 par la condition 4 du paragraphe 11.1 et le lemme 4.7) ;

11.6. Pour tout sous-tore maximal T défini sur F de G^*, on a une application naturelle

$$\mathcal{E}(T, G^*, F_v) \to \mathcal{E}(T, G^*, \mathbb{A}_F/F)$$

et donc une application duale

$$\mathcal{K}(T, G^*, F) \to \mathcal{K}(T, G^*, F_v)$$

notée $\kappa \mapsto \kappa_v$.

11.7. Soit $Y_0 \in \mathfrak{h}(F)$ tel que

(1) pour tout $v \in S - V_\infty$, on ait $Y_0 \in \Omega_v$;

(2) pour tout $v \in V - S$, on ait $Y_0 \in b_v$.

L'approximation forte assure l'existence de Y_0. Soit $X_0^* \in \mathfrak{g}_{\mathrm{reg}}^*(F)$ tel que $Y_0 \to X_0^*$. On vérifie que X_0^* est semi-simple régulier et elliptique. Soit $\mathcal{Y} \subset \Sigma_G(\mathfrak{h})$ le sous-ensemble (fini) des Y tels que $Y \to X_0^*$ qu'on identifie à un système de représentants. On suppose bien sûr que $Y_0 \in \mathcal{Y}$. On a expliqué (cf. fin du paragraphe 4.4) comment tout élément de $\mathcal{Y}$ définit un élément $\kappa_Y \in \mathcal{K}(T_{X_0^*}, G^*, F)$. On pose

$$\kappa_0 = \kappa_{Y_0}$$

et

$$\mathcal{Y}_0 = \{Y \in \mathcal{Y} \mid \kappa_Y = \kappa_0\}.$$

11.8. Choix de fonctions hors S. — Pour tout $v \notin S$, soit f_v, resp. f_v^H, la fonction caractéristique de a_v, resp. b_v. Soit $c_v \neq 0$ la constante donnée par le théorème 8.4.

11.9. Constructions de fonctions aux places finies. — Soit $v \in S$ une place finie, distincte de w. Soit $X^* \in \mathfrak{g}_{\text{reg}}(F_v)$. Posons

$$\mathfrak{g}(X^*, F) = \chi_G^{-1}(\chi_{G^*}(X^*)) \cap \mathfrak{g}(F)$$

Soit $\kappa \in \mathcal{K}(T_{X^*}, G^*, F_v)$. On appelle κ-fonction au-dessus de X^* une fonction θ de $\mathfrak{g}(X^*, F)$ dans $\mathbb{C}$ qui vérifie

- θ est invariante par $G(F)$;
- pour $X_1, X_2 \in \mathfrak{g}(X^*, F)$, on a

$$\theta(X_1) = \theta(X_2)\langle \kappa_Y, \text{inv}(X^*; X_1, X_2)\rangle$$

Pour une telle fonction θ et $f \in C_c^\infty(\mathfrak{g}(F_v))$, on pose

$$I_{X^*}^\theta(f) = \sum_X \theta(X) I_X(f)$$

où la somme est prise sur $X \in \mathfrak{g}(X^*, F)$ modulo conjugaison. Lorsque $\theta = \Delta(Y, \cdot)$ (cf. 1.(7.2) du §7.4) pour $Y \in \mathfrak{h}(F)$ tel que $Y \to X^*$, on a

$$I_{X^*}^\theta(f) = I_Y^{G,H}(f).$$

Lorsque $G = G^*$ et φ est l'identité, on peut prendre $\theta = \text{inv}(X^*; \cdot, X^*)$ et on pose alors

$$I_{X^*}^\kappa(f) = I_{X^*}^\theta(f).$$

Venons-en à la construction suivante due à Waldspurger (cf. [**Wif**] proposition 8.2). Les notations non rappelées sont celles des paragraphes précédents.

Proposition 11.2. — *Il existe des fonctions $f \in C_c^\infty(\mathfrak{g}_{\text{reg}}(F_v))$ et $f^H \in C_c^\infty(\mathfrak{h}_{G-\text{reg}}(F_v))$ telles que*

(1) *Si $X \in \text{supp}(f)$, resp. $Y \in \text{supp}(f^H)$, il existe $X' \in \mathfrak{t}_{X_0^*}(F_v)$ tel que $\chi_{G^*}(X') = \chi_G(X)$, resp. tel que $Y \to X'$;*

(2) *f^H est un transfert de f ;*

(3) *soit $Z \in \mathfrak{g}_{\text{reg}}^*(F_v)$ et $\tau \in \mathcal{K}(T_Z, G^*, F_v)$. Supposons qu'il n'existe pas de $g \in G^*$ tel que la conjugaison par g induise un F_v-isomorphisme entre $(T_{X_0^*}, \kappa_0)$ et (T_Z, τ). Alors pour toute τ-fonction θ au-dessus de Z, on a*

$$I_Z^\theta(f) = 0\ ;$$

(4) *Soit $Y \in \mathfrak{h}_{G^*-\text{reg}}(F_v)$ et $\tau \in \mathcal{K}(T_Y, H, F_v)$. Si $\tau \neq 1$ alors*

$$I_Y^\tau(f^H) = 0\ ;$$

(5) *soit $\tau \in \mathcal{K}(T_{X_0^*}, G^*, F_v)$ et θ une τ-fonction au dessus de X_0^* non nulle. Alors pour $\tau \neq \kappa_0$ on a*

$$I_{X_0^*}^\tau(\hat{f}) = 0.$$

De plus, on a

$$I_{X_0^*}^{\kappa_0}(\hat{f}) \neq 0\ ;$$

(6) *Soit* $Y \in \mathcal{Y}$ *et* $\tau \in \mathcal{K}(T_Y, H, F_v)$. *Alors*

$$J_Y^\tau(\widehat{f^H}) = 0$$

dès que $Y \notin \mathcal{Y}_0$ *ou* $\tau \neq 1$. *De plus,*

$$I_Y^{G_v,H_v}(\hat{f}) = 0$$

si $Y \notin \mathcal{Y}_0$;

(7) *Il existe une constante* $c_v \neq 0$ *qui ne dépend que des données* $(G_v, H_v, \psi_v, \langle , \rangle)$ *telle que pour tout* $Y \in \mathcal{Y}_0$

$$S_Y^{H_v}(\widehat{f^H}) = c_v\, I_Y^{G_v,H_v}(\hat{f}) \neq 0.$$

Il n'est pas question ici de donner la construction élaborée de ces fonctions. En revanche, le lecteur pourra vérifier de lui-même qu'il est facile de construire des paires de fonctions qui vérifient les conditions 1 à 4 ci-dessus. Les conditions 5 à 7 sont bien plus difficiles à satisfaire. Leurs constructions reposent sur une expression explicite à «l'infini» des fonctions $\hat{i}^G$ du théorème. Plus précisément, soit $X, Z \in \mathfrak{g}_{\mathrm{reg}}(F_v)$. Waldspurger montre qu'il existe un entier naturel N tel que pour tout $\lambda \in F_v^\times$ tel que $v(\lambda) < -N$, la quantité $\hat{i}^G(\lambda X, Z)$ admette une expression explicite (cf. proposition VIII.1 de [**Wals-trace**]).

11.10. Pour tout $v \in S$ une place finie, distincte de w, soient $f_v \in C_c^\infty(\mathfrak{g}_{\mathrm{reg}}(F_v))$ et $f_v^H \in C_c^\infty(\mathfrak{h}_{G-\mathrm{reg}}(F_v))$ des fonctions qui satisfont les conclusions de la proposition 11.2.

11.11. Soit $v \in V_\infty$ une place archimédienne. On rappelle qu'on suppose que $F_v \simeq \mathbb{C}$. Soit $f_v \in \mathcal{S}(\mathfrak{g}(F_v))$ une fonction arbitraire. Soit $f_v^H \in \mathcal{S}(\mathfrak{h}(F_v))$ un transfert de f (cf. théorème 9.2).

11.12. Posons $f = \otimes_{v\in V} f_v$ et $f^H = \otimes f_v^H$. Soit $\mathrm{Out}_G(H)$ le groupe des automorphismes de la donnée (H, κ, ξ) modulo le groupe des automorphismes intérieurs. C'est un groupe fini. la proposition suivante est la proposition 10.9 de [**Wif**]. Posons

$$c^S = \prod_{v \notin S} c_v$$

où c_v est définie au §11.8.

Proposition 11.3. — *On a l'égalité*

$$J^H(f^H) = |\,\mathrm{Out}_G(H)| c^S \tau(H) \tau(G^*)^{-1} J^G(f).$$

Démonstration. — Partons de la ligne (10.3) du paragraphe 10.5. On a donc

$$J^G(f) = \tau(G^*) \sum_{X^* \in \Sigma_{\mathrm{ell}}(\mathfrak{g}^*)} \sum_{\kappa \in \mathcal{K}(T_{X^*}, G^*, F)} I_{X^*}^{G,\kappa}(f).$$

Soit $X^* \in \Sigma_{\mathrm{ell}}(\mathfrak{g}^*)$ et $\kappa \in \mathcal{K}(T_{X^*}, G^*, F)$. Soit $(H', \kappa', \xi'$ la donnée endoscopique elliptique de G^* et $Y \in \Sigma(\mathfrak{h}')$ tel que $Y \to X^*$ et $\kappa_Y = \kappa$ (cf. lemme 10.3). On a donc (cf. §10.8)

$$I_{X^*}^{G,\kappa}(f) = \prod_{v \in V} I_Y^{G_v, H'_v}(f_v)$$

(comme d'habitude dans les produits infinis que nous considérons presque tous les facteurs sont égaux à 1.) Dans la suite, on suppose que

$$I_{X^*}^{G,\kappa}(f) \neq 0.$$

Cela implique que pour tout $v \notin S$ on $I_Y^{G_v, H'_v}(f_v) \neq 0$. Or f_v est une fonction caractéristique d'un réseau hyperspécial. On en déduit que H'_v est non-ramifié (il s'agit d'une version d'un lemme dû à Kottwitz, cf. [**Wif**] lemme 7.4). Par ailleurs pour $v \in S_2$, on sait que $I_Y^{G_v, H'_v}(f_v) \neq 0$ implique que le centralisateur de Y est déployé (par la condition 1 de la proposition 11.2 et la condition 2 du §11.5). Donc H est non ramifié hors de S_1.

On peut donc supposer que $(H', \kappa', \xi') \in \mathcal{H}$. Si $(H', \kappa', \xi') \in \mathcal{H}'$, pour $v = v_{(H', \kappa', \xi')}$, on a $I_Y^{G_v, H'_v}(f_v) \neq 0$ et donc comme ci-dessus H'_v est déployé : contradiction. Donc $F[H'] \subset F[H]$.

On utilise maintenant la place u. Par la condition 3 de la proposition 11.2 et le lemme 10.3, on voit que les données (H_u, κ, ξ) et $(H'_u, , \kappa', \xi')$ sont équivalentes sur F_u. En utilisant la condition 3 du paragraphe 11.1, on montre que cet isomorphisme se prolonge en un F-isomorphisme.

Soit $\Sigma_G(\mathfrak{h})(X^*, \kappa)$ le sous-ensemble de $\Sigma_G(\mathfrak{h})$ formé des Y tels que $Y \to X^*$ et $\kappa_Y = \kappa$. Alors lorsque $\Sigma_G(\mathfrak{h})(X^*, \kappa)$ est non vide il est de cardinal $|\,\mathrm{Out}_G(H)|$ (cf. lemme 10.3).

On en déduit que

$$J^G(f) = |\,\mathrm{Out}_G(H)|^{-1} \sum_{X^* \in \Sigma_{\mathrm{ell}}(\mathfrak{g}^*)} \; \sum_{\kappa \in \mathcal{K}(T_{X^*}, G^*, F)} \; \sum_{Y \in \Sigma_G(\mathfrak{h})(X^*, \kappa)} \; \prod_{v \in V} I_Y^{G_v, H_v}(f_v)$$

On va faire un calcul analogue pour le groupe H. Il résulte du fait que le support de f_u^H est dans les éléments G-réguliers que l'on peut écrire

$$J^H(f^H) = \tau(H) \sum_{Y \in \Sigma_{G,\mathrm{ell}}(\mathfrak{h})} \; \sum_{\kappa \in \mathcal{K}(T_Y, H, F)} I_Y^{H,\kappa}(f^H).$$

Soit $Y \in \Sigma_{G,\mathrm{ell}}(\mathfrak{h})$ et $\kappa \in \mathcal{K}(T_Y, H, F)$. On a défini au §11.6 des éléments $\kappa_v \in \mathcal{K}(T_Y, H, F_v)$ pour toute place v. On vérifie que

$$I_Y^{H,\kappa}(f^H) = \prod_{v \in V} I_Y^{H,\kappa_v}(f_v^H).$$

Donc si ce produit n'est pas nul, en particulier $I_Y^{H,\kappa_u}(f_u^H)$ ne l'est pas non plus ce qui force $\kappa_u = 1$ (cf. condition 4 de la proposition 11.2.) Comme T_Y est F-elliptique et même F_u-elliptiques (cela résulte de la construction de f_u et de la condition 3 du §11.5), on voit facilement (avec la dualité de Tate-Nakayama par exemple) que l'application

$$\mathcal{K}(T_Y, H, \mathbb{A}_F/F) \to \mathcal{K}(T_Y, H, F_u)$$

est injective. Par conséquent, on a $\kappa = 1$. On en déduit immédiatement que

$$J^H(f^H) = \tau(H) \sum_{Y \in \Sigma_{G,\mathrm{ell}}(\mathfrak{h})} \prod_{v \in V} S_Y^{H_v}(f_v^H)$$

En utilisant les propriétés de transfert entre les fonctions f_v et f_v^H et en particulier le lemme fondamental, on écrit

$$J^H(f^H) = \tau(H)\, c^S \sum_{Y \in \Sigma_{G,\mathrm{ell}}(\mathfrak{h})} \prod_{v \in V} I^{G_v,H_v}(Y, f_v).$$

Il suffit d'utiliser le fait que $\Sigma_{G,\mathrm{ell}}(\mathfrak{h})$ est la réunion disjointe des $\Sigma_{G,\mathrm{ell}}(\mathfrak{h})(X^*, \kappa)$ lorsque $X^* \in \Sigma_{\mathrm{ell}}(\mathfrak{g}^*)$ et $\kappa \in \mathcal{K}(T_{X^*}, G^*, F)$ pour conclure. □

11.13. Raffinement des conditions aux places archimédiennes. — On peut supposer de plus (cf. [**Wif**] lemme 10.7) que les fonctions archimédiennes sont choisies de sorte qu'on ait

(1) pour tout $v \in V_\infty$,

$$I_{\varphi_v^{-1}(X_0^*)}(\hat{f}_v) \neq 0\ ;$$

(2) soit $X \in \mathfrak{g}^*(F)$. Si en toute place v, $\varphi_v^{-1}(X_v)$ est conjugué par un élément de $G(\bar{F}_v)$ à un élément du support de $\hat{f}_v$, alors X est stablement conjugué à X_0^* ;

(3) soit $Y \in \mathfrak{h}(F)$. Si en toute place v, Y est conjugué par un élément de $H(\bar{F}_v)$ à un élément du support de $\widehat{f_v^H}$, alors $Y \to X_0^*$;

Bien sûr, pour tout $v \in V_\infty$, on continue de supposer que f_v^H est un transfert de f_v ce qui implique donc que $\widehat{f_v^H}$ est un transfert de $\hat{f}_v$ (cf. théorème 9.2).

11.14. Utilisation de la formule sommatoire de Poisson. — Celle-ci se traduit par l'identité suivante : pour tout $x \in G(\mathbb{A})$

$$\sum_{X \in \mathfrak{g}(F)} f(\mathrm{Ad}(x)^{-1}X) = \sum_{X \in \mathfrak{g}(F)} \hat{f}(\mathrm{Ad}(x)^{-1}X).$$

D'après la forme du support de f_u, on voit que la somme de gauche porte effectivement sur les $X \in \mathfrak{g}_{\mathrm{ell}}(F)$. Vu les considérations du paragraphe précédent, la somme de droite porte elle aussi uniquement sur les $X \in \mathfrak{g}_{\mathrm{ell}}(F)$. Par intégration sur $G(F)\backslash G(\mathbb{A})^1$ (cf. théorème 10.1), on obtient l'égalité

$$J^G(f) = J^G(\hat{f}). \tag{11.2}$$

Le même raisonnement donne aussi

(11.3) $$J^H(f^H) = J^H(\widehat{f^H}).$$

11.15. Calcul de $J^G(\hat{f})$. — On a le lemme suivant.

Lemme 11.4. — *On a l'égalité*

$$J^G(\hat{f}) = \tau(G^*) I_{X_0^*}^{\kappa_0}(\hat{f}).$$

Démonstration. — On utilise la pré-stabilisation de $J^G(\hat{f})$. Comme on a, de plus, la condition 2 du paragraphe 11.13, on aboutit à

$$J^G(\hat{f}) = \tau(G^*) \sum_{\kappa \in \mathcal{K}(T_{X_0^*}, G^*, F)} I_{X_0^*}^{\kappa}(\hat{f}).$$

Soit $\kappa \in \mathcal{K}(T_{X_0^*}, G^*, F)$ tel que $I_{X_0^*}^{\kappa}(\hat{f}) \neq 0$. Soit (H', κ', ξ) et $Y \in \mathfrak{h}'_{G-\mathrm{reg}}(F)$ tel que $Y \to X_0^*$ et $\kappa'_Y = \kappa$ (cf. lemme 10.3). On a donc

$$I_{X_0^*}^{\kappa}(\hat{f}) = \prod_{v \in V} I_Y^{G_v, H'_v}(\hat{f}_v).$$

Or puisque $I_Y^{G_u, H'_u}(\hat{f}_u) \neq 0$, on voit d'après la condition 5 de la proposition 11.2 que $\kappa_{0,u} = \kappa_u$. Comme le tore $T_{X_0^*}$ est F-elliptique et F_u-elliptique, on sait (voir plus haut) que l'application de localisation des caractères endoscopiques est injective. Par conséquent, on a

$$\kappa_0 = \kappa$$

ce qu'il fallait démontrer.□

11.16. Calcul de $J^H(\widehat{f^H})$. — On a le lemme suivant.

Lemme 11.5. — *On a l'égalité*

$$J^H(\widehat{f^H}) = \tau(H) \sum_{Y \in \mathcal{Y}_0} \prod_{v \in V} S_Y^{H_v}(\widehat{f_v^H}).$$

Démonstration. — On utilise la pré-stabilisation de $J^H(\widehat{f^H})$. Celle-ci se traduit par l'égalité

$$J^H(\widehat{f^H}) = \sum_{Y \in \mathcal{Y}} \sum_{\kappa \in \mathcal{K}(T_Y, H, F)} I_Y^{\kappa}(\widehat{f^H}).$$

Soit $Y \in \mathcal{Y}$ et $\kappa \in \mathcal{K}(T_Y, H, F)$ tels que $I_Y^{\kappa}(\widehat{f^H}) \neq 0$. On a

$$I_Y^{\kappa}(\widehat{f^H}) = \prod_{v \in V} I_Y^{\kappa_{Y,v}}(\widehat{f_v^H}).$$

Il résulte de la condition 6 de la proposition 11.2 que $Y \in \mathcal{Y}_0$ et $\kappa_u = 1$. D'où toujours par le même argument d'injectivité $\kappa = 1$. Le produit ci-dessus est donc composé d'intégrales orbitales stables.□

11.17. Conclusion. — Il résulte des lemmes 11.4 et 11.5 ci-dessus, de la proposition 11.3 et des égalités (11.2) et (11.3) qu'on a l'égalité

$$\sum_{Y\in\mathcal{Y}_0}\prod_{v\in V} S_Y^{H_v}(\widehat{f_v^H}) = c^S|\operatorname{Out}_G(H)|I_{X_0^*}^{\kappa_0}(\hat{f}) \tag{11.4}$$

Posons

$$c^w = \prod_{v\in V-V_\infty, v\neq w} c_s.$$

En utilisant le lemme fondamental, les propriétés des fonctions f_v aux places finies $v\in S$ distinctes de w (propriété 7 de la proposition 11.2) et les propriétés de f_v aux places archimédiennes, on obtient que pour tout $Y\in\mathcal{Y}_0$

$$\prod_{v\in V} S_Y^{H_v}(\widehat{f_v^H}) = c^w S_Y^{H_w}(\widehat{f_w^H})\prod_{v\in V, v\neq w} I_Y^{G_v,H_v}(\hat{f}_v).$$

Soit $v\in V$, $v\neq w$. Soit $X_0\in\mathfrak{g}(F_v)$ tel $Y\to X_0$. Il revient au même de demander que X_0 et X_0^* ont même image dans $\mathfrak{car}$. En particulier, la condition précédente ne dépend pas du choix de $Y\in\mathcal{Y}_0$. Alors pour tout $X\in\mathfrak{g}(F_v)$ tel que $Y\to X$, on a (cf. 1.(7.2) du 7.4)

$$\begin{aligned}\Delta_v(Y,X) &= \Delta_v(Y,X_0)\langle\kappa_{Y,v},\operatorname{inv}(X^*;X,X_0)\rangle\\ &= \Delta_v(Y,X_0)\langle\kappa_{0,v},\operatorname{inv}(X^*;X,X_0)\rangle\\ &= \frac{\Delta_v(Y,X_0)}{\Delta_v(Y_0,X_0)}\Delta_v(Y_0,X)\end{aligned}$$

Par conséquent, on a

$$I_Y^{G_v,H_v}(\hat{f}_v) = \frac{\Delta_v(Y,X_0)}{\Delta_v(Y_0,X_0)} I_{Y_0}^{G_v,H_v}(\hat{f}_v).$$

En utilisant la formule du produit pour les facteurs de transfert (cf. §10.7), on aboutit à

$$\prod_{v\in V, v\neq w} I_Y^{G_v,H_v}(\hat{f}_v) = \frac{\Delta_w(Y_0,X_0)}{\Delta_w(Y,X_0)}\prod_{v\in V, v\neq w} I_{Y_0}^{G_v,H_v}(\hat{f}_v)$$

Lemme 11.6. — *Pour $Y\in\mathcal{Y}_0$, on a l'égalité*

$$S_{Y_0}^{H_w}(\widehat{f_w^H}) = \frac{\Delta_w(Y_0,X_0)}{\Delta_w(Y,X_0)} S_Y^{H_w}(\widehat{f_w^H})$$

Démonstration. — Esquissons une démonstration (pour plus de détails, cf. section 3 de [**Wif**]). Puisque $\kappa_Y=\kappa_{Y_0}$, il existe un automorphisme i de (H,κ,ξ) qui envoie Y_0 sur Y (cf. lemme 10.3). On démontre (cf. lemme 2.4 de [**Wif**]) qu'il existe une constante $\delta\neq 0$ tel que pour tout $Z\in\mathfrak{h}_{G-\mathrm{reg}}(F_w)$ et tout $X\in\mathfrak{g}(F_w)$ tel que $Z\to X$ on ait

$$\Delta_w(i(Z),X) = \delta\Delta_w(Z,X). \tag{11.5}$$

En particulier, comme f_w^H est par définition un transfert de f, on obtient : pour tout $Z \in \mathfrak{h}_{G-\mathrm{reg}}(F_w)$

$$S_{i(Z)}^{H_w}(f_w^H) = \delta S_Z^{H_w}(f_w^H).$$

On remarque ensuite que l'isomorphisme i préserve la forme $\langle\cdot,\cdot\rangle$ sur $\mathfrak{h}(F_w)$. On en déduit qu'elle préserve aussi les fonctions $\hat{\imath}^{H_w}$. En utilisant le théorème 6.2, on aboutit à l'égalité

$$S_{i(Z)}^{H_w}(\widehat{f_w^H}) = \delta S_Z^{H_w}(\widehat{f_w^H}).$$

pour tout $Z \in \mathfrak{h}_{G-\mathrm{reg}}(F_w)$. Pour conclure, il suffit d'évaluer δ (prendre $(Z,X) = (Y_0, X_0)$ dans (11.5).□

En utilisant le lemme précédent, on a donc

$$\prod_{v\in V} S_Y^{H_v}(\widehat{f_v^H}) = c^w S_{Y_0}^{H_w}(\widehat{f_w^H}) \prod_{v\in V, v\neq w} I_{Y_0}^{G_v,H_v}(\hat{f}_v).$$

En particulier, ce produit ne dépend pas de Y. L'égalité (11.4) devient

$$c^w S_{Y_0}^{H_w}(\widehat{f_w^H}) \prod_{v\in V, v\neq w} I_{Y_0}^{G_v,H_v}(\hat{f}_v) = c^S I_{X_0^*}^{\kappa_0}(\hat{f}).$$

Comme on a aussi

$$I_{X_0^*}^{\kappa_0}(\hat{f}) = \prod_{v\in V} I_{Y_0}^{G_v,H_v}(\hat{f}_v)$$

on en déduit que

$$\left(c^w S_{Y_0}^{H_w}(\widehat{f_w^H}) - c^S I_{Y_0}^{G_w,H_w}(\hat{f}_w)\right) \prod_{v\in V, v\neq w} I_{Y_0}^{G_v,H_v}(\hat{f}_v) = 0.$$

On a tout fait pour que le second facteur soit non nul on a donc

$$c^w S_{Y_0}^{H_w}(\widehat{f_w^H}) = c^S I_{Y_0}^{G_w,H_w}(\hat{f}_w)$$

et cette égalité est vraie aussi en Y_w (cf. le choix de Ω_w au §11.4) d'où (11.1).

11.18. Remarque finale. — Cette technique d'isoler une classe de conjugaison globale dans la formule des traces par des choix locaux est utilisée aussi par Hales dans l'article [**Hales**] où il montre que le lemme fondamental à presque toutes les places implique le lemme fondamental à toutes les places où il fait sens et pour toutes les fonctions de l'algèbre de Hecke sphérique. En fait, le résultat de Waldspurger est un analogue pour les algèbres de Lie de celui de Hales.

Références

[Bouaziz1] A. Bouaziz, Intégrales orbitales sur les algbres de lie rductives. *Invent. math.* 115 :163–207, 1994.

[Hales] T. Hales, On the fundamental lemma for standard endoscopy : reduction to unit elements. *Canad. J. Math.* **47**(5) :974–994, 1995.

[HC-Debacker] Harish-Chandra, Admissible invariant distributions on reductive p-adic groups. Notes by S.DeBacker and P.J. Sally Jr, volume 16 of Univ. Lect. Series, *Amer. Math. Soc.,* 1999.

[Ko2] R. Kottwitz, Stable trace formula : elliptic singular terms, *Math. Ann.* 275 :365–399, 1986.

[KoTF] R. Kottwitz. Transfer factors for Lie algebras, *Represent. Theory,* 3 :127–138 (electronic), 1999.

[LabBreen] J.-P. Labesse, Cohomologie, stabilisation et changement de base, *Astérisque*, (257) :vi+161, 1999. Appendice A par L. Clozel et J.-P. Labesse, et Appendice B par L. Breen.

[Langlands-formule-stable] R. Langlands, Les débuts d'une formule des traces stable, volume 13 of *Publications Mathématiques de l'Université Paris VII*, Université de Paris VII U.E.R. de Mathématiques, Paris, 1983.

[LS] R. Langlands & D. Shelstad, On the definition of transfer factors, *Math. Ann.* 278 :219–271, 1987.

[Ngo2] B. C. Ngô, Le lemme fondamental pour les algébres de Lie, *Publ. Math. Inst. Hautes Études Sci.,* **111** (2010) :1–169.

[Wals-trace] J.-L. Waldspurger, Une formule des traces locale pour les algébres de Lie p-adiques, *J. Reine Angew. Math.* 465 :41–99, 1995.

[Wif] J.-L. Waldspurger, Le lemme fondamental implique le transfert *Compositio Math.* 105 :153–236, 1997.

[W-car] J.-L. Waldspurger, Endoscopie et changement de caractéristique, *J. Inst. Math. Jussieu*, 5(3) :423–525, 2006.

CNRS et Université Paris-Sud, UMR 8628, Mathématique, Bâtiment 425, F-91405 Orsay Cedex, France

E-mail address: Pierre-Henri.Chaudouard@math.u-psud.fr

A propos du lemme fondamental tordu
[W.II.B.2]

J.-L. Waldspurger

1. Introduction

On présente une preuve de l'assertion suivante : le lemme fondamental tordu résulte du lemme fondamental pour les algèbres de Lie et d'un lemme que nous appelons lemme fondamental non standard. Ces différents lemmes seront énoncés précisément dans la section 3. Disons tout de suite que ce que nous appelons ici lemme fondamental tordu devrait plutôt être appelé lemme fondamental tordu pour les unités des algèbres de Hecke. Par ailleurs, Ngo Bao Chau a récemment démontré à la fois le lemme fondamental pour les algèbres de Lie et le lemme fondamental non standard. Le bilan est que le lemme fondamental tordu, au sens où nous l'entendons, est vrai. Le présent article est pour l'essentiel extrait de l'article [W1]. La preuve donnée dans cette référence est presque entièrement reproduite ici. Par contre, on passe plus rapidement sur certains préliminaires (fonction exponentielle, sous-groupes hyperspéciaux, normalisation des mesures etc.). On renvoie pour les détails à [W1] ou à des articles plus généraux de théorie des groupes. De même, plusieurs propriétés ne sont vérifiées que si p est grand. On indique une borne précise en 2.7, mais on ne cherchera pas à la justifier. Dans [W1], on se préoccupait non seulement du lemme fondamental, mais aussi du transfert, ce qui conduisait à considérer une situation parfaitement générale. Se limiter au lemme fondamental simplifie sensiblement la situation : les groupes de départ sont non ramifiés et on n'a à considérer que des éléments compacts de ces groupes, c'est-à-dire des éléments qui engendrent des sous-groupes relativement compacts. Cela permet parfois de simplifier les preuves, ce que l'on a tenté de faire. On a ajouté quelques exemples. En particulier, dans la section 5, on considère en détail le cas du changement de base pour les groupes unitaires.

2. La situation

2.1. Corps locaux. Soit F un corps local non archimédien de caractéristique nulle. On note $\mathfrak{o}$ son anneau d'entiers, $\mathbb{F}_q$ son corps résiduel, q le nombre d'éléments de $\mathbb{F}_q$ et p sa caractéristique, $|.|_F$ la valeur absolue usuelle de F. On fixe une clôture algébrique $\bar{F}$ de F, on note F^{nr} le plus grand sous-corps de $\bar{F}$ non ramifié sur F, $\bar{\mathbb{F}}_q$ son corps résiduel, $\mathfrak{o}_{\bar{F}}$ et $\mathfrak{o}_{F^{nr}}$ les anneaux d'entiers de $\bar{F}$ et F^{nr}. On prolonge $|.|_F$ en une valeur absolue sur $\bar{F}$, à valeurs dans $\mathbb{Q}$. On note Γ, resp. Γ^{nr}, le groupe de Galois de l'extension $\bar{F}$ sur F, resp. F^{nr} sur F. Le groupe Γ^{nr} s'identifie au groupe de Galois de l'extension $\bar{\mathbb{F}}_q$ sur $\mathbb{F}_q$. Il est engendré topologiquement par l'élément de Frobenius, que l'on note ϕ. On note $W^{nr} \subset \Gamma^{nr}$ le sous-groupe formé des puissances entières de ϕ. Il y a une suite exacte :

$$1 \to I \to \Gamma \to \Gamma^{nr} \to 1$$

où I est le sous-groupe d'inertie. On note W le groupe de Weil de l'extension $\bar{F}$ sur F. Il y a une suite exacte :

$$1 \to I \to W \to W^{nr} \to 1.$$

On peut considérer W comme un sous-groupe de Γ, muni d'une topologie plus fine que la topologie induite. On utilisera des notations similaires pour des extensions E de F, en les affectant d'un indice E (par exemple $\mathfrak{o}_E$, Γ_E etc.)

2.2. Groupes. Soit M un groupe agissant sur un ensemble X. Pour tout sous-ensemble $X' \subset X$, on note $Norm_M(X) = \{m \in M; m(X') \subset X\}$ et $Z_M(X') = \{m \in M; \forall x \in X', m(x) = x\}$. Si $X' = \{x\}$, on pose simplement $Z_M(x) = Z_M(\{x\})$. Quand l'ensemble X ne sera pas précisé, il sera entendu que $X = M$ et que M agit par conjugaison. On note Z_M le centre de M. Si θ est un automorphisme de M, on note M^θ le sous-groupe des points fixes. Si M est abélien, on pose $(1-\theta)(M) = \{m\theta(m)^{-1}; m \in M\}$ et $M_{/\theta} = M/(1-\theta)(M)$. Si X est une variété algébrique définie sur un corps algébriquement clos k, on note encore X son groupe de points sur k, c'est-à-dire $X = X(k)$ (on fera toutefois quelques exceptions dans des cas ambigus). Soit M un groupe algébrique défini sur k. On note M^0 sa composante neutre. Supposons que M agisse algébriquement sur la variété algébrique X. Pour $x \in X$, on pose $M_x = (Z_M(x))^0$.

2.3. Groupes réductifs. Soit G un groupe réductif connexe défini sur $\bar{F}$. On note $\mathfrak{g}$ son algèbre de Lie, G_{SC} le revêtement simplement connexe de son groupe dérivé, G_{AD} son groupe adjoint. Pour $g \in G$, on note g_{ad} son image dans G_{AD}. Par contre, pour $g \in G_{SC}$, on note simplement g son image dans G. Si H est un sous-groupe de G, on note H_{sc} l'image réciproque de H dans G_{SC} et H_{ad} l'image de H dans G_{AD}. Ces groupes dépendent du groupe ambiant G. Le groupe H_{sc}, resp. H_{ad}, n'a aucune raison d'être simplement connexe, resp. adjoint. Pour $g \in G$, on note int_g l'automorphisme intérieur de G associé à g, c'est-à-dire l'automorphisme $x \to gxg^{-1}$. Cet automorphisme se relève en un automorphisme de G_{SC}, il se descend en un automorphisme de G_{AD} et définit par dérivation un automorphisme de $\mathfrak{g}$. On note encore int_g ces automorphismes. On notera aussi $X \to gXg^{-1}$ celui de $\mathfrak{g}$. Ces automorphismes ne dépendent que de g_{ad} et, en fait, on peut aussi les définir pour $g \in G_{AD}$. On appelle paire de Borel de G un couple (B,T) où B est un sous-groupe de Borel de G et T un sous-tore maximal de B. Considérons une telle paire. On note $X_*(T)$, resp. $X^*(T)$, le groupe des cocaractères, resp. des caractères, de T, $\Sigma \subset X^*(T)$ l'ensemble des racines de T dans $\mathfrak{g}$, $\check{\Sigma} \subset X_*(T)$ l'ensemble de coracines associé, $\Delta \subset \Sigma$ et $\check{\Delta} \subset \check{\Sigma}$ les ensembles de racines et coracines simples définis par B. Pour $\alpha \in \Sigma$, on note $\mathfrak{g}_\alpha \subset \mathfrak{g}$ la droite radicielle correspondante. Un épinglage relatif à (B,T) est une famille $(E_\alpha)_{\alpha\in\Delta}$ où, pour tout $\alpha \in \Delta$, E_α est un élément non nul de $\mathfrak{g}_\alpha$. On appelle paire de Borel épinglée une famille $(B,T,(E_\alpha)_{\alpha\in\Delta})$ d'objets comme ci-dessus.

Supposons G défini sur F. On dit qu'une paire de Borel (B,T) est définie sur F si B et T le sont. Si c'est le cas, Γ agit sur $X^*(T)$ en préservant Σ et Δ. On dit qu'une paire de Borel épinglée $(B,T,(E_\alpha)_{\alpha\in\Delta})$ est définie sur F si (B,T) l'est et si, de plus, $\sigma(E_\alpha) = E_{\sigma(\alpha)}$ pour tous $\sigma \in \Gamma$ et $\alpha \in \Delta$. L'ensemble des paires de Borel épinglées définies sur F n'est pas vide et forme une seule classe de conjugaison par $G_{AD}(F)$. On utilisera des notations similaires si le corps de base F est remplacé par $\mathbb{F}_q$ et la clôture algébrique $\bar{F}$ est remplacée par $\bar{\mathbb{F}}_q$.

2.4. Eléments compacts, tores non ramifiés. Soit M un groupe algébrique défini sur F, de composante neutre M^0 réductive. Pour $x \in M(F)$, on dit que x est compact si le sous-groupe qu'il engendre est d'adhérence compacte dans $M(F)$. On dit que x est topologiquement unipotent si $lim_{n\to\infty}x^{p^n} = 1$. On note $M(F)_c$, resp. $M(F)_{tu}$, le sous-ensemble des éléments compacts, resp. topologiquement unipotents, de $M(F)$. Remarquons que, si M/M^0 est d'ordre premier à p, tout élément de $M(F)_{tu}$ appartient à $M^0(F)$. On note $M(F)_{p'}$ le sous-ensemble des éléments de $M(F)$ d'ordre fini premier à p. Si M est abélien, ces sous-ensembles sont des sous-groupes. Tout élément $x \in M(F)_c$ se décompose de façon unique en produit $x = x_{tu}x_{p'}$, où $x_{tu} \in M(F)_{tu}$, $x_{p'} \in M(F)_{p'}$ et ces deux éléments commutent entre eux. De plus, ces deux éléments appartiennent à l'adhérence du sous-groupe engendré par x.

Puisque M est réunion des $M(E)$ quand E parcourt les extensions finies de F contenues dans $\bar{F}$, on n'a aucun mal à remplacer $M(F)$ par M dans les définitions et résultats ci-dessus.

Quand M est le groupe multiplicatif $\mathbb{G}_m$, on a $M(F) = F^\times$ et $M(F)_c = \mathfrak{o}^\times$. Le groupe $M(F)_{tu}$ est le sous-groupe $\mathfrak{o}^\times_{tu}$ des éléments de $\mathfrak{o}^\times$ dont la réduction dans $\mathbb{F}_q$ est égale à 1. Le groupe $M(F)_{p'} = \mathfrak{o}^\times_{p'}$ est d'ordre $q-1$

et l'application de réduction l'identifie à $\mathbb{F}_q^\times$. Soit T un tore défini sur F et non ramifié, c'est-à-dire déployé sur F^{nr}. On a $T(F^{nr}) = X_*(T) \otimes_{\mathbb{Z}} F^{nr,\times}$ et $T(F) = T(F^{nr})^{\Gamma^{nr}}$. Le tore T possède une structure de schéma en groupes lisse sur $\mathfrak{o}$, pour laquelle $T(\mathfrak{o}_{F^{nr}}) = X_*(T) \otimes_{\mathbb{Z}} \mathfrak{o}_{F^{nr}}^\times$ et $T(\mathfrak{o}) = T(\mathfrak{o}_{F^{nr}})^{\Gamma^{nr}}$. On a les égalités :

$$T(F^{nr})_c = T(\mathfrak{o}_{F^{nr}}), T(F^{nr})_{tu} = X_*(T) \otimes_{\mathbb{Z}} \mathfrak{o}_{F^{nr},tu}^\times,$$

$$T(F^{nr})_{p'} = X_*(T) \otimes_{\mathbb{Z}} \mathfrak{o}_{F^{nr},p'}^\times, T(F)_c = T(\mathfrak{o}), T(F)_{tu} = T(F^{nr})_{tu}^{\Gamma^{nr}},$$
$$T(F)_{p'} = T(F^{nr})_{p'}^{\Gamma^{nr}}.$$

Fixons une uniformisante ϖ de F et notons $T_\varpi = X_*(T) \otimes_{\mathbb{Z}} \varpi^{\mathbb{Z}}$. Les applications produits :

$$T(\mathfrak{o}_{F^{nr}}) \times T_\varpi \to T(F^{nr}), T(\mathfrak{o}) \times T_\varpi^{\Gamma^{nr}} \to T(F)$$

sont des isomorphismes. On a $T_{p'} = X_*(T) \otimes_{\mathbb{Z}} \mathfrak{o}_{\bar{F},p'}^\times$. Mais $\mathfrak{o}_{\bar{F},p'}^\times = \mathfrak{o}_{F^{nr},p'}^\times$, car extraire des racines d'ordre premier à p ne crée que des extensions non ramifiées. Donc $T_{p'} = T(F^{nr})_{p'} = T(\mathfrak{o}_{F^{nr}})_{p'}$. La fibre spéciale $\mathbf{T}$ du schéma en groupes est le tore sur $\mathbb{F}_q$ de groupe de cocaractères $X_*(T)$. L'application de réduction identifie $T(\mathfrak{o}_{F^{nr}})_{p'}$ à $\mathbf{T}(\bar{\mathbb{F}}_\mathbf{q})$ et $T(\mathfrak{o})_{p'}$ à $\mathbf{T}(\mathbb{F}_\mathbf{q})$.

2.5. Sous-groupes hyperspéciaux. Soit G un groupe réductif connexe défini sur F. On suppose G non ramifié, c'est-à-dire quasi-déployé sur F et déployé sur une extension non ramifiée de F. A la suite de Bruhat et Tits, on introduit l'immeuble $Imm(G, F)$ de G sur F et la notion de point hyperspécial de cet immeuble ([T] 1.11.2). Le groupe $G(F)$ agit sur l'immeuble. Le fixateur d'un point hyperspécial est un sous-groupe ouvert compact de $G(F)$ que l'on appelle sous-groupe hyperspécial. A un sous-groupe hyperspécial K est attaché un schéma en groupes lisse G_K sur $\mathfrak{o}$, de fibre générique G, tel que $G_K(\mathfrak{o}) = K$. Ce schéma en groupes possède une algèbre de Lie $\mathfrak{g}_K$. Le réseau $\mathfrak{k} = \mathfrak{g}_K(\mathfrak{o}) \subset \mathfrak{g}(F)$ est dit lui-aussi hyperspécial. Si E est une extension non ramifiée de F, on pose $K_E = G_K(\mathfrak{o}_E)$ et $\mathfrak{k}_E = \mathfrak{g}_K(\mathfrak{o}_E)$. Dans le cas où $E = F^{nr}$, on pose simplement $K^{nr} = K_{F^{nr}}$ et $\mathfrak{k}^{nr} = \mathfrak{k}_{F^{nr}}$. On note $\mathbf{G_K}$ et $\boldsymbol{\mathfrak{g}}_K$ les fibres spéciales de G_K et $\mathfrak{g}_K$.

Soit $(B, T, (E_\alpha)_{\alpha\in\Delta})$ une paire de Borel épinglée de G définie sur F. Elle détermine un sous-groupe hyperspécial K, caractérisé de la façon suivante. Soit $\alpha \in \Delta$. On peut identifier $\check{\alpha} \in X_*(T)$ à un élément de $\mathfrak{t}$. Il existe un unique $E_{-\alpha} \in \mathfrak{g}_{-\alpha}$ tel que le triplet $(E_\alpha, \check{\alpha}, E_{-\alpha})$ forme un $\mathfrak{sl}_2$-triplet. De E_α et $E_{-\alpha}$ se déduisent des homomorphismes e_α et $e_{-\alpha}$ du groupe additif $\mathbb{G}_a$ dans G (les dérivés de ces homomorphismes envoient 1 sur E_α et $E_{-\alpha}$). Alors K^{nr} est le sous-groupe de $G(F^{nr})$ engendré par $T(\mathfrak{o}_{F^{nr}})$ et les sous-groupes $e_\alpha(\mathfrak{o}_{F^{nr}})$ et $e_{-\alpha}(\mathfrak{o}_{F^{nr}})$ pour $\alpha \in \Delta$. On a $K = K^{nr,\Gamma^{nr}}$.

Inversement, soit K un sous-groupe hyperspécial. Soit T_0 un sous-tore déployé maximal de G tel que K fixe un point hyperspécial de l'appartement de $Imm(G, F)$ associé à T_0. Soit $T = Z_G(T_0)$. Complétons T en une paire de

Borel (B,T) de G définie sur F. Alors on peut choisir un épinglage $(E_\alpha)_{\alpha\in\Delta}$ relatif à cette paire de sorte que la paire de Borel épinglée $(B,T,(E_\alpha)_{\alpha\in\Delta})$ soit définie sur F et que le sous-groupe hyperspécial qu'on lui associe comme ci-dessus soit K. Les sous-groupes hyperspéciaux de $G(F)$ forment une seule classe de conjugaison par $G_{AD}(F)$. Soit K un tel sous-groupe. Le normalisateur de K dans $G(F)$ est égal à $Z_G(F)K$. L'image réciproque K_{sc} de K dans $G_{SC}(F)$, resp. l'image K_{ad} de K dans $G_{AD}(F)$, est un sous-groupe hyperspécial de $G_{SC}(F)$, resp. $G_{AD}(F)$.

2.6. L'exponentielle. Soit G un groupe réductif connexe défini sur F. On note $dim(G)$ sa dimension et $rang(G)$ la dimension d'un sous-tore maximal de G. Notons ϕ la fonction indicatrice d'Euler et, pour tout entier $n\geq 1$, $\psi(n)$ le plus grand entier d tel que $\phi(d)\leq n$. On pose :

$$N(G)=sup(\psi(rang(G)),dim(G)).$$

Notons e_F l'indice de ramification de F sur $\mathbb{Q}_p$. On suppose :

(1) $p>N(G)e_F+1$.

Soit $X\in\mathfrak{g}(F)$. Notons X_s sa partie semi-simple et T le plus grand tore central dans le commutant G_{X_s} de X_s. On a $X_s\in\mathfrak{t}(F)$. Tout élément de $X^*(T)$ définit une forme linéaire sur $\mathfrak{t}$, à valeurs dans $\bar{F}$. On dit que X est topologiquement nilpotent si et seulement si $|x^*(X)|_F<1$ pour tout $x^*\in X^*(T)$. On note $\mathfrak{g}(F)_{tn}$ le sous-ensemble des éléments topologiquement nilpotents de $\mathfrak{g}(F)$. On sait définir l'exponentielle sur un voisinage de 0 dans $\mathfrak{g}(F)$, à valeurs dans un voisinage de 1 dans $G(F)$. Grâce à l'hypothèse (1), on peut prendre pour voisinage de 0 l'ensemble $\mathfrak{g}(F)_{tn}$. L'application ainsi définie, notée exp, prend ses valeurs dans $G(F)_{tu}$ et est un homéomorphisme de $\mathfrak{g}(F)_{tn}$ sur $G(F)_{tu}$. Elle possède toutes les propriétés usuelles. Par exemple, si $X\in\mathfrak{g}(F)_{tn}$ et $g\in G(F)$, on a l'égalité $exp(gXg^{-1})=gexp(X)g^{-1}$. Si H est un sous-groupe algébrique de G, l'exponentielle se restreint à $\mathfrak{h}(F)_{tn}$ en l'exponentielle relative à ce groupe.

Supposons de plus G non ramifié, soient K un sous-groupe hyperspécial de $G(F)$ et $\mathfrak{k}\subset\mathfrak{g}(F)$ le réseau hyperspécial associé. Pour $X\in\mathfrak{g}(F)_{tn}$, on a $X\in\mathfrak{k}$ si et seulement si $exp(X)\in K$.

2.7. Groupes tordus. On introduit dans ce paragraphe la partie p-adique des données que nous conserverons dans la suite de l'article. Soit G un groupe réductif connexe défini sur F. On suppose :

(Hyp1) $p>N(G)e_F+1$;

(Hyp2) G est non ramifié.

On fixe un sous-groupe hyperspécial K de $G(F)$. Soit θ un automorphisme de G défini sur F. On suppose :

(Hyp3) $\theta(K)=K$.

A la suite de Labesse, on introduit l'espace tordu $\tilde{G}$. C'est une variété algébrique définie sur F, isomorphe à G par un isomorphisme noté $g\to g\theta$.

Le groupe G agit à droite et à gauche sur $\tilde{G}$, ces actions étant définies par $g_1(g\theta)g_2 = g_1g\theta(g_2)\theta$. Pour $g \in G$, on note encore int_g l'automorphisme $\delta \to g\delta g^{-1}$ de $\tilde{G}$ et on utilise les notations de 2.2 pour cette action de G sur $\tilde{G}$ (par exemple, pour $\delta \in \tilde{G}$, on définit son commutant $Z_G(\delta)$ dans G et la composante neutre G_δ de celui-ci). On peut aussi considérer que tout élément de $\tilde{G}$ agit sur G : l'élément $\delta = g\theta$ agit par $aut_\delta = int_g \circ \theta$. On définit le sous-ensemble $\tilde{K} = K\theta \subset \tilde{G}(F)$. Pour tout $\delta \in \tilde{K}$, aut_δ conserve K.

Fixons une paire de Borel épinglée $(B, T, (E_\alpha)_{\alpha\in\Delta})$ de G, définie sur F, dont K soit le sous-groupe hyperspécial associé. Son image par θ est encore une paire de Borel épinglée définie sur F. Il y a donc un élément de $G_{AD}(F)$ dont l'automorphisme intérieur associé envoie cette paire image sur la paire de départ. Relevons cet élément en un élément $g_\theta \in G_{SC}$. On pose $\theta^* = int_{g_\theta} \circ \theta$. C'est un automorphisme de G, défini sur F, qui conserve la paire de Borel épinglée. On suppose :

(Hyp4) θ^* est d'ordre fini.

Remarquons que cette hypothèse est automatiquement vérifiée si G est semi-simple puisqu'alors, tout automorphisme conservant la paire de Borel épinglée est uniquement déterminé par la permutation de Δ qu'il induit. On montre que l'hypothèse (Hyp1) entraîne que θ^* est d'ordre premier à p. Il en est de même de l'ordre de $Z_{G_{SC}}$. Cela entraîne que $Z_{G_{SC}} \subset G_{SC}(F^{nr})$, car $Z_{G_{SC}} \subset T_{sc,p'} = T_{sc}(\mathfrak{o}_{F^{nr}})_{p'}$. Puisque θ^* conserve la paire de Borel épinglée, il conserve aussi K. Donc int_{g_θ} conserve K, ce qui implique $g_{\theta,ad} \in K_{ad}$. On a :

(1) l'homomorphisme $K_{sc}^{nr} \to K_{ad}^{nr}$ est surjectif.

Fixons une uniformisante ϖ de F. Soit E une extension non ramifiée de F. Le groupe K_E est muni d'une filtration naturelle $(K_E(n))_{n\in\mathbb{N}}$ par des sous-groupes distingués. Pour $n = 0$, $K_E(0) = K_E$; pour $n \geq 1$, $K_E(n) = exp(\varpi^n\mathfrak{k}_E)$. De même, les groupes $K_{sc,E}$ et $K_{ad,E}$ sont filtrés. Appliquons cela à $E = F^{nr}$. L'application (1) est compatible aux filtrations. Il suffit de prouver que sa restriction :

(2) $K_{sc}^{nr}(1) \to K_{ad}^{nr}(1)$

est surjective et qu'il en est de même de l'homomorphisme :

$$K_{sc}^{nr}/K_{sc}^{nr}(1) \to K_{ad}^{nr}/K_{ad}^{nr}(1).$$

Ce dernier s'identifie à l'homorphisme :

$$\mathbf{G}_{K,SC}(\bar{\mathbb{F}}_q) \to \mathbf{G}_{K,AD}(\bar{\mathbb{F}}_q),$$

où $\mathbf{G}_{K,SC}$ et $\mathbf{G}_{K,AD}$ sont les fibres spéciales des schémas en groupes associés à K_{sc} et K_{ad}. L'homomorphisme est surjectif car $\mathbf{G}_{K,AD}$ est le groupe adjoint de $\mathbf{G}_{K,SC}$. Pour prouver que (2) est surjectif, il suffit de prouver que, pour toute extension E de F, finie et non ramifiée, l'homomorphisme analogue :

$$K_{sc,E}(1) \to K_{ad,E}(1)$$

est surjectif. Puisque E est une extension finie de F, les groupes en question sont complets et il suffit de prouver que, pour tout $n \geq 1$, l'homomorphisme gradué :

$$K_{sc,E}(n)/K_{sc,E}(n+1) \to K_{ad,E}(n)/K_{ad,E}(n+1)$$

est surjectif. Il s'identifie à l'homomorphisme :

$$\mathfrak{g}_{K,SC}(\mathbb{F}_{q_E}) \to \mathfrak{g}_{K,AD}(\mathbb{F}_{q_E}),$$

où $\mathbb{F}_{q_E}$ est le corps résiduel de E. Sous l'hypothèse (Hyp1), l'homomorphisme $\mathbf{G}_{K,SC} \to \mathbf{G}_{K,AD}$ est surjectif et lisse et la surjectivité cherchée s'ensuit. Cela prouve (1).

Il en résulte que $g_\theta \in K_{sc}^{nr}$. Pour $\sigma \in \Gamma$, on pose $z(\sigma) = g_\theta\sigma(g_\theta)^{-1}$. Alors z est un cocycle de Γ dans $Z_{G_{SC}}$. Il est trivial sur le groupe d'inertie I et on peut aussi bien considérer qu'il est défini sur Γ^{nr}. Notons Θ^* le groupe d'automorphismes de G engendré par θ^*, introduisons le groupe non connexe $G^+ = G \rtimes \Theta^*$. On considère G^+ comme un groupe algébrique défini sur F, les éléments de Θ^* étant tous dans $G^+(F)$. La composante $G\theta^* \subset G^+$ est un espace tordu comme ci-dessus. Les deux espaces tordus $\tilde{G}$ et $G\theta^*$ sont isomorphes sur F^{nr}, par l'isomorphisme :

$$\begin{array}{rccc} \tilde{\psi}: & \tilde{G} & \to & G\theta^* \\ & g\theta & \mapsto & gg_\theta^{-1}\theta^*. \end{array}$$

Ils ne sont en général pas isomorphes sur F. Pour $\sigma \in \Gamma$ et $\delta \in \tilde{G}$, on a la relation $\tilde{\psi} \circ \sigma(\delta) = z(\sigma)^{-1}\sigma \circ \tilde{\psi}(\delta)$. Pour alléger la notation, on va oublier $\tilde{\psi}$. On considère désormais $\tilde{G}$ comme étant égal à $G\theta^*$, mais muni d'une action galoisienne tordue par le cocycle z selon la formule précédente.

Soit $\boldsymbol{\omega}$ un caractère de $G(F)$ (c'est-à-dire un homomorphisme continu de $G(F)$ dans $\mathbb{C}^\times$). D'après Langlands, le groupe des caractères de $G(F)$ est en bijection avec $H^1(W, Z_{\hat{G}})$, où $\hat{G}$ est le groupe dual de G. Soit $\mathbf{a}_\omega$ l'élément de $H^1(W, Z_{\hat{G}})$ correspondant à $\boldsymbol{\omega}$. On suppose :

(Hyp5) $\mathbf{a}_\omega$ provient par inflation d'un élément de $H^1(W^{nr}, Z_{\hat{G}})$.

On montre en [W1] 4.1(1) que cette condition entraîne que $\boldsymbol{\omega}$ est trivial sur K. En fait, la preuve montre que ces deux conditions sont équivalentes.

2.8. Exemples. Exemple 1. On considère $G = SL_2$, $K = SL_2(\mathfrak{o})$, $\boldsymbol{\omega} = 1$. Soit $\alpha \in \mathfrak{o}_{F^{nr}}^\times$ tel que $\alpha^2 \in F^\times$. On pose $\theta = int_{g_\theta}^{-1}$, où :

$$g_\theta = \begin{pmatrix} \alpha & 0 \\ 0 & \alpha^{-1} \end{pmatrix}.$$

Ces données vérifient les conditions requises. On a $\theta^* = 1$. Remarquons que, si $\alpha \notin F^\times$, les espaces tordus $\tilde{G}$ et $G\theta^*$ ne sont pas isomorphes sur F.

Exemple 2. On considère $G = GL_n$, $K = GL_n(\mathfrak{o})$, $\theta = 1$ et $\boldsymbol{\omega} = \chi \circ det$, où χ est un caractère non ramifié de $F^\times$ d'ordre divisant n.

Exemple 3. On considère $G = GL_n$, $K = GL_n(\mathfrak{o})$, $\boldsymbol{\omega} = 1$ et θ est défini par la formule :

$$\theta(g) = \Phi {}^t g^{-1} \Phi^{-1},$$

où :

$$\Phi = \begin{pmatrix} & & & & 1 \\ & & & -1 & \\ & & \cdot & & \\ & \cdot & & & \\ (-1)^{n+1} & & & & \end{pmatrix}.$$

On prend pour B le groupe triangulaire supérieur, pour T le tore diagonal et pour épinglage l'épinglage évident, c'est-à-dire celui pour lequel tout E_α est une matrice dont tous les coefficients sont nuls sauf un seul, qui vaut 1. Alors $\theta = \theta^*$.

2.9. Données endoscopiques. Introduisons le groupe dual $\hat{G}$ et le L-groupe ${}^LG = \hat{G} \rtimes W$. On munit $\hat{G}$ d'une paire de Borel épinglée $(\hat{B}, \hat{T}, (\hat{E}_{\hat{\alpha}})_{\hat{\alpha} \in \hat{\Delta}})$ conservée par l'action de Γ. Une telle paire existe par définition de $\hat{G}$. Il y a des isomorphismes en dualité $j : X_*(T) \to X^*(\hat{T})$ et $\hat{j} : X_*(\hat{T}) \to X^*(T)$, qui sont équivariants pour les actions de Γ et identifient les racines de T avec les coracines de $\hat{T}$ et vice-versa. On introduit l'automorphisme $\hat{\theta}$ de $\hat{G}$ qui conserve la paire de Borel épinglée et vérifie les relations équivalentes $j \circ \theta = \hat{\theta}^{-1} \circ j$, $\hat{j} \circ \hat{\theta} = \theta^{-1} \circ \hat{j}$. On prolonge $\hat{\theta}$ en un automorphisme de ${}^LG$ qui agit par l'identité sur W. On peut introduire les espaces tordus $\hat{G}\hat{\theta}$ ou ${}^LG\hat{\theta}$ sur lesquels $\hat{G}$ agit à droite et à gauche. Cela permet d'utiliser les notations introduites en 2.2 et 2.7 pour l'action par conjugaison de $\hat{G}$ sur chacun de ces espaces tordus. Par exemple, pour $s \in \hat{G}$, on note $\hat{G}_{s\hat{\theta}}$ la composante neutre du groupe $Z_{\hat{G}}(s\hat{\theta}) = \{x \in \hat{G}; xs\hat{\theta}(x)^{-1} = s\}$.

Une donnée endoscopique de $(G, \theta, \boldsymbol{\omega})$ est un quadruplet $(H, \mathcal{H}, s, \hat{\xi})$ satisfaisant les conditions suivantes :

(1) H est un groupe réductif connexe quasi-déployé sur F ;

(2) $\mathcal{H}$ est une extension scindée de W par $\hat{H}$, de sorte que l'action de W sur $\hat{H}$ qui s'en déduit coïncide avec celle qui provient de la structure sur F de H ;

(3) $s \in \hat{G}$ est un élément tel que l'automorphisme $int_s \circ \hat{\theta}$ de $\hat{G}$ soit semi-simple, c'est-à-dire préserve une paire de Borel ;

(4) $\hat{\xi} : \mathcal{H} \to {}^LG$ est un homomorphisme continu et injectif, commutant aux projections sur W ;

(5) $\hat{\xi}$ se restreint en un isomorphisme de $\hat{H}$ sur $\hat{G}_{s\hat{\theta}}$;

(6) il existe un cocycle $a : W \to Z_{\hat{G}}$, dont la classe est $\mathbf{a}_{\boldsymbol{\omega}}$, tel que l'on ait l'égalité $int_s \circ \hat{\theta} \circ \hat{\xi}(\hat{h}) = a(\hat{h})\hat{\xi}(\hat{h})$ pour tout $\hat{h} \in \mathcal{H}$, où, si $w \in W$ est la projection de $\hat{h}$, on a posé $a(\hat{h}) = a(w)$.

Un isomorphisme entre deux données $(H_1, \mathcal{H}_1, s_1, \hat{\xi}_1)$ et $(H_2, \mathcal{H}_2, s_2, \hat{\xi}_2)$ est un élément $x \in \hat{G}$ tel que $int_x \circ \hat{\xi}_1(\mathcal{H}_1) = \hat{\xi}_2(\mathcal{H}_2)$ et $xs_1\hat{\theta}(x)^{-1} \in Z_{\hat{G}}s_2$.

On dit que la donnée endoscopique $(H, \mathcal{H}, s, \hat{\xi})$ est non ramifiée si elle vérifie les conditions supplémentaires :

(7) H est non ramifié sur F ;

(8) $\mathcal{H} = {}^L H$;

(9) $\hat{\xi}$ est l'identité sur le groupe d'inertie I.

On fixe pour la suite de l'article une donnée endoscopique $(H, \mathcal{H}, s, \hat{\xi})$ de $(G, \theta, \boldsymbol{\omega})$. On suppose :

(Hyp6) cette donnée est non ramifiée.

On fixe une paire de Borel épinglée $(B_H, T_H, (E_{H,\alpha})_{\alpha \in \Delta_H})$ de H, définie sur F, et une paire de Borel $(\hat{B}_H, \hat{T}_H)$ de $\hat{H}$, conservée par l'action de Γ. Il existe un élément $x \in \hat{G}$ tel que $xs\hat{\theta}(x)^{-1} \in \hat{T}$, $int_x \circ \hat{\xi}(\hat{T}_H) = \hat{T}^{\hat{\theta},0}$ et $int_x \circ \hat{\xi}(\hat{B}_H) = \hat{B} \cap int_x \circ \hat{\xi}(\hat{H})$. On peut remplacer notre donnée endoscopique par la donnée $(H, {}^L H, xs\hat{\theta}(x)^{-1}, int_x \circ \hat{\xi})$. En oubliant la donnée initiale, on est ramené au cas où sont vérifiées les hypothèses supplémentaires :

(10) $s \in \hat{T}$;

(11) $\hat{\xi}(\hat{T}_H) = \hat{T}^{\hat{\theta},0}$, $\hat{\xi}(\hat{B}_H) = \hat{B} \cap \hat{H}$.

On suppose désormais ces hypothèses vérifiées. On peut montrer que nos résultats ultérieurs ne dépendent pas du choix de l'élément x ci-dessus. L'isomorphisme $\hat{\xi}$ de $\hat{T}_H$ sur $\hat{T}^{\hat{\theta},0}$ se dualise en un isomorphisme $\xi : T_{/\theta^*} \to T_H$. On note encore $\xi : T \to T_H$ le composé de cet isomorphisme et de la projection naturelle de T sur $T_{/\theta^*}$. Notons Ω le groupe de Weyl de G relatif à T et Ω_H celui de H relatif à T_H. Ces groupes s'identifient respectivement au groupe de Weyl de $\hat{G}$ relatif à $\hat{T}$ et à celui de $\hat{H}$ relatif à $\hat{T}_H$. De l'isomorphisme $\hat{\xi}$, ou de ξ, se déduit un plongement de Ω_H dans Ω^{θ^*}. Pour tout $\sigma \in \Gamma$, il existe un unique $\omega(\sigma) \in \Omega^{\theta^*}$ tel que l'on ait l'égalité sur $\hat{T}_H$:

$$\hat{\xi} \circ \sigma = \omega(\sigma) \circ \sigma \circ \hat{\xi}.$$

On a l'égalité équivalente sur T :

$$\sigma \circ \xi = \xi \circ \omega(\sigma) \circ \sigma.$$

L'application ω se factorise en une application définie sur Γ^{nr}. On vérifie que $\xi(Z_G) \subset Z_H$. D'après la relation précédente, la restriction de ξ à Z_G est équivariante pour les actions de Γ. Pour $\sigma \in \Gamma$, posons $z_H(\sigma) = \xi \circ z(\sigma)$. On introduit une variété H_z : c'est la variété algébrique sur F telle qu'il existe un isomorphisme $\psi_z : H_z \to H$ défini sur $\bar{F}$, de sorte que, pour tous $\sigma \in \Gamma$ et $h \in H_z$, on ait l'égalité $\psi_z \circ \sigma(h) = z_H(\sigma)^{-1} \sigma \circ \psi_z(h)$. En fait, ψ_z est défini sur F^{nr}. Pour alléger la notation, on oublie ψ_z et on considère H_z comme étant égal à H, avec une structure sur F tordue par le cocycle z_H. Remarquons que H agit sur H_z par conjugaison.

De la paire de Borel épinglée de H se déduit un sous-groupe hyperspécial K_H de $H(F)$. On note simplement H_K le schéma en groupes associé. On pose $K_z = H_z(F) \cap K_H^{nr}$.

On peut interpréter H_z comme un espace tordu de la façon suivante. Rappelons que z prend ses valeurs dans $Z_{G_{SC}}$ (plus exactement, ici, dans l'image de ce groupe dans G), donc dans $Z_{G,p'}$. Alors z_H est un cocycle de Γ dans $Z_{H,p'}$, qui se factorise en un cocycle défini sur Γ^{nr}. Poussons-le en un cocycle de Γ^{nr} dans $T_{H,p'}$. Comme on l'a dit en 2.4, $T_{H,p'} = T_H(\mathfrak{o}_{F^{nr}})_{p'} \simeq \mathbf{T}_H(\bar{\mathbb{F}}_q)$. D'après le théorème de Lang, $H^1(\Gamma^{nr}, \mathbf{T}_H(\bar{\mathbb{F}}_q)) = 0$. On peut donc choisir $h_\theta \in T(\mathfrak{o}_{F^{nr}})_{p'}$ tel que $z_H(\sigma) = h_\theta\sigma(h_\theta)^{-1}$ pour tout $\sigma \in \Gamma$. Remarquons que $h_\theta \in T_{H,ad}(\mathfrak{o})$. On pose $\theta_H = int_{h_\theta}^{-1}$ et on introduit l'espace tordu $\tilde{H} = H\theta_H$ comme en 2.7. Alors l'application $h \to hh_\theta\theta_H$ identifie H_z à $\tilde{H}$. L'automorphisme θ_H conserve K_H et K_z s'identifie à $\tilde{K}_H$, ce dernier ensemble étant défini comme en 2.7. Posons $\boldsymbol{\omega}_H = 1$. Alors les données H, θ_H, $\boldsymbol{\omega}_H$ et K_H vérifient les mêmes hypothèses qu'en 2.7. La seule différence, qui n'aura pas d'incidence, est que la donnée auxiliaire h_θ n'appartient qu'à K_H^{nr} et pas à $K_{H,sc}^{nr}$.

2.10. Correspondance entre classes de conjugaison. L'homomorphisme ξ^{-1} se quotiente en une application :

$$(1) \qquad T_H/\Omega_H \to T_{/\theta^*}/\Omega^{\theta^*}.$$

Soit δ un élément semi-simple de $\tilde{G}$. On peut conjuguer δ en un élément de la forme $\nu\theta^*$, avec $\nu \in T$. Alors l'image de ν dans $T_{/\theta^*}/\Omega^{\theta^*}$ est bien déterminée. De la sorte, $T_{/\theta^*}/\Omega^{\theta^*}$ paramètre les classes de conjugaison par G dans l'ensemble des éléments semi- simples de $\tilde{G}$. Soit γ un élément semi-simple de H_z. On peut conjuguer γ en un élément $\mu \in T_H$. L'image de cet élément dans T_H/Ω_H est bien déterminée. De la sorte, T_H/Ω_H paramètre les classes de conjugaison par H dans l'ensemble des éléments semi-simples de H_z. Alors l'application (1) apparaît comme une correspondance entre classes de conjugaison d'éléments semi-simples dans $\tilde{G}$ et dans H_z. Les structures sur F de $\tilde{G}$ et H_z induisent des actions de Γ sur les ensembles de départ et d'arrivée de l'application (1). Cette application est équivariante pour ces actions (c'est pour cela que l'on a introduit H_z).

On note $\tilde{G}_{reg}$ le sous-ensemble des $\delta \in \tilde{G}$ tels que $Z_G(\delta)$ soit commutatif et $Z_G(\delta)^0$ soit un tore. On note $H_{z,\tilde{G}-reg}$ le sous-ensemble des $\gamma \in H_z$ qui sont semi-simples et dont la classe de conjugaison correspond par (1) à la classe d'un élément de $\tilde{G}_{reg}$. On montre que tout tel γ est fortement régulier, c'est-à-dire que $Z_H(\gamma)$ est un tore. On note $\mathcal{D}$ l'ensemble des couples (γ, δ) tels que $\gamma \in H_{z,\tilde{G}-reg}(F)$, $\delta \in \tilde{G}_{reg}(F)$ et les classes de conjugaison de γ par H et de δ par G se correspondent par l'application (1).

Kottwitz et Shelstad ([KS] chapitre 4) ont défini un facteur de transfert :

$$\Delta : \mathcal{D} \to \mathbb{C}^\times.$$

Il n'est défini qu'à multiplication près par une constante. Le choix de K permet de le normaliser. Soit δ un élément semi-simple de $\tilde{K}$. L'automorphisme aut_δ se réduit en un automorphisme $\mathbf{aut}_\delta$ de $\mathbf{G}_K$. Disons que δ est de réduction régulière si le sous-groupe des points fixes de $\mathbf{aut}_\delta$ est commutatif et si sa composante neutre est un tore. On peut choisir Δ de sorte que $\Delta(\gamma,\delta) = 1$ pour tout $(\gamma,\delta) \in \mathcal{D}$ tel que δ soit un élément de $\tilde{K}$ de réduction régulière. Si p est assez grand, il existe au moins un tel couple et cela détermine Δ de façon unique. Malheureusement, il n'est pas clair que la condition (Hyp1) soit suffisante pour assurer l'existence d'un couple (γ,δ) comme ci-dessus. Dans [W1] 4.6, on a indiqué une autre façon de normaliser, qui est une variante un peu plus sophistiquée de celle que l'on vient de présenter, et qui est valable sous la seule hypothèse (Hyp1).

Soit $(\gamma,\delta) \in \mathcal{D}$. Si $\gamma' \in H_{z,\tilde{G}-reg}(F)$ est conjugué à γ par un élément de H, on a l'égalité $\Delta(\gamma',\delta) = \Delta(\gamma,\delta)$. Si $g \in G(F)$, on a l'égalité $\Delta(\gamma, g\delta g^{-1}) = \boldsymbol{\omega}(g)^{-1}\Delta(\gamma,\delta)$ ([KS] théorème 5.1.D).

3. Les résultats

3.1. Intégrales orbitales. On note $C_c^\infty(\tilde{G}(F))$ l'espace des fonctions sur $\tilde{G}(F)$, à valeurs complexes, localement constantes et à support compact. On munit $G(F)$ de la mesure de Haar pour laquelle K est de mesure 1. Cette mesure sera dite canonique (elle ne dépend pas de K puisque tous les sous-groupes hyperspéciaux sont conjugués par $G_{AD}(F)$). Soit $\delta \in \tilde{G}_{reg}(F)$. Supposons choisie une mesure de Haar sur $G_\delta(F)$. Supposons que le caractère $\boldsymbol{\omega}$ soit trivial sur ce groupe. Pour $f \in C_c^\infty(\tilde{G}(F))$, on pose :

$$O_\delta^{\tilde{G},\omega}(f) = [Z_G(\delta)(F) : G_\delta(F)]^{-1} \int_{G_\delta(F)\backslash G(F)} f(g^{-1}\delta g)\boldsymbol{\omega}(g)dg.$$

Munissons $H(F)$ de la mesure canonique. Soit $\gamma \in H_{z,\tilde{G}-reg}(F)$. Supposons choisie une mesure de Haar sur $H_\gamma(F)$. Pour $f \in C_c^\infty(H_z(F))$, on pose :

$$O_\gamma^{H_z}(f) = \int_{H_\gamma(F)\backslash H(F)} f(h^{-1}\gamma h)dh.$$

Cette définition est un cas particulier de la précédente d'après l'interprétation que l'on a donnée en 2.9 de l'espace H_z. On l'a explicitée parce que sa forme est plus simple que celle du cas général.

3.2. Intégrales orbitales endoscopiques. Pour tout sous-tore maximal T' de H, défini sur F, on fixe une mesure de Haar sur $T'(F)$. On suppose que si T' et T'' sont de tels tores et si $h \in H$ est tel que int_h envoie T' sur T'' et induit un isomorphisme défini sur F entre ces deux tores, alors les mesures sur $T'(F)$ et $T''(F)$ se correspondent par int_h. Soit $\gamma \in H_{z,\tilde{G}-reg}(F)$. Considérons un élément $\delta \in \tilde{G}_{reg}(F)$ dont la classe de conjugaison corresponde à celle de γ. D'après [KS] lemme 4.4.C, le caractère $\boldsymbol{\omega}$ est trivial sur $G_\delta(F)$. On verra en 4.9 que la mesure que l'on a fixée sur $H_\gamma(F)$ en détermine une sur $G_\delta(F)$, et c'est celle que l'on utilise dans la formule ci-dessous. Pour $f \in C_c^\infty(\tilde{G}(F))$, on peut poser :

$$O_\gamma^{\tilde{G},H_z,\boldsymbol{\omega}}(f) = \sum_\delta \Delta(\gamma,\delta) O_\delta^{\tilde{G},\boldsymbol{\omega}}(f),$$

où l'on somme sur un ensemble de représentants des classes de conjugaison par $G(F)$ dans l'ensemble des $\delta \in \tilde{G}_{reg}(F)$ tels que $(\gamma,\delta) \in \mathcal{D}$. Pour $f \in C_c^\infty(H_z(F))$, on pose :

$$SO_\gamma^{H_z}(f) = \sum_{\gamma'} O_{\gamma'}^{H_z}(f),$$

où l'on somme sur un ensemble de représentants des classes de conjugaison par $H(F)$ dans l'ensemble des $\gamma' \in H_{z,\tilde{G}-reg}(F)$ qui sont conjugués à γ par un élément de H.

3.3. Enoncé du lemme fondamental. On note $\mathbf{1}_{\tilde{K}}$ la fonction caractéristique de $\tilde{K}$ dans $\tilde{G}(F)$. On note $\mathbf{1}_{K_z}$ la fonction caractéristique de K_z dans $H_z(F)$.

Lemme fondamental. *Pour tout $\gamma \in H_{z,\tilde{G}-reg}(F)$, on a l'égalité :*

$$SO_\gamma^{H_z}(\mathbf{1}_{K_z}) = O_\gamma^{\tilde{G},H_z,\boldsymbol{\omega}}(\mathbf{1}_{\tilde{K}}).$$

3.4. Le lemme fondamental pour les algèbres de Lie. Considérons le cas où $\theta = 1$ et $\boldsymbol{\omega} = 1$. Alors $\tilde{G} = G$ et $H_z = H$. On peut oublier toutes les parties tordues de nos objets. Les définitions des paragraphes 2.10, 3.1 et 3.2 se descendent aux algèbres de Lie. Il y a une correspondance entre classes de conjugaison d'éléments semi-simples dans $\mathfrak{h}$ et classes de conjugaison d'éléments semi- simples dans $\mathfrak{g}$. Notons $\mathfrak{g}_{reg}$ le sous-ensemble des $X \in \mathfrak{g}$ dont le commutant G_X est un tore. Notons $\mathfrak{h}_{G-reg}$ le sous-ensemble des éléments semi-simples de $\mathfrak{h}$ dont la classe de conjugaison correspond à celle d'un élément de $\mathfrak{g}_{reg}$. On définit un facteur de transfert Δ sur le sous-ensemble des couples $(Y,X) \in \mathfrak{h}_{G-reg}(F) \times \mathfrak{g}_{reg}(F)$ tels que les classes de conjugaison de Y et de X se correspondent. Le choix de K permet de le normaliser de façon unique. Pour $f \in C_c^\infty(\mathfrak{g}(F))$ et $X \in \mathfrak{g}_{reg}(F)$, on définit l'intégrale orbitale $O_X^G(f)$ (modulo le choix d'une mesure de Haar sur $G_X(F)$). Puis, pour $Y \in \mathfrak{h}_{G-reg}(F)$, on définit l'intégrale orbitale endoscopique $O_Y^{G,H}(f)$ (les mesures étant fixées comme en 3.2). Pour $f \in C_c^\infty(h(F))$, on définit aussi l'intégrale orbitale stable $SO_Y^H(f)$. Enfin, on note $\mathbf{1}_{\mathfrak{k}}$ la fonction caractéristique du réseau $\mathfrak{k}$ dans $\mathfrak{g}(F)$ et $\mathbf{1}_{\mathfrak{k}_H}$ la fonction caractéristique du réseau $\mathfrak{k}_H$ dans $\mathfrak{h}(F)$.

Lemme fondamental pour les algèbres de Lie. *Pour tout* $Y \in \mathfrak{h}_{G-reg}(F)$, *on a l'égalité :*

$$SO_Y^H(\mathbf{1}_{\mathfrak{k}_H}) = O_Y^{G,H}(\mathbf{1}_{\mathfrak{k}}).$$

3.5. Le lemme fondamental non standard. Considérons deux groupes connexes semi-simples et simplement connexes G_1 et G_2, définis et quasi-déployés sur F. Pour $i = 1, 2$, on fixe un tore maximal T_i de G_i, faisant partie d'une paire de Borel définie sur F. On affecte d'un indice i les notations introduites dans les paragraphes précédents. Par exemple, on note Ω_i le groupe de Weyl de G_i relatif à T_i. Pour tout $\mathbb{Z}$-module A, on pose $A_{\mathbb{Q}} = A \otimes_{\mathbb{Z}} \mathbb{Q}$. Considérons un isomorphisme $j_* : X_*(T_1)_{\mathbb{Q}} \to X_*(T_2)_{\mathbb{Q}}$. Notons $j^* : X^*(T_2)_{\mathbb{Q}} \to X^*(T_1)_{\mathbb{Q}}$ son transposé. On suppose que ces isomorphismes sont équivariants pour les actions de Γ. On suppose qu'il existe des bijections $j_{\check{\Sigma}} : \check{\Sigma}_1 \to \check{\Sigma}_2$, $j_{\Sigma} : \Sigma_2 \to \Sigma_1$ et des fonctions $\check{b} : \check{\Sigma}_1 \to \mathbb{Q}_{>0}$ et $b : \Sigma_2 \to \mathbb{Q}_{>0}$ de sorte que :

- j_{Σ} s'identifie à l'inverse de $j_{\check{\Sigma}}$ via les bijections naturelles de Σ_i sur $\check{\Sigma}_i$ pour $i = 1, 2$;
- pour tout $\check{\alpha} \in \check{\Sigma}_1$, on a l'égalité $j_*(\check{\alpha}) = \check{b}(\check{\alpha}) j_{\check{\Sigma}}(\check{\alpha})$;
- pour tout $\alpha \in \Sigma_2$, on a l'égalité $j^*(\alpha) = b(\alpha) j_{\Sigma}(\alpha)$.

A ces conditions, on dit que le triplet (G_1, G_2, j_*) est endoscopique non standard. On montre que pour $\check{\alpha}_1 \in \check{\Sigma}_1$ et $\alpha_2 \in \Sigma_2$ tels que $\check{\alpha}_2 = j_{\check{\Sigma}}(\check{\alpha}_1)$, on a nécessairement $\check{b}(\check{\alpha}_1) = b(\alpha_2)$. On montre que l'application qui, à la symétrie élémentaire relative à une coracine $\check{\alpha}_1 \in \check{\Sigma}_1$, associe la symétrie élémentaire relative à $j_{\check{\Sigma}}(\check{\alpha}_1)$, se prolonge en un isomorphisme j_{Ω} de Ω_1 sur Ω_2.

On dit que le triplet endoscopique non standard est non ramifié si :

- $p > N(G_i)e_F + 1$ pour $i = 1, 2$;
- G_1 et G_2 sont non ramifiés ;
- b et $\check{b}$ prennent leurs valeurs dans l'ensemble des éléments de $\mathbb{Q}_{>0}$ de valuation p-adique nulle.

Exemple. Prenons pour G_1 le groupe symplectique Sp_{2n} et pour G_2 le groupe spinoriel $Spin_{2n+1}$ déployé. On peut identifier $X_*(T_1)$ à $\mathbb{Z}^n$, l'ensemble Σ_1 étant l'ensemble des formes linéaires :

$$(a_1, \dots, a_n) \to \pm a_k \pm a_l$$

pour $k, l = 1, \dots, n$, $k \neq l$, ou

$$(a_1, \dots, a_n) \to \pm 2a_k$$

pour $k = 1, \dots, n$. On peut identifier $X_*(T_2)$ au sous-$\mathbb{Z}$-module des $(a_1, \dots, a_n) \in \mathbb{Z}^n$ tels que $a_1 + \dots + a_n$ est pair. L'ensemble Σ_2 est l'ensemble des formes linéaires :

$$(a_1, \dots, a_n) \to \pm a_k \pm a_l$$

pour $k,l=1,\dots,n$, $k\neq l$, ou

$$(a_1,\dots,a_n)\to \pm a_k$$

pour $k=1,\dots,n$. Les espaces $X_*(T_1)_{\mathbb{Q}}$ et $X_*(T_2)_{\mathbb{Q}}$ s'identifient tous deux à $\mathbb{Q}^n$. On note $j_*:X_*(T_1)_{\mathbb{Q}}\to X_*(T_2)_{\mathbb{Q}}$ l'isomorphisme issu de ces identifications. Alors (G_1,G_2,j_*) est un triplet endoscopique non standard, qui est non ramifié pourvu que p soit assez grand.

Considérons un triplet endoscopique non standard (G_1,G_2,j_*), non ramifié. De l'homomorphisme j_* se déduit une bijection :

$$\mathfrak{t}_1/\Omega_1\simeq \mathfrak{t}_2/\Omega_2,$$

qui est équivariante pour les actions de Γ. On peut alors, comme dans le cas endoscopique ordinaire, définir une correspondance entre classes de conjugaison d'éléments semi-simples dans $\mathfrak{g}_1$ et $\mathfrak{g}_2$. Pour $i=1,2$, soit $X_i\in\mathfrak{g}_{i,reg}(F)$. Supposons que les classes de conjugaison de X_1 et de X_2 se correspondent. Alors toute mesure de Haar sur $G_{1,X_1}(F)$ en détermine une sur $G_{2,X_2}(F)$ et on suppose dans ce qui suit que les mesures se correspondent ainsi. Fixons des sous-groupes hyperspéciaux K_i de $G_i(F)$, pour $i=1,2$.

Lemme fondamental non standard. *Pour tout couple $(X_1,X_2)\in\mathfrak{g}_{1,reg}(F)\times\mathfrak{g}_{2,reg}(F)$ formé d'éléments dont les classes de conjugaison se correspondent, on a l'égalité :*

$$SO_{X_1}^{G_1}(\mathbf{1}_{\mathfrak{k}_1})=SO_{X_2}^{G_2}(\mathbf{1}_{\mathfrak{k}_2}).$$

Remarque. Il serait tentant de formuler un énoncé analogue pour les groupes au lieu des algèbres de Lie. Il serait faux, ainsi que le montre un calcul élémentaire dans le cas de l'exemple ci-dessus, pour $n=1$.

3.6. Notre résultat. Théorème. *Supposons vérifiés le lemme fondamental pour les algèbres de Lie de 3.4 ainsi que le lemme fondamental non standard de 3.5. Alors le lemme fondamental de 3.3 est vérifié.*

3.7. Conséquence. Théorème. *Le lemme fondamental de 3.3 est vérifié.*

En effet, Ngo Bao Chau a prouvé les assertions 3.4 et 3.5 ([N]).

4. Les preuves

La situation est celle décrite au chapitre 1.

4.1. Eléments d'ordre fini premier à p. Pour tout groupe algébrique M défini sur F, dont la composante neutre est réductive, on a défini en 2.4 les notions d'éléments compacts, topologiquement unipotents ou d'ordre fini premier à p de $M(F)$. On a $H_z(F) \subset H_z(F^{nr}) = H(F^{nr})$. Un élément de $H_z(F)$ est dit compact, resp. d'ordre fini premier à p, s'il l'est dans $H(F^{nr})$. De même, introduisons le groupe G^+ de 2.7. On a $\tilde{G}(F) \subset \tilde{G}(F^{nr}) = G(F^{nr})\theta^* \subset G^+(F^{nr})$. Un élément $\eta \in \tilde{G}(F)$ est dit compact, resp. d'ordre fini premier à p, s'il l'est dans $G^+(F^{nr})$. Remarquons que ces notions dépendent du choix de l'élément g_θ de 2.7. On utilise les notations $H_z(F)_c$, $\tilde{G}(F)_c$ etc. comme en 2.4. La décomposition d'un élément compact $\gamma \in H_z(F)_c$ s'écrit $\gamma = \gamma_{tu}\gamma_{p'}$, où $\gamma_{tu} \in H(F)_{tu}$ et $\gamma_{p'} \in H_z(F)_{p'}$. En pratique, on l'écrira $\gamma = exp(Y)\epsilon$, où $\epsilon = \gamma_{p'}$ et Y est l'élément de $\mathfrak{h}_\epsilon(F)_{tn}$ tel que $exp(Y) = \gamma_{tu}$. De même, la décomposition d'un élément $\delta \in \tilde{G}(F)_c$ s'écrit $\delta = exp(X)\eta$, où $\eta = \delta_{p'} \in \tilde{G}(F)_{p'}$ et X est l'élément de $\mathfrak{g}_\eta(F)_{tn}$ tel que $exp(X) = \delta_{tu} \in G_\eta(F)_{tu}$.

Lemme. *Soit $\eta \in \tilde{G}(F)_{p'} \cap \tilde{K}$. Alors :*

(i) le groupe G_η est non ramifié et $G_\eta(F) \cap K$ est un sous-groupe hyperspécial de $G_\eta(F)$;

(ii) l'ensemble des $g \in G$ tels que $int_g(\eta) \in K^{nr}\theta^*$ est égal à $K^{nr}G_\eta$ et l'ensemble des $g \in G(F)$ tels que $int_g(\eta) \in \tilde{K}$ est égal à $KG_\eta(F)$.

Preuve. Montrons d'abord que :

(1) il existe $k \in K^{nr}$ et $\nu \in T(\mathfrak{o}_{F^{nr}})_{p'}$ tels que $int_k(\delta) = \nu\theta^*$.

Introduisons le groupe $\mathbf{G}_K^+ = \mathbf{G}_K \rtimes \Theta^*$. L'application de réduction de K^{nr} sur $\mathbf{G}_K$ se prolonge en une application du sous-ensemble $K^{nr}\theta^*$ de $G^+(F^{nr})$ sur la composante $\mathbf{G}_K\theta^*$ de $\mathbf{G}_K^+$. Notons $\boldsymbol{\eta}$ l'image de η par cette application. C'est un élément d'ordre fini premier à p dans $\mathbf{G}_K^+$. On peut donc trouver $\mathbf{k} \in \mathbf{G}_K$ et $\boldsymbol{\nu} \in \boldsymbol{T}$ de sorte que $int_{\mathbf{k}}(\boldsymbol{\eta}) = \boldsymbol{\nu}\theta^*$. Relevons $\mathbf{k}$ en un élément $k \in K^{nr}$ et $\boldsymbol{\nu}$ en un élément $\nu \in T(\mathfrak{o}_{F^{nr}})_{p'}$. Posons $\eta' = int_{k^{-1}}(\nu\theta^*)$. C'est un élément de $K^{nr}\theta^*$ d'ordre fini premier à p. Il a même réduction que η. Alors (1) résulte de l'assertion suivante :

(2) soient x et y deux éléments de $K^{nr}\theta^*$ d'ordre fini premier à p. Supposons que leurs réductions $\mathbf{x}$ et $\mathbf{y}$ dans $\mathbf{G}_K\theta^*$ sont égales. Alors il existe $k \in K^{nr}$ tel que $x = int_k(y)$.

Fixons une extension finie et non ramifiée E de F telle que x et y appartiennent à $K_E\theta^*$ et utilisons les définitions posées dans la preuve de 2.7(1). Pour $n \geq 1$, le quotient $K_E(n)/K_E(n+1)$ s'identifie à $\mathfrak{g}(\mathbb{F}_{q_E}) = \mathfrak{k}_E/\varpi\mathfrak{k}_E$, l'action aut_y de y sur le premier ensemble s'identifiant à l'action $aut_{\mathbf{y}}$ de $\mathbf{y}$ sur le second. Parce que y est d'ordre fini premier à p, et parce que deux racines distinctes de l'unité d'ordre premier à p dans $\bar{F}$ ont des réductions distinctes dans $\bar{\mathbb{F}}_q$, un raisonnement familier (essentiellement le lemme de Hensel) permet de décomposer $\mathfrak{k}_E$ en somme directe de deux sous-réseaux $\mathfrak{k}_0$ et $\mathfrak{k}_1$, qui sont le noyau et l'image de $aut_y - 1$. On a $\mathfrak{k}_0 = \mathfrak{k}_E \cap \mathfrak{g}_y(E)$. Les images de $\mathfrak{k}_0$ et $\mathfrak{k}_1$ dans $\mathfrak{g}(\mathbb{F}_{q_E})$ sont respectivement le noyau et l'image de $aut_{\mathbf{y}} - 1$. L'hypothèse de (2) signifie que xy^{-1} appartient à $K_E(1)$. Soit $n \geq 1$. Supposons qu'il existe $k_n \in K_E(1)$ et $Y_n \in \mathfrak{k}_0$ tels que $k_n x k_n^{-1} y^{-1} exp(-\varpi Y_n) \in K_E(n)$. Ecrivons cet élément sous la forme $exp(\varpi^n Z)$, avec $Z \in \mathfrak{k}_E$. Des propriétés ci-dessus résulte que l'on peut trouver $Z_0 \in \mathfrak{k}_0$ et $Z' \in \mathfrak{k}_E$ de sorte que $Z = (aut_y - 1)(Z') + Z_0$. Posons $k_{n+1} = exp(\varpi^n Z')k_n$ et $Y_{n+1} = \varpi^{n-1} Z_0 + Y_n$. On vérifie que $k_{n+1} x k_{n+1}^{-1} y^{-1} exp(-\varpi Y_{n+1})$ appartient à $K_E(n+1)$. En passant à la limite, on obtient des éléments $k \in K_E$ et $Y \in \mathfrak{k}_0$ de sorte que $kxk^{-1} = exp(\varpi Y)y$. Mais $Y \in \mathfrak{g}_y(E)$. L'égalité précédente est la décomposition de l'élément compact kxk^{-1} en produit d'un élément topologiquement unipotent et d'un élément d'ordre fini premier à p. Puisque kxk^{-1} est lui-même d'ordre fini premier à p, l'unicité de la décomposition entraîne que $Y = 0$ et $kxk^{-1} = y$. Cela prouve (2) et (1).

Soient k et ν vérifiant (1). Fixons une extension finie non ramifiée E de F telle que $k \in K_E$, $\nu \in T(\mathfrak{o}_E)$ et G soit déployé sur E. La paire $(B \cap G_{\nu\theta^*}, T^{\theta^*,0})$ est une paire de Borel de $G_{\nu\theta^*}$, définie sur E. On n'a aucun mal à construire un épinglage $(E'_{\alpha'})_{\alpha' \in \Delta'}$ relatif à cette paire, que l'on complète comme en 2.5 par un épinglage opposé $(E'_{-\alpha'})_{\alpha' \in \Delta'}$, de sorte que tous ces éléments appartiennent à $\mathfrak{k}_E$. Soit K'_E le sous-groupe hyperspécial de $G_{\nu\theta^*}(E)$ associé à la paire de Borel épinglée. Il est engendré par $T^{\theta^*,0}(\mathfrak{o}_E)$ et les groupes $exp(\mathfrak{o}_E E'_{\pm\alpha'})$ pour $\alpha' \in \Delta'$. Tous ces groupes sont inclus dans $K_E \cap G_{\nu\theta^*}(E)$. Donc $K'_E \subset K_E \cap G_{\nu\theta^*}(E)$. Puisque K'_E est un sous-groupe compact maximal de $G_{\nu\theta^*}(E)$, cette inclusion est une égalité et $K_E \cap G_{\nu\theta^*}(E)$ est un sous-groupe hyperspécial de $G_{\nu\theta^*}(E)$. Cette propriété se conserve par conjugaison par k. Donc $K_E \cap G_\eta(E)$ est un sous-groupe hyperspécial de $G_\eta(E)$. Mais ce groupe est défini sur F. Les points hyperspéciaux de l'immeuble $Imm(G_\eta, F)$ sont les points hyperspéciaux de $Imm(G_\eta, E)$ qui sont fixes par l'action du groupe de Galois $Gal(E/F)$. Il en résulte que $K \cap G_\eta(F)$ est encore un groupe hyperspécial de $G_\eta(F)$. Enfin, le groupe G_η est déployé sur E : il est isomorphe à $G_{\nu\theta^*}$ qui contient le tore $T^{\theta^*,0}$, lequel est déployé sur E. Sous l'hypothèse (Hyp1), un groupe déployé sur une extension non ramifiée et qui possède un sous-groupe hyperspécial est forcément non ramifié. Cela prouve le (i) de l'énoncé.

Passons au (ii). On veut montrer que l'ensemble des $g \in G$ tels que $int_g(\eta) \in K^{nr}\theta^*$ est égal à $K^{nr}G_\eta$. Cette propriété est invariante par conjugaison de η par un élément de K^{nr}. Grâce à (1), on peut donc remplacer η par $\nu\theta^*$ dans l'assertion ci-dessus. Soit $g \in G$, posons $\eta' = int_g(\nu\theta^*)$, supposons $\eta' \in K^{nr}\theta^*$. On peut appliquer (1) à η' : il existe $k \in K^{nr}$ et $\nu' \in T(\mathfrak{o}_{F^{nr}})_{p'}$ tels que $int_k(\eta') = \nu'\theta^*$. Quitte à remplacer g par kg, on peut supposer $\eta' = \nu'\theta^*$. Alors int_g envoie $G_{\nu\theta^*}$ sur $G_{\nu'\theta^*}$. Les deux groupes possèdent une paire de Borel définie sur F^{nr} dont le tore maximal est $T^{\theta^*,0}$. Quitte à multiplier g à droite par un élément de $G_{\nu\theta^*}$, on peut supposer que int_g conserve ce tore. Alors int_g conserve aussi son commutant T et définit un élément $\omega \in \Omega$. Un élément de Ω qui conserve $T^{\theta^*,0}$ appartient à Ω^{θ^*}. On sait que Ω^{θ^*} est le groupe de Weyl de $G^{\theta^*,0}$ et que tout élément de ce groupe peut se représenter par un élément de $K^{nr} \cap G^{\theta^*,0}(F^{nr})$. On peut donc écrire $g = kt$, avec $t \in T$ et $k \in K^{nr} \cap G^{\theta^*,0}(F^{nr})$. On a $int_{k^{-1}}(\nu'\theta^*) = k^{-1}\nu'k\theta^*$, car $k \in G^{\theta^*,0}$, et $k^{-1}\nu'k$ est encore un élément de $T(\mathfrak{o}_{F^{nr}})_{p'}$. Quitte à remplacer ν' par cet élément et g par $k^{-1}g$, on peut donc supposer $g \in T$. Fixons une extension finie E de F telle que $g \in T(E)$ et une uniformisante ϖ_E de cette extension. Décomposons g en produit $g = t_{tu}t_{p'}t_{\varpi_E}$, avec $t_{tu} \in T(\mathfrak{o}_E)_{tu}$, $t_{p'} \in T(\mathfrak{o}_E)_{p'}$ et $t_{\varpi_E} \in T_{\varpi_E}$, cf. 2.4. L'égalité $int_g(\nu\theta^*) = \nu'\theta^*$ et le fait que ν et ν' appartiennent à $T(\mathfrak{o}_{F^{nr}})_{p'}$ entrainent que $int_{t_{p'}}(\nu\theta^*) = \nu'\theta^*$, tandis que t_{tu} et t_{ϖ_E} commutent à $\nu\theta^*$. Cela entraîne que t_{tu} et t_{ϖ_E} commutent à θ^*. Le groupe $T_{\varpi_E}^{\theta^*}$ est égal à $X_*(T)^{\theta^*} \otimes_{\mathbb{Z}} \varpi_E^{\mathbb{Z}}$, donc est inclus dans $T^{\theta^*,0}(E)$. Parce que l'ordre de θ^* est premier à p, on vérifie aussi que $T(\mathfrak{o}_E)_{tu}^{\theta^*} = T^{\theta^*,0}(\mathfrak{o}_E)_{tu}$. On a donc $t_{tu}t_{\varpi_E} \in T^{\theta^*,0} \subset G_{\nu\theta^*}$. De plus, $t_{p'} \in T_{p'} = T(\mathfrak{o}_{F^{nr}})_{p'} \subset K^{nr}$. Alors $g \in K^{nr}G_{\nu\theta^*}$, ce qui achève de prouver la première assertion de (ii).

D'après (i), $G_\eta(F) \cap K$ est un sous-groupe hyperspécial de $G_\eta(F)$. Notons-le K_η et notons $G_{\eta,K}$ le schéma en groupes associé. Soit $g \in G(F)$ tel que $int_g(\eta) \in \tilde{K}$. Ainsi que l'on vient de le prouver, on peut écrire $g = kx$, avec $x \in G_\eta$ et $k \in K^{nr}$. Nécessairement, $x \in G_\eta(F^{nr})$. Pour $\sigma \in \Gamma^{nr}$, l'hypothèse $g \in G(F)$ entraîne que $\sigma(k)^{-1}k = \sigma(x)x^{-1}$. Cette expression définit un cocycle de Γ^{nr} dans $G_\eta(F^{nr}) \cap K^{nr} = K_\eta^{nr}$. Ce groupe possède une filtration dont les quotients sont, soit $\mathbf{G}_{\eta,K}(\bar{\mathbb{F}}_q)$, soit $\mathfrak{g}_{\eta,K}(\bar{\mathbb{F}}_q)$. Si le groupe était complet, il résulterait du théorème de Lang que $H^1(\Gamma^{nr}, K_\eta^{nr}) = 0$. Le groupe n'est pas complet, mais on peut contourner cette difficulté comme en 2.7(1) et prouver la nullité de ce H^1. On peut donc choisir $k_0 \in K_\eta^{nr}$ tel que $\sigma(x)x^{-1} = \sigma(k_0)k_0^{-1}$ pour tout $\sigma \in \Gamma^{nr}$. Posons $x' = k_0^{-1}x$ et $k' = kk_0$. Alors $x' \in G_\eta(F)$, $k' \in K$ et $g = k'x'$. Cela prouve la seconde assertion de (ii).

4.2. Classes de conjugaison coupant $\tilde{K}$. Soit η un élément semi-simple de $\tilde{G}(F)$. A la suite de Labesse, on introduit le groupe algébrique non connexe $I_\eta = Z_G^{\theta^*}G_\eta$. C'est un sous-groupe de $Z_G(\eta)$, défini sur F. D'après (Hyp1), l'ordre de $Z_G^{\theta^*}/Z_G^{\theta^*,0}$ est premier à p. Puisque $Z_G^{\theta^*,0}$ est divisible, on a $Z_G^{\theta^*} = Z_{G,p'}^{\theta^*}Z_G^{\theta^*,0}$. Puisque $Z_G^{\theta^*,0} \subset G_\eta$, on a :

(1) $I_\eta = Z_{G,p'}^{\theta^*}G_\eta$.

Remarquons que $Z_{G,p'}^{\theta^*} \subset K^{nr}$, puisque $Z_{G,p'}^{\theta^*} \subset T_{p'} = T(\mathfrak{o}_{F^{nr}})_{p'}$. On a :

(2) si $\eta \in \tilde{G}_{reg}(F)$, alors $I_\eta = Z_G(\eta)$.

Il suffit de prouver que $Z_G(\eta) \subset Z_G G_\eta$. Cette assertion étant géométrique, on peut conjuguer η en un élément de $T\theta^*$. Puisque $\eta \in \tilde{G}_{reg}$, G_η est le tore $T^{\theta^*,0}$ et $Z_G(\eta)$ est inclus dans son commutant T, donc est égal à T^{θ^*}. Il reste à prouver que $T^{\theta^*} = Z_G^{\theta^*} T^{\theta^*,0}$. Cela résulte du fait que $T_{ad}^{\theta^*}$ est connexe, cf. [KS] 1.1.

Rappelons que l'on note ϕ l'élément de Frobenius de Γ^{nr}. Posons :

$$\mathcal{Z} = \{z \in K^{nr}; \phi(z)^{-1} z \in Z_{G,p'}^{\theta^*}\}.$$

C'est un groupe. Pour $z \in \mathcal{Z}$, l'automorphisme int_z de G, resp. de $\tilde{G}$, est défini sur F. Il conserve K, resp. $\tilde{K}$. Le groupe $\mathcal{Z}$ contient K comme sous-groupe distingué. Soit $\eta \in \tilde{K} \cap \tilde{G}(F)_{p'}$. On fixe un ensemble de représentants $\mathcal{Z}_\eta$ de l'ensemble de doubles classes $K \backslash \mathcal{Z}/(\mathcal{Z} \cap I_\eta)$.

Soit $\delta \in \tilde{G}(F)_{reg} \cap \tilde{G}(F)_c$. On écrit $\delta = exp(X)\eta$, avec $\eta = \delta_{p'}$ et $X \in \mathfrak{g}_\eta(F)_{tn}$. Supposons $\eta \in \tilde{K}$. Appelons classe de conjugaison stable de δ l'intersection de $\tilde{G}(F)$ avec la classe de conjugaison de δ par G. De même, appelons classe de conjugaison stable de X l'intersection de $\mathfrak{g}_\eta(F)$ avec la classe de conjugaison de X par G_η. Fixons un ensemble de représentants $\mathcal{E}_X$ des classes de conjugaison par $G_\eta(F)$ dans la classe de conjugaison stable de X. Posons :

$$\mathcal{X}_\delta = \{int_z(exp(X')\eta); z \in \mathcal{Z}_\eta, X' \in \mathcal{E}_E\}.$$

Lemme. *Sous ces hypothèses, les propriétés suivantes sont vérifiées :*

(i) $\mathcal{X}_\delta$ est inclus dans la classe de conjugaison stable de δ ;

(ii) soit C une classe de conjugaison par $G(F)$ contenue dans la classe de conjugaison stable de δ. Supposons que C coupe $\tilde{K}$. Alors l'intersection $C \cap \mathcal{X}_\delta$ a pour nombre d'éléments :

$$|\mathcal{Z}_\eta|[Z_G(\delta)(F) : G_\delta(F)]^{-1}.$$

Preuve. Le (i) résulte immédiatement des propriétés des automorphismes int_z pour $z \in \mathcal{Z}$.

Soit C comme en (ii). Montrons d'abord que :

(3) C coupe $\mathcal{X}_\delta$.

On peut fixer $\delta_1 \in C \cap \tilde{K}$, que l'on décompose en $\delta_1 = exp(X_1)\eta_1$. On peut aussi fixer $g \in G$ tel que $int_g(\delta) = \delta_1$. Cela implique $int_g(\eta) = \eta_1 \in \tilde{K}$. Appliquons le lemme 4.1 (ii) pour écrire $g = kx_1$ avec $k \in K^{nr}$ et $x_1 \in G_\eta$. Considérons l'application $h \to \phi(h)^{-1}h$ de K^{nr} dans lui-même. Comme dans la preuve du lemme 4.1, une adaptation du théorème de Lang montre qu'elle se quotiente en une bijection : $K\backslash K^{nr} \to K^{nr}_{\phi-fin}$, où $K^{nr}_{\phi-fin}$ est le sous-ensemble des $h \in K^{nr}$ pour lesquels il existe un entier $n \geq 1$ de sorte que $\phi^{n-1}(h)\ldots\phi(h)h = 1$. Un résultat analogue vaut si l'on remplace K par le groupe K_η. On note $K^{nr}_{\eta,\phi-fin}$ le sous-ensemble analogue à $K^{nr}_{\phi-fin}$. Soit $\sigma \in \Gamma$. Puisque δ et δ_1 appartiennent à $\tilde{G}(F)$, on a $\sigma(g)^{-1}g \in Z_G(\delta)$. D'après (2), $Z_G(\delta) = I_\delta \subset I_\eta$. Donc $\sigma(g)^{-1}g \in I_\eta$, puis $\sigma(k)^{-1}k \in I_\eta$. Ecrivons $\phi(k)^{-1}k = \zeta u$, avec $\zeta \in Z^{\theta^*}_{G,p'}$ et $u \in G_\eta$. On a $\phi(k)^{-1}k \in K^{nr}_{\phi-fin}$ et cet ensemble est invariant par multiplication par $Z_{G,p'}$. Donc $u \in K^{nr}_{\phi-fin} \cap G_\eta = K^{nr}_{\eta,\phi-fin}$. On peut fixer $v \in K^{nr}_\eta$ de sorte que $u = \phi(v)^{-1}v$. Alors kv^{-1} appartient à $\mathcal{Z}$ et on peut fixer $h \in K$, $z \in \mathcal{Z}_\eta$ et $y \in I_\eta$ de sorte que $kv^{-1} = hzy$. Alors $g = hzx_2$, où $x_2 = yvx_1 \in I_\eta$. Posons $\delta_2 = int_{x_2}(\delta)$. On a $\delta_2 = int^{-1}_{hz}(\delta_1)$. Puisque int_{hz} est défini sur F, δ_2 appartient à $\tilde{G}(F)_c$. Puisque $x_2 \in I_\eta$, on a $\delta_{2,p'} = \eta$. Ecrivons $\delta_2 = exp(X_2)\eta$. On a $X_2 = int_{x_2}(X)$ et, d'après (1), on peut dans cette égalité remplacer x_2 par un élément de G_η. Donc X_2 appartient à la classe de conjugaison stable de X et on peut fixer $X' \in \mathcal{E}_X$ et $x' \in G_\eta(F)$ de sorte que $X_2 = int_{x'}(X')$. Posons $\delta_3 = int_z(exp(X')\eta)$. C'est un élément de $\mathcal{X}_\delta$. On a $\delta_1 = int_{hzx'z^{-1}}(\delta_3)$. Puisque int_z est défini sur F, $hzx'z^{-1}$ appartient à $G(F)$. Donc $\delta_3 \in C$, ce qui prouve (3).

Soient $z \in \mathcal{Z}_\eta$ et $X' \in \mathcal{E}_X$. Posons $\delta' = int_z(exp(X')\eta)$ et notons $c(\delta')$ le nombre d'éléments de $\mathcal{X}_\delta$ qui sont conjugués à δ' par un élément de $G(F)$. Pour achever la démonstration, il suffit de prouver que :

(4) $c(\delta') = |\mathcal{Z}_\eta|[Z_G(\delta)(F) : G_\delta(F)]^{-1}$.

Soit $g \in G(F)$, supposons $int_g(\delta') \in \mathcal{X}_\delta$, écrivons $int_g(\delta') = int_{z_0}(exp(X_0)\eta)$, avec $z_0 \in \mathcal{Z}_\eta$ et $X_0 \in \mathcal{E}_X$. Les éléments $\eta' = int_z(\eta)$ et $\eta_0 = int_{z_0}(\eta)$ appartiennent à $\tilde{K} \cap \tilde{G}(F)_{p'}$ et on a $int_g(\eta') = \eta_0$. D'après le lemme 4.1(ii), g appartient à $KG_{\eta'}(F)$. Puisque $int_{z^{-1}}$ envoie $G_{\eta'}(F)$ sur $G_\eta(F)$, on peut écrire $gz = kzy$, avec $k \in K$ et $y \in G_\eta(F)$. Les éléments X et X_0 sont stablement conjugués à X, on peut donc fixer deux éléments x' et x_0 de G_η tels que $X' = int_{x'}(X)$, $X_0 = int_{x_0}(X)$. On a alors $int_{kzyx'}(\delta) = int_{z_0x_0}(\delta)$, donc $(kzyx')^{-1}z_0x_0 \in Z_G(\delta) \subset I_\eta$. Alors $(kz)^{-1}z_0$ appartient à $I_\eta \cap \mathcal{Z}$, donc $z_0 \in Kz(I_\eta \cap \mathcal{Z})$. Par définition de $\mathcal{Z}_\eta$, cela force l'égalité $z_0 = z$. Posons $k' = z^{-1}kz$. C'est encore un élément de K et la relation $(kz)^{-1}z_0 \in I_\eta$ se simplifie en $k' \in I_\eta$. L'élément $k'y$ appartient à $I_\eta(F)$ et on voit que l'on a $int_{k'y}(X') = X_0$. Cela prouve que tous les éléments de $\mathcal{X}_\delta$ qui sont conjugués à δ' par un élément de $G(F)$ sont de la forme $int_z(exp(X_0)\eta)$, où $X_0 \in \mathcal{X}_X$ est conjugué à X' par un élément de $I_\eta(F)$. Inversement, il est immédiat que tout élément de cette forme est conjugué à δ' par un élément de $G(F)$. Donc $c(\delta')$ est le nombre d'éléments X_0 de la forme ci-dessus, autrement dit le nombre de classes de conjugaison par $G_\eta(F)$ dans la classe de conjugaison de X' par $I_\eta(F)$. Ce nombre est égal au nombre d'éléments de l'ensemble de doubles classes :

$$G_\eta(F)\backslash I_\eta(F)/Z_G(\delta')(F).$$

Parce que $G_\eta\backslash I_\eta$ est abélien, ce nombre est égal à :

$$[I_\eta(F) : G_\eta(F)][Z_G(\delta')(F) : (Z_G(\delta')(F) \cap G_\eta(F))]^{-1}.$$

Le groupe $Z_G(\delta')\cap G_\eta$ est égal à $Z_{G_\eta}(X)$, qui est connexe et égal à $G_{\delta'}$. D'autre part, les éléments δ et δ' étant réguliers et conjugués par G, leurs commutants sont isomorphes sur F. On obtient alors :

(5) $c(\delta')=[I_\eta(F):G_\eta(F)][Z_G(\delta)(F):G_\delta(F)]^{-1}$.

Considérons l'application :

$$(6)\qquad \begin{cases} \mathcal{Z} & \to & Z^{\theta^*}_{G,p'} \\ k & \to & \phi(k)^{-1}k. \end{cases}$$

Puisque $Z^{\theta^*}_{G,p'}\subset K^{nr}_{\phi-fin}$, elle est surjective. Elle se quotiente en une application injective définie sur $K\backslash\mathcal{Z}$. En la composant avec la surjection :

$$Z^{\theta^*}_{G,p'}\to I_\eta(F^{nr})/G_\eta(F^{nr}),$$

on obtient une surjection :

$$K\backslash\mathcal{Z}\to I_\eta(F^{nr})/G_\eta(F^{nr}),$$

qui se quotiente ensuite en une surjection :

$$(7)\qquad K\backslash\mathcal{Z}/(\mathcal{Z}\cap G_\eta)\to I_\eta(F^{nr})/G_\eta(F^{nr}).$$

Montrons que celle-ci est aussi injective. Soient $k,h\in\mathcal{Z}$, supposons que $\phi(k)^{-1}k$ et $\phi(h)^{-1}h$ aient même image dans $I_\eta(F^{nr})/G_\eta(F^{nr})$. Soit $u\in G_\eta(F^{nr})$ tel que $\phi(k)^{-1}k=\phi(h)^{-1}hu$. On a $u\in Z^{\theta^*}_{G,p'}\cap G_\eta(F^{nr})\subset K^{nr}_{\eta,\phi-fin}$. On peut trouver $v\in K^{nr}_\eta$ tel que $\phi(v)^{-1}v=u$. On a $v\in\mathcal{Z}\cap G_\eta$. Les éléments k et hv ont même image par l'application (6), donc ont même image dans $K\backslash\mathcal{Z}$. Alors k et h ont même image dans l'espace de départ de (7), ce qui prouve l'injectivité de cette application.

Le groupe $I_\eta(F^{nr})/G_\eta(F^{nr})$ est abélien et est muni d'un automorphisme ϕ. D'après la construction de la bijection (7), celle-ci se quotiente en une bijection de $\mathcal{Z}_\eta$ sur l'espace des coinvariants $(I_\eta(F^{nr})/G_\eta(F^{nr}))_{/\phi}$. Pour tout groupe abélien fini A muni d'un automorphisme ϕ, les groupes d'invariants A^ϕ et de coinvariants $A_{/\phi}$ ont même nombre d'éléments. De l'injection de $I_\eta(F)$ dans $I_\eta(F^{nr})$ se déduit une injection de $I_\eta(F)/G_\eta(F)$ dans $(I_\eta(F^{nr})/G_\eta(F^{nr}))^\phi$. D'après (1), la suite :

$$1\to K^{nr}_\eta\to K^{nr}\cap I_\eta\to I_\eta(F^{nr})/G_\eta(F^{nr})\to 1$$

est exacte. Comme on l'a dit dans la preuve du lemme 4.1, on a $H^1(\Gamma^{nr},K^{nr}_\eta)=0$. La suite de cohomologie associée à la suite exacte ci-dessus montre que $(K^{nr}\cap I_\eta)^{\Gamma^{nr}}=K\cap I_\eta(F)$ s'envoie surjectivement sur $(I_\eta(F^{nr})/G_\eta(F^{nr}))^{\Gamma^{nr}}=(I_\eta(F^{nr})/G_\eta(F^{nr}))^\phi$. A fortiori, l'injection de $I_\eta(F)/G_\eta(F)$ dans $(I_\eta(F^{nr})/G_\eta(F^{nr}))^\phi$ est surjective. Cela démontre les égalités :

$$|\mathcal{Z}_\eta|=|(I_\eta(F^{nr})/G_\eta(F^{nr}))_{/\phi}|=[I_\eta(F):G_\eta(F)].$$

Jointes à (5), elles conduisent à l'égalité (4), ce qui achève la démonstration.

4.3. Eléments d'ordre fini premier à p dans $H_z(F)$. Pour γ et γ' deux éléments semi-simples de $H_z(F)$, on dit que γ et γ' sont stablement conjugués s'il existe $h \in H$ tel que $int_h(\gamma) = \gamma'$ et, pour tout $\sigma \in \Gamma$, $\sigma(h)^{-1}h$ appartient à H_γ. Pour un tel élément h, int_h se restreint en un torseur intérieur de H_γ sur $H_{\gamma'}$. Pour les éléments de $H_{z,\tilde{G}-reg(F)}$, la conjugaison stable se confond avec la conjugaison par H.

Lemme. *Soit $\epsilon \in H_z(F)_{p'}$.*

(i) Il existe $\epsilon' \in H_z(F)_{p'}$ qui est stablement conjugué à ϵ et tel que $H_{\epsilon'}$ est quasi- déployé.

(ii) Soient ϵ' vérifiant les conditions de (i) et γ un élément de $H_{z,\tilde{G}-reg}(F)$ $\cap H_z(F)_c$ tel que $\gamma_{p'} = \epsilon$. Alors il existe un élément $\gamma' \in H_{z,\tilde{G}-reg}(F) \cap H_z(F)_c$ qui est stablement conjugué à γ et tel que $\gamma'_{p'} = \epsilon'$.

(iii) Supposons $\epsilon \in K_z$. Alors H_ϵ est non ramifié et $K_H \cap H_\epsilon(F)$ est un sous-groupe hyperspécial de $H_\epsilon(F)$.

(iv) Supposons H_ϵ quasi-déployé. Alors ϵ est stablement conjugué à un élément de K_z si et seulement si H_ϵ est non ramifié.

Preuve. Le (i) résulte de [K1] lemme 3.3. Evidemment, ce lemme est énoncé pour un groupe tel que H et non pour un espace tordu comme H_z mais on vérifie que la torsion par le cocycle z_H n'influe pas sur la démonstration.

Le (ii) résulte de [K1] corollaire 2.2.

Le (iii) est le (i) du lemme 4.1 appliqué à H_z au lieu de $\tilde{G}$, ce qui est loisible d'après ce que l'on a dit en 2.9.

La partie seulement si du (iv) résulte de (iii). Inversement, supposons H_ϵ non ramifié. Soit $T^{\flat *}$ un sous-tore maximal de H_ϵ, défini sur F et déployé sur F^{nr}. Les deux tores $T^{\flat *}$ et T_H sont maximaux dans H et déployés sur F^{nr}. Il existe donc $h \in H(F^{nr})$ tel que $int_h(T^{\flat *}) = T_H$. Puisque les deux tores sont définis sur F, on a $\phi(h)h^{-1} \in Norm_H(T_H)$. On sait que $Norm_H(T_H) \cap K_H^{nr}$ s'envoie surjectivement sur Ω_H. En fait, on montre, par exemple en utilisant des sections de Springer, qu'il existe un sous-groupe fini de $Norm_H(T_H) \cap K_H^{nr}$, invariant par Γ^{nr}, qui s'envoie surjectivement sur Ω_H. Un tel sous-groupe est contenu dans $K_{H,\phi-fin}^{nr}$ (cf. la preuve du lemme précédent pour la définition et les propriétés de cet ensemble). On peut donc trouver $k \in K_H^{nr}$ tel que $\phi(k)k^{-1} \in Norm_H(T_H)$ et que cet élément ait même image que $\phi(h)h^{-1}$ dans Ω_H. Posons $T' = int_{k^{-1}}(T_H)$. C'est un sous-tore maximal de H déployé sur F^{nr}. Puisque $\phi(k)k^{-1} \in Norm_H(T_H)$, T' est défini sur F. On a $int_{k^{-1}h}(T^{\flat *}) = T'$ et on vérifie que $\phi(k^{-1}h)^{-1}k^{-1}h$ appartient à $T^{\flat *}$. Posons $\epsilon' = int_{k^{-1}h}(\epsilon)$. Les conditions précédentes entraînent que $\epsilon' \in H_z(F)_{p'}$ et ϵ' est stablement conjugué à ϵ. On a $int_h(\epsilon) \in T_{H,p'} = T_H(\mathfrak{o}_{F^{nr}})_{p'} \subset K_H^{nr}$. Donc aussi $\epsilon' = int_{k^{-1}} \circ int_h(\epsilon) \in K_H^{nr}$. Puisque $\epsilon' \in H_z(F)$, on a en fait $\epsilon' \in K_z$. L'élément ϵ' vérifie donc les conditions requises, ce qui prouve (iv).

4.4. Correspondance entre classes de conjugaison, compacité, éléments d'ordre fini premier à p.

Lemme. *Soient γ, resp. δ, un élément semi-simple de $H_z(F)$, resp. $\tilde{G}(F)$. Supposons que les classes de conjugaison de γ et δ se correspondent, cf. 2.10. Alors γ est compact, resp. d'ordre fini premier à p, si et seulement si δ l'est.*

Preuve. Dire que les classes de conjugaison de γ et δ se correspondent signifie qu'il existe $h \in H$, $g \in G$ et $\nu \in T$ de sorte que $int_g(\delta) = \nu\theta^*$ et $int_{h^{-1}} \circ \xi(\nu) = \gamma$. Fixons de tels éléments. Supposons δ compact ou δ d'ordre fini premier à p. Ces propriétés se conservent par conjugaison. Donc $\nu\theta^*$ est compact ou d'ordre fini premier à p. Soit N l'ordre de θ^*. Alors $(\nu\theta^*)^N$ est compact ou d'ordre fini premier à p, donc aussi $\xi(\nu\theta^*)^N$. Mais on a l'égalité :

$$(\nu\theta^*)^N = \nu^N(\theta^*(\nu)\nu^{-1}) \dots ((\theta^*)^{N-1}(\nu)\nu^{-1}).$$

Pour tout $i \geq 1$,

$$(\theta^*)^i(\nu)\nu^{-1} = \prod_{j=1,\dots,i} (\theta^*)^j(\nu)(\theta^*)^{j-1}(\nu^{-1}).$$

Un tel élément est annulé par ξ. Donc $\xi(\nu\theta^*)^N = \xi(\nu)^N$. Puisque N est premier à p grâce à l'hypothèse (Hyp1), $\xi(\nu)$ est compact ou d'ordre fini premier à p. Par conjugaison, il en est de même de γ.

Inversement, supposons γ compact ou d'ordre fini premier à p. Posons $\mu = int_h(\gamma)$. C'est un élément de T_H qui est compact ou d'ordre fini premier à p. Donc $\mu \in T_H(\mathfrak{o}_{\bar{F}})$ ou $\mu \in T_H(\mathfrak{o}_{F^{nr}})_{p'}$. De l'homomorphisme ξ se déduit un homomorphisme surjectif de $X_*(T)$ sur $X_*(T_H)$ (pour la suite du raisonnement, il suffirait que son image soit de conoyau fini d'ordre premier à p). On n'a aucun mal à en déduire que les homomorphismes :

$$\xi : T(\mathfrak{o}_{\bar{F}}) \to T_H(\mathfrak{o}_{\bar{F}}) \text{ et } \xi : T(\mathfrak{o}_{F^{nr}})_{p'} \to T_H(\mathfrak{o}_{F^{nr}})_{p'}$$

sont surjectifs. On peut donc choisir $\nu' \in T(\mathfrak{o}_{\bar{F}})$ ou $\nu' \in T(\mathfrak{o}_{F^{nr}})_{p'}$ tel que $\xi(\nu') = \mu$. Il est clair que $\nu'\theta^*$ est compact ou d'ordre fini premier à p. On a $\xi(\nu') = \xi(\nu)$, donc il existe $t \in T$ tel que $\nu = t\theta^*(t)^{-1}\nu'$. Alors $\nu\theta^* = int_t(\nu'\theta^*)$ et $\nu\theta^*$ est lui-aussi compact ou d'ordre fini premier à p. Comme au début de la preuve, δ a les mêmes propriétés.

4.5. Diagrammes. On appelle diagramme modérément ramifié un septuplet $D = (\epsilon, T^\flat, T^\diamond, T^\natural, h, g, \eta)$ vérifiant les conditions suivantes. Le terme ϵ est un élément de $H_z(F)_{p'}$. Le terme η est un élément de $\tilde{K} \cap \tilde{G}(F)_{p'}$. Les termes $T^\flat$, resp. $T^\diamond$, $T^\natural$, sont des sous-tores maximaux et définis sur F de H, resp. G, G_η, et $T^\diamond$ est le commutant de $T^\natural$ dans G. Le terme h, resp. g, est un élément de H_{SC}, resp. G_{SC}. On suppose que $int_h(T^\flat) = T_H$, $int_g(T^\diamond) = T$ et que l'homomorphisme composé :

$$int_{h^{-1}} \circ \xi \circ int_g : T^\diamond \to T^\flat$$

est défini sur F. On suppose que $int_g(\eta)$ appartient à $T\theta^*$ et, si l'on note $\nu_D\theta^*$ cet élément, on a l'égalité $int_{h^{-1}} \circ \xi(\nu_D) = \epsilon$.

On dit que le diagramme D ci-dessus est fortement non ramifié s'il vérifie de plus les conditions suivantes : $\epsilon \in K_z \cap H_z(F)_{p'}$, $h \in K^{nr}_{H,sc}$, $g \in K^{nr}_{sc}$.

On utilisera une terminologie naturelle, par exemple on dira qu'un diagramme D comme ci-dessus joint ϵ à η, que ϵ est la source de D et η son extrémité. Il est clair que, s'il existe un diagramme modérement ramifié joignant ϵ à η, les classes de conjugaison de ces deux éléments se correspondent.

Lemme. *(i) Soit $\gamma \in H_{z,\tilde{G}-reg}(F) \cap H_z(F)_c$. Supposons qu'il existe $\delta \in \tilde{K} \cap \tilde{G}_{reg}(F)$ dont la classe de conjugaison corresponde à celle de γ. Posons $\epsilon = \gamma_{p'}$. Alors il existe un diagramme modérément ramifié de source ϵ.*

(ii) Soit $\epsilon \in K_z \cap H_z(F)_{p'}$. Alors il existe un diagramme fortement non ramifié de source ϵ.

Preuve. Fixons δ vérifiant l'hypothèse de (i). Comme on l'a dit dans la preuve du lemme précédent, dire que les classes de conjugaison de γ et δ se correspondent signifie qu'il existe $h \in H$, $g \in G$ et $\nu' \in T$ de sorte que $int_{g'}(\delta) = \nu'\theta^*$ et $int_{h'}^{-1} \circ \xi(\nu') = \gamma$. On peut remplacer h' par un élément $h \in H_{SC}$ qui a même image que h' dans H_{AD}. Ecrivons $g' = zg$, où $z \in Z_G$ et $g \in G_{SC}$. Posons $\nu = z\theta^*(z)^{-1}\nu'$. Alors $int_g(\delta) = \nu\theta^*$ et $\xi(\nu) = \xi(\nu')$, donc encore $int_{h^{-1}} \circ (\nu) = \gamma$. Posons $T^\natural = G_\delta$ et notons $T^\diamond$ le commutant de $T^\natural$ dans G. On a $int_g(T^\natural) = G_{\nu\theta^*}$ et $int_g(T^\diamond)$ est le commutant de ce groupe. Or, celui-ci contient $T^{\theta^*,0}$, donc son commutant est contenu dans T. Par égalité des dimensions, cela entraîne $int_g(T^\natural) = T^{\theta^*,0}$ et $int_g(T^\diamond) = T$. Posons $T^\flat = H_\gamma$. On a de même $int_h(T^\flat) = T_H$. Soit $\sigma \in \Gamma$. Parce que T et $T^\diamond$ sont définis sur F, $g\sigma(g^{-1})$ normalise T. Donc cet élément définit un élément $\omega_D(\sigma) \in \Omega$. Pour la même raison, cet élément conserve $T^{\theta^*,0}$, donc appartient à Ω^{θ^*}. Introduisons le groupe $G_{SC}^{\theta^*,0}$: c'est, au choix, le revêtement simplement connexe du groupe dérivé de $G^{\theta^*,0}$ ou la composante neutre du sous-groupe des points fixes par θ^* dans G_{SC}. On sait que son groupe de Weyl est Ω^{θ^*}. On peut donc écrire $\sigma(g)g^{-1} = tn$, avec $t \in T_{sc}$ et $n \in G_{SC}^{\theta^*,0}$. L'égalité $int_g(\delta) = \nu\theta^*$ et le fait que $\delta \in \tilde{G}(F)$ entraînent que :

$$\sigma(\nu)\theta^* = z(\sigma)^{-1} int_{\sigma(g)g^{-1}}(\nu\theta^*),$$

d'où :

$$\sigma(\nu) = z(\sigma)^{-1} tn\nu n^{-1}\theta^*(t)^{-1} = z(\sigma)^{-1} t\theta^*(t)^{-1}\omega_D(\sigma)^{-1}(\nu).$$

De même, l'élément $h\sigma(h)^{-1}$ normalise T_H. Il définit un élément $\omega_{D,H}(\sigma) \in \Omega^H$. En posant $\mu = int_h(\gamma)$, on a l'égalité :

$$\sigma(\mu) = z_H(\sigma)^{-1}\omega_{D,H}(\sigma)^{-1}(\mu).$$

On a dit en 2.9 que Ω_H s'identifiait à un sous-groupe de Ω^{θ^*}. De plus, il existe un élément $\omega(\sigma) \in \Omega^{\theta^*}$ de sorte que $\sigma \circ \xi = \xi \circ \omega(\sigma) \circ \sigma$. Posons $\omega_0(\sigma) = \omega_{D,H}(\sigma)\omega(\sigma)\omega_D(\sigma)^{-1}$. Puisque $\xi(\nu) = \mu$, les égalités précédentes entraînent :

$$\xi \circ \omega_0(\sigma)(\nu) = \xi(\nu).$$

Comme ci-dessus, on peut représenter $\omega_0(\sigma)$ par un élément $n_0 \in G_{SC}^{\theta^*,0}$. Il existe alors $t_0 \in T$ de sorte que $int_{n_0}(\nu) = t_0^{-1}\theta^*(t_0)\nu$. Donc $t_0 n_0$ commute à $\nu\theta^*$. On a déjà dit que le commutant de $\nu\theta^*$ était inclus dans T (il est en fait égal à T^{θ^*}). Donc $\omega_0(\sigma) = 1$. Considérons l'application :

$$(1) \qquad int_{h^{-1}} \circ \xi \circ int_g : T^{\diamond} \to T^{\flat}.$$

On calcule :

$$\sigma \circ int_{h^{-1}} \circ \xi \circ int_g = int_{h^{-1}} \circ \xi \circ \omega_0(\sigma) \circ int_g \circ \sigma.$$

L'égalité $\omega_0(\sigma) = 1$ entraîne que l'application (1) est équivariante pour les actions de Γ. Posons $\eta = \delta_{p'}$. Rappelons que cet élément appartient à l'adhérence du groupe engendré dans G^+ par δ. Donc $\eta \in \tilde{K}$. En reprenant les arguments de la preuve du lemme 4.4, on vérifie que $int_g(\eta) \in T\theta^*$ et, si l'on note $\nu_D\theta^*$ cet élément, on a $int_{h^{-1}} \circ \xi(\nu_D) = \epsilon$. Alors :

$$D = (\epsilon, T^{\flat}, T^{\diamond}, T^{\natural}, h, g, \eta)$$

est un diagramme modérément ramifié de source ϵ, ce qui prouve (i).

Soit $\epsilon \in K_z \cap H_z(F)_{p'}$. Le groupe $K_H \cap H_\epsilon(F)$ est un sous-groupe hyperspécial de $H_\epsilon(F)$ (lemme 4.3(iii)). Fixons un sous-tore de H_ϵ défini sur F et maximal tel que $K_H \cap H_\epsilon(F)$ fixe un point hyperspécial de l'appartement de $Imm(H_\epsilon, F)$ associé à ce tore. Notons $T^{\flat}$ le commutant de ce tore dans H_ϵ. C'est un sous-tore maximal de H_ϵ, défini sur F et déployé sur F^{nr}. En appliquant la relation (1) de 4.1 où on remplace $\tilde{G}$ par H_z, on voit qu'il existe $h_0 \in K_H^{nr}$ et $\mu \in T_H(\mathfrak{o}_{F^{nr}})$ tels que $int_{h_0}(\epsilon) = \mu$. Les deux tores $int_{h_0}(T^{\flat})$ et T_H sont des tores maximaux de H_μ, déployés sur F^{nr}, et $K_H^{nr} \cap H_\mu(F^{nr})$ fixe un point hyperspécial dans chacun des appartements de $Imm(H_\mu, F^{nr})$ associés à ces tores. Il en résulte que ces tores sont conjugués par un élément de $K_H^{nr} \cap H_\mu(F^{nr})$. Quitte à multiplier h_0 à gauche par cet élément, on peut supposer $int_{h_0}(T^{\flat}) = T_H$. Le groupe $K_{H,sc}^{nr}$ s'envoie surjectivement sur $K_{H,ad}^{nr}$ et on peut aussi bien remplacer h_0 par un élément $h \in K_{H,sc}^{nr}$ qui a même image que h_0 dans $K_{H,ad}^{nr}$. On a noté ϕ l'élément de Frobenius de Γ^{nr}, on le relève en un élément ϕ de Γ. Comme plus haut, on a les égalités $\phi \circ \xi = \xi \circ \omega(\phi) \circ \phi$ et $int_{h\phi(h)^{-1}} \circ \xi = \xi \circ \omega_{D,H}(\phi)$. Posons $\omega_D(\phi) = \omega_{D,H}(\phi)\omega(\phi)$. L'application $K_{sc}^{nr} \cap Norm_{G_{SC}^{\theta^*,0}}(T_{sc}^{\theta^*,0}) \to \Omega^{\theta^*}$ est surjective. Par le même argument utilisé dans la preuve du lemme 4.3(iv), on voit qu'il existe $g \in K_{sc}^{nr} \cap G_{K,SC}^{\theta^*,0}$ tel que $g\phi(g^{-1})$ appartienne à $Norm_{G_{SC}^{\theta^*,0}}(T_{sc}^{\theta^*,0})$ et ait $\omega_D(\phi)$ pour image dans Ω^{θ^*}. Posons $T^{\diamond} = int_{g^{-1}}(T)$. La définition de g entraîne :

- $T^\diamond$ est défini sur F et invariant par θ^* ;
- l'application :

$$(2) \qquad int_{h^{-1}} \circ \xi \circ int_g : T^\diamond \to T^\flat$$

est définie sur F et se quotiente en un isomorphisme défini sur F de $T^\diamond/\theta^*$ sur $T^\flat$.

Comme on l'a dit dans la preuve du lemme 4.4, ξ envoie $T(\mathfrak{o}_{F^{nr}})_{p'}$ sur $T_H(\mathfrak{o}_{F^{nr}})_{p'}$. Soit $\nu \in T(\mathfrak{o}_{F^{nr}})_{p'}$ tel que $\xi(\nu) = \mu$. Posons $\rho = int_{g^{-1}}(\nu)$. C'est un élément de $T^\diamond(\mathfrak{o}_{F^{nr}})_{p'}$. Un calcul analogue à l'un de ceux faits dans la preuve de (i) montre qu'il existe $t \in T^\diamond$ tel que $\phi(\rho) = z(\phi)t\theta^*(t)^{-1}\rho$. Comme dans la preuve du lemme 4.1(ii), on peut remplacer t par un élément de $T^\diamond(\mathfrak{o}_{F^{nr}})_{p'}$. Or $H^1(\Gamma^{nr}, T^\diamond(\mathfrak{o}_{F^{nr}})_{p'}) = 0$, puisque $T^\diamond(\mathfrak{o}_{F^{nr}})_{p'} \simeq \mathbf{T}^\diamond(\bar{\mathbb{F}}_q)$. On peut fixer $t_1 \in T^\diamond(\mathfrak{o}_{F^{nr}})_{p'}$ tel que $t = t_1\phi(t_1)^{-1}$. Posons $\eta = t_1\rho\theta^*t_1^{-1}$. C'est un élément de $G(F^{nr})\theta^*$ et on voit qu'il vérifie la relation $\phi(\eta) = z(\phi)\eta$. Donc $\eta \in \tilde{G}(F)$. Puisque ν, g et t_1 appartiennent à K^{nr}, η appartient à $\tilde{K}$. Comme ν, c'est un élément d'ordre fini premier à p. En posant $T^\natural = T^{\diamond,\theta^*,0}$, on vérifie que $(\epsilon, T^\flat, T^\diamond T^\natural, h, g, \eta)$ est un diagramme non ramifié.

4.6. Construction d'une nouvelle donnée endoscopique. Soit $D = (\epsilon, T^\flat, T^\diamond T^\natural, h, g, \eta)$ un diagramme modérément ramifié. On pose $\nu = \nu_D$ et $\bar{G} = G_\eta$. D'après le lemme 4.1(i), $\bar{G}$ est quasi-déployé sur F. On fixe une paire de Borel $(\bar{B}, \bar{T})$ définie sur F de $\bar{G}$. Le couple $(int_{g^{-1}}(B) \cap \bar{G}, T^\natural)$ est aussi une paire de Borel de $\bar{G}$. On fixe $g_0 \in \bar{G}_{SC}$ tel que int_{g_0} envoie la première paire sur la seconde.

On suppose que H_ϵ est quasi-déployé sur F. On fixe une paire de Borel $(B^{\flat*}, T^{\flat*})$ définie sur F de H_ϵ et un élément $h_0 \in H_{\epsilon,SC}$ tel que int_{h_0} envoie cette paire sur $(int_{h^{-1}}(B_H) \cap H_\epsilon, T^\flat)$.

On va identifier $\bar{T}$ à $T^{\theta^*,0}$ et $T^{\flat*}$ à $T_{/\theta^*}$. Ces identifications ne sont pas équivariantes pour les actions de Γ et on va distinguer les différentes actions dans la notation. Pour $\sigma \in \Gamma$, on note simplement σ l'action naturelle de σ sur $T^{\theta^*,0}$ ou $T_{/\theta^*}$, on note $\sigma_{\bar{T}}$ l'action sur $T^{\theta^*,0}$ transportée de l'action naturelle sur $\bar{T}$ et on note $\sigma_{T^{\flat*}}$ l'action sur $T_{/\theta^*}$ transportée de l'action naturelle sur $T^{\flat*}$. On note de la même façon les actions duales sur $\hat{T}_{/\hat{\theta}}$ ou $\hat{T}^{\hat{\theta},0}$. On identifie $\bar{T}$ à $T^{\theta^*,0}$ par l'application int_{gg_0}. Cette application identifie le groupe de Weyl de $\bar{G}$ relatif à $\bar{T}$ à un sous-groupe de Ω^{θ^*}. Par abus de notation, c'est ce dernier que nous notons $\Omega_{\bar{G}}$. Pour $\sigma \in \Gamma$, on a l'égalité d'actions sur $T^{\theta^*,0}$:

$$(1) \qquad \sigma_{\bar{T}} = \omega_{\bar{G}}(\sigma)\omega_D(\sigma) \circ \sigma,$$

où $\omega_{\bar{G}}(\sigma)$ est l'image dans $\Omega_{\bar{G}}$ de $gg_0\sigma(g_0)^{-1}g^{-1}$ et, comme dans la preuve du lemme 4.5, $\omega_D(\sigma)$ est l'image dans Ω^{θ^*} de $g\sigma(g)^{-1}$. On identifie $T^{\flat*}$ à $T_{/\theta^*}$ par $\xi^{-1} \circ int_{hh_0}$. Cette application identifie le groupe de Weyl de H_ϵ relatif à $T^{\flat*}$ à un sous-groupe de Ω_H, lui- même sous-groupe de Ω^{θ^*}. C'est ce sous-groupe que l'on note Ω_{H_ϵ}. Pour $\sigma \in \Gamma$, on a l'égalité d'actions sur $T_{/\theta^*}$:

$$(2) \qquad \sigma_{T^{\flat *}} = \omega_{H_\epsilon}(\sigma)\omega_{D,H}(\sigma)\omega(\sigma) \circ \sigma,$$

où $\omega_{H_\epsilon}(\sigma)$ est l'image dans Ω_{H_ϵ} de $hh_0\sigma(h_0^{-1})h_1$, $\omega_{D,H}(\sigma)$ est l'image dans Ω_H de $h\sigma(h^{-1})$ et $\omega(\sigma) \in \Omega^{\theta^*}$ est l'élément tel que $\sigma_{T_H} \circ \xi = \xi \circ \omega(\sigma) \circ \sigma$, σ_{T_H} étant l'action naturelle sur T_H.

Les formules (1) et (2) permettent d'étendre chacune des actions galoisiennes à T, ou à tout tore déduit naturellement de T ($T^{\theta^*,0}$, $\hat{T}$ etc.). Par le même calcul que l'on a fait dans la preuve du lemme 4.5, le fait que $int_{h^{-1}} \circ \xi \circ int_g : T^{\diamond} \to T^{\flat}$ soit défini sur F équivaut à l'égalité :

$$\omega_{D,H}(\sigma)\omega(\sigma) = \omega_D(\sigma).$$

La comparaison des formules (1) et (2) conduit à l'égalité :

$$(3) \qquad \sigma_{T^{\flat *}} = \omega_{H_\epsilon}(\sigma)\omega_{\bar{G}}(\sigma)^{-1}\sigma_{\bar{T}}.$$

Introduisons le L-groupe ${}^L\bar{G} = \hat{\bar{G}} \rtimes W$. Fixons une paire de Borel $(\hat{\bar{B}}, \hat{\bar{T}})$ de $\hat{\bar{G}}$ conservée par l'action de W. De l'identification de $\bar{T}$ à $T^{\theta^*,0}$ résulte une identification de $\hat{\bar{T}}$ à $\hat{T}_{/\hat{\theta}}$. Notons $s_\sharp$ l'image de s dans $\hat{T}_{/\hat{\theta}}$ et $\hat{H}_\sharp = \hat{\bar{G}}_{s_\sharp} = Z_{\hat{\bar{G}}}(s_\sharp)^0$. On munira plus loin ce groupe d'une action de Γ. Pour l'instant, contentons-nous d'introduire le groupe réductif connexe $H_\sharp$ sur $\bar{F}$ dont $\hat{H}_\sharp$ est le groupe dual. On fixe une paire de Borel épinglée $(B_\sharp, T_\sharp, (E_{\alpha_\sharp})_{\alpha_\sharp \in \Sigma_\sharp})$ de $H_\sharp$. Le tore $T_\sharp$ s'identifie à $\bar{T} \simeq T^{\theta^*,0}$.

On va décrire les ensembles de racines et coracines de différents groupes. Posons pour simplifier $X_* = X_*(T)$ et $X^* = X^*(T)$. Rappelons que l'on dispose de l'ensemble de racines $\Sigma \subset X^*$ et de l'ensemble de coracines $\check{\Sigma} \subset X_*$. Le groupe Θ^* agit sur ces ensembles. Pour $\alpha \in \Sigma$, on note n_α le nombre d'éléments de l'orbite de α pour cette action et $N\alpha$ la somme des éléments de cette orbite. On définit de même $n_{\check{\alpha}}$ et $N\check{\alpha}$ pour $\check{\alpha} \in \check{\Sigma}$. Posons :

$$Y^* = X^*/(X^* \cap (1-\theta^*)(X^*_{\mathbb{Q}})), Y_* = X_*/(X_* \cap (1-\theta^*)(X_{*,\mathbb{Q}})),$$

(cf. 3.5 pour les notations $X^*_{\mathbb{Q}}$ et $X_{*,\mathbb{Q}}$) et notons $p^* : X^* \to Y^*$ et $p_* : X_* \to Y_*$ les applications naturelles. Remarquons que Y^* est le dual de $X_*^{\theta^*}$ tandis que Y_* est le dual de X^{*,θ^*}. D'après [KS] 1.3, on distingue trois types de racines. Pour $\alpha \in \Sigma$, on dit que α est de type 1 si $2p^*(\alpha)$ et $\frac{1}{2}p^*(\alpha)$ n'appartiennent pas à $p^*(\Sigma)$; on dit que α est de type 2 si $2p^*(\alpha) \in p^*(\Sigma)$; on dit que α est de type 3 si $\frac{1}{2}p^*(\alpha) \in p^*(\Sigma)$. On montre que ces deux dernières conditions sont exclusives l'une de l'autre. Si α est de type 2, alors n_α est pair. L'application $\alpha \to \alpha + (\theta^*)^{n_\alpha/2}(\alpha)$ est une surjection de l'ensemble des racines de type 2 sur celui des racines de type 3, dont toutes les fibres ont deux éléments. On dit qu'un élément $\check{\alpha} \in \check{\Sigma}$ est de type 1, 2 ou 3 si la racine α associée est de ce type. Posons :

$$\Sigma^{res} = p^*(\Sigma) \subset Y^*,$$

$$\check{\Sigma}^{res} = \{N\check{\alpha}; \check{\alpha} \in \check{\Sigma}, \check{\alpha} \text{ de type 1 ou 3}\} \cup \{2N\check{\alpha}; \check{\alpha} \in \check{\Sigma}, \check{\alpha} \text{ de type 2}\} \subset X_*^{\theta^*},$$

$$\Sigma_{res}=\{N\alpha;\alpha\in\Sigma,\alpha \text{ de type 1 ou 3}\}\cup\{2N\alpha;\alpha\in\Sigma,\alpha \text{ de type 2}\}\subset X^{*,\theta^*},$$
$$\check{\Sigma}_{res}=p_*(\check{\Sigma})\subset Y_*.$$

D'après [KS] 1.3, Σ^{res} est l'ensemble des racines de $G^{\theta^*,0}$ relatif à $T^{\theta^*,0}$ et $\check{\Sigma}^{res}$ est l'ensemble de coracines associé. De même $\check{\Sigma}_{res}$ est l'ensemble des racines de $\hat{G}^{\hat{\theta},0}$ relatif à $\hat{T}^{\hat{\theta},0}$ et Σ_{res} est l'ensemble de coracines associé.

Remarquons que H_ϵ et $H_\sharp$ s'obtiennent par composition de deux opérations analogues, mais qui ne sont pas prises dans le même ordre. En effet, pour définir H_ϵ, on commence par prendre le commutant de $s\hat{\theta}$ dans $\hat{G}$, ce qui nous fournit $\hat{H}$, puis on prend le commutant de ϵ dans H. Pour définir $H_\sharp$, on commence par prendre le commutant de η dans G, ce qui nous fournit $\bar{G}$, puis on prend le commutant de $s_\sharp$ dans $\hat{\bar{G}}$, ce qui nous fournit $\hat{H}_\sharp$. En appliquant la recette de [KS] 1.3, on peut décrire les ensembles de racines et coracines de H_ϵ relatifs à $T^{\flat *}$ et ceux de $H_\sharp$ relatifs à $T_\sharp$. On note ces ensembles Σ_{H_ϵ}, $\check{\Sigma}_{H_\epsilon}$, $\Sigma_\sharp$ et $\check{\Sigma}_\sharp$. Alors :

$$\Sigma_{H_\epsilon}=\{N\alpha;\alpha\in\Sigma,\alpha \text{ de type 1}, N\check{\alpha}(s)=1,N\alpha(\nu)=1\}\cup$$
$$\{2N\alpha;\alpha\in\Sigma,\alpha \text{ de type 2}, N\check{\alpha}(s)=1,N\alpha(\nu)=\pm 1\}\cup$$
$$\{N\alpha;\alpha\in\Sigma,\alpha \text{ de type 3}, N\check{\alpha}(s)=-1,N\alpha(\nu)=1\};$$

$$\check{\Sigma}_{H_\epsilon}=\{p_*(\check{\alpha});\check{\alpha}\in\check{\Sigma},(N\check{\alpha}(s),N\alpha(\nu))=\begin{cases}(1,1), & \text{si } \check{\alpha} \text{ est de type 1,}\\ (1,\pm 1), & \text{si } \check{\alpha} \text{ est de type 2,}\\ (-1,1), & \text{si } \check{\alpha} \text{ est de type 3;}\end{cases}\}$$

$$\Sigma_\sharp=\{p^*(\alpha);\alpha\in\Sigma,(N\check{\alpha}(s),N\alpha(\nu))=\begin{cases}(1,1), & \text{si } \alpha \text{ est de type 1,}\\ (\pm 1,1), & \text{si } \alpha \text{ est de type 2,}\\ (1,-1), & \text{si } \alpha \text{ est de type 3;}\end{cases}\}$$

$$\check{\Sigma}_\sharp=\{N\check{\alpha};\check{\alpha}\in\check{\Sigma},\check{\alpha} \text{ de type 1}, N\check{\alpha}(s)=1,N\alpha(\nu)=1\}\cup$$
$$\{2N\check{\alpha};\check{\alpha}\in\check{\Sigma},\check{\alpha} \text{ de type 2}, N\check{\alpha}(s)=\pm 1,N\alpha(\nu)=1\}\cup$$
$$\{N\check{\alpha};\check{\alpha}\in\check{\Sigma},\check{\alpha} \text{ de type 3}, N\check{\alpha}(s)=1,N\alpha(\nu)=1\}.$$

De l'application naturelle de $X_*^{\theta^*}$ dans Y_* se déduit un isomorphisme de $X_{*,\mathbb{Q}}^{\theta^*}$ sur $Y_{*,\mathbb{Q}}$. On note j_* son inverse. On note $j^*: Y_{\mathbb{Q}}^* \to X_{\mathbb{Q}}^{*,\theta^*}$ l'isomorphisme transposé de j_*. On voit alors que les ensembles de racines ci-dessus vérifient une partie des hypothèses de 3.5, c'est-à-dire qu'il existe des bijections $j_{\check{\Sigma}}: \check{\Sigma}_{H_\epsilon}\to\check{\Sigma}_\sharp$ et $j_\Sigma:\Sigma_\sharp\to\Sigma_{H_\epsilon}$ et des fonctions $\check{b}:\check{\Sigma}_{H_\epsilon}\to\mathbb{Q}_{>0}$ et $b:\Sigma_\sharp\to\mathbb{Q}_{>0}$ vérifiant les conditions de ce paragraphe. Pour $p^*(\alpha)\in\Sigma_\sharp$, avec α de type 1, on a simplement $j_\Sigma(p^*(\alpha))=N\alpha$ et $b(p^*(\alpha))=n_\alpha$. Pour les racines de type 2 ou 3, la définition est plus subtile, cf. [W1] 3.3. Comme on l'a dit en 3.5, cela a pour conséquence que le groupe de Weyl Ω_{H_ϵ} s'identifie au groupe de Weyl $\Omega_\sharp$ de $H_\sharp$. Ce groupe est donc contenu dans $\Omega_H\cap\Omega_{\bar{G}}$. Soit $\sigma\in\Gamma$. Montrons que :

(4) $\sigma_{T^{\flat *}}$ conserve $\Sigma_\sharp$.

Notons $\Sigma_{\bar{G}}\subset\Sigma^{res}$ l'ensemble des racines de $\bar{G}$. Remarquons que $\Sigma_\sharp$ est égal au sous-ensemble des éléments de $\Sigma_{\bar{G}}$ qui sont proportionnels à un élément de $(j^*)^{-1}(\Sigma_{H_\epsilon})$. L'action $\sigma_{T^{\flat *}}$ conserve Σ_{H_ϵ} par définition, donc aussi le sous-ensemble des éléments de $Y^*_{\mathbb{Q}}$ qui sont proportionnels à un élément de $(j^*)^{-1}(\Sigma_{H_\epsilon})$. Il suffit de prouver qu'elle conserve aussi $\Sigma_{\bar{G}}$. Mais cet ensemble est conservé par $\sigma_{\bar{T}}$ par définition de cette dernière action, et est aussi conservé par $\Omega_{\bar{G}}$. Alors l'égalité (3) entraîne la propriété requise.

Le sous-groupe de Borel B détermine des sous-ensembles positifs dans Σ, Σ^{res} etc., en particulier dans Σ_{H_ϵ} et $\Sigma_\sharp$. Par construction, ces derniers sont aussi les sous-ensembles positifs déterminés par $B^{\flat *}$, resp. $B_\sharp$. L'action $\sigma_{T^{\flat *}}$ conserve le sous-ensemble positif de Σ_{H_ϵ}, donc aussi celui de $\Sigma_\sharp$. Cela permet de munir $H_\sharp$ de l'unique action galoisienne qui conserve la paire de Borel épinglée que l'on a fixée et dont l'action induite sur $T_\sharp$ est $\sigma \to \sigma_{T^{\flat *}}$. Ainsi, $H_\sharp$ devient un groupe réductif défini et quasi-déployé sur F.

Notons $\hat{\hat{T}}_{ad}$ l'image de $\hat{\hat{T}} \simeq \hat{T}_{/\hat{\theta}}$ dans $\hat{\hat{G}}_{AD}$. On note $\bar{s}$ l'image de $s_\sharp$ dans $\hat{\hat{T}}_{ad}$ et $\hat{\hat{H}} = \hat{\hat{G}}_{AD,\bar{s}} = Z_{\hat{\hat{G}}_{AD}}(\bar{s})^0$. On introduit le groupe réductif connexe $\bar{H}$ sur $\bar{F}$ dont $\hat{\hat{H}}$ est le groupe dual. Il y a une suite d'homomorphismes :

$$\hat{H}_\sharp \to \hat{\hat{H}} \to \hat{H}_{\sharp,AD} = \hat{\hat{H}}_{AD},$$

et dualement une suite d'homomorphismes :

$$\bar{H}_{SC} = H_{\sharp,SC} \to \bar{H} \to H_\sharp.$$

A cause de l'égalité (3), l'action $\sigma \to \sigma_{T^{\flat *}}$ conserve $Z_{\hat{\hat{G}}}$. Il en résulte que l'action galoisienne sur $\hat{H}_\sharp$ se descend en une action sur $\hat{\hat{H}}$ et, dualement, l'action galoisienne sur $H_\sharp$ se relève en une action sur $\bar{H}$. Ainsi, $\bar{H}$ est un groupe réductif connexe défini et quasi- déployé sur F. On note $(B_{\bar{H}}, T_{\bar{H}})$ la paire de Borel de $\bar{H}$ image réciproque de celle de $H_\sharp$ que l'on a fixée. Soit $\sigma \in \Gamma$. Montrons que l'on a l'égalité dans $\hat{\hat{T}}_{ad}$:

(5) $\sigma_{T^{\flat *}}(\bar{s}) = \bar{s}$.

L'action galoisienne étant continue, on peut supposer que σ appartient au sous-groupe W de Γ. Il y a une surjection de $\hat{T}$ sur $\hat{\hat{T}}_{ad}$ et $\bar{s}$ est l'image de s par cette surjection. Relevons l'élément $\omega(\sigma) \in \Omega^{\theta^*}$ en un élément $n \in Norm_{\hat{G}^{\hat{\theta},0}}(\hat{T}^{\hat{\theta},0})$. On a l'égalité $\omega(\sigma) \circ \sigma(s) = int_n \circ \sigma(s)$, d'où $\omega(\sigma) \circ \sigma(s)\hat{\theta} = int_n \circ \sigma(s\hat{\theta})$. Posons $\hat{\xi}(\sigma) = (x, \sigma)$, où $x \in \hat{G}$. Alors x normalise $\hat{T}$ et, d'après les définitions de 2.9, son image dans Ω est égale à $\omega(\sigma)$. Il existe donc $t \in \hat{T}$ tel que $n = tx$. Alors $int_n \circ \sigma(s\hat{\theta}) = t\hat{\theta}(t)^{-1} int_x \circ \sigma(s\hat{\theta})$. Ce dernier élément est égal à $t\hat{\theta}(t)^{-1} a(\sigma)^{-1} s\hat{\theta}$ d'après 2.9(6). D'où la relation :

$$\omega(\sigma) \circ \sigma(s) \in sZ_{\hat{G}}(1 - \hat{\theta})(\hat{T}).$$

L'élément s est fixé par tout élément de Ω_H. On en déduit la relation :

$$\omega_{D,H}(\sigma)\omega(\sigma) \circ \sigma(s) \in sZ_{\hat{G}}(1 - \hat{\theta})(\hat{T}).$$

On vérifie que $Z_{\hat{G}}(1-\hat{\theta})(\hat{T})$ s'envoie sur l'élément neutre de $\hat{\hat{T}}_{ad}$. On obtient :

$$\omega_{D,H}(\sigma)\omega(\sigma) \circ \sigma(\bar{s}) = \bar{s}.$$

Enfin, l'élément $\bar{s}$ est fixé par $\Omega_\sharp$ par définition de $\hat{H}_\sharp$, donc par $\omega_{H_\epsilon}(\sigma)$. Alors l'égalité (2) entraîne (5).

Pour tout $\sigma \in W$, fixons un élément $y(\sigma) \in Norm_{\hat{\bar{G}}_{AD}}(\hat{\bar{T}}_{ad})$ dont l'image dans $\Omega_{\bar{G}}$ soit $\omega_{H_\epsilon}(\sigma)\omega_{\bar{G}}(\sigma)^{-1}$. Notons $\bar{\mathcal{H}}$ le sous-groupe de ${}^L\bar{G}_{AD}$ engendré par $\hat{\bar{H}}$ et les éléments $(y(\sigma), \sigma)$ pour $\sigma \in W$. Notons $\hat{\bar{\xi}}$ l'injection naturelle de $\bar{\mathcal{H}}$ dans ${}^L\bar{G}_{AD}$. Les égalités (3) et (5) entraînent facilement que $(\bar{H}, \bar{\mathcal{H}}, \bar{s}, \hat{\bar{\xi}})$ est une donnée endoscopique de $\bar{G}_{SC}$ (il s'agit d'endoscopie ordinaire ; si l'on veut utiliser les définitions de 2.9, $\bar{G}_{SC}$ est muni de l'automorphisme et du caractère triviaux). Dans le cas où le diagramme de départ est fortement non ramifié, il est clair que toutes les applications galoisiennes que l'on a introduites se factorisent en des actions de Γ^{nr} et on voit que $(\bar{H}, \bar{\mathcal{H}}, \bar{s}, \hat{\bar{\xi}})$ est une donnée endoscopique non ramifiée de $\bar{G}_{SC}$.

4.7. Une donnée endoscopique non standard. On conserve les hypothèses du paragraphe précédent. On a introduit dans ce paragraphe des isomorphismes j_* et j^*, que l'on peut considérer comme des applications :

$$j_* : X_*(T^{\flat *})_{\mathbb{Q}} \to X_*(T_\sharp)_{\mathbb{Q}}, \quad j^* : X^*(T_\sharp)_{\mathbb{Q}} \to X^*(T^{\flat *})_{\mathbb{Q}}.$$

Elles sont équivariantes pour les actions de Γ : on a défini l'action sur $T_\sharp$ pour qu'il en soit ainsi. Introduisons les revêtements simplement connexes $H_{\epsilon,SC}$ et $\bar{H}_{SC}$ des groupes dérivés de H_ϵ et $\bar{H}$. On peut décomposer :

$$(1) \qquad X_*(T^{\flat *})_{\mathbb{Q}} = X_*(T^{\flat *}_{sc})_{\mathbb{Q}} \oplus X_*(Z^0_{H_\epsilon})_{\mathbb{Q}},$$

$$(2) \qquad X_*(T_\sharp)_{\mathbb{Q}} = X_*(T_{\bar{H},sc})_{\mathbb{Q}} \oplus X_*(Z^0_{\bar{H}})_{\mathbb{Q}} \oplus X_*(Z^0_{\bar{G}})_{\mathbb{Q}}.$$

Les correspondances que l'on a établies entre les ensembles de racines et coracines de H_ϵ et $H_\sharp$ entraînent que j_* envoie le premier facteur de (1) sur le premier facteur de (2) et le second facteur de (1) sur la somme des deux derniers de (2). Alors $(H_{\epsilon,SC}, \bar{H}_{SC}, j_*)$ a toutes les propriétés requises pour être un triplet endoscopique non standard. Si le diagramme D est fortement non ramifié, ce triplet est non ramifié.

De j_* se déduit un isomorphisme :

$$(3) \qquad \mathfrak{t}^{\flat *}_{sc} \oplus \mathfrak{z}_{H_\epsilon} = \mathfrak{t}^{\flat *} \to \mathfrak{t}_\sharp = \mathfrak{t}_{\bar{H},sc} \oplus \mathfrak{z}_{\bar{H}} \oplus \mathfrak{z}_{\bar{G}},$$

qui possède des propriétés similaires à celles de j_*.

4.8. Un triangle de correspondances. On conserve les hypothèses de 4.6. Soit γ un élément compact de $H_{z,\tilde{G}reg}(F)$ tel que $\gamma_{p'} = \epsilon$. Ecrivons $\gamma = exp(Y)\epsilon$, avec $Y \in \mathfrak{h}_\epsilon(F)_{tn}$. Décomposons Y en $Y_{sc} + Y_Z$, où $Y_{sc} \in \mathfrak{h}_{\epsilon,SC}(F)_{tn}$ et $Y_Z \in \mathfrak{z}_{H_\epsilon}(F)_{tn}$. Notons $\bar{Y}_Z + X_Z$ l'image de Y_Z par l'isomorphisme 4.7(3), avec $\bar{Y}_Z \in \mathfrak{z}_{\bar{H}}(F)_{tn}$ et $X_Z \in \mathfrak{z}_{\bar{G}}(F)_{tn}$. L'élément Y_{sc} est régulier dans $\mathfrak{h}_{\epsilon,SC}(F)$. Par la correspondance endoscopique non standard, sa classe de conjugaison correspond à celle d'un élément de $\bar{\mathfrak{h}}_{SC}(F)_{tn}$. Fixons un tel élément $\bar{Y}_{sc}$. Posons $\bar{Y} = \bar{Y}_{sc} + \bar{Y}_Z$. C'est un élément régulier de $\bar{\mathfrak{h}}(F)$.

Puisque $\bar{G}_{SC}$ est quasi-déployé, tout tel élément se transfère, par la correspondance endoscopique entre $\bar{H}$ et $\bar{G}_{SC}$, en un élément de $\bar{\mathfrak{g}}_{SC}(F)$. Fixons un tel élément X_{sc}. Il appartient à $\bar{\mathfrak{g}}_{SC}(F)_{tn}$. Posons $X = X_{sc} + X_Z$ et $\delta = exp(X)\eta$. Alors δ est un élément semi-simple et compact de $\bar{G}(F)$ et :

(1) la classe de conjugaison de δ correspond à celle de γ.

Dans le paragraphe 4.6, on a identifié $T^{\flat *}$ à $T_{/\theta^*}$ et $T_\sharp$ et $\bar{T}$ à $T^{\theta^*,0}$. Les deux membres de l'égalité (3) de 4.7 s'identifient à $\bar{\mathfrak{t}}$, l'identification n'étant pas cette fois équivariante pour les actions de Γ. Soit $Y^*_{sc} \in \mathfrak{t}^{\flat *}_{sc}$ un élément conjugué à Y_{sc} par un élément de $H_{\epsilon,SC}$. Il s'identifie par l'égalité (3) de 4.7 à un élément $\bar{Y}^*_{sc} \in \mathfrak{t}_{\bar{H},sc}$. L'élément $\bar{Y}^*_{sc} + \bar{Y}_Z$ s'identifie à un élément $X^*_{sc} \in \bar{\mathfrak{t}}_{sc}$. Dire que les classes de conjugaison de Y_{sc} et de $\bar{Y}_{sc}$ se correspondent signifie que $\bar{Y}_{sc}$ est conjugué à $\bar{Y}^*_{sc}$ par un élément de $\bar{H}_{SC}$. Donc $\bar{Y}$ est conjugué à $\bar{Y}^*_{sc} + \bar{Y}_Z$ par un élément de $\bar{H}$. Dire que les classes de conjugaison de $\bar{Y}$ et de X_{sc} se correspondent signifie donc que X_{sc} est conjugué à X^*_{sc} par un élément de $\bar{G}_{SC}$. Donc X est conjugué à $X^*_{sc} + X_Z$ par un élément de $\bar{G}$. L'élément γ est conjugué à $exp(Y^*_{sc} + Y_Z)\epsilon$ par un élément de $H_\epsilon \subset H$. L'élément δ est conjugué à $exp(X_{sc} + X_Z)\eta$ par un élément de $\bar{G} \subset G$. D'après les définitions de ces éléments et de nos identifications, on a l'égalité :

$$int^{-1}_{hh_0} \circ \xi \circ int_{gg_0}(exp(X_{sc} + X_Z)\eta) = exp(Y^*_{sc} + Y_Z)\epsilon.$$

Cela démontre (4).

Puisque $(\bar{H}, \bar{\mathcal{H}}, \bar{s}, \hat{\bar{\xi}})$ est une donnée endoscopique de $\bar{G}_{SC}$, on peut introduire un facteur de transfert $\bar{\Delta}$ défini sur un sous-ensemble adéquat de $\bar{\mathfrak{h}}(F) \times \bar{\mathfrak{g}}_{SC}(F)$ (cf. 3.4; dans ce paragraphe, on n'a considéré que des données non ramifiées, mais la définition vaut en général). Dans le cas où D est fortement non ramifié, on le normalise en choisissant pour sous-groupe hyperspécial de $\bar{G}_{SC}(F)$ l'image réciproque $\bar{K}_{sc}$ dans ce groupe du sous-groupe $\bar{K} = K \cap \bar{G}(F)$, qui est un sous-groupe hyperspécial de $\bar{G}(F)$ d'après le lemme 4.1(i).

Théorème. *Il existe $c \in \mathbb{C}^\times$ tel que, pour tout γ comme ci-dessus, on ait l'égalité $\Delta(\gamma, \delta) = c\bar{\Delta}(\bar{Y}, X_{sc})$. Dans le cas où D est fortement non ramifié, cet élément c est égal à* 1.

C'est le résultat principal de [W1]. Cf. cette référence, pages 76 à 230, pour la démonstration.

4.9. Normalisation des mesures. Pour tout groupe réductif connexe M défini sur F, il y a une bijection naturelle entre mesures de Haar sur $M(F)$ et sur $\mathfrak{m}(F)$: on demande que les mesures se correspondent par l'exponentielle au voisinage des éléments neutres. Fixons des mesures de Haar sur $M(F)$ et $Z^0_M(F)$. Elles déterminent des mesures de Haar sur $\mathfrak{m}(F)$ et $\mathfrak{z}_M(F)$. En vertu de l'égalité $\mathfrak{m}(F) = \mathfrak{m}_{SC}(F) \oplus \mathfrak{z}_M(F)$, on récupère une mesure de Haar sur $\mathfrak{m}_{SC}(F)$, puis sur $M_{SC}(F)$. Si T est un sous-tore maximal de M défini sur F et si l'on munit $T(F)$ d'une mesure de Haar, on en déduit par le même procédé une mesure de Haar sur $T_{sc}(F)$.

Soient $\gamma \in H_{z,\tilde{G}-reg}(F)$ et $\delta \in \tilde{G}_{reg}(F)$ deux éléments dont les classes de conjugaison se correspondent. Notons $T^\flat = H_\gamma$ et $T^\natural = G_\delta$. Comme dans la preuve du lemme 4.5, on peut choisir $h \in H$ et $g \in G$ de sorte que $int_{h^{-1}} \circ \xi \circ int_g$ définisse un homomorphisme défini sur F de $T^\natural$ sur $T^\flat$. C'est une isogénie, dont on déduit un isomorphisme de $\mathfrak{t}^\natural(F)$ sur $\mathfrak{t}^\flat(F)$. On a fixé en 3.2 une mesure de Haar sur $T^\flat(F)$. Elle se descend en une mesure sur $\mathfrak{t}^\flat(F)$, que l'on transporte en une mesure sur $\mathfrak{t}^\natural(F)$ et on relève celle-ci en une mesure sur $T^\natural(F)$. Cette dernière ne dépend pas des choix de h et g.

Remarque. Indiquons une façon possible de normaliser les mesures sur les tores. Soit T un tore défini sur F. Même si T n'est pas déployé sur F^{nr}, il possède une structure naturelle de schéma en groupes lisse sur $\mathfrak{o}$. La fibre spéciale $\mathbf{T}$ n'est pas connexe. Notons $T(\mathfrak{o})^0$ le sous-groupe des $t \in T(\mathfrak{o})$ dont la réduction appartient à $\mathbf{T}^0(\mathbb{F}_q)$. C'est un sous- groupe ouvert compact de $T(F)$. On normalise la mesure sur $T(F)$ en imposant que $T(\mathfrak{o})^0$ ait pour mesure 1. On vérifie que cette normalisation est compatible aux procédés de transfert de mesures décrits ci-dessus.

Soient D et ϵ comme en 4.6. On fixe des mesures de Haar sur $H_\epsilon(F)$, $\bar{H}(F)$, $\bar{G}(F)$, $Z^0_{\bar{H}}(F)$ et $Z_{\bar{G}}(F)$. Chaque fois que l'un de ces groupes est non ramifié (ce qui est toujours le cas pour $\bar{G}$ et $Z_{\bar{G}}$), on choisit la mesure canonique pour laquelle un sous-groupe hyperspécial est de mesure 1. Les différents isomorphismes établis dans les paragraphes précédents et les recettes ci-dessus munissent d'une mesure de Haar tous les groupes qui interviendront dans la suite. Il y a parfois plusieurs façons de les définir, mais qui donnent le même résultat. De plus, ces mesures sont les mesures canoniques quand les groupes sont non ramifiés.

4.10. Descente d'intégrales orbitales endoscopiques. Notre but est d'établir l'égalité énoncée en 3.3. Si $\gamma \in H_{z,\tilde{G}-reg}(F)$ n'est pas compact, sa classe de conjugaison ne coupe pas K_z et $SO_\gamma^{H_z}(\mathbf{1}_{K_z}) = 0$. En utilisant le lemme 4.4, on voit de même que $O_\gamma^{\tilde{G},H_z,\boldsymbol{\omega}}(\mathbf{1}_{\tilde{K}}) = 0$. On peut donc se limiter aux éléments de $H_{z,\tilde{G}reg}(F) \cap H_z(F)_c$.

Soit γ un tel élément, que l'on écrit $\gamma = exp(Y)\epsilon$. L'égalité à démontrer ne dépend que de la classe de conjugaison stable de γ. En utilisant le lemme 4.3(ii), on peut supposer H_ϵ quasi-déployé sur F. Dans le cas où H_ϵ est non ramifié, on peut supposer $\epsilon \in K_z$ d'après le lemme 4.3(iv).

Supposons qu'il n'existe pas de diagramme modérément ramifié de source ϵ. D'après le lemme 4.5(i), il n'existe aucun élément de $\tilde{K} \cap \tilde{G}_{reg}(F)$ dont la classe de conjugaison corresponde à celle de γ. Alors $O_\gamma^{\tilde{G},H_z,\boldsymbol{\omega}}(\mathbf{1}_{\tilde{K}}) = 0$. Supposons qu'il existe un élément $\gamma' \in K_z$ stablement conjugué à γ. Posons $\epsilon' = \gamma'_{p'}$. Alors $\epsilon' \in K_z$ (car ϵ' appartient à l'adhérence du groupe engendré par γ') et est stablement conjugué à ϵ. Alors H_ϵ est non ramifié d'après le lemme 4.3(iv), donc $\epsilon \in K_z$ d'après l'hypothèse que l'on a faite ci- dessus. Le lemme 4.5(ii) contredit notre hypothèse sur l'inexistence d'un diagramme de source ϵ. Il n'existe donc aucun élément γ' comme ci-dessus, ce qui entraîne $SO_\gamma^{H_z}(\mathbf{1}_{K_z}) = 0$. L'égalité de 3.3 est vérifiée.

On est ramené au cas où il existe un diagramme modérément ramifié de source ϵ. On fixe un tel diagramme D. Si H_ϵ est non ramifié, on le suppose fortement non ramifié. On effectue les constructions des paragraphes 4.6 à 4.8.

Lemme. *Sous ces hypothèses, les propriétés suivantes sont vérifiées.*

(i) Si H_ϵ n'est pas non ramifié, $O_\gamma^{\tilde{G},H_z,\boldsymbol{\omega}}(\mathbf{1}_{\tilde{K}}) = 0$.

(ii) Si H_ϵ est non ramifié, on a l'égalité :

$$O_\gamma^{\tilde{G},H_z,\boldsymbol{\omega}}(\mathbf{1}_{\tilde{K}}) = O_{\bar{Y}}^{\bar{G}_{SC},\bar{H}}(\mathbf{1}_{\bar{\mathfrak{k}}_{sc}}).$$

Preuve. Posons $\delta = exp(X)\eta$. D'après 4.8(1), la classe de conjugaison de δ correspond à celle de γ. L'intégrale orbitale endoscopique que l'on doit calculer est une somme sur les classes de conjugaison de δ par $G(F)$ contenues dans sa classe de conjugaison stable. Evidemment, les classes de conjugaison qui ne coupent pas $\tilde{K}$ ne contribuent pas. En utilisant le lemme 4.2, on peut remplacer la sommation par une somme sur les éléments de $\mathcal{X}_\delta$, à condition de diviser le tout par $|\mathcal{Z}_\eta|[Z_G(\delta)(F) : G_\delta(F)]^{-1}$. Remarquons que ce dernier indice va compenser le facteur de normalisation que l'on a introduit dans la définition des intégrales orbitales, cf. 3.1. Pour $z \in \mathcal{Z}_\eta$ et $X' \in \mathcal{E}_X$, posons $\delta(z, X') = int_z(exp(X')\eta)$. On obtient :

$$(1) \qquad O_\gamma^{\tilde{G},H_z,\boldsymbol{\omega}}(\mathbf{1}_{\tilde{K}}) = |\mathcal{Z}_\eta|^{-1} \sum_{z \in \mathcal{Z}_\eta} I(z),$$

où :

$$I(z) = \sum_{X' \in \mathcal{E}_X} \Delta(\gamma, \delta(z, X')) \int_{G_{\delta(z,X')}(F)\backslash G(F)} \mathbf{1}_{\tilde{K}}(g^{-1}\delta(z, X')g)\boldsymbol{\omega}(g)dg.$$

Soit $z \in \mathcal{Z}$. Rappelons que $z_{ad} \in G_{AD}(F)$, que les automorphismes int_z de G, resp. $\tilde{G}$, sont définis sur F et que le second préserve $\tilde{K}$. La définition des facteurs de transfert montre que la fonction sur $\mathcal{D}$:

$$(\gamma', \delta') \to \Delta(\gamma', int_z(\delta'))$$

est encore un facteur de transfert, donc est proportionnel à Δ. Puisque int_z préserve $\tilde{K}$, ce nouveau facteur vérifie la même condition de normalisation que Δ donc lui est égal. En particulier $\Delta(\gamma, \delta(z, X')) = \Delta(\gamma, exp(X')\eta)$. Le groupe $G_{AD}(F)$ agit trivialement sur $Hom(G(F), \mathbb{C}^\times)$. Donc $\boldsymbol{\omega} \circ int_z = \boldsymbol{\omega}$. En changeant g en $int_z(g)$ dans l'intégrale ci-dessus, on voit que $I(z)$ ne dépend pas de z. Le nombre de ces intégrales compense le facteur $|\mathcal{Z}_\eta|^{-1}$

figurant dans (1) et on obtient simplement :

$$O_\gamma^{\tilde{G},H_z,\boldsymbol{\omega}}(\mathbf{1}_{\tilde{K}})$$
$$= \sum_{X'\in\mathcal{E}_X} \Delta(\gamma, exp(X')\eta) \int_{G_{exp(X')\eta}(F)\backslash G(F)} \mathbf{1}_{\tilde{K}}(g^{-1}exp(X')\eta g)\boldsymbol{\omega}(g)dg.$$

Soit $X' \in \mathcal{E}_X$. On vérifie que, pour toute fonction $f \in C_c^\infty(G_{exp(X')\eta}(F)\backslash G(F))$, à support dans $G_{exp(X')\eta}(F)\backslash G_\eta(F)K$, on a l'égalité :

$$\int_{G_{exp(X')\eta}(F)\backslash G(F)} f(g)dg = \int_{G_{exp(X')\eta}(F)\backslash G_\eta(F)\times K} f(gk)dkdg$$
$$= \int_{\bar{G}_{X'}(F)\backslash \bar{G}(F)\times K} f(gk)dkdg.$$

La fonction $g \to \mathbf{1}_{\tilde{K}}(g^{-1}exp(X')\eta g)\boldsymbol{\omega}(g)$ vérifie la condition requise d'après le lemme 4.1(ii). Elle est invariante à droite par K (on a dit en 2.7 que $\boldsymbol{\omega}$ était trivial sur K). De plus, pour $g \in \bar{G}(F)$, on a l'égalité :

$$\mathbf{1}_{\tilde{K}}(g^{-1}exp(X')\eta g) = \mathbf{1}_{\bar{\mathfrak{k}}}(g^{-1}X'g).$$

Cela conduit à l'égalité :

$$O_\gamma^{\tilde{G},H_z,\boldsymbol{\omega}}(\mathbf{1}_{\tilde{K}}) = \sum_{X'\in\mathcal{E}_X} \Delta(\gamma, exp(X')\eta) \int_{\bar{G}_{X'}(F)\backslash \bar{G}(F)} \mathbf{1}_{\bar{\mathfrak{k}}}(g^{-1}X'g)\boldsymbol{\omega}(g)dg.$$

L'ensemble de représentants $\mathcal{E}_X$ est nécessairement de la forme $\mathcal{E}_X = \{X'_{sc}+X_Z; X'_{sc} \in \mathcal{E}_{X_{sc}}\}$, où $\mathcal{E}_{X_{sc}}$ est un ensemble de représentants des classes de conjugaison par $\bar{G}(F)$ dans la classe de conjugaison stable de X_{sc}. Soit X'_{sc} un élément de cet ensemble. Fixons un ensemble $\mathcal{A}_{X'_{sc}}$ de représentants des doubles classes $\bar{G}_{X'_{sc}}(F)\backslash\bar{G}(F)/\bar{G}_{SC}(F)$. On vérifie que, pour toute fonction $f \in C_c^\infty(\bar{G}_{X'}(F)\backslash\bar{G}(F))$, on a l'égalité :

$$\int_{\bar{G}_{X'}(F)\backslash\bar{G}(F)} f(g)dg = \sum_{a\in\mathcal{A}_{X'_{sc}}} \int_{\bar{G}_{SC,a^{-1}X'_{sc}a}(F)\backslash\bar{G}_{SC}(F)} f(ag)dg.$$

On applique cela à la fonction f définie par $f(g) = \mathbf{1}_{\bar{\mathfrak{k}}}(g^{-1}(X'_{sc}+X_Z)g)\boldsymbol{\omega}(g)$. Soient $a \in \mathcal{A}_{X'_{sc}}$ et $g \in \bar{G}_{SC}(F)$. Tout caractère est trivial sur $\bar{G}_{SC}(F)$, donc $\boldsymbol{\omega}(ag) = \boldsymbol{\omega}(a)$. L'élément X_Z appartient à $\mathfrak{z}_{\bar{G}}(F)_{tn}$. Mais $Z^0_{\bar{G}}$ est un tore non ramifié. Il possède un unique sous-groupe hyperspécial, qui est son plus grand sous-groupe compact. Ce sous-groupe contient tous les éléments topologiquement unipotents. De même, tout élément topologiquement nilpotent de $\mathfrak{z}_{\bar{G}}(F)$ appartient à l'unique réseau hyperspécial de cette algèbre, et celui-ci est contenu dans $\bar{\mathfrak{k}}$. Donc $\mathbf{1}_{\bar{\mathfrak{k}}}(g^{-1}a^{-1}(X'_{sc}+X_Z)ag) = \mathbf{1}_{\bar{\mathfrak{k}}_{sc}}(g^{-1}a^{-1}X'_{sc}ag)$. On obtient :

$$O_\gamma^{\tilde{G},H_z,\boldsymbol{\omega}}(\mathbf{1}_{\tilde{K}}) = \sum_{X'_{sc}\in\mathcal{E}_{X_{sc}}} \Delta(\gamma, exp(X'_{sc}+X_Z)\eta)$$

$$\sum_{a\in\mathcal{A}_{X'_{sc}}}\boldsymbol{\omega}(a)\int_{\bar{G}_{SC,a^{-1}X'_{sc}a}(F)\backslash\bar{G}_{SC}(F)}\mathbf{1}_{\bar{\mathfrak{k}}_{sc}}(g^{-1}a^{-1}X'_{sc}ag)dg.$$

D'après [KS] théorème 5.1.D, on a l'égalité :

$$\Delta(\gamma,exp(X'_{sc}+X_Z)\eta)\boldsymbol{\omega}(a)=\Delta(\gamma,exp(a^{-1}X'_{sc}a+X_Z)\eta).$$

Dans le théorème 4.8, X_{sc} était un élément quelconque de $\bar{\mathfrak{g}}(F)$ dont la classe de conjugaison correspondait à celle de $\bar{Y}$. On peut aussi bien le remplacer par $a^{-1}X'_{sc}a$. Le facteur Δ ci-dessus est donc égal à $c\bar{\Delta}(\bar{Y},a^{-1}X'_{sc}a)$. On obtient :

$$O_\gamma^{\tilde{G},H_z,\boldsymbol{\omega}}(\mathbf{1}_{\tilde{K}})=c\sum_{X'_{sc}\in\mathcal{E}_{X_{sc}}}\sum_{a\in\mathcal{A}_{X'_{sc}}}\bar{\Delta}(\bar{Y},a^{-1}X'_{sc}a)$$
$$\int_{\bar{G}_{SC,a^{-1}X'_{sc}a}(F)\backslash\bar{G}_{SC}(F)}\mathbf{1}_{\bar{\mathfrak{k}}_{sc}}(g^{-1}a^{-1}X'_{sc}ag)dg.$$

Mais l'ensemble :

$$\{a^{-1}X'_{sc}a;X'\in\mathcal{E}_{X_{sc}},a\in\mathcal{A}_{X'_{sc}}\}$$

est un ensemble de représentants des classes de conjugaison par $G_{SC}(F)$ dans la classe de conjugaison stable de X_{sc}. Alors :

$$O_\gamma^{\tilde{G},H_z,\boldsymbol{\omega}}(\mathbf{1}_{\tilde{K}})=cO_{\bar{Y}}^{\bar{G}_{SC},\bar{H}}(\mathbf{1}_{\bar{\mathfrak{k}}_{sc}}),$$

par définition de ce dernier terme.

Si H_ϵ est non ramifié, le diagramme D est fortement non ramifié et $c=1$ d'après le théorème 4.8. Cela démontre le (ii) de l'énoncé. Supposons que H_ϵ n'est pas non ramifié. Alors le tore $T^{\flat *}$ n'est pas déployé sur F^{nr}. En vertu de l'isomorphisme (3) de 4.8 et puisque $Z^0_{\tilde{G}}$ est, lui, déployé sur F^{nr}, le tore $T_{\bar{H}}$ ne l'est pas. Donc la donnée endoscopique $(\bar{H},\bar{\mathcal{H}},\bar{s},\hat{\bar{\xi}})$ de $\bar{G}_{SC}$ n'est pas non ramifiée. Une adaptation aux algèbres de Lie de la proposition 7.5 de [K2] montre que $O_{\bar{Y}}^{\bar{G}_{SC},\bar{H}}(\mathbf{1}_{\bar{\mathfrak{k}}_{sc}})=0$. Cela démontre le (i) de l'énoncé.

4.11. Preuve du théorème 3.6. L'élément γ est comme dans le paragraphe précédent. On a :

(1) si H_ϵ n'est pas non ramifié, $SO_\gamma^{H_z}(\mathbf{1}_{K_z})=0$;

(2) si H_ϵ est non ramifié, $SO_\gamma^{H_z}(\mathbf{1}_{K_z})=SO_{Y_{sc}}^{H_{\epsilon,SC}}(\mathbf{1}_{\mathfrak{k}_{\epsilon,sc}})$,

où on a noté $\mathfrak{k}_{\epsilon,sc}$ le réseau hyperspécial $\mathfrak{k}_H\cap\mathfrak{h}_{\epsilon,SC}(F)$ de $\mathfrak{h}_{\epsilon,SC}(F)$. Ces assertions sont celles du lemme précédent, appliqué au cas où $\tilde{G}=H_z$. Jointes à ce lemme, elles conduisent tout de suite à l'égalité :

$$(3)\qquad SO_\gamma^{H_z}(\mathbf{1}_{K_z})=O_\gamma^{\tilde{G},H_z,\boldsymbol{\omega}}(\mathbf{1}_{\tilde{K}})$$

dans le cas où H_ϵ n'est pas non ramifié. Supposons H_ϵ non ramifié. D'après 4.8(3), $\bar{H}$ est lui-aussi non ramifié. Introduisons un réseau hyperspécial $\mathfrak{k}_{\bar{H}}$ de

$\bar{\mathfrak{h}}(F)$ et posons $\mathfrak{k}_{\bar{H},sc} = \mathfrak{k}_{\bar{H}} \cap \bar{\mathfrak{h}}_{SC}(F)$.Le lemme fondamental pour les algèbres de Lie entraîne l'égalité :

$$O_{\bar{Y}}^{\bar{G}_{SC},\bar{H}}(\mathbf{1}_{\bar{k}_{sc}}) = SO_{\bar{Y}}^{\bar{H}}(\mathbf{1}_{\mathfrak{k}_{\bar{H}}}).$$

Un calcul analogue à celui fait dans la preuve du lemme 4.10 montre que :

$$SO_{\bar{Y}}^{\bar{H}}(\mathbf{1}_{\mathfrak{k}_{\bar{H}}}) = SO_{\bar{Y}_{sc}}^{\bar{H}_{SC}}(\mathbf{1}_{\mathfrak{k}_{\bar{H},sc}}).$$

Le lemme fondamental non standard entraîne l'égalité :

$$SO_{\bar{Y}_{sc}}^{\bar{H}_{SC}}(\mathbf{1}_{\mathfrak{k}_{\bar{H},sc}}) = SO_{Y_{sc}}^{H_{\epsilon},SC}(\mathbf{1}_{\mathfrak{k}_{\epsilon}}).$$

En utilisant ces égalités, celle du (ii) du lemme 4.10 et l'égalité (2) ci-dessus, on obtient de nouveau l'égalité (3), ce qui achève la démonstration.

5. L'exemple du changement de base pour un groupe unitaire non ramifié

5.1. La situation. Soit E l'extension quadratique non ramifiée de F et n un entier strictement positif. On note τ l'élément non trivial du groupe de Galois de E/F. On note U_n le groupe unitaire quasi-déployé relatif à cette extension. Pour distinguer les différentes actions galoisiennes, on les affecte de l'indice du groupe sur lequel porte l'action. On note de la même façon les actions sur les groupes duaux. Ainsi, on note $\sigma \mapsto \sigma_{U_n}$ l'action de Γ sur U_n ou sur le groupe dual $\hat{U}_n$. On fait une exception pour le groupe GL_n : on note simplement $\sigma \mapsto \sigma$ l'action naturelle sur GL_n. Dans l'exemple 3 de 2.8, on a introduit un automorphisme de GL_n, alors noté θ, et que nous noterons maintenant θ_n. Alors le groupe U_n peut se décrire ainsi : $U_n(\bar{F}) = GL_n(\bar{F})$ et, pour $\sigma \in \Gamma$ et $g \in GL_n(\bar{F})$, on a $\sigma_{U_n}(g) = \theta_n \circ \sigma(g)$. Du groupe U_n se déduit un groupe sur E puis, par restriction des scalaires, un groupe G sur F. La façon la plus naturelle de le décrire est la suivante :

- $G(\bar{F}) = U_n(\bar{F}) \times U_n(\bar{F})$;
- pour $\sigma \in \Gamma$ et $(g', g'') \in G(\bar{F})$, on a $\sigma_G(g', g'') = (\sigma_{U_n}(g'), \sigma_{U_n}(g''))$ si $\sigma \in \Gamma_E$ et $\sigma_G(g', g'') = (\sigma_{U_n}(g''), \sigma_{U_n}(g'))$ si $\sigma \notin \Gamma_E$.

Le groupe G est muni d'un automorphisme θ défini par $\theta(g', g'') = (g'', g')$. Il est plus commode et plus habituel de transformer cette description en la suivante, qui est celle que nous utiliserons (on passe de l'une à l'autre en remplaçant g'' par $\theta_n(g'')$) :

- $G(\bar{F}) = GL_n(\bar{F}) \times GL_n(\bar{F})$;
- pour $\sigma \in \Gamma$ et $(g', g'') \in G(\bar{F})$, on a $\sigma_G(g', g'') = (\sigma(g'), \sigma(g''))$ si $\sigma \in \Gamma_E$ et $\sigma_G(g', g'') = (\sigma(g''), \sigma(g'))$ si $\sigma \notin \Gamma_E$.

Alors G est muni d'un automorphisme θ défini par $\theta(g', g'') = (\theta_n(g''), \theta_n(g'))$. L'application :

$$\begin{array}{ccc} GL_n(E) & \to & G(F) \\ g & \mapsto & (g, \tau(g)) \end{array}$$

est un isomorphisme qui identifie les automorphismes $\theta_n \circ \tau$ de $GL_n(E)$ et θ de $G(F)$. On pose $\boldsymbol{\omega} = 1$. On prend pour sous-groupe hyperspécial K de $G(F)$ l'image par l'isomorphisme ci-dessus du sous-groupe $GL_n(\mathfrak{o}_E)$ de $GL_n(E)$. On prend pour paire de Borel épinglée le produit des deux paires dans GL_n décrites dans l'exemple 3 de 2.8. Alors $\theta^* = \theta$. On a $\hat{U}_n = GL_n(\mathbb{C})$. Notons $\hat{\theta}_n$ l'automorphisme de $GL_n(\mathbb{C})$ défini par la même formule que θ_n. Pour $\sigma \in \Gamma$ et $x \in GL_n(\mathbb{C})$, on a $\sigma_{U_n}(x) = x$ si $\sigma \in \Gamma_E$ et $\sigma_{U_n}(x) = \hat{\theta}_n(x)$ si $\sigma \notin \Gamma_E$. On a $\hat{G} = GL_n(\mathbb{C}) \times GL_n(\mathbb{C})$. Pour $\sigma \in \Gamma$ et $(x', x'') \in \hat{G}$, on a $\sigma_G(x', x'') = (x', x'')$ si $\sigma \in \Gamma_E$ et $\sigma_G(x', x'') = (x'', x')$ si $\sigma \notin \Gamma_E$. On a $\hat{\theta}(x', x'') = (\hat{\theta}_n(x''), \hat{\theta}_n(x'))$.

Les données endoscopiques les plus intéressantes sont les elliptiques. Rappelons qu'avec les notations de 2.9, une donnée endoscopique $(H, \mathcal{H}, s, \hat{\xi})$ est dite elliptique si $\hat{\xi}(Z_{\hat{H}}^{\Gamma,0}) \subset Z_{\hat{G}}$. A équivalence près, les données endoscopiques elliptiques de $(G, \theta, \boldsymbol{\omega})$ sont les suivantes. Soient $n_+, n_- \in \mathbb{N}$ tels que $n_+ + n_- = n$. On pose $H = U_{n_+} \times U_{n_-}$, $\mathcal{H} = {}^L H$, $s = (s', 1) \in GL_n(\mathbb{C}) \times GL_n(\mathbb{C})$, où :

$$s' = \begin{pmatrix} 1 & & & & & & \\ & \cdot & & & & & \\ & & \cdot & & & & \\ & & & 1 & & & \\ & & & & -1 & & \\ & & & & & \cdot & \\ & & & & & & \cdot \\ & & & & & & & -1 \end{pmatrix},$$

avec n_+ termes 1 et n_- termes -1. Pour $(x_+, x_-) \in GL_{n_+}(\mathbb{C}) \times GL_{n_-}(\mathbb{C}) = \hat{H}$, on pose :

$$\hat{\xi}(x_+, x_-) = \left(\begin{pmatrix} x_+ & 0 \\ 0 & x_- \end{pmatrix}, \begin{pmatrix} \hat{\theta}_{n_-}(x_-) & 0 \\ 0 & \hat{\theta}_{n_+}(x_+) \end{pmatrix} \right).$$

Pour $w \in I$, $\hat{\xi}(w) = w$. Pour un élément $\phi \in W$ ayant pour image dans W^{nr} l'élément de Frobenius, $\hat{\xi}(\phi) = (\mathbf{w}, \hat{\theta}_n(\mathbf{w})s', \phi) \in {}^L G$, où $\mathbf{w}$ est la matrice de permutation qui échange les décompositions $n = n_- + n_+$ et $n = n_+ + n_-$. Autrement dit, la matrice $\mathbf{w} = (\mathbf{w}_{i,j})_{i,j=1,\dots n}$ est définie par $\mathbf{w}_{i,j} = 1$ si $i = 1, \dots, n_+$ et $j = i + n_-$ ou si $i = n_+ + 1, \dots, n$ et $j = i - n_+$, et $\mathbf{w}_{i,j} = 0$ sinon.

5.2. Description des éléments semi-simples de $G(F)$. Fixons un espace vectoriel V sur E de dimension n, muni d'une base $(e_i)_{i=1,\dots,n}$. Alors $GL_n(E)$ s'identifie de la façon usuelle avec le groupe des automorphismes E-linéaires de V. Notons $\tilde{G}(F)$ l'ensemble des formes sesquilinéaires non dégénérées sur V (pour préciser le sens du mot sesquilinéaire, indiquons que

$Q(e'v', e''v'') = \tau(e')e''Q(v', v'')$ pour $Q \in \tilde{G}(F)$, $e', e'' \in E$ et $v', v'' \in V$). Le groupe $G(F) = GL_n(E)$ agit à gauche et à droite sur $\tilde{G}(F)$ par :

$$g_1 Q g_2(v', v'') = Q(g_1^{-1}(v'), g_2(v''))$$

pour $g_1, g_2 \in GL_n(E)$, $Q \in \tilde{G}(F)$ et $v', v'' \in V$. Notons Q_θ l'élément de $\tilde{G}(F)$ défini par :

$$Q_\theta(e_i, e_j) = (-1)^j \delta_{i,n+1-j}$$

où $\delta_{i,j}$ est le symbole de Kronecker. Alors $\tilde{G}(F)$ s'identifie avec l'ensemble tordu introduit en 2.7, c'est-à-dire avec $G(F)\theta$, par l'application :

$$\begin{array}{ccc} G(F)\theta = GL_n(E)\theta & \rightarrow & \tilde{G}(F) \\ g\theta & \mapsto & Q_{g\theta}, \end{array}$$

où $Q_{g\theta}(v', v'') = gQ_\theta(v', v'') = Q_\theta(g^{-1}(v'), v'')$.

Il y a deux procédés élémentaires pour construire des formes sesquilinéaires sur un espace vectoriel sur E. Dans le premier, on considère une extension finie K_λ de E, un élément $\lambda \in K_\lambda$ qui engendre K_λ sur E et un automorphisme τ_λ de K_λ qui conserve E et dont la restriction à E soit égale à τ. On suppose que $\lambda\tau_\lambda(\lambda) = 1$. On considère un espace vectoriel V_λ sur K_λ, de dimension finie et une forme sesquilinéaire q_λ non dégénérée sur V_λ. Sesquilinéaire signifie que $q_\lambda(k'v', k''v'') = \tau_\lambda(k')k''q_\lambda(v', v'')$ pour $k', k'' \in K_\lambda$ et $v', v'' \in V_\lambda$. On suppose que q_λ vérifie la condition de symétrie $q_\lambda(v'', v') = \lambda\tau_\lambda(q_\lambda(v', v''))$. On peut considérer V_λ comme un espace vectoriel sur E, dont on note N_λ la dimension. On munit V_λ de la forme Q_λ sesquilinéaire sur E définie par $Q_\lambda(v', v'') = trace_{K_\lambda/E}(q_\lambda(v', v''))$. Dans le second procédé, on considère une extension finie K_λ de E et un élément $\lambda \in K_\lambda$ qui engendre K_λ sur E. On suppose qu'il n'existe pas d'automorphisme τ_λ de K_λ se restreignant à E en l'automorphisme τ et tel que $\lambda\tau_\lambda(\lambda) = 1$. On considère un espace vectoriel W_λ sur K_λ de dimension finie, et son dual W_λ^*. On pose $V_\lambda = W_\lambda \oplus W_\lambda^*$, que l'on munit de la structure d'espace vectoriel sur E telle que $e(w \oplus w^*) = ew \oplus \tau(e)w^*$ pour $e \in E$, $w \in W_\lambda$ et $w^* \in W_\lambda^*$. On note N_λ la dimension de V_λ sur E. On munit V_λ de la forme Q_λ sesquilinéaire sur E définie par :

$$Q_\lambda(w_1 \oplus w_1^*, w_2 \oplus w_2^*) = trace_{K_\lambda/E}(w_1^*(w_2)) + \tau \circ trace_{K_\lambda/E}(\lambda^{-1}w_2^*(w_1)).$$

Considérons deux couples (λ, K_λ) et $(\lambda', K_{\lambda'})$ de l'une ou l'autre des deux situations ci-dessus. On dit que ces couples sont équivalents si l'une des deux conditions suivantes est vérifiée :

- il existe un isomorphisme de K_λ sur $K_{\lambda'}$ se restreignant en l'identité de E et envoyant λ sur λ' ;

- il existe un isomorphisme de K_λ sur $K_{\lambda'}$ se restreignant en l'automorphisme τ de E et envoyant λ sur λ'^{-1}.

Il est commode de fixer une fois pour toutes un plongement de E dans $\bar{F}$ et des plongements des corps K_λ dans $\bar{F}$ qui prolongent celui de E. On peut ainsi considérer les λ comme des éléments de $\bar{F}^\times$.

Considérons maintenant les données suivantes :

- un ensemble fini Λ, union disjointe de deux sous-ensembles Λ^e et Λ^d ;

- pour $\lambda \in \Lambda^e$, des objets λ, K_λ, τ_λ, V_λ, q_λ comme dans le premier procédé ci-dessus ; comme on le voit, on note de la même façon l'objet principal λ et l'élément de l'ensemble de paramètres, on pense que cela n'induit pas de confusion ;

- pour $\lambda \in \Lambda^d$, des objets λ, K_λ, W_λ comme dans le deuxième procédé.

On suppose que, pour deux éléments distincts $\lambda, \lambda' \in \Lambda$, les couples (λ, K_λ) et $(\lambda', K_{\lambda'})$ ne sont pas équivalents. On suppose $\sum_{\lambda\in\Lambda} N_\lambda = n$. On peut alors fixer un isomorphisme E-linéaire de $\oplus_{\lambda\in\Lambda} V_\lambda$ sur V. La forme sesquilinéaire $\oplus_{\lambda\in\Lambda} Q_\lambda$ s'identifie à une forme sesquilinéaire sur E, c'est-à-dire à un élément de $\tilde{G}(F)$, que l'on note Q_λ. Un peu d'algèbre linéaire élémentaire montre que cet élément est semi-simple. Inversement, pour tout élément semi-simple de $\tilde{G}(F)$, on peut construire des données comme ci-dessus de sorte que l'élément en question soit conjugué à Q_λ par un élément de $G(F)$. Ces données sont essentiellement uniques : les seules modifications que l'on peut faire sont de remplacer les objets par des objets isomorphes ou, de façon plus subtile, remplacer un couple (λ, K_λ) par un couple équivalent au sens défini ci-dessus. On peut décrire de même les intersections avec $\tilde{G}(F)$ des classes de conjugaison par $G(\bar{F})$. Il suffit d'oublier les formes sesquilinéaires q_λ : deux éléments semi-simples paramétrés par des données Λ et Λ' sont conjugués par un élément de $G(\bar{F})$ si et seulement s'il existe une bijection $\lambda \mapsto \lambda'$ de Λ sur Λ' de sorte que, pour tout $\lambda \in \Lambda$, les couples (λ, K_λ) et $(\lambda', K_{\lambda'})$ soient équivalents et $N_\lambda = N_{\lambda'}$.

Considérons des données comme ci-dessus et l'élément semi-simple $\delta = Q_\lambda \in \tilde{G}(F)$ qui leur est associé. Le commutant $Z_G(\delta)$ est connexe. Son groupe de points sur F est le produit des groupes linéaires $GL_{K_\lambda}(W_\lambda)$ pour $\lambda \in \Lambda^d$ et des groupes $U(q_\lambda)$ pour $\lambda \in \Lambda^e$, où $U(q_\lambda)$ est le groupe des automorphismes K_λ-linéaires de V_λ qui conservent la forme q_λ. Pour $\lambda \in \Lambda^e$, notons K'_λ le sous-corps des points fixes par τ_λ dans K_λ. C'est une extension de F disjointe de E et K_λ est le composé des deux extensions E et K'_λ de F. Grâce au théorème Hilbert 90, on peut fixer $\alpha_\lambda \in K_\lambda$ tel que $\lambda = \alpha_\lambda \tau_\lambda(\alpha_\lambda)^{-1}$. Alors la forme $\alpha_\lambda^{-1} q_\lambda$ est une forme hermitienne sur V_λ relative à l'extension K_λ/K'_λ. Le groupe $U(q_\lambda)$ est un groupe unitaire : il est égal au groupe $U(\alpha_\lambda^{-1} q_\lambda)$. Puisque $Z_G(\delta)$ est connexe, on a $G_\delta = I_\delta = Z_G(\delta)$. L'élément δ est compact, resp. d'ordre fini premier à p, si et seulement si, pour tout $\lambda \in \Lambda$, λ appartient à $\mathfrak{o}_{\bar{F}}^\times$, resp. λ est d'ordre fini premier à p. L'élément δ est conjugué à un élément de $\tilde{K}$ par un élément de $G(F)$ si et seulement s'il est compact et, de plus, pour tout $\lambda \in \Lambda^e$, l'espace V_λ possède un sous-$\mathfrak{o}_E$-réseau autodual pour la forme Q_λ (ce n'est pas la même chose que de demander que V_λ possède un sous-$\mathfrak{o}_{K_\lambda}$-réseau autodual pour la forme q_λ). L'élément δ appartient à $\tilde{G}_{reg}(F)$ si et seulement si $dim_{K_\lambda}(V_\lambda) = 1$ pour tout $\lambda \in \Lambda^e$ et $dim_{K_\lambda}(W_\lambda) = 1$ pour tout $\lambda \in \Lambda^d$. Supposons $\delta \in \tilde{G}_{reg}(F)$. Soit $\lambda \in \Lambda^e$. On peut supposer $V_\lambda = K_\lambda$. Toute forme sesquilinéaire sur K_λ vérifiant la condition de symétrie requise est de la forme $q(v',v'') = \alpha\tau_\lambda(v')v''$, où α est un élément de K_λ tel que $\alpha\tau_\lambda(\alpha)^{-1} = \lambda$. Ces éléments α forment un espace principal homogène sous l'action par multiplication de $K_\lambda'^\times$. La classe d'isomorphie de q est déterminée par l'orbite $\alpha Norm_{K_\lambda/K'_\lambda}(K_\lambda^\times)$. Il y a donc deux classes d'isomorphie. En particulier, à q_λ est associé un élément de $K_\lambda^\times$ que l'on note α_λ et la classe de q_λ est déterminée par la classe $\alpha_\lambda Norm_{K_\lambda/K'_\lambda}(K_\lambda^\times)$.

Exemple. Supposons $n = 1$. Alors $V = E$. Tout élément $Q \in \tilde{G}(F)$ (c'est-à-dire toute forme sesquilinéaire non dégénérée Q sur E) s'écrit $Q(v',v'') = \alpha\tau(v')v''$, pour un $\alpha \in E^\times$. Le groupe $G(F) = E^\times$ agit sur $\tilde{G}(F)$ par conjugaison : pour $g \in E^\times$ et $Q \in \tilde{G}(F)$ écrite comme ci-dessus, gQg^{-1} est définie par $gQg^{-1}(v',v'') = \alpha Norm_{E/F}(g)^{-1}\tau(v')v''$. Tout élément $Q \in \tilde{G}(F)$ est semi-simple et sa classe de conjugaison est paramétrée par les données suivantes : l'ensemble Λ se réduit à un unique élément λ, qui appartient à Λ^e ; on a $K_\lambda = E$, $\tau_\lambda = \tau$, λ est un élément du noyau E^1 de l'homomorphisme $Norm_{E/F}$, $V_\lambda = E$ et $q_\lambda = Q$. Ecrivons Q comme ci-dessus. Alors $\lambda = \alpha\tau(\alpha)^{-1}$. Le commutant de Q dans $G(F)$ est le groupe $U_1(F) = E^1$. Soit x un élément de ce groupe et soient λ', $q_{\lambda'}$ paramétrant la classe de conjugaison de xQ. Par définition, on a $q_{\lambda'}(v',v'') = (xQ)(v',v'') = \alpha x\tau(v')v''$, donc $\lambda' = \alpha x\tau(\alpha x)^{-1} = \lambda x^2$.

5.3. Description des éléments semi-simples d'un groupe unitaire. Considérons un espace vectoriel V_ι sur E de dimension finie n_ι, muni d'une forme hermitienne Q_ι non dégénérée. On suppose qu'il existe un sous-$\mathfrak{o}_E$-réseau de V_ι qui est autodual pour la forme Q_ι. On fixe un tel sous-réseau L_ι. Notons U_ι le groupe unitaire de (V_ι, Q_ι). Il est quasi-déployé. Le sous-groupe K_ι des éléments de $U_\iota(F)$ qui conservent L_ι est un sous-groupe hyperspécial.

Il y a deux procédés pour construire des éléments semi-simples d'un groupe unitaire. Dans le premier, on considère une extension finie K_λ de E, un élément $\lambda \in K_\lambda$ qui engendre K_λ sur E et un automorphisme τ_λ de K_λ qui conserve E et dont la restriction à E soit égale à τ. On suppose que $\lambda\tau_\lambda(\lambda) = 1$. On considère un espace vectoriel V_λ sur K_λ, de dimension finie et une forme hermitienne q_λ non dégénérée sur V_λ. On peut considérer V_λ comme un espace vectoriel sur E, dont on note $N_{\iota,\lambda}$ la dimension. On munit V_λ de la forme Q_λ hermitienne sur E définie par $Q_\lambda(v',v'') = trace_{K_\lambda/E}(q_\lambda(v',v''))$. L'élément semi-simple du groupe unitaire de (V_λ, Q_λ) est l'application $h_\lambda : v \mapsto \lambda v$. Dans le second procédé, on considère une extension finie K_λ de E et un élément $\lambda \in K_\lambda$ qui engendre K_λ sur E. On suppose qu'il n'existe pas d'automorphisme τ_λ de K_λ se restreignant à E en l'automorphisme τ et tel que $\lambda\tau_\lambda(\lambda) = 1$. On considère un espace vectoriel W_λ sur K_λ de dimension finie, et son dual W_λ^*. On pose $V_\lambda = W_\lambda \oplus W_\lambda^*$, que l'on munit de la structure d'espace vectoriel sur E telle que $e(w \oplus w^*) = ew \oplus \tau(e)w^*$ pour $e \in E$, $w \in W_\lambda$ et $w^* \in W_\lambda^*$. On note

$N_{\iota,\lambda}$ la dimension de V_λ sur E. On munit V_λ de la forme Q_λ hermitienne sur E définie par :

$$Q_\lambda(w_1 \oplus w_1^*, w_2 \oplus w_2^*) = trace_{K_\lambda/E}(w_1^*(w_2)) + \tau \circ trace_{K_\lambda/E}(w_2^*(w_1)).$$

L'élément semi-simple du groupe unitaire de (V_λ, Q_λ) est l'application h_λ : $w \oplus w^* \mapsto \lambda w \oplus \lambda^{-1}w^*$.

Considérons maintenant les données suivantes :

- un ensemble fini Λ_ι, union disjointe de deux sous-ensembles Λ_ι^e et Λ_ι^d ;

- pour $\lambda \in \Lambda_\iota^e$, des objets λ, K_λ, τ_λ, V_λ, q_λ comme dans le premier procédé ci-dessus ;

- pour $\lambda \in \Lambda_\iota^d$, des objets λ, K_λ, W_λ comme dans le deuxième procédé.

On suppose que, pour deux éléments distincts $\lambda, \lambda' \in \Lambda_\iota$, les couples (λ, K_λ) et $(\lambda', K_{\lambda'})$ ne sont pas équivalents. On suppose $\sum_{\lambda\in\Lambda_\iota} N_{\iota,\lambda} = n_\iota$. On suppose qu'il existe un isomorphisme E-linéaire de $\oplus_{\lambda\in\Lambda_\iota} V_\lambda$ sur V_ι qui identifie les formes hermitiennes $\oplus_{\lambda\in\Lambda_\iota} Q_\lambda$ et Q_ι. Alors cet isomorphisme identifie l'élément $\oplus_{\lambda\in\Lambda_\iota} h_\lambda$ du groupe unitaire de la première forme avec un élément $h_{\Lambda_\iota} \in U_\iota(F)$. C'est un élément semi-simple. Tout élément semi-simple de $U_\iota(F)$ est conjugué par un élément de $U_\iota(F)$ à un élément de cette forme et les données Λ_ι sont déterminées par l'élément semi-simple de façon essentiellement unique. Comme dans le paragraphe précédent, on peut lire sur les ensembles de paramètres la conjugaison stable (il suffit d'oublier les formes q_λ pour $\lambda \in \Lambda_\iota^e$), le fait que h_{Λ_ι} soit régulier, ou compact, ou d'ordre fini premier à p, ou conjugué à un élément de K_ι par un élément de $U_\iota(F)$. Soit $h = h_{\Lambda_\iota}$ un élément comme ci-dessus. Le commutant de h dans U_ι est connexe. Son groupe de points sur F est le produit des groupes $GL_{K_\lambda}(W_\lambda)$ pour $\lambda \in \Lambda_\iota^d$ et des groupes unitaires de (V_λ, q_λ) pour $\lambda \in \Lambda_\iota^e$. On retrouve de façon élémentaire le fait que h est stablement conjugué à un élément dont le commutant est quasi-déployé : il suffit de remplacer les formes q_λ par des formes de même dimension dont les groupes unitaires sont quasi-déployés. On a aussi une propriété plus forte que le lemme 4.3(iv) (et qui n'est pas toujours vraie dans le cas général) : si h est d'ordre fini premier à p, il est stablement conjugué à un élément de K_ι. Il suffit de remplacer les formes q_λ par des formes de même dimension et admettant un réseau autodual. L'hypothèse sur h assure que K_λ/E est non ramifiée, donc Q_λ admet elle-aussi un réseau autodual.

Exemple. Supposons $n_\iota = 1$, donc $V_\iota = E$. On a $U_\iota(F) = E^1$. Tout élément γ de ce groupe est semi-simple et sa classe de conjugaison (qui se réduit à lui-même puisque le groupe est commutatif) est paramétrée par les données suivantes : Λ_ι possède un unique élément qui appartient à Λ_ι^e ; cet unique élément est formé des objets $\lambda = \gamma \in E^1$, $K_\lambda = E$, $\tau_\lambda = \tau$, $V_\lambda = E$, $q_\lambda = Q_\lambda = Q_\iota$.

5.4. Correspondance entre classes de conjugaison. On considère la donnée endoscopique $(H, \mathcal{H}, s, \hat{\xi})$ de $(G, \theta, \boldsymbol{\omega})$ qui est associée à une décomposition $n = n_+ + n_-$, cf. 5.1. On a $H = U_{n_+} \times U_{n_-}$. En remplaçant l'indice ι du paragraphe précédent par $+$ ou $-$, on peut identifier H au produit des groupes unitaires de deux espaces hermitiens (V_+, Q_+) et (V_-, Q_-). Soient $\gamma = (\gamma_+, \gamma_-)$ un élément semi-simple de $H(F)$ et δ un élément semi-simple de $\tilde{G}(F)$. Les éléments γ_+, γ_- et δ sont paramétrés par des données notées Λ_{γ_+}, Λ_{γ_-} et Λ_δ. Quitte à changer ces données par des données équivalentes, on peut supposer que, si λ et λ' sont deux éléments de la réunion de ces trois ensembles, les couples (λ, K_λ) et $(\lambda', K_{\lambda'})$ ou bien sont égaux, ou bien ne sont pas équivalents. Tout élément par exemple de Λ_δ est une famille de données λ, K_λ etc. et on a identifié dans la notation cette famille avec l'unique terme $\lambda \in \bar{F}^\times$. On peut ainsi considérer Λ_δ comme un sous-ensemble de $\bar{F}^\times$, en se rappelant qu'à chaque élément est affecté des données supplémentaires. Alors les classes de conjugaison de γ et δ se correspondent si et seulement si les conditions suivantes sont vérifiées :

- $\Lambda_\delta = \{(-1)^{n+1}\lambda; \lambda \in \Lambda_{\gamma_+} \cup \Lambda_{\gamma_-}\}$;
- pour tout $\lambda \in \Lambda_\delta$, $N_\lambda = N_{+,(-1)^{n+1}\lambda} + N_{-,(-1)^{n+1}\lambda}$.

Les signes perturbateurs $(1)^{n+1}$ sont inévitablement dus à la présence de signes dans la définition de l'automorphisme θ_n. Remarquons que si ces conditions sont vérifiées et si λ appartient à $\Lambda^e_{\gamma_\pm}$, resp. $\Lambda^d_{\gamma_\pm}$, on a nécessairement $(-1)^{n+1}\lambda \in \Lambda^e_\delta$, resp. $(-1)^{n+1}\lambda \in \Lambda^d_\delta$.

Supposons que ces conditions soient vérifiées et que de plus δ appartienne à $\tilde{G}_{reg}(F)$, autrement dit $(\gamma, \delta) \in \mathcal{D}$. Dans ce cas, Λ_δ est union disjointe de $\Lambda_{\delta,+}$ et $\Lambda_{\delta,-}$, où $\Lambda_{\delta,\pm} = \{(-1)^{n+1}\lambda; \lambda \in \Lambda_{\gamma_\pm}\}$. On pose $\Lambda^e_{\delta,\pm} = \Lambda^e_\delta \cap \Lambda_{\delta,\pm}$ et $\Lambda^d_{\delta,\pm} = \Lambda^d_\delta \cap \Lambda_{\delta,\pm}$. Notons $\mathcal{P}_{\gamma_\pm}$ le sous-ensemble de $\bar{F}^\times$ dont les éléments sont :

- les conjugués distincts $\sigma(\lambda)$ pour $\sigma \in \Gamma_E$ et $\lambda \in \Lambda^e_{\gamma_\pm}$;
- les conjugués distincts $\sigma(\lambda)$ pour $\sigma \in \Gamma_E$ et $\sigma(\lambda)^{-1}$ pour $\sigma \in \Gamma \setminus \Gamma_E$, pour $\lambda \in \Lambda^d_{\gamma_\pm}$.

On peut aussi dire que $\mathcal{P}_{\gamma_\pm}$ est l'ensemble des valeurs propres de $\gamma_\pm$ vu comme un automorphisme E-linéaire de $V_\pm$. Notons $\mathcal{P}_{\gamma,\delta,1}$ l'ensemble des couples (x, y) tels que $x \in \mathcal{P}_{\gamma_-}$, $y \in \mathcal{P}_{\gamma_+} \cup \mathcal{P}_{\gamma_-}$ et $x \neq y$. Notons $\mathcal{P}_{\gamma,\delta,2}$ l'ensemble des couples (x, y) tels que $x \in \mathcal{P}_{\gamma_-}$ et $y \in \mathcal{P}_{\gamma_+}$. Pour $i = 1, 2$, posons :

$$P_{\gamma,\delta,i} = \prod_{(x,y) \in \mathcal{P}_{\gamma,\delta,i}} (x - y).$$

C'est un élément de $E^\times$. Notons val_E la valuation habituelle de E (elle vaut 1 sur une uniformisante). On peut expliciter le facteur de transfert sous la forme :

$$(1) \quad \Delta(\gamma, \delta) = \Big(\prod_{\lambda \in \Lambda^d_{\delta,-}} (-1)^{val_E \circ Norm_{K_\lambda/E}(\lambda)}\Big)\Big(\prod_{\lambda \in \Lambda^e_{\delta,-}} (-1)^{val_E \circ Norm_{K_\lambda/E}(\alpha_\lambda)}\Big)$$

$$\times(-1)^{val_E(P_{\gamma,\delta,1})} q^{-val_E(P_{\gamma,\delta,2})}.$$

Remarque. Pour $\lambda \in \Lambda^e_\delta$, l'extension K'_λ de F est disjointe de E, donc le degré résiduel de l'extension K_λ/E est impair. Il en résulte que

$val_E \circ Norm_{K_\lambda/E}(\alpha_\lambda)$ change effectivement de parité quand δ varie dans sa classe de conjugaison par $G(\bar{F})$.

Dans le cas où $n_+ = n$ et $n_- = 0$, on a simplement $\Delta(\gamma, \delta) = 1$. Dans le cas où $n_+ = 0$ et $n_- = n$, on vérifie que $\Delta(\gamma, \delta) = \chi(\delta)$, où χ est la fonction sur $\tilde{G}(F)$ définie de la façon suivante. Revenons à la définition $\tilde{G}(F) = GL_n(E)\theta$. Alors $\chi(g\theta) = (-1)^{val_E \circ det(g)}$. Plus généralement, introduisons la donnée endoscopique $(H', \mathcal{H}', s', \hat{\xi}')$ associée à la décomposition $n = n_- + n_+$, notons Δ' le facteur de transfert associé défini par la formule similaire à (1). Posons $\gamma' = (\gamma_-, \gamma_+)$. Alors $\gamma' \in H'(F)$ et les classes de conjugaison stable de γ' et δ se correspondent. On vérifie l'égalité $\Delta'(\gamma', \delta) = \chi(\delta)\Delta(\gamma, \delta)$.

Exemple. Supposons $n = n_+ = 1$. Soient $\gamma = (\gamma_+, 1)$ un élément semi-simple de $H(F)$ et δ un élément semi-simple de $\tilde{G}(F)$. Les données paramétrant la classe de conjugaison de γ_+ se résument à l'unique élément $\lambda_+ = \gamma_+ \in E^1$. Celles paramétrant la classe de conjugaison de δ sont un couple (λ, q_λ), avec en fait $q_\lambda = \delta$ quand on se rappelle que δ est une forme sesquilinéaire sur E. Les classes de conjugaison se correspondent si et seulement si $\lambda = \lambda_+$. Ecrivons q_λ sous la forme $q_\lambda(v', v'') = \alpha\tau(v')v''$. Comme on l'a dit en 5.2, on a $\lambda = \alpha\tau(\alpha)^{-1}$ et on retrouve la correspondance $\alpha \mapsto \lambda_+ = \alpha\tau(\alpha)^{-1}$ que le lecteur attend (peut-être).

5.5. Description des groupes $\tilde{G}$, $\bar{H}$ et H_ϵ. Considérons la situation qui intervient dans la preuve du théorème 3.6. La donnée endoscopique $(H, \mathcal{H}, s, \hat{\xi})$ de $(G, \theta, \boldsymbol{\omega})$ est associée à une décomposition $n = n_+ + n_-$. On considère un couple $(\gamma, \delta) \in \mathcal{D}$. On suppose γ et δ compacts et on les écrit $\gamma = exp(Y)\epsilon$, $\delta = exp(X)\eta$. On suppose $\eta \in \tilde{K}$. Les éléments γ, Y et ϵ se décomposent en $\gamma = (\gamma_+, \gamma_-)$, $Y = (Y_+, Y_-)$, $\epsilon = \epsilon_+, \epsilon_-$.

La classe de conjugaison par $G(F)$ de l'élément η est paramétrée par des données que l'on note Λ_η. Pour simplifier, on suppose que Λ_η est réduit à un unique élément, qui appartient à Λ_η^e (le cas où l'élément appartient à Λ_η^d est plus simple, il n'apparaît que des groupes linéaires ; le cas général est produit, en un sens évident, de ces cas particuliers). On note λ_η cet unique élément. C'est une famille d'objets que l'on note λ_η, K_η, τ_η, V_η, q_η. Elle est en fait uniquement déterminée par le premier élément $\lambda_\eta \in \mathfrak{o}_{\bar{F},p'}$: K_η est l'extension non ramifiée de E engendrée par λ_η, τ_η est l'unique automorphisme de ce corps qui se restreint à E en l'automorphisme τ et envoie λ_η sur λ_η^{-1}, V_η est de dimension $n/[K_\eta : E]$ sur K_η et q_η est l'unique (à isomorphisme près) forme sesquilinéaire vérifiant la condition de symétrie requise et possédant un réseau autodual.

Seules comptent pour nous les classes de conjugaison stable de $\epsilon_\pm$ et leurs paramétrages se déduisent de celui de η. Pour éliminer les $(-1)^{n+1}$ perturbateurs, on va supposer n impair. Alors la classe de conjugaison stable de $\epsilon_\pm$ est paramétrée par l'unique donnée $\lambda_{\epsilon_\pm} = \lambda_\eta$, $K_{\epsilon_\pm} = K_\eta$, $\tau_{\epsilon_\pm} = \tau_\eta$, $V_{\epsilon_\pm}$, elle-aussi entièrement déterminée par λ_η. Remarquons que $dim_{K_\eta}(V_{\epsilon_\pm}) = n_\pm/[K_\eta : E]$.

Fixons $\alpha_\eta \in K_\eta$ tel que $\lambda_\eta = \alpha_\eta \tau_\eta(\alpha_\eta)^{-1}$. Le commutant $\bar{G} = G_\eta$ est le groupe unitaire de $(V_\eta, \alpha_\eta^{-1} q_\eta)$. L'élément X appartenant à son algèbre de Lie, sa classe de conjugaison par $\bar{G}(F)$ est paramétrée comme en 5.3. Il convient d'adapter le paramétrage et la notation aux algèbres de Lie. Ainsi, on va noter $\mathcal{X}$ l'ensemble de paramètres, qui se décompose en $\mathcal{X}^e \cup \mathcal{X}^d$. Un élément $x \in \mathcal{X}^e$ est une famille d'objets x, K_x, τ_x, V_x, q_x. Le premier objet x est un élément de $\bar{F}$. Le second K_x est l'extension de K_η engendrée par x. Le troisième τ_x est l'unique automorphisme de K_x dont la restriction à K_η est l'automorphisme τ_η et qui envoie x sur $-x$. On note K'_x le sous-corps des points fixes dans K_x par l'automorphisme τ_x. Puisque X est régulier, on peut supposer $V_x = K_x$. Alors q_x est une forme hermitienne sur K_x, relativement à l'extension K_x/K'_x. On pose $Q_x(v', v'') = trace_{K_x/K_\eta}(q_x(v', v''))$. Un élément $x \in \mathcal{X}^d$ est une famille x, K_x, W_x. Les deux premiers termes sont comme ci-dessus et W_x est un espace vectoriel de dimension finie sur K_x. On pose $V_x = W_x \oplus W_x^*$, que l'on munit de la structure d'espace vectoriel sur K_η définie en 5.3, puis d'une forme hermitienne Q_x comme dans ce paragraphe. On fixe un isomorphisme de V_η sur $\oplus_{x \in \mathcal{X}} V_x$ qui identifie les formes hermitiennes $\alpha_\eta^{-1} q_\eta$ et $\sum_{x \in \mathcal{X}} Q_x$. Modulo cet isomorphisme, X est l'élément de $End_{K_\eta}(V_\eta)$ qui agit par $v \mapsto xv$ sur V_x pour $x \in \mathcal{X}^e$ et par $(w \oplus w^*) \mapsto (xw \oplus -xw^*)$ sur V_x pour $x \in \mathcal{X}^d$.

L'espace V_η est égal à V. Selon notre identification de $\tilde{G}(F)$ à un ensemble de formes sesquilinéaires sur V, l'élément η se confond avec la forme Q_η définie par $Q_\eta(v', v'') = trace_{K_\eta/E}(q_\eta(v', v''))$. L'élément $\delta = exp(X)\eta$ se confond avec la forme Q_δ définie par $Q_\delta(v', v'') = Q_\eta(exp(-X)(v'), v'')$. On en déduit l'ensemble Λ_δ qui paramètre la classe de conjugaison par $G(F)$ de δ. Il y a une bijection $x \mapsto \lambda_x$ de $\mathcal{X}$ sur Λ_δ de sorte que :

- $\lambda_x = exp(2x)\lambda_\eta$;
- pour $x \in \mathcal{X}^e$, $K_{\lambda_x} = K_x$, $\tau_{\lambda_x} = \tau_x$, $V_{\lambda_x} = V_x$, $q_{\lambda_x} = \alpha_\eta exp(x) q_x$;
- pour $x \in \mathcal{X}^d$, $K_{\lambda_x} = K_x$ et $W_{\lambda_x} = W_x$.

Remarque. Le facteur 2 de la première formule vient des formules décrivant les formes sesquilinéaires. Soit par exemple $x \in \mathcal{X}^e$. d'après la formule ci-dessus décrivant q_{λ_x}, λ_x est égal à $\alpha_\eta exp(x)\tau_x(\alpha_\eta exp(x))^{-1}$, qui est bien égal à $exp(2x)\lambda_\eta$. Cf. l'exemple de 5.2 dans le cas $n = 1$.

L'élément $Y_\pm$ appartient à l'algèbre de Lie du groupe unitaire de $V_{\epsilon_\pm}$. Sa classe de conjugaison stable est paramétrée par un ensemble $\mathcal{Y}_\pm = \mathcal{Y}_\pm^e \cup \mathcal{Y}_\pm^d$. Un élément $y \in Y_\pm^e$ est une famille d'objets y, K_y, τ_y, V_y. Un élément $y \in \mathcal{Y}_\pm^d$ est une famille d'objets y, K_y, W_y. On en déduit l'ensemble $\Lambda_{\gamma_\pm}$ qui paramètre la classe de conjugaison stable de $\gamma_\pm = exp(Y_\pm)\epsilon_\pm$. Il y a une bijection $y \mapsto \lambda_y$ de $\mathcal{Y}_\pm$ sur $\Lambda_{\gamma_\pm}$ de sorte que :

- $\lambda_y = exp(y)\lambda_\eta$;
- pour $y \in \mathcal{Y}_\pm^e$, $K_{\lambda_y} = K_y$, $\tau_{\lambda_y} = \tau_y$, $V_{\lambda_y} = V_y$;
- pour $y \in \mathcal{Y}_\pm^d$, $K_{\lambda_y} = K_y$ et $W_{\lambda_y} = W_y$.

Puisque $\delta \in \tilde{G}_{reg}(F)$, le fait que les classes de conjugaison de γ et δ se correspondent signifie que $\Lambda_\delta = \Lambda_{\gamma_+} \sqcup \Lambda_{\gamma_-}$, du moins si l'on considère ces ensembles comme des sous- ensembles de $\bar{F}^\times$. Modulo les bijections ci-dessus, il y a donc une décomposition $\mathcal{X} = \mathcal{X}_+ \sqcup \mathcal{X}_-$ et des bijections $x \mapsto y_x$ de $\mathcal{X}_\pm$ sur $\mathcal{Y}_\pm$ de sorte que :

(1) $y_x = 2x$;

- pour $x \in \mathcal{X}_\pm^e = \mathcal{X}^e \cap \mathcal{X}_\pm$, $K_{y_x} = K_x$, $\tau_{y_x} = \tau_x$, $V_{y_x} = V_x$;
- pour $x \in \mathcal{X}_\pm^d = \mathcal{X}^d \cap \mathcal{X}_\pm$, $K_{y_x} = K_x$ et $W_{y_x} = W_x$.

On a défini une donnée endoscopique $(\bar{H}, \bar{\mathcal{H}}, \bar{s}, \hat{\bar{\xi}})$ de $\bar{G}_{SC}$. En fait, le passage au groupe simplement connexe $\bar{G}_{SC}$ n'est imposé dans le cas général que par la présence du caractère $\boldsymbol{\omega}$. Il est ici trivial et on peut aussi bien définir $(\bar{H}, \bar{\mathcal{H}}, \bar{s}, \hat{\bar{\xi}})$ comme une donnée endoscopique de $\bar{G}$. On peut aussi considérer $\bar{G}$ comme un groupe unitaire sur K'_η (le corps fixé par τ_η), auquel cas son L-groupe devient simplement le produit semi- direct $GL_{\bar{n}}(\mathbb{C}) \rtimes W_{K'_\eta}$, où $\bar{n} = n/[K_\eta : E]$. Il est plus ou moins évident que $\bar{s}$ est l'élément diagonal de $GL_{\bar{n}}(\mathbb{C})$ qui a $\bar{n}_+$ valeurs propres 1 et $\bar{n}_-$ valeurs propres -1, où $\bar{n}_\pm = n_\pm/[K_\eta : E]$. Alors $\bar{H}$ est le produit de deux groupes unitaires quasi-déployés sur K'_η de rangs respectifs $\bar{n}_+$ et $\bar{n}_-$. On constate que $\bar{H}$ est isomorphe à H_ϵ. Introduisons l'élément $\bar{Y} = (\bar{Y}_+, \bar{Y}_-) \in \bar{\mathfrak{h}}(F)$ de 4.8. Puisque cet élément correspond à la fois à Y par la correspondance endoscopique non standard entre H_ϵ et $\bar{H}$ (on oublie ici les passages inutiles aux groupes simplement connexes) et à X par la correspondance endoscopique ordinaire entre $\bar{H}$ et $\bar{G}$, on voit que $\bar{Y}_\pm$ est paramétré par l'ensemble $\mathcal{X}_\pm$, en oubliant les formes hermitiennes puisque seule compte la classe de conjugaison stable de $\bar{Y}$. En identifiant les groupes H_ϵ et $\bar{H}$, la relation (1) montre que Y s'identifie à $2\bar{Y}$. Le lemme fondamental non standard que l'on a utilisé dans la preuve du théorème dit que :

$$SO_Y^{H_\epsilon}(\mathbf{1}_{\mathfrak{k}_\epsilon}) = SO_{Y/2}^{H_\epsilon}(\mathbf{1}_{\mathfrak{k}_\epsilon}).$$

C'est trivial puisque $p \neq 2$.

Remarque. On peut montrer plus généralement que le lemme fondamental non standard ne sert à rien dans toute situation de changement de base, car les deux groupes $\bar{H}$ et H_ϵ sont isomorphes. Mais, comme le montre notre exemple, même dans ce cas, la correspondance endoscopique non standard entre $\mathfrak{h}_\epsilon(F)$ et $\bar{\mathfrak{h}}(F)$ n'est pas celle qui provient de l'isomorphisme entre les deux groupes : c'en est un multiple. Cela est évidemment lié au fait que, dans le cas général, il n'y a de lemme fondamental non standard que pour les algèbres de Lie et non pas pour les groupes. En utilisant la formule (1) de 5.4 et la formule analogue de [W2] proposition X.8 (dont la preuve est beaucoup plus brève que celle du théorème 4.8), on vérifie aisément l'égalité des facteurs de transfert énoncée au théorème 4.8.

Références

[K1] R. Kottwitz, Rational conjugacy classes in reductive groups, *Duke Math. Journal* **49** (1982), 785–806.

[K2] ———, Stable trace formula : elliptic singular terms, *Math. Annalen* **275** (1986), 365-399.

[KS] ———, D. Shelstad, Foundations of twisted endoscopy, *Astérisque* **255** (1999).

[Lab] J.-P. Labesse, Stable twisted trace formula : elliptic terms, *Journal of the Inst. of Math. Jussieu* **3** (2004), 473–530.

[N] Ngo Bao Chau, Le lemme fondamental pour les algèbres de Lie, prépublication 2007.

[T] J. Tits, Reductive groups over local fields, in Automorphic forms, representations and L-functions, *Proc. Symposia in Pure Math.* XXXIII, AMS (1979).

[W1] J.-L. Waldspurger, L'endoscopie tordue n'est pas si tordue, *Memoirs AMS* **908** (2008).

[W2] ———, Intégrales orbitales nilpotentes et endoscopie pour les groupes classiques non ramifiés, *Astérisque* **269** (2001).

Subsection II.C

The fundamental lemma and the Hitchin fibration

Lemme Fondamental pour les algèbres de Lie (d'après Ngô Bao-Châu)

[DT.II.C.1]

Dat Jean-François and Ngô Dac-Tuan

La formule des traces sur les corps de nombres a été introduite par Selberg, puis développée par Arthur. Parmi les applications importantes de la formule des traces d'Arthur-Selberg, on trouve plusieurs cas du principe de fonctorialité de Langlands et l'expression de la cohomologie des variétés de Shimura en termes automorphes. Cependant, pour beaucoup de ces applications, il est nécessaire de la stabiliser. La stabilisation a été faite par Langlands et Kottwitz pour la partie elliptique modulo une conjecture connue sous le nom "*lemme fondamental de Langlands-Shelstad*". Récemment Arthur a réussi à stabiliser toute la formule des traces invariante modulo le lemme fondamental pondéré - une version élargie du lemme fondamental.

Le lemme fondamental est de nature locale. Etant donnés un groupe réductif G sur un corps local et un groupe endoscopique H de G associé à une donnée endoscopique (κ, ρ_κ), le lemme fondamental est une égalité conjecturale entre la κ-intégrale orbitale de G et l'intégrale orbitale stable de H. Plusieurs cas particuliers de cette conjecture ont été prouvés, nous renvoyons le lecteur au [**19**, Introduction] pour une liste des cas connus. Grâce aux travaux de Waldspurger [**21**] et Cluckers-Hales-Loeser [**5**], pour prouver le lemme fondamental, il suffit de démontrer sa variante pour les algèbres de Lie pour les corps locaux d'égales caractéristiques dont la caractéristique peut être supposée très grande par rapport au groupe réductif G en question. Voici l'énoncé de cette variante [**19**, Introduction]:

THÉORÈME (Lemme fondamental pour les algèbres de Lie).
Soient k un corps fini à q éléments, $\mathcal{O}$ un anneau de valuation discrète complet de corps résiduel k et F son corps des fractions. Soit G un schéma en groupes réductifs au-dessus de $\mathcal{O}$ dont le nombre de Coxeter multiplié par deux est plus petit que la caractéristique de k. Soient (κ, ρ_κ) une donnée endoscopique de G au-dessus de $\mathcal{O}$ et H le schéma en groupes endoscopiques associé. On a l'égalité entre la κ-intégrale orbitale et l'intégrale orbitale

stable

$$\Delta_G(a)\,\mathrm{O}_a^{\kappa}(1_{\mathfrak{g}}, \mathrm{d}\,t) = \Delta_H(a_H)\,\mathrm{O}_a^{\kappa}(1_{\mathfrak{h}}, \mathrm{d}\,t)$$

associées aux classes de conjugaison stable semi-simples régulières a et a_H de $\mathfrak{g}(F)$ et $\mathfrak{h}(F)$ qui se correspondent, aux fonctions caractéristiques et des compacts $\mathfrak{g}(\mathcal{O})$ et $\mathfrak{h}(\mathcal{O})$ dans $\mathfrak{g}(F)$ et $\mathfrak{h}(F)$ et où on a noté

$$\Delta_G(a) = q^{-\operatorname{val}(\mathfrak{D}_G(a))/2}, \quad \Delta_H(a) = q^{-\operatorname{val}(\mathfrak{D}_H(a_H))/2},$$

$\mathfrak{D}_G$, $\mathfrak{D}_H$ étant les fonctions discriminant de G et de H.

Kazhdan et Lusztig ont remarqué que les intégrales orbitales apparaissent naturellement dans les comptages des points des fibres de Springer affines. Ainsi, l'intégrale orbitale admet une interprétation en termes de la géométrie. Puis Goresky, Kottwitz, MacPherson ont donné une interprétation géométrique du lemme fondamental en utilisant la cohomologie équivariante des fibres de Springer affines [**9**]. Modulo une conjecture de pureté, ils ont réussi à démontrer le lemme fondamental dans un cas particulier dit nonramifié. Toujours sous l'hypothèse de la validité de la conjecture de pureté, par un argument de déformation, Laumon a généralisé cette méthode et prouvé le lemme fondamental pour les groupes unitaires dans le cas éventuellement non ramifié [**13, 14**]. Soulignons que tous ces résultats sont de nature locale.

Dans [**17**], Ngô propose d'étudier les fibrations de Hitchin comme analogues globaux des fibres de Springer affines et donne une interprétation géométrique du processus de stabilisation elliptique en termes de la cohomologie des fibrations de Hitchin. Dans un travail en commun avec Laumon [**16**], ils combinent cette nouvelle approche avec la technique de cohomologie équivariante pour démontrer le lemme fondamental pour les algèbres de Lie des groupes unitaires. Récemment, Ngô a prouvé le lemme fondamental pour les algèbres de Lie en toute généralité. Dans sa preuve, la cohomologie équivariante n'apparait plus, et l'argument clef est une version forte du théorème de support, qui peut être considéré comme l'analogue global de la conjecture de pureté. La preuve de ce théorème de support est basée essentiellement sur une technique due à Goresky et MacPherson.

Le but de ce texte est de présenter une introduction à la preuve de Ngô Bao Châu du lemme fondamental pour les algèbres de Lie. Nous nous contentons d'esquisser les grandes lignes et d'expliquer les arguments clefs de sa preuve. Nous espérons que ce texte facilitera la lecture de l'article original [**19**]. Pour d'autres articles de synthèse sur le même sujet, le lecteur pourra consulter les articles [**6, 15, 18**].

1. Préliminaires

1.1. Notations. Soit k un corps fini à q éléments. On utilise librement la notion de champs dont une excellente référence est le livre de Laumon-Moret Bailly [**12**]. Sauf mention contraire, tous les schémas et les champs considérés dans ce texte seront définis sur k.

Fixons une fois pour toutes X une courbe projective, lisse, géométriquement connexe sur k. On note $\bar{k}$ une clôture algébrique de k et $\bar{X}$ le produit fibré $X \times_k \bar{k}$. Soit F le corps de fonctions de X, d'anneau des adèles $\mathbb{A}$; on note $\bar{F} = F \times_k \bar{k}$. Pour une place finie v de F, on note $v : F^\times \longrightarrow \mathbb{Z}$ la valuation associé. Puis F_v sera la complétion de F par rapport à cette valuation, $\mathcal{O}_v$ sera l'anneau des entiers de F_v, k_v sera son corps résiduel et $\deg(v)$ le degré de l'extension k_v/k.

Soit $\mathbb{G}$ un groupe réductif déployé sur k. Fixons une fois pour toutes $\mathbb{T}$ un tore maximal déployé sur k de $\mathbb{G}$ et $\mathbb{B}$ un sous-groupe de Borel de $\mathbb{G}$ qui contient $\mathbb{T}$. Avec la donnée de $(\mathbb{T}, \mathbb{B})$, on notera $\mathbb{W}$ le groupe de Weyl de $\mathbb{G}$ et Φ (resp. Δ) l'ensemble des racines de $\mathbb{G}$ (resp. celui des racines simples). Les algèbres de Lie de $\mathbb{G}$, $\mathbb{T}$, et $\mathbb{B}$ seront notées $\mathbb{g}$, $\mathbb{t}$, et $\mathbb{b}$. On note ad l'action adjointe de $\mathbb{G}$ sur $\mathbb{g}$.

Pour chaque racine simple $\alpha \in \Delta$, on choisit un vecteur non nul x_α dans $\mathbb{g}_\alpha$. Il existe alors un unique vecteur non nul $x_{-\alpha}$ dans $\mathbb{g}_{-\alpha}$ tel que $[x_\alpha, x_{-\alpha}] = \alpha^\vee$. On pose $x_+ = \sum_{\alpha\in\Delta} x_\alpha$ et $x_- = \sum_{\alpha\in\Delta} x_{-\alpha}$. La donnée de ces vecteurs est appelée *épinglage*.

Soit D un diviseur de X qui s'écrit sous la forme $D = 2D'$ où D' est un diviseur effectif de degré plus grand que g - le genre de la courbe X. On notera $\mathbb{t}_D = \mathbb{t} \otimes_{\mathcal{O}_X} \mathcal{O}_X(D)$ et $\mathbb{g}_D = \mathbb{g} \otimes_{\mathcal{O}_X} \mathcal{O}_X(D)$; ce sont les fibrés vectoriels sur X obtenus en tordant $\mathbb{t}$ et $\mathbb{g}$ par le diviseur D.

Désormais, on suppose que la caractéristique de k est plus grand que le nombre de Coxeter multiplié par deux.

Exemple : $\mathbb{G} = \mathrm{GL}_n$. Dans ce cas, on prend pour $\mathbb{T}$ le sous-groupe des matrices diagonales et pour $\mathbb{B}$ le sous-groupe des matrices triangulaires supérieures. L'algèbre de Lie $\mathbb{g}$ (resp. $\mathbb{t}$ et $\mathbb{b}$) est l'algèbre des matrices carrées de taille n (resp. la sous-algèbre des matrices diagonales et des matrices triangulaires supérieures de taille $n \times n$). L'ensemble Φ est l'ensemble des caractères $\alpha_{i,j} : \mathrm{diag}(t_1, \cdots, t_n) \in \mathbb{T} \mapsto t_i t_j^{-1}$ pour $i \neq j$, et le sous-ensemble Δ est constitué des $\alpha_{i,i+1}$ pour $i = 1, \cdots, n-1$. L'épinglage est donné par les matrices $x_{i,j}$ ayant une seule entrée non nulle et égale à 1 à la position (i, j). Le groupe $\mathbb{W}$ est le groupe symétrique $\mathfrak{S}_n$ des permutations de $\{1, 2, \ldots, n\}$. Un élément $w \in \mathfrak{S}_n$ envoie l'élément $t = \mathrm{diag}(t_1, t_2, \ldots, t_n)$ de $\mathbb{t}$ sur l'élément $\mathrm{diag}(t_{\sigma(1)}, t_{\sigma(2)}, \ldots, t_{\sigma(n)})$.

1.2. Morphisme de Chevalley. On considère l'action adjointe de $\mathbb{G}$ sur $\mathbb{g}$ et on s'intéresse à l'algèbre $k[\mathbb{g}]^{\mathbb{G}}$ des fonctions $\mathbb{G}$-invariantes sur $\mathbb{g}$. En restreignant à $\mathbb{t}$, on obtient un homomorphisme $k[\mathbb{g}]^{\mathbb{G}} \longrightarrow k[\mathbb{t}]^W$ et on vérifie que c'est en fait un isomorphisme. De plus, on montre que $k[\mathbb{g}]^{\mathbb{G}}$ est une algèbre de polynôme, c'est-à-dire il existe des fonctions régulières homogènes $a_1, a_2, \ldots, a_r$ de degré $d_1, d_2, \ldots, d_r$ dans $k[\mathbb{g}]^{\mathbb{G}}$ telles que $k[\mathbb{g}]^{\mathbb{G}} = k[a_1, a_2, \ldots, a_r]$. Bien que le choix des fonctions homogènes $a_1, a_2, \ldots, a_r$ ne soit pas unique, les entiers $d_1, d_2, \ldots, d_r$ sont uniques à une permutation près. On les appelle les *exposants de Kostant* de $\mathbb{G}$.

On définit $\mathbb{c} = \operatorname{Spec} k[\mathbb{g}]^{\mathbb{G}} = \operatorname{Spec} k[\mathbb{t}]^W$ et une action de $\mathbb{G}_m$ sur $\mathbb{c}$ via les exposants de Kostant:

$$t \circ (a_1, a_2, \ldots, a_r) = (t^{d_1} a_1, t^{d_2} a_2, \ldots, t^{d_r} a_r).$$

On considère le morphisme naturel dit *caractéristique de Chevalley* $\chi : \mathbb{g} \longrightarrow \mathbb{c}$. Si $\mathbb{G}_m$ agit sur $\mathbb{g}$ par l'homothétie et sur $\mathbb{c}$ via ses exposants de Kostant, χ est $\mathbb{G}_m$-équivariant, de sorte qu'on a un 1-morphisme de champs :

$$\chi : [\mathbb{g}/(\mathbb{G} \times G_m)] \longrightarrow [\mathbb{c}/\mathbb{G}_m].$$

On va rappeler maintenant la construction due à Kostant d'une section de $\chi : \mathbb{g} \longrightarrow \mathbb{c}$. On a déjà construit un vecteur non nul $x_+ \in \mathbb{g}$; on note $\mathbb{g}^{x_+}$ le centralisateur de x_+ dans $\mathbb{g}$. Alors la restriction de $\mathbb{g} \longrightarrow \mathbb{c}$ au sous-espace $x_- + \mathbb{g}^{x_+}$ est un isomorphisme. L'inverse de cet isomorphisme

$$\epsilon : \mathbb{c} \longrightarrow \mathbb{g}$$

sera appelé la section de Kostant de $\mathbb{g}$. Notons que cela permet de munir $\mathbb{c}$ d'une structure d'espace affine sous $\mathbb{g}^{x_+}$, et même d'espace vectoriel avec l'origine évidente. Cette structure d'espace vectoriel est compatible à l'action de $\mathbb{G}_m$ définie ci-dessus.

Exemple : $\mathbb{G} = \mathrm{GL}_n$. Dans ce cas, $\mathbb{c}$ est l'espace affine des polynômes unitaires de degré n, et χ envoie une matrice A sur son polynôme caractéristique $\det(X.I_n - A)$. Plus explicitement, l'espace des polynômes unitaires de degré n est isomorphe à $\mathbb{A}^n$ via $(a_1, \cdots, a_n) \mapsto X^n - a_1 X^{n-1} + \cdots + (-1)^n a_n$. Identifiant donc $\mathbb{c}$ à $\mathbb{A}^n$, χ envoie une matrice carrée A de taille n sur le n-uplet

$$a = (\mathrm{tr}(A), \mathrm{tr}(\wedge^2 A), \cdots, \mathrm{tr}(\wedge^n A)).$$

En particulier les exposants de Kostant sont $d_1 = 1, d_2 = 2, \cdots, d_n = n$. Bien que ce ne soit pas tout à fait la section de Kostant, χ a une section bien connue qui envoie un polynôme sur sa matrice “compagnon”. La matrice associée à $a = (a_1, \cdots, a_n)$ est la suivante :

$$\epsilon(a) = \begin{pmatrix} 0 & 0 & \ldots & 0 & (-1)^{n-1} a_n \\ 1 & 0 & \ldots & 0 & (-1)^{n-2} a_{n-1} \\ \vdots & \vdots & \ddots & \vdots & \vdots \\ 0 & 0 & \ldots & 0 & -a_2 \\ 0 & 0 & \ldots & 1 & a_1 \end{pmatrix}.$$

1.3. Lieux réguliers. Centralisateurs réguliers. Pour toute racine $\alpha \in \Phi$, on dispose d'une flèche $\mathrm{d}\,\alpha : \mathbb{t} \longrightarrow k$. Le discriminant $\mathfrak{D} = \prod_{\alpha \in \Phi} \mathrm{d}\,\alpha$ est une fonction régulière $\mathbb{W}$-invariante sur $\mathbb{t}$, donc elle est une fonction régulière sur $\mathbb{c}$ et définit un diviseur sur $\mathbb{c}$ noté $\mathfrak{D}$. On montre aisément que c'est un diviseur réduit, cf. [**19**, Lemme 1.10.1]. Le complémentaire de ce diviseur sera appelé l'ouvert des éléments semisimples réguliers de $\mathbb{c}$ et on le note $\mathbb{c}^{\mathrm{rs}}$.

On a vu que le schéma $\mathfrak{c}$ est le schéma quotient de $\mathfrak{t}$ par l'action du groupe fini $\mathbb{W}$. Le morphisme $\mathfrak{t} \longrightarrow \mathfrak{c}$ est donc fini et plat. De plus, au-dessus de l'ouvert régulier $\mathfrak{c}^{\mathrm{rs}}$, $\mathfrak{t}_{|\mathfrak{c}^{\mathrm{rs}}}$ est étale galoisien de groupe de Galois $\mathbb{W}$. Si l'on considère l'action de $\mathbb{G}_m$ sur $\mathfrak{t}$ par l'homothétie et sur $\mathfrak{c}$ via les exposants de Kostant, ce morphisme est $\mathbb{G}_m$-équivariant.

On considère le schéma en groupes des centralisateurs I au-dessus de $\mathfrak{g}$ défini par

$$I = \{(g, x) \in \mathbb{G} \times \mathfrak{g} : \mathrm{ad}(g)x = x\}.$$

Il n'est pas lisse au-dessus de $\mathfrak{g}$, ni même plat. On définit l'ouvert régulier $\mathfrak{g}^{\mathrm{reg}}$ de $\mathfrak{g}$ comme suit. Un $\bar{k}$-point x de $\mathfrak{g}$ est dans $\mathfrak{g}^{\mathrm{reg}}(\bar{k})$ si et seulement si son centralisateur I_x est de dimension minimale, c'est-à-dire $\dim I_x = \mathrm{rang}\, G$. On montre qu'au-dessus de cet ouvert $\mathfrak{g}^{\mathrm{reg}}$, $I_{|\mathfrak{g}^{\mathrm{reg}}}$ est un schéma en groupes commutatifs lisse. Rappelons que le morphisme de Chevalley $\chi : \mathfrak{g} \longrightarrow \mathfrak{c}$ admet une section de Kostant $\epsilon : \mathfrak{c} \longrightarrow \mathfrak{g}$ dont l'image est contenue dans $x^- + \mathfrak{g}^{x+}$, donc dans $\mathfrak{g}^{\mathrm{reg}}$. Ainsi, le centralisateur régulier $J = \epsilon^* I$ est un schéma en groupes commutatifs lisse. Notons que J est un tore au-dessus du lieu semi-simple régulier $\mathfrak{c}^{\mathrm{rs}}$, mais peut avoir des fibres non connexes et une partie additive le long du diviseur $\mathfrak{D}$. Par la descente fidèlement plate, on montre qu'il existe un isomorphisme G-équivariant canonique $\chi^* J \longrightarrow I$ sur $\mathfrak{g}^{\mathrm{reg}}$. Puis, par normalité, Ngô montre que cet isomorphisme s'étend en un homomorphisme $\chi^* J \longrightarrow I$ sur tout l'espace $\mathfrak{g}$, cf. [**17**, Proposition 3.2].

Exemple : $\mathbb{G} = \mathrm{GL}_n$. Dans ce cas, $\mathfrak{c}^{\mathrm{rs}}$ est le lieu des polynômes unitaires de degré n qui sont séparables (sans racines multiples). Si $a = (a_1, a_2, \dots, a_n) \in \mathfrak{c}(\bar{k}) = \bar{k}^n$ représente le polynôme unitaire $P_a = t^n - a_1 t^{n-1} + \ldots + (-1)^n a_n$, alors J_a est le groupe des éléments inversibles de l'algèbre $\bar{k}[T]/P_a$. En particulier, la partie additive de J_a est non triviale si et seulement si le polynôme P_a a des racines multiples. Soit $A \in \mathfrak{g}(\bar{k})$ une matrice de caractéristique a; on fait agir $\bar{k}[T]$ sur $\mathfrak{g}(\bar{k})$ de manière suivante: l'action des éléments de $\bar{k}$ est par la multiplication et l'action de T est par la multiplication avec la matrice A à droite. La flèche obtenue n'est rien d'autre que l'homomorphisme $J_a \longrightarrow I_A$. On vérifie sans peine que si A est régulier, c'est un isomorphisme.

1.4. Formes quasi-déployées de $\mathbb{G}$ sur X. Le lemme fondamental concerne des groupes quasi-déployés sur un corps local. Ce sont des formes "extérieures" de groupes déployés. De même ici, on doit considérer des formes extérieures de $\mathbb{G}$ sur X. Une telle forme est donnée par un morphisme

$$\rho : \pi_1(X, x) \longrightarrow \mathrm{Out}(\mathbb{G})$$

(le groupe discret des automorphismes extérieurs de $\mathbb{G}$). Plus précisément, la forme G de $\mathbb{G}$ sur X associée à ρ s'obtient en choisissant un revêtement Galoisien (X_ρ, x_ρ) de (X, x) de groupe noté Θ_ρ et tel que ρ se factorise par $\pi_1(X, x) \longrightarrow \Theta_\rho$, puis en tordant $\mathbb{G}$ par ce torseur. On a donc $G = \mathbb{G} \times^{\Theta_\rho} X_\rho$. Comme $\mathrm{Out}(\mathbb{G})$ s'identifie au sous-groupe de $\mathrm{Aut}(\mathbb{G})$ qui préserve la paire $(\mathbb{T}, \mathbb{B})$ et l'épinglage des x_α, toutes les constructions précédentes peuvent être tordues pour donner une paire (T, B) dans G au-dessus de X, un groupe de Weyl W, des algèbres de Lie $\mathfrak{t}$, $\mathfrak{g}$, un quotient $\mathfrak{c} = \mathfrak{t}/W$, son lieu régulier $\mathfrak{c}^{\mathrm{rs}}$, le morphisme de Chevalley $\chi : \mathfrak{g} \longrightarrow \mathfrak{c}$, une section de Kostant $\varepsilon : \mathfrak{c} \longrightarrow \mathfrak{g}$ de ce morphisme, ainsi qu'un centralisateur régulier J commutatif et lisse au-dessus de $\mathfrak{c}$ et un morphisme $\chi^* J \longrightarrow I$ de schémas en groupes au-dessus de $\mathfrak{g}$. On remarque que $\mathfrak{c}$ s'identifie au quotient

$$\mathfrak{c} = (\mathbb{t} \times X_\rho)/(\mathbb{W} \rtimes \Theta_\rho).$$

Exemple : groupes unitaires. Pour $\mathbb{G} = \mathrm{GL}_n$, on a $\mathrm{Out}(\mathbb{G}) = \mathbb{Z}/2\mathbb{Z}$, donc une forme de $\mathbb{G}$ est donnée par un revêtement étale X_ρ de degré 2. L'élément non trivial de $\mathrm{Aut}(\mathbb{G})$ préservant l'épinglage qu'on a fixé plus haut est donné par $g \mapsto \Phi_n {}^t g^{-1} \Phi_n^{-1}$ où Φ_n est la matrice antidiagonale. La forme associée à X_ρ est le schéma en groupes qui à un X-schéma S associe

$$U_{X_\rho/X}(n)(S) := \{g \in \mathrm{GL}_n(\Gamma(S \times_X X_\rho, \mathcal{O}_{S \times_X X_\rho})), \Phi_n \tau({}^t g^{-1})\Phi_n^{-1} = g\}$$

où τ est l'automorphisme non-trivial de X_ρ. C'est un groupe unitaire sur X, qui devient isomorphe au groupe constant GL_n après extension à X_ρ.

1.5. Constructions sur le groupe dual. Soit $\hat{\mathbb{G}}$ le groupe dual de $\mathbb{G}$ sur $\bar{\mathbb{Q}}_\ell$. Sa donnée radicielle s'obtient à partir de celle de $\hat{\mathbb{G}}$ en échangeant racines et coracines. Il vient avec une paire $(\hat{\mathbb{T}}, \hat{\mathbb{B}})$ et même un épinglage $(\hat{x}_\alpha)_\alpha$, et on a $\mathrm{Out}(\mathbb{G}) = \mathrm{Out}(\hat{\mathbb{G}})$.

Soit κ un élément de $\hat{\mathbb{T}}$. On note $\hat{\mathbb{H}} = \hat{\mathbb{H}}(\kappa)$, resp. $\pi_0(\kappa)$, la composante neutre, resp. le groupe des composantes, du centralisateur de κ dans $\hat{\mathbb{G}} \rtimes \mathrm{Out}(\mathbb{G})$. Le groupe $\hat{\mathbb{H}}$ est réductif, contient la paire de Borel $(\hat{\mathbb{T}}, \hat{\mathbb{B}} \cap \hat{\mathbb{H}})$ et est muni de l'épinglage induit par celui de $\hat{\mathbb{G}}$. Le groupe $\pi_0(\kappa)$ est muni de morphismes évidents

$$\mathrm{Out}(\hat{\mathbb{G}}) \longleftarrow \pi_0(\kappa) \longrightarrow \mathrm{Out}(\hat{\mathbb{H}}).$$

D'après [**19**, 1.9.8], on peut alors prolonger l'inclusion $\mathbb{W}_{\hat{\mathbb{H}}} \hookrightarrow \mathbb{W}_{\hat{\mathbb{G}}}$ en un morphisme

$$\mathbb{W}_{\hat{\mathbb{H}}} \rtimes \pi_0(\kappa) \longrightarrow \mathbb{W}_{\hat{\mathbb{G}}} \rtimes \pi_0(\kappa)$$

compatible avec l'action sur $\hat{\mathbb{T}}$ et la projection sur $\pi_0(\kappa)$.

Au groupe $\hat{\mathbb{H}}$ correspond un groupe réductif déployé $\mathbb{H}$ sur k, muni d'une paire $(\mathbb{T}_\mathbb{H}, \mathbb{B}_\mathbb{H})$ et d'un épinglage, ainsi que d'un isomorphisme $\mathbb{T}_\mathbb{H} \xrightarrow{\sim} \mathbb{T}$ compatible avec le morphisme $\mathbb{W}_\mathbb{H} \rtimes \pi_0(\kappa) \longrightarrow \mathbb{W}_\mathbb{G} \rtimes \pi_0(\kappa)$ évoqué ci-dessus. En particulier, cet isomorphisme induit un morphisme de *transfert*

$$\nu : \mathbb{c}_\mathbb{H} \longrightarrow \mathbb{c}$$

compatible avec l'action de $\pi_0(\kappa)$.

Le transfert n'envoie bien-sûr pas $\mathbb{c}_\mathbb{H}^{\mathrm{rs}}$ dans $\mathbb{c}^{\mathrm{rs}}$. On note $\mathbb{c}_\mathbb{H}^{G-\mathrm{rs}}$ l'image réciproque de l'ouvert $\mathbb{c}^{\mathrm{rs}}$ dans $\mathbb{c}_\mathbb{H}^{\mathrm{rs}}$. On montre que le diviseur $\nu^*(\mathfrak{D}_\mathbb{G})$ est de la forme $\mathfrak{D}_\mathbb{H} + 2\mathfrak{R}_\mathbb{H}^\mathbb{G}$ pour un diviseur réduit effectif $\mathfrak{R}_\mathbb{H}^\mathbb{G}$ de $\mathbb{c}_\mathbb{H}$, cf. [**19**, Lemme 1.10.2].

Exemple : $\mathbb{G} = \mathrm{GL}_n$. Dans ce cas, $\mathbb{H}$ est un produit de GL_{n_i} avec $\sum_i n_i = n$, et $\pi_0(\kappa)$ est soit trivial, soit égal à $\mathbb{Z}/2\mathbb{Z}$. Dans ce dernier cas, $\mathbb{H}$ est isomorphe à $\mathrm{GL}_{m_{-k}} \times \cdots \times \mathrm{GL}_{m_{-1}} \times (\mathrm{GL}_{n_1} \times \cdots \times \mathrm{GL}_{n_r}) \times \mathrm{GL}_{m_1} \cdots \times \mathrm{GL}_{m_k}$ avec $m_{-i} = m_i$ pour $i = 1, \cdots, k$. L'élément non trivial de $\pi_0(\kappa)$ échange GL_{m_i} et $\mathrm{GL}_{m_{-i}}$ et induit $g \mapsto \Phi_{n_i}{}^t g^{-1} \Phi_{n_i}^{-1}$ sur chaque GL_{n_i}. De plus, le transfert ν est donné par le produit des polynômes unitaires correspondants.

1.6. Données endoscopiques. Une *donnée endoscopique* pour le groupe G est un couple (κ, ρ_κ) avec κ comme ci-dessus et un morphisme $\rho_\kappa : \pi_1(X, x) \longrightarrow \pi_0(\kappa)$ qui induit ρ. Le *groupe endoscopique* associé à la donnée (κ, ρ_κ) est alors la forme de $\mathbb{H}$ associée à l'homorphisme $\pi_1(X, x) \longrightarrow \mathrm{Out}(\mathbb{H})$ déduit de ρ_κ. Il est de la forme $H = \mathbb{H} \times^{\Theta_\rho} X_\rho$ si l'on a pris soin de choisir (X_ρ, x_ρ) (voir plus haut) de sorte que ρ_κ se factorise par Θ_ρ. Le morphisme de transfert $\nu : \mathbb{c}_{\mathbb{H}} \longrightarrow \mathbb{c}$ passe à la torsion par ρ_κ pour donner un transfert

$$\nu : \mathfrak{c}_H \longrightarrow \mathfrak{c}.$$

Exemple : groupes linéaires. Si $G = \mathrm{GL}_n$, les groupes endoscopiques sont isomorphes à des sous-groupes de Levi de G, c'est-à-dire de la forme $H = \prod_i \mathrm{GL}_{n_i}$ avec $\sum_i n_i = n$.

Exemple : groupes unitaires. Soit G le groupe unitaire associé au revêtement X_ρ de degré 2 de X. Pour une donnée endoscopique, on doit avoir $\pi_0(\kappa)$ non trivial, donc égal à $\mathbb{Z}/2\mathbb{Z}$. Avec les notations ci-dessus, le groupe endoscopique H est de la forme $\mathrm{GL}_{m_1} \times \cdots \times \mathrm{GL}_{m_k} \times U_{X_\rho/X}(n_1) \times \cdots \times U_{X_\rho/X}(n_r)$. Un tel groupe est isomorphe à un sous-groupe de Levi si et seulement si $r = 1$ (et éventuellement $n_1 = 0$).

2. Lemme fondamental et fibres de Springer affines

Il s'agit de la partie locale de la théorie. Soit G un groupe du type décrit au paragraphe 1.4 dont on reprend les notations. Fixons un point fermé v de X, notons $\mathcal{O}_v$ son anneau local complété, F_v le corps des fractions de celui-ci et Γ_v un groupe de décomposition en v du groupe de Galois de $F(X)$. Fixons aussi un point

$$a \in \mathfrak{c}(\mathcal{O}_v) \cap \mathfrak{c}^{\mathrm{rs}}(F_v).$$

2.1. Intégrales orbitales. Le point a définit une classe de conjugaison *stable* semi-simple régulière $\chi^{-1}(a)$ de $\mathfrak{g}(F_v)$, ainsi qu'un représentant de Kostant $\gamma_0 = \epsilon(a) \in \mathfrak{g}(F_v)$ de cette classe. Les centralisateurs des éléments de $\chi^{-1}(a)$ sont des tores, tous *canoniquement* isomorphes à celui de γ_0, que l'on note I_{γ_0}. Ils sont aussi canoniquement isomorphes au tore $J_a := \mathbb{T} \wedge^{\mathbb{W} \rtimes \Theta_\rho} \pi_\rho^{-1}(a)$ obtenu en tordant $\mathbb{T}$ par le $\mathbb{W} \rtimes \Theta_\rho$-torseur $\pi_\rho^{-1}(a)$ au-dessus de $\mathrm{Spec}(F_v)$, où $\pi_\rho : \mathfrak{t} \times X_\rho \longrightarrow \mathfrak{c}$ est la projection canonique. Notons qu'un tel torseur est donné par un morphisme

$$\pi_{\rho,a} : \Gamma_v \longrightarrow \mathbb{W} \rtimes \Theta_\rho$$

qui relève la restriction $\rho_{|\Gamma_v} : \Gamma_v \longrightarrow \Theta_\rho$ de ρ à Γ_v.

Si $\gamma \in \chi^{-1}(a)$, le transporteur de γ sur γ_0 dans G est un torseur sous le centralisateur de γ_0 donc définit un torseur sous J_a. L'application ainsi obtenue

$$\begin{aligned}\chi^{-1}(a) &\longrightarrow H^1(F_v, J_a)\\ \gamma &\mapsto \mathrm{inv}(\gamma, \gamma_0)\end{aligned}$$

induit une bijection entre l'ensemble des classes de conjugaison dans $\chi^{-1}(a)$ et le "noyau" $\ker(H^1(F_v, J_a) \longrightarrow H^1(F_v, G))$ qui est le même ensemble apparu dans la formule (3.1) de [**10**]. Le groupe $H^1(F_v, J_a)$ peut se calculer au moyen de la dualité de Tate-Nakayama. Celle-ci fournit un accouplement parfait

$$\langle\, , \rangle : H^1(F_v, J_a) \times \pi_0(\hat{\mathbb{T}}^{\pi_{\rho,a}(\Gamma_v)}) \longrightarrow \mathbb{Q}/\mathbb{Z}. \tag{2.1.1}$$

Étant donné un caractère κ de $H^1(F_v, J_a)$, on définit la κ-intégrale orbitale d'une fonction localement constante f sur $\mathfrak{g}(F_v)$

$$\mathbf{O}_a^\kappa(f, \mathrm{d}t_v) = \sum_{\gamma \in \chi^{-1}(a)/\mathrm{conj}} \langle \kappa, \mathrm{inv}(\gamma, \gamma_0)\rangle \mathbf{O}_\gamma(f, \mathrm{d}t_v),$$

où la somme porte sur les classes de conjugaison dans la classe stable $\chi^{-1}(a)$ et

$$\mathbf{O}_\gamma(f, \mathrm{d}t_v) = \int_{I_\gamma(F_v)\backslash G(F_v)} f\left(\mathrm{ad}(g_v^{-1})\gamma\right) \frac{\mathrm{d}g_v}{\mathrm{d}t_v}$$

désigne l'intégrale orbitale de f en γ pour la mesure de Haar $\mathrm{d}t_v$ sur $J_a(F_v)$ (identifié au centralisateur de γ) et la mesure de Haar $\mathrm{d}g_v$ normalisée par $G(\mathcal{O}_v)$ sur $G(F_v)$.

2.2. Lemme fondamental. Fixons maintenant une donnée endoscopique (κ, ρ_κ), ainsi qu'un élément $a_H \in \mathfrak{c}_H(\mathcal{O}_v)$ d'image a dans $\mathfrak{c}(\mathcal{O}_v)$. Comme ci-dessus, on en déduit un morphisme $\pi_{a_H,\rho} : \Gamma_v \longrightarrow \mathbb{W}_{\mathbb{H}} \rtimes \Theta_\rho$, dont le morphisme précédent $\pi_{a,\rho}$ se déduit par composition avec le morphisme naturel $\mathbb{W}_{\mathbb{H}} \rtimes \Theta_\rho \longrightarrow \mathbb{W} \rtimes \Theta_\rho$ de 1.5. Il s'ensuit que l'isomorphisme $\mathbb{T}_{\mathbb{H}} \xrightarrow{\sim} \mathbb{T}$ induit un isomorphisme $J_{H,a_H} \xrightarrow{\sim} J_a$, et par conséquent des identifications

$$H^1(F_v, J_{H,a_H})^* = H^1(F_v, J_a)^* = \pi_0(\hat{\mathbb{T}}^{\pi_{a,\rho}(\Gamma_v)}) = \pi_0(\hat{\mathbb{T}}^{\pi_{a_H,\rho}(\Gamma_v)}).$$

Par définition, l'élément κ est dans $\hat{\mathbb{T}}^{\pi_{a,\rho}(\Gamma_v)}$ et définit donc un caractère des groupes $H^1(F_v, J_{H,a_H})$ et $H^1(F_v, J_a)$. De plus, tout choix de mesure de Haar $\mathrm{d}t_v$ sur $J_a(F_v)$ en détermine un sur $J_{H,a_H}(F_v)$.

Le lemme fondamental est l'égalité suivante :

$$\mathbf{O}_a^\kappa(1_{\mathfrak{g}(\mathcal{O}_v)}, \mathrm{d}t_v) = q^{r_{H,v}^G(a_H)} \mathbf{O}_{a_H}^\kappa(1_{\mathfrak{h}(\mathcal{O}_v)}, \mathrm{d}t_v)$$

où $r_{H,v}^G(a_H) = \deg_v(a_H^* \mathfrak{R}_H^G)$.

On notera que κ est dans le centre de $\hat{\mathbb{H}}$, donc le caractère qu'il définit sur $H^1(F_v, J_{H,a_H})$ est trivial sur le noyau de $H^1(F_v, J_{H,a_H}) \longrightarrow H^1(F_v, H)$. En d'autres termes, pour tout γ_H dans $\chi_H^{-1}(a_H)$, on a

$$\langle \kappa, \mathrm{inv}(\gamma_H, \gamma_{H,0}) \rangle = 1.$$

Le membre de droite de l'égalité ci-dessus est donc l'*intégrale orbitale stable* en a_H.

2.3. Fibres de Springer affines. Notons $X_v := \mathrm{Spec}(\mathcal{O}_v)$ le disque formel en v et $X_v^\bullet := \mathrm{Spec}(F_v)$ le disque épointé. Pour un k-schéma S, on note $X_v \hat{\times} S$ la complétion v-adique de $X_v \times S$ et $X_v^\bullet \hat{\times} S$ le complémentaire de $\{v\} \times S$. Enfin, notons E_0 le G-torseur trivial sur $X_v^\bullet$.

La fibre de Springer affine $\mathcal{M}_v(a)$ associée à a peut se définir comme le champ qui à S associe le groupoïde

$$\mathcal{M}_v(a,S) := \left\{ (E,\varphi,\iota), \begin{array}{l} E \text{ un } G\text{-torseur sur } X_v\hat{\times}S, \\ \varphi \text{ une section de } \mathfrak{g} \wedge^G E \text{ sur } X_v\hat{\times}S \\ \iota \text{ un isomorphisme } (E,\varphi)_{|X_v^\bullet\hat{\times}S} \xrightarrow{\sim} (E_0,\gamma_0) = (E_0,\epsilon(a)) \end{array} \right\}.$$

Ce groupoïde est en réalité discret, *i.e.* équivalent à l'ensemble de ses classes d'isomorphisme. En particulier, on a

$$\mathcal{M}_v(a,k) \simeq \{g \in G(F_v)/G(\mathcal{O}_v),\ \mathrm{ad}(g)^{-1}(\gamma_0) \in \mathfrak{g}(\mathcal{O}_v)\}.$$

On montre que $\mathcal{M}_v(a)$ est représentable par un ind-schéma sur k.

Soit (E,φ,ι) un triplet comme ci-dessus. Le morphisme $\chi^* J \longrightarrow I$ induit un morphisme $J_a \longrightarrow \underline{\mathrm{Aut}}(E,\varphi)$, ce qui permet de tordre (E,φ) par tout J_a-torseur. On en déduit une action sur $\mathcal{M}_v(a)$ du champ de Picard (ou "champ en groupes abéliens") $\mathcal{P}_v(a)$ qui à S associe le groupoïde

$$\mathcal{P}_v(a,S) := \left\{ (F,\iota), \begin{array}{l} F \text{ un } J_a\text{-torseur sur } X_v\hat{\times}S, \\ \iota \text{ une trivialisation de } F \text{ sur } X_v^\bullet\hat{\times}S \end{array} \right\}.$$

Ici encore, ce groupoïde est équivalent à l'ensemble de ses classes d'isomorphisme, lequel est un groupe abélien. Par exemple,

$$\mathcal{P}_v(a,k) \simeq J_a(F_v)/J_a(\mathcal{O}_v) \simeq I_{\gamma_0}(F_v)/I_{\gamma_0}(\mathcal{O}_v)$$

agit de la manière évidente sur $\mathcal{M}_v(a,k)$. On montre que $\mathcal{P}_v(a)$ est représentable par un ind-schéma en groupes sur k.

Exemple : groupes linéaires et unitaires. Pour $G = \mathrm{GL}_n$, $\mathcal{M}_v(a,k)$ s'identifie à l'ensemble des $\mathcal{O}_v$-réseaux γ_0-stables de F_v^n. Pour un groupe unitaire, $\mathcal{M}_v(a,k)$ s'identifie à l'ensemble des $\mathcal{O}(X_v \times_X X_\rho)$-réseaux autoduaux dans $(F_v \otimes_{F(X)} F(X_\rho))^n$.

2.4. Quotients naïfs et intégrales orbitales. Rappelons que si M est un ensemble sur lequel agit un groupe P, on peut former le groupoïde quotient $[M/P]$ dont les objets sont les éléments de M et les morphismes sont donnés par l'action de P. Lorsque le quotient M/P est fini et les fixateurs des éléments de M dans P sont finis, alors on définit le cardinal de $[M/P]$ par

$$\sharp[M/P] := \sum_{m \in M/P} \frac{1}{\sharp \mathrm{Aut}_{[M/P]}(m)}.$$

On remarque alors d'après le paragraphe précédent que

$$\sharp[\mathcal{M}_v(a,k)/\mathcal{P}_v(a,k)] = \mathbf{O}_{\gamma_0}(1_{\mathfrak{g}(\mathcal{O}_v)}, \mathrm{d}t_v)$$

pour la mesure de Haar normalisée par $J_a(\mathcal{O}_v)$. Ainsi, *l'intégrale orbitale admet une interprétation naturelle en termes de la géométrie.*

2.5. Quotients champêtres et κ-intégrales orbitales. Supposons maintenant que M est un k-schéma sur lequel agit un k-schéma en groupes. On peut alors former le champ quotient $[M/P]$, comme le champ "associé" au pré-champ $S \mapsto [M(S)/P(S)]$. Le groupoïde $[M/P](k)$ admet deux descriptions utiles :

$$\begin{aligned}[M/P](k) &= \{(Q,j),\ Q \text{ un } P\text{-torseur sur } k \text{ et } j : Q \longrightarrow M \text{ un } k\text{-}P\text{-morphisme}\} \\ &= \{(m,p) \in M(\bar{k}) \times P(\bar{k}),\ p.\sigma(m) = m\}.\end{aligned}$$

Dans la seconde, σ désigne le Frobenius de $\bar{k}$ et les morphismes $(m,p) \longrightarrow (m',p')$ sont les éléments $h \in P(\bar{k})$ tels que $m' = hm$ et $p' = hp\sigma(h)^{-1}$. On passe à la première description en associant à (m,p) la paire (Q,j) obtenue en descendant le morphisme $\bar{j} : a \in P_{\bar{k}} \longrightarrow a.m \in M_{\bar{k}}$ pour le Frobenius tordu $\sigma_p : a \mapsto \sigma(a)p^{-1}$ sur $P_{\bar{k}}$. On voit bien sur la première description qu'on a une application

$$\mathrm{cl} : [M/P](k) \longrightarrow H^1(k,P).$$

Etant donné un caractère κ de $H^1(k,P)$, et en supposant que l'ensemble des classes d'isomorphismes et les groupes d'automorphismes sont finis, on définit alors le comptage κ-pondéré par la formule

$$\sharp[M/P](k)_\kappa = \sum_{x \in [M/P](k)/\sim} \frac{(\mathrm{cl}(x),\kappa)}{\sharp\, \mathrm{Aut}(x)}$$

L'hypothèse de finitude est vérifiée dans le cas de l'action de $\mathcal{P}_v(a)$ sur $\mathcal{M}_v(a)$, cf paragraphe 4.1. Par un théorème de Steinberg donnant le premier isomorphisme ci-dessous, on a un morphisme

$$\begin{aligned} H^1(F_v, J_a) &\simeq H^1(\mathrm{Gal}(\bar{k}/k), J_a(F_v \otimes_k \bar{k})) \\ &\longrightarrow H^1(\mathrm{Gal}(\bar{k}/k), J_a(F_v \otimes_k \bar{k})/J_a(\mathcal{O}_v \otimes_k \bar{k})) = H^1(k, \mathcal{P}_v(a)),\end{aligned}$$

qui d'ailleurs est un isomorphisme si la fibre spéciale de J_a est connexe.

Ainsi un caractère κ de $H^1(k,\mathcal{P}_v(a))$ induit un caractère encore noté κ de $H^1(F_v,J_a)$. Le *lien entre κ-intégrales orbitales et fibres de Springer affines* est alors donné par la proposition suivante, cf. [**19**, proposition 8.2.7], qui est une élaboration des travaux de Kottwitz, Goresky et MacPherson [**9**].

PROPOSITION 2.6. *Avec les notations ci-dessus on a*

$$\sharp[\mathcal{M}_v(a)/\mathcal{P}_v(a)](k)_\kappa = \mathbf{O}_a^\kappa(1_{\mathfrak{g}(\mathcal{O}_v)}, \mathrm{d}\,t_v),$$

où $\mathrm{d}\,t_v$ *la mesure de Haar normalisée par* $J_a^0(\mathcal{O}_v)$ *avec* J_a^0 *la composante neutre de* J_a. *De plus, si* κ' *est un caractère de* $\mathrm{H}^1(F_v, J_a)$ *qui ne provient pas d'un caractère de* $\mathrm{H}^1(k, \mathcal{P}_v(a))$, *alors la* κ'*-intégrale orbitale* $\mathbf{O}_a^{\kappa'}(1_{\mathfrak{g}_v}, \mathrm{d}\,t_v)$ *est nulle.*

Ce résultat est plus facile à prouver lorsque J_a est à fibres connexes. Si ce n'est pas le cas, on prouve d'abord un résultat pour un modèle connexe de J_{a,F_v} sur $\mathcal{O}_v$, *i.e.* un schéma en groupes lisse abélien à fibres connexes J'_a sur $\mathcal{O}_v$ muni d'un morphisme $J'_a \longrightarrow J_a$ qui est un isomorphisme en fibres génériques. Pour un tel J'_a, on peut considérer le champ $\mathcal{P}_v(J'_a)$ des J'_a-torseurs trivialisés sur F_v, lequel agit encore sur $\mathcal{M}_v(a)$. Cette fois on a un isomorphisme $H^1(F_v, J_a) \xrightarrow{\sim} H^1(k, \mathcal{P}_v(J'_a))$. La formule obtenue, cf. [**19**, proposition 8.2.5], devient

PROPOSITION 2.7. *Avec les notations ci-dessus, on a*

$$\sharp[\mathcal{M}_v(a)/\mathcal{P}_v(J'_a)](k)_\kappa = \mathrm{vol}(J'_a(\mathcal{O}_v), \mathrm{d}\,t_v) \cdot \mathbf{O}_a^\kappa(1_{\mathfrak{g}(\mathcal{O}_v)}, \mathrm{d}\,t_v),$$

où $\mathrm{d}\,t_v$ *est n'importe quelle mesure de Haar de* $J_a(F_v)$.

Rappelons que le discriminant $\mathfrak{D}$ est une fonction régulière sur $\mathfrak{c}$. Dans le cas où $v(\mathfrak{D}(a)) \leq 1$, on peut calculer explicitement les comptages $\sharp[\mathcal{M}_v(a)/\mathcal{P}_v(J'_a)](k)_\kappa$ et $\sharp[\mathcal{M}_v(a)/\mathcal{P}_v(J_a)](k)_\kappa$, cf. [**19**, section 8.3].

3. Fibration de Hitchin

Il s'agit de la partie globale de la théorie. Ici encore, G désigne un groupe du type décrit au paragraphe 1.4.

3.1. Champ de Hitchin. Commençons par définir la notion de paire de Higgs. Soit S un k-schéma; on appelle une paire de Higgs sur S la donnée d'un couple (E, φ) où E est un G-torseur sur le produit fibré $X \times S$ et φ est une section globale sur $X \times S$ du fibré vectoriel $\mathrm{ad}(E) \otimes_{\mathcal{O}_X} \mathcal{O}_X(D)$. Ici $\mathrm{ad}(E)$ désigne le fibré vectoriel $\mathfrak{g} \times^G E$ tordu de $\mathfrak{g}$ par le G-torseur E pour l'action adjointe de G sur $\mathfrak{g}$. Le torseur E sera appelé le fibré de Higgs et la section φ sera appelée le champ de Higgs.

Introduisons maintenant le champ de Hitchin $\mathcal{M}$ classifiant les paires de Higgs. Soit S un k-schéma; $\mathcal{M}(S)$ est la catégorie dont les objets sont des paires de Higgs sur S et les morphismes sont des isomorphismes entre ces paires. Soit $S' \longrightarrow S$ un morphisme entre deux k-schémas, le pull-back induit un foncteur $\mathcal{M}(S) \longrightarrow \mathcal{M}(S')$. Ainsi la collection des groupoïdes $\mathcal{M}(S)$ lorsque S parcourt la catégorie des k-schémas définit un champ $\mathcal{M}$ appelé le champ de Hitchin.

Dans le langage des champs algébriques, on peut réinterpréter cette définition comme suit. Soit $\mathbb{B}G$ (resp. $\mathbb{B}\mathbb{G}_m$) le champ algébrique classifiant les G-torseurs sur k (resp. les fibrés inversibles sur k). Ainsi pour tout k-schéma S, $\mathbb{B}G(S)$ est la catégorie dont les objets sont les G-torseurs sur S et les morphismes sont les isomorphismes entre de tels torseurs. On note $h_D : X \longrightarrow \mathbb{B}\mathbb{G}_m$ le morphisme qui correspond au fibré inversible $\mathcal{O}_X(D)$.

Soit S un k-schéma. La donnée d'un fibré de Higgs E sur $X \times S$ est équivalente à celle d'une flèche

$$h_E : X \times S \longrightarrow \mathbb{B}G.$$

Puis la donnée d'un champ de Higgs φ est équivalente à celle d'une flèche

$$h_{E,\varphi} : X \times S \longrightarrow [\mathfrak{g}/G \times \mathbb{G}_m]$$

au-dessus de $h_E \times h_D : X \times S \longrightarrow \mathbb{B}G \times \mathbb{B}\mathbb{G}_m$.

On définit l'espace de Hitchin $\mathcal{A}$ comme le foncteur qui associe à tout schéma S l'ensemble des flèches $X \times S \longrightarrow [\mathfrak{c}/\mathbb{G}_m]$ au-dessus de $h_D : X \times S \longrightarrow \mathbb{B}\mathbb{G}_m$. On voit aisément que $\mathcal{A}$ est l'espace des sections globales sur X du fibré vectoriel $\mathfrak{c}_D = \mathfrak{c} \otimes_{\mathcal{O}_X} \mathcal{O}_X(D)$, en particulier, il est un schéma.

Définissons maintenant le morphisme de Hitchin. La façon la plus simple est d'utiliser la description champêtre des paires de Higgs. Soit S un schéma; un S-point de $\mathcal{M}$ consiste en la donnée de deux flèches $h_E : X \times S \longrightarrow \mathbb{B}G$ et $h_{E,\varphi} : X \times S \longrightarrow [\mathfrak{g}/G \times \mathbb{G}_m]$ telles que le diagramme naturel

$$\begin{array}{ccc} X \times S & \xrightarrow{h_{E,\varphi}} & [\mathfrak{g}/G \times \mathbb{G}_m] \\ & \searrow^{h_E \times h_D} & \downarrow \\ & & \mathbb{B}G \times \mathbb{B}\mathbb{G}_m \end{array}$$

soit commutatif. En composant $h_{E,\varphi}$ avec le morphisme de Chevalley $\chi : [\mathfrak{g}/G \times \mathbb{G}_m] \longrightarrow [\mathfrak{c}/\mathbb{G}_m]$, on obtient un S-point de $\mathcal{A}$. Le morphisme que l'on vient de définir est appelé le morphisme caractéristique de Hitchin et il est noté:

$$m : \mathcal{M} \longrightarrow \mathcal{A}.$$

Pour tout point $a \in \mathcal{A}$, on notera $\mathcal{M}_a$ la fibre $m^{-1}(a)$ au dessus de a et on considérera $\mathcal{M}$ comme une famille des fibres $\mathcal{M}_a$ lorsque a varie.

Introduisons quelques sous-schémas ouverts de $\mathcal{A}$ sur lesquels on va travailler dans la suite. On a déjà identifié $\mathcal{A}$ à l'espace des sections globales sur X du fibré vectoriel $\mathfrak{c}_D = \mathfrak{c} \otimes_{\mathcal{O}_X} \mathcal{O}_X(D)$. Avec cette interprétation, on identifie $\mathcal{A}(\bar{k})$ à un sous ensemble de $\mathfrak{c}(F \otimes_k \bar{k})$ en envoyant le point a à la valeur au point générique de la section globale qui correspond à a. Sous cette identification, on définit le sous-schéma $\mathcal{A}^{\heartsuit}$ (resp. $\mathcal{A}^{\mathrm{ani}}$) de manière suivante: un $\bar{k}$-point de $\mathcal{A}$ est dans $\mathcal{A}^{\heartsuit}$ (resp. $\mathcal{A}^{\mathrm{ani}}$) si et seulement si, étant considéré comme un élément de $\mathfrak{c}(F \otimes_k \bar{k})$, il est régulier semisimple (resp. régulier semisimple et elliptique). On vérifie aisément que ce sont des sous-schémas ouverts de $\mathcal{A}$.

Pour tout point géométrique $a \in \mathcal{A}(\bar{k})$ considéré comme un morphisme $a : \bar{X} \longrightarrow \mathfrak{c}_D$, on définit U_a comme l'image inverse dans $\bar{X}$ de $\mathfrak{c}_D^{\mathrm{rs}}$; lorsque $a \in \mathcal{A}^{\heartsuit}(\bar{k})$, c'est un ouvert dense de $\bar{X}$.

Suivant la terminologie de Donagi et Gaitsgory [**7**], on introduit le revêtement caméral $\tilde{X}$ en formant le diagramme cartésien

$$\begin{array}{ccc} \tilde{X} & \longrightarrow & \mathfrak{t}_D \\ \downarrow & & \downarrow \\ X \times \mathcal{A} & \longrightarrow & \mathfrak{c}_D. \end{array}$$

Soit $a \in \mathcal{A}(\bar{k})$ un $\bar{k}$-point de $\mathcal{A}$, on obtient un revêtement dit caméral $\tilde{X}_a$ de $\bar{X}$. On vérifie facilement que si a est régulier, alors le revêtement $\tilde{X}_a \longrightarrow \bar{X}$ est génériquement un torseur sous W et la courbe $\tilde{X}_a$ est réduite.

On notera $\mathcal{M}^{\heartsuit}$ (resp. $\mathcal{M}^{\mathrm{ani}}$) le sous-champ ouvert de $\mathcal{M}$ défini comme l'image inverse de $\mathcal{A}^{\heartsuit}$ (resp. $\mathcal{A}^{\mathrm{ani}}$) via le morphisme de Hitchin m. Les propriétés importantes de ces sous-champs peuvent être résumées dans le théorème suivant qui est dû à Biswas-Ramanan [**3**] et Faltings [**8**].

THÉORÈME 3.2. *Avec les notations ci-dessus,*

i) $\mathcal{M}^{\heartsuit}$ *est un champ lisse.*

ii) Supposons que G est semisimple, alors le morphisme de Hitchin $\mathcal{M}^{\mathrm{ani}} \longrightarrow \mathcal{A}^{\mathrm{ani}}$ au-dessus de l'ouvert $\mathcal{A}^{\mathrm{ani}}$ est propre. De plus, le champ $\mathcal{M}^{\mathrm{ani}}$ est de Deligne-Mumford.

Exemple : $G = \mathrm{GL}_n$. La donnée d'une paire de Higgs (E, φ) sur un schéma S est équivalente à la donnée d'un fibré vectoriel E de rang n sur $X \times S$ muni d'un endomorphisme tordu $\varphi : E \longrightarrow E \otimes \mathcal{O}(D)$. L'espace affine de Hitchin est alors simplement

$$\mathcal{A} = \bigoplus_{1 \leq i \leq n} \mathrm{H}^0(X, \mathcal{O}_X(iD))$$

et le morphisme caractéristique de Hitchin m envoie une paire (E, φ) sur le point

$$(\mathrm{tr}(\varphi), \mathrm{tr}(\wedge^2 \varphi), \ldots, \mathrm{tr}(\wedge^n \varphi)) \in \mathcal{A}$$

où $\mathrm{tr}(\wedge^i \varphi))$ est la trace du morphisme $\wedge^i \varphi : \wedge^i E \longrightarrow \wedge^i E \otimes \mathcal{O}_X(iD)$.

Soit $a = (a_1, a_2, \ldots, a_n)$ un $\bar{k}$-point de $\mathcal{A}$ qui correspond à une collection de sections globales des fibrés vectoriels $a_i \in \mathrm{H}^0(\bar{X}, \mathcal{O}_{\bar{X}}(iD))$. En évaluant ces sections au point générique de $\bar{X}$, on obtient un polynôme $P_a(t) = t^n - a_1 t^{n-1} + \ldots + (-1)^n a_n$ à coefficients dans $F \otimes_k \bar{k}$. On voit que a est régulier (resp. régulier elliptique) si et seulement si P_a n'a que des racines simples (resp. est irréductible).

Exemple : groupes unitaires. Si G est un groupe unitaire associé à un revêtement X_ρ de degré 2, le champ de Hitchin classifie les "fibrés hermitiens" munis d'un "endomorphisme hermitien tordu". Le morphisme de Hitchin est encore donné par l'application polynôme caractéristique.

3.3. Groupes des symétries. On renvoie à la section 1.3 pour la définition du centralisateur régulier J au-dessus de $\mathfrak{c}$. Comme J est $\mathbb{G}_m$-équivariant, on peut descendre J en un schéma en groupes commutatifs lisse sur le champ quotient $[\mathfrak{c}/\mathbb{G}_m]$. On note celui-ci encore J et on l'appelle toujours le schéma en groupes des centralisateurs réguliers.

Soit a un S-point de $\mathcal{A}$; c'est équivalent à donner une flèche $a : X\times S \longrightarrow [\mathfrak{c}/\mathbb{G}_m]$ au-dessus de $h_D : X \times S \longrightarrow B\mathbb{G}_m$. En prenant l'image inverse de J, on obtient un schéma en groupes commutatifs $J_a = a^*J$ sur $X \times S$ et on introduit $\mathcal{P}_a$ le champ classifiant des J_a-torseurs sur $X\times S$. Soit $(E, \varphi) \in \mathcal{M}_a$; la flèche $\chi^*J \longrightarrow I$ sur $\mathfrak{g}$ induit une flèche

$$\chi^* J_a \longrightarrow \mathrm{Aut}(E, \varphi).$$

Ainsi, on peut tordre cette paire par tout J_a-torseur et on obtient donc une action de $\mathcal{P}_a$ sur $\mathcal{M}_a$. Lorsque l'on varie a, $\mathcal{P}_a$ s'organise en famille et définit un champ $\mathcal{P}$ au-dessus de $\mathcal{A}$. C'est un champ algébrique qui agit sur le champ de Hitchin $\mathcal{M}$. On notera $\mathcal{P}^{\heartsuit}$ (resp. $\mathcal{P}^{\mathrm{ani}}$) la restriction de $\mathcal{P}$ au-dessus de l'ouvert $\mathcal{A}^{\heartsuit}$ (resp. $\mathcal{A}^{\mathrm{ani}}$).

On montre que $\mathcal{P}^{\heartsuit}$ est un champ relativement lisse au-dessus de $\mathcal{A}^{\heartsuit}$. On peut alors définir le faisceau en groupes commutatifs $\pi_0(\mathcal{P})$ des composantes connexes de $\mathcal{P}^{\heartsuit}$ pour la topologie étale sur $\mathcal{A}^{\heartsuit}$. Sa fibre en un point géométrique $a \in \mathcal{A}^{\heartsuit}(\bar{k})$ est le groupe des composantes connexes de $\mathcal{P}_a$:

$$[\pi_0(\mathcal{P})]_a = \pi_0(\mathcal{P}_a).$$

Le lien entre l'action de $\mathcal{P}_a$ et l'endoscopie viendra du résultat suivant, dont la preuve utilise d'ailleurs une technique de Kottwitz classique pour traduire des résultats de cohomologie Galoisienne en termes de L-groupes [**11**].

Proposition 3.4. [**17**, Prop. 6.8] *Il existe un homomorphisme canonique*

$$\mathbb{X}_*T \times^W \mathrm{Irr}(\tilde{X}_a) \to \pi_0(\mathcal{P}_a)$$

et c'est un isomophisme si G est adjoint. Cet homomorphisme s'organise en famille. De plus, si G est semisimple, a est dans $\mathcal{A}^{\mathrm{ani}}(\bar{k})$ si et seulement si $\pi_0(\mathcal{P}_a)$ est fini.

Exemple : $G = \mathrm{GL}_n$. Reprenons les notations de l'exemple du paragraphe précédent. Fixons une caractéristique $a \in \mathcal{A}(\bar{k})$ régulière, autrement dit $a \in \mathcal{A}^{\heartsuit}(\bar{k})$. Suivant Beauville, Narasimhan et Ramanan [**2**], on définit la courbe spectrale Y_a associée à a; c'est une courbe tracée sur l'espace total Σ_D du fibré inversible $\mathcal{O}_{\bar{X}}(D)$ donnée par l'équation

$$t^n - a_1 t^{n-1} + \ldots + (-1)^n a_n = 0.$$

C'est un revêtement fini de $\bar{X}$ qui est génériquement étale Galoisien de groupe de Galois $W = \mathfrak{S}_n$. On voit bien que la condition que a est régulier (resp. régulier elliptique) entraîne que la courbe Y_a est réduite (resp. irréductible). Vue la forme explicite de J_a donnée dans l'exemple du paragraphe 1.3, on voit aussi que le groupe des symétries $\mathcal{P}_a$ est le champ de Picard de Y_a qui classifie les fibrés inversibles sur Y_a. Beauville, Narasimhan et Ramanan ont montré que la fibre $\mathcal{M}_a$ peut se voir comme le champ classifiant les faisceaux cohérents sans torsion de rang générique 1 sur Y_a. Avec ces identifications, l'action de $\mathcal{P}_a$ sur $\mathcal{M}_a$ est donnée par le produit tensoriel $\mathcal{L} \otimes \mathcal{F}$ d'un fibré inversible $\mathcal{L}$ sur Y_a avec un faisceau cohérent sans torsion $\mathcal{F}$ de rang générique 1. Notons que dans le cas $G = \mathrm{GL}_1$, on retrouve la théorie classique des variétés de Picard.

Ces constructions se généralisent aux groupes unitaires, et plus généralement aux groupes classiques.

3.5. Formule de produit. On veut établir un lien entre le contexte global et le contexte local. On se donne un $\bar{k}$-point a de $\mathcal{A}$; on le considère aussi comme un morphisme $a : \bar{X} \longrightarrow [\mathfrak{c}/\mathbb{G}_m]$. Rappelons qu'on a déjà défini U_a comme l'ouvert de $\bar{X}$ des points réguliers défini par a. Puis on considère le champ quotient $[\mathcal{M}_a/\mathcal{P}_a]$. A priori, c'est une 2-catégorie mais on vérifie aisément que c'est en fait une 1-catégorie. Soit S un $\bar{k}$-schéma; $[\mathcal{M}_a/\mathcal{P}_a](S)$ est le groupoide dont les objets sont des paires de Higgs (E, φ) de caractéristique a et les morphismes entre deux telles paires (E, φ) et (E', φ') consistent en un élément $p \in \mathcal{P}_a(S)$ et un isomorphisme

$$p \circ (E, \varphi) \xrightarrow{\sim} (E', \varphi').$$

PROPOSITION 3.6 (Formule de produit). *Avec les notations ci-dessus, on a une équivalence de catégories:*

$$[\mathcal{M}_a/\mathcal{P}_a] = \prod_{v \in \bar{X} - U_a} [\mathcal{M}_v(a)/\mathcal{P}_v(a)].$$

On se content d'expliquer le foncteur de la catégorie $\prod_{v\in\bar{X}-U_a}[\mathcal{M}_v(a)/\mathcal{P}_v(a)]$ vers la catégorie $[\mathcal{M}_a/\mathcal{P}_a]$. Soit $(x_v)_{v\in\bar{X}-U_a}$ une famille d'éléments de $(\mathcal{M}_v(a))_{v\in\bar{X}-U_a}$. On considère la paire de Higgs (E, φ) sur $\bar{X}$ définie par la section de Kostant au-dessus de a. L'isomorphisme sur le disque pointé $X_v^\bullet = \bar{X} \times_{\bar{k}} \bar{F}_v$ permet de recoller la restriction de (E, φ) sur U_a avec les données locales $(x_v)_{v\in\bar{X}-U_a}$ pour obtenir une nouvelle paire de Higgs (E', φ') sur $\bar{X}$. Ainsi on obtient un foncteur $\prod_{v\in\bar{X}-U_a} \mathcal{M}_v(a) \longrightarrow \mathcal{M}_a$. Il est compatible avec l'action de $\prod_{v\in\bar{X}-U_a} \mathcal{P}_v(a)$ et de $\mathcal{P}_a$, donc on obtient un foncteur $\prod_{v\in\bar{X}-U_a}[\mathcal{M}_v(a)/\mathcal{P}_v(a)] \longrightarrow [\mathcal{M}_a/\mathcal{P}_a]$. Une démonstration du fait que ce foncteur est pleinement fidèle et essentiellement surjectif se trouve dans [**17**, Théorème 4.6][1].

Rappelons que du côté de la formule des traces, pour pouvoir exprimer l'intégrale orbitale globale comme un produit des intégrales orbitales locales, on a besoin d'extraire le facteur de volume. La formule ci-dessus est l'interprétation géométrique de cette opération.

[1]On remarquera que notre notation est ici celle de [**19**]. Dans [**17**], ce que nous notons $\mathcal{M}_v(a)$ est noté $\mathcal{M}_{a,v}^\bullet$

3.7. Invariant δ. Soit $a \in \mathcal{A}^{\heartsuit}(\bar{k})$; on le considère comme un morphisme $a : \bar{X} \longrightarrow \mathfrak{c}_D$. Rappelons que U_a est l'ouvert de $\bar{X}$ des points réguliers défini par a. Au-dessus de U_a, le schéma en groupes commutatifs lisse J_a est un tore. En dehors de U_a, il a aussi un facteur unipotent. D'après Bosch, Lutkebohmert et Raynaud [**4**], il existe un modèle de Néron $J_a^{\flat}$ de J_a sur $\bar{X}$. C'est un schéma en groupes commutatifs lisse sur $\bar{X}$ muni d'un homomorphisme $J_a \longrightarrow J_a^{\flat}$ qui est un isomorphisme sur U_a et qui est maximal pour ces propriétés. Ceci signifie que pour tout homomorphisme $J_a \longrightarrow J'_a$ qui est un isomorphisme sur U_a avec J'_a un schéma en groupes commutatifs lisse sur $\bar{X}$, il existe un homomorphisme $J'_a \longrightarrow J_a^{\flat}$ rendant commutatif le diagramme évident. Le modèle de Néron $J_a^{\flat}$ peut être construit explicitement à partir de la courbe camérale $\tilde{X}_a$, cf. [**19**, corollaire 4.8.1]. On notera $\mathcal{P}_a^{\flat}$ le champ classifiant les $J_a^{\flat}$-torseurs sur $\bar{X}$. L'homomorphisme $J_a \longrightarrow J_a^{\flat}$ induit un homomorphisme $\mathcal{P}_a \longrightarrow \mathcal{P}_a^{\flat}$. Ngô montre:

Proposition 3.8. [**19**, Prop 4.8.2]
i) L'homomorphisme $\mathcal{P}_a(\bar{k}) \longrightarrow \mathcal{P}_a^{\flat}(\bar{k})$ est surjectif.
ii) Le composante connexe de $\mathcal{P}_a^{\flat}$ est un champ abélien.
iii) Le noyau $\mathcal{R}_a$ de $\mathcal{P}_a \longrightarrow \mathcal{P}_a^{\flat}$ est un groupe algébrique affine qui peut s'écrire comme un produit des groupes algébriques affines locaux.

L'invariant δ_a mesure la dimension de la partie affine de $\mathcal{P}_a$:

$$\delta_a := \dim \mathcal{R}_a.$$

Exemple : $G = \mathrm{GL}_n$. Reprenons les notations de l'exemple du paragraphe précédent. On considère $\xi : Y_a^{\flat} \longrightarrow Y_a$ la normalisation de la courbe spectrale Y_a. La flèche $\mathcal{L} \mapsto \xi^*\mathcal{L}$ qui associe à tout fibré inversible sur Y_a son pull-back sur $Y_a^{\flat}$ définit un homomorphisme $\xi^* : \operatorname{Pic} Y_a \longrightarrow \operatorname{Pic} Y_a^{\flat}$. On considère la suite exacte:

$$0 \longrightarrow \mathcal{O}_{Y_a}^{\times} \longrightarrow \xi_*\mathcal{O}_{Y_a^{\flat}}^{\times} \longrightarrow \xi_*\mathcal{O}_{Y_a^{\flat}}^{\times}/\mathcal{O}_{Y_a}^{\times} \longrightarrow 0.$$

La suite exacte longue de cohomologie implique que ξ^* est essentiellement surjectif, c'est-à-dire il est surjectif au niveau des $\bar{k}$-points. On voit bien que le noyau de ξ^* consiste en les fibrés inversibles sur Y_a dont le pull-back sur $Y_a^{\flat}$ est isomorphe au fibré trivial $\mathcal{O}_{Y_a^{\flat}}$. L'invariant δ_a est par définition l'entier $\dim \mathrm{H}^0(Y_a, \xi_*\mathcal{O}_{Y_a^{\flat}}/\mathcal{O}_{Y_a})$. Il mesure la dimension du groupe affine $\ker \xi^*$. Puis on introduit le groupe des composantes connexes $\pi_0(Y_a)$. Le morphisme $\xi^* : \operatorname{Pic} Y_a \longrightarrow \operatorname{Pic} Y_a^{\flat}$ induit un homomorphisme $\pi_0(\operatorname{Pic} Y_a) \longrightarrow \pi_0(\operatorname{Pic} Y_a^{\flat}) = \mathbb{Z}^{\pi_0(Y_a^{\flat})}$.

3.9. Normalisation en famille. On construit une stratification de $\mathcal{A}$ par normalisation en famille. Par définition, étant donné un morphisme propre, plat, de fibres réduites de dimension un $Y \longrightarrow S$, une normalisation en famille $Y' \longrightarrow Y$ est un morphisme propre birationnel qui est un isomorphisme sur un ouvert U de Y dense dans chaque fibre de Y au-dessus de S et tel que le morphisme composé $Y' \longrightarrow S$ soit un morphisme propre et lisse.

Ngô montre qu'il existe une stratification

$$\mathcal{A}^{\heartsuit} \times_k \bar{k} = \bigsqcup_{\psi \in \Psi} \mathcal{A}_\psi$$

de $\mathcal{A}^{\heartsuit} \times_k \bar{k}$ par des sous-schémas localements fermés $\mathcal{A}_\psi$ indexés par un ensemble fini d'indices Ψ qui vérifie les conditions suivantes:

i) Au-dessus de chaque strate $\mathcal{A}_\psi$, la courbe camérale $\tilde{X}_\psi$ admet une normalisation en famille après un changement de base radiciel.

ii) Pour tout $\bar{k}$-schéma S connexe réduit qui est muni d'un morphisme $s : S \longrightarrow \mathcal{A} \times_k \bar{k}$ tel que la restriction de la courbe camérale à S admette une normalisation en famille, alors le morphisme s se factorise par l'une des strates $\mathcal{A}_\psi$.

On utilise la relation d'ordre suivante sur Ψ: soient ψ et ψ' deux éléments de Ψ, on dit que $\psi \leq \psi'$ si et seulement si $\mathcal{A}_\psi$ est contenue dans l'adhérence de $\mathcal{A}_{\psi'}$. On vérifie que l'invariant δ_a reste inchangé dans une normalisation en famille de base connexe. Ainsi on dispose d'une application $\delta : \Psi \longrightarrow \mathbb{N}$; par rapport à la relation d'ordre sur Ψ, c'est une fonction décroissante. On définit alors une stratification de $\mathcal{A}^{\heartsuit} \times_k \bar{k}$ par l'invariant δ en regroupant tous les strates de même ordre:

$$\mathcal{A}_\delta = \bigsqcup_{\delta(\psi)=\delta} \mathcal{A}_\psi.$$

L'estimation suivante joue un rôle clef dans la preuve de la version forte du théorème de support.

Proposition 3.10. *Si le degré du diviseur D est très grand par rapport à δ, alors la strate $\mathcal{A}_\delta$ est de codimension supérieure ou égale à δ.*

On notera $\mathcal{A}^{\text{good}}$ l'ouvert qui contient toutes les strates $\mathcal{A}_\delta$ qui vérifient $\operatorname{codim} \mathcal{A}_\delta \geq \delta$. Lorsque D est assez grand, c'est un ouvert non vide de $\mathcal{A}^{\heartsuit}$. Son complémentaire dans $\mathcal{A}^{\heartsuit}$ est noté $\mathcal{A}^{\text{bad}}$.

Remarquons aussi que, sur chaque strate, le groupe des composantes connexes $\pi_0(\mathcal{P}_a)$ reste constant. On montre que si G est semisimple, alors l'ensemble des strates telles que, pour tout point géométrique a dans cette strate, le groupe des composantes connexes soit fini forme l'ouvert non vide $\mathcal{A}^{\text{ani}}$ de $\mathcal{A}^{\heartsuit}$.

Exemple : $G = \mathrm{GL}_n$. Reprenons les notations des exemples des paragraphes précédents. On considère le foncteur $\mathcal{B}$ qui à tout schéma S associe l'ensemble $\mathcal{B}(S)$ des triplets $(a, Y_a^\flat, \xi)$ où:

i) $a : S \longrightarrow \mathcal{A}^{\heartsuit}$ est un S-point de $\mathcal{A}^{\heartsuit}$,

ii) $\xi : Y_a^\flat \longrightarrow Y_a$ est une normalisation en famille de la courbe spectrale Y_a.

On prouve qu'il est représentable par un k-schéma de type fini et que le morphisme d'oubli $\mathcal{B} \longrightarrow \mathcal{A}^{\heartsuit}$ est une bijection au niveau des $\bar{k}$-points. On note Ψ l'ensemble des composantes connexes de $\mathcal{B}$; c'est un ensemble fini. Puis pour tout $\psi \in \Psi$, on note $\mathcal{A}_\psi$ l'image de la composante connexe $\mathcal{B}_\psi$ de $\mathcal{B}$. On obtient ainsi la stratification par normalisation en famille de $\mathcal{A}^\heartsuit$:

$$\mathcal{A}^\heartsuit \times_k \bar{k} = \bigsqcup_{\psi \in \Psi} \mathcal{A}_\psi.$$

L'intérêt de cette stratification réside dans le fait que les invariants δ_a et $\pi_0(Y_a^\flat)$ sont constants sur chaque strate de cette stratification.

4. Comptage local et global

4.1. Une κ-formule de Lefschetz. Reprenons les notations du début du paragraphe 2.5, où un k-schéma en groupes commutatifs P agit sur un k schéma M. On suppose M et P localement de type fini sur k, et que les conditions suivantes sont vérifiées:

i) Le groupe des composantes connexes $\pi_0(P)$ de P est un groupe abélien de type fini.
ii) Le stabilisateur dans P de tout point de M est un sous-groupe de type fini.
iii) Il existe un k-sous-groupe discret sans torsion Λ de P tel que P/Λ et M/Λ soient de type fini.

Sous ces hypothèses, la catégorie quotient $[M/P]$ est équivalente à la catégorie $[(M/\Lambda)/(P/\Lambda)]$. Donc il n'y a qu'un nombre fini de classes d'équivalence dans $[M/P](k)$ et pour tout $x \in [M/P](k)$, son groupe d'automorphismes $\mathrm{Aut}(x)$ est fini. Le nombre rationnel $\sharp[M/P](k)_\kappa$ associé à un caractère κ de $H^1(k,P)$ est ainsi bien défini. Si κ est le caractère trivial, on le note simplement $\sharp[M/P](k)$.

Cette somme $\sharp[M/P](k)_\kappa$ a une interprétation cohomologique. Soit P_σ le groupe des classes de σ-conjugaison dans $P(\bar{k})$. Ainsi $H^1(k,P)$ s'identifie au sous-groupe de torsion de P_σ. D'après [**19**, lemme 8.1.12], il existe toujours un relèvement de κ sur P_σ, noté encore $\kappa : P_\sigma \longrightarrow \bar{\mathbb{Q}}_\ell^\times$. Choisissons un tel relèvement. Quitte à remplacer Λ par un sous-groupe d'indice fini, on suppose que κ est trivial sur Λ. On notera $\mathrm{H}_c^n(M/\Lambda)_\kappa$ (cohomologie ℓ-adique) le sous-espace de $\mathrm{H}_c^n(M/\Lambda)$ sur lequel le groupe P/Λ agit via le caractère κ. On prouve, cf. [**19**, proposition 8.1.6]:

Proposition 4.2. *Soit P^0 le composante connexe de P. On a alors la formule:*

$$\sharp P^0(k) \cdot \sharp[M/P](k)_\kappa = \sum_n (-1)^n \,\mathrm{tr}(\sigma, \mathrm{H}_c^n(M/\Lambda)_\kappa).$$

De plus, si Λ' est un sous-groupe d'indice fini et σ-invariant de Λ, alors pour tout entier n, on a un isomorphisme canonique

$$\mathrm{H}_c^n(M/\Lambda')_\kappa \xrightarrow{\sim} \mathrm{H}_c^n(M/\Lambda)_\kappa.$$

4.3. Soit $a \in \mathcal{A}^{\mathrm{ani}}(k)$ un k-point anisotrope de $\mathcal{A}$; il s'identifie à un morphisme $a : X \longrightarrow \mathfrak{c}_D$. Soit J'_a un schéma en groupes commutatifs lisse de fibres connexes sur X muni d'un homomorphisme $J'_a \longrightarrow J_a$ qui est un isomorphisme sur U_a et soit $\mathcal{P}'_a$ le champ classifiant les J'_a-torseurs sur X. L'homomorphisme $J'_a \longrightarrow J_a$ induit un homomorphisme $\mathcal{P}'_a \longrightarrow \mathcal{P}_a$. La formule de produit se généralise:

$$[\mathcal{M}_a/\mathcal{P}'_a] = \prod_{v \in X - U_a} [\mathcal{M}_v(a)/\mathcal{P}_v(J'_a)].$$

Et cette équivalence de catégories est compatible avec l'action de $\sigma \in \mathrm{Gal}(\bar{k}/k)$.

En toute place $v \in X - U_a$, on fixe une trivialisation du fibré inversible $\mathcal{O}_X(D')_{|X_v}$ et on choisit une mesure de Haar $\mathrm{d}\, t_v$ du tore $J_a(F_v)$. Ainsi, on peut

i) identifier la restriction de a à $X_v = X \times_k F_v$ avec un élément $a_v \in \mathfrak{c}^{\heartsuit}(F_v) \cap \mathfrak{c}(\mathcal{O}_v)$,

ii) identifier J_a et J_{a_v}, ce qui permet de munir une mesure de Haar $\mathrm{d}\, t_v$ à $J_{a_v}(F_v)$,

iii) identifier la fibre de Springer affine $\mathcal{M}_v(a)$ munie d'une action de $\mathcal{P}_v(J'_a)$ avec la fibre de Springer affine $\mathcal{M}_v(a_v)$ munie d'une action de $\mathcal{P}_v(J'_{a_v})$.

Soit $\kappa : \pi(\mathcal{P}_a)_\sigma \longrightarrow \bar{\mathbb{Q}}_\ell^\times$ un caractère. Comme J'_a est de fibres connexes, on a déjà démontré le lien entre le comptage local $[\mathcal{M}_v(a_v)/\mathcal{P}_v(J'_a)](k)$ et l'intégrale orbitale:

$$\sharp[\mathcal{M}_v(a_v)/\mathcal{P}_v(J'_a)](k)_\kappa = \mathrm{vol}(J'_a(\mathcal{O}_v), \mathrm{d}\, t_v)\, \mathrm{O}^\kappa_{a_v}(1_{\mathfrak{g}_v}, \mathrm{d}\, t_v).$$

Puis la formule des produits entraîne:

$$\sharp[\mathcal{M}_a/\mathcal{P}'_a](k)_\kappa = \prod_{v \in X - U_a} \mathrm{vol}(J'_a(\mathcal{O}_v), \mathrm{d}\, t_v)\, \mathrm{O}^\kappa_{a_v}(1_{\mathfrak{g}_v}, \mathrm{d}\, t_v).$$

5. Théorème de stabilisation et Lemme fondamental

Pour simplifier l'exposition, on suppose désormais que G est un groupe déployé semisimple, simplement connexe. Alors, pour tout élément $\kappa \in \hat{T}$, le centralisateur $\hat{G}_\kappa$ de κ dans $\hat{G}$ est toujours connexe, ce qui implique que le morphisme ρ_κ est toujours trivial. Ainsi se donner une donnée endoscopique (κ, ρ_κ) est équivalent à se donner un élément $\kappa \in \hat{T}$.

5.1. L'ouvert étale $\tilde{\mathcal{A}}$. Comme Ngô a expliqué dans son article [**19**, section 1.7], dans la définition d'une κ-intégrale orbitale, il est nécéssaire de choisir un certain point base qui nous permet de relier le groupe de cohomologie $H^1(F_v, J_a)$ avec le tore dual $\hat{\mathbb{T}}$, cf. (2.1.1). On va de même construire un ouvert étale $\tilde{\mathcal{A}}$ de $\mathcal{A}$ qui consiste à faire le choix d'un point base de la courbe camérale dans le lieu où celle-ci est étale au-dessus de X. On verra dans la suite l'énoncé de stabilisation deviendra beaucoup plus agréable sur $\tilde{\mathcal{A}}$.

On fixe d'abord un point géométrique $\infty \in X(\bar{k})$. Soit $\mathcal{A}^\infty$ l'ouvert de $\mathcal{A}$ défini sur $\bar{k}$ tel qu'un point géométrique $a \in \mathcal{A}(\bar{k})$ appartienne à $\mathcal{A}^\infty(\bar{k})$ si et seulement si le revêtement caméral $\tilde{X}_a \longrightarrow \bar{X}$ est étale au-dessus de ∞. Cet ouvert est contenu dans $\mathcal{A}^\heartsuit$. Puis on introduit $\tilde{\mathcal{A}}$ dont les $\bar{k}$-points sont les couples $\tilde{a} = (a, \tilde{\infty})$ où $a \in \mathcal{A}^\infty(\bar{k})$ et $\tilde{\infty}$ est un $\bar{k}$-point de $\tilde{X}_a$ au-dessus de ∞. Alors, $\tilde{\mathcal{A}}$ est irréductible lisse et le morphisme d'oubli $\tilde{\mathcal{A}} \longrightarrow \mathcal{A}^\infty$ est un torseur sous le groupe W. Notons que si ∞ est défini sur k, alors $\mathcal{A}^\infty$ et $\tilde{\mathcal{A}}$ sont aussi définis sur k. Les homomorphismes de la proposition 3.4 pour $a \in \mathcal{A}^\infty$ proviennent simplement d'un homomorphisme $\mathbb{X}_*T \longrightarrow \pi_0(\mathcal{P})_{|\tilde{\mathcal{A}}}$, cf. [**19**, Prop 5.7.1].

5.2. La κ-décomposition perverse. On notera $\tilde{\mathcal{A}}^{\mathrm{ani}}$ l'intersection de l'ouvert anisotrope $\mathcal{A}^{\mathrm{ani}}$ avec l'ouvert étale $\tilde{\mathcal{A}}$ de $\mathcal{A}$. On considère le morphisme de Hitchin au-dessus de cet ouvert

$$\tilde{m}^{\mathrm{ani}} : \tilde{\mathcal{M}}^{\mathrm{ani}} \longrightarrow \tilde{\mathcal{A}}^{\mathrm{ani}}.$$

C'est un morphisme propre. De plus, le champ $\tilde{\mathcal{M}}^{\mathrm{ani}}$ est un champ algébrique de Deligne-Mumford lisse muni d'une action de $\tilde{\mathcal{P}}^{\mathrm{ani}}$. Par le théorème de pureté de Deligne, $K = \tilde{m}_*^{\mathrm{ani}}\bar{\mathbb{Q}}_\ell$ est un faisceau pervers pur muni d'une action de $\tilde{\mathcal{P}}^{\mathrm{ani}}$. Par le lemme d'homotopie [**16**], cette action se factorise à travers le schéma en groupes commutatifs finis $\pi_0(\tilde{\mathcal{P}}^{\mathrm{ani}})$. Ainsi, on obtient *l'interprétation géométrique de la κ-décomposition* dans le processus de stabilisation de la formule des traces:

$$K = \bigoplus_\kappa K_\kappa$$

où

i) κ parcourt l'ensemble des caractères $\mathbb{X}_*T \longrightarrow \bar{\mathbb{Q}}_\ell^\times$ qui se factorisent à travers le groupe fini $\pi_0(\tilde{\mathcal{P}}^{\mathrm{ani}})$,

ii) K_κ est le facteur direct de K sur lequel $\pi_0(\tilde{\mathcal{P}}^{\mathrm{ani}})$ agit par le caractère κ.

On appelle K_κ le κ-morceau de K et si κ appartient au centre de $\hat{\mathbb{G}}$, on l'appelle le morceau stable de K et on le note aussi K_{st}. Alors, *la stabilisation géométrique consiste à décrire les différents morceaux K_κ en termes des morceaux stables associés aux groupes endoscopiques de G.*

5.3. Transfert endoscopique. Soit κ un élément de $\hat{T}$. Rappelons que l'on dispose alors d'un morphisme fini $\mathfrak{c}_H \longrightarrow \mathfrak{c}_G$. A son tour, celui-ci induit un morphisme fini et non ramifié $\mathcal{A}_H \longrightarrow \mathcal{A}_G$ puis en restreignant à l'ouvert étale $\mathcal{A}$, on obtient une immersion fermée $\nu : \tilde{\mathcal{A}}_H \longrightarrow \tilde{\mathcal{A}}_G$, cf [**19**, 6.3].

Soit $\tilde{a}_H \in \tilde{\mathcal{A}}_H(k)$ d'image $\tilde{a} \in \tilde{\mathcal{A}}_G(k)$. On note a (resp. a_H) l'image de $\tilde{a}$ dans $\tilde{\mathcal{A}}_G(k)$ (resp. $\tilde{a}_H$ dans $\tilde{\mathcal{A}}_H(k)$). Alors, il existe un homomorphisme canonique $J_a \longrightarrow J_{H,a_H}$ dont la restriction au-dessus de U_a est un isomorphisme. On choisit une fois pour toutes un schéma en groupes commutatifs lisse de fibres connexes J'_a muni d'un homomorphisme $J'_a \longrightarrow J_a$ qui est un isomorphisme génériquement. On peut choisir J'_a de manière que

$\mathrm{H}^0(\bar{X}, J'_a) = 0$ de telle sorte que le champ $\mathcal{P}'_a$ classifiant les J'_a-torseurs sur X est de type fini. Les morphismes naturels $\mathcal{P}'_a \longrightarrow \mathcal{P}_a$ et $\mathcal{P}'_a \longrightarrow \mathcal{P}_{H,a_H}$ induisent des actions de $\mathcal{P}'_a$ sur $\mathcal{M}_a$ et $\mathcal{M}_{H,a_H}$. Dans la suite, on travaille toujours avec les quotients $[\mathcal{M}_a/\mathcal{P}'_a]$ et $[\mathcal{M}_{H,a_H}/\mathcal{P}'_a]$.

Comme on l'a déjà dit au paragraphe 1.4, il existe un diviseur effectif réduit $\mathfrak{R}^H_G$ sur $\mathfrak{c}_H$ (resp. $\mathfrak{R}^H_{G,D}$ sur $\mathfrak{c}_{H,D}$) vérifiant la relation $\nu^*\mathfrak{D}_G = \mathfrak{D}_H + 2\mathfrak{R}^H_G$.

5.4. Rappelons que κ s'identifie à un caractère de $\mathbb{X}_*T$. Via le morphisme $\mathbb{X}_*T \longrightarrow \pi_0(\mathcal{P})_{|\tilde{\mathcal{A}}}$, on peut donc considérer le facteur direct $K_{G,\kappa}$ de K_G. C'est aussi un faisceau pervers pur. L'argument clef de la preuve de Ngô réside dans le théorème de support suivant dont la preuve sera esquissée dans le chapitre de Ngô [**20**].

Théorème 5.5 (Théorème de support).

i) *(Version faible) Le support de $K_{G,\kappa}$ est contenu dans l'image de $\tilde{\mathcal{A}}^{\mathrm{ani}}_H$.*

ii) *(Version forte) Soit $\mathcal{K}$ un facteur direct de $K_{G,\kappa}$. Supposons que le support de $\mathcal{K}$ a une intersection non vide avec l'ouvert $\tilde{\mathcal{A}}^{\mathrm{good}}_H \cap \tilde{\mathcal{A}}^{\mathrm{ani}}_H$. Alors, il contient $\tilde{\mathcal{A}}^{\mathrm{good}}_H \cap \tilde{\mathcal{A}}^{\mathrm{ani}}_H$.*

Ainsi, on obtient l'interprétation géométrique de la stabilisation:

Théorème 5.6. *Avec les notations ci-dessus, après semi-simplification, on a un isomorphisme:*

$$\nu^* K_{G,\kappa}[2r^G_H(D)](r^G_H(D)) \xrightarrow{\sim} K_{H,\mathrm{st}}$$

où $r^G_H(D)$ est la codimension de $\tilde{\mathcal{A}}_H$ dans $\tilde{\mathcal{A}}_G$.

Remarquons que le facteur $r^G_H(D)$ qui apparaît dans cette formule correspond bien à la valuation du facteur de transfert dans l'énoncé du lemme fondamental. L'énoncé géométrique ci-dessus est obtenu à partir du théorème de support et à l'aide des comptages des fibres de Springer affines et de la fibration de Hitchin, au terme d'un raisonnement au cours duquel sont prouvés aussi le lemme fondamental et l'énoncé i) ci-dessous.

Théorème 5.7. *Avec les notations précédents, on a :*

i) (Stabilisation) Pour tout k-point $\tilde{a}_H \in \tilde{\mathcal{A}}^{\mathrm{ani}}_H(k)$ d'image $\tilde{a} \in \tilde{\mathcal{A}}^{\mathrm{ani}}(k)$, si a (resp. a_H) est l'image de $\tilde{a}$ (resp. $\tilde{a}_H$), on a toujours l'égalité globale:

$$\sharp[\mathcal{M}_a/\mathcal{P}'_a](k)_\kappa = q^{r^G_H(D)} \sharp[\mathcal{M}_{H,a_H}/\mathcal{P}'_a](k).$$

ii) (Lemme fondamental) Pour tout k-point $a_H \in \mathfrak{c}^{\mathrm{rs}}_H(F_v) \cap \mathfrak{c}_H(\mathcal{O}_v)$ d'image $a \in \mathfrak{c}^{\mathrm{rs}}_G(F_v) \cap \mathfrak{c}_G(\mathcal{O}_v)$, on a toujours l'égalité locale:

$$\sharp[\mathcal{M}_v(a)/\mathcal{P}_v(J'_a)](k)_\kappa = q^{\deg(v) r^G_{H,v}(a_H)} \sharp[\mathcal{M}_{H,v}(a_H)/\mathcal{P}_v(J'_a)](k)$$

où $r^G_H(a_{H,v})$ est le degré du diviseur $a^*_H \mathfrak{R}^H_G$ en v.

L'équivalence de ii) et l'énoncé du lemme fondamental résulte du comptage sur les fibres de Springer affines, cf proposition 2.7.

Expliquons pourquoi i) implique la stabilisation géométrique, i.e. le théorème 5.6. En effet, $K_{G,\kappa}$ (resp. $K_{H,\kappa}$) est un faisceau pervers pur sur $\tilde{\mathcal{A}}_H^{\mathrm{ani}}$, donc il existe un ouvert non vide $\tilde{\mathcal{U}}$ de $\tilde{\mathcal{A}}_H^{\mathrm{ani}}$ tel que

i) la restriction de $\nu^* K_{G,\kappa}[2r_H^G(D)](r_H^G(D))$ (resp. $K_{H,\kappa}$) sur cet ouvert soit un système local,

ii) le faisceau pervers $\nu^* K_{G,\kappa}[2r_H^G(D)](r_H^G(D))$ (resp. $K_{H,\kappa}$) soit le prolongement intermédiaire de sa restriction sur $\tilde{\mathcal{U}}$.

Compte tenu du comptage global, le point i) du théorème ci-dessus implique que, pour toute extension finie k' de k et pour tout k'-point $\tilde{a}_H$ de $\tilde{\mathcal{A}}_H^{\mathrm{ani}}$ d'image $\tilde{a} \in \tilde{\mathcal{A}}_G(k')$, on a l'égalité:

$$\sum_n (-1)^n \operatorname{tr}(\sigma_{k'}, (K_{G,\kappa}^n)_{\tilde{a}}) = \sum_n (-1)^n \operatorname{tr}(\sigma_{k'}, (K_{H,\kappa}^n)_{\tilde{a}_H}).$$

En particulier, pour toute extension k'_j de degrée j de k', on considère $\tilde{a}_H$ comme un k'_j-point de $\tilde{\mathcal{A}}_H$ et obtient aussi l'égalité:

$$\sum_n (-1)^n \operatorname{tr}(\sigma_{k'}^j, (K_{G,\kappa}^n)_{\tilde{a}}) = \sum_n (-1)^n \operatorname{tr}(\sigma_{k'}^j, (K_{H,\kappa}^n)_{\tilde{a}_H}).$$

Comme $K_{G,\kappa}$ et $K_{H,\kappa}$ sont purs, les valeurs propres de $\sigma_{k'}$ dans $(K_{G,\kappa}^n)_{\tilde{a}})$ et $(K_{G,\kappa}^n)_{\tilde{a}})$ sont toutes de modules $q^{n\deg(k'/k)/2}$. On en déduit que

$$\operatorname{tr}(\sigma_{k'}, (K_{G,\kappa}^n)_{\tilde{a}}) = \operatorname{tr}(\sigma_{k'}, (K_{H,\kappa}^n)_{\tilde{a}_H}).$$

Le théorème de Chebotarev appliqué aux systèmes locaux $\nu^* K_{G,\kappa}[2r_H^G(D)](r_H^G(D))$ et $K_{H,\kappa}$ entraîne qu'ils sont isomorphes après semi-simplification.

5.8. Voici les grandes lignes de la preuve du théorème 5.7 de Ngô Bao Châu. On se place d'abord sur un ouvert $\tilde{\mathcal{A}}_H^{\mathrm{trans}}$ dit transversal de $\tilde{\mathcal{A}}_H^{\mathrm{ani}}$. Par définition, un point géométrique $\tilde{a}_H \in \tilde{\mathcal{A}}_H^{\mathrm{ani}}(\bar{k})$ d'image $a_H \in \mathcal{A}_H^{\mathrm{ani}}(\bar{k})$ est dans cet ouvert si et seulement si $a \in \mathcal{A}_H^{\mathrm{good}} \cap \mathcal{A}_H^{\mathrm{ani}}(\bar{k})$ et que l'image $a(\bar{X})$ coupe transversalement le diviseur réduit $\mathfrak{D}_H + \mathfrak{R}_G^H$ dans $\mathfrak{c}_{D,H}$.

Soit $\tilde{a}_H \in \tilde{\mathcal{A}}_H^{\mathrm{trans}}(k)$ d'image $\tilde{a} \in \tilde{\mathcal{A}}^{\mathrm{ani}}(k)$; on note a_H (resp. a) l'image de $\tilde{a}_H$ dans $\mathcal{A}_H(k)$ (resp. de $\tilde{a}$ dans $\mathcal{A}(k)$). On peut calculer explicitement les nombres locaux $\sharp[\mathcal{M}_v(a)/\mathcal{P}_v(J'_a)](k)_\kappa$ et $\sharp[\mathcal{M}_{H,v}(a_H)/\mathcal{P}_v(J'_a)](k)$ et vérifie directement que

$$\sharp[\mathcal{M}_v(a)/\mathcal{P}_v(J'_a)](k)_\kappa = q^{\deg(v) r_{H,v}^G(a_H)} \sharp[\mathcal{M}_{H,v}(a_H)/\mathcal{P}_v(J'_a)](k).$$

Compte tenue du fait que $r_H^G(D) = \sum_v \deg(v) r_{H,v}^G(a_H)$, l'égalité globale s'en suit. Ainsi, le théorème de stabilisation est valide pour tout point transversal $\tilde{a}_H \in \tilde{\mathcal{A}}_H^{\mathrm{trans}}(k)$.

Puis, le théorème de support implique immédiatement que le théorème de stabilisation est encore valide pour tout point $\tilde{a}_H \in \tilde{\mathcal{A}}_H^{\mathrm{good}} \cap \tilde{\mathcal{A}}_H^{\mathrm{ani}}(k)$. Démontrons maintenant le lemme fondamental. En effet, soit $a_{H,v} \in$

$\mathfrak{c}_H^{\mathrm{rs}}(F_v)\cap\mathfrak{c}_H(\mathcal{O}_v)$. Alors, si l'on choisit le diviseur D de degré suffisamment grand, il existe $\tilde{a}_H\in\tilde{\mathcal{A}}_H^{\mathrm{good}}\cap\tilde{\mathcal{A}}_H^{\mathrm{ani}}(k)$ dont la composante locale en v est $a_{H,v}$ et que l'image $a_H(X-\{v\})$ coupe transversalement le diviseur $\mathfrak{D}_H+\mathfrak{R}_G^H$ dans $\mathfrak{c}_{D,H}$. On note $\tilde{a}$ l'image de $\tilde{a}_H\in\tilde{\mathcal{A}}_G(k)$. Puisque $\tilde{a}_H\in\tilde{\mathcal{A}}_H^{\mathrm{good}}\cap\tilde{\mathcal{A}}_H^{\mathrm{ani}}(k)$, on a l'égalité globale:

$$\sharp[\mathcal{M}_a/\mathcal{P}'_a](k)_\kappa=q^{\deg(v)r_H^G(a_H)}\sharp[\mathcal{M}_{H,a_H}/\mathcal{P}'_a](k).$$

Comme $a_H(X-\{v\})$ coupe transversalement le diviseur $\mathfrak{D}_H+\mathfrak{R}_G^H$ dans $\mathfrak{c}_{D,H}$, pour toute place $w\neq v$, on a:

$$\sharp[\mathcal{M}_w(a)/\mathcal{P}_w(J'_a)](k)_\kappa=q^{\deg(w)r_{H,w}^G(a_H)}\sharp[\mathcal{M}_{H,w}(a_H)/\mathcal{P}_w(J'_a)](k)$$

et ils ne s'annulent pas. On obtient ainsi la même égalité en la place v, autrement dit le lemme fondamental:

$$\sharp[\mathcal{M}_v(a)/\mathcal{P}_v(J'_a)](k)_\kappa=q^{\deg(v)r_{H,v}^G(a_H)}\sharp[\mathcal{M}_{H,v}(a_H)/\mathcal{P}_v(J'_a)](k).$$

Enfin, supposons qu'on a déjà démontré le lemme fondamental. Démontrons le théorème de stabilisation pour tout point $\tilde{a}_H\in\tilde{\mathcal{A}}_H^{\mathrm{ani}}(k)$. En effet, le lemme fondamental implique que, pour toute place v de X, on a:

$$\sharp[\mathcal{M}_v(a)/\mathcal{P}_v(J'_a)](k)_\kappa=q^{\deg(v)r_{H,v}^G(a_H)}\sharp[\mathcal{M}_{H,v}(a_H)/\mathcal{P}_v(J'_a)](k).$$

On fait le produit sur toutes les places pour obtenir l'égalité globale:

$$\sharp[\mathcal{M}_a/\mathcal{P}'_a](k)_\kappa=q^{\deg(v)r_H^G(a_H)}\sharp[\mathcal{M}_{H,a_H}/\mathcal{P}'_a](k).$$

La preuve est donc terminée.

Références

[1] J. Arthur, An introduction to the trace formula, *Harmonic analysis, the trace formula, and Shimura varieties, Clay Math. Proc.* **4**, Amer. Math. Soc., Providence, RI (2005), 1–263.

[2] A. Beauville, M. Narasimhan & S. Ramanan, *Spectral curves and generalized theta divisor, J. Reine Angew. Math.* **398** (1989), 169–179.

[3] I. Biswas, S. Ramanan, Infinitesimal study of Hitchin pairs, *J. London Math. Soc.* **49** (1994), 219–231.

[4] S. Bosch, W. Lutkebohmert, M. Raynaud, Neron models, *Ergeb. der Math.* **21**, Springer Verlag (1990).

[5] R. Cluckers, T. Hales & F. Loeser, Chapitre dans ce livre.

[6] J-F. Dat, Lemme fondamental et endoscopie, une approche géométrique, *Séminaire Bourbaki* **940**, novembre 2004.

[7] R. Donagi & D. Gaitsgory, The gerb of Higgs bundles, *Transform. Groups* **7** (2002), 109–153.

[8] G. Faltings, Stable G-bundles and projective connections, *J. Alg. Geom.* **2** (1993), 507–568.

[9] M. Goresky, R. Kottwitz & R. MacPherson, Homology of affine Springer fiber in the unramified case, *Duke Math. J.* **121** (2004), 509–561.

[10] M. Harris, Introduction, Chapitre dans ce livre.

[11] R. Kottwitz, Isocristal with additionnal structures, *Compos. Math.* **56** (1985), 201–220.

[12] G. Laumon & L. Moret-Bailly, Champs algébriques, *Ergebnisse der Mathematik* **39**, Springer-Verlag, Berlin (2000).

[13] G. Laumon, Fibres de Springer et Jacobiennes compactifiées, in Algebraic geometry and number theory, *Progr. Math.* **253**, Birkhäuser Boston, Boston, MA, (2006), 515–563.

[14] G. Laumon, Sur le lemme fondamental pour les groupes unitaires, arxiv.org/abs/math/0212245.

[15] G. Laumon, Aspects géométriques du Lemme Fondamental de Langlands-Shelstad, *International Congress of Mathematicians Vol. II, Eur. Math. Soc.*, Zürich, (2006), 401–419.

[16] G. Laumon & B-C. Ngô, Le lemme fondamental pour les groupes unitaires, *Annals of Math.* **168**(2) (2008), 477–573.

[17] B-C. Ngô, Fibration de Hitchin et endoscopie, *Invent. Math.* **164**(2) (2006), 399–453.

[18] B-C. Ngô, Fibration de Hitchin et structure endoscopique de la formule des traces, *International Congress of Mathematicians Vol. II, Eur. Math. Soc.*, Zürich, (2006), 1213–1225.

[19] B-C. Ngô, Le lemme fondamental pour les algèbres de Lie, arXiv:math/0801.0446v3".

[20] B-C. Ngô, Chapitre dans ce livre.

[21] J-L. Waldspurger, Endoscopie et changement de caractéristique, *J. Inst. Math. Jussieu* **5**(3) (2006), 423–525.

Institut Mathématique de Jussieu, Université Pierre et Marie Curie (Paris 6), 175 rue du Chevaleret, 75013 Paris, France

E-mail address: `dat@math.jussieu.fr`

CNRS — Université de Paris Nord (Paris 13), LAGA - Département de Mathématiques, 99 avenue Jean-Baptiste Clément, 93430 Villetaneuse, France

E-mail address: `ngodac@math.univ-paris13.fr`

Decomposition theorem and abelian fibration

[N.II.C.2]

Ngô Bao Châu

The text is the account of a talk given in Bonn in a conference in honor of M. Rapoport. We will review the proof of the support theorem which is the main geometric ingredient in the proof of the fundamental lemma. The original source is the chapter 7 of [**16**].

1. Support in the decomposition of a pure direct image

Let X be a smooth algebraic variety over a field k, S a connected scheme of finite type of k and $f : X \to S$ a proper k-morphism. By Deligne's theorem [**5**], the direct image $f_* \mathbb{Q}_\ell$ is a pure complex i.e. for the perverse t-structure, the cohomology ${}^p\mathrm{H}^j(f_*\mathbb{Q}_\ell)$ are pure perverse sheaves. After a base change to $S \otimes_k \bar{k}$, ${}^p\mathrm{H}^j(f_*\mathbb{Q}_\ell)$ is direct sum of simple perverse sheaves according to Beilinson, Bernstein, and Deligne [**1**].

Let us recall the description of simple perverse sheaves over $S \otimes_k \bar{k}$. Let Z be a closed irreducible subscheme of $S \otimes_k \bar{k}$, $i : Z \hookrightarrow S \otimes_k \bar{k}$ denoting the closed immersion. Let U be a smooth open dense subscheme of Z, $j : U \hookrightarrow Z$ denoting the open immersion. Let $\mathcal{K}$ be an irreducible local system on U. Then

$$K = i_* j_{!*} \mathcal{K}[\dim(Z)]$$

is a simple perverse sheaf on $S \otimes_k \bar{k}$, $j_{!*}$ being Goresky-MacPherson's intermediate extension. Moreover, it is known that all simple perverse sheaves over $S \otimes_k \bar{k}$ are of this form, see [**1**]. In particular the closed reduced subscheme Z of $S \otimes_k \bar{k}$ is completely determined by K and is called the support of K.

One of the basic invariants that one can attach to a proper morphism f with smooth source is the set of supports of the simple perverse sheaves occurring in the perverse cohomology of $f_*\mathbb{Q}_\ell$. A question of a particular interest is in which circumstances the only present support is $S \otimes_k \bar{k}$.

The knowledge of the set of supports is useful in proving certain equality between numbers of points. Let k be a finite field. Let $f_1 : X_1 \to S$ and $f_2 : X_2 \to S$ be morphisms as above. Assume that the only support in

$f_{i,*}\mathbb{Q}_\ell$ is $S \otimes_k \bar{k}$ for $i \in \{1,2\}$. Assume that there exists a dense open subset $U \subset S$ such that for all extension k'/k, for all $s \in U(k')$ we have the equality

$$\sharp X_{1,s}(k') = \sharp X_{2,s}(k').$$

Then the above equality is also true for all $s \in S(k)$. This is a simple application of Deligne's purity theorem, the Grothendieck-Lefschetz fixed points formula and Chebotarev's theorem. Since the only present support is $S \otimes_k \bar{k}$, the class of ${}^p\mathrm{H}^j(f_*\mathbb{Q}_\ell)$ in the appropriate Grothendieck group is determined by its restriction to U which is in turn determined by the traces of the Frobenius operator σ_s attached to closed points s of U. In [**13**], Laumon explained an induction process for the above strategy in a more general case where the set of supports might contain more than $S \otimes_k \bar{k}$. In any case, the control on the set of supports allows a principle of continuation of an equality between the numbers of points in the fibers of f_1 and f_2.

In general however, it is difficult to determine the supports of the simple perverse constituents of the direct image. Here are a few examples where such a determination is possible. First assume $f : X \to S$ is a proper and smooth morphism. Since the source X is already supposed to be smooth, so is the target S. In this case, the ordinary cohomology sheaves $\mathrm{H}^j(f_*\mathbb{Q}_\ell)$ are local systems on S and they coincide with the perverse cohomology sheaves. Thus the only present support is $S \otimes_k \bar{k}$.

Less obvious examples consist in small and semi-small morphisms in Goresky-MacPherson's sense. A proper morphism $f : X \to S$ is called semi-small if $\dim(X \times_S X) \leq \dim(X)$ and small if $\dim(X \times_S X - \Delta X) < \dim(X)$. Here ΔX denotes the diagonal of X in $X \times_S X$. Assume as always X smooth. A simple dimension computation shows that if f is small then $f_*\mathbb{Q}_\ell[\dim X]$ is a perverse sheaf and the only present support is $S \otimes_k \bar{k}$. If f is semi-small then $f_*\mathbb{Q}_\ell[\dim X]$ is a perverse sheaf and the support of its constituents are the images of maximal dimensional components of $X \times_S X$ in S. The Springer resolution of the nilpotent cone in a semi-simple Lie algebra is a semi-small morphism and the Grothendieck-Springer simultaneous resolution is a small morphism. These examples had quite an impact in representation theory, see [**14**] for instant.

Let us consider another simple but important example which was pointed out to me by Goresky. Let $f : X \to S$ be a relative proper curve. Assume that X/k is smooth of dimension 2, S/k is a smooth connected base of dimension one and the fibers X_s are irreducible curves. Then one can prove that the only support in $f_*\mathbb{Q}_\ell$ is S. The proof is an application of Poincaré duality. We are going to explain it in some details since this argument will be important in the sequel. Consider the decomposition

$$f_*\mathbb{Q}_\ell[2] = \bigoplus K[n]$$

where K are simple perverse sheaves on $S \otimes_k \bar{k}$ and n are integers. By Poincaré duality, we know if $K[n]$ occurs in the above direct sum then so does $K^\vee[-n]$. Let Z be a support of certain simple summand $K[n]$.

Assume $\dim(Z) = 0$. Then $K[n]$ must contribute in ordinary cohomological degree $-n$ and $K^\vee[-n]$ in degree n. But $f_*\mathbb{Q}_\ell[2]$ is concentrated in ordinary cohomological degrees $0, -1, -2$. It follows that $n = 0$. But $\mathrm{H}^0(f_*\mathbb{Q}_\ell[2]) = \mathrm{H}^2(f_*\mathbb{Q}_\ell)$ is a constant sheaf of rank one in which a punctual sheaf cannot contribute as a direct factor. This yields a contradiction.

This argument provides in fact a general upper bound for the codimension of the support.

Proposition 1 (Goresky-MacPherson). Assume X/k smooth and S connected. Let $f : X \to S$ be a proper morphism of relative dimension d. Let Z be the support of a simple perverse sheaf occurring in $f_*\mathbb{Q}_\ell$. Then we have the inequality

$$\mathrm{codim}(Z) \leq d.$$

In this degree of generality, the above inequality seems to be of little use. In concrete situations, we have to do a little bit more to extract precise information about the set of supports. The good point is that the above Poincaré duality argument is very malleable to be adapted to different specific situations.

2. Algeraic abelian fibration

Algebraic abelian fibration is a somewhat vague terminology for a denegerating family of abelian varieties. It is however difficult to coin exactly what an abelian fibration is. We are going to introduce instead a loose notion of weak abelian fibration by keeping the properties that are conserved by arbitrary base change and a more restrictive notion of δ-regular abelian fibration. A good notion of algebraic abelian fibration must be somewhere in between.

A *weak abelian fibration* will consist in a proper morphism $f : M \to S$ equipped with an action of a smooth commutative group scheme $g : P \to S$ i.e. we have an action of P_s on M_s depending algebraically on $s \in S$. In this section, it is convenient to assume P have connected fibers. In general, we can replace P by the open sub-group schemes of neutral components. We will require the following three properties to be satisfied.

(1) The morphism f and g have the same relative dimension d.
(2) The action has *affine stabilizers* : for all geometric points $s \in s$, $m \in M_s$, the stabilizer $P_{s,m}$ of m is an affine subgroup of P_s. We can rephrase this property as follows. According to Chevalley, for all geometric point $s \in S$, there exists an exact sequence

$$1 \to R_s \to P_s \to A_s \to 1$$

where A_s is an abelian variety and R_s is a connected affine commutative group. Then for all geometric points $s \in s$, $m \in M_s$, we require that the stabilizer $P_{s,m}$ is a subgroup of R_s.

(3) The group scheme P has a *polarizable Tate module*. Let $\mathrm{H}_1(P/S) = \mathrm{H}^{2g-1}(g_!\mathbb{Q}_\ell)$ with fiber $\mathrm{H}_1(P/S)_s = T_{\mathbb{Q}_\ell}(P_s)$. This is a sheaf for the étale topology of S whose stalk over a geometric point $s \in S$ is the $\mathbb{Q}_\ell$-Tate module of P_s. The Chevalley exact sequence induces

$$0 \to T_{\mathbb{Q}_\ell}(R_s) \to T_{\mathbb{Q}_\ell}(P_s) \to T_{\mathbb{Q}_\ell}(A_s) \to 0.$$

We require that locally for the étale topology there exists an alternating form ψ on $\mathrm{H}_1(P/S)$ such that over any geometric point $s \in S$, ψ is null on $T_{\mathbb{Q}_\ell}(R_s)$ and induces a non-degenerating form on $T_{\mathbb{Q}_\ell}(A_s)$.

We observe that the notion of weak abelian fibration is conserved by arbitrary base change. In particular, the generic fiber of P is not necessarily an abelian variety. We are going now to introduce a strong restriction called *δ-regularity* which implies in particular that the generic P is an abelian variety.

Let us stratify S by the dimension $\delta(s) = \dim(R_s)$ of the affine part of P_s. We know that δ is an upper semi-continuous function. Let us denote

$$S_\delta = \{s \in S | \delta(s) = \delta\}$$

which is a locally closed subset of S. The group scheme $g : P \to S$ is δ-regular if

$$\mathrm{codim}(S_\delta) \geq \delta.$$

A *δ-regular abelian fibration* is a weak abelian fibration $f : M \to S$ equipped with an action of a δ-regular group scheme $g : P \to S$.

We observe that δ-regularity is conserved by flat base change.

For a δ-regular abelian fibration, the open subset S_0 is a non empty open subset i.e. generically P is an abelian variety. Combined with the affineness of stabilizer and with the assumption f and g having the same relative dimension, it follows that the generic fiber of f is a finite union of abelian varieties.

A typical example is the following one. Let $X \to S$ be a family of reduced irreducible curves with only plane singularities. Let $P = \mathrm{Jac}_{X/S}$ be the relative Jacobian. Let $M = \overline{\mathrm{Jac}}_{X/S}$ be the compactified relative Jacobian. For every $s \in S$, P_s classifies invertible sheaves of degree 0 on X_s, M_s classifies rank one torsion-free sheaves of degree 0 on X_s and P_s acts on M_s by tensor product. The Weil pairing defines a polarization of the Tate module $\mathrm{H}_1(P/S)$. For every geometric point $s \in S$, we can check that $\delta(s)$ is Serre's δ-invariant

$$\delta(s) = \dim \mathrm{H}^0(X_s, c_*\mathcal{O}_{\tilde{X}_s}/\mathcal{O}_{X_s})$$

of X_s. Here $c : \tilde{X}_s \to X_s$ denote the normalization of X_s. It is well known that the δ-regularity is true for a versal deformation of curve with plane singularities, see [**6**], and thus is true in the neighborhood of any point s of S where the family $X \to S$ is versal. But it is not true in general.

It is not obvious to prove the δ-regularity of a given weak abelian fibration.

One family of examples is given by algebraic integrable systems over the field of complex numbers. As we will see, in this case the existence of the symplectic form implies the δ-regularity. Let $f : M \to S$ and $g : P \to S$ form a weak abelian fibration. Assume that M is a complex smooth algebraic variety of dimension $2d$ equipped with a symplectic form and that S is smooth of dimension $\dim(S) = \dim(M)/2$. Assume that for every $m \in M$ over $s \in S$, the tangent space $T_m M_s$ to the fiber is coisotropic i.e. its orthogonal $(T_m M_s)^\perp$ with respect to the symplectic form is contained into itself. The tangent application $T_m M \to T_s S$ defines by duality a linear map

$$T_s^* S \to T_m^* M \cong T_m M$$

by identifying $T_m^* M$ with $T_m M$ using the symplectic form. Let $\mathrm{Lie}(P/S)$ be the relative Lie algebra of P whose stalk at s is $\mathrm{Lie}(P_s)$. Assume that we have an isomorphism $\mathrm{Lie}(P/S) \cong T^* S$ of vector bundles on S such that for each point s, the infinitesimal action of P_s on M_s at the point $m \in M_s$ is given by the above linear map. Consider the Chevalley exact sequence

$$1 \to R_s \to P_s \to A_s \to 1$$

of P_s. The connected affine subgroup R_s acting on the proper scheme M_s must have a fixed point according to Borel. Denote m a fixed point. The map $P_s \to M$ given by $p \mapsto pm$ factors through A_s so that on the infinitesimal level, the map $\mathrm{Lie}(P_s) \to T_m M$ factors through $\mathrm{Lie}(A_s)$. By duality, for every point $m \in M_s$ fixed under the action of the affine part R_s, the image of the tangent application

$$T_m M \to T_s S$$

is contained in $\mathrm{Lie}(A_s)^*$ which is a subspace of codimension $\delta(s)$ independent of m. In characteristic zero, the δ-regularity follows. Roughly speaking when s moves in such a way that $\delta(s)$ remains constant, the tangent direction of the motion of s cannot get away from the fixed subvector space $\mathrm{Lie}(A_s)^*$ of $T_s S$ which has codimension $\delta(s)$.

Unfortunately, this argument does not work well in positive characteristic. In the case of Hitchin fibration, we can use some kind of global-local argument. One can define a local variant of the δ-invariant. A computation of the codimension of δ-constant strata can be derived from Goresky-Kottwitz-MacPherson's result [**8**]. One can use Riemann-Roch's type argument to obtain a global estimate from the local estimates in certain circumstance as in [**16**]. In *loc. cit,* we proved a weaker form of δ-regularity which is good enough to prove local statements as the fundamental lemma but unsatisfying from the point of view of Hitchin fibration. We hope to be able to remove this caveat in future works.

3. Support in the direct image by an abelian fibration

We are now summarizing chapter 7 of [**16**].

THEOREM 2 (Support). Let $f : M \to S$ and $g : P \to S$ be a δ-regular abelian fibration of relative dimension d with the total space M smooth over k. Assume moreover S connected and f projective.

Let K be a simple perverse sheaf occuring in $f_*\mathbb{Q}_\ell$ and let Z be its support. There exists an open subset U of $S \otimes_k \bar{k}$ such that $U \cap Z \neq \emptyset$ and a non trivial local system L on $U \cap Z$ such that the constructible sheaf i_*L is a direct factor of $\mathrm{R}^{2d} f_*\mathbb{Q}_\ell|_U$. Here i is the inclusion of $U \cap Z$ in Z.

In particular, if the geometric fibers of f are all irreducible then $Z = S \otimes_k \bar{k}$.

For any weak abelian fibration, we prove in fact an estimate on the codimension of Z which can be looked as an improvement of Goresky-MacPherson inequality.

PROPOSITION 3 (δ-Inequality). Let $f : M \to S$ equipped with $g : P \to S$ be a weak abelian fibration of relative dimension d with total space M smooth over the base field k. Assume moreover S connected and f projective. Let K be a simple perverse sheaf occurring in $f_*\mathbb{Q}_\ell$.

Let Z be the support of K. Let δ_Z be the minimal value of δ on Z. Then we have the inequality

$$\mathrm{codim}(Z) \leq \delta_Z.$$

If equality occurs, there exists an open subset U of $S \otimes_k \bar{k}$ such that $U \cap Z \neq \emptyset$ and a non trivial local system L on $U \cap Z$ such that i_*L is a direct factor of $\mathrm{R}^{2d} f_*\mathbb{Q}_\ell|_U$. In particular, if the geometric fibers of f are irreducible then $Z = S \otimes_k \bar{k}$.

It is immediate that the above δ-inequality implies the support theorem.

What follows now is a false start which gives however an intuitive idea about the δ-inequality.

The problem is local arround any point of Z. Let us fix such a point s in Z. The δ-inequality is an improvement of Goresky-MacPherson's inequality $\mathrm{codim}(Z) \leq d$ in the case of abelian fibrations. It can be even reduced to this inequality if we make the following lifting assumptions on an neighborhood around s:

- there exists a lift of A_s to an abelian scheme $A_{S'}$ over an étale neighborhood S' of s,
- there exists a homomorphism $A_{S'} \to P_{S'} = P \times_S S'$ such that over the point s, its composition with the projection $P_s \to A_s$ is an isogeny of the abelian variety A_s.

Under these assumptions, we have an action of the abelian scheme $A_{S'}$ on $M_{S'} = M \times_S S'$ with finite stabilizers. Consider the quotient $[M_{S'}/A_{S'}]$ which is an algebraic stack proper and smooth over S' of relative dimension δ_Z. The δ-inequality follows from the fact that the morphism $M_{S'} \to [M_{S'}/A_{S'}]$ is proper and smooth and from Goresky-MacPherson's inequality for the morphism $[M_{S'}/A_{S'}] \to S'$.

In practice, the above lifting assumptions almost never happen because in most of the cases, the generic fiber of P is an irreducible abelian variety. Our strategy is in fact to imitate the above proof at the homological level instead of the geometry level.

Consider the homology complex $\Lambda_P = g_!\mathbb{Q}_\ell[2d](d)$ of P which is concentrated in ordinary cohomological degrees $-2d, -2d+1, \ldots, 0$. For all geometric point $s \in S$, $\mathrm{H}^{-1}(\Lambda_P)_s$ is the $\mathbb{Q}_\ell$-Tate module $T_{\mathbb{Q}_\ell}(P_s)$. For all $i = 1, \ldots, d$, $\mathrm{H}^{-i}(\Lambda_P)_s = \wedge^i T_{\mathbb{Q}_\ell}(P_s)$.

As the S-group scheme $g : P \to S$ acts on the S-scheme $f : M \to S$, the trace map induces a canonical map $\Lambda_P \otimes f_!\mathbb{Q}_\ell \to f_!\mathbb{Q}_\ell$. When $f = g$, $\Lambda_P \otimes \Lambda_P \to \Lambda_P$ is just the Pontryagin product fiberwise. Thus Λ_P has an algebra structure for which $f_!\mathbb{Q}_\ell$ is a Λ_P-module. In our case, f is proper thus $f_*\mathbb{Q}_\ell$ is a Λ_P-module on the homology of a commutative algebraic group. We have to be aware however that Λ_P and $f_*\mathbb{Q}_\ell$ are objects of a triangular category.

The hypothetical geometric lifting assumptions can be replaced by the following homological property : *$f_!\mathbb{Q}_\ell$ is free over the abelian part of Λ_P*. It is not immediately clear what we mean by the abelian part of Λ_P and by the freeness of objects in derived category. In fact, it requires quite a few preparations to formulate this freeness property in the right way.

Consider the decomposition of the n-th perverse cohomology ${}^p\mathrm{H}^n(f_*\mathbb{Q}_\ell)$ into a direct sum of simple perverse sheaves over $S \otimes_k \bar{k}$. We can put together those simple perverse sheaves having the same support so that we have a canonical decomposition

$$ {}^p\mathrm{H}^n(f_*\mathbb{Q}_\ell) = \bigoplus_{\alpha \in \mathfrak{A}} K_\alpha^n $$

where $(Z_\alpha)_{\mathfrak{A}}$ is a finite set of distinct closed irreducible subschemes of $S \otimes_k \bar{k}$ and where K_α^n is the sum of simple perverse sheaves with support Z_α occuring in ${}^p\mathrm{H}^n(f_*\mathbb{Q}_\ell)$. For all $\alpha \in \mathfrak{A}$, let denote

$$ K_\alpha = \bigoplus_n K_\alpha^n[-n]. $$

Let x_α be a geometric generic point of Z_α and let denote S_α the strict henselization of $S \otimes_k \bar{k}$ at x_α. The restriction of K_α to S_α is just a graded vector space supported by the closed point x_α of S_α. We have a canonical structure of graded $\Lambda_{P_{x_\alpha}}$-module on K_{α,x_α}. Let

$$ 1 \to R_\alpha \to P_{x_\alpha} \to A_\alpha \to 1 $$

be Chevalley canonical exact sequence attached to P_{x_α}. For any homological lifting $T_{\mathbb{Q}_\ell}(A_\alpha) \to T_{\mathbb{Q}_\ell}(T_{P_\alpha})$, we have a homomorphism of graded algebras $\Lambda_{A_\alpha} \to \Lambda_{P_{x_\alpha}}$ and thus a structure of Λ_{A_α}-module on K_{α,x_α}.

PROPOSITION 4 (Freeness). For all $\alpha \in \mathfrak{A}$, for all homological lifting $T_{\mathbb{Q}_\ell}(A_\alpha) \to T_{\mathbb{Q}_\ell}(T_{P_\alpha})$, K_{α,x_α} is a free Λ_{A_α}-module.

We can derive the δ-inequality from this freeness property as follows. It amounts to prove that for all $\alpha \in \mathfrak{A}$, $\mathrm{codim}(Z_\alpha) \leq \delta_{x_\alpha}$ and that if the equality occurs then there exists a direct factor of $\mathrm{R}^{2d} f_* \mathbb{Q}_\ell[-2d]|_{S_\alpha}$ with support x_α. The freeness property implies that the set

$$\{n \in \mathbb{Z} \mid K^n_{\alpha,x_\alpha} \neq 0\}$$

is an union of integral intervals of length $2(d - \delta_{x_\alpha})$. By Poincaré duality, this set is symmetric with respect to the integer $\dim(M)$. Thus there exist $n \geq \dim(M) + d - \delta_\alpha$ such that $K^n_{\alpha,x_\alpha} \neq 0$. But $K^n_{\alpha,x_\alpha}[-n]$ is a vector space supported by x_α and placed on the degree $n - \dim(Z_\alpha)$ so that

$$n - \dim(Z_\alpha) \leq 2d$$

and

$$\dim(M) + d - \delta_\alpha - \dim(Z_\alpha) \leq n - \dim(Z_\alpha) \leq 2d$$

which implies eventually $\mathrm{codim}(Z_\alpha) \leq \delta_\alpha$. If equality occurs then the punctual sheaf $K^n_{\alpha,x_\alpha}[n - \dim(Z_\alpha)]$ is a direct summand of $\mathrm{R}^{2d} f_* \mathbb{Q}_\ell|_{S_\alpha}$. In particular, if f has only irreducible fibers around x_α, $\mathrm{R}^{2d} f_* \mathbb{Q}_\ell|_{S_\alpha}$ is the constant sheaf which can not admit the punctual sheaf as a direct summand unless $S_\alpha = \{x_\alpha\}$. In this case, we have $Z_\alpha = S \otimes_k \bar{k}$.

Before sketching the proof of the freeness property, let us have a look at the freeness property fiberwise. Let P be a smooth connected commutative group over an algebraically closed field k which acts on a projective scheme M/k with affine stabilizers. Then Λ_P acts on $\mathrm{H}^*(M)$. Let $1 \to R \to P \to A \to 1$ be the canonical Chevalley exact sequence. Any homological lifting $T_{\mathbb{Q}_\ell}(A) \to T_{\mathbb{Q}_\ell}(P)$ induces a homomorphism of algebras $\Lambda_A \to \Lambda_P$. Via any of such homomorphisms, $\mathrm{H}^*(M)$ is a free module over Λ_A.

If $P = A$ is an abelian variety, then A acts on M with finite stabilizers. The quotient $[M/A]$ is an algebraic stack which is proper over k. Using the projectivity assumption on M, we can prove that the Leray spectral sequence associated to the composite map $M \to [M/A] \to \mathrm{Spec}(k)$ degenerate. It follows that there exists an isomorphism

$$\mathrm{H}^*(M) = \mathrm{H}^*([M/A]) \otimes \Lambda_A$$

that makes $\mathrm{H}^*(M)$ a free Λ_A-module.

If P is a smooth connected commutative group defined over a finite field k, there exists a quasi-lifting $A \to P$ i.e. such that the composite map $A \to P \to A$ is the multiplication by a certain non-zero integer N on the abelian variety A. Indeed, the group $\mathrm{Ext}^1(A, R)$ of extensions of A by R which are defined over k is finite and thus it is annihilated by a divisible-enough non-zero integer. That quasi-lifting induces a $\mathrm{Gal}(\bar{k}/k)$-equivariant homological lifting $T_{\mathbb{Q}_\ell}(A) \to T_{\mathbb{Q}_\ell}(P)$ with respect to which the freeness $\mathrm{H}^*_c(M)$ on Λ_A can be proved by reducing to the case where P is an abelian variety. For an arbitrary homological lifting $T_{\mathbb{Q}_\ell}(A) \to T_{\mathbb{Q}_\ell}(P)$, the freeness can be established by the following homogeneity argument due to Deligne. Taking as a base point the linear lifting $T_{\mathbb{Q}_\ell}(A) \to T_{\mathbb{Q}_\ell}(P)$ induced by the quasi-lifting $A \to P$, we can identify the space of all linear liftings with the $\mathbb{Q}_\ell$-vector space $\mathrm{Hom}(T_{\mathbb{Q}_\ell}(A), T_{\mathbb{Q}_\ell}(R))$, this identification being compatible with the natural actions of $\mathrm{Gal}(\bar{k}/k)$. The space of all linear liftings $T_{\mathbb{Q}_\ell}(A) \to T_{\mathbb{Q}_\ell}(P)$ for which $\mathrm{H}^*(M)$ is a free module over Λ_A is an open subset of that vector space which is stable under $\mathrm{Gal}(\bar{k}/k)$ and which contains 0. As the eigenvalues of Frobenius on $T_{\mathbb{Q}_\ell}(A)$ and $T_{\mathbb{Q}_\ell}(R)$ are of different weights, this open subset is then necessarily the whole vector space $\mathrm{Hom}(T_{\mathbb{Q}_\ell}(A), T_{\mathbb{Q}_\ell}(R))$. The case of an arbitrary field can be reduced to finite fields by the standard spreading out argument.

We come now to the proof of the freeness property over a strict henselian base. Consider the restriction of the summands $K_{\alpha'}$ to S_α. Only those $K_{\alpha'}$ with $x_\alpha \in Z_{\alpha'}$ survive to this restriction. In that case, we have $\dim(Z_{\alpha'}) > \dim(Z_\alpha)$. Let us prove the proposition by a downward induction with respect to the integer $\dim(Z_\alpha)$. By induction hypothesis, we know the freeness property for $\alpha' \neq \alpha$ such that $x_\alpha \in Z_{\alpha'}$. The induction hypothesis is however concerned with a freeness property over $S_{\alpha'}$ instead of S_α so that we will need to migrate it to S_α. This follows essentially from the fact that the agebra $\Lambda_{A_{\alpha'}}$ is a free module over Λ_{A_α}. This in turns follows from the assumption $T_{\mathbb{Q}_\ell}(P)$ polarizable.

Except $K_\alpha|_{S_\alpha}$, all other perverse summands $K_{\alpha'}|_{S_\alpha}$ of $f_*\mathbb{Q}_\ell|_{S_\alpha}$ are now known to be free Λ_{A_α}-modules. We want to prove $K_\alpha|_{S_\alpha}$ is also a free Λ_{A_α}-module. For this, we look at the spectral sequence associated with the perverse filtration of $f_*\mathbb{Q}_\ell$ with respect to the fiber functor at x_α. The spectral sequence degenerates at E_2 by an application of the decomposition theorem. The abutment is the cohomology of the special fiber $(f_*\mathbb{Q}_\ell)_{x_\alpha}$ which is known to be a free Λ_{A_α}-module. Now, except the summands which are simple perverse sheaves supported by s, all other direct summands contribute a free Λ_{A_s}-module to the abutment. We can prove that the remaining summands also contribute a free Λ_{A_s}-module to the abutment by using a peculiar property of the ring Λ_{A_α}, its injective modules are also projective, combined with the fact that the contribution $K_\alpha|_{S_\alpha}$ occurs as one subquotient in a filtration of the abutment with only three layers.

4. Application to Hitchin fibration

Fix a proper smooth irreducible curve X defined over a field k and a line bundle D of large degree on X. Let G be a semi-simple group. We put semi-simple instead of reductive only for simplicity. We can construct from these data a Hitchin fibration $f : \mathcal{M} \to \mathcal{A}$ and its anisotropic part $f^{\mathrm{ani}} : \mathcal{M}^{\mathrm{ani}} \to \mathcal{A}^{\mathrm{ani}}$ which is a proper morphism. This construction has been recalled in [**4**]. In the sequel, we will write $f : \mathcal{M} \to \mathcal{A}$ instead of $f^{\mathrm{ani}} : \mathcal{M}^{\mathrm{ani}} \to \mathcal{A}^{\mathrm{ani}}$. As recalled in loc. cit. there exists a group scheme $g : \mathcal{P} \to \mathcal{A}$ which acts on $f : \mathcal{M} \to \mathcal{A}$ and which satisfies the axioms of a weak abelian fibration.

When k is the field of complex numbers and D is the canonical sheaf of X, $\mathcal{M}$ is equipped with a sympectic form and $f : \mathcal{M} \to \mathcal{A}$ is an algebraic completely integrable system. In that case our weak abelian fibration is δ-regular. Thus we can apply theorem 2. In order to get the most satisfying result, we need to take care of the action of sheaf $\pi_0(\mathcal{P})$ of the connected components of the fibers of $\mathcal{P}$ with the interpolating property $\pi_0(\mathcal{P})_a = \pi_0(\mathcal{P}_a)$ for all $a \in \mathcal{A}$. Over our anisotropic base, $\pi_0(\mathcal{P})$ is a sheaf of finite abelian groups. It is known that for every $a \in \mathcal{A}$, $\pi_0(\mathcal{P}_a)$ acts simply transitively on the set of irreducible components of $\mathcal{M}_a$. The action of $\mathcal{P}$ in the perverse sheaves ${}^p\mathrm{H}^n(f_*\mathbb{Q}_\ell)$ factors through the sheaf of connected components $\pi_0(\mathcal{P})$. Let us denote ${}^p\mathrm{H}^j(f_*\mathbb{Q}_\ell)_{st}$ the largest direct factor on which $\pi_0(\mathcal{P})$ acts trivially.

Corollary 5. *Under the above assumptions, if a simple perverse sheaf K is a direct factor of ${}^p\mathrm{H}^j(f_*\mathbb{Q}_\ell)_{st}$ then the support of K is $\mathcal{A} \otimes_k \bar{k}$.*

By theorem 2, it suffices to prove that the stable part $\mathrm{H}^{2d}(f_*\mathbb{Q}_\ell)_{st}$ of the top ordinary cohomology is the constant sheaf $\mathbb{Q}_\ell$. But this follows from the fact that for all geometric point $a \in \mathcal{A}$, $\pi_0(\mathcal{P}_a)$ acts simply transitively on the set of irreducible components of $\mathcal{M}_a$.

If G_1 and G_2 are dual semi-simple groups in Langlands' sense, or more generally having isogenous root data as in [**16**], generic fibers of $f_1 : \mathcal{M}_1 \to \mathcal{A}_1$ and $f_2 : \mathcal{M}_2 \to \mathcal{A}_2$ are roughly speaking isogenous abelian schemes. It follows that ${}^p\mathrm{H}^j(f_{1,*}\mathbb{Q}_\ell)_{st}$ and ${}^p\mathrm{H}^j(f_{2,*}\mathbb{Q}_\ell)_{st}$ are isomorphic local system over an open part of $\mathcal{A}$. It follows from support theorem that there exists an isomorphism between those two perverse sheaves.

If k is a finite field, by counting points in corresponding fibers of f_1 and f_2 we obtain Waldspurger's non-standard variant of the fundamental lemma. However, in positive characteristic, we have not been able to prove the δ-regularity of the fibration. We can only prove that after cutting off a closed subset of $\mathcal{A}^{\mathrm{bad}}$, $g : \mathcal{P} \to \mathcal{A}$ becomes a δ-regular over the remaining part $\mathcal{A}^{\mathrm{good}} = \mathcal{A} - \mathcal{A}^{\mathrm{bad}}$ and moreover that $\mathcal{A}^{\mathrm{good}}$ is still big enough to contain the informations about local orbital integrals which are needed for proving Waldspurger's non-standard fundamental lemma.

As in [**15**], endoscopy theory is closely related to the sheaf $\pi_0(\mathcal{P})$ interpolating the groups of connected components of the fibers of $\mathcal{P}$. It can be useful to understand first the case of SL(2) exposed in the last section of [**15**]. In that case, by direct computation one sees that $\pi_0(\mathcal{P})$ is a quotient of the constant sheaf $\mathbb{Z}/2\mathbb{Z}$ which is non trivial over the locus consisting of the parameter $a \in \mathcal{A}$ coming from a parameter $a_H \in \mathcal{A}_H$ for certain one-dimensional unramified twisted torus H. This locus is a disjoint union of smooth irreducible subschemes of $\mathcal{B}_H$ indexed by those one-dimensional unramified twisted tori H. The action of $\mathbb{Z}/2\mathbb{Z}$ on ${}^p\mathrm{H}^j(f_*\mathbb{Q}_\ell)$ induces a decomposition

$${}^p\mathrm{H}^j(f_*\mathbb{Q}_\ell) = {}^p\mathrm{H}^j(f_*\mathbb{Q}_\ell)_+ \oplus {}^p\mathrm{H}^j(f_*\mathbb{Q}_\ell)_-$$

where ${}^p\mathrm{H}^j(f_*\mathbb{Q}_\ell)_+$ is the stable part which was denoted above by the subscript st and the supplement ${}^p\mathrm{H}^j(f_*\mathbb{Q}_\ell)_-$ will be called the endoscopic part. It is obvious that the restriction of ${}^p\mathrm{H}^j(f_*\mathbb{Q}_\ell)_-$ to the open complement of $\bigcup \mathcal{B}_H$ is trivial. It can be proved that up to twisting ${}^p\mathrm{H}^j(f_*\mathbb{Q}_\ell)_-$ are local system supported on the union of $\mathcal{B}_H$ which are comparable with the cohomology of Hitchin fibration for those tori.

In the general case, it is agreeable to base change to a certain étale covering $\tilde{\mathcal{A}} \to \mathcal{A}$, see [**4**, 5.1] so that over $\tilde{\mathcal{A}}$, we have a canonical surjective homomorphism $\mathbf{X}_* \to \pi_0(\mathcal{P})$ from the constant sheaf of values the group $\mathbf{X}_*$ of cocharacters of a fixed maximal torus T. For every character of finite order $\kappa : \mathbf{X}_* \to \mathbb{Q}_\ell{}^\times$, the locus $\tilde{a} \in \tilde{\mathcal{A}}$ such that κ factor through $\mathbf{X}_* \to \pi_0(\mathcal{P}_a)$ can be identified canonically with disjoint union of $\tilde{\mathcal{A}}_H$ for a certain unramified endoscopic groups H attached to κ. We have a finite action of $\mathbf{X}_*$ on ${}^p\mathrm{H}^j(f_*\mathbb{Q}_\ell)$ which induces a direct decomposition following characters of finite order. The κ-part ${}^p\mathrm{H}^j(f_*\mathbb{Q}_\ell)_\kappa$ have a trivial restriction to the open complement of union of images of $\tilde{\mathcal{A}}_H$ attached to certain unramified endoscopic groups H associated with κ.

Over a base field k characteristic zero, since δ-regularity is available, the support theorem 2 implies :

COROLLARY 6. *Any simple perverse constituent of ${}^p\mathrm{H}^j(f_*\mathbb{Q}_\ell)_\kappa$ of $f_*\mathbb{Q}_\ell|_{\tilde{\mathcal{A}}}$ must have support $\tilde{\mathcal{A}}_H$ where H is one of the unramified endoscopic groups attached to κ .*

This makes possible a comparison between the restriction of ${}^p\mathrm{H}^j(f_*\mathbb{Q}_\ell)_\kappa$ to $\tilde{\mathcal{A}}_H$ and the stable part of the cohomology of Hitchin fibration for H.

Over a base field k of positive characteristic, we will have to throw away a closed subset of $\tilde{\mathcal{A}}_H$ which might violate the δ-regularity before applying the δ-inequality 3. We obtain in that way the theorem [**4**, 5.5]. As in the non-standard case, the remaining part $\tilde{\mathcal{A}}_H^{\mathrm{good}}$ of $\tilde{\mathcal{A}}_H$ still contains all informations about local orbital integrals. Thus after completing the exercices of counting points on Hitchin fibers and affine Springer fibers, as well as the approximation of local conjugacy class by global one, we obtain a proof of the fundamental lemma for Lie algebra conjectured by Langlands and Shelstad in positive characteristic case. In virtue of the theorem of Waldspurger [**17**], and of Cluckers-Hales-Loeser [**2**], the fundamental lemma for p-adic fields is also proved at least large prime p. By Hales [**10**], we know also the fundamental lemma for p-adic fields for all p. However, the fundamental lemma for formal series fields in small characteristic is not known so far.

5. Acknowledgement

The author expresses his gratitude to G. Laumon who carefully read this chapter for errors. This chapter was written during the author's stay at the

Institute for Advanced Study, with the support of the Simonyi Fund and the Monell Foundation.

References

[1] A. Beilinson, J. Bernstein& P. Deligne, Faisceaux pervers, *Astérisque* **100** (1982).
[2] R. Cluckers, T. Hales& F. Loeser, Chapter in this volume.
[3] J.-F. Dat, Lemme fondamental et endoscopie, une approche géométrique, *Séminaire Bourbaki* **940** novembre 2004.
[4] J.F. Dat & Ngo Dac Tuan, Chapter in this volume.
[5] P. Deligne, La conjecture de Weil II, *Publ. Math. de l'I.H.E.S.* **52** (1980) 137–252.
[6] B. Fantechi, L. Göttsche & D. Van Straten, Euler number of the compactified Jacobian and multiplicity of rational curves, *J. Alg. Geom.* **8** (1999), 115–133.
[7] N. Goresky, R. Kottwitz& R. MacPherson, Homology of affine Springer fiber in the unramified case. *Duke Math. J.* **121** (2004) 509–561.
[8] M. Goresky, R. Kottwitz& R. MacPherson, Codimensions of root valuation strata, Preprint arXiv.
[9] M. Goresky, R. Kottwitz & R. MacPherson, Purity of equivalued affine Springer fibers. *Represent. Theory* **10** (2006), 130–146.
[10] T. Hales, On the fundamental lemma for standard endoscopy: reduction to unit elements, *Canad. J. Math.* **47** (1995), no. 5, 974–994.
[11] N. Hitchin, Stable bundles and integrable connections. *Duke Math. J.* **54** (1987) 91–114.
[12] R. Langlands, Les débuts d'une formule des traces stables, *Publications de l'Université Paris 7*, **13** (1983).
[13] G. Laumon,Transformation de Fourier, constantes d'equations fonctionnelles et conjecture de Weil, *Inst. Hautes Études Sci. Publ. Math.* No. 65 (1987), 131–210.
[14] G. Lusztig, Intersection cohomology methods in representation theory. *Proceedings of the International Congress of Mathematicians,* Vol. I, II (Kyoto, 1990), 155–174.
[15] B.C. Ngô, Fibration de Hitchin et endoscopie. *Inv. Math.* **164** (2006) 399–453.
[16] B.C. Ngô, Le lemme fondamental pour les algèbres de Lie. Preprint arXiv.
[17] J.-L. Waldspurger, Endoscopie et changement de caractéristique, prépublication.

Endoscopie et Changement de Caractéristique D'après J.-L. Waldspurger
[L.II.D]

Bertrand Lemaire

Dans son article "Endoscopie et changement de caractéristique" [W2], Waldspurger montre qu'une certaine *conjecture de transfert* pour les algèbres de Lie en caractéristique positive implique la même conjecture en caractéristique nulle, pourvu que la caractéristique résiduelle des corps locaux soit suffisamment grande par rapport au rang du groupe. Précisément, il montre que la conjecture de transfert pour les algèbres de Lie ne dépend pas vraiment du corps local en question mais seulement de son corps résiduel. Comme conséquence de ce résultat, il obtient que le *lemme fondamental* pour les algèbres de Lie en caractéristique positive implique le lemme fondamental pour les algèbres de Lie en caractéristique nulle (en supposant toujours la caractéristique résiduelle assez grande). Ce sont ces résultats que nous décrivons ici.

Rappelons qu'en caractéristique nulle, Waldspurger a montré dans un travail antérieur [W1], par une méthode locale-globale, que le lemme fondamental pour les algèbres de Lie (en presque toute caractéristique résiduelle) implique la conjecture de transfert pour les groupes (en toute caractéristique résiduelle). On sait aussi, d'après des travaux de Clozel, Labesse et Hales, que le lemme fondamental pour les algèbres de Lie implique le lemme fondamental pour les groupes.

En caractéristique positive assez grande, le lemme fondamental pour les algèbres de Lie a été démontré par Ngô Bao Châu [N] — la méthode est géométrique. Il s'ensuit que la conjecture de transfert et le lemme fondamental pour les groupes sont désormais établis en caractéristique nulle. D'ailleurs, le même raisonnement s'applique à l'endoscopie tordue [N, W2, W3].

1. Le principe de la démonstration (introduction)

1.1. Dans tout ce papier, F désigne un corps commutatif localement compact non archimédien, de caractéristique résiduelle p. On note $\mathfrak{o}_F$ son anneau d'entiers, $\mathfrak{p}_F$ l'idéal maximal de $\mathfrak{o}_F$, et $\kappa_F = \mathfrak{o}_F/\mathfrak{p}_F$ le corps résiduel. On note ω_F la valuation sur F normalisée par $\omega_F(F^\times) = \mathbb{Z}$, et l'on pose $|x|_F = q^{-\omega_F(x)}$ $(x \in F)$ où q est le cardinal de κ_F. Pour toute extension algébrique E de F, on définit de la même manière $\mathfrak{o}_E$, $\mathfrak{p}_E$ et κ_E, et l'on note encore ω_F l'unique valuation de E prolongeant ω_F.

Si X est un espace topologique totalement discontinu, on note $C_c^\infty(X)$ l'espace des fonctions complexes sur X qui sont localement constantes et à support compact. Si de plus X est compact (resp. fini), on écrit simplement $C^\infty(X)$ (resp. $C(X)$). Le dual algébrique $\mathrm{Hom}_{\mathbb{C}}(C_c^\infty(X), \mathbb{C})$ est l'espace des distributions sur X ; on le note $\mathcal{D}(X)$.

Si Γ est un groupe opérant sur un ensemble X, on note $X^\Gamma \subset X$ le sous-ensemble formé des éléments fixés par Γ, et $\Gamma_X \subset \Gamma$ le sous-groupe $\{\gamma \in \Gamma : \gamma(x) = x, \forall x \in X\}$.

Pour tout groupe algébrique H défini sur un corps κ, on note $\mathfrak{h}$ (même lettre gothique) son algèbre de Lie, vue comme espace affine défini sur κ.

1.2. Soit F comme en 1.1, et soit G un groupe algébrique réductif connexe défini sur F (fixé jusqu'à la fin du numéro 1.3). On pose $\mathfrak{g} = \mathrm{Lie}(G)$, et pour $r \in \mathbb{R}$, on note $\Phi_F(r)$ l'ensemble des *r-facettes généralisées* [D, 3.2.2] de l'immeuble étendu $\mathfrak{I}(G, F)$. On rappelle que deux points x, $y \in \mathfrak{I}(G, F)$ sont dans la même r-facette généralisée ϕ si et seulement si $\mathfrak{g}(F)_{x,r} = \mathfrak{g}(F)_{y,r}$ et $\mathfrak{g}(F)_{x,r^+} = \mathfrak{g}(F)_{y,r^+}$, où $\mathfrak{g}(F)_{x,r}$ et $\mathfrak{g}(F)_{x,r^+}$ sont les réseaux dans $\mathfrak{g}(F)$ définis par Moy et Prasad; on pose alors $\mathfrak{g}(F)_\phi = \mathfrak{g}(F)_{x,r}$ et $\mathfrak{g}(F)_{\phi^+} = \mathfrak{g}(F)_{x,r^+}$. Soient

$$\mathcal{S}_F = \bigoplus_{r \in \mathbb{R}} \mathcal{S}_F(r), \qquad \mathcal{S}_F(r) = \bigoplus_{\phi \in \Phi_F(r)} C(\mathfrak{g}(F)_\phi/\mathfrak{g}(F)_{\phi^+}).$$

Les inclusions $C(\mathfrak{g}(F)_\phi/\mathfrak{g}(F)_{\phi^+}) \subset C^\infty(\mathfrak{g}(F)_\phi) \subset C_c^\infty(\mathfrak{g}(F))$ définissent une application linéaire $\mathcal{S}_F \to C_c^\infty(\mathfrak{g}(F))$, $\varphi \mapsto \tilde{\varphi}$.

Si T est un tore défini sur une extension algébrique E de F, on note $\omega_{\mathfrak{t},F} : \mathfrak{t}(E) \to \mathbb{Q} \cup \{+\infty\}$ la valuation sur $\mathfrak{t}(E)$ définie par $\omega_{\mathfrak{t},F}(0) = +\infty$ et

$$\omega_{\mathfrak{t},F}(X) = \inf\{\omega_F(\alpha(X)) : \alpha \in \mathrm{X}^*(T)\} \qquad (X \in \mathfrak{t}(E) \smallsetminus \{0\}).$$

Et pour $s \in \mathbb{R}$, on pose $\mathfrak{t}(E)_s = \{X \in \mathfrak{t}(E) : \omega_{\mathfrak{t},F}(X) \geq s\}$ et $\mathfrak{t}(E)_{s^+} = \bigcup_{s'>s} \mathfrak{t}(E)_{s'}$.

1.3. Soit un élément $X \in \mathfrak{g}(F)$ semisimple régulier. Notons $T = G_X^\circ$ le centralisateur connexe de X dans $\mathfrak{g}$; c'est un tore maximal de G défini sur F. Via le choix d'une mesure $G(F)$-invariante $d\bar{g}_X$ sur l'espace homogène $T(F)\backslash G(F)$, pour $f \in C_c^\infty(\mathfrak{g}(F))$, on définit l'intégrale orbitale

$$\Lambda^{G(F)}(f, X) = \Delta(X) \int_{T(F)\backslash G(F)} f(\mathrm{Ad}\, g^{-1}(X)) d\bar{g}_X$$

où $\Delta(X)$ désigne le facteur de normalisation habituel :

$$\Delta(X) = |\det_F(\mathrm{ad}(X); \mathfrak{t}(F)\backslash\mathfrak{g}(F))|_F^{\frac{1}{2}}.$$

L'intégrale ci-dessus est absolument convergente, et l'application $\varphi \mapsto \Lambda^{G(F)}(\tilde{\varphi}, X)$ est une forme linéaire sur $\mathcal{S}_F$.

On suppose que "p est grand par rapport au rang de G" ; en particulier, on suppose que T se déploie sur une extension modérément ramifiée de F. Posons $r(X) = \omega_{\mathfrak{t},F}(X)$, et soit une fonction $\varphi \in C(\mathfrak{g}(F)_\phi/\mathfrak{g}(F)_{\phi^+})$ pour un $r \in \mathbb{R}$ et une r-facette généralisée ϕ. Si $r(X) < r$, alors la $G(F)$-orbite de X n'intersecte pas le support de $\tilde{\varphi}$; par conséquent $\Lambda^{G(F)}(X, \tilde{\varphi}) = 0$. On peut donc supposer $r(X) \geq r$. On distingue alors deux cas :

Cas 1 : $r(X) > r$. D'après les travaux de DeBacker [D], il existe une fonction $\varphi_+ \in \oplus_{s>r} \mathcal{S}_F(s)$ telle que $\Lambda^{G(F)}(\tilde{\varphi}, X) = \Lambda^{G(F)}(\tilde{\varphi}_+, X)$.

Cas 2 : $r(X) = r$. Notons $G' \subset G$ le "r-centralisateur" de X dans G, c'est-à-dire le sous-groupe engendré par T et par les sous-groupes radiciels de G associés aux racines $\alpha \in \Sigma(G,T)$ telles que $\omega_{\mathfrak{t},F}(\alpha(X)) > r$. Le groupe G' est réductif connexe et défini sur F. D'après les travaux de Kim-Murnaghan [KM], il existe un élément $X' \in \mathfrak{g}'(F)$ semisimple régulier et une fonction $\varphi' \in \mathcal{S}'_F(r)$, tels que $\Lambda^{G(F)}(\tilde{\varphi}, X) = \Lambda^{G'(F)}(\tilde{\varphi}', X')$; où $\mathcal{S}'_F(r)$ désigne l'analogue de $\mathcal{S}_F(r)$ pour G'.

On peut donc ramener le "calcul" de $\Lambda^{G(F)}(X, \tilde{\varphi})$ à celui d'une intégrale orbitale semisimple régulière sur un groupe plus petit. L'idée est de montrer (par récurrence sur $r(X)-r$ et sur le rang de G) que les applications linéaires $\mathcal{S}_F(r) \to \oplus_{s>r}\mathcal{S}_F(s)$, $\varphi \mapsto \varphi_+$ et $\mathcal{S}_F(r) \to \mathcal{S}'_F(r)$, $\varphi \mapsto \varphi'$ ci-dessus, ne dépendent pas vraiment des données F, G, X, mais seulement de $\overline{\mathbb{F}}_q$, d'une donnée radicielle sur $\overline{\mathbb{F}}_q$ (etc.). Le même principe s'applique aux intégrales orbitales stables et aux intégrales orbitales endoscopiques.

1.4. Décrivons brièvement comment sont organisés les résultats.

Dans la section 2, le groupe de Galois Γ de l'extension modérément ramifiée maximale F^{mod} de F, est décrit en termes indépendants de F. Via le choix d'uniformisantes, on a une décomposition $\Gamma = I \rtimes \Theta$. On note $F^{\mathrm{nr}} = (F^{\mathrm{mod}})^I$ la sous-extension non ramifiée maximale de F^{mod}.

Dans la section 3, au lieu de se donner le groupe G défini sur F, on fixe une donnée de racines $\mathcal{D} = (\mathrm{X}^*, \Sigma, \Delta, \mathrm{X}_*, \check{\Sigma}, \check{\Delta})$ munie d'une action de Γ. On suppose que "p est grand par rapport au rang de $\mathcal{D}$" (i.e. le rang commun de X^* et de X_* sur $\mathbb{Z}$). On note W le groupe de Weyl de Σ. À cette donnée $\mathcal{D}$ est associée un groupe réductif connexe G, défini sur F et quasi-déployé sur F. On note V l'espace $\mathrm{X}_* \otimes_{\mathbb{Z}} \mathbb{R}$. Le groupe Γ opère sur V via son action sur X_*, et le sous-espace $V^{\mathrm{nr}} = V^I \subset V$ s'identifie à un appartement de l'immeuble étendu $\mathfrak{I}(G, F^{\mathrm{nr}})$. L'action du stabilisateur $N(F^{\mathrm{nr}})$ de V^{nr} dans $G(F^{\mathrm{nr}})$ est décrite en termes indépendants de F ; en particulier, elle se factorise à travers un groupe $\mathbf{N}^{\mathrm{nr}}$ indépendant de F.

Dans la section 4, grâce aux travaux de Kottwitz, on attache à $\mathcal{D}$ un ensemble D de cocycles de Θ à valeurs dans $\mathbf{N}^{\mathrm{nr}}$, qui paramétrise l'ensemble $\mathrm{H}^1(\Theta, G(F^{\mathrm{nr}})) = \mathrm{H}^1(\Gamma, G(F^{\mathrm{mod}}))$. À chaque $d \in D$ est associé un cocycle $d_F \in \mathrm{Z}^1(\Theta, G(F^{\mathrm{nr}}))$, lequel définit un torseur intérieur ${}_d\xi : {}_dG \to G$ défini sur F^{nr}. Soit $d \in D$, vu comme un cocycle de Γ trivial sur I. L'espace V muni de l'action tordue $\Gamma \times V \to V$, $(\theta, v) \mapsto d(\gamma) \cdot \gamma(v)$, est noté ${}_dV$. Le sous-espace ${}_dW = ({}_dV)^{\Gamma} \subset V^{\mathrm{nr}}$ s'identifie à un appartement de l'immeuble étendu $\mathfrak{I}({}_dG, F)$.

Dans les sections 5 et 6, on décrit les quotients de la filtration de Moy-Prasad $\{{}_d\mathfrak{g}(F)_{v,r} : r \in \mathbb{R}\}$ dans ${}_d\mathfrak{g}(F) = \mathrm{Lie}({}_dG)(F)$ associée à un point $v \in {}_dW$, en termes indépendants de F. Notons $\mathbb{Z}_{(p)}$ le localisé de $\mathbb{Z}$ en p, et posons ${}_dW_{(p)} = {}_dW \cap (\mathrm{X}_* \otimes_{\mathbb{Z}} \mathbb{Z}_{(p)})$; c'est un sous-ensemble dense de ${}_dW$. Pour $(v, r) \in {}_dW_{(p)} \times \mathbb{Z}_{(p)}$, on définit un $\overline{\mathbb{F}}_q$-espace vectoriel ${}_d\underline{\mathfrak{g}}_{v,r}$ muni d'une action de Θ, tel que le $\mathbb{F}_q$-espace vectoriel ${}_d\underline{\mathfrak{g}}_{v,r}^{\Theta}$ s'identifie canoniquement à ${}_d\mathfrak{g}(F)_{v,r}/{}_d\mathfrak{g}(F)_{v,r^+}$. L'espace $V^{\mathrm{nr}} \times \mathbb{R}$ est muni d'une décomposition en facettes, indépendante de F, vérifiant les propriétés suivantes : notons ${}_d\Phi$ l'ensemble de ces facettes qui sont invariantes pour l'action de Θ sur V^{nr} tordue par d. Pour toute facette $\phi \in {}_d\Phi$, l'intersection $\phi \cap ({}_dW_{(p)} \times \mathbb{Z}_{(p)})$ est dense dans $\phi \cap ({}_dW \times \mathbb{R})$; et pour tous (v, r), $(v', r') \in \phi \cap ({}_dW_{(p)} \times \mathbb{Z}_{(p)})$, on a ${}_d\mathfrak{g}(F)_{v,r} = {}_d\mathfrak{g}(F)_{v',r'}$, ${}_d\mathfrak{g}(F)_{v,r^+} = {}_d\mathfrak{g}(F)_{v',r'^+}$ et ${}_d\underline{\mathfrak{g}}_{v,r}^{\Theta} = {}_d\underline{\mathfrak{g}}_{v',r'}^{\Theta}$. On peut donc remplacer la paire d'indices "v, r" par un indice "ϕ" dans les notations. Posons ${}_d\mathcal{S} = \oplus_{\phi \in {}_d\Phi}\, {}_d\mathcal{S}_{\phi}$ avec ${}_d\mathcal{S}_{\phi} = C({}_d\underline{\mathfrak{g}}_{\phi}^{\Theta})$. Via les identifications $C({}_d\underline{\mathfrak{g}}_{\phi}^{\Theta}) = C({}_d\mathfrak{g}(F)_{\phi}/{}_d\mathfrak{g}(F)_{\phi^+})$ pour $\phi \in {}_d\Phi$, on obtient une application linéaire ${}_d\mathrm{rea}_F : {}_d\mathcal{S} \to C_{\mathrm{c}}^{\infty}({}_d\mathfrak{g}(F))$.

Dans la section 7, on montre la version indépendante de F du **cas 1** de 1.3. Pour $X \in {}_d\mathfrak{g}(F)$ semisimple régulier, le centralisateur T_X de X dans ${}_dG$ est un tore maximal défini sur F, et l'espace homogène $T_X(F)\backslash{}_dG(F)$ est muni d'une mesure ${}_dG(F)$-invariante canonique ; la distribution $\Lambda^{{}_dG(F)}(\cdot, X)$ sur ${}_d\mathfrak{g}(F)$ est donc bien définie. À toute facette $\phi \subset V^{\mathrm{nr}} \times \mathbb{R}$ est associé un invariant $r(\phi) \in \mathbb{Z}_{(p)}$. Pour $r \in \mathbb{Z}_{(p)}$, posant ${}_d\mathcal{S}_r = \oplus_{\phi \in {}_d\Phi, r(\phi) \geq r}\, {}_d\mathcal{S}_{\phi} \subset {}_d\mathcal{S}$ et ${}_d\mathcal{S}_{r^+} = \oplus_{\phi \in {}_d\Phi, r(\phi) > r}\, {}_d\mathcal{S}_{\phi} \subset {}_d\mathcal{S}_r$, on montre qu'il existe une application linéaire ${}_dl_r^+ : {}_d\mathcal{S}_r \to {}_d\mathcal{S}_{r^+}$ telle que l'application ${}_d\mathrm{rea}_F \circ {}_dl_r^+$ "réalise" l'application $\varphi \mapsto \varphi_+$ de 1.3 (proposition 7.4).

Dans les sections 8 et 9, on montre la version indépendante de F du **cas 2** de 1.3. On commence par décrire le centralisateur d'un "bon élément" (au sens de [KM]) de ${}_d\mathfrak{g}(F)$, en termes indépendants de F. On considère des quadruplets $(\mathcal{D}', \psi, r, \underline{Z}')$ où $\mathcal{D}' = (\mathrm{X}'^*, \Sigma', \Delta', \mathrm{X}'_*, \check{\Sigma}', \check{\Delta}')$ est une donnée de racines munie d'une action de Γ ; ψ est une paire d'isomorphismes de $\mathbb{Z}$-modules $(\mathrm{X}'^* \to \mathrm{X}^*, \mathrm{X}'_* \to \mathrm{X}_*)$ en dualité ; $r \in \mathbb{Z}_{(p)}$; et $\underline{Z}' \in \underline{\mathfrak{t}}'(r)^{\Gamma}$ où $\underline{\mathfrak{t}}'(r)$ désigne le $\overline{\mathbb{F}}_q$-espace vectoriel $\mathrm{X}'_* \otimes_{\mathbb{Z}} \overline{\mathbb{F}}_q$ munie d'une certaine action tordue de Γ définie par r. On demande que $\mathcal{D}'$ s'identifie via ψ à un sous-système de Levi standard de $\mathcal{D}$; que pour tout $\gamma \in \Gamma$, il existe un $w_{\psi}(\gamma) \in W$ tel que $\psi \circ \gamma = w_{\psi}(\gamma) \circ \gamma \circ \psi$; et que l'ensemble $\{\alpha \in \Sigma : \alpha \circ \psi(\underline{Z}') = 0\}$ coïncide avec $\psi(\Sigma')$. Affectons d'un exposant "$'$" les objets attachés à $\mathcal{D}'$ comme précédemment. On a une application naturelle $\psi_{D'} : D' \to D$. Pour $d' \in D'$, posant $d = \psi_{D'}(d')$, on montre qu'il existe une application linéaire

${}_{d'}l'_r : {}_d\mathcal{S}_r \to {}_{d'}\mathcal{S}'_r$ telle que l'application ${}_{d'}\mathrm{rea}'_F \circ {}_{d'}l'_r$ "réalise" l'application $\varphi \mapsto \varphi'$ de 1.3 (proposition 9.3).

Dans la section 10 est montré le théorème principal de [W2]. On considère simultanément toutes les formes tordues ${}_dG$ pour $d \in D$. On sait que les classes de conjugaison stable semisimples régulières dans ${}_D\mathfrak{g}(F) = \coprod_{d\in D} {}_d\mathfrak{g}(F)$ sont paramétrées par l'ensemble $\mathcal{Z}_F = (\mathfrak{t}^{\mathrm{mod}}_{\mathrm{reg}}/W)^\Gamma$ où $\mathfrak{t}^{\mathrm{mod}}_{\mathrm{reg}}$ désigne l'ensemble des éléments W-réguliers de $\mathfrak{t}^{\mathrm{mod}} = \mathrm{X}_* \otimes_{\mathbb{Z}} F^{\mathrm{mod}}$. Pour $z_F \in \mathcal{Z}_F$, on note $\mathcal{O}_{\mathrm{st}}(z_F) \subset {}_D\mathfrak{g}(F)$ la classe de conjugaison stable associée à z_F. On construit (de manière indépendante de F) un ensemble $\mathcal{Z}$, et pour chaque $z \in \mathcal{Z}$, un groupe fini D_z, vérifiant les propriétés suivantes : il existe une application surjective $\zeta_F : \mathcal{Z}_F \to \mathcal{Z}$, et pour tout $z_F \in \mathcal{Z}_F$, il existe une application surjective $\delta_{z_F} : \mathcal{O}_{\mathrm{st}}(z_F) \to D_{\zeta_F(z_F)}$ dont les fibres sont exactement les classes de conjugaison ordinaire dans $\mathcal{O}_{\mathrm{st}}(z_F)$. Posons $\mathcal{S} = \oplus_{d\in D}\, {}_d\mathcal{S}$, $\mathcal{C}_F = \oplus_{d\in D} C^\infty_{\mathrm{c}}({}_d\mathfrak{g}(F))$, et notons rea_F l'application linéaire $\oplus_{d\in D}\, {}_d\mathrm{rea}_F : \mathcal{S} \to \mathcal{C}_F$. Pour $f \in \mathcal{C}_F$ et $X \in {}_D\mathfrak{g}(F)$ semisimple régulier, on pose $\Lambda^{{}_DG(F)}(f, X) = \Lambda^{{}_dG(F)}(f_d, X)$ où d est l'élément de D tel que $X \in {}_d\mathfrak{g}(F)$, et f_d est la composante de f sur ${}_d\mathfrak{g}(F)$. Pour $z \in \mathcal{Z}$ et $\delta \in D_z$, on montre qu'il existe une forme linéaire $\Lambda^{\mathcal{D}}(\cdot, z, \delta)$ sur $\mathcal{S}$ telle que pour tous $z_F \in \zeta_F^{-1}(z)$ et $X \in \delta_{z_F}^{-1}(\delta)$, on ait l'égalité $\Lambda^{{}_DG(F)}(\mathrm{rea}_F(\cdot), X) = \Lambda^{\mathcal{D}}(\cdot, z, \delta)$ (théorème 10.7).

Dans la section 11, on généralise le théorème principal au transfert endoscopique. On introduit la notion (indépendante de F) de *donnée endoscopique de $\mathcal{D}$*, c'est-à-dire un triplet $(\mathcal{D}_\natural, \eta, s)$ où $\mathcal{D}_\natural = (\mathrm{X}^*_\natural, \Sigma_\natural, \Delta_\natural, \mathrm{X}_{*,\natural}, \check{\Sigma}_\natural, \check{\Delta}_\natural)$ est une donnée de racines munie d'une action de Γ, etc. (cf. 11.1). Un tel triplet détermine un triplet endoscopique $(G_\natural, \hat{\eta}, s)$ de G. Mieux : $(G_\natural, \hat{\eta}, s)$ est un triplet endoscopique de ${}_dG$ pour tout $d \in D$. Affectons d'un indice "$\natural$" les objets relatifs à la donnée $\mathcal{D}_\natural$. De η se déduit un isomorphisme $\mathfrak{t}^{\mathrm{mod}}_\natural \to \mathfrak{t}^{\mathrm{mod}}$, qui n'est en général pas Γ-équivariant. Notons $\mathfrak{t}^{\mathrm{mod}}_{\natural,\mathcal{D}-\mathrm{reg}}$ l'image réciproque de $\mathfrak{t}^{\mathrm{mod}}_{\mathrm{reg}}$, et posons $\mathcal{Z}_{\natural,\mathcal{D}-\mathrm{reg},F} = (\mathfrak{t}^{\mathrm{mod}}_{\natural,\mathcal{D}-\mathrm{reg}}/W_\natural)^\Gamma$. On a une application naturelle $\eta_F : \mathcal{Z}_{\natural,\mathcal{D}-\mathrm{reg},F} \to \mathcal{Z}_F$, mais il n'existe en général pas d'application $\mathcal{Z}_\natural \to \mathcal{Z}$ qui lui corresponde. Pour résoudre cette difficulté, on construit (de manière indépendante de F) un ensemble $\mathcal{Y}$ muni d'applications $\eta_{\mathcal{Y}} : \mathcal{Y} \to \mathcal{Z}$ et $\epsilon : \mathcal{Y} \to \mathcal{Z}_\natural$, et pour chaque $y \in \mathcal{Y}$, un groupe fini D_y muni d'identifications canoniques $D_{\eta_{\mathcal{Y}}(y)} = D_y = D_{\epsilon(y)}$, vérifiant les propriétés suivantes : il existe une application surjective $\tau_F : \mathcal{Z}_{\natural,\mathcal{D}-\mathrm{reg},F} \to \mathcal{Y}$ telle que $\eta_{\mathcal{Y}} \circ \tau_F = \zeta_F \circ \eta_F$ et $\epsilon \circ \tau_F = \zeta_{\natural,F}|_{\mathcal{Z}_{\natural,\mathcal{D}-\mathrm{reg},F}}$; et pour tout $z_{\natural,F} \in \mathcal{Z}_{\natural,\mathcal{D}-\mathrm{reg},F}$, posant $z_F = \eta_F(z_{\natural,F})$, $y = \tau_F(z_{\natural,F})$ et $z = \eta_{\mathcal{Y}}(y) = \zeta_F(z_F)$, il existe un $s_y \in \mathrm{Hom}(D_y, \mathbb{C}^\times)$ tel que pour $X \in \mathcal{O}_{\mathrm{st}}(z_F)$, le facteur de transfert de Langlands-Shelstad $\Delta(z_{\natural,F}, X)$ coïncide avec $s_y(\delta_{z_F}(X)^{-1})$. Pour $f \in \mathcal{C}_F$ et $z_{\natural,F} \in \mathcal{Z}_{\natural,\mathcal{D}-\mathrm{reg},F}$, on définit l'intégrale orbitale endoscopique

$$\Lambda^{{}_DG(F),G_\natural(F)}(f, z_{\natural,F}) = \sum_X \Delta(z_{\natural,F}, X)\Lambda^{{}_DG(F)}(f, X)$$

où X parcourt un système de représentants dans $\mathcal{O}_{\mathrm{st}}(X)$ des classes de conjugaison ordinaire. Pour $y \in \mathcal{Y}$, le théorème 10.7 implique qu'il existe une forme linéaire $\Lambda^{\mathcal{D},\mathcal{D}^\natural}(\cdot, y)$ sur $\mathcal{S}$ telle que pour tout $z_{\natural,F} \in \tau_F^{-1}(y)$, on ait l'égalité $\Lambda^{D^{G(F)},G_\natural(F)}(\mathrm{rea}_F(\cdot), z_{\natural,F}) = \Lambda^{\mathcal{D},\mathcal{D}_\natural}(\cdot, y)$ (proposition 11.4). On en déduit que la conjecture de transfert pour les fonctions dans l'image des applications rea_F et $\mathrm{rea}_{\natural,F}$, est indépendante de F (corollaire 11.4). Si maintenant les données $\mathcal{D}$ et $\mathcal{D}_\natural$ sont non ramifiées, en supposant toujours que p est grand par rapport au rang de $\mathcal{D}$, on obtient que le lemme fondamental pour les algèbres de Lie est une conjecture indépendante de F (corollaire 11.5).

2. L'extension modérément ramifiée maximale d'un corps local

2.1. Fixons un nombre premier p et une puissance $q = p^r$ de p. Fixons aussi une clôture algébrique $\overline{\mathbb{F}}_q$ de $\mathbb{F}_q$. Posons $\Theta = \hat{\mathbb{Z}}$ et notons θ_1 le générateur topologique canonique de Θ ($\theta_1 = 1$). Le groupe Θ s'identifie au groupe de Galois de $\overline{\mathbb{F}}_q/\mathbb{F}_q$, θ_1 s'identifiant au Frobénius.

Notons $\mathcal{P}$ l'ensemble des entiers $e \geq 1$ qui sont premiers à p. Posons $I = \varprojlim_{e\in\mathcal{P}} \zeta_e(\overline{\mathbb{F}}_q)$ où, pour tout corps κ, $\zeta_e(\kappa)$ désigne le groupe des racines e-ièmes de l'unité dans κ; la limite projective étant prise par rapport aux morphismes de transition $\zeta_{ee'}(\overline{\mathbb{F}}_q) \to \zeta_e(\overline{\mathbb{F}}_q)$, $\mu \mapsto \mu^{e'}$ (e, $e' \in \mathcal{P}$). Pour $\sigma \in I$, on note σ_e l'image de σ dans $\zeta_e(\overline{\mathbb{F}}_q)$. Et pour $e \in \mathcal{P}$, on note $I(e)$ l'unique sous-groupe de I d'indice e. On munit I de l'action de Θ donnée par $\theta_1(\sigma) = \sigma^q$, et l'on pose $\Gamma = \Theta \ltimes I$.

Notons $\mathbb{Z}_{(p)}$ le localisé de $\mathbb{Z}$ en p, et

$$I \times \mathbb{Z}_{(p)} \to \overline{\mathbb{F}}_q^\times,\ (\sigma, r) \mapsto r(\sigma)$$

la forme bilinéaire définie comme suit : pour $(\sigma, r) \in I \times \mathbb{Z}_{(p)}$, on écrit $r = \frac{n}{e}$ avec $n \in \mathbb{Z}$ et $e \in \mathcal{P}$, et l'on pose $r(\sigma) = \sigma_e^{-n}$. Enfin, posons $\mathbf{F}^* = \mathbb{Z}_{(p)} \times \overline{\mathbb{F}}_q^\times$ et munissons $\mathbf{F}^*$ de l'action de Γ définie comme suit : pour $(r, x) \in \mathbb{Z}_{(p)} \times \overline{\mathbb{F}}_q^\times$ et $(\theta, \sigma) \in \Theta \ltimes I$, on pose $\theta(r, x) = (r, \theta(x))$ et $\sigma(r, x) = (r, r(\sigma)^{-1}x)$.

2.2 On suppose que κ_F est de cardinal q. Fixons une clôture séparable $\overline{F}$ de F, et notons F^{nr} (resp. F^{mod}) la sous-extension non ramifiée (resp. modérément ramifiée) maximale de $\overline{F}/F$. On a donc les inclusions $F \subset F^{\mathrm{nr}} \subset F^{\mathrm{mod}} \subset \overline{F}$. Posons $\overline{\kappa}_F = \kappa_{F^{\mathrm{nr}}} = \kappa_{\overline{F}}$. On a une identification canonique $\mathrm{Gal}(F^{\mathrm{nr}}/F) = \Theta$. Fixons un isomorphisme $\iota : \overline{\kappa}_F \to \overline{\mathbb{F}}_q$. Pour $e \in \mathcal{P}$, la composée de ι et de la réduction modulo $\mathfrak{p}_{F^{\mathrm{nr}}}$, induit une identification $\zeta_e(F^{\mathrm{nr}}) = \zeta_e(\overline{\mathbb{F}}_q)$. Fixons une uniformisante ϖ_1 de F, et pour tout $e \in \mathcal{P}$, fixons un élément $\varpi_{\frac{1}{e}} \in F^{\mathrm{mod}}$ de sorte que pour e, $e' \in \mathcal{P}$, on ait l'égalité $(\varpi_{\frac{1}{ee'}})^{e'} = \varpi_{\frac{1}{e}}$. Alors on a

$$F^{\mathrm{mod}} = \bigcup_{e\in\mathcal{P}} F^{\mathrm{nr}}(\varpi_{\frac{1}{e}}).$$

Pour $r \in \mathbb{Z}_{(r)}$, on écrit $r = \frac{n}{e}$ avec $n \in \mathbb{Z}$ et $e \in \mathcal{P}$, et l'on pose $\varpi_r = (\varpi_{\frac{1}{e}})^n$.

On identifie Γ au groupe de Galois $\mathrm{Gal}(F^{\mathrm{mod}}/F)$ de la manière suivante : le groupe Θ opère naturellement sur F^{nr} (via l'identification $\Theta = \mathrm{Gal}(F^{\mathrm{nr}}/F)$) ; le groupe I fixe tout élément de F^{nr} ; et pour $\sigma \in I$ et $e \in \mathcal{P}$, on a $\sigma(\varpi_{\frac{1}{e}}) = \sigma_e \varpi_{\frac{1}{e}}$. Pour $(\sigma, r) \in I \times \mathbb{Z}_{(p)}$, on a donc

$$r(\sigma) = \sigma(\varpi_{-r})\varpi_r.$$

On définit un homomorphisme $\mathbf{r}_F : F^{\mathrm{mod},\times} \to \mathbf{F}^*$ de la manière suivante : pour $e \in \mathcal{P}$, on pose $\mathbf{r}_F(\varpi_{\frac{1}{e}}) = (\frac{1}{e}, 1)$; et pour $x \in F^{\mathrm{mod}}$ tel que $\omega_F(x) = 0$, notant $\bar{x}$ l'image de x dans $\overline{\kappa}_F$, on pose $\mathbf{r}_F(x) = (0, \iota(\bar{x}))$. Notons $U^+_{F^{\mathrm{mod}}}$ le sous-groupe de $U_{F^{\mathrm{mod}}} = \mathfrak{o}^\times_{F^{\mathrm{mod}}}$ formé des x tels que $\omega_F(1 - x) > 0$. On a la suite exacte courte

$$1 \to U^+_{F^{\mathrm{mod}}} \to F^{\mathrm{mod},\times} \xrightarrow{\mathbf{r}_F} \mathbf{F}^* \to 1.$$

De plus, l'homomorphisme $\mathbf{r}_F$ est Γ-équivariant. On définit une section $s_F : \mathbb{Z}_{(p)} \to F^{\mathrm{mod},\times}$ de $\mathbf{r}$ au-dessus de $\mathbb{Z}_{(p)}$: pour $r \in \mathbb{Z}_{(p)}$, on pose $s_F(r) = \varpi_r$.

Pour $e \in \mathcal{P}$, on note $F^e \subset F^{\mathrm{mod}}$ le sous-corps fixé par $I(e)$; on a $F^e = F(\varpi_{\frac{1}{e}})$. Pour alléger l'écriture, on écrit $\mathfrak{o}^e$, $\mathfrak{p}^e$ au lieu de $\mathfrak{o}_{F^e}$, $\mathfrak{p}_{F^e}$. Et si $e = 1$, on remplace l'exposant "e" par un exposant "nr" : $\mathfrak{o}^{\mathrm{nr}}$, $\mathfrak{p}^{\mathrm{nr}}$.

2.3. On a adjoint à F (tel que κ_F soit de cardinal q) des données $\overline{F}$, ι et $\{\varpi_{\frac{1}{e}}\}_{e \in \mathcal{P}}$, obtenant ainsi un quadruplet $\underline{F} = (F, \overline{F}, \iota, \{\varpi_{\frac{1}{e}}\})$. Deux quadruplets $\underline{F} = (F, \overline{F}, \iota, \{\varpi_{\frac{1}{e}}\})$ et $\underline{F}' = (F', \overline{F}', \iota', \{\varpi'_{\frac{1}{e}}\})$ sont dits *isomorphes* s'il existe un isomorphisme de corps $\lambda : \overline{F} \to \overline{F}'$ tel que $\lambda(F) = F'$ (ce qui entraîne $\lambda(F^{\mathrm{nr}}) = F'^{\mathrm{nr}}$ et $\lambda(F^{\mathrm{mod}}) = F'^{\mathrm{mod}}$), $\iota' \circ \bar{\lambda} = \iota$ où $\bar{\lambda} : \overline{\kappa}_F \to \overline{\kappa}_{F'}$ désigne l'isomorphisme de corps déduit de λ par restriction et passage aux quotients, et $\lambda(\varpi_{\frac{1}{e}}) = \varpi'_{\frac{1}{e}}$ pour tout $e \in \mathcal{P}$. On fixe un ensemble CL_q de représentants des classes d'isomorphisme de tels quadruplets. Pour alléger l'écriture, on considérera les éléments de CL_q comme des corps locaux F munis de "données supplémentaires occultes" $\overline{F}$, ι et $\{\varpi_{\frac{1}{e}}\}$.

3. Groupes et données radicielles

3.1. Soit une donnée de racines $\mathcal{D} = (\mathrm{X}^*, \Sigma, \Delta, \mathrm{X}_*, \check{\Sigma}, \check{\Delta})$ munie d'une action de Γ. C'est-à-dire que X^* et X_* sont des $\mathbb{Z}$-modules libres de type fini en dualité, Σ est un système de racines *réduit* dans X^*, $\Delta \subset \Sigma$ est une base de racines simples, et $\check{\Sigma}$ (resp. $\check{\Delta}$) est l'ensemble des coracines associé à Σ (resp. Δ) dans X_*. La base Δ détermine un sous-ensemble de racines positives $\Sigma^+ \subset \Sigma$. On note W le groupe de Weyl de Σ. Le rang de $\mathcal{D}$, noté $\mathrm{rg}(\mathcal{D})$, est par définition le rang commun de X^* et de X_* sur $\mathbb{Z}$.

De $\mathcal{D}$ se déduisent de la manière habituelle des données de racines $\mathcal{D}_{\mathrm{sc}} = (\mathrm{X}^*_{\mathrm{sc}}, \Sigma, \Delta, \mathrm{X}_{*,\mathrm{sc}}, \check{\Sigma}, \check{\Delta})$ et $\mathcal{D}_{\mathrm{der}} = (\mathrm{X}^*_{\mathrm{der}}, \Sigma, \Delta, \mathrm{X}_{*,\mathrm{der}}, \check{\Sigma}, \check{\Delta})$ munies d'une action de Γ. Ici, $\mathrm{X}_{*,\mathrm{sc}} \subset \mathrm{X}_*$ désigne le réseau engendré par $\check{\Delta}$, et

$X^*_{\rm sc} \subset X^* \otimes_{\mathbb{Z}} \mathbb{Q}$ celui engendré par les poids fondamentaux; et l'on pose $X_{*,\rm der} = X_* \cap (X_{*,\rm sc} \otimes_{\mathbb{Z}} \mathbb{Q})$ et $X^*_{\rm der} = \mathrm{Hom}_{\mathbb{Z}}(X_{*,\rm der}, \mathbb{Z})$.

On suppose que p est "grand par rapport au rang de $\mathcal{D}$". Pour tout système de racines Σ' dans un espace vectoriel réel V' qu'il engendre, introduisons l'ensemble des coracines $\check{\Sigma}'$ dans l'espace dual de V', et le réseau $\check{X}'$ engendré par $\check{\Sigma}'$. Si Σ' est réduit et irréductible, on sait définir le nombre de Coxeter $h(\Sigma')$ de Σ'. Si Σ' est seulement réduit, on écrit $\Sigma' = \coprod_{i=1}^n \Sigma'_i$ où Σ'_i est une composante irréductible de Σ', et l'on pose $h(\Sigma') = \max\{h(\Sigma'_i) : i = 1, \ldots, n\}$. On définit les conditions :

[$(P^1_{\Sigma'})$] pour tout sous-ensemble linéairement indépendant $A \subset \Sigma'$, l'indice du réseau $\mathbb{Z}[A]$ dans le réseau $\mathbb{Q}[A] \cap \mathrm{Hom}_{\mathbb{Z}}(\check{X}', \mathbb{Z})$ est premier à p;

[$(P^2_{\Sigma'})$] pour tout sous-ensemble linéairement indépendant $A \subset \Sigma'$ et toute famille $\{c_\alpha\}_{\alpha \in A}$ de nombres rationnels telle que $\sum_{\alpha \in A} c_\alpha \alpha \in \Sigma'$, on a $\omega_{\mathbb{Q}_p}(1 - \sum_{\alpha \in A} c_\alpha) \in \{0, +\infty\}$;

[$(P_{\Sigma'})$] si Σ' est réduit, on a $p > 3(h(\Sigma') - 1)$.

Pour $N \in \mathbb{N}$, on définit la condition :

[(P_N)] $p \geq N + 2$, et pour tout système de racines Σ' de rang $\leq N$, les conditions $(P^1_{\Sigma'})$, $(P^2_{\Sigma'})$, $(P_{\Sigma'})$ sont vérifiées.

L'hypothèse "p est grand par rapport au rang de $\mathcal{D}$" est précisément la suivante : la condition $(P_{\mathrm{rg}(\mathcal{D})})$ est vérifiée.

3.2. Soit k un corps commutatif de caractéristique 0 ou p. Fixons une clôture séparable $\overline{k}$ de k, et supposons qu'il existe un homomorphisme $\Gamma_k = \mathrm{Gal}(\overline{k}/k) \to \Gamma$. Ainsi Γ_k opère sur $\mathcal{D}$, et il existe un groupe réductif connexe G, un sous groupe de Borel B de G et un tore maximal $T \subset B$ de G, tous trois définis sur F et vérifiant la condition suivante : il existe un couple d'isomorphismes de $\mathbb{Z}[\Gamma_k]$-modules $j^* : X^* \to X^*(T)$ et $j_* : X_* \to X_*(T)$ en dualité, tel que $j^*(\Sigma) = \Sigma(G, T)$, $j_*(\check{\Sigma}) = \check{\Sigma}(G, T)$ et $j^*(\Delta) = \Delta(G, B, T)$. Posant $j = (j^*, j_*)$, le quadruplet (G, B, T, j) est unique à isomorphisme près. Via j, on identifie X^* à $X^*(T)$ et X_* à $X_*(T)$.

Pour $\alpha \in \Sigma$, on note $\mathfrak{g}_\alpha$ le sous-espace radiciel de $\mathfrak{g}$ associé à α. Fixons un épinglage $\{E_\alpha\}_{\alpha \in \Delta}$ de $\mathfrak{g}$ défini sur k : pour $(\alpha, \gamma) \in \Delta \times \Gamma$, on a $E_\alpha \in \mathfrak{g}_\alpha(\overline{k}) \smallsetminus \{0\}$ et $\gamma(E_\alpha) = E_{\gamma(\alpha)}$. Le quintuplet $(G, B, T, j, \{E_\alpha\}_{\alpha \in \Delta})$ est unique à isomorphisme *unique* près.

Pour $\alpha \in \Delta$, notons $E_{-\alpha}$ l'unique élément de $\mathfrak{g}_{-\alpha}(\overline{k})$ tel que $[E_\alpha, E_{-\alpha}] = \check{\alpha}$. Posons $N = \mathrm{N}_G(T)$. Le quotient $N/T = N(\overline{k})/T(\overline{k})$ s'identifie à W. Puisque $p \gg \mathrm{rg}(\mathcal{D})$, on peut définir l'exponentielle $\exp(X) \in G(\overline{k})$ d'un élément nilpotent $X \in \mathfrak{g}(\overline{k})$. On définit une section ensembliste de la projection canonique

$$n : W \to N(\overline{k}),$$

caractérisée par les propriétés suivantes [Sp, 11.2.9] :

[-] pour $\alpha \in \Delta$, on a $n(s_\alpha) = \exp(E_\alpha)\exp(-E_{-\alpha})\exp(E_\alpha)$, où $s_\alpha \in W$ désigne la symmétrie associée à la racine α;

[-] pour tous $w, w' \in W$ tels que $l(ww') = l(w) + l(w')$, on a $n(ww') = n(w)n(w')$, où $l : W \to \mathbb{N}$ désigne l'application longueur.

Plus généralement [LS, 2.1.A], pour tous w, $w' \in W$, posant

$$\Sigma^+_{w,w'} = \{\alpha \in \Sigma^+ : w^{-1}(\alpha) < 0,\ w'^{-1}w^{-1}(\alpha) > 0\},$$

on a $n(w)n(w') = \nu(w,w')n(ww')$ avec $\nu(w,w') = \prod_\alpha \check{\alpha}(-1)$ où α parcourt les éléments de $\Sigma^+_{w,w'}$. Notons que par construction, la section n est Γ-équivariante. On sait que l'on peut prolonger la famille $\{E_\alpha\}_{\alpha\in\pm\Delta}$ en une famille $\{E_\alpha\}_{\alpha\in\Sigma}$ vérifiant les conditions suivantes [Sp, 11.3.6] :

[-] pour $\alpha \in \Sigma$, $E_\alpha \in \mathfrak{g}_\alpha(\bar{k})$;

[-] pour $(\alpha,\gamma) \in \Sigma \times \Gamma$, il existe un $\epsilon \in \{\pm 1\}$ tel que $\gamma(E_\alpha) = \epsilon E_{\gamma(\alpha)}$;

[-] pour $(\alpha, w) \in \Sigma \times W$, il existe un $\epsilon \in \{\pm 1\}$ tel que $\mathrm{Ad}(n(w))(E_\alpha) = \epsilon E_{w(\alpha)}$.

Le prolongement est unique au remplacement près de certains E_α par leurs opposés.

On note $\rho : G_{\mathrm{sc}} \to G$ le revêtement simplement connexe du groupe dérivé G_{der}. Notons que G_{sc} est "le" groupe associé à la donnée de racines $\mathcal{D}_{\mathrm{sc}}$. On note B_{sc}, T_{sc} (etc.) les images réciproques de B, T (etc.) dans G_{sc}, et l'on définit comme ci-dessus une section ensembliste $n_{\mathrm{sc}} : W \to N_{\mathrm{sc}}$ de la projection canonique $N_{\mathrm{sc}} \to W$. On a l'égalité $n = \rho \circ n_{\mathrm{sc}}$.

3.3. Appliquons la construction de 3.2 au corps $k = \mathbb{F}_q$ et à l'homomorphisme $\Theta \hookrightarrow \Gamma$. On obtient un groupe réductif connexe défini sur $\mathbb{F}_q$, que l'on note $\underline{G}$. Plus généralement, on souligne par un trait "_" tous les objets associés à $\mathcal{D}$ en 3.2 : $\underline{B}$, $\underline{T}$ (etc.). Et comme en 3.2, on fixe un épinglage $\{\underline{E}_\alpha\}_{\alpha\in\Delta}$ de $\underline{\mathfrak{g}}$ défini sur $\mathbb{F}_q$, que l'on prolonge en une famille $\{\underline{E}_\alpha\}_{\alpha\in\Sigma}$.

Si $\underline{H}$ est un groupe algébrique défini sur une sous-extension de $\overline{\mathbb{F}}_q/\mathbb{F}_q$, on l'identifie (comme d'habitude) au groupe de ses points $\overline{\mathbb{F}}_q$-rationnels $\underline{H}(\overline{\mathbb{F}}_q)$.

Puisque le groupe I opère sur $\mathcal{D}$, il opère aussi sur $\underline{G}$. Soit $\sigma \in I$. De l'action de σ sur X_* se déduit une action sur $\underline{T} = \mathrm{X}_* \otimes_{\mathbb{Z}} \overline{\mathbb{F}}_q^\times$. Cette action se prolonge à $\underline{G}$ de sorte que pour $\alpha \in \Delta$, on ait $\sigma(\underline{E}_\alpha) = \underline{E}_{\sigma(\alpha)}$. Plus généralement, pour $\alpha \in \Sigma$, il existe un $\epsilon \in \{\pm 1\}$ tel que $\sigma(\underline{E}_\alpha) = \epsilon\underline{E}_{\sigma(\alpha)}$. Puisque I ne commute pas à Θ, l'action de I sur $\underline{G}$ n'est en général pas définie sur $\mathbb{F}_q$; mais les actions de I et de Θ se combinent néanmoins en une action de Γ sur $\underline{G}$. Notons en particulier que la section ensembliste $\underline{n} : W \to \underline{N}$ est Γ-équivariante.

Posons $T_\varpi = \mathrm{X}_* \otimes_{\mathbb{Z}} \mathbb{Z}_{(p)}$ et $\mathbf{T} = \mathrm{X}_* \otimes_{\mathbb{Z}} \mathbf{F}^*$. Le groupe $\underline{N}$ opère sur X_* via sa projection sur W ; il opère donc aussi sur T_ϖ. On pose $\mathbf{N} = \underline{N} \ltimes T_\varpi$ et l'on note $\underline{\pi} : \mathbf{N} \to \underline{N}$ la projection sur le premier facteur. On a les égalités $\mathbf{T} = T_\varpi \underline{T}$ et $\mathbf{N} = T_\varpi \underline{N} = \mathbf{T}\underline{N} = \mathbf{T}\underline{n}(W)$.

Des actions de Γ sur X_* et sur $\mathbf{F}^*$, on déduit une action sur $\mathbf{T}$. D'autre part l'action de Γ sur $\underline{G}$ se restreint en une action sur $\underline{N}$. Les deux actions coïncident sur $\underline{T} = \mathbf{T} \cap \underline{N}$. On en déduit donc une action de Γ sur $\mathbf{N}$. Notons que T_ϖ n'est pas Γ-stable pour cette action : pour $\lambda \otimes r \in T_w$ et $\sigma \in I$, on a $\sigma(\lambda \otimes r) = \sigma(\lambda) \otimes r(\sigma)^{-1} r \in \mathbf{T}$. L'application $\underline{\pi}$ n'est donc pas Γ-équivariante.

Posons $V = X_* \otimes_{\mathbb{Z}} \mathbb{R}$ et $V_{(p)} = X_* \otimes_{\mathbb{Z}} \mathbb{Z}_{(p)} \subset V$. On note

$$\mathbf{N} \times V \to V,\ (n, v) \mapsto n_{\mathbb{R}}(v)$$

l'action de $\mathbf{N}$ sur V définie comme suit : pour $t \in T_w$ $(= V_{(p)})$, on pose $t_{\mathbb{R}}(v) = v - t$; et le groupe $\underline{N}$ opère via l'action de son quotient W sur X_*. De l'action de Γ sur X_* se déduit une action sur V qui stabilise $V_{(p)}$. Et pour $(n, v) \in \mathbf{N} \times V$ et $\gamma \in \Gamma$, on a $\gamma(n_{\mathbb{R}}(v)) = \gamma(n)_{\mathbb{R}}(\gamma(v))$.

On peut faire les mêmes constructions pour le groupe $\underline{G}_{\rm sc}$; on définit en particulier $\mathbf{N}_{\rm sc}$ et $V_{\rm sc}$. Posons $V_{\rm cent} = \{v \in V : \alpha(v)$ pour tout $\alpha \in \Sigma\}$; c'est un sous-espace vectoriel Γ-stable de V, et l'on a la décomposition $V = V_{\rm cent} \oplus V_{\rm sc}$. Si u est un automorphisme affine de $V_{\rm sc}$, on identifie u à un automorphisme affine de V en le prolongeant par l'identité sur $V_{\rm cent}$. Alors pour $n \in \mathbf{N}_{\rm sc}$, $n_{\mathbb{R}}$ s'identifie à $\rho(n)_{\mathbb{R}}$.

Soit $e \in \mathcal{P}$. On note Σ^e l'ensemble des orbites de $I(e)$ dans Σ, et $W^e \subset W$, $V^e \subset V$, $\mathbf{N}^e \subset \mathbf{N}$ (etc.) les sous-ensembles formés des points fixés par $I(e)$. L'action de $\mathbf{N}^e$ sur V stabilise V^e, et l'on note $\mathbf{W}^e$ le groupe d'automorphismes affines de V^e image de $\mathbf{N}^e$ par cette action. On définit $\mathbf{W}^e_{\rm sc}$ de la même manière. Comme ci-dessus, on identifie $\mathbf{W}^e_{\rm sc}$ à un groupe d'automorphismes affines de V^e. Toute paire $(\alpha, r) \in \Sigma^e \times \mathbb{R}$ définit une fonction affine $\alpha + r$ sur V^e, et l'on pose $H^e_{\alpha+r} = \{v \in V : \alpha(v) + r = 0\}$. On note $\Sigma^e_{\rm aff}$ l'ensemble de ces fonctions $\alpha + r$ pour lesquelles il existe un élément $w \in \mathbf{W}^e_{\rm sc}$ tel que $H^e_{\alpha+r} = \{v \in V^e : w(v) = v\}$. Les hyperplans H^e_ψ pour $\psi \in \Sigma^e_{\rm aff}$ définissent une décomposition en facettes. L'action de $\mathbf{W}^e$ sur V^e conserve $\Sigma^e_{\rm aff}$ et la décomposition en facettes. On note $\mathcal{C}^e$ la chambre qui contient $\epsilon\check{\rho}$ pour $\epsilon > 0$ assez petit, avec $\check{\rho} = \frac{1}{2}\sum_{\alpha \in \Sigma^e_+} \check{\alpha}$ où $\Sigma^e_+ \subset \Sigma^e$ désigne l'ensemble des orbites de $I(e)$ dans Σ_+. Cette chambre est Γ-stable. On note $\mathbf{W}^e_{\mathcal{C}}$ le stabilisateur de $\mathcal{C}^e$ dans $\mathbf{W}^e$, et $\mathbf{N}^e_{\mathcal{C}}$ son image réciproque dans $\mathbf{N}^e$.

Si $e = 1$, on remplace l'exposant "e" par un exposant "nr" : $\Sigma^{\rm nr}$, $W^{\rm nr}$, $V^{\rm nr}$ (etc.). On utilisera souvent cette convention d'écriture dans la suite du papier, sans la réintroduire explicitement.

3.4. Soit $F \in {\rm CL}_q$. Appliquons la construction de 3.2 au corps $k = F$ et à l'homorphisme ${\rm Gal}(\overline{F}/F) \twoheadrightarrow {\rm Gal}(F^{\rm mod}/F) = \Gamma$. On obtient un groupe réductif connexe G défini sur F et déployé sur $F^{\rm mod}$; un sous-groupe de Borel B de G et un tore maximal $T \subset B$ de G, tous deux définis sur F. On fixe un épinglage $\{E_\alpha\}_{\alpha \in \Delta}$ de $\mathfrak{g}$ défini sur F, que l'on prolonge en une famille $\{E_\alpha\}_{\alpha \in \Sigma}$. Puisque G est déployé sur $F^{\rm mod}$, on peut supposer que $\{E_\alpha\}_{\alpha \in \Delta} \subset \mathfrak{g}(F^{\rm mod})$; alors pour $\alpha \in \Sigma$, on a $E_\alpha \in \mathfrak{g}(F^{\rm mod})$. On pose $G^{\rm mod} = G(F^{\rm mod})$, $B^{\rm mod} = B(F^{\rm mod})$ (etc.). Pour $e \in \mathcal{P}$, on pose $G^e = G(F^e)$, $B^e = B(F^e)$ (etc.).

Pour toute extension algébrique E de F, on pose

$$T(E)_+ = \{x \in T(E) : \omega_F(\chi(x) - 1) > 0,\ \forall \chi \in X^*\}.$$

Si $E = F^{\rm mod}$ (resp. F^e pour $e \in \mathcal{P}$, $F^{\rm nr}$), on écrit $T^{\rm mod}_+$ (resp. T^e_+, $T^{\rm nr}_+$). De l'homomorphisme $\mathbf{r}_F : F^{{\rm mod},\times} \to \mathbf{F}^*$ se déduit un homomorphisme

$\mathbf{r}_{T,F} : T^{\mathrm{mod}} = \mathrm{X}_* \otimes_{\mathbb{Z}} F^{\mathrm{mod},\times} \to \mathrm{X}_* \otimes_{\mathbb{Z}} \mathbf{F}^* = \mathbf{T}$. On a la suite exacte courte :

$$1 \to T_+^{\mathrm{mod}} \to T^{\mathrm{mod}} \xrightarrow{\mathbf{r}_{T,F}} \mathbf{T} \to 1.$$

D'après 3.2, l'homomorphisme $\mathbf{r}_{T,F}$ se prolonge en un homomorphisme $\mathbf{r}_{N,F} : N^{\mathrm{mod}} \to \mathbf{N}$ qui, pour tout $w \in W$, vérifie $\mathbf{r}_{T,F}(n(w)) = \underline{n}(w)$. On a $\ker \mathbf{r}_{N,F} = \ker \mathbf{r}_{T,F}$, et $\mathbf{r}_{N,F}$ est Γ-équivariant. Pour tout sous-groupe fermé Γ' de Γ, on a $\mathrm{H}^i(\Gamma', T_+^{\mathrm{mod}}) = 0$ ($i \in \mathbb{Z}_{\geq 1}$). Par conséquent pour $e \in \mathcal{P}$, on a la suite exacte courte :

$$1 \to T_+^e \to N^e \xrightarrow{\mathbf{r}_{N,F}} \mathbf{N}^e \to 1.$$

De la section s_F de $\mathbf{r}_F$ au-dessus de $\mathbb{Z}_{(p)}$ définie en 2.2, se déduit une section $s_{T,F} : T_\varpi \to T^{\mathrm{mod}}$ de $\mathbf{r}_{T,F}$ au-dessus de T_ϖ.

Soit $e \in \mathcal{P}$. On note $\mathfrak{I}_F^e$ l'immeuble étendu $\mathfrak{I}(G, F^e)$; c'est un $G^e \rtimes \mathrm{Gal}(F^e/F)$-ensemble. Pour $v \in \mathfrak{I}_F^e$, il existe un $\mathfrak{o}^e$-schéma en groupes affine lisse $\mathcal{G}_v^e$ de fibre générique $G \times_F F^e$, unique à isomorphisme unique près, tel que le groupe de ses points $\mathfrak{o}^e$-rationnels $G_v^e = \mathcal{G}_v^e(\mathfrak{o}^e)$ coïncide avec le stabilisateur de v dans G^e. On note $\overline{\mathcal{G}}_v^e$ sa fibre spéciale $\mathcal{G}_v^e \times_{\mathfrak{o}^e} \overline{\kappa}_F$, et $\mathbb{G}_v^e$ le quotient réductif maximal de $\overline{\mathcal{G}}_v^e$. Ainsi $\mathbb{G}_v^e$ est un groupe réductif (en général non connexe) défini sur une sous-extension finie de $\overline{\kappa}_F/\kappa_F$. Comme en 3.3, on identifie $\overline{\mathcal{G}}_v^e$, $\mathbb{G}_v^e$ (etc.) à $\overline{\mathcal{G}}_v^e(\overline{\kappa}_F)$, $\mathbb{G}_v^e(\overline{\kappa}_F)$ (etc.).

Notons S_e le plus grand sous-tore de T déployé sur F^e ; c'est un sous-tore F^e-déployé maximal de $G \times_F F^e$. On a $\mathrm{X}_*(S_e) = X_*^{I(e)}$, $\mathrm{Z}_G(S_e) = T$ et $\mathrm{N}_G(S_e) = N$. Au tore S_e est associé un appartement $A(G, S_e, F^e)$ de $\mathfrak{I}_F^e$. Le stabilisateur de $A(G, S_e, F^e)$ dans G^e est le groupe N^e ($= N(F^e)$). On vérifie que l'on peut identifier les N^e-ensembles $A(G, S_e, F^e)$ et V^e de sorte que l'action d'un élément $n \in N^e$ soit $\mathbf{r}_{N,F}(n)_{\mathbb{R}}$. L'immeuble $\mathfrak{I}_F^e$ s'identifie donc au quotient de $G^e \times V^e$ par la relation d'équivalence : $(g, v) \sim (g', v')$ si et seulement s'il existe un $n \in N^e$ et un $h \in G_v^e$ tels que $v' = \mathbf{r}_{N,F}(n)_{\mathbb{R}}(v)$ et $g' = ghn^{-1}$. Au tore S_e sont associés un ensemble de racines affines et un groupe de Weyl affine étendu, qui ne sont autres que Σ_{aff}^e et $\mathbf{W}^e$. De plus, $\mathbf{W}_{\mathrm{sc}}^e$ est le groupe de Weyl affine associé à Σ_{aff}^e, et l'on a la décomposition $\mathbf{W}^e = \mathbf{W}_{\mathrm{sc}}^e \rtimes \mathbf{W}_{\mathcal{C}}^e$.

Soient maintenant $e, e' \in \mathcal{P}$ tels que e divise e'. Puisque l'extension $F^{e'}/F^e$ est modérément ramifiée, l'immeuble $\mathfrak{I}_F^e$ coïncide avec l'ensemble des points fixes $(\mathfrak{I}_F^{e'})^{\mathrm{Gal}(F^{e'}/F^e)} = (\mathfrak{I}_F^{e'})^{I(e)}$. Et pour $v \in I(e)$, le groupe G_v^e n'est autre que le groupe des points fixes $(G_v^{e'})^{I(e)} = G_v^{e'} \cap G(F^e)$.

4. Torsion (d'après des constructions de Kottwitz)

Dans toute cette section, on fixe un corps $F \in \mathrm{CL}_q$.

4.1. Soit $e \in \mathcal{P}$. Appliquons la construction de 3.4 au cas $G = T$: pour tout $v \in V^e$, il existe un $\mathfrak{o}^e$-schéma en groupes affine lisse $\mathcal{T}_v^e$ de fibre générique $T \times_F F^e$, unique à isomorphisme unique près, tel que $T_v^e = \mathcal{T}_v^e(\mathfrak{o}^e)$ coïncide avec le stabilisateur de v dans T^e. Puisque $T_v^e = (X_* \otimes_{\mathbb{Z}} U_{F^{\mathrm{mod}}})^{I(e)}$,

$\mathcal{T}_v^e$ est indépendant de v. Soit $\mathcal{T}^{e,\circ}$ la composante neutre de $\mathcal{T}_v^e$; c'est aussi la composante neutre du modèle de Néron de $T\times_F F^e$. Posons $T^{e,\circ}=\mathcal{T}^{e,\circ}(\mathfrak{o}^e)$. On a l'inclusion $T^{e,\circ}\subset T_v^e$. Si de plus $I(e)$ opère trivialement sur $\mathcal{D}$, alors on a $T_v^e=\mathrm{X}_*\otimes_{\mathbb{Z}}U_{F^e}=T^{e,\circ}$; de la valuation ω_F sur F^e se déduit un homomorphisme $\omega_{T,F}^e : T^e\rightarrow\frac{1}{e}X_*$, et l'on a la suite exacte courte :

$$1\rightarrow T^{e,\circ}\rightarrow T^e\xrightarrow{\omega_{T,F}^e}\tfrac{1}{e}\mathrm{X}_*\rightarrow 1.$$

Dans le cas général, on construit de la manière suivante un homomorphisme $\omega_{T,F}^e : T^e\rightarrow\frac{1}{e}\mathrm{X}_{*,I(e)}$ où $\mathrm{X}_{*,I(e)}$ désigne le groupe des coinvariants de X_* pour l'action de $I(e)$: on choisit un $e'\in\mathcal{P}$ tel que e divise e' et $I(e')$ opère trivialement sur $\mathcal{D}$. L'application norme $T^{e'}\rightarrow T^e$ est surjective, et par fonctorialité, elle induit une application $T^{e',\circ}\rightarrow T^{e,\circ}$. La suite exacte ci-dessus se prolonge en un diagramme à carrés commutatifs dont les deux suites horizontales sont exactes :

$$\begin{array}{ccccccccc}
1 & \longrightarrow & T^{e',\circ} & \longrightarrow & T^{e'} & \xrightarrow{\omega_{T,F}^{e'}} & \frac{1}{e'}\mathrm{X}_* & \longrightarrow & 1\\
 & & \downarrow & & \downarrow & & \downarrow & & \\
1 & \longrightarrow & T^{e,\circ} & \longrightarrow & T^{e} & \xrightarrow{\omega_{T,F}^{e}} & \frac{1}{e}\mathrm{X}_{*,I(e)} & \longrightarrow & 1
\end{array}.$$

Les deux première flêches verticales sont les normes, la troisième est la composée de la surjection canonique $\frac{1}{e}\mathrm{X}_*\rightarrow\frac{1}{e}\mathrm{X}_{*,I(e)}$ et de la multiplication par $\frac{e'}{e}$. L'application $\omega_{T,F}^e$ est bien définie, i.e. elle ne dépend pas du choix de e'.

Puisque l'extension $F^{e'}/F^e$ est modérément ramifiée, l'application norme $T^{e'}\rightarrow T^e$ induit une application surjective $T_+^{e'}\rightarrow T_+^e$. Or $T_+^{e'}=\mathrm{X}_*\otimes_{\mathbb{Z}}U_{F^e}^+\subset T^{e',\circ}$; d'où l'inclusion $T_+^e\subset T^{e,\circ}$. Soit $\underline{T}^{e,\circ}=\mathrm{X}_*^{I(e)}\otimes_{\mathbb{Z}}\overline{\mathbb{F}}_q^\times$ la composante neutre du groupe algébrique $\underline{T}^e=\underline{T}^{I(e)}$. Via l'homorphisme $\mathbf{r}_{T,F}$, le diagramme précédent induit le diagramme à carrés commutatifs et suites horizontales exactes suivant :

$$\begin{array}{ccccccccc}
1 & \longrightarrow & \underline{T} & \longrightarrow & \mathbf{T}^{e'} & \xrightarrow{\omega_{\mathbf{T}}^{e'}} & \frac{1}{e'}\mathrm{X}_* & \longrightarrow & 1\\
 & & \downarrow & & \downarrow & & \downarrow & & \\
1 & \longrightarrow & \underline{T}^{e,\circ} & \longrightarrow & \mathbf{T}^{e} & \xrightarrow{\omega_{\mathbf{T}}^{e}} & \frac{1}{e}\mathrm{X}_{*,I(e)} & \longrightarrow & 1
\end{array}.$$

Tout élément $n\in\mathbf{N}^e$ s'écrit de manière unique $n=t\underline{n}(w)$ avec $t\in\mathbf{T}^e$ et $w\in W^e$. En envoyant n sur $(\omega_{\mathbf{T}}^e(t),w)$, on définit un homomorphisme $\omega_{\mathbf{N}}^e : \mathbf{N}^e\rightarrow\frac{1}{e}\mathrm{X}_{*,I(e)}\rtimes W^e$. Et de la deuxième suite exacte du diagramme ci-dessus, on déduit la suite exacte courte :

$$1\rightarrow\underline{T}^{e,\circ}\rightarrow\mathbf{N}^e\xrightarrow{\omega_{\mathbf{N}}^e}\tfrac{1}{e}\mathrm{X}_{*,I(e)}\rtimes W^e\rightarrow 1.$$

4.2. Prenons $e=1$, et considérons la surjection canonique $\mathrm{X}_{*,I}\rightarrow(\mathrm{X}_*/\mathrm{X}_{*,\mathrm{sc}})_I$. Elle est W^{nr}-équivariante pour les actions naturelles de W^{nr}. Or le groupe W opère trivialement sur $\mathrm{X}_*/\mathrm{X}_{*,\mathrm{sc}}$: pour $w\in W$ et $\lambda\in\mathrm{X}_*$, on a $w(\lambda)-\lambda\in\mathrm{X}_{*,\mathrm{sc}}$. Par conséquent la surjection précédente se prolonge en un homomorphisme surjectif $X_{*,I} \rtimes W^{nr} \to (X_*/X_{*,sc})_I$. En composant ce dernier avec $\omega_{\mathbf{N}}^{nr}$, on obtient un homomorphisme surjectif $\tilde{\omega}_{\mathbf{N}}^{nr} : \mathbf{N}^{nr} \to (X_*/X_{*,sc})_I$. Le sous-groupe $\mathbf{N}_C^{nr} \subset \mathbf{N}^{nr}$ défini en 3.3 contient $\underline{T}^{nr} = \underline{T}^I$; à fortiori, il contient la composante neutre $\underline{T}^{nr,\circ}$. D'après le lemme 1.8.1 de [W2], on a la suite exacte courte

$$1 \to \underline{T}^{nr,\circ} \to \mathbf{N}_C^{nr} \xrightarrow{\tilde{\omega}_{\mathbf{N}}^{nr}} (X_*/X_{*,sc})_I \to 1.$$

D'autre part, on a l'égalité $H^1(\Theta, (X_*/X_{*,sc})_I) = (X_*/X_{*,sc})_{\Gamma,tor}$ où $(X_*/X_{*,sc})_{\Gamma,tor}$ désigne le sous-groupe de torsion de $(X_*/X_{*,sc})_\Gamma$: à un cocycle $\delta \in Z^1(\Theta, (X_*/X_{*,sc})_I)$, on associe l'image de $\delta(\theta_1)$ dans $[(X_*/X_{*,sc})_I]_\Theta = (X_*/X_{*,sc})_\Gamma$. L'homomorphisme $\tilde{\omega}_{\mathbf{N}}^{nr}$ est Θ-équivariant. Il induit donc une application $H^1(\Theta, \mathbf{N}_C^{nr}) \to (X_*/X_{*,sc})_{\Gamma,tor}$. D'après le lemme 1.8.2 de [W2], cette application est *bijective*.

Fixons un sous-ensemble $D \subset Z^1(\Theta, \mathbf{N}_C^{nr})$ tel que l'application naturelle $D \to H^1(\Theta, \mathbf{N}_C^{nr})$ soit bijective. On munit D de la structure de groupe déduite de celle de $(X_*/X_{*,sc})_{\Gamma,tor}$ via la bijection $D \to (X_*/X_{*,sc})_{\Gamma,tor}$. On considère aussi les éléments de D comme des cocycles de Γ triviaux sur I. Notons N_C^{nr} l'image réciproque de $\mathbf{N}_C^{nr}$ par l'homomorphisme $\mathbf{r}_{N,F}^{nr}$. Ce dernier définit une application $H^1(\Theta, N_C^{nr}) \to H^1(\Theta, \mathbf{N}_C^{nr})$ qui, puisque $H^i(\Theta, T_+^{nr}) = 0$ $(i \in \mathbb{Z}_{\geq 1})$, est *bijective*. Plus précisément, tout cocycle de Θ à valeurs dans $\mathbf{N}_C^{nr}$ se relève en un cocycle de Θ à valeur dans N_C^{nr}. Pour chaque $d \in D$, on fixe un tel relèvement que l'on note d_F.

4.3. Dans [K2, §7], Kottwitz a défini un homomorphisme $\omega_G^{nr} : G^{nr} \to (X_*/X_{*,sc})_I$ qui possède les propriétés suivantes :

[-] la restriction de ω_G^{nr} à T^{nr} coïncide avec la composée de ω_T^{nr} et de la surjection canonique $X_{*,I} \to (X_*/X_{*,sc})_I$;

[-] $\rho(G_{sc}^{nr}) \subset \ker \omega_G^{nr}$.

On en déduit que la restriction de ω_G^{nr} à N^{nr} coïncide avec l'application composée $\tilde{\omega}_{\mathbf{N}}^{nr} \circ \mathbf{r}_{N,F}^{nr}$. D'autre part, on a les égalités

$$H^1(\mathrm{Gal}(\overline{F}/F), G(\overline{F})) = H^1(\Gamma, G^{mod}) = H^1(\Theta, G^{nr});$$

et ω_G^{nr} induit une application bijective

$$H^1(\Theta, G^{nr}) \to H^1(\Theta, (X_*/X_{*,sc})_I) = (X_*/X_{*,sc})_{\Gamma,tor}.$$

D'après 4.2, l'inclusion $N_C^{nr} \subset G^{nr}$ induit une application bijective $H^1(\Theta, N_C^{nr}) \to H^1(\Theta, G^{nr})$. Ainsi l'application $d \mapsto d_F$ identifie D à un ensemble de cocycles représentant l'ensemble $H^1(\Theta, G^{nr})$.

Tout élément $d \in D$ définit une nouvelle structure de G sur F, notée ${}_dG$. Précisément, ${}_dG$ est muni d'un isomorphisme de groupes algébriques ${}_d\xi : {}_dG \to G$ défini sur F^{nr} et vérifiant ${}_d\xi \circ \theta = \mathrm{Ad}(d_F(\theta)) \circ \theta \circ {}_d\xi$ pour tout $\theta \in \Theta$; i.e. vérifiant ${}_d\xi \circ \gamma = \mathrm{Ad}(d_F(\gamma)) \circ \gamma \circ {}_d\xi$ pour tout $\gamma \in \Gamma$. Notons ${}_dT$ l'image réciproque de T par ${}_d\xi$. C'est un tore de ${}_dG$ défini sur F. Soit ${}_dS$ le plus grand sous-tore de ${}_dT$ déployé sur F.

LEMME [W2, lemme 1.9]. — *Le tore ${}_dS$ est un sous-tore F-déployé maximal de ${}_dG$.*

Notons ${}_dV$ l'espace V muni de l'action de Γ suivante : $(\gamma, v) \mapsto d(\gamma)_{\mathbb{R}}(\gamma(v))$ pour $\gamma \in \Gamma$ et $v \in V$. Soit $e \in \mathcal{P}$ tel que $I(e)$ opère trivialement sur $\mathcal{D}$. De même que l'on a construit $\mathfrak{I}_F^e$ comme quotient de $G^e \times V$ (cf. 3.3), on construit l'immeuble étendu ${}_d\mathfrak{I}_F^e = \mathfrak{I}({}_dG, F^e)$ comme quotient de ${}_dG(F^e) \times {}_dV$. D'ailleurs, via l'isomorphisme de groupes ${}_dG(F^e) \to G^e$ induit par ${}_d\xi$, les G^e-ensembles ${}_d\mathfrak{I}_F^e$ et $\mathfrak{I}_F^e$ sont isomorphes. L'action de Γ sur ${}_dG(F^e) \times {}_dV$ passe au quotient et définit une action sur ${}_d\mathfrak{I}_F^e$. Grâce au lemme précédent, l'immeuble ${}_d\mathfrak{I}_F = \mathfrak{I}({}_dG, F)$ s'identifie au sous-ensemble $({}_d\mathfrak{I}_F^e)^\Gamma \subset {}_d\mathfrak{I}_F^e$ des points fixés par Γ. Au tore ${}_dS$ est associé un appartement $A({}_dG, {}_dS, F)$ dans ${}_d\mathfrak{I}_F$, qui n'est autre que le sous-ensemble ${}_dW = ({}_dV)^\Gamma \subset {}_dV$ des points fixés par Γ.

Pour $v \in {}_dW$, il existe un $\mathfrak{o}_F$-schéma en groupes affine lisse ${}_d\mathcal{G}_v$ de fibre générique ${}_dG$, unique à isomorphisme unique près, tel que le groupe des points $\mathfrak{o}^{\mathrm{nr}}$-rationnels ${}_dG_v^{\mathrm{nr}} = {}_d\mathcal{G}_v(\mathfrak{o}^{\mathrm{nr}})$ coïncide avec le stabilisateur de v dans ${}_dG(F^{\mathrm{nr}})$. L'application ${}_d\xi$ induit un isomorphisme ${}_dG_v^{\mathrm{nr}} \xrightarrow{\sim} G_v^{\mathrm{nr}}$. Par fonctorialité, on en déduit qu'il existe un unique isomorphisme de $\mathfrak{o}^{\mathrm{nr}}$-schémas en groupes ${}_d\xi_v : {}_d\mathcal{G}_v \times_{\mathfrak{o}_F} \mathfrak{o}^{\mathrm{nr}} \to \mathcal{G}_v^{\mathrm{nr}}$ prolongeant ${}_d\xi$; i.e. tel que ${}_d\xi_v \times_{\mathfrak{o}^{\mathrm{nr}}} F^{\mathrm{nr}} = {}_d\xi$.

On pose ${}_d\mathfrak{g} = \mathrm{Lie}({}_dG)$ et ${}_d\mathfrak{g}^{\mathrm{nr}} = {}_d\mathfrak{g}(F^{\mathrm{nr}})$. Le groupe Θ opère naturellement sur ${}_d\mathfrak{g}^{\mathrm{nr}}$, et l'on a ${}_d\mathfrak{g}^{\mathrm{nr}} = ({}_d\mathfrak{g}^{\mathrm{nr}})^\Theta \otimes_F F^{\mathrm{nr}}$ avec $({}_d\mathfrak{g}^{\mathrm{nr}})^\Theta = {}_d\mathfrak{g}(F)$.

5. Filtrations de Moy-Prasad (sur F^{nr})

5.1. Posons $V_{(p)}^{\mathrm{nr}} = V_{(p)}^I$ ($= \mathrm{X}_*^I \otimes_{\mathbb{Z}} \mathbb{Z}_{(p)}$). De la forme bilinéaire $I \times \mathbb{Z}_{(p)} \to \overline{\mathbb{F}}_q^\times$, $(\sigma, r) \mapsto r(\sigma)$ définie en 2.1, on déduit une forme bilinéaire

$$I \times V_{(p)}^{\mathrm{nr}} \to \mathrm{X}_*^I \otimes_{\mathbb{Z}} \overline{\mathbb{F}}_q^\times = \underline{T}^{\mathrm{nr},\circ},\ (\sigma, v) \mapsto v(\sigma).$$

On peut aussi la définir autrement : pour $v \in V_{(p)}$, il existe un unique élément $t_v \in T_\varpi$ tel que $(t_v)_{\mathbb{R}}(0) = v$ (avec l'identification $T_\varpi = V_{(p)}$, on a $t_v = -v$). Rappelons qu'on a défini en 3.3 une action de Γ sur $\mathbf{T}$, pour laquelle T_ϖ n'est pas Γ-stable. Pour $(\sigma, v) \in I \times V_{(p)}^{\mathrm{nr}}$, on a

$$v(\sigma) = \sigma(t_v)t_v^{-1}.$$

Soit $v \in V_{(p)}^{\mathrm{nr}}$. Le groupe I opère sur $\underline{G}$, et pour $\sigma \in I$, on note $\underline{\rho}_v(\sigma)$ le $\overline{\mathbb{F}}_q$-automorphisme $\mathrm{Ad}(v(\sigma)) \circ \sigma$ de $\underline{G}$. L'application $I \to \mathrm{Aut}_{\overline{\kappa}_F\text{-gr}}(\underline{G})$, $\sigma \mapsto \underline{\rho}_v$ est un homomorphisme. Soit $\underline{G}_v \subset \underline{G}$ le sous-groupe (en général non connexe) formé des éléments fixés par $\underline{\rho}_v(\sigma)$ pour tout $\sigma \in I$.

Pour $(v, r) \in V_{(p)}^{\mathrm{nr}} \times \mathbb{Z}_{(p)}$, on pose

$$\underline{\mathfrak{g}}_{v,r} = \{X \in \underline{\mathfrak{g}} : \mathrm{d}(\underline{\rho}_v(\sigma))(X) = r(\sigma)X \text{ pour tout } \sigma \in I\}.$$

Cet espace est stable sous l'action de $\underline{G}_v$. Et pour $r = 0$, on a $\underline{\mathfrak{g}}_{v,0} = \mathrm{Lie}(\underline{G}_v)$.

Soient (v,r), $(v',r') \in V^{\mathrm{nr}}_{(p)} \times \mathbb{Z}_{(p)}$. Choisissons un entier $e \in \mathcal{P}$ tel que ev, $ev' \in \mathrm{X}_*$ et er, $er' \in \mathbb{Z}$. Posons $\lambda = ev' - ev$. Pour $x \in \overline{\mathbb{F}}_q^\times$, on a donc $\lambda(x) \in \underline{T}$. Pour $i \in \mathbb{Z}$, posons

$$\begin{aligned}
&\underline{\mathfrak{g}}[i] = \{X \in \underline{\mathfrak{g}} : \mathrm{Ad}(\lambda(x))(X) = x^i X \text{ pour tout } x \in \overline{\mathbb{F}}_q^\times\},\\
&\underline{\mathfrak{g}}[\geq i] = \oplus_{j\geq i}\, \underline{\mathfrak{g}}[j],\\
&\underline{\mathfrak{g}}[> i] = \oplus_{j> i}\, \underline{\mathfrak{g}}[j].
\end{aligned}$$

On vérifie que pour $i = er' - er$, les espaces $\underline{\mathfrak{g}}[i]$, $\underline{\mathfrak{g}}[\geq i]$ et $\underline{\mathfrak{g}}[> i]$, ne dépendant pas du choix de e ; de plus on a $\underline{\mathfrak{g}}_{v,r} = \oplus_{i\in\mathbb{Z}}\, (\underline{\mathfrak{g}}_{v,r} \cap \underline{\mathfrak{g}}[i])$. On pose

$$\begin{aligned}
\underline{\mathfrak{m}}_{v,r;v',r'} &= \underline{\mathfrak{g}}_{v,r} \cap \underline{\mathfrak{g}}[er' - er],\\
\underline{\mathfrak{p}}_{v,r;v',r'} &= \underline{\mathfrak{g}}_{v,r} \cap \underline{\mathfrak{g}}[\geq er' - er],\\
\underline{\mathfrak{u}}_{v,r;v',r'} &= \underline{\mathfrak{g}}_{v,r} \cap \underline{\mathfrak{g}}[> er' - er].
\end{aligned}$$

On a $\underline{\mathfrak{p}}_{v,r;v',r'} = \underline{\mathfrak{m}}_{v,r;v',r'} \oplus \underline{\mathfrak{m}}_{v,r;v',r'}$. Si de plus $r = r' = 0$, alors $\underline{\mathfrak{p}}_{v,r;v',r'}$ est une sous-$\overline{\mathbb{F}}_q$-algèbre parabolique de $\underline{\mathfrak{g}}_{v,r}$ de radical nilpotent $\underline{\mathfrak{u}}_{v,r;v',r'}$, et $\underline{\mathfrak{m}}_{v,r;v',r'}$ est une composante de Levi de $\underline{\mathfrak{p}}_{v,r;v',r'}$.

5.2. Pour $\alpha \in \Sigma^{\mathrm{nr}}$, on note $\underline{\mathfrak{g}}_\alpha$ l'espace $\oplus_\beta\, \underline{\mathfrak{g}}_\beta$ où β parcourt les éléments de Σ appartenant à l'orbite α. On pose $\tilde{\Sigma}^{\mathrm{nr}} = \Sigma^{\mathrm{nr}} \cup \{0\}$ et $\underline{\mathfrak{g}}_0 = \underline{\mathfrak{t}}$ avec (rappel) $\underline{\mathfrak{t}} = \mathrm{Lie}(\underline{T})$. Pour $\alpha \in \tilde{\Sigma}^{\mathrm{nr}}$, on note R_α l'ensemble des $r \in \mathbb{Z}_{(p)}$ pour lesquels il existe un $X \in \underline{\mathfrak{g}}_\alpha \smallsetminus \{0\}$ tel que $\sigma(X) = r(\sigma)X$ pour tout $\sigma \in I$. L'ensemble R_α est stable par translation par $\mathbb{Z}$ et son image dans $\mathbb{Z}_{(p)}/\mathbb{Z}$ est *finie*. La collection des hyperplans $\mathcal{H}_{\alpha,r} = \{(v,r') \in V^{\mathrm{nr}} \times \mathbb{R} : r' - \alpha(v) = r\} \subset V^{\mathrm{nr}} \times \mathbb{R}$ pour $\alpha \in \tilde{\Sigma}^{\mathrm{nr}}$ et $r \in R_\alpha$, définit une *décomposition en facettes*. On note Φ l'ensemble de ces facettes. Pour $(v,r) \in V^{\mathrm{nr}} \times \mathbb{R}$, on note $\phi_{v,r} \in \Phi$ la facette à laquelle appartient (v,r). Chaque facette est stable par translation par l'espace $V^{\mathrm{nr}}_{\mathrm{cent}} = (V_{\mathrm{cent}})^I$ (cf. 3.3). Posons $\overline{V}^{\mathrm{nr}} = V^{\mathrm{nr}}/V^{\mathrm{nr}}_{\mathrm{cent}}$. Pour $\phi \in \Phi$, le sous-ensemble $\bar{\phi} = \phi/V^{\mathrm{nr}}_{\mathrm{cent}}$ de $\overline{V}^{\mathrm{nr}} \times \mathbb{R}$ est relativement compact, et sa projection sur $\mathbb{R}$ est un intervalle borné de $\mathbb{R}$ dont on note $r(\phi)$ la borne supérieure. La condition $(P_{\mathrm{rg}(\mathcal{D})})$ implique le

LEMME [W2, lemmes 2.4.1 et 2.4.2]. — *Pour $\phi \in \Phi$, l'intersection $\phi \cap (V^{\mathrm{nr}}_{(p)} \times \mathbb{Z}_{(p)})$ est dense dans ϕ, et l'on a $r(\phi) \in \mathbb{Z}_{(p)}$. De plus, il existe un $e \in \mathcal{P}$ tel que $er(\phi) \in \mathbb{Z}$ pour tout $\phi \in \Phi$.*

Pour $\phi \in \Phi$, on note ω_ϕ l'ensemble (fini) des facettes $\phi' \in \Phi$ dont l'adhérence contient ϕ ; et pour $(v,r) \in V^{\mathrm{nr}} \times \mathbb{R}$, on pose $\omega_{v,r} = \omega_{\phi_{v,r}}$.

5.3. Soit $F \in \mathrm{CL}_q$ (fixé jusqu'à la fin de la section 5), et soit $e \in \mathcal{P}$ tel que $I(e)$ opère trivialement sur $\mathcal{D}$. Le point $v = 0 \in V = V^e$ est *hyperspécial*, par conséquent la fibre spéciale $\overline{\mathcal{G}}^e_0$ du $\mathfrak{o}^e$-schéma en groupes $\mathcal{G}^e_0$ associé à 0 en 3.4, est *réductive* et *connexe* ; i.e. on a $\mathbb{G}^e_0 = \overline{\mathcal{G}}^e_0$ et $\overline{\mathcal{G}}^e_0$ est un groupe algébrique connexe. D'autre part, puisque T est déployé sur F^e, on a $\mathcal{T}^e_{v'} = \mathcal{T}^{e,\circ}$ pour tout $v' \in V$. Notons $\pi^e_0 : G^e_0 = \mathcal{G}^e_0(\mathfrak{o}^e) \to \mathbb{G}^e_0 = \mathbb{G}^e_0(\overline{\kappa}_F)$ la projection canonique (réduction modulo $\mathfrak{p}^e$). Alors $\mathbb{B}^e_0 = \pi^e_0(B^e \cap G^e_0)$ et

$\mathbb{T}_0^e = \pi_0^e(T^e \cap G_0^e)$ sont respectivement un sous-groupe de Borel et un tore maximal de $\mathbb{G}_0^e$. On a une identification canonique $\mathrm{X}_*(\mathbb{T}_0^e) = \mathrm{X}_*$; et via cette identification, $\Sigma(\mathbb{G}_0^e, \mathbb{T}_0^e)$ s'identifie à Σ. La famille $\{\mathrm{d}(\pi_0^e)(E_\alpha)\}_{\alpha\in\Delta}$ est un épinglage de $\mathrm{Lie}(\mathbb{G}_0^e)$. On peut donc identifier les groupes $\mathbb{G}_0^e$ et $\underline{G}$ de sorte que :

[-] $\mathbb{T}_0^e$ s'identifie à $\underline{T}$ de manière compatible avec les identifications $\mathrm{X}_*(\mathbb{T}_0^e) = \mathrm{X}_* = \mathrm{X}_*(\underline{T})$;

[-] pour $\alpha \in \Delta$, $\mathrm{d}(\pi_0^e)(E_\alpha)$ s'identifie à $\underline{E}_\alpha$.

Alors on vérifie que :

[-] l'intersection $N^e \cap G^e$ est l'image réciproque de $\underline{N}$ par l'application $N^e \xrightarrow{\mathbf{r}_{N,F}} \mathbf{N}^e$;

[-] les applications $\mathbf{r}_{N,F}$ et π_0^e coïncident sur cette intersection.

La famille $\{\mathrm{d}(\pi_0^e)(E_\alpha)\}_{\alpha\in\Sigma}$ prolonge l'épinglage $\{\underline{E}_\alpha\}_{\alpha\in\Delta}$ de $\underline{\mathfrak{g}}$ au sens de 3.2, c'est-à-dire qu'elle vérifie les trois conditions imposées à la fin de 3.2. Pour $\alpha \in \Delta$, on a nécessairement $\mathrm{d}(\pi_0^e)(E_{-\alpha}) = \underline{E}_{-\alpha}$. Et d'après 3.2, quitte à remplacer la famille de départ $\{E_\alpha\}_{\alpha\in\Sigma}$ en la multipliant par certains signes, on peut supposer que $\mathrm{d}(\pi_0^e)(E_\alpha) = \underline{E}_\alpha$ pour tout $\alpha \in \Sigma$.

5.4. Soit $v \in V_{(p)}^{\mathrm{nr}}$. Choisissons un entier $e \in \mathcal{P}$ tel que $ev \in \mathrm{X}_*$ et $I(e)$ opère trivialement sur $\mathcal{D}$. On a vu en 3.4 que $(G_v^e)^I = G_v^{\mathrm{nr}}$. L'action de I sur G_v^e se descend en une action (algébrique) de I sur la fibre spéciale $\mathbb{G}_v^e$, et puisque l'extension F^e/F^{nr} est modérément ramifiée, l'égalité $(G_v^e)^I = G_v^{\mathrm{nr}}$ se complète en le diagramme commutatif suivant (cf. [L, 3.5]) :

$$\begin{array}{ccc} G_v^{\mathrm{nr}} & \xrightarrow{\pi_v^{\mathrm{nr}}} & \mathbb{G}_v^{\mathrm{nr}} \\ \| & & \| \\ (G_v^e)^I & \xrightarrow{\pi_v^e} & (\mathbb{G}_v^e)^I \end{array}$$

où π_v^{nr} est la composée des projections canoniques $G_v^{\mathrm{nr}} \to \overline{\mathcal{G}}_v^{\mathrm{nr}} = \overline{\mathcal{G}}_v^{\mathrm{nr}}(\overline{\kappa}_F)$ (réduction modulo $\mathfrak{p}^{\mathrm{nr}}$) et $\overline{\mathcal{G}}_v^{\mathrm{nr}} \to \mathbb{G}_v^{\mathrm{nr}}$.

Posons $t_{F,v} = \mathbf{s}_{T,F}(t_v) \in T^{\mathrm{mod}}$ où t_v est l'élément de T_ϖ défini en 5.1, et $\mathbf{s}_{T,F}$ est la section de $\mathbf{r}_{T,F}$ au-dessus de T_ϖ définie en 3.4. On vérifie que $t_{F,v} \in T^e$, puis que $G_v^e = \mathrm{Ad}(t_{F,v})(G_0^e)$. Notons $\tilde{\underline{\iota}}_v : G_v^{\mathrm{nr}} \to \underline{G}$ l'homomorphisme composé de l'application $\pi_0^e \circ \mathrm{Ad}(t_{F,v}^{-1})$ et de l'inclusion $G_v^{\mathrm{nr}} \subset G_v^e$. D'après le lemme 2.3.1 de [W2], on a $\tilde{\underline{\iota}}_v(G_v^{\mathrm{nr}}) = \underline{G}_v$ et $\tilde{\underline{\iota}}_v$ se quotiente en un isomorphisme

$$\underline{\iota}_v : \mathbb{G}_v^{\mathrm{nr}} \xrightarrow{\sim} \underline{G}_v$$

qui ne dépend pas du choix de e. On pose $\underline{\pi}_v = \underline{\iota}_v \circ \pi_v^{\mathrm{nr}} : G_v^{\mathrm{nr}} \twoheadrightarrow \underline{G}_v$.

Notons que pour $n \in N^{\mathrm{mod}} \cap G_v^e = N^e \cap G_v^e$, on a

$$\pi_0^e \circ \mathrm{Ad}(t_{F,v}^{-1})(n) = \mathrm{Ad}(t_v^{-1})(\mathbf{r}_{N,F}(n)) = \underline{\pi} \circ \mathbf{r}_{N,F}(n).$$

La première égalité résulte de ce que π_0^e et $\mathbf{r}_{N,F}$ coïncident sur l'intersection $N^e \cap G_0^e$; et puisque $N^e \cap G_0^e$ coïncide avec l'image réciproque de $\underline{N}$ par la surjection $N^e \xrightarrow{\mathbf{r}_{N,F}} \mathbf{N}^e$, on a $\mathrm{Ad}(t_v^{-1})(\mathbf{r}_{N,F}(n)) \in \underline{N}$. Or $\underline{\pi}$ se restreint en l'identité sur $\underline{N}$, et $\underline{\pi}(t_v) = 1$, d'où la seconde égalité.

5.5. Soit $e \in \mathcal{P}$. Pour $(v,r) \in V^e \times \mathbb{R}$, Moy et Prasad ont défini un $\mathfrak{o}^e$-réseau $\mathfrak{g}^e_{v,r} = \mathfrak{g}(F^e)_{v,r}$ dans $\mathfrak{g}(F^e)$. Fixé v, l'application $r \mapsto \mathfrak{g}^e_{v,r}$ est décroissante, et l'on pose $\mathfrak{g}^e_{v,r+} = \bigcup_{s>r} \mathfrak{g}^e_{v,s}$. Si $e' \in \mathcal{P}$ divise e, alors pour $(v,r) \in V^{e'} \times \mathbb{R}$, on a l'égalité $\mathfrak{g}^{e'}_{v,r} = (\mathfrak{g}^e_{v,r})^{I(e')}$.

Supposons de plus que $I(e)$ opère trivialement sur $\mathcal{D}$. Alors pour $(v,r) = (0,0)$, posant $\mathfrak{g}^e_0 = \mathrm{Lie}(\mathcal{G}^e_0)(\mathfrak{o}^e)$, on a $\mathfrak{g}^e_{0,0} = \mathfrak{g}^e_0$; et pour $(v,r) = (0,r)$ avec $r \in \frac{1}{e}\mathbb{Z}$, on a $\mathfrak{g}^e_{0,r} = \varpi_r \mathfrak{g}^e_0$.

Soit maintenant $(v,r) \in V^{\mathrm{nr}}_{(p)} \times \mathbb{Z}_{(p)}$. Choisissons un $e \in \mathcal{P}$ tel que $ev \in \mathrm{X}_*$, $er \in \mathbb{Z}$ et $I(e)$ opère trivialement sur $\mathcal{D}$. Alors on a l'égalité $\mathfrak{g}^e_{v,r} = \mathrm{Ad}(t_{F,v})(\varpi_r \mathfrak{g}^e_0)$. Notons $\underline{\pi}_{v,r} : \mathfrak{g}^{\mathrm{nr}}_{v,r} \to \underline{\mathfrak{g}}$ l'application linéaire définie comme la composée de la suite d'applications suivante :

$$\mathfrak{g}^{\mathrm{nr}}_{v,r} = (\mathfrak{g}^e_{v,r})^I \subset \mathfrak{g}^e_{v,r} \xrightarrow{\mathrm{Ad}(t^{-1}_{F,v})} \varpi_r \mathfrak{g}^e_0 \xrightarrow{\times \varpi_{-r}} \mathfrak{g}^e_0 \xrightarrow{\mathrm{d}(\pi^e_0)} \underline{\mathfrak{g}}.$$

D'après le lemme 2.3.2 de [W2], on a $\underline{\pi}_{v,r}(\mathfrak{g}^{\mathrm{nr}}_{v,r}) = \underline{\mathfrak{g}}_{v,r}$ et $\underline{\pi}_{v,r}$ se quotiente en un isomorphisme

$$\mathfrak{g}^{\mathrm{nr}}_{v,r} / \mathfrak{g}^{\mathrm{nr}}_{v,r} \xrightarrow{\sim} \underline{\mathfrak{g}}_{v,r}$$

qui ne dépend pas du choix de e.

5.6. D'après le lemme 2.5.1 de [W2], pour (v,r), $(v',r') \in V^{\mathrm{nr}} \times \mathbb{R}$, on a $\phi_{v,r} = \phi_{v',r'}$ si et seulement si $\mathfrak{g}^{\mathrm{nr}}_{v,r} = \mathfrak{g}^{\mathrm{nr}}_{v',r'}$ et $\mathfrak{g}^{\mathrm{nr}}_{v,r+} = \mathfrak{g}^{\mathrm{nr}}_{v',r'+}$. Cela signifie que pour $r \in \mathbb{R}$, la décomposition en facettes de V^{nr} obtenue en coupant les éléments de Φ avec $V^{\mathrm{nr}} \times \{r\}$ puis en les projetant sur V^{nr}, coïncide avec la décomposition de V^{nr} en r-facettes généralisées définie par DeBacker (cf. 1.1). En particulier pour $r = 0$, on retrouve la décomposition en facettes de V^{nr} définie en 3.3.

Pour $\phi \in \Phi$ tel que $\phi = \phi_{v,r}$ pour un $(v,r) \in V^{\mathrm{nr}} \times \mathbb{R}$, on peut donc poser $\mathfrak{g}^{\mathrm{nr}}_\phi = \mathfrak{g}^{\mathrm{nr}}_{v,r}$ et $\mathfrak{g}^{\mathrm{nr}}_{\phi+} = \mathfrak{g}^{\mathrm{nr}}_{v,r+}$. Pour $\phi, \phi' \in \Phi$ tels que $\phi' \in \omega_\phi$, on a les inclusions

$$\mathfrak{g}^{\mathrm{nr}}_{\phi+} \subset \mathfrak{g}^{\mathrm{nr}}_{\phi'+} \subset \mathfrak{g}^{\mathrm{nr}}_{\phi'} \subset \mathfrak{g}^{\mathrm{nr}}_{\phi}.$$

Soient (v,r), $(v',r') \in V^{\mathrm{nr}}_{(p)} \times \mathbb{Z}_{(p)}$ tels que $\phi_{v',r'} \in \omega_{v,r}$. On a deux façons d'envoyer $\mathfrak{g}^{\mathrm{nr}}_{v',r'}$ dans $\underline{\mathfrak{g}}$: la première est l'application $\underline{\pi}_{v',r'}$ (cf. 5.5) ; la seconde est la composée de l'application $\underline{\pi}_{v,r}$ et de l'inclusion $\mathfrak{g}^{\mathrm{nr}}_{v',r'} \subset \mathfrak{g}^{\mathrm{nr}}_{v,r}$. D'après le lemme 2.5.2 de [W2], on a $\underline{\pi}_{v,r}(\mathfrak{g}^{\mathrm{nr}}_{v',r'}) = \underline{\mathfrak{p}}_{v,r;v',r'}$, $\underline{\pi}_{v,r}(\mathfrak{g}^{\mathrm{nr}}_{v',r'+}) = \underline{\mathfrak{u}}_{v,r;v',r'}$, $\underline{\mathfrak{g}}_{v',r'} = \underline{\mathfrak{m}}_{v,r;v',r'}$, et l'application $\underline{\pi}_{v',r'}$ coïncide avec la composée de la projection canonique $\underline{\mathfrak{p}}_{v,r;v',r'} \to \underline{\mathfrak{m}}_{v,r;v',r'}$ et de $\underline{\pi}_{v,r}$. Cela implique que l'espace $\underline{\mathfrak{g}}_{v',r'}$ et la projection $\underline{\pi}_{v',r'} : \mathfrak{g}^{\mathrm{nr}}_{v',r'} \to \underline{\mathfrak{g}}_{v',r'}$ ne dépendent pas vraiment de (v',r'), mais seulement de la facette $\phi' = \phi_{v',r'}$ contenant (v',r') ; on les note respectivement $\underline{\mathfrak{g}}_{\phi'}$ et $\underline{\pi}_{\phi'}$. Posant $\phi = \phi_{v,r}$, on définit de la même manière $\underline{\mathfrak{g}}_\phi$ et $\underline{\pi}_\phi$. Alors les espaces $\underline{\mathfrak{m}}_{v,r;v',r'}$, $\underline{\mathfrak{p}}_{v,r;v',r'}$, $\underline{\mathfrak{u}}_{v,r;v',r'}$ et la restriction de $\underline{\pi}_{v,r}$ à $\mathfrak{g}^{\mathrm{nr}}_{v',r'}$ ne dépendent eux aussi que des facettes ϕ et ϕ' ; on les note respectivement $\underline{\mathfrak{m}}_{\phi;\phi'}$, $\underline{\mathfrak{p}}_{\phi;\phi'}$, $\underline{\mathfrak{u}}_{\phi;\phi'}$ et $\underline{\pi}_{\phi;\phi'}$. On a donc $\underline{\mathfrak{g}}_{\phi'} = \underline{\mathfrak{m}}_{\phi;\phi'}$.

6. Filtrations de Moy-Prasad et torsion

Dans toute cette section, on fixe un élément $d \in D$. On rappelle que d est un cocycle de Θ (considéré aussi comme un cocycle de Γ trivial sur I) à valeurs dans $\mathbf{N}_{\mathcal{C}}^{\mathrm{nr}}$.

6.1. Notons $\Theta_d \subset \Theta$ le sous-groupe formé des θ opèrant trivialement sur ${}_dV$; i.e. tels que $d(\theta)_{\mathbb{R}} \circ \theta$ opère trivialement sur V. Puisque $\mathrm{rg}(\mathcal{D}) < p-1$, pour $\theta \in \Theta$, l'automorphisme affine $d(\theta)_{\mathbb{R}} \circ \theta$ de V est d'ordre fini premier à p [W2, 2.8]. On en déduit que l'indice $(\Theta : \Theta_d)$ est fini et premier à p. Pour $\theta \in \Theta$, $d(\theta)_{\mathbb{R}} \circ \theta$ stabilise le sous-espace $V^{\mathrm{nr}} \subset V$, et induit une permutation de l'ensemble Φ (défini en 5.2) des facettes de $V^{\mathrm{nr}} \times \mathbb{R}$. On note ${}_d\Phi \subset \Phi$ le sous-ensemble formé des facettes Θ-stables pour cette action de Θ. Par convexité, tout élément de ${}_d\Phi$ coupe ${}_dW \times \mathbb{R}$.

Le sous-espace $V_{(p)} \subset V$ est Γ-stable pour l'action de Γ tordue par d définie en 4.3; i.e. il définit un sous-Γ-ensemble ${}_dV_{(p)} \subset {}_dV$. On pose ${}_dW_{(p)} = ({}_dV_{(p)})^{\Gamma} \subset {}_dV_{(p)}$; on a donc ${}_dW_{(p)} = {}_dW \cap V_{(p)}$. D'après le lemme 5.2, pour tout $\phi \in {}_d\Phi$, l'intersection $\phi \cap ({}_dW_{(p)} \times \mathbb{Z}_{(p)})$ est dense dans $\phi \cap ({}_dW \times \mathbb{R})$. En effet, la projection de $V^{\mathrm{nr}} \times \mathbb{R}$ sur ${}_dW \times \mathbb{R}$ définie par

$$(v, r) \mapsto \frac{1}{(\Theta : \Theta_d)} \sum_{\theta \in \Theta/\Theta_d} (d(\theta)_{\mathbb{R}} \circ \theta(v), r)$$

envoie $V_{(p)}^{\mathrm{nr}} \times \mathbb{Z}_{(p)}$ sur ${}_dW_{(p)} \times \mathbb{Z}_{(p)}$. On en déduit en particulier que pour tout $\phi \in {}_d\Phi$, l'adhérence de ϕ dans V^{nr} coupe ${}_dW_{(p)} \times \{r(\phi)\}$.

Pour $\phi \in {}_d\Phi$ et $r \in \mathbb{R}$, on pose ${}_d\omega_\phi = \omega_\phi \cap {}_d\Phi$, et l'on note ${}_d\omega_\phi(> r)$ l'ensemble des $\phi' \in {}_d\omega_\phi$ tels que $r(\phi') > r$.

6.2. Puisque Θ fixe le groupe T_ϖ, l'application $\underline{d} = \underline{\pi} \circ d : \Theta \to \underline{N}$ est encore un cocycle. Comme en 4.3, on peut définir un $\mathbb{F}_q$-groupe algébrique ${}_d\underline{G}$, muni d'un $\overline{\mathbb{F}}_q$-isomorphisme ${}_d\underline{\xi} : {}_d\underline{G} \to \underline{G}$ vérifiant ${}_d\underline{\xi} \circ \theta = \mathrm{Ad}(\underline{d}(\theta)) \circ \theta \circ {}_d\underline{\xi}$ pour tout $\theta \in \Theta$.

Soit $v \in {}_dW_{(p)}$. Pour $\theta \in \Theta$ et $\sigma \in I$, les automorphismes $\mathrm{Ad}(\underline{d}(\theta)) \circ \theta$ et $\underline{\rho}_v(\sigma)$ de $\underline{G}$ ne commutent pas, mais on a l'égalité :

$$\mathrm{Ad}(\underline{d}(\theta)) \circ \theta \circ \underline{\rho}_v(\sigma) = \underline{\rho}_v(\theta\sigma\theta^{-1}) \circ \mathrm{Ad}(\underline{d}(\theta)) \circ \theta.$$

Par conséquent $\mathrm{Ad}(\underline{d}(\theta)) \circ \theta$ stabilise le sous-groupe $\underline{G}_v \subset \underline{G}$, et le sous-groupe ${}_d\underline{G}_v = {}_d\underline{\xi}^{-1}(\underline{G}_v)$ de ${}_d\underline{G}$ est défini sur $\mathbb{F}_q$. On peut aussi dire les choses autrement : puisque $v \in {}_dW$, on a $t_v^{-1}d(\gamma)\gamma(t_v) \in \underline{N}$ pour tout $\gamma \in \Gamma$. L'application $\gamma \mapsto t_v^{-1}d(\gamma)\gamma(t_v)$ est un cocycle, et pour $\gamma \in \Gamma$, on note ${}_d\underline{\rho}_v(\gamma)$ le $\overline{\mathbb{F}}_q$-automorphisme $\mathrm{Ad}(t_v^{-1}d(\gamma)\gamma(t_v)) \circ \gamma$ de $\underline{G}$. On munit ${}_d\underline{G}$ d'une action de Γ de sorte que ${}_d\underline{\xi} \circ \gamma = {}_d\underline{\rho}_v(\gamma) \circ {}_d\underline{\xi}$ pour tout $\gamma \in \Gamma$. Cette action prolonge celle de Θ définie plus haut, et ${}_d\underline{G}_v$ coïncide avec le sous-groupe ${}_d\underline{G}^I \subset {}_d\underline{G}$ des points fixés par I.

On note ${}_dG_v^{\Theta} = {}_d\underline{G}_v(\mathbb{F}_q)$ le groupe des points $\mathbb{F}_q$-rationnels de ${}_d\underline{G}_v$.

Posons ${}_d\underline{\mathfrak{g}} = \mathrm{Lie}({}_d\underline{G})$. Comme en 3.3, on identifie ${}_d\underline{\mathfrak{g}}$ à ${}_d\underline{\mathfrak{g}}(\overline{\mathbb{F}}_q)$. Pour $(v,r) \in {}_dW_{(p)} \times \mathbb{Z}_{(p)}$, notons ${}_d\underline{\mathfrak{g}}_{v,r} \subset {}_d\underline{\mathfrak{g}}$ le sous-espace formé des X tels que $\sigma(X) = r(\sigma)X$ pour tout $\sigma \in I$. Ce sous-espace est défini sur $\mathbb{F}_q$, i.e. il est Θ-stable. Et il est ${}_d\underline{G}_v$-stable pour l'action adjointe de ${}_d\underline{G}_v$ sur ${}_d\underline{\mathfrak{g}}$. Par construction on a l'égalité $\mathrm{d}({}_d\xi)({}_d\underline{\mathfrak{g}}_{v,r}) = \underline{\mathfrak{g}}_{v,r}$.

Pour $\phi \in {}_d\Phi$, on pose ${}_d\underline{\mathfrak{g}}_\phi = \mathrm{d}({}_d\xi)^{-1}(\underline{\mathfrak{g}}_\phi)$; i.e. ${}_d\underline{\mathfrak{g}}_\phi = {}_d\underline{\mathfrak{g}}_{v,r}$ pour tout $(v,r) \in \phi \cap ({}_dW_{(p)} \cap \mathbb{Z}_{(p)})$. Pour $\phi \in {}_d\Phi$ et $\phi' \in {}_d\omega_\phi$, on a défini en 5.6 les sous-espaces $\underline{\mathfrak{m}}_{\phi;\phi'}$, $\underline{\mathfrak{p}}_{\phi;\phi'}$ et $\underline{\mathfrak{u}}_{\phi;\phi'}$ de $\underline{\mathfrak{g}}_\phi$. Leurs images réciproques par $\mathrm{d}({}_d\xi)$, notées respectivement ${}_d\underline{\mathfrak{m}}_{\phi;\phi'}$, ${}_d\underline{\mathfrak{p}}_{\phi;\phi'}$ et ${}_d\underline{\mathfrak{u}}_{\phi;\phi'}$, sont des sous-espaces Θ-stables de ${}_d\underline{\mathfrak{g}}_\phi$. Puisque $\underline{\mathfrak{g}}_{\phi'} = \underline{\mathfrak{m}}_{\phi;\phi'}$, on a ${}_d\underline{\mathfrak{g}}_{\phi'} = {}_d\underline{\mathfrak{m}}_{\phi;\phi'}$. On note ${}_d\underline{\mathfrak{g}}_\phi^\Theta \subset {}_d\underline{\mathfrak{g}}_\phi$, ${}_d\underline{\mathfrak{m}}_{\phi,\phi'}^\Theta \subset {}_d\underline{\mathfrak{m}}_{\phi,\phi'}$, (etc.), les sous-espaces formés des points fixés par Θ. Et pour tous (v,r), $(v',r') \in {}_dW_{(p)} \cap \mathbb{Z}_{(p)}$ tels que $(v,r) \in \phi$ et $(v',r') \in \phi'$, on pose ${}_d\underline{\mathfrak{g}}_{v,r}^\Theta = {}_d\underline{\mathfrak{g}}_\phi^\Theta$, ${}_d\underline{\mathfrak{m}}_{v,r;v',r'}^\Theta = {}_d\underline{\mathfrak{g}}_{\phi,\phi'}^\Theta$, (etc.).

6.3. Dans ce numéro et dans le suivant, le terme "nilpotent" signifie nilpotent dans la $\overline{\mathbb{F}}_q$-algèbre ${}_d\underline{\mathfrak{g}}$. Soit $v \in {}_dW_{(p)}$. L'application qui à $r \in \mathbb{Z}_{(p)}$ associe le caractère $\sigma \mapsto r(\sigma)$ de I se quotiente en un isomorphisme de $\mathbb{Z}_{(p)}/\mathbb{Z}$ sur le groupe des caractères lisses d'ordre fini de I. On en déduit que pour $r \in \mathbb{Z}_{(p)}$, le sous-espace ${}_d\underline{\mathfrak{g}}_{v,r}$ ne dépend que de l'image de r dans $\mathbb{Z}_{(p)}/\mathbb{Z}$. L'ensemble des $r \in \mathbb{Z}_{(p)}/\mathbb{Z}$ tels que ${}_d\underline{\mathfrak{g}}_{v,r} \neq \{0\}$ est fini, et l'on a la décomposition

$$ {}_d\underline{\mathfrak{g}} = \oplus_r \, {}_d\underline{\mathfrak{g}}_{v,r} $$

où r parcourt les éléments de $\mathbb{Z}_{(p)}/\mathbb{Z}$.

Notons que pour r, $s \in \mathbb{Z}_{(p)}$, l'espace $[{}_d\underline{\mathfrak{g}}_{v,r} : {}_d\underline{\mathfrak{g}}_{v,s}] = \{[X,Y] : X \in {}_d\underline{\mathfrak{g}}_{v,r},\ Y \in {}_d\underline{\mathfrak{g}}_{v,s}\}$ est contenu dans ${}_d\underline{\mathfrak{g}}_{v,r+s}$. DeBacker a remarqué que la théorie des $\mathfrak{sl}_2$-triplets s'adaptait a cette situation [D, Appendix A]. Soient $r \in \mathbb{Z}_{(p)}$ et $X \in {}_d\underline{\mathfrak{g}}_{v,r}^\Theta$ *nilpotent*. Grâce à la condition (P_Σ) de 3.1, on montre qu'il existe un élément $Y \in {}_d\underline{\mathfrak{g}}_{v,-r}^\Theta$ et un élément *semisimple* $H \in {}_d\underline{\mathfrak{g}}_{v,0}^\Theta$ tels que (X,H,Y) soit un $\mathfrak{sl}_2$-triplet [W2, lemme 2.9.1]. On gradue alors l'espace ${}_d\underline{\mathfrak{g}}$ de la manière habituelle suivante : pour $i \in \mathbb{Z}$, on pose ${}_d\underline{\mathfrak{g}}(i) = \{Z \in {}_d\underline{\mathfrak{g}} : [H,Z] = iZ\}$. Cette graduation est définie sur $\mathbb{F}_q$ au sens où les sous-espaces ${}_d\underline{\mathfrak{g}}(i)$ sont Θ-stables. Pour $i \in \mathbb{Z}$, on pose ${}_d\underline{\mathfrak{g}}(\geq i) = \oplus_{j \geq i}\, {}_d\underline{\mathfrak{g}}(j)$. À nouveau grâce à la condition (P_Σ), on montre que pour $Z \in {}_d\underline{\mathfrak{g}}_{v,r}^\Theta \cap {}_d\underline{\mathfrak{g}}(\geq 3)$, il existe un élément $g \in {}_d\underline{G}_v^\Theta$ tel que $\mathrm{Ad}(g)(X) = X + Z$ [W2, lemme 2.9.2].

6.4. Pour $\phi \in {}_d\Phi$, on pose ${}_d\mathcal{S}_\phi = C({}_d\underline{\mathfrak{g}}_\phi^\Theta)$ et ${}_dS_\phi^* = \mathcal{D}({}_d\underline{\mathfrak{g}}_\phi^\Theta)$. Puisque ${}_d\underline{\mathfrak{g}}_\phi^\Theta$ est un espace vectoriel de dimension finie sur $\mathbb{F}_q$, c'est un ensemble fini ; et $\dim_{\mathbb{C}}({}_d\mathcal{S}_\phi) < +\infty$.

Soient $\phi \in {}_d\Phi$ et $\phi' \in {}_d\omega_\phi$. D'après 6.2, on a ${}_d\underline{\mathfrak{g}}_{\phi'}^\Theta = {}_d\underline{\mathfrak{m}}_{\phi;\phi'}^\Theta$ et ${}_d\underline{\mathfrak{p}}_{\phi;\phi'}^\Theta = {}_d\underline{\mathfrak{m}}_{\phi;\phi'}^\Theta \oplus {}_d\underline{\mathfrak{u}}_{\phi;\phi'}^\Theta$. Cela permet de définir une application linéaire injective ${}_ds_{\phi;\phi'} : {}_d\mathcal{S}_{\phi'} \to {}_d\mathcal{S}_\phi$. Précisément, à toute fonction $\varphi' \in {}_d\mathcal{S}_{\phi'}$, ${}_ds_{\phi;\phi'}$ associe la fonction $\varphi \in {}_d\mathcal{S}_\phi$ à support dans ${}_d\underline{\mathfrak{p}}_{\phi;\phi'}^\Theta$ donnée par $\varphi(X+Y) = \varphi'(X)$ pour tout $X \in {}_d\underline{\mathfrak{m}}_{\phi;\phi'}^\Theta$ et tout $Y \in {}_d\underline{\mathfrak{u}}_{\phi;\phi'}^\Theta$. Supposons de plus que $r(\phi') > r$

pour un $r \in \mathbb{Z}_{(p)}$ tel que ϕ coupe ${}_dW_{(p)} \times \{r\}$. Alors il existe deux éléments $v, v' \in {}_dW_{(p)}$ et un $r' \in \mathbb{Z}_{(p)}$ tels que $r' > r$, $(v,r) \in \phi$ et $(v',r') \in \phi'$. Choisissons un $e \in \mathcal{P}$ tel que ev, $ev' \in \mathrm{X}_*$ et er, $er' \in \mathbb{Z}$. Posons $\lambda = ev' - ev$ et graduons l'espace $\underline{\mathfrak{g}}$ comme en 5.1 : pour $i \in \mathbb{Z}$, posons

$$\underline{\mathfrak{g}}[i] = \{X \in \underline{\mathfrak{g}} : \mathrm{Ad}(\lambda(x))(X) = x^i X \text{ pour tout } x \in \overline{\mathbb{F}}_q^\times\}.$$

L'espace $\underline{\mathfrak{g}}[\geq 1]$ est le radical nilpotent d'une sous-algèbre parabolique de $\underline{\mathfrak{g}}$; il est donc formé d'éléments nilpotents. Puisque $r' > r$, l'espace ${}_d\underline{\mathfrak{p}}_{\phi;\phi'} = \mathrm{d}({}_d\underline{\xi})^{-1}(\underline{\mathfrak{g}}[er' - er])$ est contenu dans $\mathrm{d}({}_d\underline{\xi})^{-1}(\underline{\mathfrak{g}}[\geq 1])$, donc est lui aussi formé d'éléments nilpotents. En particulier, toute fonction dans l'image de ${}_ds_{\phi;\phi'}$ est à support nilpotent.

Pour $\phi \in {}_d\Phi$, les injections ${}_ds_{\phi;\phi'}$ définissent une application linéaire ${}_ds_\phi : \oplus_{\phi' \in {}_d\omega_\phi} {}_d\mathcal{S}_{\phi'} \to {}_d\mathcal{S}_\phi$. Grâce à aux résultats de 6.3, on montre le

LEMME [W2, lemme 2.10.2]. — *Soit $\phi \in {}_d\Phi$ et soit $(v,r) \in \phi \cap ({}_dW_{(p)} \times \mathbb{Z}_{(p)})$. Il existe une application linéaire ${}_dl_\phi^+ : {}_d\mathcal{S}_\phi \to \oplus_{\phi' \in {}_d\omega_\phi(>r)} {}_d\mathcal{S}_{\phi'}$ vérifiant la propriété suivante : pour toute paire $(\varphi, J) \in {}_dS_\phi \times {}_d\mathcal{S}_\phi^*$ telle que J soit ${}_d\underline{G}_v^\Theta$-invariante (pour l'action adjointe de ${}_d\underline{G}_v^\Theta$ sur ${}_d\underline{\mathfrak{g}}_\phi^\Theta$) et à support nilpotent, on a l'égalité $J({}_ds_\phi \circ {}_dl_\phi^+(\varphi)) = J(\varphi)$.*

6.5. Soit $F \in \mathrm{CL}_q$, et soit $v \in {}_dW_{(p)}$. On a défini en 4.3 le $\mathfrak{o}_F$-schéma en groupes ${}_d\mathcal{G}_v$. Notons ${}_d\mathbb{G}_v$ le plus grand quotient réductif de la fibre spéciale ${}_d\overline{\mathcal{G}}_v = {}_d\mathcal{G}_v \times_{\mathfrak{o}_F} \kappa_F$ de ${}_d\mathcal{G}_v$. Ainsi ${}_d\mathbb{G}_v$ est un groupe réductif défini sur κ_F, en général non connexe. Notons ${}_d\pi_v^{\mathrm{nr}} : {}_dG_v^{\mathrm{nr}} \to {}_d\mathbb{G}_v = {}_d\mathbb{G}_v(\overline{\kappa}_F)$ la composée des projections canoniques ${}_dG_v^{\mathrm{nr}} \to {}_d\overline{\mathcal{G}}_v = {}_d\overline{\mathcal{G}}_v(\overline{\kappa}_F)$ (réduction modulo $\mathfrak{p}^{\mathrm{nr}}$) et ${}_d\overline{\mathcal{G}}_v \to {}_d\mathbb{G}_v$. Pour $\theta \in \Theta$, on a l'égalité $\underline{\pi}_v \circ \mathrm{Ad}(d_F(\theta)) \circ \theta = {}_d\rho_v(\theta) \circ \underline{\pi}_v$. On en déduit qu'il existe un (unique) isomorphisme de groupes algébriques ${}_d\iota_v : {}_d\mathbb{G}_v \to {}_d\underline{G}_v$ *défini sur* κ_F (identifié à $\mathbb{F}_q$ via ι), tel que posant ${}_d\underline{\pi}_v = {}_d\iota_v \circ {}_d\pi_v^{\mathrm{nr}} : {}_dG_v^{\mathrm{nr}} \twoheadrightarrow {}_d\underline{G}_v$, le diagramme

$$\begin{array}{ccc} {}_dG_v^{\mathrm{nr}} & \xrightarrow{{}_d\underline{\pi}_v} & {}_d\underline{G}_v \\ {\scriptstyle {}_d\xi}\downarrow & & \downarrow{\scriptstyle {}_d\underline{\xi}} \\ G_v^{\mathrm{nr}} & \xrightarrow{\underline{\pi}_v} & \underline{G}_v \end{array}$$

soit commutatif. Puisque ${}_d\iota_v$ est défini sur κ_F, ${}_d\underline{\pi}_v$ est Θ-équivariant.

Posons ${}_d\mathrm{X}_* = \{\lambda \in \mathrm{X}_*^{\mathrm{nr}} : d(\theta) \circ \theta(\lambda) = \lambda, \forall \theta \in \Theta\}$. Puisque le point v appartient à l'appartement ${}_dW = A({}_dG, {}_dS, F)$ de ${}_d\mathfrak{I}_F$, la théorie de Bruhat-Tits nous dit que l'image de ${}_dS(F^{\mathrm{nr}}) \cap {}_dG_v^{\mathrm{nr}}$ par ${}_d\underline{\pi}_v$ est le groupe des points $\overline{\mathbb{F}}_q$-rationnels d'un sous-tore maximal ${}_d\underline{S}_v$ de ${}_d\underline{G}_v$ défini sur $\mathbb{F}_q$. Ce tore est nécessairement déployé sur $\overline{\mathbb{F}}_q$. Et par construction, on a ${}_d\underline{\xi}({}_d\underline{S}_v) \subset \underline{T}$ et $\mathrm{X}_*({}_d\underline{\xi}({}_d\underline{S}_v)) = {}_d\mathrm{X}_*$.

Soit une facette $\phi \in {}_d\Phi$. On note ${}_d\mathfrak{g}_\phi^{\mathrm{nr}}$ le $\mathfrak{o}^{\mathrm{nr}}$-réseau dans ${}_d\mathfrak{g}^{\mathrm{nr}}$ défini par ${}_d\mathfrak{g}_\phi^{\mathrm{nr}} = \mathrm{d}({}_d\xi)^{-1}(\mathfrak{g}_\phi^{\mathrm{nr}})$. Il est Θ-stable. Posons ${}_d\mathfrak{g}(F)_\phi = [{}_d\mathfrak{g}_\phi^{\mathrm{nr}}]^\Theta$; c'est un $\mathfrak{o}_F$-réseau dans ${}_d\mathfrak{g}(F)$, et l'on a ${}_d\mathfrak{g}_\phi^{\mathrm{nr}} = {}_d\mathfrak{g}(F)_\phi \otimes_{\mathfrak{o}_F} \mathfrak{o}^{\mathrm{nr}}$. Il existe une (unique) projection Θ-équivariante ${}_d\underline{\pi}_\phi : {}_d\mathfrak{g}_\phi^{\mathrm{nr}} \to {}_d\underline{\mathfrak{g}}_\phi$ rendant le diagramme

$$\begin{array}{ccc} {}_d\mathfrak{g}_\phi^{\mathrm{nr}} & \xrightarrow{{}_d\underline{\pi}_\phi} & {}_d\underline{\mathfrak{g}}_\phi \\ \mathrm{d}({}_d\xi)\downarrow & & \downarrow \mathrm{d}({}_d\underline{\xi}) \\ \mathfrak{g}_\phi^{\mathrm{nr}} & \xrightarrow{\underline{\pi}_\phi} & \underline{\mathfrak{g}}_\phi \end{array}$$

commutatif. Pour $(v,r) \in \phi \cap ({}_dW_{(p)} \times \mathbb{Z}_{(p)})$, on écrit aussi ${}_d\underline{\pi}_{v,r} = {}_d\underline{\pi}_\phi$. Soit ${}_d\mathfrak{g}_{\phi+}^{\mathrm{nr}} = \mathrm{d}({}_d\xi)^{-1}(\mathfrak{g}_{\phi+}^{\mathrm{nr}})$. D'après 5.5, ${}_d\underline{\pi}_\phi$ se quotiente en un isomorphisme

$${}_d\mathfrak{g}_\phi^{\mathrm{nr}} / {}_d\mathfrak{g}_{\phi+}^{\mathrm{nr}} \xrightarrow{\sim} {}_d\underline{\mathfrak{g}}_\phi .$$

En particulier, ${}_d\mathfrak{g}_{\phi+}^{\mathrm{nr}}$ est un $\mathfrak{o}^{\mathrm{nr}}$-réseau Θ-stable dans ${}_d\mathfrak{g}^{\mathrm{nr}}$. Posons ${}_d\mathfrak{g}(F)_{\phi+} = [{}_d\mathfrak{g}_{\phi+}^{\mathrm{nr}}]^\Theta$; c'est un $\mathfrak{o}_F$-réseau dans ${}_d\mathfrak{g}(F)$, et l'on a ${}_d\mathfrak{g}_{\phi+}^{\mathrm{nr}} = {}_d\mathfrak{g}(F)_{\phi+} \otimes_{\mathfrak{o}_F} \mathfrak{o}^{\mathrm{nr}}$. Puisque $\mathrm{H}^1(\Theta, {}_d\mathfrak{g}_{\phi+}^{\mathrm{nr}}) = 0$, l'application ${}_d\underline{\pi}_\phi$ induit un isomorphisme

$${}_d\mathfrak{g}(F)_\phi / {}_d\mathfrak{g}(F)_{\phi+} \xrightarrow{\sim} {}_d\underline{\mathfrak{g}}_\phi^{\mathrm{O}} .$$

Pour $(v,r) \in \phi \cap ({}_dW \times \mathbb{R})$, on pose ${}_d\mathfrak{g}(F)_{v,r} = {}_d\mathfrak{g}(F)_\phi$ et ${}_d\mathfrak{g}(F)_{v,r^+} = {}_d\mathfrak{g}(F)_{\phi+}$; par construction, on a donc ${}_d\mathfrak{g}(F)_{v,r^+} = \bigcup_{s>r} {}_d\mathfrak{g}(F)_{v,s}$. Notons que ${}_d\mathfrak{g}(F)_{v,r}$ est bien le $\mathfrak{o}_F$-réseau dans ${}_d\mathfrak{g}(F)$ associé à (v,r) par Moy et Prasad.

7. Analyse harmonique : une première réduction

Dans toute cette section, on fixe un élément $d \in D$. Dans les trois premiers numéros, on fixe aussi un corps $F \in \mathrm{CL}_q$.

7.1. Pour $v \in {}_dW_{(p)}$, on a défini en 6.5 un homomorphisme Θ-équivariant ${}_d\underline{\pi}_v : {}_dG_v^{\mathrm{nr}} \twoheadrightarrow {}_d\underline{G}_v$. Posons ${}_dK_v^+ = (\ker {}_d\pi_v)^\Theta$; c'est un sous-groupe ouvert distingué de $({}_dG_v^{\mathrm{nr}})^\Theta = {}_d\mathcal{G}_v(\mathfrak{o}_F)$. Puisque $\mathrm{H}^1(\Theta, \ker {}_d\pi_v) = 0$, l'application ${}_d\underline{\pi}_v$ induit un isomorphisme ${}_d\mathcal{G}_v(\mathfrak{o}_F)/{}_dK_v^+ \xrightarrow{\sim} {}_d\underline{G}_v^\Theta$.

La chambre $\mathcal{C}^{\mathrm{nr}}$ définie en 3.3 appartient à ${}_d\Phi$: pour $\theta \in \Theta$, $d(\theta) \in \mathbf{N}_{\mathcal{C}}^{\mathrm{nr}}$ par conséquent $d(\theta)_{\mathbb{R}} \circ \theta$ stabilise $\mathcal{C}^{\mathrm{nr}}$. Soit $v \in {}_dW_{(p)} \cap \mathcal{C}^{\mathrm{nr}}$. Considérons la mesure de Haar ${}_ddg$ sur ${}_dG(F)$ définie par $\mathrm{vol}({}_d\mathcal{G}_v(\mathfrak{o}_F), {}_ddg) = |{}_d\underline{G}_v^\Theta| q^{-\frac{1}{2}\dim({}_d\underline{G}_v)}$. Alors pour tout $v' \in {}_dW_{(p)}$, on a

$$\mathrm{vol}({}_d\mathcal{G}_{v'}(\mathfrak{o}_F), {}_ddg) = |{}_d\underline{G}_{v'}^\Theta| q^{-\frac{1}{2}\dim({}_d\underline{G}_{v'})}.$$

En effet, d'après le paragraphe précédent, l'égalité ci-dessus équivaut à

$$\frac{\mathrm{vol}({}_dK_v^+, {}_ddg)}{\mathrm{vol}({}_dK_{v'}^+, {}_ddg)} = q^{\frac{1}{2}[\dim({}_d\underline{G}_{v'}) - \dim({}_d\underline{G}_v)]}.$$

Quitte à remplacer v' par gv' pour un $g \in {}_dG(F)$, on peut supposer que v' appartient à l'adhérence de la chambre $\mathcal{C}^{\mathrm{nr}}$. Alors on a ${}_dK_{v'}^+ \subset {}_dK_v^+ \subset {}_d\mathcal{G}_{v'}(\mathfrak{o}_F)$, et l'application ${}_d\underline{\pi}_{v'}$ identifie le quotient ${}_dK_v^+/{}_dK_{v'}^+$ au groupe des point $\mathbb{F}_q$-rationnels du radical unipotent $\underline{U}$ d'un sous-groupe parabolique $\underline{P}$ de ${}_d\underline{G}_v^\circ$ (la composante neutre de ${}_d\underline{G}_v$) défini sur $\mathbb{F}_q$, tel que $\underline{P}/\underline{U}$ soit $\mathbb{F}_q$-isomorphe à ${}_d\underline{G}_{v'}^\circ$. D'où l'égalité cherchée.

Le groupe G lui-même est défini sur F, et l'on peut comme ci-dessus définir une mesure de Haar dg sur $G(F)$; cela revient à prendre pour d le cocycle représentant la classe de cohomologie triviale dans $\mathrm{H}^1(\Theta, \mathbf{N}_C^{\mathrm{nr}})$. Le F^{nr}-isomorphisme ${}_d\xi$ est un torseur intérieur de ${}_dG$ vers sa forme quasi-déployée G. On sait qu'un tel torseur établit une correspondance entre mesures de Haar sur $G(F)$ et mesures de Haar sur ${}_dG(F)$. D'après Kottwitz [K1, §1], les mesures dg et ${}_ddg$ se correspondent (cf. [W2, lemme 3.1.2]).

7.2. Pour $r \in \mathbb{R}$, on pose

$$
{}_d\mathfrak{g}(F)_r = \bigcup_{g \in {}_dG(F)} \bigcup_{v \in {}_dW} \mathrm{Ad}(g)({}_d\mathfrak{g}(F)_{v,r}), \qquad {}_d\mathfrak{g}(F)_{r^+} = \bigcup_{s>r} {}_d\mathfrak{g}(F)_s.
$$

Pour $X \in {}_d\mathfrak{g}(F) \smallsetminus \{0\}$ semisimple, on pose $r(X) = \sup\{r \in \mathbb{R} : X \in {}_d\mathfrak{g}(F)_r\}$; et pour $X = 0$, on pose $r(0) = +\infty$.

Choisissons une sous-extension galoisienne finie E_1/F de F^{mod}/F tel que le sous-groupe $\mathrm{Gal}(F^{\mathrm{mod}}/E_1) \subset \Gamma$ opère trivialement sur $\mathcal{D}$ et que le cocycle d se factorise à travers $\mathrm{Gal}(E_1/F)$. Puisque $|W|$ est fini, il n'y a qu'un nombre fini de sous-extensions E_2/E_1 de $\overline{F}/E_1$ de degré divisant $|W|$. Grâce à la condition $(P_{\mathrm{rg}(\mathcal{D})})$, $|W|$ est premier à p et une telle extension E_2 est contenue dans F^{mod}. Soit E la composée sur F de ces extensions E_2. C'est une extension galoisienne finie de F, qui vérifie la propriété suivante [W2, lemme 3.2.(i)] : pour tout $X \in {}_d\mathfrak{g}(F)$ semisimple, il existe un $g \in {}_d\mathfrak{g}(E)$ tel que $\mathrm{Ad}(g) \circ \mathrm{d}({}_d\xi)(X) \in \mathfrak{t}(E)$. Alors pour $r \in \mathbb{R}$, on a $X \in {}_d\mathfrak{g}(F)_r$ si et seulement si $\mathrm{Ad}(g) \circ \mathrm{d}({}_d\xi)(X) \in \mathfrak{t}(E)_r$. Cela résulte du lemme 2.2.1 de [KM] et du fait que la définition de la filtration $r \mapsto {}_d\mathfrak{g}(F)_r$ est compatible aux extensions modérément ramifiées de F [KM, 2.1]. Notons $e \in \mathcal{P}$ l'indice de ramification de E/F. Puisque T est déployé sur E, pour $X' \in \mathfrak{t}(E) \smallsetminus \{0\}$, on a $\sup\{r \in \mathbb{R} : X' \in \mathfrak{t}(E)_r\} \in \omega(K^\times) = \frac{1}{e}\mathbb{Z}$. On en déduit que pour $X \in {}_d\mathfrak{g}(F) \smallsetminus \{0\}$ semisimple, on a $r(X) \in \frac{1}{e}\mathbb{Z}$.

7.3. Posons ${}_d\mathcal{S} = \oplus_{\phi \in {}_d\Phi}\, {}_d\mathcal{S}_\phi$ et notons ${}_d\mathcal{C}_F = C_{\mathrm{c}}^\infty({}_d\mathfrak{g}(F))$. Pour $\phi \in {}_d\Phi$ et $\varphi \in {}_d\mathcal{S}_\phi$, on note $\mathrm{rea}_F(\varphi) \in C({}_d\mathfrak{g}(F)_\phi/{}_d\mathfrak{g}(F)_{\phi^+})$ la composée de l'isomorphisme ${}_d\mathfrak{g}(F)_\phi/{}_d\mathfrak{g}(F)_{\phi^+} \to {}_d\underline{\mathfrak{g}}_\phi^\Theta$ de 6.5 et de φ. Via les inclusions $C({}_d\mathfrak{g}(F)_\phi/{}_d\mathfrak{g}(F)_{\phi^+}) \subset C^\infty({}_d\mathfrak{g}(F)_\phi) \subset {}_d\mathcal{C}_F$, cela définit une application linéaire ${}_d\mathrm{rea}_F : {}_d\mathcal{S} \to {}_d\mathcal{C}_F$, qui n'est en général ni injective ni surjective.

Soient $\phi \in {}_d\Phi$ et $\phi' \in {}_d\omega_\phi$. On a défini en 6.4 une application injective ${}_ds_{\phi;\phi'} : {}_d\mathcal{S}_{\phi'} \to {}_d\mathcal{S}_\phi$. D'après 6.2, pour $\varphi' \in {}_d\mathcal{S}_{\phi'}$, on a l'égalité ${}_d\mathrm{rea}_F(\varphi') = {}_d\mathrm{rea}_F \circ {}_ds_{\phi;\phi'}(\varphi')$.

Pour $r \in \mathbb{R}$, on pose ${}_d\Phi_r = \{\phi \in {}_d\Phi : r(\phi) \geq r\}$ et ${}_d\Phi_{r^+} = \{\phi \in {}_d\Phi : r(\phi) > r\}$; on pose aussi ${}_d\mathcal{S}_r = \oplus_{\phi \in {}_d\Phi_r}\, {}_d\mathcal{S}_\phi$ et ${}_d\mathcal{S}_{r^+} = \oplus_{\phi \in {}_d\Phi_{r^+}}\, {}_d\mathcal{S}_\phi$. D'après 6.1, pour $\phi' \in {}_d\Phi$, l'adhérence de ϕ' coupe ${}_dW_{(p)} \times \{r(\phi')\}$. En d'autres termes, il existe un point $v \in {}_dW_{(p)}$ tel que posant $\phi = \phi_{v,r(\phi')}$, on ait $\phi' \in {}_d\omega_\phi$. On en déduit que pour $r \in \mathbb{R}$ et $\varphi \in {}_d\mathcal{S}$, on a $\varphi \in {}_d\mathcal{S}_r$ si et seulement si le support de la fonction ${}_d\mathrm{rea}_F(\varphi)$ est contenu dans ${}_d\mathfrak{g}(F)_r$.

Le résultat suivant est bien connu :

LEMME [W2, lemme 3.3]. — *Soit* $\phi \in {}_d\Phi$ *et soit* $(v, r) \in \phi \cap ({}_dW_{(p)} \times \mathbb{Z}_{(p)})$. *Tout élément de* ${}_d\mathfrak{g}_{\underline{\phi}}^{\Theta}$ *appartenant à l'image de* ${}_d\mathfrak{g}(F)_{r+} \cap {}_d\mathfrak{g}_{\phi}^{\mathrm{nr}}$ *par l'application* ${}_d\underline{\pi}_{\phi}$, *est nilpotent (dans* ${}_d\underline{\mathfrak{g}}$*).*

Notons ${}_d\mathfrak{g}(F)_{\mathrm{reg}} \subset {}_d\mathfrak{g}(F)$ le sous-ensemble formé des éléments semisimples réguliers. D'après la condition (P_Σ), pour $X \in {}_d\mathfrak{g}(F)_{\mathrm{reg}}$, le centralisateur T_X de X dans ${}_dG$ est un groupe algébrique *connexe*; c'est donc un tore maximal de ${}_dG$, défini sur F et (d'après 7.2) déployé sur F^{mod}. Il correspond à une donnée $\mathcal{D}_X = (\mathrm{X}^*(T_X), \emptyset, \emptyset, \mathrm{X}_*(T_X), \emptyset, \emptyset)$ munie d'une action de Γ comme en 3.1. On peut lui associer, comme en 7.1, une mesure de Haar dt. Soit ${}_dd\bar{g}_X$ la mesure $\frac{{}_ddg}{dt}$ sur l'espace homogène $T_X(F)\backslash{}_dG(F)$. Posons $\mathfrak{t}_X = \mathrm{Lie}(T_X)$, et notons $\Lambda^{{}_dG(F)}(\cdot, X)$ la distribution sur ${}_d\mathfrak{g}(F)$ définie par

$$\Lambda^{{}_dG(F)}(f, X) = \Delta(X) \int_{T_X(F)\backslash{}_dG(F)} f(\mathrm{Ad}(g^{-1})(X))\,{}_dd\bar{g}_X$$

avec

$$\Delta(X) = |\det{}_F(\mathrm{ad}(X); \mathfrak{t}_X(F)\backslash{}_d\mathfrak{g}(F))|_F^{\frac{1}{2}}.$$

7.4. Jusqu'à présent, le corps F était fixé. En combinant les lemmes 6.4 et 7.3, on peut le supprimer des données :

PROPOSITION [W2, proposition 3.4]. — *Soit* $r \in \mathbb{Z}_{(p)}$. *Il existe une application linéaire* ${}_dl_r^+ : {}_d\mathcal{S}_r \to {}_d\mathcal{S}_{r+}$ *telle que pour tous* $\varphi \in {}_d\mathcal{S}_r$, $F \in \mathrm{CL}_q$ *et* $X \in {}_d\mathfrak{g}(F)_{\mathrm{reg}} \cap {}_d\mathfrak{g}(F)_{r+}$, *on ait l'égalité*

$$\Lambda^{{}_dG(F)}({}_d\mathrm{rea}_F(\varphi), X) = \Lambda^{{}_dG(F)}({}_d\mathrm{rea}_F \circ {}_dl_r^+(\varphi), X).$$

Démonstration : Par linéarité, il suffit de définir ${}_dl_r(\varphi)$ pour $\varphi \in {}_d\mathcal{S}_{\phi'}$ avec $\phi' \in {}_d\Phi_r$. Si $r(\phi') > r$, alors $\varphi \in {}_d\mathcal{S}_{r+}$ et l'on pose ${}_dl_r^+(\varphi) = \varphi$. Supposons donc $r(\phi') = r$. D'après 6.1 (cf. aussi 7.3), il existe un point $v \in {}_dW_{(p)}$ tel que, posant $\phi = \phi_{v,r}$, on ait $\phi' \in {}_d\omega_\phi$. Supposons l'application ${}_dl_r^+$ définie sur ${}_d\mathcal{S}_\phi$. Alors on définit ${}_dl_r^+$ sur ${}_d\mathcal{S}_{\phi'}$ comme la composée de l'injection ${}_ds_{\phi;\phi'} : {}_d\mathcal{S}_{\phi'} \to {}_d\mathcal{S}_\phi$ et de l'application précédente. D'après le second paragraphe de 7.3, cette application convient. Reste à définir ${}_dl_r^+$ sur $\mathcal{S}_\phi$: puisque $\oplus_{\phi'' \in {}_d\omega_\phi(>r)}\, {}_d\mathcal{S}_{\phi''}$ est contenu dans ${}_d\mathcal{S}_{r+}$, on peut prendre l'application ${}_dl_\phi^+$ du lemme 6.4. Montrons qu'elle convient. Soient $F \in \mathrm{CL}_q$, $X \in {}_d\mathfrak{g}(F)_{\mathrm{reg}} \cap {}_d\mathfrak{g}(F)_{r+}$ et $\varphi \in {}_d\mathcal{S}_\phi$. Posons $f = {}_d\mathrm{rea}_F(\varphi)$, $f^+ = {}_d\mathrm{rea}_F \circ {}_dl_\phi^+(\varphi)$, et notons Ω l'ensemble $\{g \in T_X(F)\backslash{}_dG(F) : \mathrm{Ad}(g^{-1})(X) \in {}_d\mathfrak{g}(F)_\phi\}$; il est stable par multiplication à droite par ${}_d\mathcal{G}_v(\mathfrak{o}_F)$, et pour $g \in {}_dG(F) \smallsetminus \Omega$, on a $f(\mathrm{Ad}(g^{-1}(X)) = 0$. Posant $c = \mathrm{vol}({}_d\mathcal{G}_v(\mathfrak{o}_F), {}_ddg)$, on a donc

$$(*) \qquad \Lambda^{{}_dG(F)}(f, X) = \Delta(X) \int_\Omega \int_{{}_d\mathcal{G}_v(\mathfrak{o}_F)} c^{-1} f(\mathrm{Ad}(k^{-1}g^{-1})(X)\,{}_ddk\,{}_dd\bar{g}_X$$

où ${}_d dk$ désigne la restriction de la mesure ${}_d dg$ à ${}_d\mathcal{G}_v(\mathfrak{o}_F)$. Pour $g \in \Omega$, on définit $J_g \in {}_d\mathcal{S}_\phi^*$ par :

$$J_g(\varphi_1) = |{}_d\underline{G}_v^\Theta|^{-1} \sum_{x \in {}_d\underline{G}_v^\Theta} \varphi_1\left(\mathrm{Ad}(x^{-1}) \circ {}_d\pi_\phi \circ \mathrm{Ad}(g^{-1})(X)\right).$$

Par définition de l'application ${}_d\mathrm{rea}_F$, l'intégrale sur ${}_d\mathcal{G}_v(\mathfrak{o}_F)$ dans la formule $(*)$ n'est autre que $J_g(\varphi)$. On a donc $\Lambda^{{}_dG(F)}(f, X) = \Delta(X) \int_\Omega J_g(\varphi){}_d d\bar{g}_X$. Posons $\varphi'' = {}_d s_\phi \circ {}_d l_\phi^+(\varphi)$ et $f'' = {}_d\mathrm{rea}_F(\varphi'')$. Le calcul ci-dessus conduit à l'égalité $\Lambda^{{}_dG(F)}(f'', X) = \Delta(X) \int_\Omega J_g(\varphi''){}_d d\bar{g}_X$. Mais d'après le lemme 7.3, pour $g \in \Omega$, la distribution J_g sur ${}_d\underline{\mathfrak{g}}_\phi^\Theta$ vérifie les hypothèses du lemme 6.4 : on a donc $J_g(\varphi) = J_g(\varphi'')$. D'où l'égalité $\Lambda^{{}_dG(F)}(f, X) = \Lambda^{{}_dG(F)}(f'', X)$. Or d'après le second paragraphe de 7.3, on a $f'' = f^+$. □

8. Centralisateurs d'éléments semisimples

Dans toute cette section, on fixe un corps $F \in \mathrm{CL}_q$.

8.1. Soit δ un cocycle de Γ à valeurs dans $\mathbf{N}$. On définit les espaces ${}_\delta W = ({}_\delta V)^\Gamma \subset {}_\delta V$ comme en 4.3, et les espaces ${}_\delta W_{(p)} = ({}_\delta V_{(p)})^\Gamma \subset {}_\delta V_{(p)}$ comme en 6.1. Pour $v \in V_{(p)}$, on a $v \in {}_\delta W$ si et seulement si $t_v^{-1}\delta(\gamma)\gamma(t_v) \in \underline{N}$ pour tout $\gamma \in \Gamma$. Si $v \in {}_\delta W_{(p)}$, l'application $\gamma \mapsto t_v^{-1}\delta(\gamma)\gamma(t_v)$ est un cocycle à valeurs dans $\underline{N}$, et pour $\gamma \in \Gamma$, on note ${}_\delta\underline{\rho}_v(\gamma)$ le $\overline{\mathbb{F}}_q$-automorphisme de $\underline{G}$ défini comme en 6.2.

Soit $A \subset V$ un sous-espace affine tel que $A_{(p)} = A \cap V_{(p)}$ soit dense dans A. On note $\Sigma(A)$ l'ensemble des racines $\alpha \in \Sigma$ qui prennent une valeur constante, notée $\alpha(A)$, sur A. Puisque $A_{(p)}$ est dense dans A, pour $\alpha \in \Sigma(A)$, on a $\alpha(A) \in \mathbb{Z}_{(p)}$. L'ensemble $\Sigma(A)$ est un sous-ensemble de Levi de Σ. Il lui correspond un sous-groupe connexe $\underline{H}(A) \subset \underline{G}$ dont l'algèbre de Lie $\underline{\mathfrak{h}}(A)$ est engendré par $\mathfrak{t}$ et les $\underline{\mathfrak{g}}_\alpha$ pour $\alpha \in \Sigma(A)$. Pour $n \in \mathbf{N}$, on a l'égalité $\underline{H}(n_\mathbb{R}(A)) = \mathrm{Ad}(\underline{\pi}(n))(\underline{H}(A))$; et si de plus $\underline{\pi}(n) \in \underline{H}(A)$ et $n_\mathbb{R}$ fixe un élément de $A_{(p)}$, alors $n_\mathbb{R}$ fixe tout élément de A [W2, lemme 4.2.1]. Supposons de plus que $A \subset {}_\delta W$. Alors $A_{(p)} = A \cap {}_\delta W_{(p)}$; et d'après le lemme 4.2.2 de [W2], pour tout $v \in A_{(p)}$ et tout $\gamma \in \Gamma$, l'automorphisme ${}_\delta\underline{\rho}_v(\gamma)$ stabilise le sous-groupe $\underline{H}(A) \subset \underline{G}$ et sa restriction à $\underline{H}(A)$ est indépendante de v.

8.2. Soit $e \in \mathcal{P}$. Pour $v \in V^e$, on note $\underline{G}_v^e \subset \underline{G}$ le sous-groupe formé des éléments fixés par $\underline{\rho}_v(\sigma)$ pour tout $\sigma \in I(e)$ (cf. 5.1); et l'on définit comme en 5.4 un homomorphisme surjectif $\underline{\pi}_v : G_v^e \to \underline{G}_v^e$. Rappelons la construction : on choisit un entier $e' \in \mathcal{P}$ tel que e divise e', $e'v \in \mathrm{X}_*$, et $I(e')$ opère trivialement sur $\mathcal{D}$. L'égalité $(G_v^{e'})^{I(e)} = G_v^e$ se complète en le diagramme commutatif suivant :

$$\begin{array}{ccc} G_v^e & \xrightarrow{\pi_v^e} & \mathbb{G}_v^e \\ \| & & \| \\ (G_v^{e'})^{I(e)} & \xrightarrow{\pi_v^{e'}} & (\mathbb{G}_v^{e'})^{I(e)} \end{array}$$

où π_v^e est la composée des projections canoniques $G_v^e \to \overline{\mathcal{G}}_v^e$ (réduction modulo $\mathfrak{p}^e$) et $\overline{\mathcal{G}}_v^e \to \mathbb{G}_v^e$. L'homomorphisme $\underline{\tilde{\iota}}_v : G_v^e \to \underline{G}$ composé de l'application $\pi_0^{e'} \circ \mathrm{Ad}(t_{F,v}^{-1})$ et de l'inclusion $G_v^e \subset G_v^{e'}$, se quotiente en un isomorphisme $\underline{\iota}_v : \mathbb{G}_v^e \to \underline{G}_v^e$. On pose alors $\underline{\pi}_v = \underline{\iota}_v \circ \pi_v^e : G_v^e \to \underline{G}_v^e$. D'après 5.4 et 5.3, les applications $\pi_0^{e'} \circ \mathrm{Ad}(t_{F,v}^{-1})$ et $\underline{\pi} \circ \mathbf{r}_{N,F}$ coïncident sur l'intersection $N^{e'} \cap G_v^{e'}$, et induisent une application surjective $N^{e'} \cap G_v^{e'} \to \underline{N}$. Le noyau de cette surjection est cohomologiquement trivial. On en déduit que $\underline{\pi}_v$ se restreint en une application surjective $N^e \cap G_v^e \to \underline{N} \cap \underline{G}_v^e$, qui coïncide avec $\underline{\pi} \circ \mathbf{r}_{N,F}$. Notons que pour $v \in V^{\mathrm{nr}}$, le sous-groupe $\underline{G}_v^e \subset \underline{G}$ est stabilisé par $\underline{\rho}_v(\sigma)$ pour tout $\sigma \in I$, et le diagramme

$$\begin{array}{ccc} G_v^{\mathrm{nr}} & \xrightarrow{\underline{\pi}_v} & \underline{G}_v \\ \cap & & \cap \\ G_v^e & \xrightarrow[\underline{\pi}_v]{} & \underline{G}_v^e \end{array}$$

est commutatif (les notations sont donc cohérentes).

Soit un cocycle δ_F de Γ à valeur dans N^{mod}, et soit $e \in \mathcal{P}$ tel que δ_F soit trivial sur $I(e)$. Posons $\delta = \mathbf{r}_{N,F} \circ \delta_F : \Gamma \to \mathbf{N}$, et soit $v \in {}_\delta W_{(p)}$. Puisque δ est trivial sur $\mathbf{N}^e$, on a ${}_\delta W_{(p)} \subset V_{(p)}^e$. Pour tout $\gamma \in \Gamma$, l'automorphisme $\mathrm{Ad}(\delta_F(\gamma)) \circ \gamma$ de G^{mod} stabilise G^e ; et puisque $v \in {}_\delta W$, il stabilise aussi G_v^e. Pour $g \in G_v^e$ et $\gamma \in \Gamma$, on a l'égalité $\underline{\pi}_v \circ \mathrm{Ad}(\delta_F(\gamma)) \circ \gamma(g) = {}_\delta\underline{\rho}_v \circ \underline{\pi}_v(g)$. On en déduit que pour tout $\gamma \in \Gamma$, l'automorphisme ${}_\delta\underline{\rho}_v(\gamma)$ de $\underline{G}$ stabilise le groupe $\underline{\pi}_v(G_v^e) = \underline{G}_v^e$.

Soit A un sous-espace affine de V tel que $A_{(p)}$ soit dense dans A. Choisissons un $e \in \mathcal{P}$ tel que $e\alpha(A) \in \mathbb{Z}$ pour tout $\alpha \in \Sigma(A)$, et $I(e)$ opère trivialement sur $\mathcal{D}$. On a définit en 3.3 une décomposition en facettes de $V = V^e$. Parmi les facettes qui coupent A, on en choisit une de dimension maximale, que l'on note $\mathcal{F}$. D'après le lemme 4.4.1 de [W2], pour tout $v \in A_{(p)} \cap \mathcal{F}$, la composante neutre $\underline{G}_v^{e,\circ}$ de $\underline{G}_v^e = \underline{\pi}_v(G_v^e)$ coïncide avec $\underline{H}(A)$; et pour tout $g \in G_v^e$, il existe un $n \in N^e \cap G_v^e$ tel que $\underline{\pi}_v(n^{-1}g) \in \underline{H}(A)$. Au sous-ensemble de Levi $\Sigma(A) \subset \Sigma$ correspond, comme en 8.1, un sous-groupe réductif connexe $H(A) \subset G$. Posons $H_A^e = H(A)(F^e) \subset G^e$. D'après le lemme 4.4.2 de [W2], pour tout $v \in A_{(p)}$, le groupe $H_A^e \cap G_v^e$ est indépendant de v ; et le groupe $\underline{\pi}_v(H_A^e \cap G_v^e)$ coïncide avec $\underline{H}(A)$.

8.3. Pour $r \in \mathbb{Z}_{(p)}$, on note $\underline{\mathfrak{t}}(r)$ le $\overline{\mathbb{F}}_q$-espace vectoriel $\underline{\mathfrak{t}} = \mathrm{X}_* \otimes_{\mathbb{Z}} \overline{\mathbb{F}}_q$ muni de l'action de Γ définie comme suit : pour $\lambda \in \mathrm{X}_*$, $x \in \overline{\mathbb{F}}_q$, $\theta \in \Theta$ et $\sigma \in I$, on pose $\theta(\lambda \otimes x) = \theta(\lambda) \otimes \theta(x)$ et $\sigma(\lambda \otimes x) = \sigma(\lambda) \otimes r(\sigma)^{-1}x$.

Soit $\mathcal{L}(\mathcal{D})$ l'ensemble des quadruplets $(\mathcal{D}', \psi, r, \underline{Z}')$ de la forme suivante :

[-] $\mathcal{D}' = (\mathrm{X}'^*, \Sigma', \Delta', \mathrm{X}'_*, \check{\Sigma}', \check{\Delta}')$ est une donnée de racines munie d'une action de Γ comme en 3.1. On affecte les objets relatifs à $\mathcal{D}'$, introduits précédemment pour la donnée $\mathcal{D}$, d'un exposant "$'$" : W', $\underline{G}'$, $\underline{T}'$ (etc.). On fait la même chose pour les objets dépendant de F : G', T', (etc.).

[-] $\psi = (\mathrm{X}'^* \to \mathrm{X}^*, \mathrm{X}'_* \to \mathrm{X}_*)$ est un couple d'isomorphismes de $\mathbb{Z}$-modules en dualité. Pour simplifier, on note encore ψ chacun de ces isomorphismes, ainsi que tout isomorphisme qui s'en déduit par

fonctorialité. On suppose que $\psi(\Delta') \subset \Delta$ et $\psi(\check{\Delta}') \subset \check{\Delta}$, et que $\psi(\Sigma')$ (resp. $\psi(\check{\Sigma}')$) est le sous-système de racines de Σ engendré par $\psi(\Delta')$ (resp. par $\psi(\check{\Delta}')$). En d'autres termes, on suppose que ψ identifie $\mathcal{D}'$ à un sous-système de Levi standard de $\mathcal{D}$. On suppose aussi que pour tout $\gamma \in \Gamma$, il existe un $w \in W$ tel que $\psi \circ \gamma = w \circ \gamma \circ \psi$. Cet élément w est unique ; on le note $w_\psi(\gamma)$.

[-] $r \in \mathbb{Z}_{(p)}$.

[-] $\underline{Z}' \in \underline{\mathfrak{t}}'_{\mathrm{cent}}(r)^\Gamma$ avec $\underline{\mathfrak{t}}'_{\mathrm{cent}}(r) = \{X \in \underline{\mathfrak{t}}'(r) : \alpha'(X) = 0, \forall \alpha' \in \Sigma'\}$. Ici, $\underline{\mathfrak{t}}'(r)$ désigne le $\overline{\mathbb{F}}_q$-espace vectoriel $\underline{\mathfrak{t}}' = \mathrm{X}'_* \otimes_{\mathbb{Z}} \overline{\mathbb{F}}_q$ muni de l'action de Γ définie plus haut. On suppose que pour tout $\alpha \in \Sigma \smallsetminus \psi(\Sigma')$, on a $\alpha \circ \psi(\underline{Z}') \neq 0$.

Deux éléments $(\mathcal{D}'_1, \psi_1, r_1, \underline{Z}'_1)$ et $(\mathcal{D}'_2, \psi_2, r_2, \underline{Z}'_2)$ de $\mathcal{L}(\mathcal{D})$ sont dits isomorphes si $r_1 = r_2$ et s'il existe un isomorphisme Γ-équivariant $f : \mathcal{D}'_1 \to \mathcal{D}'_2$ (i.e. un couple d'isomorphismes de $\mathbb{Z}[\Gamma]$-modules $(\mathrm{X}'^*_1 \to \mathrm{X}'^*_2, \mathrm{X}'_{*,1} \to \mathrm{X}'_{*,2})$ en dualité, tel que $f(\Sigma'_1) = \Sigma'_2$, $f(\Delta'_1) = \Delta'_2$, $f(\check{\Sigma}'_1) = \check{\Sigma}'_2$ et $f(\check{\Delta}'_1) = \check{\Delta}'_2$) et un élément $w \in W$ tels que $\psi_2 \circ f = w \circ \psi_1$ et $f(\underline{Z}'_1) = \underline{Z}'_2$. Grâce à l'hypothèse (P_Σ), on vérifie que le groupe d'automorphismes d'un élément de $\mathcal{L}(\mathcal{D})$ est trivial.

8.4. Soit un élément $(\mathcal{D}', \psi, r, \underline{Z}') \in \mathcal{L}(\mathcal{D})$, fixé jusqu'à la fin de la section 8. On définit un plongement $\underline{i}_\psi : \underline{G}' \to \underline{G}$, caractérisé comme suit : $\underline{i}_\psi$ se restreint en l'isomorphisme $\underline{T}' \xrightarrow{\sim} \underline{T}$ déduit de ψ ; et pour tout $\alpha' \in \Delta'$, on a $\mathrm{d}(\underline{i}_\psi)(\underline{E}'_{\alpha'}) = \underline{E}_{\psi(\alpha')}$. Alors $\underline{i}_\psi$ se restreint en un plongement $\underline{N}' \to \underline{N}$ qui, par construction, est compatible aux sections $\underline{n}' : W' \to \underline{N}'$ et $\underline{n} : W \to \underline{N}$: pour $w' \in W'$, on a $\underline{i}_\psi \circ \underline{n}'(w') = \underline{n} \circ \psi(w')$. D'autre part, de ψ se déduit un isomorphisme de groupes $T'_\varpi \xrightarrow{\sim} T_\varpi$ qui permet de prolonger le plongement $\underline{i}_\psi : \underline{N}' \to \underline{N}$ en un plongement $\mathbf{i}_\psi : \mathbf{N}' \to \mathbf{N}$.

Notons $\mathbf{Z}_\psi \subset \mathbf{T}$ le sous-groupe formé des t tels que $\alpha(t) = 0$ pour tout $\alpha \in \psi(\Sigma')$, et posons $\underline{Z}_\psi = \mathbf{Z}_\psi \cap \underline{T}$. Les éléments de $\mathbf{Z}_\psi$ commutent à ceux de $\mathbf{i}_\psi(\mathbf{N}')$, et $\underline{Z}_\psi$ est le centre de $\underline{i}_\psi(\underline{G}')$.

Posons $\underline{Z} = \psi(\underline{Z}') \in \underline{\mathfrak{t}}$. Pour γ, on pose $\sum^+_{\psi,\gamma} = \{\alpha \in \Sigma^+ : w_\psi(\gamma)^{-1}(\alpha) < 0\}$ et

$$z_{\varpi,\psi}(\gamma) = \left(\textstyle\sum_{\alpha \in \Sigma^+_{\psi,\gamma}} \check{\alpha}\right) \otimes r \in T_\varpi,$$

$$\underline{z}_\psi(\gamma) = \textstyle\prod_{\alpha \in \Sigma^+_{\psi,\gamma}} \check{\alpha} \circ \alpha(\underline{Z}) \in \underline{T},$$

$$\mathbf{n}_\psi(\gamma) = z_{\varpi,\psi}(\gamma)\underline{z}_\psi(\gamma)\underline{n}(w_\psi(\gamma)) \in \mathbf{N},$$

$$\underline{n}_\psi(\gamma) = \underline{\pi}(\mathbf{n}_\psi(\gamma)) = \underline{z}_\psi(\gamma)\underline{n}(w_\psi(\gamma)) \in \underline{N}.$$

D'après le lemme 4.6 de [W2], l'application $\gamma \mapsto \mathbf{n}_\psi(\gamma)$ est un cocycle (cela est une conséquence de la preuve du lemme 2.2.A de [LS]), et pour tout $\gamma \in \Gamma$, on a $z_{\varpi,\psi}(\gamma) \in \mathbf{Z}_\psi$ et $\underline{z}_\psi(\gamma) \in \underline{Z}_\psi$; de plus, on a les égalités :

$$\underline{i}_\psi \circ \gamma = \mathrm{Ad}(\underline{n}_\psi(\gamma)) \circ \gamma \circ \underline{i}_\psi,$$

$$\mathbf{i}_\psi \circ \gamma = \mathrm{Ad}(\mathbf{n}_\psi(\gamma)) \circ \gamma \circ \mathbf{i}_\psi(\gamma).$$

8.5. On construit comme en 8.4 un plongement $i_\psi : G' \to G$, qui est défini sur F^{mod}.

Pour $s \in \mathbb{R}$, posant $\mathfrak{t}_s^{\mathrm{mod}} = \mathfrak{t}(F^{\mathrm{mod}})_s$ et $\mathfrak{t}_{s+}^{\mathrm{mod}} = \mathfrak{t}(F^{\mathrm{mod}})_{s+}$ (cf. 1.2), on a $\mathfrak{t}_s^{\mathrm{mod}} \subset \mathfrak{t}_{s+}^{\mathrm{mod}}$ avec égalité si et seulement si $s \in \mathbb{R} \smallsetminus \mathbb{Z}_{(p)}$. Pour $s \in \mathbb{Z}_{(p)}$, on définit comme suit un isomorphisme de Γ-modules $\mathfrak{t}_s^{\mathrm{mod}}/\mathfrak{t}_{s+}^{\mathrm{mod}} \to \underline{\mathfrak{t}}(s)$: pour $\lambda \in X_*$ et $x \in F^{\mathrm{mod}}$ avec $\omega_F(x) \geq s$, il envoie $\lambda \otimes x$ (en fait sa classe modulo $\mathfrak{t}_{s+}^{\mathrm{mod}}$) sur $\lambda \otimes \iota(\overline{\varpi_{-s}x})$ où $\overline{\varpi_{-s}x} \in \overline{\kappa}_F$ désigne la réduction de $\varpi_{-s}x \in \mathfrak{o}_{F^{\mathrm{mod}}}$ modulo $\mathfrak{p}_{F^{\mathrm{mod}}}$. Puisque le Γ-module $\mathfrak{t}_{s+}^{\mathrm{mod}}$ est cohomologiquement trivial, par passage aux points fixes sous Γ, on obtient un isomorphisme $\mathfrak{t}(F)_s/\mathfrak{t}(F)_{s+} \to \underline{\mathfrak{t}}(s)^\Gamma$.

On pose $\mathfrak{t}'_{\mathrm{cent}} = \{X \in \mathfrak{t} : \alpha'(X) = 0, \forall \alpha' \in \Sigma'\}$; c'est un sous-espace affine de $\mathfrak{t}'$ défini sur F. Pour $s \in \mathbb{R}$, on pose $\mathfrak{t}'^{\mathrm{mod}}_{\mathrm{cent},s} = \mathfrak{t}'_{\mathrm{cent}}(F^{\mathrm{mod}})_s$ et $\mathfrak{t}'^{\mathrm{mod}}_{\mathrm{cent},s+} = \mathfrak{t}'_{\mathrm{cent}}(F^{\mathrm{mod}})_{s+}$. Et pour $s \in \mathbb{Z}_{(p)}$, on définit comme ci-dessus un isomorphisme de Γ-modules $\mathfrak{t}'^{\mathrm{mod}}_{\mathrm{cent},s}/\mathfrak{t}'^{\mathrm{mod}}_{\mathrm{cent},s+} \to \underline{\mathfrak{t}}'_{\mathrm{cent}}(s)$, qui par passage aux points fixes sous Γ, induit un isomorphisme $\mathfrak{t}'_{\mathrm{cent}}(F)_s/\mathfrak{t}'_{\mathrm{cent}}(F)_{s+} \to \underline{\mathfrak{t}}'_{\mathrm{cent}}(s)^\Gamma$.

Prenons $s = r$. Puisque $\underline{Z}' \in \underline{\mathfrak{t}}'(r)^\Gamma$, on peut fixer un élément $Z' \in \mathfrak{t}'_{\mathrm{cent}}(F)_r$ dont la classe modulo $\mathfrak{t}'_{\mathrm{cent}}(F)_{r+}$ s'envoie sur $\underline{Z}'$ par l'isomorphisme précédent. À l'aide de Z', on construit comme en 8.4 un cocycle $n_{\psi,F} \in Z^1(\Gamma, N^{\mathrm{mod}})$: pour $\gamma \in \Gamma$, on pose $z_{\psi,F}(\gamma) = \prod_{\alpha \in \Sigma^+_{\psi,\gamma}} \check{\alpha} \circ \alpha(\psi(Z'))$ et $n_{\psi,F}(\gamma) = s_{T,F}(z_{\varpi,F}(\gamma)) z_{\psi,F}(\gamma) n(w_\psi(\gamma))$. Par construction, on a $\mathbf{n}_\psi = \mathbf{r}_{N,F} \circ n_{\psi,F}$; et d'après la démonstration du lemme 4.6 de [W2], on a l'égalité

$$i_\psi \circ \gamma = \mathrm{Ad}(n_{\psi,F}(\gamma)) \circ \gamma \circ i_\psi.$$

8.6. Comme pour la donnée $\mathcal{D}$, on fixe un ensemble D' de cocycles de Θ à valeurs dans $\mathbf{N}'^{\mathrm{nr}}_{\mathcal{C}'}$ (l'analogue de $\mathbf{N}^{\mathrm{nr}}_{\mathcal{C}}$ pour $\mathcal{D}'$) tel que l'application naturelle $D' \to H^1(\Theta, \mathbf{N}'^{\mathrm{nr}}_{\mathcal{C}'})$ soit bijective. De ψ se déduit un homomorphisme surjectif $X'_*/X'_{*,\mathrm{sc}} \to X_*/X_{*,\mathrm{sc}}$, qui est Γ-équivariant puisque W opère trivialement sur l'espace d'arrivée. D'où un homomorphisme surjectif $(X'_*/X'_{*,\mathrm{sc}})_\Gamma \to (X_*/X_{*,\mathrm{sc}})_\Gamma$. En passant aux sous-groupes de torsion, on obtient un homomorphisme $\psi_{D'} : D' \to D$, qui n'est en général pas surjectif.

Soit un cocycle $d' \in D'$, fixé jusqu'à la fin de la section 8, et soit $d = \psi_{D'}(d)$. Pour $\gamma \in \Gamma$, posons $d_0(\gamma) = \mathbf{i}_\psi \circ d'(\gamma)\mathbf{n}_\psi(\gamma)$. D'après 8.4, d_0 est un cocycle de Γ dans $\mathbf{N}$. De même, on définit un cocycle $d_{0,F}$ de Γ dans N^{mod} en posant $d_{0,F} = i_\psi \circ d'_F(\gamma) n_{\psi,F}(\gamma)$.

LEMME [W2, lemme 4.8]. — *Les cocycles $d_{0,F}$ et d_F définissent le même élément de* $H^1(\Gamma, G^{\mathrm{mod}})$.

Fixons un sous-groupe ouvert $\Gamma_0 \subset \Gamma$ opérant trivialement sur $\mathcal{D}$ et sur $\mathcal{D}'$. D'après 6.1, on peut supposer que l'indice $(\Gamma : \Gamma_0)$ est premier à p. On peut donc remplacer Γ par Γ/Γ_0 dans les constructions précédentes ; en particulier, l'application $\gamma \mapsto z_{\varpi,\psi}$ se factorise à travers Γ/Γ_0. On note

$z_0 \in T_\varpi$ l'élément défini par :

$$z_0 = \frac{1}{(\Gamma : \Gamma_0)} \sum_{\gamma \in \Gamma/\Gamma_0} z_{\varpi,\psi}(\gamma).$$

Notons $\psi_{V'} : V' \to V$ l'isomorphisme déduit de ψ. D'après le lemme 4.9 de [W2], l'application

$$_{d'}V' \to {}_{d_0}V,\ v' \mapsto z_{0,\mathbb{R}} \circ \psi_{V'}(v')$$

est Γ-équivariante. Par passage aux points fixes sous Γ, on obtient un isomorphisme $_{d'}W' \to {}_{d_0}W$. D'après le lemme 4.10 de [W2], il existe un couple $(n, g) \in \mathbf{N} \times \underline{G}$ vérifiant les conditions suivantes :

[-] $n_\mathbb{R}({}_{d_0}W) \subset {}_dW$;

[-] posant $A = n_\mathbb{R}({}_{d_0}W)$, $A_{(p)}$ est dense dans A, et g appartient à $\underline{H}(A)$;

[-] posant $m(\gamma) = nd_0(\gamma)\gamma(n)^{-1}d(\gamma)^{-1}$ $(\gamma \in \Gamma)$, on a $\underline{\pi}(m(\gamma)) = \mathrm{Ad}(g) \circ {}_d\underline{\rho}_v(\gamma)$ pour tout $\gamma \in \Gamma$ et tout $v \in A_{(p)}$.

8.7. Fixons un couple $(n, g) \in \mathbf{N} \times \underline{G}$ vérifiant les conditions de 8.6, et posons $\underline{n} = \underline{\pi}(n)$. On définit une application affine :

$$_{d'}\psi_{V'} : {}_{d'}W' \to {}_dW,\ v' \mapsto n_\mathbb{R} \circ z_{0,\mathbb{R}} \circ \psi_{V'}(v').$$

Elle envoie $_{d'}W'_{(p)}$ dans $_dW_{(p)}$. Soit $v' \in {}_{d'}W'_{(p)}$, et posons $v = {}_{d'}\psi_{V'}(v')$.

D'après la preuve du lemme 4.11 de [W2], pour $\gamma \in \Gamma$, on a l'égalité

$$_d\underline{\rho}_v(\gamma) \circ \mathrm{Ad}(g^{-1}\underline{n}) \circ \underline{i}_\psi = \mathrm{Ad}(g^{-1}\underline{n}) \circ \underline{i}_\psi \circ {}_{d'}\underline{\rho}'_{v'}(\gamma).$$

Par conséquent l'homomorphisme $\mathrm{Ad}(g^{-1}\underline{n}) \circ \underline{i}_\psi : \underline{G}' \to \underline{G}$ envoie $\underline{G}'_{v'}$ dans $\underline{G}_v$, et il existe un unique homomorphisme $_{d'}\underline{i}_{\psi,v'} : {}_{d'}\underline{G}'_{v'} \to {}_d\underline{G}_v$, défini sur $\mathbb{F}_q$, tel que

$$\mathrm{Ad}(g^{-1}\underline{n}) \circ \underline{i}_\psi \circ {}_{d'}\underline{\xi}' = {}_d\underline{\xi} \circ {}_{d'}\underline{i}_{\psi,v'}.$$

De même, pour $s \in \mathbb{Z}_{(p)}$, l'homomorphisme $\mathrm{Ad}(g^{-1}\underline{n}) \circ \mathrm{d}(\underline{i}_\psi) : \underline{\mathfrak{g}}' \to \underline{\mathfrak{g}}$ envoie $\underline{\mathfrak{g}}'_{v',s}$ dans $\underline{\mathfrak{g}}_{v,s}$, et il existe un unique homomorphisme $_{d'}\underline{i}_{\psi,v',s} : {}_{d'}\underline{\mathfrak{g}}'_{v',s} \to {}_d\underline{\mathfrak{g}}_{v,s}$, défini sur $\mathbb{F}_q$, tel que

$$\mathrm{Ad}(g^{-1}\underline{n}) \circ \mathrm{d}(\underline{i}_\psi) \circ \mathrm{d}({}_{d'}\underline{\xi}') = \mathrm{d}({}_d\underline{\xi}) \circ {}_{d'}\underline{i}_{\psi,v',s}.$$

Bien sûr, $_{d'}\underline{i}_{\psi,v',s}$ ne dépend pas vraiment de (v', s) mais seulement des facettes $\phi' \in {}_{d'}\Phi'$ et $\phi \in {}_d\Phi$ contenant respectivement (v', s) et (v, s).

Choisissons un élément $n_F \in N^{\mathrm{mod}}$ tel que $\mathbf{r}_{N,F}(n_F) = n$. Posons $A = {}_{d'}\psi_{V'}({}_{d'}W') = n_\mathbb{R}({}_{d_0}W)$; on a donc $v \in A_{(p)}$. Choisissons un $e \in \mathcal{P}$ tel que $I(e)$ opère trivialement sur $\mathcal{D}$ et sur $\mathcal{D}'$, $n_F \in N^e$, et $e\alpha(A) \in \mathbb{Z}$ pour tout $\alpha \in \Sigma(A)$. D'après 8.2, on a $\underline{\pi}_v(H_A^e \cap G_v^e) = \underline{H}(A)$. Soit un élément $g_F \in H_A^e \cap G_v^e$ tel que $\underline{\pi}_v(g_F) = g$. D'après [W2, 4.12], on peut choisir g_F de telle manière que pour tout $\gamma \in \Gamma$, on ait l'égalité :

$$d_F(\gamma) = g_F^{-1} n_F\, \iota_\psi \circ d'_F(\gamma) n_{\psi,F}(\gamma)\, \gamma(n_F^{-1} g_F).$$

Alors il existe un unique homomorphisme ${}_{d'}i_{\psi} : {}_{d'}G' \to {}_{d}G$, défini sur F, tel que

$$\mathrm{Ad}(g_F^{-1} n_F) \circ i_{\psi} \circ {}_{d'}\xi' = {}_{d}\xi \circ {}_{d'}i_{\psi}.$$

LEMME [W2, lemme 4.12.1]. — *L'application* ${}_{d'}\psi_{V'} : {}_{d'}W' \to {}_{d}W$ *se prolonge en une application* ${}_{d'}\psi : \mathfrak{I}({}_{d'}G', F) \to \mathfrak{I}({}_{d}G, F)$ *telle que pour tout* $g' \in {}_{d'}G'(F)$ *et tout* $x' \in \mathfrak{I}({}_{d'}G, F))$, *on ait l'égalité* ${}_{d'}\psi(g' \cdot x') = {}_{d'}i_{\psi}(g') \cdot {}_{d'}\psi(x')$. *De plus, ce prolongement est unique.*

Par construction, ${}_{d'}i_{\psi}$ se restreint en un plongement ${}_{d'}G_{v'}^{\prime\mathrm{nr}} \to {}_{d}G_{v}^{\mathrm{nr}}$ défini sur $\mathfrak{o}_F$, et pour $s \in \mathbb{Z}_{(p)}$, $\mathrm{d}({}_{d'}i_{\psi})$ se restreint en un plongement ${}_{d'}\mathfrak{g}_{v',s}^{\prime\mathrm{nr}} \to {}_{d}\mathfrak{g}_{v,s}^{\mathrm{nr}}$ défini sur $\mathfrak{o}_F$; de plus, les deux diagrammes suivants sont commutatifs :

$$\begin{array}{ccc} {}_{d'}G_{v'}^{\prime\mathrm{nr}} & \xrightarrow{{}_{d'}i_{\psi}} & {}_{d}G_{v}^{\mathrm{nr}} \\ {\scriptstyle {}_{d'}\underline{\pi}'_{v'}}\downarrow & & \downarrow{\scriptstyle {}_{d}\underline{\pi}_{v}} \\ {}_{d'}\underline{G}'_{v'} & \xrightarrow[{}_{d'}\underline{i}_{\psi,v'}]{} & {}_{d}\underline{G}_{v} \end{array}, \qquad \begin{array}{ccc} {}_{d'}\mathfrak{g}_{v',s}^{\prime\mathrm{nr}} & \xrightarrow{\mathrm{d}({}_{d'}i_{\psi})} & {}_{d}\mathfrak{g}_{v,s}^{\mathrm{nr}} \\ {\scriptstyle {}_{d'}\underline{\pi}'_{v',s}}\downarrow & & \downarrow{\scriptstyle {}_{d}\underline{\pi}_{v,s}} \\ {}_{d'}\underline{\mathfrak{g}}'_{v',s} & \xrightarrow[{}_{d'}\underline{i}_{\psi,v',s}]{} & {}_{d}\underline{\mathfrak{g}}_{v,s} \end{array}.$$

9. Analyse harmonique : une deuxième réduction (descente centrale)

9.1. Soit $d \in D$. Pour $v \in {}_{d}W_{(p)}$, notons ${}_{d}\mathcal{M}_v$ l'ensemble des couples $\mu = (m, h) \in \mathbf{N} \times \underline{G}$ vérifiant les deux conditions suivantes (pour tout $\gamma \in \Gamma$) :

[-] $t_v^{-1} d(\gamma)\gamma(m) d(\gamma)^{-1} t_v$ appartient à $\underline{N}$,

[-] $t_v^{-1} d(\gamma)\gamma(m) d(\gamma)^{-1} t_v = h\, {}_{d}\rho_v(\gamma) h^{-1}$.

Pour un tel couple μ, on pose $\mu(v) = m_{\mathbb{R}}(v)$. Si $e \in \mathcal{P}$ est tel que $I(e)$ opère trivialement sur $\mathcal{D}$ et $ev \in \mathrm{X}_*$, alors on a $m \in \mathbf{N}^e$ et $e\mu(v) \in \mathrm{X}_*$; donc $\mu(v) \in {}_{d}W_{(p)}$. L'application $\mathrm{Ad}(\underline{\pi}(m)h)$ envoie $\underline{G}_v$ sur $\underline{G}_{\mu(v)}$ et il existe un unique isomorphisme $\underline{\mu} : {}_{d}\underline{G}_v \to {}_{d}\underline{G}_{\mu(v)}$, défini sur $\mathbb{F}_q$, tel que

$$\mathrm{Ad}(\underline{\pi}(m)h) \circ {}_{d}\underline{\xi} = {}_{d}\underline{\xi} \circ \underline{\mu}.$$

De même, pour $s \in \mathbb{Z}_{(p)}$, l'application $\mathrm{Ad}(\underline{\pi}(m)h)$ envoie $\underline{\mathfrak{g}}_{v,s}$ sur $\underline{\mathfrak{g}}_{\mu(v),s}$ et il existe un unique isomorphisme $\underline{\mu}_s : {}_{d}\underline{\mathfrak{g}}_{v,s} \to {}_{d}\underline{\mathfrak{g}}_{\mu(v),s}$, défini sur $\mathbb{F}_q$, tel que

$$\mathrm{Ad}(\underline{\pi}(m)h) \circ \mathrm{d}({}_{d}\underline{\xi}) = \mathrm{d}({}_{d}\underline{\xi}) \circ \underline{\mu}_s.$$

Soit $\underset{d}{\sim}$ la relation binaire sur ${}_{d}W_{(p)}$ définie par : $v \underset{d}{\sim} v_1$ si et seulement s'il existe un $\mu \in {}_{d}\mathcal{M}_v$ tel que $v_1 = \mu(v)$. Alors $\underset{d}{\sim}$ est une relation d'équivalence, et pour tout $v \in {}_{d}W_{(p)}$, on a l'inclusion ${}_{d}\mathrm{X}_* \times \{1\} \subset {}_{d}\mathcal{M}_v$ où le groupe ${}_{d}\mathrm{X}_*$ (cf. 6.5) est identifié à un sous-groupe de $T_{\varpi} = \mathrm{X}_* \otimes \mathbb{Z}_{(p)}$.

Soit $F \in \mathrm{CL}_q$, et soient $v, v_1 \in {}_{d}W_{(p)}$. D'après le lemme 5.1 de [W2], on a $v \underset{d}{\sim} v_1$ si et seulement s'il existe un $g \in {}_{d}G(F)$ tel que $g \cdot v = v_1$ (dans $\mathfrak{I}({}_{d}G, F)$). Plus précisément : si $v_1 = \mu(v)$ pour un $\mu \in {}_{d}\mathcal{M}_v$, alors il existe un $g \in {}_{d}G(F)$ tel que :

[-] $g \cdot v = v_1$;

[-] pour tout $s \in \mathbb{Z}_{(p)}$, les diagrammes suivants sont commutatifs :

$$\begin{array}{ccc} {}_d\mathcal{G}_v(\mathfrak{o}_F) & \xrightarrow{\mathrm{Ad}(g)} & {}_d\mathcal{G}_{v_1}(\mathfrak{o}_F) \\ {\scriptstyle {}_d\underline{\pi}_v}\downarrow & & \downarrow{\scriptstyle {}_d\underline{\pi}_{v_1}} \\ {}_d\underline{G}_v^{\Theta} & \xrightarrow[\underline{\mu}]{} & {}_d\underline{G}_{v_1}^{\Theta} \end{array}, \qquad \begin{array}{ccc} {}_d\mathfrak{g}(F)_{v,s} & \xrightarrow{\mathrm{Ad}(g)} & {}_d\mathfrak{g}(F)_{v_1,s} \\ {\scriptstyle {}_d\underline{\pi}_{v,s}}\downarrow & & \downarrow{\scriptstyle {}_d\underline{\pi}_{v_1,s}} \\ {}_d\underline{\mathfrak{g}}_{v,s}^{\Theta} & \xrightarrow[\underline{\mu}_s]{} & {}_d\underline{\mathfrak{g}}_{v_1,s}^{\Theta} \end{array}.$$

9.2. Soit un quadruplet $(\mathcal{D}', \psi, r, \underline{Z}') \in \mathcal{L}(\mathcal{D})$, et soit $d' \in D'$. Posons $d = \psi_{D'}(d')$. Comme en 9.1, on définit les ensembles ${}_{d'}\mathcal{M}'_{v'}$ pour $v' \in {}_{d'}W'_{(p)}$, et la relation d'équivalence $\underset{d'}{\sim}$ dans ${}_{d'}W'_{(p)}$. Soit $v \in {}_dW_{(p)}$. Notons $\tilde{\Omega}'_v$ l'ensemble des $v' \in {}_{d'}W'_{(p)}$ tels que ${}_{d'}\psi_{V'}(v') \underset{d}{\sim} v$. Cet ensemble est clos pour la relation d'équivalence $\underset{d'}{\sim}$. D'après 9.1, il existe un $e \in \mathcal{P}$ tel que la classe de $\underset{d}{\sim}$-équivalence de v dans ${}_dW_{(p)}$ soit contenue dans $\frac{1}{e}({}_d\mathrm{X}_*)$. On peut ensuite choisir un $e' \in \mathcal{P}$ tel que l'image réciproque de $\frac{1}{e}({}_d\mathrm{X}_*)$ par ${}_{d'}\psi_{V'}$ soit contenue dans $\frac{1}{e'}({}_{d'}\mathrm{X}'_*)$. Alors on a l'inclusion $\tilde{\Omega}'_v \subset \frac{1}{e'}({}_{d'}\mathrm{X}'_*)$. Pour tout $v' \in {}_{d'}W'_{(p)}$, on a l'inclusion ${}_{d'}\mathrm{X}'_* \times \{1\} \subset {}_{d'}\mathcal{M}'_{v'}$ (cf. 9.1). Puisque le groupe ${}_{d'}\mathrm{X}'_*$ est d'indice fini dans $\frac{1}{e'}{}_{d'}\mathrm{X}'_*$, on en déduit que $\tilde{\Omega}'_v$ est réunion d'un nombre *fini* de classes de $\underset{d'}{\sim}$-équivalence. Fixons un ensemble $\Omega'_v \subset \tilde{\Omega}'_v$ de représentants de ces classes, et pour tout $v' \in \Omega'_v$, fixons un élément $\mu_{v',v} \in {}_d\mathcal{M}_v$ tel que $\mu_{v',v}(v) = {}_{d'}\psi_{V'}(v')$.

Soit $F \in \mathrm{CL}_q$. Pour tout $v' \in \Omega'_{v'}$, fixons un élément $g_{v',v} \in {}_dG(F)$ vérifiant les deux conditions de la fin de 9.1 relativement à l'élément $\mu_{v',v} \in {}_d\mathcal{M}_v$. Alors on a les diagrammes commutatifs suivants (pour tout $s \in \mathbb{Z}_{(p)}$) :

$$\begin{array}{ccc} {}_{d'}\mathcal{G}'_{v'}(\mathfrak{o}_F) & \xrightarrow{\mathrm{Ad}(g_{v',v})^{-1}\circ {}_{d'}\iota_\psi} & {}_d\mathcal{G}_v(\mathfrak{o}_F) \\ {\scriptstyle {}_{d'}\underline{\pi}'_{v'}}\downarrow & & \downarrow{\scriptstyle {}_d\underline{\pi}_v} \\ {}_{d'}\underline{G}'^{\Theta}_{v'} & \xrightarrow[\underline{\mu}_{v',v}^{-1}\circ {}_{d'}\underline{\iota}_{\psi,v'}]{} & {}_d\underline{G}_{v_1}^{\Theta} \end{array}, \qquad \begin{array}{ccc} {}_{d'}\mathfrak{g}'(F)_{v',s} & \xrightarrow{\mathrm{Ad}(g_{v',v})^{-1}\circ \mathrm{d}({}_{d'}\iota_\psi)} & {}_d\mathfrak{g}(F)_{v,s} \\ {\scriptstyle {}_{d'}\underline{\pi}'_{v',s}}\downarrow & & \downarrow{\scriptstyle {}_d\underline{\pi}_{v,s}} \\ {}_{d'}\underline{\mathfrak{g}}'^{\Theta}_{v',s} & \xrightarrow[(\underline{\mu}_{v',v})_s^{-1}\circ {}_{d'}\underline{\iota}_{\psi,v',s}]{} & {}_d\underline{\mathfrak{g}}_{v,s}^{\Theta} \end{array}.$$

L'ensemble Ω'_v s'interprète de la manière suivante : soit $v' \in {}_{d'}W'_{(p)}$ tel que ${}_{d'}\psi_{V'}(v') = g \cdot v$ pour un $g \in {}_dG(F)$. Alors il existe un *unique* $v'' \in \Omega'_v$ tel que $v' = g' \cdot v''$ pour un $g' \in {}_{d'}G'(F)$.

9.3. Continuons avec les notations de 9.2. Supposons de plus que $\underline{Z}' \neq 0$.

PROPOSITION [W2, proposition 5.3]. — *Il existe une application linéaire* ${}_{d'}l'_r : {}_d\mathcal{S}_r \to {}_{d'}\mathcal{S}'_r$ *telle que pour tous* $\varphi \in {}_d\mathcal{S}_r$, $F \in \mathrm{CL}_q$, $Z' \in {}_{d'}\mathfrak{g}'(F)$ *tel que* $\mathrm{d}({}_{d'}\xi')(Z') \in \mathfrak{t}'_{\mathrm{cent}}(F)_r$ *et son image dans* $\underline{\mathfrak{t}}'_{\mathrm{cent}}(r)$ *soit égale à* $\underline{Z}'$, $X' \in {}_{d'}\mathfrak{g}'(F)_{\mathrm{reg}} \cap {}_{d'}\mathfrak{g}'(F)_{r+}$, *on ait l'égalité*

$$\Lambda^{{}_dG(F)}({}_d\mathrm{rea}_F(\varphi), \mathrm{d}({}_{d'}\iota_\psi)(Z' + X')) = \Lambda^{{}_{d'}G'(F)}({}_{d'}\mathrm{rea}'_F \circ {}_{d'}l'_r(\varphi), X').$$

(Remarque : l'élément $\mathrm{d}({}_{d'}\iota_\psi)(Z' + X')$ *appartient à* ${}_d\mathfrak{g}(F)_{\mathrm{reg}} \cap ({}_d\mathfrak{g}(F)_r \smallsetminus {}_d\mathfrak{g}(F)_{r+})$.*)*

Démonstration : Posons $X = \mathrm{d}({}_{d'}i_\psi)(Z'+X')$. Choisissons un $g' \in G'^{\mathrm{mod}}$ tel que l'élément $X'_1 = \mathrm{Ad}(g') \circ \mathrm{d}({}_{d'}\xi')(X')$ appartienne à $\mathfrak{t}'^{\mathrm{mod}}$, et posons $Z'_1 = \mathrm{Ad}(g') \circ \mathrm{d}({}_{d'}\xi')(Z') = \mathrm{d}({}_{d'}\xi')(Z')$. Il existe un $g \in G^{\mathrm{mod}}$ tel que l'élément $X_1 = \mathrm{Ad}(g) \circ \mathrm{d}({}_d\xi)(X)$ coïncide avec $\mathrm{d}(i_\psi)(Z'_1 + X'_1)$. Montrons que X_1 est régulier. Soit $\alpha \in \Sigma$, et posons $\alpha' = \psi^{-1}(\alpha) \in \mathrm{X}'^*$. On a donc $\alpha(X_1) = \alpha'(Z'_1 + X'_1)$. Si $\alpha' \in \Sigma'$, alors $\alpha'(Z'_1) = 0$ et $\alpha'(X'_1) \neq 0$ car X' est régulier, donc $\alpha(X_1) \neq 0$. Supposons que $\alpha' \in \mathrm{X}'^* \smallsetminus \Sigma'$. Puisque $X' \in {}_{d'}\mathfrak{g}'(F)_{r+}$, on a $\omega_F(\alpha'(X'_1)) > r$. D'autre part on a $\omega_F(\alpha'(Z'_1)) \geq r$, et la réduction modulo $\mathfrak{p}_{F^{\mathrm{mod}}}$ de $\varpi_r^{-1}\alpha'(Z'_1)$ coïncide avec $\alpha'(\underline{Z}')$. Or $\alpha'(\underline{Z}') \neq 0$ (cf. 8.3), par conséquent $\omega_F(\alpha'(Z'_1)) = r = \omega_F(\alpha'(Z'_1 + X'_1)) = \omega_F(\alpha(X_1))$. En particulier $\alpha(X_1) \neq 0$. Donc X_1 est régulier, et $X_1 \in \mathfrak{t}_r^{\mathrm{mod}} \smallsetminus \mathfrak{t}_{r+}^{\mathrm{mod}}$. Par conséquent $X \in {}_d\mathfrak{g}(F)_{\mathrm{reg}}$ et (d'après 7.2) $X \in {}_d\mathfrak{g}(F)_r \smallsetminus {}_d\mathfrak{g}(F)_{r+}$.

Par linéarité, il suffit de définir ${}_{d'}l'_r(\varphi)$ pour $\varphi \in {}_d\mathcal{S}_{\phi_1}$ avec $\phi_1 \in {}_d\Phi_r$. Si $r(\phi_1) > r$, alors pour $\varphi \in {}_d\mathcal{S}_{\phi_1}$, le support de ${}_d\mathrm{rea}_F(\varphi)$ est contenu dans ${}_d\mathfrak{g}(F)_{r+}$, et puisque $X \in {}_d\mathfrak{g}(F)_r \smallsetminus {}_d\mathfrak{g}(F)_{r+}$, on a $\Lambda^{{}_dG(F)}({}_d\mathrm{rea}_F(\varphi), X) = 0$; on pose donc ${}_{d'}l'_r(\varphi) = 0$. Supposons $r(\phi_1) = r$. Comme dans la démonstration de la proposition 7.4, on peut remplacer ϕ_1 par une facette ϕ qui coupe ${}_dW_{(p)} \times \{r\}$. Soit un point $v \in {}_dW_{(p)}$ tel que $(v, r) \in \phi$. Fixons une fonction $\varphi \in {}_d\mathcal{S}_\phi$, et posons $f = {}_d\mathrm{rea}_F(\varphi)$.

Un calcul facile donne l'égalité $\Delta(X) = q^{\frac{r}{2}(|\Sigma'|-|\Sigma|)}\Delta'(X)$.

Pour $v_1 \in \mathfrak{I}({}_dG, F)$ est défini (par transport de structure) le $\mathfrak{o}_F$-réseau ${}_d\mathfrak{g}(F)_{v_1,r}$ dans ${}_d\mathfrak{g}(F)$; et pour $g \in {}_dG(F)$, on a $\mathrm{Ad}(g^{-1})(X) \in {}_d\mathfrak{g}(F)_\phi$ si et seulement si $X \in {}_d\mathfrak{g}(F)_{g\cdot v,r}$. D'après le lemme 2.2.6 de [KM], on a l'inclusion

$$\{v_1 \in \mathfrak{I}({}_dG, F) : X \in {}_d\mathfrak{g}(F)_{v_1,r}\} \subset {}_{d'}\psi(\mathfrak{I}({}_{d'}G', F));$$

en effet, $\mathrm{d}({}_{d'}i_\psi)(Z')$ est un "bon élément" de ${}_d\mathfrak{g}(F)$, et ${}_{d'}i_\psi({}_{d'}G'(F))$ en est le centralisateur dans ${}_dG(F)$. Soit donc $\mathcal{Y} = \{g \in {}_dG(F) : g \cdot v \in {}_{d'}\psi(\mathfrak{I}({}_{d'}G, F))\}$. D'après le numéro 9.2, dont on reprend ici les notations, on a la décomposition $\Omega = \coprod_{v' \in \Omega'_v} \mathcal{Y}_{v'}$ avec $\mathcal{Y}_{v'} = {}_{d'}G'(F) g_{v',v}\, {}_d\mathcal{G}_v(\mathfrak{o}_F)$. Pour $v' \in \Omega'_v$, posons

$$J_{v'} = \int_{T_X(F)\backslash \mathcal{Y}_{v'}} f(\mathrm{Ad}(g^{-1})(X))_d d\bar{g}_X;$$

on a donc $\Lambda^{{}_dG(F)}(f, X) = \Delta(X) \sum_{v' \in \Omega'_v} J_{v'}$. Soit $v' \in \Omega'_{v'}$. Notons $\phi' \in {}_{d'}\Phi'$ la facette contenant (v', r), et $\underline{j}_{\phi,\phi'} : {}_{d'}\underline{\mathfrak{g}}'^{\Theta}_{\phi'} \to {}_d\underline{\mathfrak{g}}^{\Theta}_{\phi}$ le plongement $(\underline{\mu}_{v',v})_r^{-1} \circ {}_{d'}\underline{i}_{\psi,v',r}$ (cf. 9.2). On définit comme suit les fonctions $\tilde{\varphi} \in {}_d\mathcal{S}_\phi$, $\tilde{\varphi}'_{v'} \in {}_{d'}\mathcal{S}'_{\phi'}$ et $\varphi'_{v'} \in {}_{d'}\mathcal{S}'_{\phi'}$:

[-] $\tilde{\varphi}(Y) = q^{-\frac{1}{2}\dim({}_d\underline{G}_v)} \sum_{x \in {}_d\underline{G}_v^\Theta} \varphi(\mathrm{Ad}(x)(Y))$;

[-] $\tilde{\varphi}'_{v'}(Y') = |{}_{d'}\underline{G}'^{\Theta}_{v'}|^{-1} q^{\frac{1}{2}\dim({}_{d'}\underline{G}'_v)} \tilde{\varphi} \circ \underline{j}_{\phi,\phi'}(Y')$;

[-] $\varphi'_{v'}(Y') = \tilde{\varphi}'_{v'}(\mathrm{d}({}_{d'}\underline{\xi}')^{-1}(\underline{Z}') + Y')$.

Cette dernière définition a un sens : pour tout $v'' \in V'^{\mathrm{nr}}$, on a $\mathrm{d}({}_{d'}\underline{\xi}')^{-1}(\underline{Z}') \in {}_{d'}\mathfrak{g}(F)_{v'',r}$. Posons $f'_{v'} = {}_{d'}\mathrm{rea}'_F(\varphi'_{v'})$. Un calcul facile (mais pénible !) conduit à l'égalité $J_{v'} = \Lambda^{{}_{d'}G'(F)}(f'_{v'}, X')$.

Soit $\varphi' = \oplus_{v'\in\Omega'_v} q^{\frac{r}{2}(|\Sigma'|-|\Sigma|)}\varphi'_{v'} \in {}_{d'}\mathcal{S}'_r$, et posons $f' = {}_{d'}\mathrm{rea}_F(\varphi')$. D'après les paragraphes précédents, on a l'égalité $\Lambda^{dG(F)}(f,X) = \Lambda^{d'G'(F)}(f',X')$. Reste à poser ${}_{d'}l'_r(\varphi) = \varphi'$. Cela achève la démonstration de la proposition.□

10. Classes de conjugaison et théorème principal

10.1. Notons $\tilde{\mathcal{B}}$ l'ensemble des couples $(r,\underline{Z})$ tels que $r \in \mathbb{Z}_{(p)}$ et $\underline{Z} \in \underline{\mathfrak{t}}(r) \smallsetminus \{0\}$. Le groupe $W \rtimes \Gamma$ opère sur $\tilde{\mathcal{B}}$ via son action sur $\underline{\mathfrak{t}}(r)$, et l'on pose $\mathcal{B} = (\tilde{\mathcal{B}}/W)^\Gamma$. Comme en 7.2, on montre qu'il existe un $e \in \mathcal{P}$ vérifiant la propriété : pour tout $(r,\underline{Z}) \in \tilde{\mathcal{B}}$ tel que l'image de $(r,\underline{Z})$ dans $\tilde{\mathcal{B}}/W$ soit fixe par Γ, on a $er \in \mathbb{Z}$.

À tout élément $b \in \mathcal{B}$, on associe comme suit un élément de $\mathcal{L}(\mathcal{D})$ dont la classe d'isomorphisme est uniquement déterminée par b. Choisissons un relèvement $(r,\underline{Z})$ de b dans $\tilde{\mathcal{B}}$, et notons Σ_1 l'ensemble des $\alpha \in \Sigma$ tels que $\alpha(\underline{Z}) = 0$. D'après (P_Σ), Σ_1 est un sous-ensemble de Levi de Σ. Il existe donc un élément $w \in W$ tels que $w(\Sigma_1)$ soit standard, c'est-à-dire engendré par $(w(\Sigma_1) \cap \Delta) \cup -(w(\Sigma_1) \cap \Delta)$. Quitte à remplacer $(r,\underline{Z})$ par $(r, w(\underline{Z}))$, on peut supposer Σ_1 standard. Soit alors $\mathcal{D}_1 = (\mathrm{X}^*, \Sigma_1, \Delta_1, \mathrm{X}_*, \check{\Sigma}_1, \check{\Delta}_1)$ avec $\Delta_1 = \Sigma_1 \cap \Delta$, $\check{\Sigma}_1 = \{\check{\alpha} : \alpha \in \Sigma_1\}$ et $\check{\Delta}_1 = \check{\Sigma}_1 \cap \check{\Delta}$. Posons $W_1 = W(\Sigma_1) \subset W$. À nouveau grâce à (P_Σ), on a $W_1 = \{w_1 \in W : w_1(\underline{Z}) = \underline{Z}\}$. Puisque l'image de $(r,\underline{Z})$ dans $\tilde{\mathcal{B}}/W$ est fixe par Γ, pour tout $\gamma \in \Gamma$, il existe un $w(\gamma) \in W$ tel que $w(\gamma) \circ \gamma(\underline{Z}) = \underline{Z}$; l'élément $w(\gamma)$ n'est pas unique, mais la classe $W_1 w(\gamma)$ est unique. On a $w(\gamma) \circ \gamma(\Sigma_1) = \Sigma_1$, et quitte à remplacer $\omega(\gamma)$ par $w_1 w(\gamma)$ pour un (unique) $w_1 \in W_1$, on peut supposer que $w(\gamma) \circ \gamma(\Delta_1) = \Delta_1$. Cela détermine $w(\gamma)$ de manière unique. Posons $\mathcal{D}' = \mathcal{D}_1$, notons $\psi = (\mathrm{X}'^* \to \mathrm{X}^*, \mathrm{X}'_* \to \mathrm{X}_*)$ les morphismes identité, et munissons $\mathcal{D}'$ de l'action de Γ donnée par $\psi \circ \gamma = w(\gamma) \circ \gamma \circ \psi$. Soit $\underline{Z}' = \psi^{-1}(\underline{Z})$. Alors $(\mathcal{D}', \psi, r, \underline{Z}') \in \mathcal{L}(\mathcal{D})$, et c'est l'élément cherché. Cet élément n'est pas uniquement determiné par b car il n'y a pas unicité du relèvement $(r,\underline{Z})$ tel que Σ_1 soit standard. Mais deux tels relèvements donnent lieu à deux éléments de $\mathcal{L}(\mathcal{D})$ isomorphes au sens de 8.3. Inversement, si $(\mathcal{D}'_2, \psi_2, r, \underline{Z}'_2) \in \mathcal{L}(\mathcal{D})$ est isomorphe à $(\mathcal{D}', \psi, r, \underline{Z}')$, alors $(r, \psi_2(\underline{Z}'_2)) \in \tilde{\mathcal{B}}$ est un relèvement de b.

10.2. Notons $\tilde{\mathcal{Z}}_\bullet$ l'ensemble des suites $(r_1, \dots, r_k; \underline{Z}_1, \dots, \underline{Z}_k)$ telles que (pour $i = 1, \dots, k$) :

[-] $k \in \mathbb{Z}_{\geq 1}$;

[-] $r_i \in \mathbb{Z}_{(p)}$ et $r_1 < r_2 < \dots < r_k$;

[-] $\underline{Z}_i \in \underline{\mathfrak{t}}(r_i) \smallsetminus \{0\}$;

[-] posant $\Sigma_i = \{\alpha \in \Sigma : \alpha(\underline{Z}_j) = 0, \forall j \in \{1, \dots, i\}\}$, $\mathrm{X}_{*,i} = \mathrm{X}_* \cap \mathbb{Q}\langle\check{\alpha} : \alpha \in \Sigma_i\rangle$ et $\underline{\mathfrak{t}}_i = \mathrm{X}_{*,i} \otimes_{\mathbb{Z}} \overline{\mathbb{F}}_q \subset \underline{\mathfrak{t}}(r_i)$, on demande que $\underline{\mathfrak{t}} \supsetneq \underline{\mathfrak{t}}_1 \supsetneq \dots \supsetneq \underline{\mathfrak{t}}_k = \{0\}$ et $\underline{Z}_{j+1} \in \underline{\mathfrak{t}}_j$ $(j = 1, \dots, k-1)$.

Notons que pour $i = 1, \dots, k$, Σ_i est un sous-ensemble de Levi de Σ ; et que $\underline{\mathfrak{t}}_i$, considéré comme un sous-espace de $\underline{\mathfrak{t}}(r_i)$, n'est en général pas Γ-stable. Le groupe $W \rtimes \Gamma$ opère naturellement sur $\tilde{\mathcal{Z}}_\bullet$, et l'on pose $\mathcal{Z}_\bullet = (\tilde{\mathcal{Z}}_\bullet/W)^\Gamma$. Tout élément de $\tilde{\mathcal{Z}}_\bullet$ est *régulier* au sens où son stabilisateur dans W est trivial. En effet (avec les notations ci-dessus) posant $W_i = W(\Sigma_i)$, le fixateur de $\underline{Z}_1$ dans W est W_1 (cf. 10.1) et pour $j = 2, \ldots k$, le fixateur de $\underline{Z}_j$ dans W_{j-1} est W_j. Or $W_k = \{1\}$ car $\underline{\mathfrak{t}}_k = \{0\}$, par conséquent le fixateur de $(r_1, \ldots, r_k; \underline{Z}_1, \ldots, \underline{Z}_k)$ dans W est $\{1\}$.

À tout élément $z \in \mathcal{Z}_\bullet$, on associe comme suit un groupe D_z. On choisit un relèvement $\tilde{z}$ de z dans $\tilde{\mathcal{Z}}_\bullet$. Puisque $\tilde{z}$ est régulier et que son image dans $\tilde{\mathcal{Z}}_\bullet/W$ est fixe par Γ, pour tout $\gamma \in \Gamma$, il existe un unique $w_{\tilde{z}}(\gamma) \in W$ tel que $w_{\tilde{z}}(\gamma) \circ \gamma(\tilde{z}) = \tilde{z}$. L'application $\gamma \mapsto w_{\tilde{z}}(\gamma)$ est un cocycle. Posons ${}_{\tilde{z}}\mathrm{X}_* = \mathrm{X}_*$, notons ${}_{\tilde{z}}\varphi : {}_{\tilde{z}}\mathrm{X}_* \to \mathrm{X}_*$ le morphisme identité, et munissons ${}_{\tilde{z}}\mathrm{X}_*$ de l'action de Γ donnée par ${}_{\tilde{z}}\varphi \circ \gamma = w_{\tilde{z}}(\gamma) \circ {}_{\tilde{z}}\varphi$. On pose $D_{\tilde{z}} = ({}_{\tilde{z}}\mathrm{X}_*)_{\Gamma,\mathrm{tor}}$. Si maintenant $\tilde{z}'$ est un autre relèvement de z dans $\tilde{\mathcal{Z}}_\bullet$, alors il existe un unique $w \in W$ tel que $w(\tilde{z}) = \tilde{z}'$, et un unique isomorphisme de $\mathbb{Z}$-modules $\iota_{\tilde{z}',\tilde{z}} : {}_{\tilde{z}}\mathrm{X}_* \to {}_{\tilde{z}'}\mathrm{X}_*$ tel que ${}_{\tilde{z}'}\varphi \circ \iota_{\tilde{z}',\tilde{z}} = w \circ {}_{\tilde{z}}\varphi$. De plus, $\iota_{\tilde{z}',\tilde{z}}$ est Γ-équivariant et définit un isomorphisme $\bar{\iota}_{\tilde{z}',\tilde{z}} : D_{\tilde{z}} \to D_{\tilde{z}'}$. On note D_z la limite inductive des $D_{\tilde{z}}$ pour $\tilde{z}$ parcourant l'ensemble des relèvements de z dans $\tilde{\mathcal{Z}}_\bullet$, les morphismes de transition étant ces isomorphismes $\bar{\iota}_{\tilde{z}',\tilde{z}}$.

Continuons avec les notations ci-dessus. De l'isomomorphisme ${}_{\tilde{z}}\varphi$ se déduit un homomorphisme surjectif ${}_{\tilde{z}}\mathrm{X}_* \to \mathrm{X}_*/\mathrm{X}_{*,\mathrm{sc}}$. Puisque W opère trivialement sur $\mathrm{X}_*/\mathrm{X}_{*,\mathrm{sc}}$, ce dernier est Γ-équivariant. Il induit donc un homomorphisme $D_{\tilde{z}} \to (\mathrm{X}_*/\mathrm{X}_{*,\mathrm{sc}})_{\Gamma,\mathrm{tor}}$, qui est compatible avec les isomorphismes $\bar{\iota}_{\tilde{z}',\tilde{z}}$. D'où un homomorphisme $\mu_z : D_z \to D$.

Si $\Sigma \neq \emptyset$, on pose $\tilde{\mathcal{Z}} = \tilde{\mathcal{Z}}_\bullet$; et si $\Sigma = \emptyset$ (i.e. si $\mathcal{D}$ est la donnée de racines d'un tore), on pose $\tilde{\mathcal{Z}} = \tilde{\mathcal{Z}}_\bullet \cup \{z_\infty\}$ et l'on considère que z_∞ est fixe par Γ. Soit $\mathcal{Z} = (\tilde{\mathcal{Z}}/W)^\Gamma$; si $\Sigma = \emptyset$, on a donc $\mathcal{Z} = \mathcal{Z}_\bullet \cup \{z_\infty\}$. Toujours si $\Sigma = \emptyset$, on pose $D_{z_\infty} = (\mathrm{X}_*)_{\Gamma,\mathrm{tor}}$ et l'on note $\mu_{z_\infty} : D_{z_\infty} \to D$ l'isomorphisme canonique.

Pour $z \in \mathcal{Z}$, on note $r(z) \in \mathbb{Z}_{(p)} \cup \{+\infty\}$ et $k(z) \in \mathbb{Z}_{\geq 0}$ les invariants définis comme suit : si $\Sigma \neq \emptyset$, ou si $\Sigma = \emptyset$ et $z \neq z_\infty$, on choisit un relèvement $(r_1, \ldots, r_k; \underline{Z}_1, \ldots, \underline{Z}_k)$ de z dans $\tilde{\mathcal{Z}}_\bullet$, et l'on pose $r(z) = r_1$ et $k(z) = k$; et si $\Sigma = \emptyset$ et $z = z_\infty$, on pose $r(z) = +\infty$ et $k(z) = 0$.

10.3. L'application $\tilde{\mathcal{Z}}_\bullet \to \tilde{\mathcal{B}}$, $(r_1, \ldots, r_k; \underline{Z}_1, \ldots, \underline{Z}_k) \mapsto (r_1, \underline{Z}_1)$ est surjective et $W \rtimes \Gamma$-équivariante. Elle induit une application surjective $\mathcal{Z}_\bullet \to \mathcal{B}$ dont la fibre au-dessus d'un élément $b \in \mathcal{B}$ est notée $\mathcal{Z}_b$.

Soient $b \in \mathcal{B}$ et $z \in \mathcal{Z}_b$. Comme en 10.1, on associe à b un élément $(\mathcal{D}', \psi, r, \underline{Z}') \in \mathcal{L}(\mathcal{D})$. Comme dans la section 8, on affecte les objets relatifs à $\mathcal{D}'$, introduits précédemment pour la donnée $\mathcal{D}$, d'un exposant "′". On peut choisir un relèvement $\tilde{z}$ de z dans $\tilde{\mathcal{Z}}_\bullet$ de la forme $\tilde{z} = (r, r_2, \ldots, r_k; \psi(\underline{Z}'), \underline{Z}_2, \ldots, \underline{Z}_k)$. La suite $(r, r_2, \ldots, r_k; \underline{Z}', \psi^{-1}(\underline{Z}_2), \ldots, \psi^{-1}(\underline{Z}_k))$ appartient à $\tilde{\mathcal{Z}}'_\bullet$ et son image dans $\tilde{\mathcal{Z}}'_\bullet/W'$ est fixe par Γ. Cette image ne dépend pas du choix du relèvement $\tilde{z}$ de z ; on la note $\beta_\bullet(z) \in \mathcal{Z}'_\bullet$. D'autre part, la suite $(r_2, \ldots, r_k; \psi^{-1}(\underline{Z}_2), \ldots, \psi^{-1}(\underline{Z}_k))$ appartient à $\tilde{\mathcal{Z}}'$, et son image dans $\tilde{\mathcal{Z}}'/W'$ est fixe par Γ. À nouveau, cette image ne dépend pas du choix du relèvement $\tilde{z}$ de z ; on la note $\beta(z) \in \mathcal{Z}'$. On a donc défini deux applications

$$\mathcal{Z}_b \xrightarrow{\beta_\bullet} \mathcal{Z}'_\bullet, \qquad \mathcal{Z}_b \xrightarrow{\beta} \mathcal{Z}'.$$

Ces deux applications sont injectives. L'image de $\beta_\bullet$ est la fibre $\mathcal{Z}'_{b'}$ de la surjection $\mathcal{Z}'_\bullet \to \mathcal{B}'$ au-dessus de b', où b' est l'image de $(r, \underline{Z}')$ dans $\mathcal{B}'$. Notons $\mathcal{Z}'_{\mathrm{der}}$ l'ensemble obtenu en remplaçant $\mathcal{D}'$ par $\mathcal{D}'_{\mathrm{der}}$ (cf. 3.1) dans la définition de $\mathcal{Z}'$. On a une identification naturelle $\mathcal{Z}'_{\mathrm{der}} \subset \mathcal{Z}'$, et l'image de β est l'ensemble des $z' \in \mathcal{Z}'_{\mathrm{der}}$ tels que $r(z') > r$. Par construction, on a $D'_{\beta_\bullet(z)} = D'_{\beta(z)}$, et ψ induit un isomorphisme $\psi_{z,\beta(z)} : D'_{\beta(z)} \to D_z$. De plus, le diagramme suivant est commutatif :

$$\begin{array}{ccc} D'_{\beta(z)} & \xrightarrow{\psi_{z,\beta(z)}} & D_z \\ {\scriptstyle \mu'_{\beta(z)}}\downarrow & & \downarrow{\scriptstyle \mu_z} \\ D' & \xrightarrow[\psi_{D'}]{} & D. \end{array}$$

10.4. Soit un corps $F \in \mathrm{CL}_q$, fixé jusqu'à la fin du numéro 10.6. Pour $X \in \mathfrak{t}^{\mathrm{mod}}$, on pose $r(X) = \omega_{\mathfrak{t},F}(X)$ (cf. 1.2 et 7.2).

Pour tout sous-système de racines Σ' de Σ, on pose $\mathrm{X}_{*,\Sigma'-\mathrm{cent}} = \{X \in \mathrm{X}_* : \alpha(X) = 0, \forall \alpha \in \Sigma'\}$ et $\mathfrak{t}^{\mathrm{mod}}_{\Sigma'-\mathrm{cent}} = \mathrm{X}_{*,\Sigma'-\mathrm{cent}} \otimes_{\mathbb{Z}} F^{\mathrm{mod}}$, $\mathrm{X}_{*,\Sigma'-\mathrm{der}} = \mathrm{X}_* \cap \mathbb{Q}\langle \check{\alpha} : \alpha \in \Sigma' \rangle$ et $\mathfrak{t}^{\mathrm{mod}}_{\Sigma'-\mathrm{der}} = \mathrm{X}_{*,\Sigma'-\mathrm{der}} \otimes_{\mathbb{Z}} F^{\mathrm{mod}}$. D'après (P_Σ), le réseau $\mathrm{X}_{*,\Sigma'-\mathrm{cent}} \oplus \mathrm{X}_{*,\Sigma'-\mathrm{der}}$ est d'indice premier à p dans X_*, par conséquent on a la décomposition $\mathfrak{t}^{\mathrm{mod}} = \mathfrak{t}^{\mathrm{mod}}_{\Sigma'-\mathrm{cent}} \oplus \mathfrak{t}^{\mathrm{mod}}_{\Sigma'-\mathrm{der}}$.

Posons $\mathfrak{t}^{\mathrm{mod}}_{\mathrm{reg}} = \{X \in \mathfrak{t}^{\mathrm{mod}} : w(X) \neq X, \forall w \in W\}$ et $\mathcal{Z}_F = (\mathfrak{t}^{\mathrm{mod}}_{\mathrm{reg}}/W)^\Gamma$. On définit comme suit une application $\zeta_F : \mathcal{Z}_F \to \mathcal{Z}$. Soit $X \in \mathfrak{t}^{\mathrm{mod}}_{\mathrm{reg}}$. Si $X = 0$ (ce qui implique $\Sigma = \emptyset$ puisque X est régulier), on pose $\tilde{\zeta}_F(X) = z_\infty$. Supposons maintenant $X \neq 0$, et notons $r_1, \ldots, r_k$ les éléments du sous-ensemble $\{r(X)\} \cup \{\omega_F(\alpha(X)) : \alpha \in \Sigma\}$ de $\mathbb{Z}_{(p)}$, ordonnés de manière croissante. On a donc $r_1 = r(X)$. Pour $i = 1, \ldots, k$, posons $\Sigma_i = \{\alpha \in \Sigma : \omega_F(\alpha(X)) > r_i\}$. Les ensembles Σ_i sont des sous-systèmes de racines de Σ, et l'on a $\Sigma \supset \Sigma_1 \supsetneq \Sigma_2 \supsetneq \cdots \supsetneq \Sigma_k$. Notons que si $\mathfrak{t}^{\mathrm{mod}} = \mathfrak{t}^{\mathrm{mod}}_{\Sigma-\mathrm{der}}$, alors l'inclusion $\Sigma \supset \Sigma_1$ est stricte. Par conséquent $\mathfrak{t}^{\mathrm{mod}} \supsetneq \mathfrak{t}^{\mathrm{mod}}_{\Sigma_1-\mathrm{cent}}$. D'après le paragraphe précédent, on a la décomposition :

$$\begin{aligned} \mathfrak{t}^{\mathrm{mod}} = \mathfrak{t}^{\mathrm{mod}}_{\Sigma_1-\mathrm{cent}} &\oplus (\mathfrak{t}^{\mathrm{mod}}_{\Sigma_1-\mathrm{der}} \cap \mathfrak{t}^{\mathrm{mod}}_{\Sigma_2-\mathrm{cent}}) \oplus (\mathfrak{t}^{\mathrm{mod}}_{\Sigma_2-\mathrm{der}} \cap \mathfrak{t}^{\mathrm{mod}}_{\Sigma_3-\mathrm{cent}}) \\ &\oplus \cdots \oplus (\mathfrak{t}^{\mathrm{mod}}_{\Sigma_{k-1}-\mathrm{der}} \cap \mathfrak{t}^{\mathrm{mod}}_{\Sigma_k-\mathrm{cent}}). \end{aligned}$$

Écrivons $X = X_1 + X_2 + \cdots + X_k$ suivant cette décomposition. Par construction, pour $i = 1, \ldots, k$, on a $X_i \in \mathfrak{t}^{\mathrm{mod}}_{r_i} \smallsetminus \mathfrak{t}^{\mathrm{mod}}_{(r_i)+}$. Notons $\underline{X}_i \in \underline{\mathfrak{t}}(r_i)$ l'image de X_i par la projection canonique $\mathfrak{t}^{\mathrm{mod}}_{r_i} \to \underline{\mathfrak{t}}(r_i)$ (cf. 8.5). Alors la suite $(r_1, \ldots, r_k; \underline{X}_1, \ldots, \underline{X}_k)$ appartient à $\tilde{\mathcal{Z}}_\bullet$; on la note $\tilde{\zeta}_F(X)$. L'application $\tilde{\zeta}_F : \mathfrak{t}^{\mathrm{mod}}_{\mathrm{reg}} \to \tilde{\mathcal{Z}}$ ainsi définie est $W \rtimes \Gamma$-équivariante. Elle induit donc une application $\zeta_F : \mathcal{Z}_F \to \mathcal{Z}$. Pour $z \in \mathcal{Z}$, on pose $\mathcal{Z}_F(z) = \zeta_F^{-1}(z)$

Soient $b \in \mathcal{B}$ et $z \in \mathcal{Z}_b$. Comme en 10.1, on associe à b un élément $(\mathcal{D}', \psi, r, \underline{Z}') \in \mathcal{L}(\mathcal{D})$. On pose $\mathfrak{t}'^{\mathrm{mod}}_{\mathrm{reg}} = \{X' \in \mathfrak{t}'^{\mathrm{mod}} : w(X') \neq X', \forall w \in W'\}$, $\mathcal{Z}'_F = (\mathfrak{t}'^{\mathrm{mod}}_{\mathrm{reg}}/W')^\Gamma$, et l'on définit comme ci-dessus une application $\zeta'_F : \mathcal{Z}'_F \to \mathcal{Z}'$. Soit $z'_\bullet = \beta_\bullet(z) \in \mathcal{Z}'_\bullet$, et posons $\mathcal{Z}'_F(z'_\bullet) = \zeta'^{-1}_F(z'_\bullet) \subset \mathcal{Z}'_F$. Rappelons que l'isomorphisme $i_\psi : \mathfrak{t}'^{\mathrm{mod}} \to \mathfrak{t}^{\mathrm{mod}}$ n'est pas Γ-équivariant; mais l'application $\mathfrak{t}'^{\mathrm{mod}}/W' \to \mathfrak{t}^{\mathrm{mod}}/W$ qui s'en déduit est Γ-équivariante. Cette dernière induit une application *bijective* (cf. [W2] 6.6)

(1) $$\mathcal{Z}'_F(z'_\bullet) \to \mathcal{Z}_F(z).$$

D'autre part, à la donnée $\mathcal{D}'_{\mathrm{der}}$ sont associés comme ci-dessus un ensemble $\mathcal{Z}'_{\mathrm{der},F}$ et une application $\zeta'_{\mathrm{der},F} : \mathcal{Z}'_{\mathrm{der},F} \to \mathcal{Z}'_{\mathrm{der}}$. Comme pour $\mathcal{Z}'_{\mathrm{der}}$, on a une identifcation naturelle $\mathcal{Z}'_{\mathrm{der},F} \subset \mathcal{Z}'_F$. Soit $z' = \beta(z) \in \mathcal{Z}'_{\mathrm{der}}$, et posons $\mathcal{Z}'_{\mathrm{der},F}(z') = \zeta'^{-1}_{\mathrm{der},F}(z')$. Notons $\mathfrak{t}'_{\mathrm{cent}}(F)(\underline{Z}')$ l'ensemble des éléments de $\mathfrak{t}'_{\mathrm{cent}}(F)_r$ qui s'envoient sur $\underline{Z}'$ par la projection canonique $\mathfrak{t}'_{\mathrm{cent}}(F)_r \to \underline{\mathfrak{t}}'_{\mathrm{cent}}(r)^\Gamma$. De la décomposition $\mathfrak{t}'^{\mathrm{mod}} = \mathfrak{t}'^{\mathrm{mod}}_{\mathrm{cent}} \oplus \mathfrak{t}'^{\mathrm{mod}}_{\mathrm{der}}$, on déduit une application *bijective* (cf. [W2] 6.6)

(2) $$\mathcal{Z}'_F(z'_\bullet) \to \mathfrak{t}'_{\mathrm{cent}}(F)(\underline{Z}') \times \mathcal{Z}'_{\mathrm{der},F}(z').$$

Des bijections (1) et (2) ci-dessus, par récurrence sur le rang de $\mathcal{D}$, on déduit le

LEMME [W2, lemme 6.6.1]. — *L'application ζ_F est surjective.*

10.5. On note ${}_D\mathfrak{g}$ la variété $\coprod_{d\in D} {}_d\mathfrak{g}$ sur F, et l'on pose ${}_D\mathfrak{g}^{\mathrm{mod}} = \coprod_{d\in D} {}_d\mathfrak{g}^{\mathrm{mod}}$ et ${}_D\mathfrak{g}(F)_{\mathrm{reg}} = \coprod_{d\in D} {}_d\mathfrak{g}(F)_{\mathrm{reg}}$. Soient d, $d' \in D$, $X \in {}_d\mathfrak{g}(F)_{\mathrm{reg}}$ et $X' \in {}_{d'}\mathfrak{g}(F)_{\mathrm{reg}}$. On dit que X et X' sont :

[-] *conjugués* si $d = d'$ et s'il existe un $g \in {}_dG(F)$ tel que $\mathrm{Ad}(g)(X) = X'$;

[-] *stablement conjugués* s'il existe un $\overline{g} \in G(\overline{F})$ tel que $\mathrm{Ad}(\overline{g}) \circ {}_d\xi(X) = {}_{d'}\xi(X')$.

D'après 7.2, si X et X' sont stablement conjugués, alors on peut choisir $\overline{g}$ dans G^{mod}.

Soit $X \in {}_d\mathfrak{g}(F)_{\mathrm{reg}}$ pour un $d \in D$. D'après 7.2, il existe un $g \in G^{\mathrm{mod}}$ tel que $\mathrm{Ad}(g) \circ {}_d\xi(X) \in \mathfrak{t}^{\mathrm{mod}}_{\mathrm{reg}}$. L'image de $\mathrm{Ad}(g) \circ {}_d\xi(X)$ dans $\mathfrak{t}^{\mathrm{mod}}_{\mathrm{reg}}/W$ est fixe par Γ, et ne dépend pas du choix de g; on la note $t(X)$. L'application ${}_D\mathfrak{g}(F)_{\mathrm{reg}} \to \mathcal{Z}_F$, $X \mapsto t(X)$ est surjective, et ses fibres sont les classes de conjugaison stable dans ${}_D\mathfrak{g}(F)_{\mathrm{reg}}$. Pour $z_F \in \mathcal{Z}_F$, on pose $\mathcal{O}_{\mathrm{st}}(z_F) = t^{-1}(z_F)$.

Soient $z \in \mathcal{Z}$ et $z_F \in \mathcal{Z}_F(z)$. On définit comme suit une application $\delta = \delta_{z_F} : \mathcal{O}_{\mathrm{st}}(z_F) \to D_z$. Choisissons un relèvement $\tilde{z}$ de z dans $\tilde{Z}$. D'après la définition de ζ, il existe un $Z \in \mathfrak{t}^{\mathrm{mod}}_{\mathrm{reg}}$ relevant z_F et tel que $\tilde{\zeta}_F(Z) = \tilde{z}$. Puisque $\tilde{z}$ est régulier, la condition $\tilde{\zeta}_F(Z) = \tilde{z}$ détermine Z de manière unique. Puisque l'application $\tilde{\zeta}_F$ est Γ-équivariante, pour $\gamma \in \Gamma$, $w_{\tilde{z}}(\gamma)$ (cf. 10.2) est l'unique élément de W vérifiant $w_{\tilde{z}}(\gamma) \circ \gamma(Z) = Z$. Notons ${}_{\tilde{z}}T$ le tore défini sur F dont le groupe des cocaractères est ${}_{\tilde{z}}\mathrm{X}_*$. On a un isomorphisme ${}_{\tilde{z}}T \to T$ défini sur F^{mod}, que l'on note encore ${}_{\tilde{z}}\varphi$, tel que ${}_{\tilde{z}}\varphi \circ \gamma = w_{\tilde{z}}(\gamma) \circ {}_{\tilde{z}}\varphi$

pour tout $\gamma \in \Gamma$. D'après 4.3, l'homomorphisme de Kottwitz $\omega^{\rm nr}_{\tilde{z}T}$ induit une application bijective $\mathrm{H}^1(\Gamma, ({}_{\tilde{z}}T)^{\rm mod}) \to ({}_{\tilde{z}}\mathrm{X}_*)_{\Gamma,\rm tor} = D_{\tilde{z}}$. On identifie $\mathrm{H}^1(\Gamma, ({}_{\tilde{z}}T)^{\rm mod})$ et $D_{\tilde{z}}$ via cette application. Pour $\gamma \in \Gamma$, on pose $\Sigma^+_{\tilde{z},\gamma} = \{\alpha \in \Sigma^+ : w_{\tilde{z}}(\gamma)^{-1}(\alpha) < 0\}$ et

$$t_{\tilde{z}}(\gamma) = \textstyle\prod_{\alpha\in\Sigma^+_{\tilde{z},\gamma}} \check{\alpha} \circ \alpha(Z) \in T^{\rm mod},$$

$$n_{\tilde{z}}(\gamma) = n(w_{\tilde{z}}(\gamma)) \in N^{\rm mod}.$$

Soit $X \in {}_d\mathfrak{g}(F)_{\rm reg} \cap \mathcal{O}_{\rm st}(z_F)$ pour un $d \in D$. Choisissons un $g \in G^{\rm mod}$ tel que $\mathrm{Ad}(g) \circ {}_d\xi(X) = Z$. Pour $\gamma \in \Gamma$, l'élément $gd_F(\gamma)\gamma(g)^{-1}$ appartient à $N^{\rm mod}$, et son image dans W est $w_{\tilde{z}}(\gamma)$. Alors l'élément $gd_F(\gamma)\gamma(g)^{-1}n_{\tilde{z}}(\gamma)^{-1}t_{\tilde{z}}(\gamma)^{-1}$ appartient à $T^{\rm mod}$, et son image par l'isomorphisme ${}_{\tilde{z}}\varphi^{-1} : T \to {}_{\tilde{z}}T$ appartient à $({}_{\tilde{z}}T)^{\rm mod}$; on la note $\delta_{\tilde{z},X,g}$. Grâce au lemme 2.2.A de [LS], on vérifie que l'application $\gamma \mapsto \delta_{\tilde{z},X,g}$ est un cocycle. Elle définit un élément $\delta_{\tilde{z}}(X) \in \mathrm{H}^1(\Gamma, ({}_{\tilde{z}}T)^{\rm mod}) = D_{\tilde{z}}$, qui ne dépend pas du choix de g. D'après le lemme 2.3.A de [LS], l'application $\delta_{\tilde{z}} : \mathcal{O}_{\rm st}(z_F) \to D_{\tilde{z}}$, est compatible avec les isomorphismes $\bar{\iota}_{\tilde{z}',\tilde{z}}$. D'où une application $\delta : \mathcal{O}_{\rm st}(z_F) \to D_z$.

Pour $z_F \in \mathcal{Z}_F$, notons $\mathrm{cl}(z_F)$ l'ensemble des classes de conjugaison (ordinaire) contenues dans $\mathcal{O}_{\rm st}(z_F)$. Par construction, δ_{z_F} se factorise en une application $\overline{\delta}_{z_F} : \mathrm{cl}(z_F) \to D_{\zeta_F(z_F)}$.

LEMME [W2, lemme 6.5]. — *Soient $z \in \mathcal{Z}$ et $z_F \in \mathcal{Z}_F(z)$.*
[(1)] Pour $d \in D$ et $X \in {}_d\mathfrak{g}(F)_{\rm reg} \cap \mathcal{O}_{\rm st}(z_F)$, on a $\mu_z \circ \delta_{z_F}(X) = d$.
[(2)] L'application $\overline{\delta}_{z_F}$ est bijective.

10.6. Soient $b \in \mathcal{B}$, $z \in \mathcal{Z}_b$ et $z_F \in \mathcal{Z}_F(z)$. Posons $z'_\bullet = \beta_\bullet(z)$ et $z' = \beta(z)$. Notons $z'_{\bullet,F} \in \mathcal{Z}'_F(z_\bullet)$ l'image réciproque de z_F par la bijection (1) de 10.4, et $(Z', z'_F) \in \mathfrak{t}'_{\rm cent}(F)(\underline{Z}') \times \mathcal{Z}'_{{\rm der},F}(z')$ l'image de $z'_{\bullet,F}$ par la bijection (2) de 10.4. Pour définir cette bijection (2), on a choisi un élément $(\mathcal{D}', \psi, r, \underline{Z}') \in \mathcal{L}(\mathcal{D})$ associé à b comme en 10.1. On définit comme en 10.5 les classes de conjugaison stable $\mathcal{O}_{\rm st}(z'_{\bullet,F})$, $\mathcal{O}_{\rm st}(z'_F) \subset {}_{D'}\mathfrak{g}'(F)_{\rm reg}$, c'est-à-dire les fibres de la surjection $\zeta'_F : \mathcal{Z}'_F \to \mathcal{Z}'$ au-dessus de $z'_{\bullet,F}$ et de z'_F (on considère ici z'_F comme un élément de $\mathcal{Z}'_F$, et non de $\mathcal{Z}'_{{\rm der},F}$). De même, on définit comme en 10.5 les ensembles $\mathrm{cl}(z'_{\bullet,F})$ et $\mathrm{cl}(z'_F)$.

Pour tout $d' \in D$, on a défini en 8.7 un F-plongement ${}_{d'}i_\psi : {}_{d'}G' \to {}_dG$ avec $d = \psi_{D'}(d')$; d'où un F-plongement $\mathrm{d}({}_{d'}i_\psi) : {}_{d'}\mathfrak{g}' \to {}_d\mathfrak{g}$. Les $\mathrm{d}({}_{d'}i_\psi)$ pour $d' \in D'$, se regroupent en un F-plongement $I_\psi : {}_{D'}\mathfrak{g}' \to {}_D\mathfrak{g}$. Ce dernier est compatible avec les conjugaisons stable et ordinaire. On vérifie que si $X'_\bullet \in \mathcal{O}_{\rm st}(z'_{\bullet,F})$, alors $I_\psi(X'_\bullet) \in \mathcal{O}_{\rm st}(z_F)$. Et la restriction de I_ψ à $\mathcal{O}_{\rm st}(z'_{\bullet,F})$ se factorise en une application $\overline{I}_\psi : \mathrm{cl}(z'_{\bullet,F}) \to \mathrm{cl}(z_F)$.

L'élément Z' est central pour la donnée $\mathcal{D}'$ et l'on peut décrire son action sur les classes de conjugaison : pour $d' \in D'$, la restriction à $\mathfrak{t}'_{\rm cent}$ de l'isomorphisme $\mathrm{d}({}_{d'}\xi')^{-1} : \mathfrak{g}' \to {}_{d'}\mathfrak{g}'$ est définie sur F, par conséquent $Y' = \mathrm{d}({}_{d'}\xi')^{-1}(Z')$ appartient à ${}_{d'}\mathfrak{g}'(F)$. Et l'application $X' \mapsto X' + Y'$

induit une application bijective ${}_{d'}\mathfrak{g}'(F)\cap\mathcal{O}_{\rm st}(z'_F) \to {}_{d'}\mathfrak{g}'(F)\cap\mathcal{O}_{\rm st}(z'_{\bullet,F})$. D'où une application bijective $\mathrm{cl}(z'_{\bullet,F}) \to \mathrm{cl}(z'_F)$, qui permet d'identifier $\mathrm{cl}(z'_{\bullet,F})$ et $\mathrm{cl}(z'_F)$. D'après le lemme 6.6.2 de [W2], les deux carrés du diagramme suivant sont commutatifs :

$$\begin{array}{ccccc} \mathrm{cl}(z'_{\bullet,F}) & = & \mathrm{cl}(z'_F) & \xrightarrow{\overline{I}_\psi} & \mathrm{cl}(z_F) \\ \downarrow \overline{\delta}'_{z'_{\bullet,F}} & & \downarrow \overline{\delta}'_{z'_F} & & \downarrow \overline{\delta}_{z_F} \\ D'_{z'_\bullet} & = & D'_{z'} & \xrightarrow[\psi_{z,z'}]{} & D_z. \end{array}$$

(La commutativité du carré de gauche est immédiate par construction.) La commutativité du carré de droite implique en particulier que l'application $\overline{I}_\psi$ est bijective.

10.7. Pour $d \in D$ et $F \in \mathrm{CL}_q$, on a défini en 7.3 une application linéaire ${}_d\mathrm{rea}_F : {}_d\mathcal{S} \to {}_d\mathcal{C}_F$. On pose $\mathcal{S} = \oplus_{d\in D}{}_d\mathcal{S}$ et $\mathcal{C}_F = \oplus_{d\in D}\,{}_d\mathcal{C}_F$, et l'on note $\mathrm{rea}_F : \mathcal{S} \to \mathcal{C}_F$ l'application linéaire $\oplus_{d\in D}\,{}_d\mathrm{rea}_F$. On considère les éléments de $\mathcal{C}_F$ comme des fonctions sur ${}_D\mathfrak{g}(F)$. Pour $X \in {}_D\mathfrak{g}(F)_{\rm reg}$ et $f \in \mathcal{C}_F$, on pose $\Lambda^{{}_DG(F)}(f,X) = \Lambda^{{}_dG(F)}(f_d,X)$ où d est l'unique élément de D tel que $X \in {}_d\mathfrak{g}(F)$ et f_d est la restriction de f à ${}_d\mathfrak{g}(F)$.

Le résultat suivant est le théorème principal de [W2] :

THÉORÈME [W2, théorème 7.2]. — *Soient $z \in \mathcal{Z}$ et $\delta \in D_z$. Il existe une forme linéaire $\Lambda^{\mathcal{D}}(\cdot,z,\delta)$ sur $\mathcal{S}$ telle que pour tous $\varphi \in \mathcal{S}$, $F \in \mathrm{CL}_q$, $z_F \in \mathcal{Z}_F(z)$ et $X \in \delta_{z_F}^{-1}(\delta)$, on ait l'égalité*

$$\Lambda^{{}_DG(F)}(\mathrm{rea}_F(\varphi),X) = \Lambda^{\mathcal{D}}(\varphi,z,\delta).$$

Démonstration : À z sont associés deux invariants $r(z) \in \mathbb{Z}_{(p)} \cup \{+\infty\}$ et $k(z) \in \mathbb{Z}_{\geq 0}$, cf. le dernier paragraphe de 10.2. On raisonne par récurrence sur l'entier $k(z)$. Précisément, fixé un entier $k \geq 0$, on suppose le théorème vrai pour toutes les données $(\mathcal{D}_1,z_1,\delta_1)$ telles que $\mathrm{rg}(\mathcal{D}_1) \leq \mathrm{rg}(\mathcal{D})$ et $k(z_1) < k(z)$. On suppose désormais $k(z) = k$.

Commençons par traiter le cas $k = 0$. On a $\Sigma = \emptyset$, $z = z_\infty$, et $D_z = (\mathrm{X}_*)_{\Gamma,\rm tor}$ s'identifie canoniquement à D. Soit $F \in \mathrm{CL}_q$. On a $\mathcal{Z}_F = \mathfrak{g}(F) = \mathfrak{t}(F)$, $\mathcal{Z}_F(z) = \{0\}$ et $\delta_0^{-1}(\delta) = \{X\}$ où X est l'élément 0 de ${}_\delta\mathfrak{g}(F)$. Soit $\varphi \in \mathcal{S}$. Écrivons $\varphi = \sum_{d\in D}\varphi_d$ avec $\varphi_d \in {}_d\mathcal{S}$, et pour $d \in D$, posons $f_d = {}_d\mathrm{rea}_F(\varphi_d) \in {}_d\mathcal{C}_F$. Alors on a les égalités $\Lambda^{{}_DG(F)}(\mathrm{rea}_F(\varphi),X) = f_\delta(0) = \varphi_\delta(0)$. Il suffit donc de poser $\Lambda^{\mathcal{D}}(\varphi,z,\delta) = \varphi_\delta(0)$.

On suppose désormais $k \geq 1$, i.e. $z \in \mathcal{Z}_\bullet$. Par linéarité, il suffit de définir $\Lambda^{\mathcal{D}}(\cdot,z,\delta)$ sur ${}_d\mathcal{S}_\phi$ avec $d \in D$ et $\phi \in {}_d\Phi$ fixés. Soient $\varphi \in {}_d\mathcal{S}_\phi$, et F, z_F, X comme dans l'énoncé. Posons $f = {}_d\mathrm{rea}_F(\varphi) \in {}_d\mathcal{C}_F$ et $d_1 = \psi_{D,z}(\delta)$. D'après le lemme 10.5, on a $X \in {}_{d_1}\mathfrak{g}(F)$. Si $d \neq d_1$, on a donc $\Lambda^{{}_DG(F)}(f,X) = 0$; on pose alors $\Lambda^{\mathcal{D}}(\varphi,z,\delta) = 0$.

Supposons $d = d_1$. Si $r(z) < r(\phi)$, puisque tout conjugué de X dans ${}_dG(F)$ appartient à ${}_d\mathfrak{g}(F)_{r(z)} \smallsetminus {}_d\mathfrak{g}(F)_{r(z)^+}$, le support de f ne coupe pas la ${}_dG(F)$-orbite de X, donc $\Lambda^{{}_DG(F)}(f,X) = 0$; et l'on pose à nouveau $\Lambda^{\mathcal{D}}(\varphi,z,\delta)=0$. On suppose désormais $r(z)\geq r(\phi)$. D'après 7.2 et le premier paragraphe de 10.1, il existe un $e\in\mathcal{P}$ indépendant des données $r(z)$ et $r(\phi)$, tel que $r(z)$, $r(\phi)\in\frac{1}{e}\mathbb{Z}$. On peut donc raisonner par récurrence sur l'entier $e(r(z)-r(\phi))$. Précisément, on peut supposer que la forme linéaire $\Lambda^{\mathcal{D}}(\cdot,z,\delta)$ est définie sur ${}_d\mathcal{S}_{r(\phi)+}$. Posons $r=r(\phi)$. On distingue deux cas :

Cas 1 : $r(z)>r$. Alors on pose $\Lambda^{\mathcal{D}}(\varphi,z,\delta)=\Lambda^{\mathcal{D}}({}_dl_r^+(\varphi),z,\delta)$ où ${}_dl_r^+ : {}_d\mathcal{S}_r\to{}_d\mathcal{S}_{r+}$ est l'application linéaire de la proposition 7.4.

Cas 2 : $r(z)=r$. Notons b l'image de z dans $\mathcal{B}$ (i.e. l'élément tel que $z\in\mathcal{Z}_b$), et associons à b un élément $(\mathcal{D}',\psi,r,\underline{Z}')$ comme en 10.1. Posons $z'=\beta(z)\in\mathcal{Z}'$, $\delta'=\psi_{z,z'}^{-1}(\delta)\in D'_{z'}$ et $d'=\mu'_{z'}(\delta')\in D'$. D'après 10.3, on a $d=\psi_{D'}(d')$. Puisque $k(z')=k-1$, on peut poser $\Lambda^{\mathcal{D}}(\varphi,z,\delta)=\Lambda^{\mathcal{D}'}({}_{d'}l'_r(\varphi),z',\delta')$ où ${}_{d'}l'_r : {}_d\mathcal{S}_r\to{}_{d'}\mathcal{S}'_r$ est l'application linéaire de la proposition 9.3. Montrons que cette définition convient. Soient $F\in\mathrm{CL}_q$, $z_F\in\mathcal{Z}_F(z)$ et $X\in\delta_{z_F}^{-1}(\delta)$. Soit $(Z',z'_F)\in\mathfrak{t}'_{\mathrm{cent}}(F)(\underline{Z}')\times\mathcal{Z}'_{\mathrm{der},F}(z')$ l'image de z_F par la composée des bijections (1) et (2) de 10.4. Soit $X'\in\delta'^{-1}_{z'_F}(\delta')\subset\mathcal{O}_{\mathrm{st}}(z'_F)$. D'après le lemme 10.5, on a $X\in{}_{d'}\mathfrak{g}'(F)$, et d'après 10.6, les éléments X et $\mathrm{d}({}_{d'}i_\psi)(\mathrm{d}({}_{d'}\xi'^{-1})(Z')+X')$ sont conjugués dans ${}_dG(F)$. La proposition 9.3 appliquée au couple $(\mathrm{d}({}_{d'}\xi'^{-1})(Z'),X')$ entraîne donc l'égalité : $\Lambda^{{}_dG(F)}(f,X)=\Lambda^{{}_{d'}G'(F)}({}_{d'}\mathrm{rea}_F\circ{}_{d'}l'_r(\varphi),X')$. Mais d'après l'hypothèse de récurrence, on a $\Lambda^{{}_{d'}G'(F)}({}_{d'}\mathrm{rea}_F\circ{}_{d'}l'_r(\varphi),X')=\Lambda^{\mathcal{D}'}({}_{d'}l'_r(\varphi),z',\delta')$. □

Le théorème précédent permet de comparer des intégrales orbitales relatives à des groupes définis sur des corps de base différents. Mais il affirme aussi, pour un corps $F\in\mathrm{CL}_q$ fixé, la propriété de constance locale suivante des intégrales orbitales des fonctions dans l'image de rea_F : pour tous z_F, $t_F\in\mathcal{Z}_F$ tels que $\zeta_F(z_F)=\zeta_F(t_F)$, et tous $X\in\mathcal{O}_{\mathrm{st}}(z_F)$, $Y\in\mathcal{O}_{\mathrm{st}}(t_F)$ tels que $\delta_{z_F}(X)=\delta_{t_F}(Y)$, on a l'égalité

$$\Lambda^{{}_DG(F)}(\mathrm{rea}_F(\varphi),X)=\Lambda^{{}_DG(F)}(\mathrm{rea}_F(\varphi),Y)\qquad(\varphi\in\mathcal{S}).$$

11. Endoscopie, transfert et lemme fondamental

11.1. On appelle *donnée endoscopique de* $\mathcal{D}$ un triplet $(\mathcal{D}_\natural,\eta,s)$ de la forme suivante :

[-] $\mathcal{D}_\natural=(\mathrm{X}_\natural^*,\Sigma_\natural,\Delta_\natural,\mathrm{X}_{*,\natural},\check{\Sigma}_{*,\natural},\check{\Delta}_{*,\natural})$ est une donnée de racines munie d'une action de Γ comme en 3.1. Comme en 8.3, on affecte d'un indice "$\natural$" les objets relatifs à $\mathcal{D}_\natural$: $W_\natural$, $\underline{G}_\natural$, $\underline{T}_\natural$ (etc.);

[-] $\eta=(\mathrm{X}_\natural^*\to\mathrm{X}^*,\mathrm{X}_{*,\natural}\to\mathrm{X}_*)$ est un couple d'isomorphismes de $\mathbb{Z}$-modules en dualité. On note encore η chacun de ces isomorphismes, ainsi que tout isomorphisme qui s'en déduit par fonctorialité. On suppose que $\eta(\Sigma_\natural)\subset\Sigma$ et $\eta(\check{\Sigma}_\natural)\subset\check{\Sigma}$, et que pour tout $\gamma\in\Gamma$, il existe un élément $w\in W$ tel que $\eta\circ\gamma=w\circ\gamma\circ\eta$. Cet élément w est unique, et noté $w_\eta(\gamma)$.

[-] $s \in \hat{Z}_\natural^\Gamma$ où l'on a posé $\hat{T}_\natural = X_\natural^* \otimes_{\mathbb{Z}} \mathbb{C}^\times$ et $\hat{Z}_\natural = \{t \in \hat{T}_\natural : \check{\alpha}(t) = 1, \forall \alpha \in \Sigma_\natural\}$. De l'action de Γ sur $X_\natural^*$ se déduisent des actions sur $\hat{T}_\natural$ et sur $\hat{Z}_\natural$.

On suppose aussi que pour tout $\alpha \in \Sigma \smallsetminus \eta(\Sigma_\natural)$, on a $\eta^{-1}(\check{\alpha})(s) \neq 1$.

Deux données endoscopiques $(\mathcal{D}_{1,\natural}, \eta_1, s_1)$ et $(\mathcal{D}_{2,\natural}, \eta_2, s_2)$ de $\mathcal{D}$ sont dites *isomorphes* s'il existe un isomorphisme Γ-équivariant $f : \mathcal{D}_{1,\natural} \to \mathcal{D}_{1,\natural}$ (cf. 8.3) et un élément $w \in W$ tels que

$$\eta_2 \circ f = w \circ \eta_1, \qquad \hat{f}(s_1) \in s_2 \hat{Z}_{2,\natural}^{\Gamma,\circ}.$$

Ici, $\hat{f}$ désigne l'isomorphisme $\hat{T}_{1,\natural} \to \hat{T}_{2,\natural}$ déduit de $X_{1,\natural}^* \xrightarrow{f} X_{2,\natural}^*$, et $\hat{Z}_{2,\natural}^{\Gamma,\circ}$ la composante neutre de $\hat{Z}_{2,\natural}^\Gamma$.

Soit $(\mathcal{D}_\natural, \eta, s)$ une donnée endoscopique de $\mathcal{D}$. Notons $\check{\mathcal{D}}$ la donnée duale $(X_*, \check{\Sigma}_*, \check{\Delta}_*, X^*, \Sigma, \Delta)$. À cette donnée $\check{\mathcal{D}}$, on associe comme en 3.3 un groupe complexe $\hat{G}$ muni d'une action de Γ préservant un épinglage. De même, à la donnée duale $\check{\mathcal{D}}_\natural$, on associe un groupe complexe $\hat{G}_\natural$ muni d'une action de Γ préservant un épinglage. De η se déduit un plongement $\hat{\eta} : \hat{G}_\natural \to \hat{G}$, et $\hat{Z}_\natural$ est le centre de $\hat{G}_\natural$. Soit $F \in \mathrm{CL}_q$. À $\mathcal{D}$ et $\mathcal{D}_\natural$, on associe comme en 3.4 des groupes réductifs connexes G et $G_\natural$ définis sur F. Le triplet $(G_\natural, \hat{\eta}, s)$ est un *triplet endoscopique de* G au sens habituel. Plus généralement, c'est un triplet endoscopique de ${}_dG$ pour tout $d \in D$. En revanche, la notion d'isomorphisme ci-dessus est plus fine que celle habituellement utilisée (d'habitude, on demande que $\hat{f}(s_1) \in s_2 \hat{Z}_{2,\natural}^{\Gamma,\circ} \hat{\eta}_2^{-1}(\hat{Z}^\Gamma)$ où $\hat{Z}$ est le centre de $\hat{G}$) et mieux adaptée à la considération simultanée de tous les ${}_dG$.

11.2. Soit $(\mathcal{D}_\natural, \eta, s)$ une donnée endoscopique de $\mathcal{D}$ fixée jusqu'à la fin du numéro 11.4. Pour $r \in \mathbb{Z}_{(p)}$ l'isomorphisme $\eta : \underline{\mathfrak{t}}_\natural(r) \to \underline{\mathfrak{t}}(r)$ n'est en général pas Γ-équivariant. Soit $\tilde{\mathcal{Y}}$ l'ensemble des suites $(r_1, \dots, r_k; \underline{Y}_1, \dots, \underline{Y}_k)$ telles que :

[-] pour $i = 1, \dots, k$, on a $r_i \in \mathbb{Z}_{(p)}$ et $\underline{Y}_i \in \underline{\mathfrak{t}}_\natural(r_i)$;

[-] la suite $(r_1, \dots, r_k; \eta(\underline{Y}_1), \dots, \eta(\underline{Y}_k))$ appartient à $\tilde{\mathcal{Z}}$.

Le groupe $W_\natural \rtimes \Gamma$ opère naturellement sur $\tilde{\mathcal{Y}}$, et l'on pose $\mathcal{Y} = (\tilde{\mathcal{Y}}/W_\natural)^\Gamma$. L'application

$$\eta_{\tilde{\mathcal{Y}}} : \tilde{\mathcal{Y}} \to \tilde{\mathcal{Z}},\ (r_1, \dots, r_k; \underline{Y}_1, \dots, \underline{Y}_k) \mapsto (r_1, \dots, r_k; \eta(\underline{Y}_1), \dots, \eta(\underline{Y}_k))$$

est bijective (par définition) mais n'est en général pas Γ-équivariante. En revanche, l'application $\tilde{\mathcal{Y}}/W_\natural \to \tilde{\mathcal{Z}}/W$ qui s'en déduit est Γ-équivariante. D'où une application $\eta_{\mathcal{Y}} : \mathcal{Y} \to \mathcal{Z}$.

On définit comme suit une application $\epsilon : \mathcal{Y} \to \mathcal{Z}_\natural$ avec $\mathcal{Z}_\natural = \mathcal{Z}_{\bullet,\natural} \cup \{z_{\infty,\natural}\}$ (cf. 10.2). Soit un élément $\tilde{y} = (r_1, \dots, r_k; \underline{Y}_1, \dots, \underline{Y}_k) \in \tilde{\mathcal{Y}}$. Si $k = 0$, on pose $\tilde{\epsilon}(\tilde{y}) = z_{\infty,\natural}$. Supposons $k \geq 1$. Pour $i = 1, \dots, k$, posons $\Sigma_\natural(i) = \{\alpha \in \Sigma_\natural : \alpha(\underline{Y}_j) = 0, \forall j \in \{1, \dots, i\}\}$; c'est un sous-système de racines de $\Sigma_\natural$. On a les inclusions $\Sigma_\natural \supset \Sigma_\natural(1) \supset \cdots \supset \Sigma_\natural(k) = \emptyset$. Considérons l'ensemble des $i \in \{2, \dots, k\}$ tels que $\Sigma_\natural(i-1) \supsetneq \Sigma_\natural(i)$. Notons E cet ensemble, et posons $E \cup \{1\} = \{i_1, \ldots, i_{k'}\}$ avec $i_1 < i_2 < \cdots < i_{k'}$. On a donc $i_1 = 1$. Pour $j = 1, \ldots, k'$, posons $\Sigma_{\natural,j} = \Sigma_\natural(i_j)$. Pour $j \in \{1, \ldots, k'\}$, on a la décomposition (cf. 10.4) $\mathfrak{t}_\natural^{\mathrm{mod}} = \mathfrak{t}_{\natural,\Sigma_{j,\natural}-\mathrm{cent}}^{\mathrm{mod}} \oplus \mathfrak{t}_{\natural,\Sigma_{j,\natural}-\mathrm{der}}^{\mathrm{mod}}$; et pour $r \in \mathbb{Z}_{(p)}$, on en déduit une décomposition $\underline{\mathfrak{t}}_\natural(r) = \underline{\mathfrak{t}}_{\natural,\Sigma_{j,\natural}-\mathrm{cent}}(r) \oplus \underline{\mathfrak{t}}_{\natural,\Sigma_{j,\natural}-\mathrm{der}}(r)$. Posons $\underline{Z}_{1,\natural} = \underline{Y}_1$, et pour $j = 2, \ldots, k'$, notons $\underline{Z}_{j,\natural}$ la projection de $\underline{Y}_j$ sur $\underline{\mathfrak{t}}_{\natural,\Sigma_{j,\natural}-\mathrm{der}}(r_{i_j})$. Par construction, la famille $(r_1, \ldots, r_{i_{k'}}; \underline{Z}_1, \ldots, \underline{Z}_{i_{k'}})$ appartient à $\tilde{\mathcal{Z}}_{\bullet,\natural}$ [W2, 8.2] ; on la note $\tilde{\epsilon}(\tilde{y})$. L'application $\tilde{\epsilon} : \tilde{\mathcal{Y}} \to \tilde{\mathcal{Z}}_\natural$ ainsi définie est $W \rtimes \Gamma$-équivariante. Elle induit donc une application $\epsilon : \mathcal{Y} \to \mathcal{Z}_\natural$.

Soit $y \in \mathcal{Y}$. Posons $z = \eta_{\mathcal{Y}}(y) \in \mathcal{Z}$ et $z_\natural = \epsilon(y) \in \mathcal{Z}_\natural$. On a une identification canonique $D_{z_\natural} = D_z$. Si $z = z_\infty$, alors $z_\natural = z_{\infty,\natural}$ et l'identification $(\mathrm{X}_{*,\natural})_{\Gamma,\mathrm{tor}} = (\mathrm{X}_*)_{\Gamma,\mathrm{tor}}$ est celle induite par η. Supposons donc $z \neq z_\natural$, et choisissons une relèvement $\tilde{y}$ de y dans $\tilde{\mathcal{Y}}$. Posons $\tilde{z} = \eta_{\tilde{\mathcal{Y}}}(\tilde{y})$ et $\tilde{z}_\natural = \tilde{\epsilon}(\tilde{y})$. Pour $\gamma \in \Gamma$, il existe un $w_{\tilde{y}}(\gamma) \in W_\natural$ tel que $w_{\tilde{y}}(\gamma) \circ \gamma(\tilde{y}) = \tilde{y}$. Puisque l'application $\tilde{\epsilon}$ est $W \rtimes \Gamma$-équivariante, on a $w_{\tilde{y}}(\gamma) = w_{\tilde{z}_\natural}(\gamma)$ $(\gamma \in \Gamma)$. D'autre part, pour $\gamma \in \Gamma$, il existe un $w_{\eta_{\tilde{\mathcal{Y}}}}(\gamma) \in W$ tel que $\eta_{\tilde{\mathcal{Y}}} \circ \gamma = w_{\eta_{\tilde{\mathcal{Y}}}}(\gamma) \circ \gamma \circ \eta_{\tilde{\mathcal{Y}}}$. On en déduit que $w_{\tilde{z}}(\gamma) = \eta(w_{\tilde{y}}(\gamma)) w_{\eta_{\tilde{\mathcal{Y}}}}(\gamma)$ $(\gamma \in \Gamma)$. Par conséquent l'application ${}_{\tilde{z}}\varphi^{-1} \circ \eta \circ {}_{\tilde{z}_\natural}\varphi_\natural : {}_{\tilde{z}_\natural}\mathrm{X}_{*,\natural} \to {}_z\mathrm{X}_*$ est Γ-équivariante ; où ${}_{\tilde{z}_\natural}\varphi_\natural : {}_{\tilde{z}_\natural}\mathrm{X}_{*,\natural} \to \mathrm{X}_{*,\natural}$ est défini comme en 10.2. Elle induit donc un isomorphisme $D_{\tilde{z}_\natural} \to D_{\tilde{z}}$, qui est compatible aux morphismes de transition (cf. 10.2) ; d'où un isomorphisme $D_{z_\natural} \to D_z$. On verifie que ce dernier ne dépend pas du choix du relèvement $\tilde{y}$.

Soit $z_\natural \in \mathcal{Z}_\natural$. À la paire $(s, z_\natural)$, on associe comme suit un homomorphisme $s_{z_\natural} : D_{z_\natural} \to \mathbb{C}^\times$. On choisit un relèvement $\tilde{z}_\natural$ de $z_\natural$ dans $\tilde{\mathcal{Z}}_\natural$, et l'on note ${}_{\tilde{z}_\natural}\mathrm{X}_\natural^*$ le Γ-module dual de ${}_{\tilde{z}_\natural}\mathrm{X}_{*,\natural}$. De l'isomorphisme ${}_{\tilde{z}_\natural}\varphi_\natural : {}_{\tilde{z}_\natural}\mathrm{X}_{*,\natural} \to \mathrm{X}_{*,\natural}$ se déduit un isomorphisme ${}_{\tilde{z}_\natural}\hat{\varphi}_\natural : \hat{T}_\natural \to {}_{\tilde{z}_\natural}\mathrm{X}_\natural^* \otimes_{\mathbb{Z}} \mathbb{C}^\times$. La restriction de ce dernier à $\hat{Z}_\natural$ est Γ-équivariant. En particulier, on a ${}_{\tilde{z}_\natural}\hat{\varphi}_\natural(s) \in [{}_{\tilde{z}_\natural}\mathrm{X}_\natural^* \otimes_{\mathbb{Z}} \mathbb{C}^\times]^\Gamma = \mathrm{Hom}(D_{\tilde{z}_\natural}, \mathbb{C}^\times)$. Si maintenant $\tilde{z}'_\natural$ est un autre relèvement de $z_\natural$ dans $\tilde{\mathcal{Z}}_\natural$, à nouveau parceque $s \in \hat{Z}_\natural$, on a ${}_{\tilde{z}_\natural}\hat{\varphi}_\natural(s) = {}_{\tilde{z}'_\natural}\hat{\varphi}_\natural(s) \circ (D_{\tilde{z}_\natural} \to D_{\tilde{z}'_\natural})$ où $D_{\tilde{z}_\natural} \to D_{\tilde{z}'_\natural}$ est l'isomorphisme défini en 10.2 (i.e. le morphisme de transition). D'où un homorphisme $D_{z_\natural} \to \mathbb{C}^\times$, que l'on note $s_{z_\natural}$.

Pour $y \in \mathcal{Y}$, on pose $s_y = s_{\epsilon(y)} \in \mathrm{Hom}(D_{\epsilon(y)}, \mathbb{C}^\times) = \mathrm{Hom}(D_{\eta_{\mathcal{Y}}(y)}, \mathbb{C}^\times)$.

11.3. Soit $F \in \mathrm{CL}_q$. De η se déduit un isomorphisme $\mathfrak{t}_\natural^{\mathrm{mod}} \to \mathfrak{t}^{\mathrm{mod}}$. Notons $\mathfrak{t}_{\natural,\mathcal{D}-\mathrm{reg}}^{\mathrm{mod}}$ l'image réciproque de $\mathfrak{t}_{\mathrm{reg}}^{\mathrm{mod}}$ par cet isomomorphisme ; c'est un sous-ensemble $W_\natural \rtimes \Gamma$-stable de $\mathfrak{t}_{\natural,\mathrm{reg}}^{\mathrm{mod}}$. Posons $\mathcal{Z}_{\natural,\mathcal{D}-\mathrm{reg},F} = (\mathfrak{t}_{\natural,\mathcal{D}-\mathrm{reg}}^{\mathrm{mod}}/W_\natural)^\Gamma$. L'application $\tilde{\eta}_F : \mathfrak{t}_{\natural,\mathcal{D}-\mathrm{reg}}^{\mathrm{mod}} \to \mathfrak{t}_{\mathrm{reg}}^{\mathrm{mod}}$ n'est en général pas Γ-équivariante, mais l'application $\mathfrak{t}_{\natural,\mathcal{D}-\mathrm{reg},F}^{\mathrm{mod}}/W_\natural \to \mathfrak{t}_{\mathrm{reg}}^{\mathrm{mod}}/W$ qui s'en déduit l'est. D'où une application $\eta_F : \mathcal{Z}_{\natural,\mathcal{D}-\mathrm{reg},F} \to \mathcal{Z}_F$.

L'application $\tilde{\tau}_F = \eta_{\tilde{\mathcal{Y}}}^{-1} \circ \tilde{\zeta}_F \circ \tilde{\eta}_F : \mathfrak{t}_{\natural,\mathcal{D}-\mathrm{reg},F}^{\mathrm{mod}} \to \tilde{\mathcal{Y}}$ est $W_\natural \rtimes \Gamma$-équivariante. Elle induit donc une application $\tau_F : \mathcal{Z}_{\natural,\mathcal{D}-\mathrm{reg},F} \to \mathcal{Y}$. D'après le lemme 8.3.1

de [W2], les diagrammes

$$\begin{array}{ccc} \mathcal{Z}_{\natural,\mathcal{D}-\mathrm{reg},F} & \xrightarrow{\tau_F} & \mathcal{Y} \\ \downarrow \eta_F & & \downarrow \eta_{\mathcal{Y}} \\ \mathcal{Z}_F & \xrightarrow{\zeta_F} & \mathcal{Z} \end{array}, \qquad \begin{array}{ccc} \mathcal{Z}_{\natural,\mathcal{D}-\mathrm{reg},F} & \xrightarrow{\tau_F} & \mathcal{Y} \\ \cap & & \downarrow \epsilon \\ \mathcal{Z}_{\natural,F} & \xrightarrow{\zeta_{\natural,F}} & \mathcal{Z}_\natural \end{array}$$

sont commutatifs (la commutativité du diagramme de gauche est immédiate). Et de la surjectivité de l'application ζ_F, on déduit que l'application τ_F est surjective [W2, lemme 8.3.2].

Soit $z_{\natural,F} \in \mathcal{Z}_{\natural,\mathcal{D}-\mathrm{reg},F}$, et posons $z_F = \eta_F(z_{\natural,F})$. À $z_{\natural,F}$ est associée comme en 10.5 une classe de conjugaison stable $\mathcal{O}_{\mathrm{st}}(z_{\natural,F}) \subset {}_{D_\natural}\mathfrak{g}_\natural(F)_{\mathrm{reg}}$. Soient $Y \in \mathcal{O}_{\mathrm{st}}(z_{\natural,F}) \cap \mathfrak{g}_\natural(F)$ et $d \in D$. Langlands et Shelstad ont défini un *facteur de transfert* $\Delta_{{}_dG(F),G_\natural(F)}(Y,\cdot)$, qui est une fonction sur ${}_d\mathfrak{g}(F)$ à support dans $\mathcal{O}_{\mathrm{st}}(z_F) \cap {}_d\mathfrak{g}(F)$. Plus précisément :

[-] Langlands et Shelstad ont défini le facteur de transfert pour les groupes [LS]. Il se descend aux algèbres de Lie (il y est d'ailleurs beaucoup plus simple que sur les groupes).

[-] Le facteur Δ_{IV} est supprimé ici puisqu'il a été incorporé dans la définition des intégrales orbitales.

[-] Le facteur de transfert n'est en général défini qu'à une constante près. Il est canoniquement défini dans le cas des groupes quasi-déployés moyennant le choix d'épinglages. Cette définition canonique s'étend à notre situation, puisque pour chaque $d \in D$, on dispose d'un cocycle $d_F \in \mathrm{Z}^1(\Theta, N_{\mathcal{C}}^{\mathrm{nr}})$ définissant le torseur intérieur ${}_d\xi : {}_dG \to G$.

On note simplement $\Delta(z_{\natural,F},\cdot)$ la fonction sur $\mathcal{O}_{\mathrm{st}}(z_F)$ telle que pour chaque $d \in D$, la restriction de $\Delta(z_{\natural,F},\cdot)$ à $\mathcal{O}_{\mathrm{st}}(z_F) \cap {}_d\mathfrak{g}(F)$ coïncide avec $\Delta_{{}_dG(F),G_\natural(F)}(Y,\cdot)$. Posons $y = \tau_F(z_{\natural,F})$ et $z = \zeta_F(z_F) = \eta_{\mathcal{Y}}(y)$. Alors le facteur de transfert $\Delta(z_{\natural,F},\cdot)$ est donné par la formule suivante :

$$\Delta(z_{\natural,F}, X) = s_y(\delta_{z_F}(X)^{-1}) \qquad (X \in \mathcal{O}_{\mathrm{st}}(z_F)).$$

11.4. Pour $F \in \mathrm{CL}_q$, $z_{\natural,F} \in \mathcal{Z}_{\natural,\mathcal{D}-\mathrm{reg},F}$ et $f \in C_{\mathrm{c}}^\infty({}_D\mathfrak{g}(F))$, on définit l'*intégrale orbitale endoscopique* :

$$\Lambda^{{}_DG(F),G_\natural(F)}(f, z_{\natural,F}) = \sum_X \Delta(z_{\natural,F}, X)\Lambda^{G_\natural(F)}(f, X)$$

où X parcourt un système de représentants de $\mathrm{cl}(\eta_F(z_{\natural,F}))$ dans ${}_D\mathfrak{g}(F)$.

PROPOSITION [W2, proposition 9.1]. — *Soit $y \in \mathcal{Y}$. Il existe une forme linéaire $\Lambda^{\mathcal{D},\mathcal{D}_\natural}(\cdot, y)$ sur $\mathcal{S}$ telle que pour tous $\varphi \in \mathcal{S}$, $F \in \mathrm{CL}_q$, et $z_{\natural,F} \in \mathcal{Z}_{\natural,\mathcal{D}-\mathrm{reg},F}$ tel que $\tau_F(z_{\natural,F}) = y$, on ait l'égalité*

$$\Lambda^{{}_DG(F),G_\natural(F)}(\mathrm{rea}_F(\varphi), z_{\natural,F}) = \Lambda^{\mathcal{D},\mathcal{D}_\natural}(\varphi, y).$$

Démonstration : Soit $z = \eta_{\mathcal{Y}}(y)$, et posons

$$\Lambda^{\mathcal{D},\mathcal{D}_\natural}(\cdot, y) = \sum_{\delta \in D_z} s_y(\delta^{-1})\Lambda^{\mathcal{D}}(\cdot, z, \delta).$$

D'après le théorème 10.7 et la formule pour le facteur de transfert donnée en 11.3, cette forme linéaire convient. □

Le triplet $(\mathcal{D}_\natural, \mathrm{id}, s)$ est une donnée endoscopique de $\mathcal{D}_\natural$. On peut donc, pour $F \in \mathrm{CL}_q$, $z_{\natural,F} \in \mathcal{Z}_{\natural,F}$ et $f_\natural \in C_{\mathrm{c}}^\infty({}_{D_\natural}\mathfrak{g}_\natural(F))$, définir comme ci-dessus l'intégrale orbitale endoscopique $\Lambda^{{}_{D_\natural}G_\natural(F),G_\natural(F)}(f_\natural, z_{\natural,F})$. Notons que la forme linéaire $\Lambda^{{}_{D_\natural}G_\natural(F),G_\natural(F)}(\cdot, z_{\natural,F})$ sur $C_{\mathrm{c}}^\infty({}_{D_\natural}\mathfrak{g}_\natural(F))$ dépend de l'élément s (même si ce dernier n'apparaît pas dans la notation). En fait, remplacer s par 1 revient à multiplier cette forme linéaire par un scalaire non nul sur chaque composante $C_{\mathrm{c}}^\infty({}_{d_\natural}G_\natural(F))$ pour $d_\natural \in D_\natural$.

Soient $f \in C_{\mathrm{c}}^\infty({}_D\mathfrak{g}(F))$ et $f_\natural \in C_{\mathrm{c}}^\infty({}_{D_\natural}\mathfrak{g}_\natural(F))$. On dit que $f_\natural$ *est un transfert de* f si pour tout $z_{\natural,F} \in \mathcal{Z}_{\natural,\mathcal{D}-\mathrm{reg},F}$, on a l'égalité $\Lambda^{{}_{D_\natural}G_\natural(F),G_\natural(F)}(f_\natural, z_{\natural,F}) = \Lambda^{{}_DG(F),G_\natural(F)}(f, z_{\natural,F})$.

COROLLAIRE [W2, corollaire 9.2]. — *Soient* $\varphi \in \mathcal{S}$, $\varphi_\natural \in \mathcal{S}_\natural$, *et* $F, F' \in \mathrm{CL}_q$. *Alors* $\mathrm{rea}_{\natural,F}(\varphi_\natural)$ *est un transfert de* $\mathrm{rea}_F(\varphi)$ *si et seulement si* $\mathrm{rea}_{\natural,F'}(\varphi_\natural)$ *est un transfert de* $\mathrm{rea}_{F'}(\varphi)$.

Démonstration : Remplaçons $\mathcal{D}$ par $\mathcal{D}_\natural$ dans les constructions précédentes : on a $\mathcal{Z}_{\natural,\mathcal{D}_\natural-\mathrm{reg},F} = \mathcal{Z}_{\natural,F}$, $\mathcal{Y}_\natural = \mathcal{Z}_\natural$, et l'application $\tau_{\natural,F} : \mathcal{Z}_{\natural,F} \to \mathcal{Z}_\natural$ coïncide avec $\zeta_{\natural,F}$. Pour $z_\natural \in \mathcal{Z}_\natural$, on dispose donc d'une forme linéaire $\Lambda^{\mathcal{D}_\natural,\mathcal{D}_\natural}(\cdot, z_\natural)$ sur $\mathcal{S}_\natural$. Soit $z_{\natural,F} \in \mathcal{Z}_{\natural,\mathcal{D}-\mathrm{reg},F}$, et posons $y = \tau_F(z_{\natural,F})$. On a les égalités

$$\Lambda^{{}_DG(F),G_\natural(F)}(\mathrm{rea}_F(\varphi), z_{\natural,F}) = \Lambda^{\mathcal{D},\mathcal{D}_\natural}(\varphi, y)$$

et

$$\Lambda^{{}_{D_\natural}G_\natural(F),G_\natural(F)}(\mathrm{rea}_F(\varphi_\natural), z_{\natural,F}) = \Lambda^{\mathcal{D}_\natural,\mathcal{D}_\natural}(\varphi_\natural, \zeta_{\natural,F}(z_{\natural,F}))$$

avec $\zeta_{\natural,F}(z_{\natural,F}) = \epsilon(y)$ (cf. 11.3). Puisque l'application τ_F est surjective, $\mathrm{rea}_{\natural,F}(\varphi_\natural)$ est un transfert de $\mathrm{rea}_F(\varphi)$ si et seulement si pour tout $y \in \mathcal{Y}$, on a l'égalité $\Lambda^{\mathcal{D}_\natural,\mathcal{D}_\natural}(\varphi_\natural, \epsilon(y)) = \Lambda^{\mathcal{D},\mathcal{D}_\natural}(\varphi, y)$. D'où le corollaire, puisque cette seconde condition ne dépend pas de F. □

11.5. Supposons les données $\mathcal{D}$ et $\mathcal{D}_\natural$ non ramifiées, c'est-à-dire telles que I opère trivialement. Le groupe $(\mathrm{X}_*/\mathrm{X}_{*,\mathrm{sc}})_{\Gamma,\mathrm{tor}}$ contient l'élément 0, représenté par le cocycle trivial $d_0 \in D$. Le point $v = 0$ appartient à ${}_{d_0}W$, et est hyperspécial. Notons $\phi_0 \in {}_{d_0}\Phi$ la facette contenant $(0,0)$, et $\varphi_0 \in {}_{d_0}\mathcal{S}_{\phi_0}$ la fonction constante égale à $|{}_{d_0}\underline{G}_0^\Theta|^{-1} q^{\frac{1}{2}\dim({}_{d_0}\underline{G}_0)}$ sur ${}_{d_0}\underline{\mathfrak{g}}_{\phi_0}^\Theta$. Pour $F \in \mathrm{CL}_q$, on pose $f_{F,0} = \mathrm{rea}_F(\varphi_0)$; c'est la fonction caractéristique du réseau hyperspécial ${}_{d_0}\mathfrak{g}(F)_{\phi_0} \subset {}_{d_0}\mathfrak{g}(F)$ multipliée par une constante > 0. On considère $f_{F,0}$ comme un élément de ${}_D\mathfrak{g}(F)$ dont la restriction à ${}_d\mathfrak{g}(F)$ pour $d \neq d_0$, est nulle. On définit de la même manière la fonction $\varphi_{\natural,0} \in {}_{d_{\natural,0}}\mathcal{S}_{\natural,\phi_{\natural,0}}$, et pour $F \in \mathrm{CL}_q$, la fonction $f_{\natural,F,0} \in C_{\mathrm{c}}^\infty({}_{D_\natural}\mathfrak{g}_\natural(F))$.

Pour $F \in \mathrm{CL}_q$, le “lemme fondamental pour les algèbres de Lie” est l'assertion suivante : $f_{\natural,F,0}$ est un transfert de $f_{F,0}$. En appliquant le corollaire 11.4 au couple $(\varphi_{\natural,F,0}, \varphi_{F,0})$, on obtient le

COROLLAIRE [W2, corollaire 9.3]. — *Soient F, $F' \in \mathrm{CL}_q$. Alors le lemme fondamental pour les algèbres de Lie est vrai pour F si et seulement s'il est vrai pour F'.*

Références

[D] S. DeBacker, Parametrizing nilpotent orbits via Bruhat-Tits theory, *Annals of Math.* **156** (2002), pp. 295–332.

[KM] J. Kim & F. Murnaghan, Character expansions and unrefined minimal K-types, *Amer. J. Math.* **125** (2003), pp. 409–481.

[K1] R. Kottwitz R., *Tamagawa numbers*, Annals Math. **127** (1988), pp. 629–646.

[K2] R. Kottwitz, Isocrystals with additional structures II, *Compositio Math.* **109** (1997), pp. 255–339.

[L] B. Lemaire, Sous-groupes parahoriques d'un groupe réductif p-adique et descente galoisienne ramifiée, manuscrit.

[N] Ngô Bao Châu, Le lemme fondamental pour les algèbres de Lie, prépublication arXiv : 0801.0446v3.

[LS] R. Langlands & D. Shelstad, On the definition of transfert factors, *Math. Ann.* **278** (1987), pp. 219–271.

[Sp] T. Springer, Linear algebraic groups, *Progress in Math.* **9**, Birkhauser, 1981.

[W1] J.-L. Waldspurger, Le lemme fondamental implique le transfert, *Compositio Math.* **105** (1997), pp. 153–236.

[W2] J.-L. Waldspurger, Endoscopie et changement de caractéristique, *Journal Inst. Math. Jussieu.* **5** (2006), pp. 423–525.

[W3] J.-L. Waldspurger, L'endoscopie tordue n'est pas si tordue, *Memoirs Amer. Math. Soc.* **194**, n° 908 (2008), pp. 1–261.

Institut de Mathématiques de Luminy et UMR 6206 du CNRS, Université Aix-Marseille II, Case Postale 907, 163 Av. de Luminy, 13288 Marseille Cedex 9, France

E-mail address: `lemaire@iml.univ-mrs.fr`

Transfer principle for the fundamental lemma
[CHL.II.E]

Raf Cluckers, Thomas Hales, and François Loeser

Introduction

The purpose of this paper is to explain how the general transfer principle of Cluckers and Loeser [**9**][**11**] may be used in the study of the fundamental lemma. We use here the word "transfer" in a sense originating with the Ax-Kochen-Eršov transfer principle in logic. Transfer principles in model theory are results that transfer theorems from one field to another. The transfer principle of [**9**][**11**] is a general result that transfers theorems about identities of p-adic integrals from one collection of fields to others. These general transfer principles are reviewed in Theorems 2.6 and 10.3. The main purpose of this article is to explain how the identities of various fundamental lemmas fall within the scope of these general transfer principles. Consequently, once the fundamental lemma has been established for one collection of fields (for example, fields of positive characteristic), it is also valid for others (fields of characteristic zero). Precise statements appear in Theorems 9.1 (for the fundamental lemma), 9.2 (for the weighted fundamental lemma), and Section 10.3 (for the Jacquet-Ye relative fundamental lemma).

In general terms, the aim of this paper is to show that certain theorems about fields in positive characteristic imply that the same theorems hold in characteristic zero. To make this aim precise, we need to state the theorems in such a way that the field is not fixed from the outset, but appears as a parameter that can be supplied at a later stage. Stated in these terms, it is natural to turn to model theory for an answer, because model theory gives a separation of language from structure; a first-order language expresses the theorems, and structures later supply the fields. The particular form of each

During the preparation of this paper, R. Cluckers was a postdoctoral fellow of the Fund for Scientific Research, Flanders (Belgium) (F.W.O.).

The research of T. Hales was supported in part by NSF grant 0503447. He would also like to thank E.N.S. for its hospitality.

The project work of F. Loeser was partially supported by ANR grant 06-BLAN-0183.

theorem is an identity of integrals. Thus, a model theoretic account of p-adic integration is needed. This is precisely what motivic integration provides.

In an unfortunate clash of terminology, the word "transfer" in the context of the fundamental lemma has come to mean the matching of smooth functions on a reductive group with those on an endoscopic group. We have nothing to say about transfer in that sense. For example, Waldspurger's 1997 article "Le lemme fondamental implique le transfert" is completely unrelated (insofar as it is possible for two articles on the fundamental lemma to be unrelated).

The intended audience for this paper being that of mathematicians working in the areas of automorphic forms, representation theory, arithmetic geometry, and Galois representations, we tried our best to make all definitions and statements from other fields that are used in this paper understandable without any prerequisite. In particular, we start the paper by giving a quick presentation of first-order languages and the Denef-Pas language and an overview on motivic constructible functions and their integration according to [**8**], before stating the general transfer principle. The bulk of the paper consists in proving the definability of the various data occurring in the fundamental lemma. Once this is achieved, it is not difficult to deduce our main result in 9.3, stating that the transfer principle holds for the integrals occurring in the fundamental lemma, which is of special interest in view of the recent advances by Laumon and Ngô [**28**] and Ngô [**30**].

Other results concerning the transfer principle for the fundamental lemma appear in [**12**], [**35**], [**36**].

We thank Michael Harris for inviting us to write an expository paper on this topic.

1. First order languages and the Denef-Pas language

1.1. Languages. A (first order) language L consists of an enumerable infinite set of symbols of variables $\mathcal{V} = \{v_0, \cdots, v_n, \cdots\}$, logical symbols $\neg$ (negation), $\wedge$ (or), $\vee$ (and), $\Longrightarrow$, $\Longleftrightarrow$, $\forall$ and $\exists$, together with two suites of sets $\mathcal{F}_n$ and $\mathcal{R}_n$, $n \in \mathbb{N}$. Elements of $\mathcal{F}_n$ will be symbols of n-ary functions, elements of $\mathcal{R}_n$ symbols of n-ary relations. A 0-ary function symbol will be called a constant symbol. The language L consists of the union of these sets of symbols.

1.2. Terms. The set $T(L)$ of terms of the language L is defined in the following way: variable symbols and constant symbols are terms and if f belongs to $\mathcal{F}_n$ and $t_1, \ldots, t_n$ are terms, then $f(t_1, \cdots, t_n)$ is also a term. A more formal definition of $T(L)$ is to view it as a subset of the set of finite words on L. For instance, to the word $ft_1 \ldots t_n$ corresponds the term $f(t_1, \cdots, t_n)$. One defines the weight of a function as its arity -1 and we give to symbols of variables the weight -1. The weight of a finite word on L is the sum of the weights of its symbols. Then terms correspond exactly to finite words on variable and function symbols of total weight -1 whose strict initial segments are of weight ≥ 0, when nonempty. If t is a term one writes $t = t[w_0, \ldots, w_n]$ to mean that all variables occurring in t belong to the w_i's.

1.3. Formulas. An atomic formula is an expression of the form $R(t_1, \ldots, t_n)$ with R an n-ary relation symbol and t_i terms. The set of formulas in L is the smallest set containing atomic formulas and such that if M and N are formulas then $\neg M$, $(M \wedge N)$, $(M \vee N)$, $(M \implies N)$, $(M \iff N)$, $\forall v_n M$ et $\exists v_n M$ are formulas. Formulas may also be defined as certain finite words on L. (Parentheses are just a way to rewrite terms and formulas in a more handy way, as opposed to writing them as finite words on L).

Let v be a variable symbol occurring in a formula F. If F is atomic we say all occurrences of v in F are free. If $F = \neg G$ the free occurrences of v in F are those in G. Free occurrences of v in $(F \alpha G)$ are those in F and those in G where α is either $\neg$, $\wedge$, $\vee$, $\implies$, or $\iff$. If $F = \forall w G$ or $\exists w G$ with $w \neq v$, free occurrences of v in F are those in G. When $v = w$, no occurrence of v in F free. Non free occurrences of a variable are said to be bound. Free variables in a formula F are those having at least one free occurrence. A sentence is a formula with no free variable.

We write $F[w_0, \ldots, w_n]$ if all free variables in F belong to the w_i's (supposed to be distinct).

1.4. Interpretation in a structure. Let L be a language. An L-structure $\mathfrak{M}$ is a set M endowed for every n-ary function symbol f in L with a function $f^{\mathfrak{M}} : M^n \to M$ and for every n-ary relation R with a subset $R^{\mathfrak{M}}$ of M^n.

If $t[w_0, \ldots, w_n]$ is a term and $a_0, \ldots, a_n$ belong to M, we denote by $t[a_0, \ldots, a_n]$ the interpretation of t in M defined by interpreting w_i by a_i. Namely, the interpretation of the term w_i is a_i, that of the constant symbol c is $c^{\mathfrak{M}}$, and that of the term $f(t_1, \ldots t_r)$ is $f^{\mathfrak{M}}(t_1[a_0, \ldots, a_n], \ldots, t_r[a_0, \ldots, a_n])$.

Similarly, if $F[w_1, \ldots, w_n]$ is a formula and $a_1, \ldots, a_n$ belong to M, there is a natural way to interpret w_i as a_i in M in the formula F, yielding a statement $F[a_1, \ldots, a_n]$ about the tuple $(a_1, \ldots, a_n)$ in M which is either true or false. One says that $(a_1, \ldots, a_n)$ satisfies F in $\mathfrak{M}$, and writes $\mathfrak{M} \models F[a_1, \ldots, a_n]$, if the statement $F[a_1, \ldots, a_n]$ obtained by interpreting w_i as a_i is satisfied (true) in M. For instance, when $F = R(t_1, \ldots, t_r)$, then $\mathfrak{M} \models F[a_1, \ldots, a_n]$ if and only if $R^{\mathfrak{M}}(t_1[a_1, \ldots, a_n], \ldots, t_r[a_1, \ldots, a_n])$ holds, and for a formula $F[w_0, \ldots, w_n]$, one has $\mathfrak{M} \models (\forall w_0 F)[a_1, \ldots, a_n]$, if and only if for every a in M, $\mathfrak{M} \models F[a, a_1, \ldots, a_n]$.

When the language contains the binary relation symbol of equality $=$, one usually assumes L-structures to be equalitarian, that is, that the relation $=^{\mathfrak{M}}$ coincides with the equality relation on the set M. From now on we shall denote in the same way symbols f and R and their interpretation $f^{\mathfrak{M}}$ and

$R^{\mathfrak{M}}$. In particular we may identify constant symbols and their interpretation in M which are elements of M.

We shall also be lax with the names of variables and allow other names, like x_i, x, y.

1.5. Some examples. Let us give some examples of languages we shall use in this paper. It is enough to give the symbols which are not variables nor logical.

For the language of abelian groups these symbols consist of the constant symbol 0, the two binary function symbols $+$, $-$ and equality. The language of ordered abelian groups is obtained by adding a binary relation symbol $<$, the language of rings by adding symbols 1 and $\cdot$ (with the obvious arity). Hence a structure for the ring language is just a set with interpretations for the symbols 0, 1, $+$, $-$ and $\cdot$. This set does not have to be a ring (but it will be in all cases we shall consider).

If S is a set, then by the ring language with coefficients in S, we mean that we add S to the set of constants in the language. For instance any ring containing S will be a structure for that language.

Note that there is a sentence φ in the ring language such that a structure $\mathfrak{M}$ satisfies φ if and only if $\mathfrak{M}$ is a field, namely the conjunction of the field axioms (there is a finite number of such axioms, each expressible by a sentence in the ring language). On the other hand, one can show there is no sentence in the ring language expressing for a field to be algebraically closed. Of course, given a natural number $n > 0$, there is a sentence expressing that every degree n polynomial has a root namely

$$\forall a_0 \forall a_1 \cdots \forall a_n \exists x (a_0 = 0 \vee a_0 x^n + a_1 x^{n-1} + \cdots + a_n = 0). \tag{1.5.1}$$

Note that here x^i is an abbreviation for $x \cdot x \cdot x \cdots x$ (i times). It is important to notice that we are not allowed to quantify over n here, since it does not correspond to a variable in the structure we are considering.

1.6. The Denef-Pas language. We shall need a slight generalization of the notion of language, that of many sorted languages (in fact 3-sorted language). In a 3-sorted language we have 3 sorts of variables symbols, and for relation and function symbols one should specify the type of the variables involved and for functions also the type of the value of f. A structure for a 3-sorted language will consist of 3 sets M_1, M_2 and M_3 together with interpretations of the non logical, non variable symbols. For instance if f is a binary function symbol, with first variable of type 2, second variable of type 3, and value of type 3, its interpretation will be a function $M_2 \times M_3 \to M_3$.

Let us fix a field k of characteristic 0 and consider the following 3-sorted language, the Denef-Pas language $\mathcal{L}_{\mathrm{DP}}$. The 3 sorts are respectively called the valued field sort, the residue field sort, and the value group sort. The language will consist of the disjoint union of the language of rings with coefficients in $k((t))$ restricted to the valued field sort, of the language of rings with coefficients in k restricted to the residue field sort and of the language of ordered groups restricted to the value group sort, together with two additional symbols of unary functions $\overline{\mathrm{ac}}$ and ord from the valued field sort to the residue field and valued groups sort, respectively. [In fact, the definition we give here is different from that in [**8**], where for the value group sort in $\mathcal{L}_{\mathrm{DP}}$ symbols $\equiv_n$ for equivalence relation modulo n, $n > 1$ in $\mathbb{N}$, are added, but since this does not change the category of definable objects, this change has no consequence on our statements.]

An example of an $\mathcal{L}_{\mathrm{DP}}$-structure is $(k((t)), k, \mathbb{Z})$ with $\overline{\mathrm{ac}}$ interpreted as the function $\overline{\mathrm{ac}} : k((t)) \to k$ assigning to a series its first nonzero coefficient if not zero, zero otherwise, and ord interpreted as the valuation function $\mathrm{ord} : k((t)) \setminus \{0\} \to \mathbb{Z}$. (There is a minor divergence here, easily fixed, since $\mathrm{ord}\, 0$ is not defined.) More generally, for any field K containing k, $(K((t)), K, \mathbb{Z})$ is naturally an $\mathcal{L}_{\mathrm{DP}}$-structure. For instance $\overline{\mathrm{ac}}(x^2 + (1 + t^3)y) - 5z^3$ and $\mathrm{ord}(x^2 + (1 + t^3)y) - 2w + 1$ are terms in $\mathcal{L}_{\mathrm{DP}}$, $\forall x \exists z \neg(\overline{\mathrm{ac}}(x^2 + (1 + t^3)y) - 5z^3 = 0)$ and $\forall x \exists w (\mathrm{ord}(x^2 + (1 + t^3)y) = 2w + 1)$ are formulas.

2. Integration of constructible motivic functions

2.1. The category of definable objects. Let φ be a formula in the language $\mathcal{L}_{\mathrm{DP}}$ having respectively m, n, and r free variables in the various sorts. To such a formula φ we assign, for every field K containing k, the subset $h_\varphi(K)$ of $K((t))^m \times K^n \times \mathbb{Z}^r$ consisting of all points satisfying φ, that is,

(2.1.1)
$$h_\varphi(K) := \Big\{(x, \xi, \eta) \in K((t))^m \times K^n \times \mathbb{Z}^r ; (K((t), K, \mathbb{Z}) \models \varphi(x, \xi, \eta)\Big\}.$$

We shall call the datum of such subsets for all K definable (sub)assignments. In analogy with algebraic geometry, where the emphasis is not put anymore on equations but on the functors they define, we consider instead of formulas the corresponding subassignments (note $K \mapsto h_\varphi(K)$ is in general not a functor).

More precisely, let $F : \mathcal{C} \to \mathrm{Set}$ be a functor from a category $\mathcal{C}$ to the category of sets. By a subassignment h of F we mean the datum, for every object C of $\mathcal{C}$, of a subset $h(C)$ of $F(C)$. Most of the standard operations of elementary set theory extend trivially to subassignments. For instance, given subassignments h and h' of the same functor, one defines subassignments $h \cup h'$, $h \cap h'$ and the relation $h \subset h'$, etc. When $h \subset h'$ we say h is a subassignment of h'. A morphism $f : h \to h'$ between subassignments of functors F_1 and F_2 consists of the datum for every object C of a map

$$f(C) : h(C) \to h'(C). \tag{2.1.2}$$

The graph of f is the subassignment

$$C \mapsto \mathrm{graph}(f(C)) \tag{2.1.3}$$

of $F_1 \times F_2$. Let k be a field and consider the category F_k of fields containing k. (To avoid any set-theoretical issues, we fix a Grothendieck universe $\mathcal{U}$ containing k and we define F_k as the small category of all fields in $\mathcal{U}$ containing k.)

We denote by $h[m, n, r]$ the functor $F_k \to \mathrm{Set}$ given by

$$h[m, n, r](K) = K((t))^m \times K^n \times \mathbb{Z}^r. \tag{2.1.4}$$

In particular, $h[0, 0, 0]$ assigns the one point set to every K. We sometimes write $\mathbb{Z}^r$ for $h[0, 0, r]$. Thus, to any formula φ in $\mathcal{L}_{\mathrm{DP}}$ having respectively m, n, and r free variables in the various sorts, corresponds a subassignment h_φ of $h[m, n, r]$ by (2.1.1). Such subassignments are called definable subassignments.

We denote by Def_k the category whose objects are definable subassignments of some $h[m, n, r]$, morphisms in Def_k being morphisms of subassignments $f : h \to h'$ with h and h' definable subassignments of $h[m, n, r]$ and $h[m', n', r']$ respectively such that the graph of f is a definable subassignment. Note that $h[0, 0, 0]$ is the final object in this category.

For example, for each $m \geq 2$, there is a definable subassignment of $h[m^2, 0, 0]$ that assigns to each field K containing k the set of elliptic elements of $GL(m, K((t)))$. This subassignment is not a functor, because there is no inclusion of elliptic elements corresponding to the inclusion of groups $GL(m, K((t))) \subset GL(m, K'((t)))$ under a general field extension K'/K.

For an example of a subassignment of $h[1, 0, 0]$ that fails to be definable, consider the functor that assigns to each field K the set of all roots of unity in $K((t))$. Another simple example of a subassignment of $h[0, 1, 0]$ that is not definable is the one that assigns to a field K the subset of K consisting of elements that are transcendental over k.

2.2. First sketch of construction of the motivic measure. The construction in [**8**] relies in an essential way on a cell decomposition theorem due to Denef and Pas [**31**]. Let us introduce the notion of cells. Fix coordinates $x = (x', z)$ on $h[m + 1, n, r]$ with x' running over $h[m, n, r]$ and z over $h[1, 0, 0]$.

A 0-cell in $h[m + 1, n, r]$ is a definable subassignment Z_A^0 defined by

$$x' \in A \quad \text{and} \quad z = c(x') \tag{2.2.1}$$

with A a definable subassignment of $h[m, n, r]$ and c a morphism $A \to h[1, 0, 0]$.

A 1-cell in $h[m + 1, n, r]$ is a definable subassignment Z_A^1 defined by

$$x' \in A, \quad \overline{\mathrm{ac}}(z - c(x')) = \xi(x') \quad \text{and} \quad \mathrm{ord}(z - c(x')) = \alpha(x') \tag{2.2.2}$$

with A a definable subassignment of $h[m, n, r]$, c, ξ and α morphisms from A to $h[1, 0, 0]$, $h[0, 1, 0] \setminus \{0\}$ and $h[0, 0, 1]$, respectively.

The Denef-Pas Cell Decomposition Theorem states that, after adding a finite number of auxiliary parameters in the residue field and value group sorts, every definable subassignment becomes a finite disjoint union of cells:

THEOREM 2.1 (Denef-Pas Cell Decomposition [**31**]). *Let A be a definable subassignment $h[m+1,n,r]$. After adding a finite number of auxiliary parameters in the residue field and value group sorts, A is a finite disjoint union of cells, that is, there exists an embedding*

$$\lambda : h[m+1,n,r] \longrightarrow h[m+1,n+n',r+r'] \tag{2.2.3}$$

such that the composition of λ with the projection to $h[m+1,n,r]$ is the identity on A and such that $\lambda(A)$ is a finite disjoint union of cells.

An example of a subassignment that can be transformed into a 1-cell by adding auxiliary parameters appears in Section 2.7.

The construction of the motivic measure $\mu(A)$ for a definable subassignment A of $h[m,n,r]$ goes roughly as follows (more details, including a description of the semi-ring in which the measure takes values, will be given in 2.6). The cell decomposition theorem expresses a definable subassignment (in $h[m,n,r]$, $m>0$) as a disjoint union of cells. The measure of a definable subassignment is defined to be the sum of the measures of its cells. In turn, a cell in $h[m,n,r]$ is expressed in terms of a definable subassignment B in $h[m-1,n',r']$ and auxiliary data. The measure of a cell can then be defined recursively in terms of the "smaller" definable subassignment B. The base case of the recursive definition is $m=0$ (with larger values of n and r). When $m=0$, one may consider the counting measure on the $\mathbb{Z}^r$-factor and the tautological measure on the $h[0,n,0]$-factor, assigning to a definable subassignment of $h[0,n,0]$ its class in $\mathrm{C}_+(\text{point})$ (a semi-ring defined in the next section). The whole point is to check that the construction is invariant under permutations of valued field coordinates. This is the more difficult part of the proof and is essentially equivalent to a form of the motivic Fubini Theorem.

2.3. Constructible functions. For X in Def_k we now define the semi-ring $\mathrm{C}_+(X)$, resp ring $\mathrm{C}(X)$, of non negative constructible motivic functions, resp. constructible motivic functions.

One considers the category Def_X whose objects are morphisms $Y \to X$ in Def_k, morphisms being morphisms $Y \to Y'$ compatible with the projections to X. We write $X[m,n,r]$ for $X \times h[m,n,r]$. Of interest to us will be the subcategory RDef_X of Def_X whose objects are definable subassignments of $X \times h[0,n,0]$, for variable n. We shall denote by $SK_0(\mathrm{RDef}_X)$ the free abelian semigroup on isomorphism classes of objects of RDef_X modulo the additivity relation

$$[Y]+[Y'] = [Y \cup Y'] + [Y \cap Y']. \tag{2.3.1}$$

It is endowed with a natural semi-ring structure. One defines similarly the Grothendieck ring $K_0(\mathrm{RDef}_X)$, which is the ring associated to the semi-ring $SK_0(\mathrm{RDef}_X)$. Proceeding this way, we only defined "half" of $\mathrm{C}_+(X)$ and $\mathrm{C}(X)$.

To get the remaining "half" one considers the ring

$$\mathbb{A} := \mathbb{Z}\Big[\mathbb{L}, \mathbb{L}^{-1}, \Big(\frac{1}{1-\mathbb{L}^{-i}}\Big)_{i>0}\Big]. \tag{2.3.2}$$

For q a real number > 1, we denote by ϑ_q the ring morphism

$$\vartheta_q : \mathbb{A} \longrightarrow \mathbb{R} \tag{2.3.3}$$

sending $\mathbb{L}$ to q and we consider the semi-ring

$$\mathbb{A}_+ := \Big\{x \in \mathbb{A} \mid \vartheta_q(x) \geq 0, \forall q > 1\Big\}. \tag{2.3.4}$$

We denote by $|X|$ the set of points of X, that is, the set of pairs (x_0, K) with K in F_k and $x_0 \in X(K)$, and we consider the subring $\mathcal{P}(X)$ of the ring of functions $|X| \to \mathbb{A}$ generated by constants in $\mathbb{A}$ and by all functions α and $\mathbb{L}^\alpha$ with $\alpha : X \to \mathbb{Z}$ definable morphisms. We define $\mathcal{P}_+(X)$ as the semi-ring of functions in $\mathcal{P}(X)$ taking their values in $\mathbb{A}_+$. These are the second "halves".

To glue the two "halves", one proceed as follows. One denotes by $\mathbb{L} - 1$ the class of the subassignment $x \neq 0$ of $X \times h[0, 1, 0]$ in $SK_0(\mathrm{RDef}_X)$, resp $K_0(\mathrm{RDef}_X)$. One considers the subring $\mathcal{P}^0(X)$ of $\mathcal{P}(X)$, resp. the subsemi-ring $\mathcal{P}^0_+(X)$ of $\mathcal{P}_+(X)$, generated by functions of the form $\mathbf{1}_Y$ with Y a definable subassignment of X (that is, $\mathbf{1}_Y$ is the characteristic function of Y), and by the constant function $\mathbb{L} - 1$. We have canonical morphisms $\mathcal{P}^0(X) \to K_0(\mathrm{RDef}_X)$ and $\mathcal{P}^0_+(X) \to SK_0(\mathrm{RDef}_X)$. We may now set

$$\mathrm{C}_+(X) = SK_0(\mathrm{RDef}_X) \otimes_{\mathcal{P}^0_+(X)} \mathcal{P}_+(X)$$

and

$$\mathrm{C}(X) = K_0(\mathrm{RDef}_X) \otimes_{\mathcal{P}^0(X)} \mathcal{P}(X).$$

There are some easy functorialities. For every morphism $f : S \to S'$, there is a natural pullback by $f^* : SK_0(\mathrm{RDef}_{S'}) \to SK_0(\mathrm{RDef}_S)$ induced by the fiber product. If $f : S \to S'$ is a morphism in $\mathrm{RDef}_{S'}$, composition with f induces a morphism $f_! : SK_0(\mathrm{RDef}_S) \to SK_0(\mathrm{RDef}_{S'})$. Similar constructions apply to K_0. If $f : S \to S'$ is a morphism in Def_k, one shows in [**8**] that the morphism f^* may naturally be extended to a morphism

$$f^* : \mathrm{C}_+(S') \longrightarrow \mathrm{C}_+(S). \tag{2.3.5}$$

If, furthermore, f is a morphism in $\mathrm{RDef}_{S'}$, one shows that the morphism $f_!$ may naturally be extended to

$$f_! : \mathrm{C}_+(S) \longrightarrow \mathrm{C}_+(S'). \tag{2.3.6}$$

Similar functorialities exist for C.

The semi-ring $\mathrm{C}_+(X)$ can be understood heuristically as follows. The right-hand side of the semi-ring consists of "raw integrands." Experience with p-adic integration has guided the choice of integrands on the right-hand side. We decided what functions we wanted to integrate and then built a semi-ring generated by precisely those functions. The variable $\mathbb{L}$ replaces the cardinality q of the residue field that appears in p-adic integrals. For example, the functions $\mathbb{L}^\alpha$ give a constructible counterpart to the common p-adic integrand consisting of the absolute value of a multivariate polynomial $|p(x_1,\ldots,x_m)|^k = q^{-k\,\mathrm{ord}(p(x_1,\ldots,x_m))}$. The left-hand side of the semi-ring (together with the ring $\mathbb{A}$) can be viewed as the "storage registers" for evaluated integrals. Imprecisely stated, integration consists of moving raw integrands from the right-hand and storing the answers on the left-hand side or in $\mathbb{A}_+$. The subsemi-ring $\mathcal{P}^0_+(X)$ and its morphisms into the two halves encode a table of integrals, matching each raw integrand with its evaluated integral. The two halves are combined into a single semi-ring to allow for multiple integrals; the value of the first integral becomes the integrand of the next integral. The denominators $(1-\mathbb{L}^{-i})$ in the semi-ring $\mathbb{A}_+$ correspond to the denominators $(1-q^{-i})$ that frequently appear in p-adic integrals when geometric series are summed.

2.4. Taking care of dimensions. In fact, we shall need to consider not only functions as we just defined, but functions defined almost everywhere in a given dimension, that we call Functions. (Note the calligraphic capital in Functions.)

To motivate why dimension should matter, it helps to recall that in the case of a local field F, there is a separate Haar measure (that is, translation-invariant regular Borel measure) on F^d, for each d. A set $X \subset F^d$ of positive measure with respect to the Haar measure on F^d will have measure zero with respect to the Haar measure on F^{d+1} under an embedding $F^d \subset F^{d+1}$. Hence, the appropriate measure for X depends on its dimension. This suggests that we should filter the semi-ring of constructible functions by dimension.

We start by defining a good notion of dimension for objects of Def_k. Heuristically, that dimension corresponds to counting the dimension only in the valued field variables, without taking in account the remaining variables. More precisely, to any algebraic subvariety Z of $\mathbb{A}^m_{k((t))}$ we assign the definable subassignment h_Z of $h[m,0,0]$ given by $h_Z(K) = Z(K((t)))$. The Zariski closure of a subassignment S of $h[m,0,0]$ is the intersection W of all algebraic subvarieties Z of $\mathbb{A}^m_{k((t))}$ such that $S \subset h_Z$. We define the dimension of S as $\dim S := \dim W$. In the general case, when S is a subassignment of $h[m,n,r]$, we define $\dim S$ as the dimension of the image of S under the projection $h[m,n,r] \to h[m,0,0]$. One can prove that isomorphic objects of Def_k have the same dimension.

For every non negative integer d, we denote by $\mathrm{C}^{\leq d}_+(S)$ the ideal of $\mathrm{C}_+(S)$ generated by functions $\mathbf{1}_Z$ with Z definable subassignments of S with $\dim Z \leq d$. We set $C_+(S) = \oplus_d C^d_+(S)$ with $C^d_+(S) := \mathrm{C}^{\leq d}_+(S)/\mathrm{C}^{\leq d-1}_+(S)$. It is a graded abelian semigroup, and also a $\mathrm{C}_+(S)$-semimodule. Elements of $C_+(S)$ are called positive constructible Functions on S. If φ is a function lying in $\mathrm{C}^{\leq d}_+(S)$ but not in $\mathrm{C}^{\leq d-1}_+(S)$, we denote by $[\varphi]$ its image in $C^d_+(S)$. One defines similarly $C(S)$ from $\mathrm{C}(S)$.

One of the reasons why we consider functions which are defined almost everywhere originates in the differentiation of functions with respect to the valued field variables: one may show that a definable function $c : S \subset h[m,n,r] \to h[1,0,0]$ is differentiable (in fact even analytic) outside a definable subassignment of S of dimension $<$ dim$S$. In particular, if $f : S \to S'$ is an isomorphism in Def_k, one may define a function ordjac$f$, the order of the jacobian of f, which is defined almost everywhere and is equal almost everywhere to a definable function, so we may define $\mathbb{L}^{-\mathrm{ordjac} f}$ in $C^d_+(S)$ when S is of dimension d.

2.5. Push-forward. Let k be a field of characteristic zero. Given S in Def_k, we define S-integrable Functions and construct pushforward morphisms for these Functions by means of Theorem 2.2. Roughly, the notion of being S-integrable for a constructible Function means that one can integrate all variables out until one is left with a function on S, where this resulting function on S need not be further integrable over S itself. The pushforward morphisms for the S-integrable Functions in idea correspond to taking integrals while taking care of dimensions of the support: if one integrates over $\mathbb{Z}$ as in (A6), the idea is summation over $\mathbb{Z}$ and the dimension of the support is left unchanged (recall that the dimension lives in the valued field); if one integrates over an open in the valued field as in (A7), one perfoms an actual motivic integral analogous to a p-adic integral and the dimension of the support goes down by 1; if one integrates so to say on a curve as in (A8), one performs a basic change of variables which of course preserves the dimension of the support; and, if one integrates over the residue field as in (A5) the idea is not to lose any relevant information and one just takes the whole class in a Grothendieck semi-ring of the part one wants to measure, the dimension of the support again being preserved. Having the pushforward morphisms after this theorem allows one to use them to integrate variables out, to send an S-integrable Function to one living on another domain by change of variables, or, as a combination of these, (possibly mimicking the usage of Leray-differential forms) to transform an S-integrable Function φ on Z into an S-integrable Function $f_!(\varphi)$ on Y, where Z and Y are the domains seen as objects in Def_S, and $f : Z \to Y$ is merely required to make a commutative diagram over S. The crucial property that relates φ and $f_!(\varphi)$ is the ability to integrate all variables out up to arriving to a constructible Function on S (by S-integrability) which is the same when calculated from φ or from $f_!(\varphi)$ by the functoriality properties.

Thus, Theorem 2.2 defines at the same time the notion of S-integrability for constructible Functions on domains Z over S, their integrals, and more generally all their transforms through the pushforward morphisms. Although all the axioms interact in the proof of the theorem, one can specify that property (A1a) and the Fubini-Tonelli property (A1b) are crucial for the definition of S-integrability, and that property (A3) can be used to pull a factor out of an integral, as is done in the example in 2.7.

THEOREM 2.2 (Cluckers-Loeser [**8**]). *Let k be a field of characteristic zero and let S be in Def_k. There exists a unique functor $Z \mapsto \mathrm{I}_S C_+(Z)$ from Def_S to the category of abelian semigroups, the functor of S-integrable Functions, assigning to every morphism $f : Z \to Y$ in Def_S a morphism $f_! : \mathrm{I}_S C_+(Z) \to \mathrm{I}_S C_+(Y)$ such that for every Z in Def_S, $\mathrm{I}_S C_+(Z)$ is a graded subsemigroup of $C_+(Z)$ and $\mathrm{I}_S C_+(S) = C_+(S)$, satisfying the following list of axioms (A1)-(A8).*

(A1a) **(Naturality)**
If $S \to S'$ is a morphism in Def_k and Z is an object in Def_S, then any S'-integrable Function φ in $C_+(Z)$ is S-integrable and $f_!(\varphi)$ is the same, considered in $\mathrm{I}_{S'}$ or in I_S.

(A1b) **(Fubini)**
A positive Function φ on Z is S-integrable if and only if it is Y-integrable and $f_!(\varphi)$ is S-integrable.

(A2) **(Disjoint union)**
If Z is the disjoint union of two definable subassignments Z_1 and Z_2, then the isomorphism $C_+(Z) \simeq C_+(Z_1) \oplus C_+(Z_2)$ induces an isomorphism $\mathrm{I}_S C_+(Z) \simeq \mathrm{I}_S C_+(Z_1) \oplus \mathrm{I}_S C_+(Z_2)$, under which $f_! = f_{|Z_1 !} \oplus f_{|Z_2 !}$.

(A3) **(Projection formula)**
For every α in $\mathrm{C}_+(Y)$ and every β in $\mathrm{I}_S C_+(Z)$, $\alpha f_!(\beta)$ is S-integrable if and only if $f^*(\alpha)\beta$ is, and then $f_!(f^*(\alpha)\beta) = \alpha f_!(\beta)$.

(A4) **(Inclusions)**
If $i : Z \hookrightarrow Z'$ is the inclusion of definable subassignments of the same object of Def_S, then $i_!$ is induced by extension by zero outside Z and sends $\mathrm{I}_S C_+(Z)$ injectively to $\mathrm{I}_S C_+(Z')$.

(A5) **(Integration along residue field variables)**
Let Y be an object of Def_S and denote by π the projection $Y[0, n, 0] \to Y$. A Function $[\varphi]$ in $C_+(Y[0, n, 0])$ is S-integrable if and only if, with notations of (2.3.6), $[\pi_!(\varphi)]$ is S-integrable and then $\pi_!([\varphi]) = [\pi_!(\varphi)]$.

Basically this axiom means that integrating with respect to variables in the residue field just amounts to taking the pushforward induced by composition at the level of Grothendieck semi-rings.

(A6) **(Integration along $\mathbb{Z}$-variables)** Basically, integration along $\mathbb{Z}$-variables corresponds to summing over the integers, but to state precisely (A6), we need to perform some preliminary constructions.

Let us consider a function in φ in $\mathcal{P}(S[0, 0, r])$, hence φ is a function $|S| \times \mathbb{Z}^r \to A$. We shall say φ is S-integrable if for every $q > 1$ and every x in $|S|$, the series $\sum_{i \in \mathbb{Z}^r} \vartheta_q(\varphi(x, i))$ is summable. One proves that if φ is S-integrable there exists a unique function $\mu_S(\varphi)$ in $\mathcal{P}(S)$ such that $\vartheta_q(\mu_S(\varphi)(x))$ is equal to the sum of the previous series for all $q > 1$ and all x in $|S|$. We denote by $\mathrm{I}_S\mathcal{P}_+(S[0,0,r])$ the set of S-integrable functions in $\mathcal{P}_+(S[0,0,r])$ and we set

$$\mathrm{I}_S\mathrm{C}_+(S[0,0,r]) = \mathrm{C}_+(S) \otimes_{\mathcal{P}_+(S)} \mathrm{I}_S\mathcal{P}_+(S[0,0,r]). \tag{2.5.1}$$

Hence $\mathrm{I}_S\mathcal{P}_+(S[0,0,r])$ is a sub-$\mathrm{C}_+(S)$-semimodule of $\mathrm{C}_+(S[0,0,r])$ and μ_S may be extended by tensoring to

$$\mu_S : \mathrm{I}_S\mathrm{C}_+(S[0,0,r]) \to \mathrm{C}_+(S). \tag{2.5.2}$$

Now we can state (A6):
Let Y be an object of Def_S and denote by π the projection $Y[0,0,r] \to Y$. A Function $[\varphi]$ in $C_+(Y[0,0,r])$ is S-integrable if and only if there exists φ' in $\mathrm{C}_+(Y[0,0,r])$ with $[\varphi'] = [\varphi]$ which is Y-integrable in the previous sense and such that $[\mu_Y(\varphi')]$ is S-integrable. We then have $\pi_!([\varphi]) = [\mu_Y(\varphi')]$.

(A7) **(Volume of balls)** It is natural to require (by analogy with the p-adic case) that the volume of a ball $\{z \in h[1,0,0] \mid \mathrm{ord}(z-c) = \alpha, \overline{\mathrm{ac}}(z-c) = \xi\}$, with α in $\mathbb{Z}$, c in $k((t))$ and ξ non zero in k, should be $\mathbb{L}^{-\alpha-1}$. (A7) is a relative version of that statement:
Let Y be an object in Def_S and let Z be the definable subassignment of $Y[1,0,0]$ defined by $\mathrm{ord}(z - c(y)) = \alpha(y)$ and $\overline{\mathrm{ac}}(z - c(y)) = \xi(y)$, with z the coordinate on the $\mathbb{A}^1_{k((t))}$-factor and α, ξ, c definable functions on Y with values respectively in $\mathbb{Z}$, $h[0,1,0] \setminus \{0\}$, and $h[1,0,0]$. We denote by $f : Z \to Y$ the morphism induced by projection. Then $[\mathbf{1}_Z]$ is S-integrable if and only if $\mathbb{L}^{-\alpha-1}[\mathbf{1}_Y]$ is, and then $f_!([\mathbf{1}_Z]) = \mathbb{L}^{-\alpha-1}[\mathbf{1}_Y]$.

(A8) **(Graphs)** This last axiom expresses the pushforward for graph projections. It relates volume and differentials and is a special case of the change of variables Theorem 2.3.
Let Y be in Def_S and let Z be the definable subassignment of $Y[1,0,0]$ defined by $z - c(y) = 0$ with z the coordinate on the $\mathbb{A}^1_{k((t))}$-factor and c a morphism $Y \to h[1,0,0]$. We denote by $f : Z \to Y$ the morphism induced by projection. Then $[\mathbf{1}_Z]$ is S-integrable if and only if $\mathbb{L}^{(\mathrm{ordjac} f)\circ f^{-1}}$ is, and then $f_!([\mathbf{1}_Z]) = \mathbb{L}^{(\mathrm{ordjac} f)\circ f^{-1}}$.

Once Theorem 2.2 is proved, one may proceed as follows to extend the constructions from C_+ to C. One defines $\mathrm{I}_S C(Z)$ as the subgroup of $C(Z)$ generated by the image of $\mathrm{I}_S C_+(Z)$. One shows that if $f : Z \to Y$ is a morphism in Def_S, the morphism $f_! : \mathrm{I}_S C_+(Z) \to \mathrm{I}_S C_+(Y)$ has a natural extension

$$f_! : \mathrm{I}_S C(Z) \to \mathrm{I}_S C(Y). \tag{2.5.3}$$

The proof of Theorem 2.2 is quite long and involved. In a nutshell, the basic idea is the following. Integration along residue field variables is controlled by (A5) and integration along $\mathbb{Z}$-variables by (A6). Integration along valued field variables is constructed one variable after the other. To integrate with respect to one valued field variable, one may, using (a variant of) the cell decomposition Theorem 2.1 (at the cost of introducing additional new residue field and $\mathbb{Z}$-variables), reduce to the case of cells which is covered by (A7) and (A8). An important step is to show that this is independent of the choice of a cell decomposition. When one integrates with respect to more than one valued field variable (one after the other) it is crucial to show that it is independent of the order of the variables, for which we use a notion of bicells.

2.6. Motivic measure. The relation of Theorem 2.2 with motivic integration is the following. When S is equal to $h[0,0,0]$, the final object of Def_k, one writes $\mathrm{I}C_+(Z)$ for $\mathrm{I}_S C_+(Z)$ and we shall say integrable for S-integrable, and similarly for C. Note that $\mathrm{I}C_+(h[0,0,0]) = C_+(h[0,0,0]) = SK_0(\mathrm{RDef}_k) \otimes_{\mathbb{N}[\mathbb{L}-1]} \mathbb{A}_+$ and that $\mathrm{I}C(h[0,0,0]) = K_0(\mathrm{RDef}_k) \otimes_{\mathbb{Z}[\mathbb{L}]} \mathbb{A}$. For φ in $\mathrm{I}C_+(Z)$, or in $\mathrm{I}C(Z)$, one defines the motivic integral $\mu(\varphi)$ by $\mu(\varphi) = f_!(\varphi)$ with f the morphism $Z \to h[0,0,0]$.

Let X be in Def_k of dimension d. Let φ be a function in $\mathrm{C}_+(X)$, or in $\mathrm{C}(X)$. We shall say φ is integrable if its class $[\varphi]_d$ in $C_+^d(X)$, resp. in $C^d(X)$, is integrable, and we shall set

$$\mu(\varphi) = \int_X \varphi \, d\mu = \mu([\varphi]_d).$$

Similarly as in the p-adic case, cf. [**19**] p.112, one may develop the integration on global (non affine) objects endowed with a differential form of top degree once the following Change of Variables Theorem 2.3 is established:

THEOREM 2.3 (Cluckers-Loeser [**8**]). *Let $f : Y \to X$ be an isomorphism in Def_k. For any integrable function φ in $\mathrm{C}_+(X)$ or $\mathrm{C}(X)$,*

$$\int_X \varphi d\mu = \int_Y \mathbb{L}^{-\operatorname{ord}\operatorname{jac}(f)} f^*(\varphi) d\mu.$$

Also, the construction we outlined of the motivic measure carries over almost literally to a relative setting: one can develop a relative theory of motivic integration: integrals depending on parameters of functions in C_+ or C still belong to C_+ or C as functions of these parameters.

More specifically, if $f : X \to \Lambda$ is a morphism and φ is a function in $\mathrm{C}_+(X)$ or $\mathrm{C}(X)$ that is relatively integrable (a notion defined in [**8**]), one constructs in [**8**] a function

$$\mu_\Lambda(\varphi) \tag{2.6.1}$$

in $\mathrm{C}_+(\Lambda)$, resp. $\mathrm{C}(\Lambda)$, whose restriction to every fiber of f coincides with the integral of φ restricted to that fiber.

2.7. An extended example. Let us give an example of a definable subassignment that can be transformed into a 1-cell by adding auxiliary parameters. Let φ be the formula with a free variable z of the valued field sort:

$$z \neq 0 \ \wedge \ \mathrm{ord}(z) \geq 0 \ \wedge \ \exists t \ (z = t^2).$$

$Z = h_\varphi \subset h[1,0,0]$ is the subassignment of nonzero squares in the valued field with nonnegative valuation. Under the embedding

$$\lambda : h[1,0,0] \to h[1,1,1], \quad z \mapsto (z, \overline{\mathrm{ac}}(z), \mathrm{ord}(z)),$$

$W = \lambda(h_\varphi)$ is a 1-cell with defining functions

$$c(\zeta, \eta) = 0, \quad \xi(\zeta,\eta) = \zeta, \quad \alpha(\zeta,\eta) = \eta.$$

on $S \times N$, where $S \subset h[0,1,0]$ and $N \subset h[0,0,1]$ are defined by the conditions

$$\zeta \neq 0 \ \wedge \ \exists \zeta_1 (\zeta_1^2 = \zeta) \quad \text{and} \quad \eta \geq 0 \ \wedge \ \exists \eta_1 (2\eta_1 = \eta), \tag{2.7.1}$$

respectively. That is, W is defined by the conditions 2.7.1 and

$$\mathrm{ord}(z) = \eta \ \wedge \ \overline{\mathrm{ac}}(z) = \zeta.$$

To continue with this example, let us compute the motivic integral over Z of a particular constructible function $\mathbb{L}^{-3\mathrm{ord}(z)}$:

$$\begin{aligned}
\int_Z \mathbb{L}^{-3\mathrm{ord}(z)} d\mu(z) &= \int_W \mathbb{L}^{-3\eta} d\mu(z,\zeta,\eta) && \text{(A5), (A6)}\\
&= \int_{S\times N} \int_{W \to S \times N} \mathbb{L}^{-3\eta} d\mu(z)\ d\mu(\zeta,\eta) && \text{(A1)}\\
&= \int_{S\times N} \mathbb{L}^{-3\eta} \int_{W \to S\times N} \mathbf{1}_W \, d\mu(z)\ d\mu(\zeta,\eta) && \text{(A3)}\\
&= \int_{S\times N} \mathbb{L}^{-3\eta}\mathbb{L}^{-\eta-1}\ d\mu(\zeta,\eta) && \text{(A7)}\\
&= [\mathbf{1}_S] \int_N \mathbb{L}^{-4\eta-1} d\mu(\eta) && \text{(A3), (A5)}\\
&= \frac{[\mathbf{1}_S]}{\mathbb{L}(1-\mathbb{L}^{-8})} && \text{(A6)}
\end{aligned}$$

This integral should be compared with the corresponding p-adic integral over the set of integer squares with respect to the Haar measure, normalized so that the ring of integers has volume 1:

$$\int_{|z|\leq 1 \ z=t^2} |z|^4 \frac{dz}{|z|} = \frac{\sigma}{q(1-q^{-8})},$$

where q is the cardinality of the residue field of the local field F, and σ is the cardinality of the set of nonzero squares in the residue field. (We have assumed that the residual characteristic is not 2.) The result of (A6) in the motivic calculation should be compared with the same infinite sum of real numbers that appears in the p-adic integral:

$$\sum_{i=0,\ i=2j}^{\infty} q^{-4i-1} = \frac{1}{q(1-q^{-8})}, \quad q > 1.$$

2.8. The transfer principle. We are now in the position of explaining how motivic integrals specialize to p-adic integrals and may be used to obtain a general transfer principle allowing to transfer relations between integrals from $\mathbb{Q}_p$ to $\mathbb{F}_p((t))$ and vice-versa.

We shall assume from now on that k is a number field with ring of integers $\mathcal{O}$. We denote by $\mathcal{A}_{\mathcal{O}}$ the set of p-adic completions of all finite extensions of k and by $\mathcal{B}_{\mathcal{O}}$ the set of all local fields of characteristic > 0 which are $\mathcal{O}$-algebras.

For F in $\mathrm{C}_{\mathcal{O}} := \mathcal{A}_{\mathcal{O}} \cup \mathcal{B}_{\mathcal{O}}$, we denote by

- R_F the valuation ring
- M_F the maximal ideal
- k_F the residue field
- $q(F)$ the cardinal of k_F
- ϖ_F a uniformizing parameter of R_F.

There exists a unique morphism $\overline{\mathrm{ac}} : F^\times \to k_F^\times$ extending $R_F^\times \to k_F^\times$ and sending ϖ_F to 1. We set $\overline{\mathrm{ac}}(0) = 0$. For $N > 0$, we denote by $\mathcal{A}_{\mathcal{O},N}$ the set of fields F in $\mathcal{A}_{\mathcal{O}}$ such that k_F has characteristic $> N$, and similarly for $\mathcal{B}_{\mathcal{O},N}$ and $\mathrm{C}_{\mathcal{O},N}$. To be able to interpret our formulas to fields in $\mathrm{C}_{\mathcal{O}}$, we restrict the language $\mathcal{L}_{\mathrm{DP}}$ to the sub-language $\mathcal{L}_{\mathcal{O}}$ for which coefficients in the valued field sort are assumed to belong to the subring $\mathcal{O}[[t]]$ of $k((t))$. We denote by $\mathrm{Def}(\mathcal{L}_{\mathcal{O}})$ the sub-category of Def_k of objects definable in $\mathcal{L}_{\mathcal{O}}$, and similarly for functions, etc. For instance, for S in $\mathrm{Def}(\mathcal{L}_{\mathcal{O}})$, we denote by $\mathrm{C}(S, \mathcal{L}_{\mathcal{O}})$ the ring of constructible functions on S definable in $\mathcal{L}_{\mathcal{O}}$.

We consider F as a $\mathcal{O}[[t]]$-algebra via

$$\lambda_{\mathcal{O},F} : \sum_{i \in \mathbb{N}} a_i t^i \longmapsto \sum_{i \in \mathbb{N}} a_i \varpi_F^i. \tag{2.8.1}$$

Hence, if we interpret a in $\mathcal{O}[[t]]$ by $\lambda_{\mathcal{O},F}(a)$, every $\mathcal{L}_{\mathcal{O}}$-formula φ defines for F in $\mathrm{C}_{\mathcal{O}}$ a subset φ_F of some $F^m \times k_F^n \times \mathbb{Z}^r$. One proves that if two $\mathcal{L}_{\mathcal{O}}$-formulas φ and φ' define the same subassignment X of $h[m, n, r]$, then $\varphi_F = \varphi'_F$ for F in $\mathrm{C}_{\mathcal{O},N}$ when $N \gg 0$. This allows us to denote by X_F the subset defined by φ_F, for F in $\mathrm{C}_{\mathcal{O},N}$ when $N \gg 0$. Similarly, every $\mathcal{L}_{\mathcal{O}}$-definable morphism $f : X \to Y$ specializes to a function $f_F : X_F \to Y_F$ for F in $\mathrm{C}_{\mathcal{O},N}$ when $N \gg 0$.

We now explain how φ in $\mathrm{C}(X, \mathcal{L}_{\mathcal{O}})$ can be specialized to $\varphi_F : X_F \to \mathbb{Q}$ for F in $\mathrm{C}_{\mathcal{O},N}$ when $N \gg 0$. Let us consider φ in $K_0(\mathrm{RDef}_X(\mathcal{L}_{\mathcal{O}}))$ of the form $[\pi : W \to X]$ with W in $\mathrm{RDef}_X(\mathcal{L}_{\mathcal{O}})$. For F in $\mathrm{C}_{\mathcal{O},N}$ with $N \gg 0$, we have $\pi_F : W_F \to X_F$, so we may define $\varphi_F : X_F \to \mathbb{Q}$ by

$$x \longmapsto \mathrm{card}\left(\pi_F^{-1}(x)\right). \tag{2.8.2}$$

For φ in $\mathcal{P}(X)$, we specialize $\mathbb{L}$ into q_F and $\alpha : X \to \mathbb{Z}$ into $\alpha_F : X_F \to \mathbb{Z}$. By tensor product we get $\varphi \mapsto \varphi_F$ for φ in $\mathrm{C}(X, \mathcal{L}_{\mathcal{O}})$. Note that, under that construction, functions in $\mathrm{C}_+(X, \mathcal{L}_{\mathcal{O}})$ specialize into non negative functions.

Let F be in $\mathrm{C}_{\mathcal{O}}$ and A be a subset of $F^m \times k_F^n \times \mathbb{Z}^r$. We consider the Zariski closure $\bar{A}$ of the projection of A into $\mathbb{A}_F^m$. One defines a measure μ

on A by restriction of the product of the canonical (Serre-Oesterlé) measure on $\bar{A}(F)$ with the counting measure on $k_F^n \times \mathbb{Z}^r$.

Fix a morphism $f : X \to \Lambda$ in $\mathrm{Def}(\mathcal{L}_{\mathcal{O}})$ and consider φ in $\mathrm{C}(X, \mathcal{L}_{\mathcal{O}})$. One can show that if φ is relatively integrable, then, for $N \gg 0$, for every F in $\mathrm{C}_{\mathcal{O},N}$, and for every λ in Λ_F, the restriction $\varphi_{F,\lambda}$ of φ_F to $f_F^{-1}(\lambda)$ is integrable.

We denote by $\mu_{\Lambda_F}(\varphi_F)$ the function on Λ_F defined by

$$\lambda \longmapsto \mu(\varphi_{F,\lambda}). \tag{2.8.3}$$

The following theorem says that motivic integrals specialize to the corresponding integrals over local fields of high enough residue field characteristic.

THEOREM 2.4 (Specialization, Cluckers-Loeser [**10**] [**11**]). *Let $f : S \to \Lambda$ be a morphism in* $\mathrm{Def}(\mathcal{L}_{\mathcal{O}})$. *Let φ be in* $\mathrm{C}(S, \mathcal{L}_{\mathcal{O}})$ *and relatively integrable with respect to f. For $N \gg 0$, for every F in* $\mathrm{C}_{\mathcal{O},N}$, *we have*

$$(\mu_\Lambda(\varphi))_F = \mu_{\Lambda_F}(\varphi_F). \tag{2.8.4}$$

We are now ready to state the following abstract transfer principle:

THEOREM 2.5 (Abstract transfer principle, Cluckers-Loeser [**10**] [**11**]). *Let φ be in* $\mathrm{C}(\Lambda, \mathcal{L}_{\mathcal{O}})$. *There exists N such that for every F_1, F_2 in* $\mathrm{C}_{\mathcal{O},N}$ *with $k_{F_1} \simeq k_{F_2}$,*

$$\varphi_{F_1} = 0 \quad \textit{if and only if} \quad \varphi_{F_2} = 0. \tag{2.8.5}$$

Putting together the two previous theorems, one immediately gets:

THEOREM 2.6 (Transfer principle for integrals with parameters, Cluckers-Loeser [**10**] [**11**]). *Let $S \to \Lambda$ and $S' \to \Lambda$ be morphisms in* $\mathrm{Def}(\mathcal{L}_{\mathcal{O}})$. *Let φ and φ' be relatively integrable functions in* $\mathrm{C}(S, \mathcal{L}_{\mathcal{O}})$ *and* $\mathrm{C}(S', \mathcal{L}_{\mathcal{O}})$, *respectively. There exists N such that for every F_1, F_2 in* $\mathrm{C}_{\mathcal{O},N}$ *with $k_{F_1} \simeq k_{F_2}$,*

$$\mu_{\Lambda_{F_1}}(\varphi_{F_1}) = \mu_{\Lambda_{F_1}}(\varphi'_{F_1}) \quad \textit{if and only if} \quad \mu_{\Lambda_{F_2}}(\varphi_{F_2}) = \mu_{\Lambda_{F_2}}(\varphi'_{F_2}).$$

In the special case where $\Lambda = h[0,0,0]$ and φ and φ' are in $\mathrm{C}(S, \mathcal{L}_{\mathcal{O}})$ and $\mathrm{C}(S', \mathcal{L}_{\mathcal{O}})$, respectively, this follows from previous results of Denef-Loeser [**13**].

REMARK. The previous constructions and statements may be extended directly - with similar proofs - to the global (non affine) setting.

Note that when $S = S' = \Lambda = h[0,0,0]$, one recovers the classical

THEOREM 2.7 (Ax-Kochen-Eršov [**6**] [**14**]). *Let φ be a first order sentence (that is, a formula with no free variables) in the language of rings. For almost all prime number p, the sentence φ is true in $\mathbb{Q}_p$ if and only if it is true in $\mathbb{F}_p((t))$.*

A major triumph of model theory is the application of this theorem to Artin's conjecture about forms over $\mathbb{Q}_p$ [**5**]. Artin conjectured that for every positive natural number d and every prime p, if $n > d^2$, then every homogeneous polynomial of degree d over $\mathbb{Q}_p$ in n variables has a nontrivial zero in $\mathbb{Q}_p$. Ax and Kochen used their result to transfer known results from positive characteristic to characteristic zero and thus to prove that Artin's conjecture holds asymptotically. That is, for each d, Artin's conjecture holds over $\mathbb{Q}_p$, except for possible failure as the prime p runs over a finite set $A(d)$.

2.9. In view of Theorem 2.6, since we are interested in the behavior of integrals for sufficiently large primes p, we shall assume throughout this paper, whenever it is useful to do so, that p is sufficiently large. *This remains a standing assumption throughout this paper.*

REMARK. Let $N > 0$ be an integer and let $\mathcal{L}_{\mathcal{O}}(1/N)$ be the language $\mathcal{L}_{\mathcal{O}}$ with one extra constant symbol to denote the rational number $1/N$. Then the above statements in section 2 remain valid if one works with $\mathcal{L}_{\mathcal{O}}(1/N)$ instead of with $\mathcal{L}_{\mathcal{O}}$, where now the conditions of big enough residue field characteristic mean in particular that the residue field characteristic is bigger than N. Indeed, $\mathcal{L}_{\mathcal{O}}(1/N)$ is a definitional expansion of $\mathcal{L}_{\mathcal{O}}$ in the sense that both languages give exactly the same definable subassignments and definable morphisms, hence they also yield the same rings and semi-rings resp. the same groups and semigroups of constructible functions and constructible Functions.

2.10. Tensoring with $\mathbb{Q}$, with $\mathbb{R}$, or with $\mathbb{C}$. Since Arthur's weight function involves volumes (see below), it is useful to work with Functions in $C(X) \otimes \mathbb{R}$ instead of in $C(X)$ and so on, for definable subassignments X. One can work similarly to tensor with $\mathbb{Q}$ or $\mathbb{C}$. Once we have the direct image operators and integration operators of Theorem 2.2, of (2.5.3), and of (2.6.1), this is easily done as follows. Let $f : Z \to Y$ be a morphism in Def_S. Clearly $\mathbb{Q}$, $\mathbb{R}$ and $\mathbb{C}$ are flat $\mathbb{Z}$-modules (as localizations, resp. direct limits of $\mathbb{Z}$, resp. of flat modules). This allows us to view $\mathrm{I_SC(Z)} \otimes_{\mathbb{Z}} \mathbb{R}$ as a submodule of $C(Z) \otimes_{\mathbb{Z}} \mathbb{R}$, and similarly for relative integrable functions. Naturally, $f_!$ extends to a homomorphism $f_! \otimes \mathbb{R} : \mathrm{I_SC(Z)} \otimes \mathbb{R} \to \mathrm{I_SC(Y)} \otimes \mathbb{R}$, and similarly for relative integrable functions. Of course, the study of semi-rings of constructible functions tensored with $\mathbb{R}$ does not have any additional value since $\mathbb{R}$ contains -1. On the other hand, $f^* \otimes \mathbb{R}$ has a natural meaning as a homomorphism from $C(Y) \otimes \mathbb{R}$ to $C(Z) \otimes \mathbb{R}$. In this setting, the analogue of the Change of Variables Theorem 2.3 clearly remains true. A function φ in $\mathrm{C}(X, \mathcal{L}_{\mathcal{O}}) \otimes \mathbb{R}$ can be specialized to $\varphi_F : X_F \to \mathbb{R}$ for F in $\mathrm{C}_{\mathcal{O},N}$ when $N \gg 0$ as in section 2.8, by tensoring with $\mathbb{R}$, where the values of φ_F now lie in $\mathbb{R}$ instead of in $\mathbb{Q}$. With this notation, the analogues in the $\otimes\mathbb{R}$-setting of Theorems 2.4, 2.5, and 2.6 and the analogues of Remarks 2.8 and 2.9 naturally hold.

In the applications to the Fundamental Lemma, the space S will be the Chevalley space of the endoscopic group and the S-Function will be the difference of a stable orbital integral and a κ-orbital integral.

3. Definability of Field Extensions

3.1. Translations into the Denef-Pas language. The Denef-Pas language has noteworthy limits to its expressive power. For example, as previously noted, there is no sentence in the language of rings that expresses that a field is algebraically closed. There is no formula φ with one free variable of the residue field sort such that $h_\varphi(K) \subset K$ is the set of roots of unity in K. There is no formula φ with two free variables of the value group sort that consists of pairs (i, j) with $i^2 = j$, because although the language of ordered groups includes a binary addition symbol, it does not include a multiplication symbol or power symbol i^2.

Some mathematical notions cannot be expressed directly in the language, but an adequate translation or proxy can be found within the language. For example, even though there is no cubic power symbol ξ^3 in the language of rings, the expansion as a product $\xi\cdot\xi\cdot\xi$ is a perfectly adequate representation of the cube within the language of rings. Similarly, if we encounter an abstract vector space of dimension n over the residue field sort, it will be represented concretely by the subassignment $h[0, n, 0]$. The endomorphism ring of the vector space will be represented concretely as the subassignment $h[0, n^2, 0]$ with free variables ξ_{ij}, $i, j = 1, \ldots, n$.

Much of the discussion that follows consists of translating mathematical concepts related to the fundamental lemma into basic terms that can be expressed in the Denef-Pas language.

3.2. Fields extensions. As an example of translating mathematical concepts into more basic terms, we consider the problem of expressing the arithmetic of a field extension E/F in the Denef-Pas language. Because of limitations of the language, we cannot work directly with a field extension E/F. However, a finite field extension E/F has a proxy in the Denef-Pas language, which is obtained by working through the minimal polynomial m of $e \in E$ that generates E/F, and an explicit basis $\{1, x, \ldots, x^{r-1}\}$ of $F[x]/(m) = E$. The minimal polynomial

$$x^r + a_{r-1}x^{r-1} + \cdots a_0 \tag{3.2.1}$$

can be identified with its list of coefficients

$$(a_{r-1}, \ldots, a_0). \tag{3.2.2}$$

In this paper, general finite field extensions will only appear as bound variables in formulas. Moreover, the degree of the extensions will always be fixed. Thus, a statements "there exists a field extension of degree r such that ..." can be translated into "there exist $a_{r-1}, \ldots, a_0$ such that $x^r + a_{r-1}x^{r-1} + \cdots + a_0$ is irreducible and such that ...." When $r > 1$, the statement "$x^r + a_{r-1}x^{r-1} + \cdots + a_0$ is irreducible" can be further expanded into the language of rings as "there do not exist $b_0, \ldots, b_{2r}$ and $c_0, \ldots, c_{2r}$ such that

$$a_j = \sum_{i=0}^{j} b_i c_{j-i}, \text{ for } j = 0, \ldots, 2r$$

where $a_r = 1$, $b_r = c_r = 0$, and $a_j = b_j = c_j = 0$ for $j > r$."

After identifying the field extension with F^r through (3.2.2), we can define field automorphisms by linear maps on F^r that respect the field operations. Thus, field automorphisms are definable, and the condition that E/F is a Galois extension is definable.

3.3. A free parameter. We will see that all of the constructions in this paper can be arranged so that the only field extensions that are "free" in a formula are unramified of fixed degree r. These can be described in a field-independent way by a minimal polynomial $x^r - a$, where a satisfies a Denef-Pas condition that it is a unit such that $x^r - a$ is irreducible. This construction introduces a parameter a to whatever formulas involve an unramified field extension. The free parameter a will be used for instance when we describe unramified unitary groups.

If $m(x) = x^r - a$ defines a field extension $F_{a,r}$ of degree r of F, then we may represent a field extension E of F of degree k containing $F_{a,r}$ by data $(b_{k-1}, \ldots, b_0, \phi)$, where $b_i \in F$ are the coefficients of a minimal polynomial $b(y)$ of an element generating E/F, and $\phi \in M_{rk}(F)$ is a matrix giving the embedding of $F^r \to F^k$, corresponding to

$$F_{a,r} = F[x]/(m(x)) \to F[y]/(b(y)). \tag{3.3.1}$$

4. Definability of Unramified Reductive Groups

4.1. Split groups. The classification of split connected reductive groups is independent of the field F. Isomorphism classes of split reductive groups are in bijective correspondence with root data:

$$D = (X^*, \Phi, X_*, \Phi^\vee), \tag{4.1.1}$$

consisting of the character group of a Cartan subgroup, the set of roots, the cocharacter group, and the set of coroots. The root data are considered up to an obvious equivalence.

In particular, we may realize each split reductive group over $\mathbb{Q}$. Fix once and for all, a rational faithful representation $\rho = \rho_D$

$$\rho : G \to GL(V) \tag{4.1.2}$$

of each reductive group over $\mathbb{Q}$, attached to root data D. Fix a basis of V over $\mathbb{Q}$. There exists a formula $\varphi_{\rho,D}$ in the Denef-Pas language (in fact a formula in the language of rings) that describes the zero set of $\rho(G)$.

4.2. Quasi-split groups. An unramified reductive group over a local field F is a quasi-split connected reductive group that splits over an unramified extension of F. The classification of unramified reductive groups is independent of the field F. Isomorphism classes of unramified reductive groups are in bijective correspondence with pairs (D, θ), where D is the root data for the corresponding split group and θ is an automorphism of finite order of D that preserves a set of simple roots in Φ.

The quasi-split group G is obtained by an outer twist of the corresponding split group G^* as follows. Suppose that θ has order r. Let $F_{a,r}$ be the unramified extension of F of degree r defined by the polynomial $x^r - a$. Let $(B, T, \{X_\alpha\})$ be a splitting of G^*, consisting of a Borel subgroup B, Cartan T, and root vectors $\{X_\alpha\}$ all defined over $\mathbb{Q}$. The automorphism θ of the root data determines a unique automorphism of G^* preserving the splitting. We let θ denote this automorphism of G^* as well. The automorphism θ acts on root vectors by $\theta(X_\alpha) = X_{\theta\alpha}$.

For any A-algebra F, we have that $G(A)$ is the set of fixed points of $G(A \otimes F_{a,r})$ under the map $\theta \circ \tau$, where

$$\tau : G^*(A \otimes F_{a,r}) \to G^*(A \otimes F_{a,r}) \tag{4.2.1}$$

extends the Frobenius automorphism of $F_{a,r}/F$. Through the representation ρ of G^*, the fixed-point condition can be expressed on the corresponding matrix groups. In particular, $G(F)$ can be explicitly realized as the fixed points of a map $\theta \circ \tau$ on $\rho(G^*(F_{a,r})) \subset M_n(F_{a,r})$, the set of n by n matrices with coefficients in $F_{a,r}$.

To express the group G by a formula in the Denef-Pas language, we introduce a free parameter a as described in Section 3, which is the proxy in the Denef-Pas language for an unramified extension of degree r. Under the identification $F_{a,r} \to F^r$, the fixed point condition becomes a ring condition on $M_n(F) \otimes F^r$.

We find that for any pair (D, θ) (and fixed ρ), there exists a formula $\varphi = \varphi_{D,\theta}$ in the Denef-Pas language with $1 + r^2 + n^2 r$ free variables such that for any p-adic field F $\varphi(a, \tau, g)$ is true exactly when a is a unit such that $x^r - a$ is irreducible over F, $F_{a,r} = F[x]/(x^r - a)$, $\tau \in M_r(F)$ is a generator of the Galois group of the field extension $F_{a,r}/F$ (under the identification $F_{a,r} = F^r$), and $g \in M_n(F) \otimes F^r$ is identified with an element $g \in G(F) \subset M_n(F_{a,r})$ (under the identification of F^r with $F_{a,r}$ determined by a).

We stress that this construction differs from the usual global to local specialization. If we take a unitary group defined over $\mathbb{Q}$ with conjugation from a quadratic extension $\mathbb{Q}(\sqrt{d})/\mathbb{Q}$, then the corresponding p-adic group at completions of $\mathbb{Q}$ can be split or inert depending on the prime. By making a (or d) a parameter to the formula, we can insure that the formula $\varphi_{D,\theta}$ defines the unramified unitary group *at every finite place*, and not merely at the inert primes.

By a similar construction, we obtain formulas in the Denef-Pas language for unramified reductive Lie algebras. If the group splits over a non-trivial unramified extension, there will be corresponding parameters, a and τ.

5. Galois Cohomology

5.1. Cocycles. Once we have expressed field extensions and Galois groups within the Denef-Pas language, we may do some rudimentary Galois cohomology.

We follow various conventions when working with Galois cohomology groups. We never work directly with the cohomology groups. Rather, we represent each class in a cohomology group $H^r(\mathrm{Gal}(E/F), A)$ explicitly as a cocycle

$$b \in Z^r(\mathrm{Gal}(E/F), A),$$

viewed as a tuple $b = (b_1, b_2, \ldots)$ of elements of A, indexed by a natural number $i = 1, \ldots, \mathrm{card}(\mathrm{Gal}(E/F))$, for a given enumeration $\sigma_1, \sigma_2, \ldots$ of the elements of the Galois group. The group law of the Galois group is not given; it is to be encoded into the formula for the cocycle condition:

$$(\sigma_i\sigma_j = \sigma_k) \implies (b_i\sigma_i(b_j) = b_k) \text{ for } i, j, k = 1, \ldots, \mathrm{card}(\mathrm{Gal}(E/F)).$$

We express that two cocycles b, b' are cohomologous by means of an existential quantifier: there exists a coboundary c such that $b = b'c$ (as tuples with component-wise multiplication).

The module A will always be one that can be expressed directly in the Denef-Pas language. For example, A may be a free $\mathbb{Z}$-module of finite rank, a definable subgroup of $\mathrm{GL}_n(E)$, or a set of elements in the coordinate ring $\mathbb{Q}[M_n]$ of the space of $n \times n$ matrices. Similarly, we restrict ourselves to actions of $\mathrm{Gal}(E/F)$ on modules A that can be expressed in our first-order language.

For example, if T is a torus that splits over an extension E/F of p-adic fields, instead of the group $H^1(F, T)$, we work with the group of cocycles $Z^1(\mathrm{Gal}(E/F), T(E))$, where $T(E)$ is represented explicitly as an affine algebraic group of invertible n by n matrices. Instead of $H^1(F, X^*(T))$, we work with $Z^1(\mathrm{Gal}(E/F), X^*(T))$, where $X^*(T)$ is viewed as a subset of the coordinate ring of T. Even more concretely, we may view T as a closed subset of M_n for some n, and represent elements of the coordinate ring of T as polynomials in n^2 variables.[1] The cup-product of $b \in Z^1(\mathrm{Gal}(E/F), T(E))$ with $p \in Z^1(\mathrm{Gal}(E/F), X^*(T))$ into $Z^2(\mathrm{Gal}(E/F), E^\times)$ is the tuple whose coordinates are $p_{\sigma'}(b_\sigma)$, where the pairing is obtained by evaluating polynomials $p_{\sigma'}$ at values b_σ.

[1] Since $X^*(T)$ is a free $\mathbb{Z}$-module of finite rank, it might be tempting to represent $X^*(T)$ within the value group sort of the Denef-Pas language. This is the wrong way to proceed!

5.2. Tate-Nakayama pairing. The canonical isomorphism of the Brauer group

$$H^2(F, \mathbb{G}_m) = \mathbb{Q}/\mathbb{Z} \tag{5.2.1}$$

relies on the Frobenius automorphism of unramified field extensions of F. We avoid the Frobenius automorphism and work with an explicitly chosen generator τ of an unramified extension.

The Tate-Nakayama pairing

$$H^1(F, T) \times H^1(F, X^*(T)) \to \mathbb{Q}/\mathbb{Z} \tag{5.2.2}$$

can be translated into the Denef-Pas language as a collection of predicates

$$\mathrm{TN}_{\ell,k}(a, \tau, b, p)$$

with free variables

$$b \in Z^1(\mathrm{Gal}(E/F), T(E)), \quad p \in Z^1(\mathrm{Gal}(E/F), X^*(T)),$$

and a generator τ of an unramified field extensions $F_{a,r}/F$, defined by parameter a. The extension E/F is assumed to be Galois, and r must be sufficiently large with respect to the degree of the extension E/F. The collection of predicates are indexed by $\ell, k \in \mathbb{N}$, with $k \neq 0$.

The predicate $\mathrm{TN}_{\ell,k}(a, \tau, b, p)$ asserts that the Tate-Nakayama cup product pairing of b and p is the class $\ell/k \in \mathbb{Q}/\mathbb{Z}$ in the Brauer group. In more detail, we construct the predicate by a rather literal translation of the Brauer group isomorphism into the language of Denef and Pas. We review Tate-Nakayama, to make it evident that the Denef-Pas language is all that is needed. There is no harm in assuming that $F_{a,r} \subset E$. Let $c \in Z^2(\mathrm{Gal}(E/F), E^\times)$ be the cup-product of b and p. There exists a coboundary d and a cocycle $c' \in Z^2(\mathrm{Gal}(F_{a,r}/F), F_{a,r}^\times)$ such that $c = c''d$, where c'' is the inflation of c' to $\mathrm{Gal}(E/F)$. Set $A = Z^2(\mathrm{Gal}(F_{a,r}/F), \mathbb{Z})$. Let $\mathrm{ord}(c') \in A$ be the tuple obtained by applying the valuation to each coordinate. Associated with the short exact sequence

$$1 \to \mathbb{Z} \to \mathbb{Q} \to \mathbb{Q}/\mathbb{Z} \to 1 \tag{5.2.3}$$

is a connecting homomorphism

$$\mathrm{Hom}(\mathrm{Gal}(F_{a,r}/F), \mathbb{Q}/\mathbb{Z}) = Z^1(\mathrm{Gal}(F_{a,r}/F), \mathbb{Q}/\mathbb{Z}) \to A. \tag{5.2.4}$$

Let $e(\ell, k)$ be the image in A of the homomorphism from $\mathrm{Gal}(F_{a,r}/F)$ to $\mathbb{Q}/\mathbb{Z}$ that sends the chosen generator τ to ℓ/k. (The predicate $\mathrm{TN}_{\ell,k}$ is defined to be false if k does not divide $r\ell$.) Finally, the predicate $\mathrm{TN}_{\ell,k}$ asserts that $e(\ell, k) = \mathrm{ord}(c') \in A$.

6. Weights

6.1. The weight function. Weighted orbital integrals can also be brought into the framework of constructible functions.

Let G be a connected reductive group over a p-adic field F and let M be a Levi subgroup of G. Let $\mathcal{P}(M)$ be the set of parabolic subgroups P of G that have a Levi decomposition $P = M_P N_P$ with $M = M_P$.

Arthur defines a real-valued weight function $w_M(x) = w_M^G(x)$ on the group $G(F)$. We recall the general form of this function. The function is defined as the value at $\lambda = 0$ of a smooth function

$$v_M(x,\lambda) = \sum_{P\in\mathcal{P}(M)} v_P(x,\lambda)\theta_P(\lambda)^{-1}, \tag{6.1.1}$$

whose terms are indexed by $P \in \mathcal{P}(M)$.

The parameter λ lies in a finite-dimensional real vector space $i\mathbf{a}_M^*$, where $\mathbf{a}_M^*$ is the dual of

$$\mathbf{a}_M = \mathrm{Hom}(X(M)_{rat}, \mathbb{R}), \tag{6.1.2}$$

and where $X(M)_{rat}$ is the group of F-rational characters of M. The function

$$\theta_P(\lambda) \in S^p[\mathbf{a}_{M,\mathbb{C}}] \tag{6.1.3}$$

is a nonzero homogeneous polynomial of some degree p. The degree p of this form is independent of $P \in \mathcal{P}(M)$.

The factor $v_P(x,\lambda)$ is defined as follows. We may assume a choice of a hyperspecial maximal compact subgroup K of $G(F)$ such that is admissible in the sense of Arthur [**3**, p.9]. Then the Iwasawa decomposition takes the form

$$G(F) = N_P(F)M_P(F)K. \tag{6.1.4}$$

Then for $x = nmk \in G(F)$, set $H_P(x) = H_M(m)$, where $H_M(m)$ is defined by the condition

$$\langle H_M(m), \chi\rangle = -\mathrm{ord}\chi(m) \tag{6.1.5}$$

for all $\chi \in X(M)_{rat}$. The function $v_P(\lambda, x)$ is then defined to be

$$v_P(\lambda, x) = e^{-\lambda(H_P(x))}. \tag{6.1.6}$$

The function $v_M(\lambda, x)$ defined by Equation 6.1.1 is not obviously a smooth function of $\lambda \in i\mathbf{a}_M^*$, because the individual summands $v_P(\lambda,x)/\theta_P(\lambda)$ do not extend continuously to $\lambda = 0$. But according to a theorem of Arthur, it is smooth.

Arthur gives the function in the following alternative form. Fix any generic λ. Let t be a real parameter. The denominator is homogeneous of degree p, so $\theta_P(t\lambda) = t^p\theta_P(\lambda)$. We compute the limit of $v_M(t\lambda, x)$ as t tends to zero by applying l'Hôpital's rule p times. The result is

$$v_M(x) = \sum_{P\in\mathcal{P}} \frac{(-1)^p(\lambda(H_P(x)))^p}{p!\theta_P(\lambda)}. \tag{6.1.7}$$

The right-hand side appears to depend on λ, but in fact it is constant as function of λ.

6.2. Weights and constructible functions. Let us recall our context for unramified groups. Associated to root data D and an automorphism θ of D that preserves positive roots, there is a formula $\varphi_{D,\theta}(a,\tau,g)$ in $1+r^2+n^2r$ variables where a determines an unramified field extension $F_{a,r}/F$ of degree r, τ is a generator of the Galois group of $F_{a,r}/F$, and g is an element of the quasi-split group determined by the root data D, automorphism θ, field extension $F_{a,r}/F$, and τ. This is a definable subassignment $\tilde{G} = \tilde{G}_{D,\theta}$.

For each p-adic field F, a and τ, the definable subassignment $\tilde{G} = \tilde{G}_{D,\theta}$, gives a reductive group G and a hyperspecial maximal compact subgroup K (by taking the integer points of G).

We pick Levi factors in standard position in the usual way. We fix a splitting $(B, T, \{X_\alpha\})$ of the split group G as in Section 4. We may take parabolic subgroups P containing B and Levi factors M generated by T and by a subset of the roots vectors $\{X_\alpha \mid \alpha \in S\}$, with S a subset of the set of simple roots. For each subset S of simple roots, we have a definable subassignment $\tilde{M}_S \subset \tilde{G}$, defined by a formula $\varphi_{S,D,\theta}(a,\tau,m)$, with a and τ as before, and m constrained to be an element of the Levi factor $M_{S,a,\tau}(F)$ of the reductive group $G_{a,\tau}$ attached to D, θ, a, τ. If S is the set of all simple roots, then $\tilde{M}_S = \tilde{G}$.

LEMMA 6.1. *Let (D,θ) be root data and an automorphism as above. For each subset S of the simple roots, There is a constructible function u_S on $\tilde{G}_{D,\theta}$ such that for for any p-adic field F,*

$$u_S(a,\tau,g) = v_M(g), \tag{6.2.1}$$

where $M = M_{S,a,\tau}$.

Proof. Constructible functions form a ring. Therefore if we express u_S as a polynomial in functions that are known to be constructible, then it follows that u_S itself is constructible.

The character group $X(M)_{rat}$ is independent of the field F. The coefficients of the form θ_P are independent of the field F. We may compute v_M from Equation 6.1.7 with respect to any sufficiently generic λ. In particular, we may choose $\lambda \in X(M)_{rat}$, once for all fields F. Then $\theta_P(\lambda)$ is a non-zero number that does not depend on F.

By definition, the function H_M on $M(F)$ is a vector of valuations of polynomial expressions in the matrix coefficients of m. The linear form $\lambda(H_M(m))$ is then clearly a constructible function.

Consider the function $\lambda(H_P(g)) = \lambda(H_M(m))$, where $g = nmk$. If we add the parameters, a, τ, its graph is

(6.2.2)

$$\{((a,\tau,g),\ell) \in \tilde{G}\times\mathbb{Z} \mid \exists k \in K,\ m \in M, n \in N_P,\ g = nmk,\ \ell = \lambda(H_M(m))\}.$$

This is a definable subassignment. Hence,

$$(a,\tau,g) \mapsto \lambda(H_P(g)), \quad P = P_{a,\tau} \tag{6.2.3}$$

is a constructible function.

As the function $v_M(g)$ is constructed as a polynomial in $\lambda(H_P(g))$ with rational coefficients, as P ranges over $\mathcal{P}(M)$, it is now clear that it can be lifted to a constructible function $u_S(a,\tau,g)$, according to the description of constructible functions recalled in Section 2.3. □

7. Measures

7.1. Differential forms. Let G be a split reductive group over $\mathbb{Q}$. Let B be a Borel subgroup of G and T a Cartan subgroup contained in B. Let N be the unipotent radical of B and Let N' be the unipotent radical of the Borel subgroup opposite to B through T. The big cell TNN' is a Zariski open subset of G. The Haar measure of G is the measure attached to the differential form, expressed on the open cell as

(7.1.1) $$\omega_G = d^*t \wedge dn \wedge dn'.$$

where d^*t, dn, dn' are differential forms of top degree on T, N, and N', which are bi-invariant by the actions of T, N, and N', respectively. The radical N (and N') can be identified with affine spaces and differential forms dn come from a choice of root vectors for the algebra N.

(7.1.2) $$dn = dx_1 \wedge \cdots \wedge dx_k.$$

We pick the root vectors to respect the rational structure of G. If we identify T with a split torus $\mathbb{G}_m^r$, then d^*t takes the form of a multiplicatively invariant form

(7.1.3) $$\frac{dt_1}{t_1} \wedge \cdots \wedge \frac{dt_r}{t_r}.$$

For any p-adic field, we obtain a Haar measure $|\omega_G|$ on $G(F)$ as the measure attached to the form ω_G.

Similarly, for an endoscopic group H, there is a Haar measure on $H(F)$ obtained from a similarly constructed differential form ω_H. According to calculations of Langlands and Shelstad of the Shalika germ associated with regular unipotent conjugacy classes ([**27**], [**33**]), the invariant measures on cosets $T\backslash G$ and $T\backslash H$ that are be used for the fundamental lemma have the form

(7.1.4) $$|d_T t|\backslash|\omega_G|, \quad |d_T t|\backslash|\omega_H|,$$

where $|d_T t|$ is a Haar measure on T, considered as a Cartan subgroup of both G and H.

LEMMA 7.1. *There is a definable subassignment $\tilde{G}_{ell}$ of regular semisimple elliptic elements of $\tilde{G}_{D,\theta}$. More generally, for each S subset of simple roots, defining a standard Levi subgroup, there is a definable subassignment $\tilde{G}_{S,ell}$ of regular semisimple elements of $\tilde{G}_{G,\theta}$ that are conjugate to an element of $\tilde{M}_S$ and that are elliptic in $\tilde{M}_S$. Similarly, there are definable subassignments $\tilde{\mathfrak{g}}_{S,ell}$ of regular semisimple elements of $\tilde{\mathfrak{g}}$.*

Proof. For each a, τ, the constraints on g are that it is not conjugate to a standard Levi subgroup of $G_{a,\tau}$, that there is a field extension (of some fixed degree k) over which g can be diagonalized, and that the Weyl discriminant is nonzero. These conditions are all readily expressed as a formula in the Denef-Pas language. The proof for the other statements are similar. □

A conjugacy class is said to be bounded if each element γ in that conjugacy class belongs to a compact subgroup of $G(F)$. Arthur states the weighted fundamental lemma (at the group level) in terms of bounded conjugacy classes. Similarly, we have the following lemma:

LEMMA 7.2. *There is a definable subassignment $\tilde{G}_{bd}$ of bounded semisimple conjugacy classes of $\tilde{G}_{D,\theta}$.*

7.2. Weyl integration formula. The Weyl integration formula can be used to fix normalizations of measures. By [**34**, page 36], the Weyl integration formula takes the form

$$\int_{\mathfrak{g}} f(X)dX = \sum_T |W(G,T)|^{-1} \int_t |D^{\mathfrak{g}}(X)| \times \int_{G/T} f(\mathrm{Ad}\, xX)dx\, dX. \tag{7.2.1}$$

The measure dX is the Serre-Oesterlé measure, which is an additive Haar measure on $\mathbf{g}$. This is compatible with the invariant form ω_G on G in the sense of [**12**]. The sum runs over conjugacy classes of Cartan subgroups. The factor $|W(G,T)|$ is the order of the Weyl group for T; and $|D^{\mathfrak{g}}(X)|$ is the usual discriminant factor. The measure dx is the quotient measure normalized by ω_G on G and dX on $\mathbf{t}$.

Let $\mathbf{c}_G = \mathrm{spec}\,(R^G)$, where R is the coordinate ring of $\mathbf{g}$, and R^G is the subring of G-invariants. We have a morphism $\mathbf{g} \to \mathbf{c}_G$, coming from the inclusion $R^G \subset R$. For $\mathbf{t} \subset \mathbf{g}$, the morphism $\mathbf{t} \to \mathbf{g} \to \mathbf{c}_G$ is $W(G,T)$-to-1 on the set of regular semisimple elements. The fiber over a regular semisimple element $\gamma \in \mathbf{c}_G$ is the stable orbit of γ in $\mathbf{g}$. Comparing this to the Weyl integration formula, we see that the choice of measure on fibers of the morphism $\mathbf{g} \to \mathbf{c}_G$ determined by the Serre-Oesterlé measures on $\mathbf{g}$ and on $\mathbf{c}_G$ is an invariant measure dx. Similar comments apply to endoscopic groups H of G.

These observations allow us to conclude that invariant measures on stable conjugacy classes are compatible with the general framework of [**11**].

8. The Langlands-Shelstad Transfer Factor

8.1. Langlands dual group. In this section we present a few facts related to the Langlands-Shelstad transfer factor. The Langlands-Shelstad transfer factor was originally defined for pairs of elements in a group G and a fixed endoscopic group. By considering the limiting behavior of this transfer factor near the identity element of the group, we obtain a transfer factor on the Lie algebra.

The fundamental lemma for the Lie algebra and the Lie group are closely related. See [**16**] for the relation in the unweighted case and the Appendix to

[**36**] for a sketch of the relation in the full weighted case of the fundamental lemma.

We work with the Lie algebra version of the transfer factor. The Lie algebra transfer factor avoids the additional complications of multiplicative characters that occur in the Lie group transfer factor.

Let F be a p-adic field. We fix an unramified connected reductive group G over F with Lie algebra $\mathbf{g}$. The L-group of G may be written in the form

$$ {}^{L}G = \hat{G} \rtimes \Gamma, \tag{8.1.1}$$

where Γ is the Galois group of a splitting field of G. The group Γ acts by automorphisms of $\hat{G}$ that fix a splitting $(\hat{T}, \hat{B}, \{X_\alpha\})$.

8.2. The parameter s **and endoscopy.** Let (s, ρ) be an unramified endoscopic datum for G. The element s belongs to $\hat{G}/Z(\hat{G})$. Taking its preimage in $\hat{G}$ and replacing the endoscopic datum by another isomorphic to it, we may take $s \in \hat{T}$ to be a semisimple element, whose connected centralizer is defined to be $\hat{H}$. We may assume that s has finite order.

By definition $\rho : \mathrm{Gal}(F_{a,r}/F) \to \mathrm{Out}(\hat{H})$ is a homomorphism for some unramified extension $F_{a,r}/F$ into the group of outer automorphisms of H. See [**24**, sec.7] for a review of endoscopic data. The expository paper [**17**] gives some additional details in the context of the fundamental lemma.

Let T be a maximally split Cartan subgroup of H and $\hat{T}$ its dual, which we identify with the Cartan subgroup containing s. We may assume further that $s \in \hat{T}^{\mathrm{Gal}(F_{a,r}/F)}$. Let $\mathbf{t}$ be the Lie algebra of $\hat{T}$. The exponential short exact sequence

$$1 \to X_*(\hat{T}) \to \mathbf{t} \to \hat{T} \to 1, \tag{8.2.1}$$

where $\mathbf{t} \to \hat{T}$ is the exponential map $\lambda \mapsto \exp(2\pi i \lambda)$, gives a connecting homomorphism

$$\hat{T}^{\mathrm{Gal}(F_{a,r}/F)} \to H^1(F, X^*(T)). \tag{8.2.2}$$

The image of s is obtained explicitly as follows. Write $s = \exp(2\pi i\lambda/k)$ for some $\lambda \in X_*(T)$ and $k > 0$. Then the cocycle $\mu_\sigma \in Z^1(F, X_*(\hat{T}))$ is given by the equation

$$k\mu_\sigma = (\sigma(\lambda) - \lambda), \quad \sigma \in \mathrm{Gal}(F_{a,r}/F). \tag{8.2.3}$$

The action of $\mathrm{Gal}(F_{a,r}/F)$ on $X_*(T)$ comes through the action of $W_H \rtimes \langle\theta\rangle$ on $X_*(T)$ by means of a homomorphism

$$\mathrm{Gal}(F_{a,r}/F) \to W_H \rtimes \langle\theta\rangle \tag{8.2.4}$$

(where W_H is the Weyl group of H). Instead of the cocycle μ_σ, we prefer to work with the corresponding cocycle

$$\mu_w \in Z^1(W_H \rtimes \langle\theta\rangle, X_*(\hat{T})) \tag{8.2.5}$$

$$k\mu_w = (w(\lambda) - \lambda), \quad w \in W_H \rtimes \langle\theta\rangle. \tag{8.2.6}$$

From the point of view of definability in the Denef-Pas language, a complex parameter s is problematical. However, the parameter $\lambda \in X^*(T)$ presents no difficulties, so we discard s and work directly with λ and the fixed natural number k. We consider $X^*(T)$ as within the coordinate ring of T. Through an explicit matrix representation of H, for any $\mu \in X^*(T)$, we may choose a polynomial P_μ representing μ in the coordinate ring of $n \times n$ matrices. The polynomials satisfy $P_\mu P_{\mu'} = P_{\mu+\mu'} \mod I$, where I is the ideal of T.

Equation 8.2.6 can be rewritten as a collection of polynomial identities

$$P_w^k = P_{w\lambda} P_{-\lambda} \mod I, \quad w \in W_H \rtimes \langle \theta \rangle, \tag{8.2.7}$$

where $P_w = P_{\mu_w}$. The collection of polynomials $\{P_w\}$ then serve as the proxy for the complex parameter s.

8.3. The normalization of transfer factors. Langlands and Shelstad defined the transfer factor, up to a scalar factor. There is a unique choice of scalar factor for the transfer factor that is compatible with the fundamental lemma. Following Kottwitz, this normalization of the transfer factor is based on a Kostant section associated to a regular nilpotent element. Let $\mathbf{g}$ be the Lie algebra of a split reductive group G. Let $\mathbf{c}_G$ be the space defined in Section 7. There is a natural morphism $\mathbf{g} \to \mathbf{c}_G$. Kostant defines a section $\mathbf{c}_G \to \mathbf{g}$ of this morphism [**22**]. The construction of this section depends on a choice X of regular nilpotent element. Kostant's construction, which is based on $\mathbf{sl}_2$ triples, can easily be carried out in the context of the Denef-Pas language. In fact, once the element X is given, the construction requires only the elementary theory of rings.

We fix a regular nilpotent element as follows. Fix a splitting $(\mathbf{b}, \mathbf{t}, \{X_\alpha\})$ of $\mathbf{g}$ defined over $\mathbb{Q}$. Pick $X \in \mathbf{b}$ such that

$$X = \sum x_\alpha X_\alpha, \tag{8.3.1}$$

where x_α is a unit for every simple root α.

8.4. The pairing. The Lie algebra version of the Langlands-Shelstad transfer factor $\Delta(\gamma_H, \gamma)$ is defined for pairs $\gamma \in \mathbf{g}$ and $\gamma_H \in \mathbf{c}_H$. It has the form $q^{m(\gamma_H,\gamma)}\Delta_0(\gamma_H, \gamma)$, where Δ_0 is a root of unity, or zero. (It is defined to be zero on the set where γ_H, γ are not matching elements.)

Normalize the transfer factor so that $\Delta_0(\gamma_H, \gamma) = 1$ if γ lies in the Kostant section associated to X. By [**16**], this normalization is independent of the choice of such X (for sufficiently large primes, as usual). Thus, we may take the quantifier over X to be either an existential or a universal quantifier, ranging over all such nilpotent elements.

Let $\tilde{\mathbf{c}}_H \times \tilde{\mathbf{g}}_{D,\theta}$ be the definable subassignment with free variables $(a, \tau, \gamma_H, \gamma)$ with defining condition that γ belongs to the Lie algebra of $\tilde{G}_{D,\theta,a,\tau}$ and γ_H belongs to the quotient $\mathbf{c}_H$ for the corresponding Lie algebra of its endoscopic group $H_{a,\tau}$.

LEMMA 8.1. *There is a constructible function on $\tilde{\mathbf{c}}_H \times \tilde{\mathbf{g}}_{D,\theta}$ that specializes to the function*

$$q^{m(\gamma_H,\gamma)}. \tag{8.4.1}$$

Proof. This is trivial. The constructible function $\mathbb{L}$ specializes to q. The ring of constructible functions contains functions of the form $\mathbb{L}^m$, with

$$m : \tilde{\mathbf{h}} \times \tilde{\mathbf{g}}_{D,\theta} \to \mathbb{Z} \tag{8.4.2}$$

definable. The function m is definable, because it is the valuation of a polynomial in the matrix coefficients of γ and γ_H. □

Let $\mathbf{k}$ be a lattice in $\mathbf{g}$ corresponding to a hyperspecial maximal compact subgroup of G. Let $\mathbf{k}_H$ be such a lattice in H. Let γ be a regular semi-simple element in $\mathbf{g}$. Assume that it is the image of some $\gamma_H \in \mathbf{k}_H$. Let $\gamma_0 \in \mathbf{g}$ be an element in the Kostant section with the same image as γ in $\mathbf{c}$. Let $\mathrm{inv}(\gamma_0,\gamma)$ be the invariant attached to γ and κ the character defined by the endoscopic data so that $\Delta_0(\gamma_H,\gamma) = \langle \mathrm{inv}(\gamma_H,\gamma),\kappa\rangle$. The construction depends on an element $g \in G(\bar{F})$ such that $\mathrm{Ad}\ g(\gamma) = \gamma_0$.

8.5. Description of the transfer factor. We arrive at the following statements that summarize the value of the transfer factor for elliptic unramified transfer factors. The description we give is rather verbose. In particular, we write out the coboundaries explicitly rather than taking classes in cohomology. This is intentional to prepare us for the proof that the transfer factor can be expressed in the Denef-Pas language.

Recall that a natural number k is used to define the complex parameter s in terms of λ. The value of the transfer factor is a kth root of unity. For $\ell \in \mathbb{N}$, let $D_\ell(a,\tau,\gamma_H,\gamma_G)$ be the set of elements in $\mathbf{h} \times \mathbf{g}$ such that γ_H is a G-regular semisimple element of $\mathbf{c}_H$, γ_G is a regular semisimple element of $\mathbf{g}$, and $\Delta_0(\gamma_H,\gamma_G) = \exp(2\pi\ell/k)$.

We fix a natural number r that is large enough that the unramified extension $F_{a,r}$ of F of degree r splits G and H, etc.

We fix a natural number $N = N_r$ that is large enough that for every Cartan subgroup in G there exists a field extension of degree dividing N containing $F_{a,r}$ that splits the Cartan subgroup. As we are only interested in the tame situation, we may assume that the residue characteristic p does not divides N.

Let $T_H \subset G$ be the fixed Cartan subgroup of G, obtained by transfer of the maximally split Cartan subgroup of H. $\Delta_0(\gamma_H,\gamma) = 0$ if and only if γ and γ_H do not have the same image in $\mathbf{c}_G$.

The set $D_\ell(a,\tau,\gamma_H,\gamma)$ is described by the following list of conditions. The parameters a and τ are as above.

(1) Let γ_0 be the element in $\mathbf{g}$, constructed as the transfer of γ_H to $\mathbf{g}$, lying in the Kostant section we have fixed.
(2) There exists a Galois field extension of degree E/F of degree N that contains a subfield isomorphic to $F_{a,r}$.

(3) There exists $g \in G(E)$ such that Ad $g(\gamma) = \gamma_0$. Let $t_\sigma = g^{-1}\sigma(g) \in T_0(E)$, for $\sigma \in \mathrm{Gal}(E/F)$, where T_0 is the centralizer of γ_0.
(4) There exists $h \in G(E)$ such that $\mathrm{ad}(h)T_0 = T_H$. Let $t'_\sigma \in Z^1(E/F, T_H^*)$ be the cocycle $\mathrm{ad}(h)t_\sigma$, with a twisted action of $\mathrm{Gal}(E/F)$ on T_H obtained by transporting the action of $\mathrm{Gal}(E/F)$ to T_H via $\mathrm{ad}(h)$. The general properties of endoscopy imply that the cocycle $Z^1(F_{a,r}/F, X^*(T_H))$, defined by the collection of polynomials P_w, also yields a cocycle $\mu_\sigma \in Z^1(E/F, X^*(T_H^*))$ for this twisted action.
(5) The predicate $\mathrm{TN}_{\ell,k}(a, \tau, t'_*, \mu_*)$ of Section 5 expressing the Tate-Nakayama duality holds.

This description of the transfer factor is equivalent to descriptions found elsewhere. We have changed the presentation slightly by working with the polynomials P_w rather than the complex parameter. Beyond that, our description is essentially the standard one. We have written the transfer factor in this form to make it apparent that nothing beyond first-order logic, basic ring arithmetic, and a valuation map are required.

We have provided justification for the following statement. It relies on the fixed natural number k that is part of our setup.

LEMMA 8.2. *There is a definable subassignment $D_\ell(a, \tau, \gamma_H, \gamma_G)$ with the following interpretation. The parameter a defines an unramified field extension $F_{a,r}$ of degree r, and τ is a generator of* $\mathrm{Gal}(F_{a,r}/F)$. *The parameter γ_H ranges over G-regular elements in $\mathbf{c}_H = \mathbf{h}^H$ of the Lie algebra $\mathbf{h}$ of the endoscopic group $H_{a,\tau}$. The parameter γ_G is a regular semisimple element in the Lie algebra $\mathbf{g}$ of the reductive group $G_{a,\tau}$. The elements γ_H and γ_G are matching elements, and the transfer factor takes the value* $\Delta_0(\gamma_H, \gamma_G) = \exp(2\pi i\ell/k)$.

We remark that the definition of D_ℓ can be described in a smaller language with two sorts, one for the valued field and another for the value group. The residue field sort and the angular component map do not enter into the description of D_ℓ.

Moreover, we have only made light use of the valuation. It is used once to fix the choice of nilpotent element that is used to construct the Kostant section. The valuation is used once again in the predicate $\mathrm{TN}_{\ell,k}$ when working with the Brauer group.

Proof. This follows directly from the description of the transfer factor in terms of ring operations and quantifiers, as provided. □

We have not made explicit estimates of the complexity of the formulas in the Denef-Pas language that define the subassignments D_ℓ. For any n, we may define $q(n)$ to be the smallest integer q such that every subassignment D_ℓ of every unramified group of rank at most n can be defined by a formula in the Denef-Pas language that contains at most q quantifiers. How rapidly does $q(n)$ grow with n?

8.6. Weighted orbital integrals. Let G be an unramified reductive group with hyperspecial maximal compact subgroup K. Let M be a Levi subgroup that is in good position with respect to K. Let M' be an unramified elliptic endoscopic group of M. Write $\mathbf{g}$, $\mathbf{k}$, $\mathbf{m}$, and so forth, for the corresponding Lie algebras. Let 1_K be the normalized unit of the Hecke algebra. The normalization appropriate for our choice of measures is discussed in [**17**].

Let $\ell' \in \mathbf{c}_{M'}$ be a G-regular element. The image ℓ_M of ℓ' in $\mathbf{c}_M$ determines a stable conjugacy class $C_M(\ell')$ in $\mathbf{m}$, given as the fiber in $\mathbf{m}$ over $\ell_M \in \mathbf{c}_M$. The image ℓ_G of ℓ' in $\mathbf{c}_G$ gives determines a stable conjugacy class $C_G(\ell')$ in $\mathbf{g}$. Any element $x \in C_G(\ell')$ permits a representation

$$x = \operatorname{Ad} g\,\gamma \tag{8.6.1}$$

with $g \in G(F)$ and $\gamma \in C_M(\ell')$. The map

$$x \mapsto (\gamma_x, g_x) = (\gamma, g) \tag{8.6.2}$$

is well-defined up to conjugacy in $M(F)$:

$$(\gamma_x, g_x) \mapsto (\operatorname{Ad} m\,\gamma_x, g_x m^{-1}). \tag{8.6.3}$$

In particular, the function

$$x \mapsto \Delta_{M',M}(\ell', \gamma_x) v_M(g_x) 1_K(x) \tag{8.6.4}$$

is well-defined, where $\Delta_{M',M}$ is the normalized transfer factor for reductive group M and endoscopic group M'. We have already established the constructibility of Δ and v_M. The Function (8.6.4) involves no more than Δ, v_M and additional existential quantifiers of the valued field sort, corresponding to the existence of the representation of Equation (8.6.1). In particular, Function (8.6.4) is constructible. (Here and elsewhere, we are slightly loose in our use of the term constructible. The precise sense of constructibility is stated in Lemmas 6.1 and 8.2.)

Define the weighted orbital integral to be

$$J^G_{M,M'}(\ell') = \int_{C_G(\ell')} \Delta_{M',M}(\ell', \gamma_x) v_M(g_x) 1_K(x), \tag{8.6.5}$$

where the integral is with respect to the quotient of the Serre-Oesterlé measures on $\mathbf{m}$ and $\mathbf{c}_M$.

8.7. The example $SL(n)$. As an example, we describe the constructible function for the transfer factor of $SL(n)$, when n factors as $n = mr$. Let E/F be an unramified extension of p-adic fields of degree r. The group $SL(n)$ over F has an endoscopic group H, where $H(F)$ is the subgroup of $GL(m, E)$ of elements whose determinant has norm 1. A basis of E/F gives an embedding of H into $SL(n)$. For the fundamental lemma, we are concerned with the transfer factor evaluated at an element in $SL(n)$ that is the image of some $\gamma \in GL(m, O_E)$, under this embedding.

Suppose that f and g are monic polynomials with roots $x_1, \ldots, x_m$ and $y_1, \ldots, y_{m'}$, respectively in an algebraic closure. The resultant $R(f, g)$ of f

and g is the product $\prod_{i,j}^{m,m'}(x_i - y_j)$. If γ_1, γ_2 are elements of Mat_ℓ, then let $R(\gamma_1, \gamma_2) = R(f_{\gamma_1}, f_{\gamma_2})$, where f_γ is the characteristic polynomial of γ. For $\gamma \in \mathrm{Mat}_m(O_E)$, define

$$\Delta_{E/F}^{m,1}(\gamma) = \Big| \prod_{\sigma \neq \tau \in \mathrm{Gal}(E/F)} R(\sigma(\gamma), \tau(\gamma)) \Big|^{\frac{1}{2}}.$$

When r is odd, define $\Delta_{E/F}^{m,2}(\gamma) = 1$. When r is even, let σ_+ denote the element of order 2 in $\mathrm{Gal}(E/F)$. Let η_E denote the unramified character of $E^\times$ of order 2. Define

$$\Delta_{E/F}^{m,2}(\gamma) = \eta_E\Big(R(\gamma, \sigma_+(\gamma))\Big).$$

Let $\Delta_{E/F}^{m}(\gamma) = \Delta_{E/F}^{m,1}(\gamma)\,\Delta_{E/F}^{m,2}(\gamma)$. By [**15**], this is the transfer factor (on the group).

To express the set

$$D_\pm = \{\gamma \mid \Delta_{E/F}^{m,2}(\gamma) = \pm 1\}$$

(when r is even) as a formula in the Denef-Pas language, pick a unit a such that $x^r - a$ is irreducible, and write elements of an unramified extension of degree r as linear combinations of $\{1, x, \ldots, x^{r-1}\}$. The automorphism σ_+ is given by $x \mapsto -x$. The element γ is a matrix

$$\gamma = (\gamma_{ij}), \quad \gamma_{ij} = \sum_{k=0}^{r-1} \gamma_{ijk} x^k.$$

The resolvant is a determinant formed from the coefficients γ_{ij} of the characteristic polynomial of γ. Since E/F is unramified, $\eta_E(x)$ is 1, exactly when the valuation of the norm $N_{E/F}x \in F^\times$ is a multiple of $2r$. Furthermore, the norm of the resolvant is a polynomial $n(\gamma_{ijk})$ in the coefficients γ_{ijk}. We obtain a formula in the Denef-Pas language of the form

$$D_+ = \{\gamma = (\gamma_{ijk}) \mid \exists l \in \mathbb{Z}.\quad \mathrm{ord}\,(n(\gamma_{ijk})) = 2rl\}$$

The formula for D_- is obtained similarly. The function

$$\Delta_{E/F}^{m,1}(\gamma) = q^{e(\gamma)}$$

may appear to have a fractional exponent, but in fact, it is an integer. Our description of $\Delta_{E/F}^{m,1}$ leads to a similar expression for $e(\gamma)$ as the valuation of a polynomial in the coefficients γ_{ijk}.

9. The statement of the Fundamental Lemma

9.1. Weighted fundamental lemma. In [**4**, Conj. 5.1], Arthur conjectures a weighted form of the fundamental lemma. In this section, we review his formulation of the conjecture. The weighted case includes the standard fundamental lemma of Langlands and Shelstad as a special case.

(For the formulation of the fundamental lemma in the twisted weighted context, see Arthur's appendix to [**37**].)

We continue to work with the Lie algebras of endoscopic groups rather than the endoscopic groups themselves, although this makes very little difference for our purposes.

Our work is now almost complete. We have already described all of the constituents of the fundamental lemma.

Let G be an unramified reductive group with hyperspecial maximal compact subgroup K. Let M be a Levi subgroup that is in good position with respect to K. Let M' be an unramified elliptic endoscopic group of M. Write $\mathbf{g}$, $\mathbf{k}$, $\mathbf{m}$, and so forth, for the corresponding Lie algebras.

To state the weighted fundamental lemma, we need the following additional data. For each G, M, M', there is a set of endoscopic data $\mathcal{E}_{M'}(G)$, defined in [**2**, Sec. 4]. The set $\mathcal{E}_{M'}$ is defined by data in the dual group that is independent of the p-adic field. (The action of the Galois group Γ on the dual group data can be replaced with the action of the automorphism θ.) For each, $G' \in \mathcal{E}_{M'}(G)$, there is a rational number number $\iota_{M'}(G, G') \in \mathbb{Q}$, that is independent of the p-adic field.

We define a function $s_M^G(\ell)$ recursively. Assume that $s_M^{G'}$ has been defined for all G' (with Levi subgroup M) such that dim $G' <$ dim G. Then, set

$$s_M^G(\ell) = J_{M,M}^G(\ell) - \sum_{G' \neq G} \iota_M(G, G') s_M^{G'}(\ell). \tag{9.1.1}$$

The sum runs over $\mathcal{E}_M(G) \setminus \{G\}$. This definition is coherent, because each group $G' \in \mathcal{E}_M(G)$ has M as a Levi subgroup, so that $s_M^{G'}$ is defined.

The conjecture of the weighted fundamental lemma is then that for all G, M, M' as above, we have

$$J_{M,M'}^G(\ell') = \sum_{G'} \iota_{M'}(G, G') s_{M'}^{G'}(\ell'). \tag{9.1.2}$$

for all G-regular elements ℓ' in $\mathbf{c}_{M'}$. The sum on the right runs over $\mathcal{E}_{M'}(G)$.

9.2. Constructibility. By our preceding discussion, we see that the integrand (Equation 8.6.4) of $J_{M,M'}^G(\ell')$ comes as specialization of a constructible function on the definable subassignment

$$Z = \tilde{\mathbf{c}}_H \times_{\tilde{\mathbf{c}}_G} \tilde{\mathbf{g}}_{D,\theta}. \tag{9.2.1}$$

This constructible function depends on parameters a, τ, $\gamma_H \in \tilde{\mathbf{c}}_H$, and $\gamma \in g_{D,\theta,a,\tau}$. If we interpret this p-adically, as we vary the parameter a (under the restriction that it is a unit), the situation specializes to isomorphic groups and Lie algebras. In particular, the fundamental lemma holds for one specialization of a if and only if it holds for all specializations of a. As we vary the generator τ of the Galois group of the unramified field extension $F_{a,r}/F$, we may obtain non-isomorphic data. Different choices of τ correspond to the fundamental lemma for various Lie algebras

$$\mathfrak{g}_{D,\theta'} \tag{9.2.2}$$

where θ' and θ generate the same group $\langle\theta'\rangle = \langle\theta\rangle$ of automorphisms of the root data. In particular, for each τ, the constructible version specializes to a version of the p-adic fundamental lemma for Lie algebras.

9.3. The main theorem. We state the transfer principle for the fundamental lemma as two theorems, once in the unweighted case and again in the weighted case. In fact, there is no needed for us to treat these two cases separately; they are both a consequence of the general transfer principle for the motivic integrals of constructible functions given in Theorem 2.6. We state them as separate theorems, only because of preprint of Ngô [**30**], which applies directly to the unweighted case of the fundamental lemma.

THEOREM 9.1 (Transfer Principle for the Fundamental Lemma). *Let (D, θ) be given. Suppose that the fundamental lemma holds for all p-adic fields of sufficiently large positive characteristic for the endoscopic groups attached to (D, θ'), as θ' ranges over automorphisms of the root data such that $\langle\theta'\rangle = \langle\theta\rangle$. Then, the fundamental lemma holds for all p-adic fields of characteristic zero with sufficiently large residual characteristic p (in the same context of all endoscopic groups attached to (D, θ')).*

THEOREM 9.2 (Transfer Principle for the weighted Fundamental Lemma). *Let (D, θ) be given. Suppose that the weighted fundamental lemma holds for all p-adic fields of sufficiently large positive characteristic for the endoscopic groups attached to (D, θ'), as θ' ranges over automorphisms of the root data such that $\langle\theta'\rangle = \langle\theta\rangle$. Then, the weighted fundamental lemma holds for all p-adic fields of characteristic zero with sufficiently large residual characteristic p (in the same context of all endoscopic groups attached to (D, θ')).*

Proof. We have successfully represented all the data entering into the fundamental lemma within the general framework of identities of motivic integrals of constructible functions. By the transfer principle given in Theorem 2.6, the fundamental lemma holds for all p-adic fields of characteristic zero, for sufficiently large primes p. □

By the main result of [**18**], the unweighted fundamental lemma holds for all elements of the Hecke algebra for all p, once it holds for all sufficiently large p (for a collection of endoscopic data obtained by descent from the original data (D, θ)). Thus, in the unweighted situation, we can derive the fundamental lemma for all local fields of characteristic zero, without restriction on p, once the fundamental lemma is known for a suitable collection of cases in positive characteristic.

10. Additive characters and the relative fundamental lemma

10.1. Adding exponentials. It is also possible to enlarge $\mathrm{C}(X)$ to a ring $\mathrm{C}(X)^{\exp}$ also containing motivic analogues of exponential functions and to construct a natural extension of the previous theory to $\mathrm{C}^{\exp}$.

This is performed as follows in [**10**] [**11**]. Let X be in Def_k. We consider the category $\mathrm{RDef}_X^{\exp}$ whose objects are triples $(Y \to X, \xi, g)$ with Y in RDef_X and $\xi : Y \to h[0,1,0]$ and $g : Y \to h[1,0,0]$ morphisms in Def_k. A morphism $(Y' \to X, \xi', g') \to (Y \to X, \xi, g)$ in $\mathrm{RDef}_X^{\exp}$ is a morphism $h : Y' \to Y$ in Def_X such that $\xi' = \xi \circ h$ and $g' = g \circ h$. The functor sending Y in RDef_X to $(Y, 0, 0)$, with 0 denoting the constant morphism with value 0 in $h[0,1,0]$, resp. $h[1,0,0]$ being fully faithful, we may consider RDef_X as a full subcategory of $\mathrm{RDef}_X^{\exp}$. To the category $\mathrm{RDef}_X^{\exp}$ one assigns a Grothendieck ring $K_0(\mathrm{RDef}_X^{\exp})$ defined as follows. As an abelian group it is the quotient of the free abelian group over symbols $[Y \to X, \xi, g]$ with $(Y \to X, \xi, g)$ in $\mathrm{RDef}_X^{\exp}$ by the following four relations

$$[Y \to X, \xi, g] = [Y' \to X, \xi', g'] \tag{10.1.1}$$

for $(Y \to X, \xi, g)$ isomorphic to $(Y' \to X, \xi', g')$,

$$\begin{aligned}[(Y \cup Y') \to X, \xi, g] + [(Y \cap Y') \to X, \xi_{|Y \cap Y'}, g_{|Y\cap Y'}] \\ = [Y \to X, \xi_{|Y}, g_{|Y}] + [Y' \to X, \xi_{|Y'}, g_{|Y'}]\end{aligned} \tag{10.1.2}$$

for Y and Y' definable subassignments of some W in RDef_X and ξ, g defined on $Y \cup Y'$,

$$[Y \to X, \xi, g + h] = [Y \to X, \xi + \bar{h}, g] \tag{10.1.3}$$

for $h : Y \to h[1,0,0]$ a definable morphism with $\mathrm{ord}(h(y)) \geq 0$ for all y in Y and $\bar{h}$ the reduction of h modulo t, and

$$[Y[0,1,0] \to X, \xi + p, g] = 0 \tag{10.1.4}$$

when $p : Y[0,1,0] \to h[0,1,0]$ is the projection and when $Y[0,1,0] \to X$, g, and ξ factorize through the projection $Y[0,1,0] \to Y$. Fiber product endows $K_0(\mathrm{RDef}_X^{\exp})$ with a ring structure.

Finally, one defines the ring $\mathrm{C}(X)^{\exp}$ of exponential constructible functions as $\mathrm{C}(X)^{\exp} := \mathrm{C}(X) \otimes_{K_0(\mathrm{RDef}_X)} K_0(\mathrm{RDef}_X^{\exp})$. One defines similarly $C(X)^{\exp}$ and $\mathrm{I}_S C(X)^{\exp}$.

In [**10**] [**11**], given S in Def_k, we construct for a morphism $f : X \to Y$ in Def_S a push-forward $f_! : \mathrm{I}_S C(X)^{\exp} \to \mathrm{I}_S C(Y)^{\exp}$ extending $f_! : \mathrm{I}_S C(X) \to \mathrm{I}_S C(Y)$ and characterized by certain natural axioms. In particular, the construction of the measure μ and its relative version μ_Λ extend to the exponential setting.

10.2. Specialization and transfer principle. We denote by $\mathrm{C}(S, \mathcal{L}_\mathcal{O})^{\exp}$ the ring of constructible functions on X definable in $\mathcal{L}_\mathcal{O}$. We denote by $\mathcal{D}_F$ the set of additive characters $\psi : F \to \mathbb{C}^\times$ such that

$\psi(x) = \exp((2\pi i/p)\mathrm{Tr}_{k_F}(\bar{x}))$ for $x \in R_F$, with p the characteristic of k_F, Tr_{k_F} the trace of k_F relatively to its prime field and $\bar{x}$ the class of x in k_F.

The construction of specialization explained in 2.8 extends as follows to the exponential case. Let φ be in $K_0(\mathrm{RDef}_X(\mathcal{L}_\mathcal{O}))^{\exp}$ of the form $[W, g, \xi]$. For ψ_F in $\mathcal{D}_F$, one specializes φ into the function $\varphi_{F,\psi_F} : X_F \to \mathbb{C}$ given by

$$x \mapsto \sum_{y \in \pi_F^{-1}(x)} \psi_F(g_F(y)) \exp((2\pi i/p)\mathrm{Tr}_{k_F}(\xi_F(y)))$$

for F in $\mathrm{C}_{\mathcal{O},N}$ with $N \gg 0$. One defines the specialization $\varphi \mapsto \varphi_{F,\psi_F}$ for φ in $\mathrm{C}(X, \mathcal{L}_\mathcal{O})^{\exp}$ by tensor product.

One can show that if φ is relatively integrable, then, for $N \gg 0$ and every F in $\mathrm{C}_{\mathcal{O},N}$, for every λ in Λ_F and every ψ_F in $\mathcal{D}_F$, the restriction $\varphi_{F,\psi_F,\lambda}$ of φ_{F,ψ_F} to $f_F^{-1}(\lambda)$ is integrable.

We denote by $\mu_{\Lambda_F}(\varphi_{F,\psi_F})$ the function on Λ_F defined by

$$\lambda \longmapsto \mu(\varphi_{F,,\psi_F,\lambda}). \tag{10.2.1}$$

The results in 2.8 generalize as follows to the exponential setting:

THEOREM 10.1 (Exponential specialization, Cluckers-Loeser [**10**] [**11**]). *Let $f : S \to \Lambda$ be a morphism in* $\mathrm{Def}(\mathcal{L}_\mathcal{O})$. *Let φ be in* $\mathrm{C}(S, \mathcal{L}_\mathcal{O})^{\exp}$ *relatively integrable with respect to f. For $N \gg 0$, for every F in* $\mathrm{C}_{\mathcal{O},N}$ *and every ψ_F in $\mathcal{D}_F$, we have*

$$(\mu_\Lambda(\varphi))_{F,\psi_F} = \mu_{\Lambda_F}(\varphi_{F,\psi_F}). \tag{10.2.2}$$

THEOREM 10.2 (Exponential abstract transfer principle, Cluckers-Loeser [**10**] [**11**]). *Let φ be in* $\mathrm{C}(\Lambda, \mathcal{L}_\mathcal{O})^{\exp}$. *There exists N such that for every F_1, F_2 in* $\mathrm{C}_{\mathcal{O},N}$ *with $k_{F_1} \simeq k_{F_2}$,*

$$\begin{gathered} \varphi_{F_1,\psi_{F_1}} = 0 \quad \textit{for all} \quad \psi_{F_1} \in \mathcal{D}_{F_1} \\ \textit{if and only if} \\ \varphi_{F_2,\psi_{F_2}} = 0 \quad \textit{for all} \quad \psi_{F_2} \in \mathcal{D}_{F_2}. \end{gathered} \tag{10.2.3}$$

THEOREM 10.3 (Exponential transfer principle for integrals with parameters, Cluckers-Loeser [**10**] [**11**]). *Let $S \to \Lambda$ and $S' \to \Lambda$ be morphisms in* $\mathrm{Def}(\mathcal{L}_\mathcal{O})$. *Let φ and φ' be relatively integrable functions in* $\mathrm{C}(S, \mathcal{L}_\mathcal{O})^{\exp}$ *and* $\mathrm{C}(S', \mathcal{L}_\mathcal{O})^{\exp}$, *respectively. There exists N such that for every F_1, F_2 in* $\mathrm{C}_{\mathcal{O},N}$ *with $k_{F_1} \simeq k_{F_2}$,*

$$\begin{gathered} \mu_{\Lambda_{F_1}}(\varphi_{F_1,\psi_{F_1}}) = \mu_{\Lambda_{F_1}}(\varphi'_{F_1,\psi_{F_1}}) \quad \textit{for all} \quad \psi_{F_1} \in \mathcal{D}_{F_1} \\ \textit{if and only if} \\ \mu_{\Lambda_{F_2}}(\varphi_{F_2,\psi_{F_2}}) = \mu_{\Lambda_{F_2}}(\varphi'_{F_2,\psi_{F_2}}) \quad \textit{for all} \quad \psi_{F_2} \in \mathcal{D}_{F_2}. \end{gathered}$$

10.3. Jacquet-Ye integrals and the relative fundamental lemma. A specific situation where Theorem 2.6 applies is that of Jacquet-Ye integrals. Let E/F be a unramified degree two extension of non archimedean local fields of residue characteristic $\neq 2$ and let ψ be a non trivial additive character of F of conductor $\mathcal{O}_F$. Let N_n be the group of upper triangular matrices with 1's on the diagonal and consider the character $\theta : N_n(F) \to \mathbb{C}^\times$ given by

$$\theta(u) := \psi(\sum_i u_{i,i+1}). \tag{10.3.1}$$

For a the diagonal matrix $(a_1, \cdots, a_n)$ with a_i in $F^\times$, we consider the integral

$$I(a) := \int_{N_n(F)\times N_n(F)} \mathbf{1}_{M_n(\mathcal{O}_F)}({}^t u_1 a u_2)\, \theta(u_1 u_2)\, du_1 du_2. \tag{10.3.2}$$

Here du denote the Haar measure on $N_n(F)$ with the normalization $\int_{N_n(\mathcal{O}_F)} du = 1$.

Similarly, one defines

$$J(a) := \int_{N_n(E)} \mathbf{1}_{M_n(\mathcal{O}_E)\cap H_n}({}^t \bar{u} a u)\, \theta(u\bar{u})\, du, \tag{10.3.3}$$

with H_n the set of Hermitian matrices.

The Jacquet-Ye Conjecture [**20**], proved by Ngô [**29**] over function fields and by Jacquet [**21**] in general, asserts that

$$I(a) = \gamma(a)\, J(a) \tag{10.3.4}$$

with

$$\gamma(a) := \prod_{1\leq i\leq n-1} \eta(a_1 \cdots a_i),$$

and η the unramified multiplicative character of order 2 on $F^\times$.

It should be clear by now to the reader that the exponential version of the Transfer Theorem 10.3 applies to (10.3.4) using the proxies for field extensions explained in section 3 and viewing a as a parameter. Note that the discrepancy between the conditions on conductors in 10.2 and 10.3 is handled by performing an homothety of ratio t. Also it is most likely that Theorem 10.3 can be used for other versions of the relative fundamental lemma.

References

[1] J. Arthur, The local behaviour of weighted orbital integrals, *Duke Math. J.* **56** (1988), 223–293.

[2] J. Arthur, Canonical normalization of weighted characters and a transfer conjecture, *C. R. Math. Acad. Sci. Soc. R. Can.* **20** (1998), 33=-52.

[3] J. Arthur, The trace formula in invariant form, *Annals of Math.* **114** (1981), 1–74.

[4] J. Arthur, A Stable Trace Formula. I. General Expansions, *J. Inst. Math. Jussieu* **1** (2002), 175–277.

[5] J. Ax & S. Kochen, Diophantine problems over local fields. I, *Amer. J. Math.* **873** (1965), 605–630.

[6] J. Ax & S. Kochen, Diophantine problems over local fields. II.: A complete set of axioms for p-adic number theory, *Amer. J. Math.* **87** (1965), 631–648.

[7] H. Chaudouard & G. Laumon, Sur l'homologie des fibres de Springer affines tronquées, *Duke Math. J.* **145** (2008), 443–535.

[8] R. Cluckers & F. Loeser, Constructible motivic functions and motivic integration, *Invent. Math.* **173** (2008), 23–121.

[9] R. Cluckers & F. Loeser, Ax-Kochen-Ersov Theorems for p-adic integrals and motivic integration, in Geometric methods in algebra and number theory, edited by F. Bogomolov and Y. Tschinkel, *Progress in Mathematics* **235**, 109–137 (2005), Birkhäuser.

[10] R. Cluckers & F. Loeser, Fonctions constructibles exponentielles, transformation de Fourier motivique et principe de transfert, *C. R. Math. Acad. Sci. Paris,* **341** (2005), 741–746.

[11] R. Cluckers & F. Loeser, Constructible exponential functions, motivic Fourier transform and transfer principle, *Ann. of Math.* **171** (2010), 1011–1065.

[12] C. Cunningham & T. Hales, Good Orbital Integrals, *Represent. Theory* **8** (2004), 414–457.

[13] J. Denef & F. Loeser, Definable sets, motives and p-adic integrals, *J. Amer. Math. Soc.* **14** (2001), 429–469.

[14] J. Eršov, On the elementary theory of maximal normed fields, *Dokl. Akad. Nauk SSSR* **165** (1965), 21–23.

[15] T. Hales, Unipotent representations and unipotent classes in $SL(n)$, *Amer. J. Math.* **1156** (1993), 1347–1383.

[16] T. Hales, A simple definition of transfer factors for unramified groups, in Representation theory of groups and algebras, 109–134, *Contemp. Math.* **145**, Amer. Math. Soc., Providence, RI, 1993.

[17] T. Hales, A statement of the fundamental lemma, in Harmonic analysis, the trace formula, and Shimura varieties, 643–658, *Clay Math. Proc.* **4**, Amer. Math. Soc., Providence, RI, 2005.

[18] T. Hales, On the fundamental lemma for standard endoscopy: reduction to unit elements, *Canad. J. Math.* **47** (1995), no. 5, 974–994.

[19] J. Igusa, An introduction to the theory of local zeta functions, *AMS/IP Studies in Advanced Mathematics*, **14**, American Mathematical Society, Providence, RI; International Press, Cambridge, MA, 2000.

[20] H. Jacquet & Y. Ye, Relative Kloosterman integrals for GL(3), *Bull. Soc. Math. France* **120** (1992), 263–295.

[21] H. Jacquet, Kloosterman identities over a quadratic extension, *Ann. of Math.* **160** (2004), 755–779.

[22] B. Kostant, Lie group representations on polynomial rings, *Am. J. Math.* **85** (1963), 327–404.

[23] R. Kottwitz, Stable trace formula: elliptic singular terms, *Math. Ann.* **275** (1986), 365–399.

[24] R. Kottwitz, Stable trace formula: cuspidal tempered terms, *Duke Math. J.* **51** (1984), 611–650.

[25] R. Kottwitz & D. Shelstad, Foundations of Twisted Endoscopy, *Astérisque* **255**, 1999.

[26] R. Kottwitz, Harmonic analysis on reductive p-adic groups and Lie algebras, Harmonic analysis, the trace formula, and Shimura varieties, 393–522, *Clay Math. Proc.* **4**, Amer. Math. Soc., Providence, RI, 2005.

[27] R.P. Langlands & D. Shelstad, On the definition of transfer factors, *Math. Ann.* **278** (1987), 219–271.

[28] G. Laumon, B. C. Ngô, Le lemme fondamental pour les groupes unitaires, *Ann. of Math.* **168** (2008), 477–573.

[29] B. C. Ngô, Faisceaux pervers, homomorphisme de changement de base et lemme fondamental de Jacquet et Ye, *Ann. Sci. École Norm. Sup.* (4) **32** (1999), 619–679.

[30] B. C. Ngô, Le lemme fondamental pour les algèbres de Lie, *Publ. Math. Inst. Hautes Études Sci.* **11** (2010), 1–169.

[31] J. Pas, Uniform p-adic cell decomposition and local zeta functions, *J. Reine Angew. Math.* **399** (1989), 137–172.

[32] J.-P. Serre, Corps Locaux, Hermann, Paris, 1968.
[33] D. Shelstad, A formula for regular unipotent germs, *Astérisque* 171–172 (1989).
[34] J.-L. Waldspurger, Une formule des traces locale pour les algèbres de Lie p-adiques, *J. Reine Angew. Math.* **465** (1995), 41–99.
[35] J.-L. Waldspurger, Endoscopie et changement de caractéristique, *J. Inst. Math. Jussieu* **5** (2006), 423–525.
[36] J.-L. Waldspurger, Endoscopie et changement de caractéristique: intégrales orbitales pondérées, *Ann. Inst. Fourier* **59** (2009), 1753–1818.
[37] D. Whitehouse, The twisted weighted fundamental lemma for the transfer of automorphic forms from $GSp(4)$ to $GL(4)$, *Astérisque* **302** (2005), 291–436.

Université Lille 1, Laboratoire Painlevé, CNRS — UMR 8524, Cité Scientifique, 59655, Villeneuve d'Ascq cedex, France, and, Katholieke Universiteit Leuven, Departement wiskunde, Celestijnenlaan 200B, B-3001 Leuven, Belgium
E-mail address: Raf.Cluckers@math.univ-lille1.fr
URL: http://wis.kuleuven.be/algebra/Raf/

Department of Mathematics, University of Pittsburgh, Pittsburgh, PA 15217
E-mail address: hales@pitt.edu
URL: www.math.pitt.edu/~thales/

École Normale Supérieure, Département de mathématiques et applications, 45 rue d'Ulm, 75230 Paris cedex 05, France (UMR 8553 du CNRS)
E-mail address: Francois.Loeser@ens.fr
URL: www.dma.ens.fr/~loeser/

Section III

Brief review of representation theory

Identités de caractères en la place archimédienne
[C.III.A]

Laurent Clozel

1. Le but de ce chapitre est d'expliciter les identités du titre, dans deux cas qui nous seront utiles. Le groupe de base est toujours un groupe unitaire réel, soit

$$G = U(p,q) \qquad (p+q=n)\,.$$

Dans le premier cas G est quasi–déployé (soit $p=q$, ou $p=q+1$). On considère le changement de base entre $G(\mathbb{R})$ et $G(\mathbb{C})$. La conjugaison complexe σ opère sur $G(\mathbb{C})$ et sur ses représentations. Si Π est une représentation irréductible tempérée σ–stable de $G(\mathbb{C})$:

$$\Pi \cong \Pi \circ \sigma\,,$$

le choix d'un opérateur d'entrelacement A_σ entre Π et $\Pi\circ\sigma$ définit un caractère tordu $\Theta_{\Pi,\sigma}$. On normalise A_σ par $A_\sigma^2 = 1$; pour $\varphi \in C_c^\infty(G)$,

$$\Theta_{\Pi,\sigma}(\varphi) = \mathrm{trace}(\Pi(\varphi)A_\sigma)\,.$$

Noter que la définition du membre de droite nécessite le choix d'une mesure de Haar dg sur G ; $\varphi\,dg$ est alors une densité de sorte que $\Theta_{\Pi,\sigma}$ est donc une fonction généralisée sur $G(\mathbb{C})$, qui ne dépend plus d'aucun choix. On sait que $\Theta_{\Pi,\sigma}$ est une fonction C^∞ en les éléments de norme régulière.

Soit ${}^LG$ le groupe dual de G sur $\mathbb{R}$, ${}^LG = \widehat{G}\rtimes W_\mathbb{R}$, et soit ${}^LG/\mathbb{C} = G\times W_\mathbb{C}$ le groupe dual de $G/\mathbb{C}$. Soit $\varphi_0 : W_\mathbb{C}^\times \longrightarrow \widehat{G}$ un paramètre de Langlands **tempéré** (= dont l'image est d'adhérence compacte). Si φ_0 provient de $\varphi : W_\mathbb{R} \longrightarrow^L G$ par restriction, φ définit un L–paquet $\Pi_\mathbb{R}$ (tempéré) de représentations de $G(\mathbb{R})$; de même φ_0 définit une représentation irréductible Π de $G(\mathbb{C}) = GL(n,\mathbb{C})$, qui est σ–stable.

Soit $g \in G(\mathbb{C})$, dont la norme $\mathcal{N}g \in G(\mathbb{R})$ (définie à conjugaison stable près) est fortement régulière. On sait depuis longtemps [**6**] que

$$\Theta_{\Pi,\sigma}(g) = \varepsilon\,\Theta_{\Pi_\mathbb{R}}(\mathcal{N}g)$$

où $\Theta_{\Pi_\mathbb{R}} = \sum_{\pi\in\Pi_\mathbb{R}} \Theta_\pi$ (la somme étant d'ailleurs sans multiplicité dans le cas réel) et $\varepsilon = \pm 1$ est un signe qui dépend du choix de A_σ. Ce signe est explicité dans [**6**] – pour un choix de A_σ dépendant de la théorie du K–type minimal – et dans [**7**] – pour des représentations Π cohomologiques, et à l'aide d'une définition cohomologique de A_σ.

Rappelons que G est quasi–déployé. Il existe alors une normalisation naturelle A_σ^W de A_σ, qui résulte de l'existence d'un modèle de Whittaker (§1). Nous allons démontrer :

THÉORÈME 1.— *Si $A_\sigma = A_\sigma^W$ est normalisé par le modèle de Whittaker,*

$$\Theta_{\Pi,\sigma}(g) = \Theta_{\Pi_{\mathbb{R}}}(\mathcal{N}g)$$

pour tout $g \in G(\mathbb{C})$ de norme fortement régulière.

Si le groupe unitaire G est remplacé par $GL(n)/\mathbb{R}$, ceci avait été démontré dans [**1**].

Le second cas qui nous intéresse est l'endoscopie. Ici G est (unitaire) arbitraire et H en est un groupe endoscopique elliptique. On sait que

$$H = U^*(a) \times U^*(b)\,, \qquad a + b = n$$

où $U^*(a)$ désigne le groupe unitaire quasi–déployé de rang (absolu) a. On suppose fixé un plongement admissible ${}^LH \longrightarrow {}^LG$ (§3.1). Soit $\varphi : W_{\mathbb{R}} \longrightarrow {}^LH$ le paramètre de Langlands d'un L–paquet de séries discrètes de $H(\mathbb{R})$. On suppose que le composé $\varphi_G : W_{\mathbb{R}} \longrightarrow^L G$ est encore discret, i.e., définit un L–paquet de séries discrètes $\Pi_{\mathbb{R}}(\varphi_G)$. Shelstad a alors démontré une identité liant, en des éléments associés, le caractère stable (sur $H(\mathbb{R})$) défini par φ et une certaine combinaison linéaire de caractères du L–paquet $\Pi_{\mathbb{R}}(\varphi_G)$. Notre travail, modeste, est de l'expliciter. Dans ce cas, à la différence du changement de base, l'identité dépend d'une application (en fait, une correspondance) $f_G \rightsquigarrow f_H$ entre fonctions C_c^∞ sur G et H : celle–ci dépend à son tour d'un choix de facteurs de transfert locaux. Nous espérons que le choix fait ici sera compatible avec les arguments globaux. Notre exposition suit entièrement Kottwitz [**13**] : pour les groupes unitaires, on peut le paraphraser de façon simple et explicite.

2. L'identité de changement de base.

Tout d'abord, le Théorème 1 se réduit aisément au cas où $\varphi : W_{\mathbb{R}} \longrightarrow {}^LG$ définit un L–paquet de séries discrètes pour $G(\mathbb{R})$. Dans le cas contraire, en effet, φ définit un L–paquet de séries discrètes pour $M(\mathbb{R})$ où $P(\mathbb{R}) = M(\mathbb{R})N(\mathbb{R})$ est un parabolique cuspidal réel. L'opérateur d'entrelacement "s'induit" de façon évidente ([**6**], Ch. 8), ainsi que le modèle de Whittaker ; on est ramené au cas d'un groupe unitaire de rang inférieur ou bien de $GL(1,\mathbb{C})$ plongé dans $GL(1,\mathbb{C})^2$, son complexifié, cas évident ...

Si φ est un tel paramètre "discret", rappelons que φ est donné sur $\mathbb{C}^\times$ par

$$z \longmapsto \operatorname{diag}((z/\overline{z})^{p_1}, \ldots, (z/\overline{z})^{p_n})$$

où $p_i \in \frac{n-1}{2} + \mathbb{Z}$ et $p_i \neq p_j$ (cf. e.g. [**2**], Ch. 5). La représentation Π est la série principale unitaire induite de $(z_1, \ldots z_n) \longmapsto (z_1/\overline{z_1})^{p_1} \ldots (z_1/\overline{z_1})^{p_n}$.

Nous allons utiliser un argument global. Soit E/F une extension quadratique CM d'un corps totalement réel. Le groupe unitaire G/F sera défini par

$$G(F) = \{g \in GL(n, E) : J_0 \, {}^t\overline{g}^{-1} J_0^{-1} = g\}$$

où J_0 est la matrice

$$\begin{pmatrix} & & & & & \iddots \\ & & & & \iddots & \\ & & & -1 & & \\ & & 1 & & & \\ & -1 & & & & \\ 1 & & & & & \end{pmatrix}$$

Il est quasi–déployé ; de plus, l'automorphisme

$$\sigma : g \longmapsto J_0 \, {}^t\overline{g}^{-1} J_0^{-1}$$

de $G(E)$ stabilise le sous–groupe de Borel des matrices triangulaires supérieures, ainsi qu'un épinglage de celui–ci. En particulier, le caractère

$$\psi : n = \begin{pmatrix} 1 & x_1 & & * \\ & \ddots & \ddots & \\ & & \ddots & x_{n-1} \\ & & & 1 \end{pmatrix} \longmapsto \alpha(x_1 + \cdots + x_{n-1})$$

de $N(\mathbb{A}_E)$, où N est le sous–groupe unipotent supérieur de $GL(n)$ et $\alpha : \mathbb{A}_E \longrightarrow \mathbb{C}^\times$ est un caractère additif non trivial stable par $\mathrm{Gal}(E/F)$, et invariant par σ. Des considérations analogues s'appliquent évidemment en toutes les places, finies ou infinies.

Notons par ailleurs λ le paramètre $(p_1, \dots p_m)$, qui paramètre donc aussi un L–paquet de représentations unitaires de $G(\mathbb{R})$, ainsi qu'une représentation irréductible de dimension finie de la forme compacte $G_c(\mathbb{R}) = U(n)$. (λ est le paramètre d'Harish-Chandra du caractère infinitésimal de la représentation). On note θ_λ son caractère.

On sait ([**7**], Prop. 3.6) qu'il existe une fonction $\varphi \in C_c^\infty(G(\mathbb{C}))$ qui est un pseudo–coefficient de $\Pi = \Pi_\lambda$, i.e., telle que pour toute représentation tempérée σ–stable $\mathbf{P}$ irréductible de $G(\mathbb{C})$, munie de $A_\sigma^{\mathbf{P}}$

$$\mathrm{trace}(\mathbf{P}(\varphi) A_\sigma^{\mathbf{P}}) = 0 \quad (\mathbf{P} \ncong \Pi)$$

$$\mathrm{trace}(\Pi(\varphi) A_\sigma) = 1$$

Dans [**8**] on démontre par voie cohomologique le résultat suivant. Rappelons que pour $\sigma \in G(\mathbb{C})$, de norme **semi–simple** $\mathcal{N}\delta \in G(\mathbb{R})$, on peut définir l'intégrale orbitale tordue stable

$$STO_\delta(\varphi, dg_\mathbb{C}, di) = \sum_{\delta'} e(\delta') TO_{\delta'}(\varphi, dg_\mathbb{C}, di')$$

où la somme de droite porte sur les classes de conjugaison (tordue) dans la classe stable de δ, di' est une mesure de Haar (positive) sur le centralisateur tordu $I(\delta')$ de δ', associée de façon unique à une mesure di sur $I = I(\delta)$, et les $e(\delta')$ sont des signes définis par Kottwitz. On renvoie à [**8**] pour les définitions de $I(\delta), e(\delta)$.

Si $\mathcal{N}\delta$ est un élément elliptique de $G(\mathbb{R})$, on peut le considérer comme une classe de conjugaison dans $G_c(\mathbb{R})$. Le centralisateur tordu I est alors un produit de groupes unitaires. Sur un groupe unitaire, il existe une mesure positive distinguée, telle que le degré formel de chaque série discrète de caractère infinitésimal égal à celui de la représentation triviale soit égal à 1. On calculera STO_δ à l'aide de cette mesure, notée di (et des di' associées). Soit par ailleurs I_c la forme compacte de I.

THÉORÈME 2.1.— *Soit $A_\sigma : \Pi \cong \Pi \circ \sigma$ un entrelacement involutif, et φ un pseudo–coefficient associé (pour la mesure $dg_{\mathbb{C}}$). Si $\delta \in G(\mathbb{C})$ est de norme semi–simple,*

$$(i) \qquad STO_\delta(\varphi) = 0 \qquad (\mathcal{N}\delta \quad \textit{non elliptique})$$

$$(ii) \qquad STO_\delta(\varphi, dg_{\mathbb{C}}, di) = \varepsilon \; e(I_c)\theta_\lambda(\mathcal{N}\delta)$$

$$(\mathcal{N}\delta \; \textit{elliptique})$$

où le signe $\varepsilon = \pm 1$ dépend de A_σ.

Ce résultat est démontré dans [**8**]. La comparaison avec cette référence appelle deux remarques. Le résultat de [**8**] ne comporte pas de signe, mais un facteur $d(G) = 2^n$. Dans [**8**], φ est remplacée par une fonction d'Euler-Poincaré, asssociée à un choix arbitraire (involutif) A d'opérateur d'entrelacement. La comparaison entre celle-ci et notre pseudo-coefficient introduit un signe, ainsi que le scalaire $d(G)$, égal à la caractéristique d'Euler-Poincaré de σ opérant dans la cohomologie de Π, égale à 2^n comme le montre le calcul de [**7**].

On précisera ce résultat après la démonstration du Théorème 1.

Nous appliquons ce théorème à toutes les représentations Π_λ ; chacune étant générique admet un vecteur de Whittaker, i.e., un vecteur–distribution ν dans l'espace dual de $\mathfrak{H} = \mathfrak{H}(\Pi)$ tel que

$$< \nu, nv >= \psi(n) < \nu, v >$$

($v \in \mathfrak{H}$, $n \in N(\mathbb{C})$). Puisque ν est unique à un scalaire près, on peut choisir uniquement A_σ tel que $A_\sigma^* \nu = \nu$. C'est l'opérateur **Whittaker–normalisé** A_σ^W.

On aura besoin d'une construction analogue en une place p–adique v de F. On suppose celle–ci décomposée, donc $G(E_v)$ est isomorphe à $GL(n, F_v) \times GL(n, F_v)$, σ opérant par

$$(g_1, g_2) \longmapsto (\theta_0(g_2), \; \theta_0(g_1))$$

où $\theta_0(g) = J_0 \; {}^t g^{-1} J_0^{-1}$. Alors σ stabilise le sous–groupe de Borel (produit), ainsi que ψ.

Soit τ une représentation supercuspidale de $GL(n, F_v)$. Alors $\tau \otimes (\tau \circ \theta_0)$ est stable par σ, l'opérateur d'entrelacement étant simplement

$$A_\sigma : (v, w) \longmapsto (w, v) .$$

Il opère par 1 sur l'espace des vecteurs de Whittaker de $\tau \otimes (\tau \circ \theta_0)$. Soit φ_1, φ_2 deux fonctions C_c^∞ sur $GL(n, F_v)$. On vérifie aussitôt que pour $\varphi = \varphi_1 \otimes \varphi_2$:

$$tr(A_\sigma(\tau \otimes \tau^{\theta_0})(\varphi)) = tr(\tau(\xi))$$

où

$$\xi(g) = \int \varphi_1(g_1)\varphi_2(\theta_0(g_1^{-1}g))dg_1 .$$

En particulier cette trace est ≥ 0 si $\varphi_2(g) = \overline{\varphi}_1(\theta_0(g^{-1}))$; elle est > 0 si de plus $\tau(\varphi_1) \neq 0$. Choisissons ainsi φ, on a

$$STO_1(\varphi) = TO_1(\varphi) = \int_{G^2/I} \varphi(g_1 g_2^{-\theta_0}, g_2 g_1^{-\theta_0}) d\bar{g}$$

où on a écrit pour un instant $G = GL(n, F_v)$, et $I = \{g, g^{\theta_0}\} \subset G^2$. Si $\varphi_2(g) = \overline{\varphi}_1(\theta_0 g^{-1})$, cette intégrale s'écrit aussi (en prenant $g_2 = 1$ puisque de tels éléments donnent des représentants de G^2/I) :

$$\int_G \varphi_1 \otimes \varphi_2(g_1, g_1^{-\theta_0}) = \int_G \varphi_1(g_1)\overline{\varphi}_1(g_1)dg_1 > 0 .$$

Soit A_σ l'endomorphisme de $L^2_{cusp}(G(E)\backslash G(\mathbb{A}_E))$ donné par $h(x) \mapsto h(\sigma x)$.

Sur $G(\mathbb{A}_E) = GL(n, \mathbb{A}_E)$ considérons une fonction φ de la forme

$$\bigotimes_{v|\infty} \varphi_{\lambda_v} \otimes \varphi_v \otimes \bigotimes_{\substack{w \neq v \\ w \nmid \infty}} \varphi_w$$

où les φ_w sont les fonctions unité de l'algèbre de Hecke. On suppose que pour $v \mid \infty, \phi_v$ est un pseudo-coefficient pour Π_λ et pour l'opérateur d'entrelacement Wh-normalisé. On a choisi F et E de façon que $\dim(F)$ soit assez grand – dim ≥ 2 suffit, voir l'article de Labesse [**14**] – et que E/F soit partout non ramifié aux places finies. D'après Labesse [**14**], la formule des traces tordue pour $G(\mathbb{A}_E)$, A_σ se stabilise. Pour $\underline{\lambda} = (\lambda_v)$ donné, elle donne simplement

$$\text{trace}(R(\varphi)A_\sigma) = \sum_\pi \text{trace}(\pi(\varphi)A_\sigma) = \sum_\delta STO_\delta(\varphi)$$

où la somme porte sur un ensemble fini de classes de conjugaison (stable) elliptiques dans $G(E)$. La représentation R est celle de $G(\mathbb{A}_E)$ sur les formes cuspidales sur $A_G G(E)\backslash G(\mathbb{A}_E)$, $A_G \cong \mathbb{R}_+^\times$ étant la composante neutre du centre déployé...Bien sûr, les termes discrets non cuspidaux de la formule des traces d'Arthur sont annulés par le choix de φ_v.(Voir par exemple [**4**], §3.15.1 pour une situation analogue). Les π sont cuspidales et σ–stables. Pour disposer des résultats d'Arthur utilisés par Labesse, il faut savoir en fait que toutes les intégrales orbitales tordues des $\varphi_v(v \mid \infty)$ en des éléments non (σ)-semi-simples s'annulent. Ceci a été démontré par Chenevier et Renard [**5**]. (On pourrait aussi rajouter une place finie et introduire une représentation de Steinberg en cette place, cf. [**4**]).

L'opérateur A_σ préserve $N(\mathbb{A}_E)$ ainsi que la fonctionnelle de Whittaker,

$$h \longmapsto \int_{N(E)\backslash N(\mathbb{A}_E)} h(n)\psi(n)dn\,.$$

Soit π l'une des représentations telles que trace$(\pi(\varphi)A_\sigma) \neq 0$. Alors, π étant générique, on a nécessairement

$$\pi = \bigotimes_{v|\infty} \Pi_{\lambda_v} \otimes (\tau \otimes \tau^{\theta_0}) \otimes \bigotimes_{w} \pi_w$$

où les π_w sont non–ramifiées (et génériques). Pour celles–ci, l'opérateur d'entrelacement Whittaker–normalisé opère par 1 sur le vecteur non–ramifié. L'opérateur $A_\pi = A_\sigma\,|_\pi$ global est donc simplement un produit fini

$$\bigotimes_{v|\infty} A_v \otimes A_{\tau\otimes\tau^{\theta_0}}\,.$$

d'opérateurs Whittaker–normalisés. D'après notre choix des φ_v, la trace tordue, pour tout π, est (strictement) positive.

La somme $\sum_\delta$ contient la classe unité :

$$\begin{aligned}\sum_\delta &= STO_1(\varphi) + \sum_{\delta'} STO_\delta(\varphi)\\ &= c\; e(G_c)^r \prod_{i=1}^{r} \varepsilon(\lambda_i) \prod_{i=1}^{r} \theta_{\lambda_i}(1)\; STO_1(\varphi_v)\\ &\quad + \sum_{\delta'} c(\delta') \prod_{i=1}^{r} \varepsilon(\lambda_i)\theta_{\lambda_i}(\mathcal{N}\delta'_i)\end{aligned} \tag{2.1}$$

Les signes $\epsilon(\lambda_i)$ sont définis par le Théorème 2.1(ii) et l'opérateur Wh-normalisé. Les constantes c, $STO_1(\varphi_v)$ sont > 0. Elles ne dépendent pas de $\underline{\lambda}$, non plus que les $c(\delta')$ (rappelons que les φ_{λ_i} peuvent être prises à support dans un compact **fixe**, de sorte que les sommes sont uniformes pour $\underline{\lambda}$ variable). Enfin, r est le degré $[F : \mathbb{Q}]$.

Supposons $r = [F : \mathbb{Q}]$ impair, ce qui est loisible ; pour simplifier on supposera $r = 3$. Fixons λ_1. Si λ_2, λ_3 tendent vers l'infini "loin des murs", l'argument de ([**4**], §3) montre que la somme (2.1) est équivalente à son "terme constant" indexé par $\delta = 1$. Puisqu'elle est positive,

$$\eta(\lambda_1)\eta(\lambda_2)\eta(\lambda_3) > 0 \qquad (\lambda_2, \lambda_3 \gg 0)$$

où on a posé $\eta(\lambda) = e(G_c)\varepsilon(\lambda)$. Prenant $\lambda_2 = \lambda_3$ on voit que $\eta(\lambda_1) = 1$.

Revenons au Théorème 2.1. Si A_σ est Wh–normalisé, on a donc $\varepsilon = e(G_c) = (-1)^{q_G}$ où $2q_G$ est la dimension réelle de l'espace symétrique de $G(\mathbb{R})$ – donc $q_G = a^2$ ou $a(a+1)$ selon la parité de n. Par ailleurs, pour δ

régulier, $STO_\delta(\varphi) = \varepsilon\theta_\lambda(\mathcal{N}\delta) = \varepsilon$ si λ est associé à la représentation triviale de $U(n)$.

Fixons ainsi λ. Soit A'_σ l'opérateur d'entrelacement tel que l'identité de caractères du Théorème 1 soit vraie, **sans signe**. Si $\varphi \in C_c(G(\mathbb{C}))$, trace$(\Pi(\varphi)A'_\sigma$ s'obtient, par intégration de Weyl – nous n'écrirons pas la formule explicite – des intégrales orbitales tordues de φ contre le caractère tordu $\Theta_{\Pi,\sigma}$ (défini par A'_σ). Celui–ci est égal à $\Theta_{\Pi_\mathbb{R}} \circ \mathcal{N}g$ (notation du Théorème 1) donc, d'après Harish–Chandra, à $(-1)^{q_G}$ (cf. [**8**]). Il est stable, donc trace$(\Pi(\varphi)A'_\sigma)$ s'écrit encore comme l'intégrale contre $(-1)^{q_G}$ des intégrales tordues stables. Si maintenant φ est le pseudo–coefficient associé à A'_σ, celles–ci sont égales d'après le Théorème 2.1 à ε où $\varepsilon = \varepsilon(A'_\sigma)$. Puisque la trace tordue est égale à 1, il en résulte que $\varepsilon = (-1)^{q_G}$. Par conséquent $\varepsilon(A'_\sigma) = \varepsilon(A^W_\sigma)$, donc $A'_\sigma = A^W_\sigma$ ce qui démontre le Théorème 1.

COROLLAIRE 2.2.— *Si* $A_\sigma = A^W_\sigma$, *les identités du Théorème 2.1 sont vraies, avec* $\varepsilon = (-1)^{q_G}$.

Remarque : L'identité obtenue est compatible avec les faits suivants :

(1) L'opérateur d'entrelacement global, $h(x) \longmapsto h(\sigma(x))$ (h = fonction sur $G(E)A_G\backslash G(\mathbb{A}_E)$) est Whittaker–normalisé.

Si φ_λ est un pseudo–coefficient de Π_λ pour A^W (en toutes les places) et f_λ un pseudo–coefficient de la représentation associée (en fait un L–paquet) Π_λ de $G(F_\infty)$, l'identité de formules des traces de Labesse ([**14**], § 9) montre que la somme (dans l'espace des formes automorphes sur G)

$$\sum_\pi \text{trace } \pi(f)$$

est positive s'il n'intervient que des représentations qui sont des séries discrètes à l'infini. C'est le cas si λ est suffisamment régulier. Dans les autres cas, l'analyse est beaucoup plus délicate et elle est ici remplacée par l'argument de [**4**].

(2) En une place réelle, $\varphi = \varphi_\lambda$ et f_λ sont alors associées au sens de Shelstad [**21**] et Clozel–Delorme (i.e., vérifient les bonnes identités d'intégrales orbitales, sans signe). Alors $STO_1(\varphi) = f_\lambda(1) = \sum_{\pi'_\lambda} \deg(\pi'_\lambda) = \theta_\lambda(1)$ (où le L–paquet $\pi_\lambda = \{\pi'_\lambda\}$). Cette formule avait été démontrée par Bouaziz [**3**].

3. Une identité endoscopique (cas tempéré)

3.1. Dans ce paragraphe $G = U(p,q)$, $p+q = n$. Le groupe dual de G est

$${}^LG = \widehat{G} \rtimes W_\mathbb{R}$$

où $\widehat{G} = GL(n,\mathbb{C})$, $W_\mathbb{R}$ opère via $\Gamma_\mathbb{R} = \mathrm{Gal}(\mathbb{C}/\mathbb{R})$, la conjugaison complexe c opérant par

$$g \longmapsto J_0^n\, {}^tg^{-1}(J_0^n)^{-1}$$

où l'on a noté J_0^n la matrice J_0 du § 2.

Rappelons la description des groupes endoscopiques elliptiques (cf. Rogawski [**17**], § 4.6). Ce sont les groupes

$$H = U^*(a) \times U^*(b)$$

où U^* est le groupe unitaire **quasi–déployé** et $a + b = n$. On a donc de nouveau

$${}^L H = GL(a, \mathbb{C}) \times GL(b, \mathbb{C}) \rtimes W_{\mathbb{R}} = \widehat{H} \rtimes W_{\mathbb{R}}$$

où $W_{\mathbb{R}}$ opère comme ci–dessus, par les deux matrices J_0^a et J_0^b. On définit un "plongement" $\xi : {}^L H \longrightarrow {}^L G$, i.e., un homomorphisme de L–groupes, de la façon suivante :

$$\text{(3.1)} \qquad \begin{array}{c} \widehat{H} \longrightarrow \widehat{G} \text{ est simplement le plongement de} \\ GL(a) \times GL(b) \text{ dans } GL(n) \text{ par des matrices diagonales par blocs} \end{array}$$

Soit $W_{\mathbb{R}} = \mathbb{C}^\times \coprod j\mathbb{C}^\times$, où $j^2 = -1$ et $jzj^{-1} = \overline{z}$ $(z \in \mathbb{C}^\times)$. Pour $w \in W_{\mathbb{R}}$, soit

$$\xi(w) = (\psi(w), w) \in \widehat{G} \times W_{\mathbb{R}} \,.$$

Si $w = z \in \mathbb{C}^\times$, w opère trivialement sur $\widehat{G}$ et $\widehat{H}$ et on doit donc avoir

$$\text{(3.2)} \qquad \xi(z) = (\mu_1(z), \mu_2(z), z)$$

où μ_i est un caractère de $\mathbb{C}^\times$, identifié à une matrice scalaire de taille a (resp. b). Vu l'action de j sur $\widehat{H}$ et $\widehat{G}$, on doit avoir

$$\xi(j) = (z(J_0^a \times J_0^b)(J_0^n)^{-1},\ j)$$

où $z \in \mathbb{C}^\times$ et $J_0^a \times J_0^b$ est la matrice diagonale par blocs. Modulo la notion naturelle d'équivalence des données endoscopiques on peut supposer

$$\text{(3.3)} \qquad \xi(j) = ((J_0^a \times J_0^b)(J_0^n)^{-1},\ j)\,.$$

Sur l'image de (3.2), $\xi(j)$ opère par conjugaison par

$$(\mu_1, \mu_2, z) \longmapsto (\mu_1^{-1}, \mu_2^{-1}, \overline{z})$$

donc les caractères μ_i doivent vérifier $\mu_i(z\overline{z}) = 1$. Un tel caractère s'écrit $z \longmapsto z^{p_i}(\overline{z})^{-p_i}$, avec $p_i \in \frac{1}{2}\mathbb{Z}$. Par ailleurs on vérifie que

$$\xi(-1) = \xi(j)^2 = ((-1)^{a+n}, (-1)^{b+n}, -1)$$

où les deux premiers éléments représentent des matrices scalaires par blocs. On a donc enfin $p_1 = \alpha$, $p_2 = \beta$ avec

$$\text{(3.4)} \qquad 2\alpha \equiv a + n\ [2], \qquad 2\beta \equiv b + n\ [2]\,.$$

Un choix naturel est $p_1 = \frac{a+n}{2}$, $p_2 = \frac{b+n}{2}$. En prévision du cas global, on évitera cependant de restreindre le choix.

Il y a une dernière donnée associée à H. En général un groupe endoscopique est défini par $s \in \widehat{G}$ (un élément semi–simple), $\widehat{H} = \mathrm{Cent}(s, \widehat{G})$, et une structure de L–groupe ${}^L H$ (étendant $\widehat{H}$) vérifiant certaines conditions [**11**] et [**17**]. Ici on a pris

$$s = \begin{pmatrix} 1_a & \\ & -1_b \end{pmatrix} \in \widehat{G}\,. \tag{3.5}$$

3.2. Rappelons, pour les groupes unitaires, la paramétrisation de Langlands pour les L–paquets de séries discrètes. Soit donc $G = U(p,q)$ – ceci s'appliquera aussi aux groupes endoscopiques $U^*(a) \times U^*(b)$, de sorte que le L–groupe a été décrit en 3.1. On considère ${}^L G$ construit à partir de $\widehat{T}$, $\widehat{B}$ où $\widehat{T}$ est le tore diagonal et $\widehat{B}$ le Borel standard ; noter que l'action de $W_{\mathbb{R}}$ préserve $(\widehat{T}, \widehat{B})$; $j \in W_{\mathbb{R}}$ opère par conjugaison sur $\widehat{T}$ par

$$(t_1, \dots t_n) \longmapsto (t_n^{-1}, \dots t_1^{-1})\,.$$

On considère des homomorphismes continus

$$\varphi = \varphi_G : W_{\mathbb{R}} \longrightarrow {}^L G$$

commutant avec la projection sur $W_{\mathbb{R}}$, et dont l'image n'est contenue dans aucun parabolique (paramètres elliptiques). L'image de $W_{\mathbb{C}} = \mathbb{C}^\times$ est contenue dans un tore maximal de $\widehat{G}$, donc à conjugaison près dans $\widehat{T}$. Langlands démontre alors (voir [**2**], Ch. 5) que $\varphi\mid_{\mathbb{C}^\times}$ est de la forme

$$z \longmapsto ((z/\overline{z})^{p_1}, \dots, (z/\overline{z})^{p_n}\ ;\ z) \tag{3.6}$$

avec $p_i \in \frac{n-1}{2} + \mathbb{Z}$, p_i distincts,

$$j \longmapsto (J_0^{-1}\ ;\ j)$$

A conjugaison près dans ${}^L G$, on peut supposer

$$p_1 > p_2 > \cdots > p_n\,. \tag{3.7}$$ [1]

Nos groupes unitaires, $G = U(p,q)$, sont tous définis par la matrice hermitienne

$$\begin{pmatrix} 1_p & \\ & -1_q \end{pmatrix}$$

et G contient donc le tore compact $T_G \cong U(1)^n$ des matrices diagonales.

Soit H un groupe endoscopique (elliptique) de G, muni d'un plongement

$$\xi : {}^L H \longrightarrow {}^L G\,.$$

Soit $\varphi_H : W_{\mathbb{R}} \longrightarrow {}^L H$ le paramètre d'un L–paquet de séries discrètes pour H. Par composition φ_H définit $\varphi_G = \xi \circ \varphi_H : W_{\mathbb{R}} \longrightarrow {}^L G$. Ce paramètre n'est pas nécessairement elliptique : si les paramètres pour $H = U^*(a) \times U^*(b)$ sont $(p_1, \dots p_a, q_1, \dots q_b)$, les paramètres pour G sont $(p_i + \alpha,\ q_j + \beta)$. Noter toutefois que d'après (3.6) $p_i + \alpha,\ q_j + \beta \equiv \frac{n-1}{2}$ [1], et que le paramètre image est elliptique quitte à changer α ou β, ce qui revient à tordre les représentations de H considérées par un caractère abélien de $H(\mathbb{R})$.

[1] On associe à φ le L–paquet $\Pi_G(\varphi)$ de séries discrètes déjà considéré dans le § 2 (là pour le groupe quasi–déployé) ; cf. Harris [**9**].

Supposons donc φ_G elliptique, il lui est donc associé un L–paquet $\Pi(\varphi_G)$ de séries discrètes.

Dans cette situation, par ailleurs, il existe une correspondance

$$f \rightsquigarrow f_H$$

envoyant les fonctions C_c^∞ sur $G(\mathbb{R})$ – disons, K–finies – vers celles sur $H(\mathbb{R})$. L'image de f n'est pas unique, mais ses intégrales orbitales stables sont spécifiées ; notons $\gamma \sim \gamma_H$ un couple d'éléments semi-simples reguliers, $\gamma \in G(\mathbb{R})$, $\gamma_H \in H(\mathbb{R})$, associés par la conjugaison stable. Alors

$$SO_{\gamma_H}(f_H) = \sum_{\gamma \sim \gamma_H} \Delta_\infty(\gamma_H, \gamma) O_\gamma(f).$$

On a choisi des mesures de Haar sur $G(\mathbb{R})$ et $H(\mathbb{R})$; les "dénominateurs" des intégrales orbitales, i.e., les mesures de Haar sur les centralisateurs des γ et des conjugués stables de γ_H , sont choisis de façon compatible : [**22**], §4. La somme porte sur les élḿents γ á conjugaison près.

On dit que Δ_∞ est un **facteur de transfert**.

L'existence des facteurs de transfert, et de f_H, a été démontrée par Shelstad. Le facteur de transfert n'est pas unique ; le point essentiel pour la formule des traces est l'existence d'une famille de facteurs locaux (en toutes les places) satisfaisant une formule du produit (Langlands–Shelstad [**15**]). Voir aussi [**16**].

Supposons donnée la correspondance, et revenons à nos paramètres $\varphi_H \rightsquigarrow \varphi_G$. Soit Θ_{φ_H} le caractère stable associé :

$$\Theta_{\varphi_H}(h) = \sum_{\pi \in \Pi(\varphi_H)} \Theta_\pi(h)$$

($h \in H(\mathbb{R})$ régulier), $\Pi(\varphi_H)$ étant un L–paquet de séries discrètes – donc ici pour $U^*(a) \times U^*(b)$. On sait que Θ_{φ_H} est invariant par conjugaison stable (Shelstad [**19**]). Donc $(\Theta_{\varphi_H}, f^H)$ est une fonction bien définie de $f \in C_c^\infty(G(\mathbb{R}))$ – disons, K–finie. Nous allons énoncer (suivant Shelstad) une identité explicite

$$(3.8) \qquad (\Theta_{\varphi_H}, f^H) = \sum_{\pi \in \Pi_G(\varphi)} \Delta_\infty(\varphi_H, \pi)(\Theta_\pi, f)\,.$$

(Les notations sont celles de Kottwitz [**11**]). L'identité dépend évidemment du choix des facteurs de transfert, qu'il nous faudra spécifier.

Revenons sur la condition (3.7). Si elle est vérifiée pour les deux facteurs de φ_H (i.e., $p_1, \dots p_a$ et $q_1, \dots q_b$) elle ne l'est pas toujours pour φ_G. On verra que la forme de (3.8) dépend en fait des positions relatives des paramètres (p_i, q_i).

Revenons à notre groupe unitaire $U(p, q) = G$, muni de son tore diagonal T. L'ensemble $\mathcal{B}$ des sous–groupes de Borel de $G(\mathbb{C}) = GL(n, \mathbb{C})$ contenant T est en bijection avec le groupe de Weyl $\Omega = \mathfrak{S}_n$, l'élément neutre correspondant au Borel standard B_0. Par ailleurs

$$^LT = (\mathbb{C}^\times)^n \rtimes W_\mathbb{R} = \widehat{T} \rtimes W_\mathbb{R}$$

$c \in \Gamma_\mathbb{R}$ opérant par $(t_i) \longmapsto (t_i^{-1})$. On construit un plongement

$$\eta_0 : {}^L T \longrightarrow {}^L G$$

en envoyant $\widehat{T}$ diagonalement dans $GL(n, \mathbb{C})$, et en posant

$$\eta_0(z) = ((z/\overline{z})^{\frac{n-1}{2}}, (z/\overline{z})^{\frac{n-3}{2}}, \dots (z/\overline{z})^{\frac{1-n}{2}}, z) \in \widehat{G} \times \mathbb{C}^\times, \qquad (3.9)$$
$$\eta_0(j) = (J_0^n, j)\,.$$

On pourra remarquer que $\eta_0|_{W_\mathbb{R}}$ est défini par le paramètre de Langlands de la "plus petite série discrète", cf. (3.6). Plus généralement, si $B = \omega B_0 \in \mathcal{B}$, on définit η_B par permutation des coordonnées (de $\widehat{T}$).

Soit H un groupe endoscopique pour G. Alors $\mathcal{B}_H$ est formé des paires de Borel dans $GL(a) \times GL(b)$, et on obtient de même des plongements

$$\eta^H_{B',B''} : {}^LT \longrightarrow {}^L H\,,$$

en particulier η_0^H ($\neq \eta_0^G$!). La différence entre η_0^H et η_0^G définit un caractère de $T(\mathbb{R})$, de la façon suivante. Ils coïncident sur $\widehat{T}$, donc si l'on pose

$$\xi \circ \eta_0^H(w) = \eta_0^G(w) a(w) \qquad (w \in W_\mathbb{R})\,,$$

$a(w)$ est un élément de $\widehat{T}$ qui par construction est un 1–cocycle. On a en fait

$$a(w) = ((z/\overline{z})^\alpha (z/\overline{z})^{\frac{a-n}{2}}\,,\ (z/\overline{z})^\beta (z/\overline{z})^{\frac{n-b}{2}}) \quad (w = z \in \mathbb{C}^\times)$$

où le premier terme est répété a fois et le second b fois. Il n'est pas nécessaire de calculer $a(j)$: par dualité de Langlands, cet élément de $H^1(W_\mathbb{R}, \widehat{T})$ définit le caractère

$$\chi_0 : (t_1, \dots t_n) \longmapsto t_1^{\mathfrak{a}} \dots t_a^{\mathfrak{a}}\ t_{a+1}^{\mathfrak{b}} \dots t_{a+b}^{\mathfrak{b}} \quad (t_i \in U(1))$$

où $\mathfrak{a} = \alpha + \frac{a-n}{2}$, $\mathfrak{b} = \beta + \frac{n-b}{2}$. Noter que $\mathfrak{a}, \mathfrak{b}$ sont entiers d'après (3.6). Si $\alpha = \frac{a+n}{2}$, $\beta = \frac{b+n}{2}$ on a

$$\mathfrak{a} = a\,, \qquad \mathfrak{b} = n\,.$$

Noter que le choix de $B = B_0 \in \mathcal{B}$ détermine B_H (comme $B \cap H(\mathbb{C})$). A partir des données ci–dessus on peut alors donner une première expression pour le facteur de transfert $\Delta = \Delta_\infty$.

DÉFINITION 3.1.— *On pose, pour $\gamma_H = \gamma \in T$ (identifié à un tore de H ou G) :*

$$\Delta_0(\gamma_H, \gamma) = (-1)^{q(G)+q(H)} \chi_0(\gamma) \Delta_{B_0}(\gamma^{-1}) \Delta_{B_0^H}(\gamma_H^{-1})^{-1}\,.$$

Les coefficients $q(G)$, $q(H)$ sont comme dans le §2 les demi–dimensions des espaces symétriques réels. Les discriminants sont définis, pour tout $B \in \mathcal{B}$, par

$$\Delta_B(\gamma) = \prod_\alpha (1 - \alpha(\gamma)^{-1}) \qquad (3.10)$$

où α parcourt les racines de (B,T), et de même pour B^H.

Cette expression est due à Kottwitz ([**11**], p. 184). Il est peut-être charitable de faire (contrairement à Kottwitz) remarquer au lecteur le point suivant. On a bien écrit "$\gamma_H = \gamma$". La condition peut paraître peu naturelle puisque le facteur de transfert est défini pour tout couple d'éléments stablement conjugués. Le facteur de transfert est bien défini par la formule indiquée (modulo les identifications spécifiées) *pour* $\gamma_H = \gamma$. Si on change γ_H dans sa classe stable, le facteur se transforme de façon compatible avec le formalisme de Langlands-Shelstad.

Nous pouvons maintenant expliciter l'égalité (3.8) dans le cas le plus simple. Disons que le paramètre φ_G est **en position dominante** s'il vérifie (3.7) :

$$z \longmapsto \operatorname{diag}((z/\overline{z})^{p_1}, \dots (z/\overline{z})^{p_n}) \qquad (z \in \mathbb{C}^\times)\,,$$
$$p_1 > \cdots > p_n\,.$$

Dans les discussions qui suivent, il sera nécessaire de distinguer le dual $\widehat{T}$ du tore anisotrope (de G ou H) et le tore de référence de $\widehat{G}$ et de $\widehat{H}$, que l'on note $\widehat{S}$.

Rappelons la paramétrisation des séries discrètes de G associées à un paramètre elliptique $\varphi = \varphi_G$. Notons $\Omega = \mathfrak{S}_n$ le groupe de Weyl de $(G_\mathbb{C}, T_\mathbb{C})$ et $\Omega_\mathbb{R} = \mathfrak{S}_p \times \mathfrak{S}_q$ le groupe de Weyl de $(G_\mathbb{R}, T_\mathbb{R})$. Soit $B \in \mathcal{B}$.

Il existe un unique isomorphisme $\widehat{T} \longrightarrow \widehat{S}$ (admissible du point de vue du L–groupe, i.e. ici réalisé par un élément $\omega \in \mathfrak{S}_n$, opérant sur $(\mathbb{C}^\times)^n = \widehat{T} = \widehat{S}$) envoyant les racines positives de B sur les racines positives de $(\widehat{S}, \widehat{B})$. Pour tout B, on a construit

$$\eta_B : {}^L T \longrightarrow {}^L G\,;$$

η_0 est donné par (3.9) et, si $B = \omega B_0$ $(\omega \in \mathfrak{S}_n)$,

$$\eta_B(z) = \eta_0(\omega^{-1} z)\,.$$

On définit un paramètre $\varphi_B : W_\mathbb{R} \longrightarrow {}^L T$ par

$$\varphi = \eta_B \circ \varphi_B\,. \tag{3.11}$$

Si $B = B_0$ et si φ vérifie (3.7) et est un paramètre pour la "plus petite série discrète", on a donc simplement

$$\varphi_B(z) = 1 \qquad (z \in \mathbb{C}^\times)\,.$$

Soit χ_B le caractère de $T(\mathbb{R})$ associé à φ_B.

Pour tout $B \in \mathcal{B}$, il existe alors une unique représentation $\pi(\varphi, B)$ dans le L–paquet associé à φ dont le caractère est donné pour $\gamma \in T(\mathbb{R})$ régulier par

$$(-1)^{q(G)} \sum_{\omega \in \Omega_\mathbb{R}} \frac{\chi_{\omega B}(\gamma)}{\Delta_{\omega B}(\gamma)}$$

où le dénominateur a été défini en (3.10).

Il est clair que $\pi(\varphi, \omega B) = \pi(\varphi, B)$ si $\omega \in \Omega_\mathbb{R}$. On vérifie que les caractères associés aux éléments de $\Omega_\mathbb{R} \backslash \Omega$ sont distincts. On obtient ainsi la paramétrisation usuelle du L–paquet par $\mathfrak{S}_n/\mathfrak{S}_p \times \mathfrak{S}_q$ (prendre $\pi(\varphi, \omega^{-1}B)$), c'est–à–dire classiquement par les orbites du groupe de Weyl complexe dans l'ensemble des orbites du groupe de Weyl compact sur les (demi)–caractères réguliers de T.

Remarque.- Si $G = U(1, q)$, le L-paquet est paramétré par $\mathfrak{S}_{q+1}/\mathfrak{S}_1 \times \mathfrak{S}_q = \{1, ..., q+1\}$ via $\sigma \mapsto \sigma.1$; notons $\pi_1, ..., \pi_{q+1}$ les représentations associées. On sait alors que, pour un choix convenable d'une structure holomorphe sur l'espace symétrique de G, π_1 est une série discrète holomorphe, π_{q+1} est antiholomorphe. Cf. [**10**], Chapter VI, [**18**].

Nous devons enfin définir l'ingrédient essentiel de la formule (3.8). Considérons de nouveau $T \subset G$; soit T_{sc} l'intersection de T avec $G_{der} = SU(p, q)$. Les groupes de Weyl Ω et $\Omega_{\mathbb{R}}$ coïncident pour G et G_{der}. Si $\Omega \in \Omega$, soit $g \in G_{der}(\mathbb{C})$ un représentant de ω (donc g appartient au normalisateur de $T_{sc}(\mathbb{C})$). Puisque $\mathrm{Gal}(\mathbb{C}/\mathbb{R})$ laisse T invariant,

$$\sigma \longmapsto g\sigma(g)^{-1} \qquad (\sigma \in \Gamma_{\mathbb{R}})$$

définit un 1–cocycle de $\Gamma_{\mathbb{R}}$ à valeurs dans $T_{sc}(\mathbb{C})$. Soit $a(\omega) \in H^1(\mathbb{R}, T)$ l'élément qui s'en déduit. On vérifie que $\omega \longmapsto a(\omega)$ est de nouveau un 1–cocycle, i.e.

$$a(\omega_1\omega_2) = a(\omega_1) + \omega_1 a(\omega_2)\,. \tag{3.12}$$

Rappelons par ailleurs que $H^1(\mathbb{R}, T)$ est dual du groupe $\pi_0(\widehat{T}^{\Gamma_{\mathbb{R}}}) = \widehat{T}^{\Gamma_{\mathbb{R}}} = (\pm 1)^n$ puisque $\Gamma_{\mathbb{C}/\mathbb{R}}$ opère sur $\widehat{T}$ par $t \longmapsto t^{-1}$. Le groupe dual est donc $(\mathbb{Z}/2\mathbb{Z})^n$.

On calcule aisément $a(\omega)$ pour les transpositions $(i, j) \in \mathfrak{S}_n$. Si (i, j) appartient à $\mathfrak{S}_p \times \mathfrak{S}_q$, c'est–à–dire si la racine associée est compacte, ω est réalisée dans $SU(p)$ ou $SU(q)$ et $a(\omega) = 0$. Si (i, j) est non compacte, $a(\omega)$ est l'élément

$$(0, \ldots \underset{i}{1}, \ 0 \ldots, \underset{j}{1}, \ldots 0) \in (\mathbb{Z}/2\mathbb{Z})^n\,.$$

En particulier $a : \Omega \longrightarrow (\mathbb{Z}/2\mathbb{Z})^n$ est d'après (3.12) définie sur $\mathfrak{S}_n/\mathfrak{S}_p \times \mathfrak{S}_q$[2]. On peut donc voir a comme fonction des sous–ensembles I de cardinal p de $\{1, \ldots n\}$, en prenant $I_0 = \{1, \ldots p\}$ comme point de base. Soit $I' = I_0 - I = \{i_1, \ldots i_r\}$ et $I'' = I \cap \{p+1, \ldots n\} = \{j_1, \ldots j_r\}$. Alors $I = \omega_1 \ldots \omega_r I_0$ où $\omega_\alpha = (i_\alpha, j_\alpha)$. Si $\omega = \omega_1 \ldots \omega_r$, $a(I) = a(\omega) = a(\omega_1) + \cdots + a(\omega_r)$ d'après (3.12), puisque $\omega_i a(\omega_j) = a(\omega_j)$, les transpositions étant disjointes. On a donc enfin :

LEMME 3.2.— *Si $\omega \in \Omega$ et $I = \omega I_0$,*

$$a(\omega) = (\varepsilon_1, \ldots \varepsilon_n)$$

[2]Si $p = q$, le groupe naturel $GU(p, q)$ qui interviendra dans les problèmes de modules n'est pas connexe. On a alors $\Omega \supset \Omega_1 \supset \Omega_{\mathbb{R}}$ où Ω_1 est le groupe de Weyl du compact maximal $K_\infty \supsetneq U(p) \times U(p)$; $\Omega_1 = \mathfrak{S}_p \times \mathfrak{S}_p \rtimes (\mathbb{Z}/2)$, $\mathbb{Z}/2$ opérant par $(\omega, \eta) \longmapsto (\eta, \omega)$. L'application passe au quotient **par** Ω_1.

où

$$\begin{aligned} \varepsilon_i = 0 \quad & si \quad i \in (I \cap I_0) \cup (\{p+1, \dots, n\} - I) \\ \varepsilon_i = 1 \quad & si \quad i \in I_0 - I \cup (\{p+1, \dots, n\} \cap I) \end{aligned}$$

Rappelons que l'on a identifié $\widehat{T}$ à $\widehat{S}$. Pour $\omega \in \Omega$, on peut donc définir $< a_\omega, s >$ puisque $s \in (\widehat{T})^{\Gamma_{\mathbb{R}}}$. Explicitement

$$< a_\omega, s >= (-1)^{\#(I' \cup I'') \cap \{a+1, \dots, n\}} . \tag{3.13}$$

On a alors :

THÉORÈME 3.3. (Shelstad, Kottwitz).— *Supposons* $\varphi_G = \xi \circ \varphi_H$ *en position dominante. Alors*

$$(\Theta_{\varphi_H}, f^H) = \sum_{\pi \in \Pi_G(\varphi)} \Delta_\infty(\varphi_H, \pi)(\Theta_\pi, f)$$

où

$$\Delta_\infty(\varphi_H, \pi) =< a_\sigma, s > \qquad si\ \pi = \pi(\varphi, \sigma^{-1} B_0)$$

C'est le cas particulier où φ_G est en position dominante de la formule donnée par Kottwitz [**13**], p.185.

Décrivons le cas général. La donnée endoscopique $\widehat{H} \subset \widehat{G}$ permet d'identifier (à conjugaison stable près) des tores elliptiques T_H et T_G de H et G. Nous supposons que T_H est le tore diagonal dans $U^*(a) \times U^*(b)$; l'isomorphisme $j : T_H \longrightarrow T_G$ ([**13**], p. 184) est simplement l'application identique. On se donne $\varphi_H : W_{\mathbb{R}} \longrightarrow {}^L H$ de la forme (3.6) (pour les deux facteurs), et $\varphi_G = \xi \circ \varphi_H$, donc

$$\varphi_G : z \longmapsto ((z/\overline{z})^{p_i + \alpha}, (z/\overline{z})^{q_j + \beta}) = ((z/\overline{z})^{r_1}, \dots, (z/\overline{z})^{r_n})$$

pour $z \in \mathbb{C}^\times$. On ne peut pas supposer en général que $r = (r_1, \dots r_n)$ – supposé régulier – est dominant. On supposera cependant que

$$r_1 > \dots > r_a, \qquad r_{a+1} > \dots > r_b . \tag{3.14}$$

La construction des L–groupes fournit un isomorphisme $\widehat{T}_H \longrightarrow \widehat{S}_H$, associé au choix d'un sous–groupe de Borel B_H relatif à H. On suppose que celui–ci est associé aux racines positives usuelles de $T_H(\mathbb{C}) = (\mathbb{C}^\times)^a \times (\mathbb{C}^\times)^b$; on le note B_H^0 (racines dans $GL(a) \times GL(b)$) de sorte que

$$\widehat{T}_H = (\mathbb{C}^\times)^n \longrightarrow \widehat{S}_H = (\mathbb{C}^\times)^n$$

est l'application identique. Rappelons que par construction $\widehat{S}_H \longrightarrow \widehat{S}_G$ est aussi l'identité. Enfin $T_H = T_G$ donne dualement $\widehat{T}_H = \widehat{T}_G$.

La donnée de φ_H et φ détermine deux sous–groupes de Borel dans $\widehat{H}$ et $\widehat{G}$. Pour $\widehat{H}$, φ_H étant dominant, $\widehat{B}_H$ est le sous–groupe de Borel standard.

Notons $\alpha_{ij} : t = (t_i) \longmapsto t_i t_j^{-1}$ $(i \neq j)$ une racine. Alors φ détermine un groupe de Borel $\widehat{B} \subset \widehat{G}$ par ses racines :

$$R(\widehat{B}, \widehat{S}) = \{\alpha_{ij} : r_i > r_j\}.$$

Noter que $R(\widehat{B}_H, \widehat{S}) \subset R(\widehat{B}, \widehat{S})$.

Dans les applications nécessitées par ce livre, on supposera que $b = 1$, donc les autres racines de $R(\widehat{B}, \widehat{S})$ sont données par

$$\begin{array}{lcl} \alpha_{i,n} & : & i < n, \ r_i > r_n \\ -\alpha_{i,n} & : & i < n, \ r_i < r_n \,. \end{array}$$

L'isomorphisme $\widehat{T} = \widehat{S}_G$ implique une correspondance entre sous–groupes de Borel pour G et $\widehat{G}$. Vu le choix, déterminé par φ, de $\widehat{B}$, on définit donc un sous–groupe de Borel B de $G_{\mathbb{C}} = GL(n)/\mathbb{C}$ par

$$R(B, T) = (\alpha_{ij} : r_i > r_j)\,. \tag{3.15}$$

Pour écrire la généralisation du Théorème 3.3, nous devons définir les objets suivants.

Tout d'abord, le facteur de transfert Δ_0 de la Définition 3.1 va être remplacé par un facteur Δ_B associé à B (Δ_0 est Δ_{B_0}). Il est donné par la même formule :

$$\Delta_B(\gamma_H, \gamma) = (-1)^{q(h)+q(H)} \chi_B(\gamma) \Delta_B(\gamma^{-1}) / \Delta_{B_H}(\gamma^{-1})$$

où $B_H = B \cap H$ et χ_B est un caractère, dépendant de B, que nous ne préciserons pas ([**13**], p. 184). Puisque les facteurs de transfert doivent être intrinsèques "à une constante près", il peut s'écrire en fonction de Δ_0. Soit $\sigma \in \Omega$. Alors il admet une décomposition de Kostant (B étant fixé)

$$\sigma = \sigma_H \ \sigma_*$$

où $\sigma \in \Omega_H = \mathfrak{S}_a \times \mathfrak{S}_b$, et $\sigma_*(B) \cap H = B \cap H$ (points complexes, bien sûr). Alors

$$\Delta_{\sigma(B)} = \det(\sigma_*) \Delta_B$$

où le déterminant est le signe usuel sur Ω. En particulier, si $\sigma(B) = B_0$,

$$\Delta_B = \det(\sigma_*) \Delta_0 \tag{3.16}$$

([**13**], p. 184). L'analogue dans le cas général du Théorème (3.3) est

$$(\Theta_{\varphi_H}, f^H) = \sum_{\pi} \Delta_\infty(\varphi_H, \pi)(\Theta_\pi, f)$$

où f et f^H **sont maintenant reliées par le facteur** Δ_B, et où il nous faut expliciter $\Delta_\infty(\varphi_H, \pi)$:

$$\Delta_\infty(\varphi_H, \pi(\varphi, \sigma^{-1} B)) = < a_\sigma, s >$$

où $\sigma \in \Omega/\Omega_{\mathbb{R}}$. Noter que, si ce facteur est apparemment identique au précédent, nous avons changé la paramétrisation de $\Pi_G(\varphi)$. Soit $\omega \in \mathfrak{S}_n$ l'unique élément tel que

$$\omega B_0 = B\,.$$

Puisque $R(B,T)$ est donné par (3.15), ω est l'unique élément tel que $\omega^{-1}r$ est dominant. Alors

$$\Delta_\infty(\varphi_H, \pi(\varphi, \sigma^{-1}B_0)) = \Delta_\infty(\varphi_H, \pi(\varphi, \sigma^{-1}\omega^{-1}B)) =< a_{\omega\sigma}, s > .$$

Par ailleurs, si $\sigma = \omega^{-1}$, $\sigma B = B_0$, et $\sigma B \cap H = B_0 \cap H = B \cap H$ puisque r était en position dominante pour H. Donc $\sigma = \sigma_*$ et, d'après (3.16),

$$\Delta_B = \det(\omega)\Delta_0 .$$

On a donc enfin "démontré" l'identité générale (cf. Kottwitz, référence ci-dessus) :

Théorème 3.4 (Shelstad, Kottwitz).— *Soit φ_H en position dominante, $\varphi_G = \xi \circ \varphi_H$; soit $\omega \in \Omega$ l'unique élément tel que $\omega^{-1}r$ soit dominant. Alors*

$$(\Theta_{\varphi_H}, f^H) = \sum_\pi \Delta_\infty(\varphi_H, \pi)(\Theta_\pi, f)$$

où $\Delta_\infty(\varphi_H, \pi) =< a_{\omega\sigma}, s > \det(\omega)$

$$si \qquad \pi = \pi(\varphi, \sigma^{-1}B_0) . \qquad (\sigma \in \Omega/\Omega_{\mathbb{R}})$$

Rappelons que $< a_{\omega\sigma}, s >$ est donné par (3.13). Cela ne rend pas son calcul évident. Il se simplifie si $b = 1$ et p ou $q = 1$. Par exemple, si $p = 1$ (Cf. ci-dessus, Remarque), de sorte que les représentations π sont paramétrées par $\{1, ..., q+1\}$, si $\omega = 1$ et $b = 1$, le Lemme 3.2 donne

$$\Delta_\infty(\phi_H, \pi_i) = 1(i \neq q+1), -1(i = q+1).$$

Références

[1] J. Arthur & L. Clozel, Simple algebras, base change and the advanced theory of the trace formule, *Ann. of Math. Studies,* Princeton University Press, 1989.

[2] N. Bergeron & L. Clozel, Spectre automorphe des variétés hyperboliques et applications topologiques, *Astérisque* **303**, S.M.F., 2005.

[3] A. Bouaziz, Formule d'inversion d'intégrales orbitales tordues, *Compositio Math.* **81** (1992), 261–290.

[4] G. Chenevier & L. Clozel, Corps de nombres peu ramifiés et formes automorphes autoduales, *J. A.M.S.* **22** (2008), 467–518.

[5] G. Chenevie &, D. Renard, On the vanishing of some non semisimple orbital integrals, preprint.

[6] L. Clozel, Changement de base pour les représentations tempérées des groupes réductifs réels, *Ann. Sc. É.N.S.* **15** (1982), 45–115.

[7] L. Clozel, Représentations galoisiennes associées aux représentations automorphes autoduales de $GL(n)$, *Publ. IHES* **73** (1991), 97–145.

[8] L. Clozel & J.–P. Labesse, Orbital integrals and distributions, in On Certain L-functions : A Volume in Honor of Freydoon Shahidi on the Occasion of his 60th birthday, *Clay Math. Proc.* **13** (2011), 109–115.

[9] M. Harris, An introduction to the stable trace formula, ce volume.

[10] A. Knapp, Representation theory of semisimple groups, Princeton University Press, 1986.

[11] R. Kottwitz, Stable trace formula : elliptic singular terms, *Math. Ann.* **275** (1986), 365–399.

[12] R. Kottwitz, Tamagawa numbers, *Ann. of Math.* **127** (1988), 629–646.

[13] R. Kottwitz, Shimura varieties and λ–adic representations, in Clozel, Milne eds., Automorphic forms, Shimura varieties, and L–functions (vol. I), *Perspectives in Math.,* vol. 10, Academic Press, 1988, p. 161–209.

[14] J.–P. Labesse, Changement de base et séries discrètes, ce volume.

[15] R.P. Langlands & D. Shelstad, On the definition of transfer factors, *Math. Ann.* **278** (1987), 219–271.

[16] D. Renard, Endoscopy for real reductive groups, ce volume.

[17] J. Rogawski, Automorphic representations of unitary groups in three variables, *Annals of Math. Studies,* Princeton University Press, 1996.

[18] W. Schmid, On the characters of the discrete series. The Hermitian symmetric case, *Inv. Math.* **30** (1975), 47–144.

[19] D. Shelstad, Characters and inner forms of a quasi–split group over $\mathbb{R}$, *Compositio Math.* **39** (1979), 11–45.

[20] D. Shelstad, L–indistinguishability for real groups, *Math. Ann.* **259** (1982), 385–430.

[21] D. Shelstad, Base change and a matching theorem for real groups, in Non commutative Harmonic analysis and Lie groups, Springer LN 880 (1981), 425–482.

[22] D. Shelstad, Orbital integrals and a family of groups attached to a real reductive group, *Ann. Sc. E. N. S.* **12** (1979), 1–31.

Discrete Series and Characters of the Component Group

[A.III.B]

Jeffrey Adams

Suppose $G(\mathbb{R})$ is a real reductive group, with L-group ${}^\vee G^\Gamma$, and $\phi : W_\mathbb{R} \to {}^\vee G^\Gamma$ is an L-homomorphism. There is a close relationship between the L-packet associated to ϕ and characters the component group of the centralizer of the image of ϕ. This was originally observed by Shelstad [**10**], and further refined by Arthur, Kottwitz, Langlands and Shelstad. This is reinterpreted in [**3**], to (among other things) make it a canonical bijection. This involves a number of changes, including using the notion of strong real form, several strong real forms at once, and taking a cover of the component group. This cover is not necessarily a two-group, so the values of the character may not be just signs.

An important special case is that of discrete series L-packets. It is worthwhile to make the correspondence between characters of (covers of) the centralizer and the (generalized) L-packet in this case. While this is a special case of [**3**], it is not so easy to extract it from there, or from the more elementary reference [**1**].

These notes consist of a mostly self-contained treatment of this special case, with some details in the case of unitary groups. We also address a question raised by Michael Harris about endoscopy.

While the primary reference is *The Langlands Classification and Irreducible Characters* [**3**], for what is in this paper other more accessible references are sufficient. *Lifting of Characters* [**1**], which concentrates on the case of regular integral infinitesimal character has most of what is needed. Some of the approach here follows that of *Algorithms for Representation Theory of Real Reductive Groups* [**4**], for which *Guide to the Atlas Software* [**2**] is a good introduction.

1. Basic Setup

Our starting point is a connected, complex, reductive group G. We begin with some definitions independent of any real structure.

We fix Cartan and Borel subgroups $H \subset B$ of G. Let $G^\vee$ be the (connected, complex) dual group, and fix Cartan and Borel subgroups $H^\vee \subset B^\vee$ of $G^\vee$. By construction we have canonical identifications

$$X^*(H) = X_*(H^\vee), \quad X_*(H) = X^*(H^\vee) \tag{1.1}$$

where X^* and X_* denote the character and co-character lattices, respectively. Let R and $R^\vee$ be the root and coroot lattices, respectively. We write $R(G,H), R^\vee(G,H)$ etc. if it is necessary to specify the extra data. Define the weight and coweight lattices $P, P^\vee$:

$$\begin{aligned} P &= \{\lambda \in X^*(H) \otimes \mathbb{R} \mid \langle \lambda, \alpha^\vee \rangle \in \mathbb{Z} \quad \text{for all } \alpha^\vee \in R^\vee\} \\ P^\vee &= \{\gamma^\vee \in X_*(H) \otimes \mathbb{R} \mid \langle \alpha, \gamma^\vee \rangle \in \mathbb{Z} \quad \text{for all } \alpha \in R\}. \end{aligned} \tag{1.2}$$

These are lattices if G is semisimple, and in general contain real vector spaces of dimension $\dim(Z)$, where $Z = Z(G)$ is the center of G. For the choice of $\mathbb{R}$ here (versus $\mathbb{Q}$ or $\mathbb{C}$) see Definition 2.1. Let W be the Weyl group. We have:

$$\begin{aligned} P^\vee &= \{\gamma^\vee \in X_*(H) \otimes \mathbb{R} \mid \exp(2\pi i \gamma^\vee) \in Z\} \\ X_*(H) &= \{\gamma^\vee \in X_*(H) \otimes \mathbb{R} \mid \exp(2\pi i \gamma^\vee) = 1\}. \end{aligned} \tag{1.3}$$

We are interested in representations of real forms of G. A *real form* of G is the fixed points $G(\mathbb{R})$ of an anti-holomorphic involution σ of G. There is a *holomorphic* involution θ of G commuting with σ, such that $G(\mathbb{R})^\theta$ is a maximal compact subgroup of $G(\mathbb{R})$. This is the *Cartan involution* of G corresponding to $G(\mathbb{R})$. The complexification of $G(\mathbb{R})^\theta$ is G^θ, and is denoted K. This is a complex reductive group (possibly disconnected).

For example if $G = \mathrm{GL}(n,\mathbb{C})$ and $\sigma(g) = \overline{g}$ (complex conjugation of coordinates), then $G(\mathbb{R}) = \mathrm{GL}(n,\mathbb{R})$. The maximal compact subgroup of $\mathrm{GL}(n,\mathbb{R})$ is $K(\mathbb{R}) = O(n)$. This is the fixed points of $\theta(g) = {}^t g^{-1}$, and $K = G^\theta = O(n,\mathbb{C})$.

The classification of real forms of G can be stated entirely in terms of the (holomorphic) Cartan involutions θ. For $g \in G$ write $\mathrm{int}(g)$ for the inner automorphism $\mathrm{int}(g)(h) = ghg^{-1}$.

Definition 1.1. A *real form* of G is a holomorphic involution θ of G. Two real forms θ_1, θ_2 are said to be *equivalent* if $\theta_2 = \mathrm{int}(g) \circ \theta_1 \circ \mathrm{int}(g^{-1})$ for some $g \in G$.

For the relation with the standard definition of real forms see [**4**, Section 3] and the references there.

For example $\theta = 1$ is the *compact* real form of G: $K = G$, so $G(\mathbb{R}) = K(\mathbb{R})$ is compact.

It is natural to consider multiple real forms simultaneously; not all real forms but those in a given *inner class*. Write $\mathrm{Aut}(G), \mathrm{Int}(G)$ for the groups of (holomorphic) automorphisms of G and inner automorphisms, respectively.

DEFINITION 1.2. Two real forms θ_1, θ_2 are in the same inner class if θ_1, θ_2 have the same image in $\mathrm{Out}(G) = \mathrm{Aut}(G)/\mathrm{Int}(G)$, i.e. $\theta_2 = \mathrm{int}(g) \circ \theta_1$ for some $g \in G$.

Fix an involution $\gamma \in \mathrm{Out}(G)$. The associated inner class of real forms of G is the set of involutions $\theta \in \mathrm{Aut}(G)$ whose image in $\mathrm{Out}(G)$ is γ.

This is equivalent to the usual notion of inner class. For more details see [**4**, Section 3].

The collection of real forms which have discrete series representations forms a single inner class. The next result is standard; the only hard part is (4), which is a basic result of Harish-Chandra.

LEMMA 1.3. *Suppose $G(\mathbb{R})$ is a real form of G. Let θ be a corresponding Cartan involution of G, and let $K = G^\theta$. The following conditions are equivalent:*

(1) *$G(\mathbb{R})$ contains a compact Cartan subgroup;*
(2) *$\mathrm{rank}(G) = \mathrm{rank}(K)$;*
(3) *θ is an inner involution;*
(4) *$G(\mathbb{R})$ has discrete series representations.*

This is the *equal rank* inner class of real forms of G. It is also known as the *compact* inner class, since it is the inner class of the compact real form $\theta = 1$.

Associated to an inner class of real forms of G is the *extended group* $G^\Gamma = G \rtimes \Gamma$, where $\Gamma = \mathrm{Gal}(\mathbb{C}/\mathbb{R})$. For example see [**4**, Section 5]. The action of Γ is by the Cartan involution of the most compact real form in the inner class. In the case of the compact inner class this action is trivial, so $G^\Gamma = G \times \Gamma$ and we may safely drop Γ from the notation. Many constructions of [**1**], [**3**] and [**4**] involving the "twist" simplify in this setting.

2. Strong real forms

We turn next to a refinement of the notion of real form which plays an important role. We emphasize that we work entirely in the equal rank inner class, i.e. the Cartan involutions are inner (cf. Lemma 1.3).

There is one technical point which arises only if G is not semisimple. An element g of G is said to be *elliptic* if the closure in the analytic topology of the cyclic subgroup generated by g is compact [**3**, Definition 22.1]. If g has finite order then it is elliptic.

DEFINITION 2.1. A *strong involution* of G (in the equal rank inner class) is an elliptic element $x \in G$ satisfying $x^2 \in Z$. A *strong real form* of G (in the equal rank inner class) is a conjugacy class of strong involutions. If x is a strong involution let $\theta_x = \mathrm{int}(x)$ and $K_x = G^{\theta_x} = \mathrm{Cent}_G(x)$.

The map $x \to \theta_x$ is a surjection

(2.1)(a) $$\{\text{strong involutions}\} \twoheadrightarrow \{\text{involutions}\}$$

and factors to a surjection

$$\text{(2.1)(b)} \qquad \{\text{strong real forms of } G\} \twoheadrightarrow \{\text{real forms of } G\}$$

(the involutions in (a) are inner, and the real forms in (b) are equal rank). If G is adjoint (b) is a bijection.

In [**3**] and elsewhere we make the stronger assumption that x^2 has finite order. The elliptic condition in Definition 2.1 makes the construction of Section 6 easier. If G is semisimple this condition is empty since Z is finite.

To emphasize the distinction with strong real forms, we sometimes refer to the real forms of Definition 1.1 as *weak real forms.*

From (1.3) we have

LEMMA 2.2. *Let* $\mathcal{X}_1 = \{x \in H \,|\, x^2 \in Z, x \text{ is elliptic}\}$. *There are canonical bijections*

$$\text{(2.2)(a)} \qquad \mathcal{X}_1 \overset{1-1}{\longleftrightarrow} P^\vee/2X_*(H)$$

and

$$\text{(2.2)(b)} \qquad \{\textit{strong real forms of } G\} \overset{1-1}{\longleftrightarrow} \mathcal{X}_1/W.$$

The set $\mathcal{X}_1$ is a special case of the set $\mathcal{X}_\tau$ (with $\tau = 1$) of [**4**, Section 11].

Proof. First assume G is semisimple. The map

$$\text{(2.3)} \qquad P^\vee \ni \gamma^\vee \to \exp(\pi i \gamma^\vee) \in \mathcal{X}_1$$

factors to the bijection (a). Every strong involution is conjugate to an element of H, i.e. an element of $\mathcal{X}_1$; two such elements are conjugate if and only if they are conjugate by W. Together with (a) this gives (b).

The proof in general is the same, using the fact [**3**, Lemma 22.2] that $x \in H$ is elliptic if and only if $x = \exp(X)$ for $X \in X_*(H) \otimes \mathbb{R}$ (cf. 1.2). □

It is often convenient to consider only those x with x^2 fixed; this is always a finite set. So fix $z \in Z$ and let

$$\text{(2.4)} \qquad \mathcal{X}_1[z] = \{x \in H \,|\, x^2 = z\}.$$

We refer to $\mathcal{X}_1[z]$ (resp. $\mathcal{X}_1[z]/W$) as the as the strong involutions (resp. strong real forms) of type z.

For $w \in Z$ the map $x \to wx$ is a bijection between $\mathcal{X}_1[z]$ and $\mathcal{X}_1[w^2z]$. In particular these two spaces map to the same set of weak real forms via (2.1). For this reason it is useful to cut down the space $\mathcal{X}_1$ to the *reduced* parameter space $\mathcal{X}_1^r$: choose a set of representative $\{z_1, \ldots, z_n\}$ of Z/Z^2, and define

$$\text{(2.5)} \qquad \mathcal{X}_1^r = \bigcup_i \mathcal{X}_1[z_i].$$

The map from X_1^r to weak real forms is surjective. See [**4**, Section 13].

An important special class of strong real forms is the following. Let $z(\rho^\vee) = \exp(2\pi i \rho^\vee)$ where $\rho^\vee = \frac{1}{2}\sum \alpha^\vee \in P(G^\vee, H^\vee)$; the sum is over any set of positive roots, and is independent of this choice. Following [**15**, Definition 2.6] we say a strong involution or real form is *pure* if $x \in \mathcal{X}_1[z(\rho^\vee)]$.

EXAMPLE 2.6. Suppose $x \in Z$. Then $\theta_x = 1$ and $K = G$, so this maps the compact weak real form of G. Thus the fiber of the map from strong real forms to the compact weak real form is Z.

At the other extreme let $x = \exp(\pi i \rho^\vee)$. We will see later in (5.7) that θ_x is the Cartan involution of a quasisplit real form.

3. Representations of strong real forms

Suppose for the moment that $G(\mathbb{R})$ is a real form of G, with Cartan involution θ and complexified maximal compact subgroup $K = G^\theta$. Let $\mathfrak{g}$ be the (complex) Lie algebra of G. A $(\mathfrak{g}, K)$-module is a vector space V, equipped with compatible actions of $\mathfrak{g}$ and K. For more details, and the relation with representations of $G(\mathbb{R})$, see [**14**].

Return now to the setting of the previous section. A *representation of a strong involution* is a pair (x, π) where x is a strong involution and π is a $(\mathfrak{g}, K_x)$-module.

Some care is required to to allow a notion of equivalence which allows conjugation by G. We say (x, π) is equivalent to (x', π') if there exists $g \in G$ such that $gxg^{-1} = x'$ and $\pi' \simeq \pi^g$. Write $[x, \pi]$ for the equivalence class of (x, π). A *representation of a strong real form* of G is an equivalence class of representations of strong involutions. Here is a key example.

EXAMPLE 3.1. Let $G = \mathrm{SL}(2, \mathbb{C})$ and $x = \mathrm{diag}(i, -i)$. Then $K_x = H = \mathbb{C}^\times$ and the corresponding real group is $SU(1,1) \simeq \mathrm{SL}(2, \mathbb{R})$. There is a unique irreducible $(\mathfrak{g}, K_x)$-module whose restriction to K_x is the direct sum of the characters z^k $(k = 2, 4, 6, \dots)$. This is the $(\mathfrak{g}, K_x)$-module of a discrete series representation of $\mathrm{SL}(2, \mathbb{R})$.

Suppose g is a representative of the nontrivial element of the Weyl group. Then $\mathrm{int}(g)$ is an automorphism of $(\mathfrak{g}, K_x)$, which acts by $h \to h^{-1}$ on K_x, and therefore takes π to the contragredient $\overline{\pi}$. However we do *not* want to consider π and $\overline{\pi}$ to be equivalent.

This is taken care of by the fact that we keep track not only of K_x but of the strong involution x. Replace π with $[x, \pi]$, the equivalence class of (x, π). Then $\mathrm{int}(g)(x) = -x$, so $[x, \pi] = [int(g)x, \pi^g] = [-x, \overline{\pi}]$, but $[x, \pi] \neq [x, \overline{\pi}]$.

Note that $\{\pi, \overline{\pi}\}$ is an L-packet of discrete series representations. Using strong real forms this is $\{[x, \pi], [x, \overline{\pi}]\}$. We prefer to think of this as

$$\{[x, \pi], [-x, \pi]\}, \tag{3.2}$$

which is parametrized by the set $\{x, -x\}$ of strong involutions. This is a special case of the general situation; see Proposition 5.1.

4. L-parameters

We continue to work with the equal rank inner class of real forms of G. The L-group of a real form only depends on its inner class; let ${}^\vee G^\Gamma$ be the

L-group of the equal rank inner class. Basic references for this section are [**5**] and [**8**]; also see [**4**, Section 7].

Let $W_{\mathbb{R}}$ be the Weil group of $\mathbb{R}$, and suppose ϕ is an admissible homomorphism of $W_{\mathbb{R}}$ into ${}^{\vee}G^{\Gamma}$. Fix a strong involution x. Langlands associates to x and ϕ a (possibly empty) L-packet denoted $\Pi(x,\phi)$. This is a finite set of $(\mathfrak{g},K_x)$-modules, all having the same infinitesimal character.

Define

$$\Pi(\phi)=\{[x,\pi]\,|\,x\in\mathcal{X}_1,\pi\in\Pi(x,\phi)\}. \tag{4.1}$$

We embed $\Pi(x,\phi)$ in $\Pi(\phi)$ by the map $\pi\to[x,\pi]$. This map is injective, since $[x,\pi]=[x,\pi']$ if and only if there exists $g\in G$ such that $gxg^{-1}=x$ and $\pi^g\simeq\pi'$. This holds if and only if $g\in K_x$ in which case $\pi^g\simeq\pi$.

It is helpful to make some choices to make this more explicit. Choose a set $\mathcal{X}_1'$ of representatives of $\mathcal{X}_1/W$; by (2.2)(b) $\mathcal{X}_1'$ parametrizes the strong real forms. Then there is a canonical bijection

$$\Pi(\phi)\overset{1-1}{\longleftrightarrow}\coprod_{x\in\mathcal{X}_1'}\Pi(x,\phi). \tag{4.2}$$

Each $\Pi(x,\phi)$ is an L-packet of $(\mathfrak{g},K_x)$-modules. Suppose two strong real forms map to the same weak real form in (2.1)(a), i.e. $x,x'\in\mathcal{X}_1'$ satisfy: x' is conjugate to zx for some $z\in Z$. Then $K_x\simeq K_{x'}$, $\Pi(x,\phi)$ and $\Pi(x',\phi)$ are isomorphic, and this L-packet occurs twice.

Thus, unless G is adjoint these sets are often larger than necessary. (In particular if G is not semisimple they are infinite). For these reasons it is sometimes helpful to fix an element $z\in Z$, and define (cf. (2.4))

$$\Pi_z(\phi)=\{[x,\pi]\,|\,x^2=z\}\subset\Pi(\phi). \tag{4.3}$$

These sets are finite, and the number of strong real forms mapping to a single weak real form is small, and often 1. On the other hand for a fixed z some real forms may fail to occur. For an example see Section 8.1.

5. Discrete Series L-packets

We specialize to the case of discrete series. Let $\phi:W_{\mathbb{R}}\to{}^{\vee}G^{\Gamma}$ be a discrete series parameter. These are characterized by any of several conditions, for example the centralizer of the image being finite. See [**5**, 10.3 and 10.5]. This determines an infinitesimal and central character, and $\Pi(x,\phi)$ is the set of discrete series $(\mathfrak{g},K_x)$-modules with this infinitesimal and central character.

Here is a little more detail. Write $W_{\mathbb{R}}=\langle\mathbb{C}^*,j\rangle$ with relations $jzj^{-1}=\overline{z}$ and $j^2=-1$. After conjugating by $G^\vee$ we may assume assume $\phi(\mathbb{C}^\times)\in H^\vee$. The fact that this is a discrete series parameter implies $\phi(j)h\phi(j)^{-1}=h^{-1}$ for $h\in H^\vee$, and $\phi(z)=(z/\overline{z})^\lambda\quad(z\in\mathbb{C}^*)$, with λ a regular element of

$\rho + X^*(H)$ (where ρ is one-half the sum of any set of positive roots). We may assume λ is B-dominant. See [**4**, Section 7]. This implies

$$S_\phi = \{h \in H^\vee \mid h^2 = 1\}. \tag{5.1}$$

For $x \in \mathcal{X}_1$, $\gamma \in \rho + X^*(H)$ regular, write $\pi_x(\gamma)$ for the discrete series $(\mathfrak{g}, K_x)$-module with Harish-Chandra parameter γ. Then $\pi_x(\lambda) \simeq \pi_x(w\lambda)$ if and only if $w \in W(K_x)$, the Weyl group of H in K_x. Thus

$$\Pi(x,\phi) = \{\pi_x(w^{-1}\lambda) \mid w \in W/W(K_x)\}. \tag{5.2)(a}$$

This embeds in $\Pi(\phi)$ as

$$\Pi(x,\phi) = \{[x, \pi_x(w^{-1}\lambda)] \mid w \in W/W(K_x)\}. \tag{5.2)(b}$$

A key point is that we may identify this subset in a different way, using $[x, \pi_x(w^{-1}\lambda)] = w[x, \pi_x(w^{-1}\lambda)] = [wx, \pi_{wx}(\lambda)]$ for all $w \in W$, i.e.

$$\Pi(x,\phi) = \{[wx, \pi_{wx}(\lambda)] \mid w \in W/W(K_x)\}. \tag{5.2)(c}$$

Recall ϕ determines λ, so the second coordinate is determined by the first, so there is a bijection

$$\Pi(x,\phi) \overset{1-1}{\longleftrightarrow} \{wx \mid w \in W/W(K_x)\} = \{y \in \mathcal{X}_1 \mid y \sim x\} \tag{5.2)(d}$$

where $\sim$ denotes G-conjugacy. Thus we obtain a map

$$\mathcal{X}_1 \to \Pi(\phi), \quad x \to [x, \pi_x(\lambda)]. \tag{5.2)(e}$$

By the preceding discussion this is a bijection. This proves:

PROPOSITION 5.1. *Suppose ϕ is a discrete series L-parameter for G. There is a canonical bijection*

$$\mathcal{X}_1 \overset{1-1}{\longleftrightarrow} \Pi(\phi). \tag{5.3}$$

Given $x \in \mathcal{X}_1$ this restricts to a bijection

$$\{y \in \mathcal{X}_1 \mid y \sim x\} \overset{1-1}{\longleftrightarrow} \Pi(x,\phi). \tag{5.4}$$

Given $z \in Z$ it restricts to a bijection (cf. (2.4) *and* (4.3)*)*

$$\mathcal{X}_1[z] \overset{1-1}{\longleftrightarrow} \Pi_z(\phi). \tag{5.5}$$

We say a $(\mathfrak{g}, K_x)$-module π is *generic* if the corresponding representation $\pi_\mathbb{R}$ of $G(\mathbb{R})$ admits a Whittaker model. This is equivalent to: π is *large*, i.e. has maximal Gelfand-Kirillov dimension. This implies $G(\mathbb{R})$ is quasisplit. See [**13**, Section 6].

It is important to know which elements of $\mathcal{X}_1$ correspond via (5.3) to generic discrete series representations. A discrete series representation $\pi(\lambda)$ is large if and only if every simple root in the chamber defined by λ is noncompact (for example see [**13**, Theorem 6.2(f)]). In our setting a root α is compact with respect to θ_x if $\alpha(x) = 1$, and is noncompact if $\alpha(x) = -1$. It is enough to consider simple roots, so:

LEMMA 5.2. *In the bijection* (5.3) $x \in \mathcal{X}_1$ *corresponds to a generic discrete series representation if and only if*

$$\alpha(x) = -1 \quad \textit{for all simple roots } \alpha. \tag{5.6}$$

Let $\rho^\vee = \frac{1}{2}\sum \alpha^\vee$ where the sum is over the roots of $H^\vee$ in $B^\vee$. Then

$$x_b = \exp(\pi i \rho^\vee) \tag{5.7}$$

is a canonical element satisfying $\alpha(x_b) = (-1)^{\langle \alpha, \rho^\vee \rangle} = -1$ for all simple roots α. This provides a canonical basepoint in $\mathcal{X}_1$, corresponding to a generic discrete series representation of a quasisplit strong involution of G. This plays an important role in what follows.

6. The groups S_ϕ and $\widetilde{S_\phi}$

Fix a discrete series L-homomorphism ϕ. Let

$$S_\phi = \mathrm{Cent}_{G^\vee}(\mathrm{Image}(\phi)). \tag{6.1}$$

We want to relate characters of S_ϕ to elements of $\Pi(\phi)$.

Recall (5.1) $S_\phi = \{h \in H^\vee \mid h^2 = 1\}$, and by (1.3) this is isomorphic to $X_*(H^\vee)/2X_*(H^\vee) = X^*(H)/2X^*(H)$.

Let $p : G^\vee_{sc} \to G^\vee$ be the *topologically* simply connected cover of $G^\vee$. Thus $G^\vee_{sc} = \mathbb{C}^n \times G^\vee_{d,sc}$ where $n = \dim(Z)$ and $G^\vee_{d,sc}$ is the simply connected cover of the derived group $G^\vee_d$. Define

$$\widetilde{S}_\phi = p^{-1}(S_\phi) \subset G^\vee_{sc}. \tag{6.2}$$

This is a variant of the *algebraic* cover $G^{\vee,alg}$ of [**3**, (1.16) and (5.10)], which is the projective limit of the finite covers of $G^\vee$. The difference is due to the fact that we define strong involutions to be elliptic elements, rather than having finite order (Definition 2.1).

The inverse image of $H^\vee$ in $G^\vee_{sc}$ is isomorphic to $X_*(H^\vee) \otimes \mathbb{C}/R^\vee(G^\vee, H^\vee) \simeq X^*(H) \otimes \mathbb{C}/R(G, H)$. From this it follows easily that there is a natural isomorphism

$$X^*(H)/2R \simeq \widetilde{S}_\phi. \tag{6.3}$$

This contains a lattice of rank $\dim(Z)$. For $\lambda \in X^*(H)$ write $s(\lambda)$ for the corresponding element of $\widetilde{S_\phi}$. The map $p : \widetilde{S_\phi} \to S_\phi$ is then the natural map

$$\widetilde{S_\phi} \simeq X^*(H)/2R \to X^*(H)/2X^*(H) \simeq S_\phi. \tag{6.4}$$

Use $\widehat{\ }$ to denote Pontriagin dual.

LEMMA 6.1. *There is a canonical group isomorphism* $\mathcal{X}_1 \simeq \widehat{\widetilde{S_\phi}}$.

Proof. Consider the map from $X^*(H)$ to the Pontriagin dual of $\mathcal{X}_1$ given by restricting to $\mathcal{X}_1$. This is surjective, since any character of the maximal compact subgroup H_c of H extends to an algebraic character of H, and $\mathcal{X}_1 \subset H_c$ by definition. The kernel is $2R$, so $\widehat{\mathcal{X}_1} \simeq X^*(H)/2R \simeq \widetilde{S_\phi}$ by (6.3). Apply Pontriagin duality. □

Explicitly, using the map s of (6.3), the character χ of $\widetilde{S_\phi}$ corresponding to $x \in \mathcal{X}_1$ is given by

$$\chi(s(\lambda)) = \lambda(x) \quad (\lambda \in X^*(H)). \tag{6.5}$$

Together with (5.3) we obtain bijections $\widetilde{S_\phi}\widehat{\ } \overset{1-1}{\longleftrightarrow} \mathcal{X}_1 \overset{1-1}{\longleftrightarrow} \Pi(\phi)$, taking the trivial character of $\widetilde{S_\phi}$ to the identity element of $\mathcal{X}_1$, and hence to a finite dimensional representation of a compact strong real form G (cf. Example 2.6). This is not the right normalization: we prefer the basepoint to be a generic discrete series of a quasisplit strong real form. This is provided (canonically) by the element x_b of 5.7.

DEFINITION 6.2. For $x \in \mathcal{X}_1$ define $\tau_x \in \widetilde{S_\phi}\widehat{\ }$ by

$$\tau_x(s(\lambda)) = \lambda(x x_b^{-1}). \tag{6.6}$$

In other words if $x = \exp(\pi i \gamma^\vee)$ $(\gamma^\vee \in X_*(H))$ then

$$\tau_x(s(\lambda)) = e^{\pi i \langle \gamma^\vee - \rho^\vee, \lambda\rangle}. \tag{6.7}$$

This defines a canonical bijection of sets

$$\widetilde{S_\phi}\widehat{\ } \overset{1-1}{\longleftrightarrow} \mathcal{X}_1 \tag{6.8}$$

taking the trivial character to x_b. Composing with (5.3) we obtain:

PROPOSITION 6.3. *There is a a canonical bijection*

$$\widetilde{S_\phi}\widehat{\ } \overset{1-1}{\longleftrightarrow} \Pi(\phi). \tag{6.9}$$

taking the trivial character of $\widetilde{S_\phi}$ to $[x_b, \pi_{x_b}(\lambda)]$, a generic discrete series representation of a quasisplit strong real form of G.

DEFINITION 6.4. Given $\tau \in \widetilde{S_\phi}\widehat{\ }$ let $x_\tau \in \mathcal{X}_1$ be the corresponding element via (6.8), and let $\pi(\tau) = [x_\tau, \pi_{x_\tau}(\lambda)] \in \Pi(\phi)$ be the corresponding element of $\Pi(\phi)$ via (6.9).

This bijection has the advantage that it is canonical. On the other hand it differs from more familiar constructions in several respects. It involves the group $\widetilde{S_\phi}$; in applications (and over other fields) it is the more natural group S_ϕ which plays a role. Furthermore $\widetilde{S_\phi}$ is often larger than necessary (even infinite), due in part to the failure of injectivity in the map (2.1)(b).

In any event for $z \in Z$ it is helpful to identify the image of $\Pi_z(\phi)$ under the bijection (6.9). The basic issue is to identify $\widehat{S_\phi}$ as a subset of $\widetilde{S_\phi}\widehat{\ }$:

$$\widehat{S_\phi} = \{\tau_{x x_b} \,|\, x^2 = 1\} = \{\tau_x \,|\, x^2 = z(\rho^\vee)\} \overset{1-1}{\longleftrightarrow} \Pi_{z(\rho^\vee)}(\phi). \tag{6.10}$$

The last set is the representations of pure strong real forms.

Now fix $y \in \mathcal{X}_1[z]$. Then $\tau_y \widehat{S_\phi} = \{\tau_y \tau_x \,|\, x^2 = z(\rho^\vee)\}$. Using the elementary identity $\tau_y \tau_x = \tau_{y x x_b^{-1}}$, and noting that $(y x x_b^{-1})^2 = y^2 = z$, we have the following result.

LEMMA 6.5. *Fix $z \in Z$ and choose $y \in \mathcal{X}_1[z]$. Then $\tau_y\widehat{S_\phi} \subset \widetilde{S_\phi}\widehat{\ }$ is independent of the choice of y, and the bijection of Proposition 6.3 restricts to a canonical bijection*

$$\text{(6.11)(a)} \qquad \tau_y\widehat{S_\phi} \overset{1-1}{\longleftrightarrow} \Pi_z(\phi).$$

In particular taking $y = x_b$ we obtain a canonical bijection

$$\text{(6.11)(b)} \qquad \widehat{S_\phi} \overset{1-1}{\longleftrightarrow} \Pi_{z(\rho^\vee)}(\phi)$$

taking the identity to $[x_b, \pi_{x_b}(\lambda)]$, a generic discrete series representation of a quasisplit strong real form of G.

Proposition 6.3 is not perfectly suited to the classical theory, and neither is the Lemma: the map from strong real forms of type z to weak real forms is not necessarily surjective or injective. Also note that, except in the setting of (6.11)(b), a further choice (of y) is required to give a bijection $\widehat{S_\phi} \overset{1-1}{\longleftrightarrow} \Pi_z(\phi)$. See Section 9.

7. Endoscopic Lifting

We continue in the setting of the previous Section, with a discrete series parameter ϕ, and the bijections $\widetilde{S_\phi}\widehat{\ } \ni \tau \to x_\tau \in \mathcal{X}_1$ and $\tau \to \pi(\tau) \in \Pi(\phi)$ of Definition 6.4.

For $x \in \mathcal{X}_1$ let

$$\text{(7.1)} \qquad e(x) = (-1)^{\dim(G/K_x) - \dim(G/K_{x_b})}.$$

Recall from (5.7) that K_{x_b} is the complexified maximal compact of the quasisplit real form of G; $e(x)$ is the *Kottwitz invariant* [**6**] of a real form of G defined by x.

DEFINITION 7.1. Fix $\tilde{s} \in \widetilde{S_\phi}$. Let

$$\text{(7.2)(a)} \qquad \widetilde{\eta}(\tilde{s}) = \sum_{\tau \in \widetilde{S_\phi}\widehat{\ }} e(x_\tau)\tau(\tilde{s})\pi(\tau).$$

Fix $x \in \mathcal{X}_1$ and let

$$\text{(7.2)(b)} \qquad \widetilde{\eta}_x(\tilde{s}) = e(x) \sum_{\{\tau \,|\, x_\tau \sim x\}} \tau(\tilde{s})\pi(\tau).$$

This only depends on the image of x in $\mathcal{X}_1/W$, and is a linear combination of representations of this strong real form of G (cf. (2.2)(b)).

This is a special case of the discussion on pages 20-21 of [**3**]; see [**3**, Theorem 26.8] for details. This is a collection of virtual representations of strong real forms of G, parametrized by $\widetilde{S_\phi}$. Now [**3**, Theorem 1.39] relates these virtual characters to characters of an endoscopic group, and is the version of endoscopic lifting of [**3**]. For the purposes of these notes we use only Definition 7.1.

In order to compare this with the lifting defined by Shelstad we write (7.2)(b) in more familiar terms. First of all we may write

$$\widetilde{\eta}_x(\widetilde{s}) = e(x)\tau_x(\widetilde{s}) \sum_{\{\tau \mid x_\tau \sim x\}} (\tau(\widetilde{s})/\tau_x(\widetilde{s}))\pi(\tau). \tag{7.3}$$

Note that $\{\tau \mid x_\tau \sim x\} = \{\tau_{wx} \mid w \in W/W(K_x)\}$. Recall $\pi(\tau_{wx}) = [wx, \pi_{wx}(\lambda)]$, and (cf. 5.2) this equals $[x, \pi_x(w^{-1}\lambda)]$, where $\pi_x(w^{-1}\lambda)$ is the discrete series $(\mathfrak{g}, K_x)$-module with Harish-Chandra parameter $w^{-1}\lambda$. We will see momentarily that τ_{wx}/τ_x factors to S_ϕ, so for $x \in \mathcal{X}_1, s \in S_\phi$ define a function $\kappa(x, s)$ on W:

$$\kappa(x, s)(w) = (\tau_{wx}/\tau_x)(s). \tag{7.4}$$

Then:

Proposition 7.2. *Given $x \in \mathcal{X}_1, \widetilde{s} \in \widetilde{S_\phi}$, let $s = p(\widetilde{s}) \in S_\phi$. Then*

$$\widetilde{\eta}_x(\widetilde{s}) = e(x)\tau_x(\widetilde{s}) \sum_{w \in W/W(K_x)} \kappa(x, s)(w)\pi_x(w^{-1}\lambda). \tag{7.5}$$

Here are some elementary properties of the constants $\kappa(x, s)$.

Lemma 7.3. (1) *$\kappa(x, s) : W \to \pm 1$;*
(2) *$\kappa(x, s)(w) = 1$ for $w \in W(K_x)$;*
(3) *$\kappa(x, s)(w_1 w_2) = \kappa(x, s)(w_1)$ for $w_2 \in W(K_x)$;*
(4) *$\kappa(x, s)(w_1 w_2) = \kappa(w_2 x, s)(w_1)\kappa(x, s)(w_2)$;*

Proof. If $x = \exp(\pi i \gamma^\vee)$ $(\gamma^\vee \in X_*(H^\vee))$ then (cf. 6.7)

$$(\tau_{wx}/\tau_x)(s(\lambda)) = e^{\pi i \langle w\gamma^\vee - \gamma^\vee, \lambda \rangle} \quad (\lambda \in X^*(H)). \tag{7.6}$$

This equals 1 if $\lambda \in 2X_*(H)$. Since the kernel of $\widetilde{S_\phi} \to S_\phi$ is $2X_*(H^\vee)/2R(H^\vee)$ (6.3) $\kappa(x, s)$ is well defined for $s \in S_\phi$. It is easy to see (7.6) implies (1) and (4), and (2) and (3) are special cases of (4). □

The functions $\kappa(x, s)$ appear in [**12**, Section 3]. It follows from Lemma 7.3 that $\widetilde{\eta}_x(\widetilde{s})$ agrees, up to a constant of absolute value 1, with the lifting of Shelstad defined in [**10**, Section 4]. This is also discussed in [**3**, pg. 289]. See [**11**] for an update of [**10**], using the transfer factors of Langlands and Shelstad [**9**], which makes some use of a cover of S_ϕ and multiple real forms.

Definition 7.1 is entirely canonical, but involves the group $\widetilde{S_\phi}$. The part of the sum (7.2)(a) over $\widehat{S_\phi} \subset \widehat{\widetilde{S_\phi}}$ involves just the pure strong real forms. Thus define

$$\eta_{z(\rho^\vee)}(s) = \sum_{\tau \in \widehat{S_\phi}} e(x_\tau)\tau(s)\pi(\tau). \tag{7.7)(a}$$

This is a linear combination of the representations of pure strong real forms.

More generally, (cf. (6.10) and (6.11)(b)), for $z \in Z$, choose $x \in \mathcal{X}_1[z]$. The part of the sum (7.2)(a) giving a linear combination of the elements of $\Pi_z(\phi)$ is:

$$\eta_z(\tilde{s}) = \tau_x(\tilde{s}) \sum_{\tau \in \widehat{S_\phi}} e(x_{\tau\tau_x})\tau(s)\pi(\tau\tau_x) \tag{7.7(b)}$$

where $s = p(\tilde{s}) \in S_\phi$. Note that (a) is obtained from (b) by making the canonical choice $x = x_b$ (so $\tau_x = 1$); in general there is no canonical choice of a basepoint x.

We return to (7.2)(b). Note that $x_\tau \sim x$ implies $\tau \in \tau_x\widehat{S_\phi}$ (cf. (6.11)(a)), so we can write the sum (7.2)(b) as being over $\{\tau\tau_x \,|\, \tau \in \widehat{S_\phi}, x_{\tau\tau_x} \sim x\}$. So: given $x \in \mathcal{X}_1, s \in S_\phi$ and $\epsilon = \pm 1$ let

$$\widehat{S_\phi}(x,s)_\epsilon = \{\tau \in \widehat{S_\phi} \,|\, x_{\tau\tau_x} \sim x, \tau(s) = \epsilon\}. \tag{7.8}$$

Then the sum (7.2)(b) becomes

$$\widetilde{\eta}_x(\tilde{s}) = e(x)\tau_x(\tilde{s})[\sum_{\widehat{S_\phi}(x,s)_+} \pi(\tau\tau_x) - \sum_{\widehat{S_\phi}(x,s)_-} \pi(\tau\tau_x)]. \tag{7.9}$$

The right hand side is a linear combination of representations of the strong real form x; the coefficients are signs times $\tau_x(\tilde{s})$, which is a complex number of absolute value 1. (If we chose our strong involutions to have finite order as in [**3**] this would be a root of unity.) The question of the signs, and precisely how they depend on the choices involved, was brought to our attention by Michael Harris; see ([**7**], pages 200-201). The case of $U(p,q)$ is discussed in Section 10.

8. Example: Strong Real forms of GL$(n, \mathbb{C})$

Let (,) be a non-degenerate Hermitian form on $\mathbb{C}^n$. It is determined up to equivalence by its signature (p,q) with $p+q = n$, and the symmetry group of this form is denoted $U(p,q)$. These are the weak real forms of GL$(n,\mathbb{C})$ of equal rank.

The map from $\mathcal{X}_1$ to weak real forms is surjective, with infinite fibers. For example

$$x = \operatorname{diag}(\overbrace{\alpha,\dots,\alpha}^{p},\overbrace{-\alpha,\dots,-\alpha}^{q}) \tag{8.1}$$

maps to $U(p,q)$ for any $\alpha \in S^1$. It is much better to fix $z \in Z$ and consider $\mathcal{X}_1[z]$, which is finite. Note that $Z/Z^2 = 1$, so $\mathcal{X}_1[z]$ is the reduced parameter space $\mathcal{X}_1^r$ (2.5).

The Hermitian form $-(,)$ is not equivalent to (,) (unless $p = q$) but their symmetry groups $U(p,q)$ and $U(q,p)$ are isomorphic. Thus the map from equivalence classes of Hermitian forms to weak real forms is two-to-one (unless $p = q$). It turns out we can identify strong real forms of type z with equivalence classes of Hermitian forms (although not canonically). Thus we can think of strong real forms of type z as $\{U(p,q)\,|\,p+q=n\}$ where we *distinguish* $U(p,q)$ and $U(q,p)$. We proceed to make this precise.

Let $G=\mathrm{GL}(n,\mathbb{C})$ and fix $z=\beta I\in Z$ with $|\beta|=1$. Furthermore fix α with $\alpha^2=\beta$. Then

$$\mathcal{X}_1[z]=\{\alpha\mathrm{diag}(\pm1,\dots,\pm1)\}. \tag{8.2}$$

The elements conjugate to

$$x=\mathrm{diag}(\overbrace{\alpha,\dots,\alpha}^{p},\overbrace{-\alpha,\dots,-\alpha}^{q}) \tag{8.3}$$

constitute a single strong real form, which we think of as $U(p,q)$. Those conjugate to $-x$ also constitute a single strong real form, distinct from the previous one unless $p=q$, which we think of as $U(q,p)$. Note that this labelling depends on a choice of $\sqrt{\beta}$. In any event the elements conjugate to $\pm x$ all map to the same weak real form $U(p,q)=U(q,p)$.

The pure strong real forms are an important special case. Note that $z(\rho^\vee)=(-1)^{n-1}I$, so $\beta=(-1)^{n-1}$ and we can take $\alpha=i^{n-1}$. Thus with this choice $x\in\mathcal{X}_1[z(\rho^\vee)]$ gives the strong real form $U(p,q)$ where p is the dimension of the i^{n-1} eigenspace of x.

8.1. Strong real forms of SL$(n,\mathbb{C})$. The equal rank real forms of $\mathrm{SL}(n,\mathbb{C})$ are the special unitary groups $SU(p,q)$ with $p+q=n$.

If n is odd then the map from strong real forms to weak real forms is bijective. If n is even the picture is essentially the same as for $\mathrm{GL}(n,\mathbb{C})$. In both cases we work with the reduced parameter space. Here are the details.

First assume n is odd. Since $Z/Z^2=1$ we can choose arbitrary $z\in Z$ and take the reduced parameter space $\mathcal{X}_1^r$ to be $\mathcal{X}_1[z]$ (cf. 2.5), and the map from strong real forms of type z to weak real forms is surjective. In fact it is a bijection in this case. For example take $z=z(\rho^\vee)=I$. Then $\mathcal{X}_1[I]\ni x=\mathrm{diag}(\pm1,\dots,\pm1)$ maps to the real form $SU(p,q)$ where q is the number of minus signs, which is even. For example if $n=5$ the weak real forms are $SU(5,0),SU(3,2)$ and $SU(1,4)$.

Now suppose $n=2m$ is even. In this case $z(\rho^\vee)=-I$, and the pure strong involutions are

$$\mathcal{X}_1[z(\rho^\vee)]=\{i\mathrm{diag}(\epsilon_1,\dots,\epsilon_n)\} \tag{8.4}$$

with $\epsilon_i=\pm1$ and $\prod\epsilon_i=(-1)^m$.

As in (8.3) the elements conjugate to

$$x=\mathrm{diag}(\overbrace{i,\dots,i}^{p},\overbrace{-i,\dots,-i}^{q}) \tag{8.5}$$

map to the weak real form $SU(p,q)$, where now $p\equiv q\equiv m \pmod 2$. The map from pure strong real forms to weak real forms is not surjective.

Since $n = 2m$ is even $|Z/Z^2| = 2$, so we choose another element of Z. Take

$$z' = \begin{cases} I & m \text{ is odd} \\ e^{2\pi i/n} I & m \text{ is even.} \end{cases} \tag{8.6}$$

Then

$$\mathcal{X}_1[z'] = \{\mathrm{diag}(\epsilon_1, \dots, \epsilon_n) \mid \prod \epsilon_i = 1\} \quad (m \text{ odd}) \tag{8.7}$$

or

$$\mathcal{X}_1[z'] = \{e^{\pi i/n}\mathrm{diag}(\epsilon_1, \dots, \epsilon_n) \mid \prod \epsilon_i = -1\} \quad (m \text{ even}). \tag{8.8}$$

There is a two-to-one map from strong real forms of type z' to the weak real forms $SU(2p, 2q)$ (m odd) or $SU(2p+1, 2q+1)$ (m even).

Thus we see the map from strong real forms of type $z(\rho^\vee) = -I$ or z' to weak real forms two-to-one, except the fiber of the weak real form $SU(m, m)$ is a singleton. This is just like the case of $U(p, q)$ except that Z/Z^2 now has two elements.

For example if $n = 6$, the pure real forms are $SU(5,1), SU(3,3)$ and $SU(1,5)$, with $z = z(\rho^\vee) = -I$. Taking $z' = I$ gives the other strong real forms $SU(6,0), SU(4,2), SU(2,4)$ and $SU(0,6)$.

Take $n = 8$. In this case the pure real forms, given by $z(\rho^\vee) = -I$, are $SU(8,0), SU(6,2), SU(4,4), SU(2,6)$ and $SU(0,8)$. On the other hand $z' = e^{2\pi i/8}$ gives the strong real forms $SU(7,1), SU(5,3), SU(3,5)$ and $SU(1,7)$.

9. Discrete series of $U(p, q)$

Let $G = \mathrm{GL}(n, \mathbb{C})$ and suppose ϕ is a discrete series L-parameter. Fix $z \in Z$ and $y \in \mathcal{X}_1[z]$. We make the bijection $\tau_y \widehat{S_\phi} \overset{1-1}{\longleftrightarrow} \Pi_z(\phi)$ of (6.11)(a) explicit.

Fix $z = \beta I$ for some $\beta \in \mathbb{C}^\times$. In this case there is a natural choice of base-point y (or rather two such). Recall (5.7) $x_b = i^{n-1}(1, -1, \dots, (-1)^{n-1}) \in \mathcal{X}[z(\rho^\vee)]$ defines a generic discrete series representation of a quasisplit strong real form. Choose $\alpha^2 = \beta$ and let

$$y = \alpha i^{-n+1} x_b = \alpha(1, -1, \dots, (-1)^{n-1}) \in X_1[z]. \tag{9.1}$$

We have $S_\phi \simeq \mathbb{Z}/2\mathbb{Z}^n$, embedded diagonally in H. For $\delta_i = \pm 1$ write $\{\delta_1, \dots, \delta_n\}$ for the corresponding character of S_ϕ.

Lemma 9.1. *With y as in (9.1) the bijection $\tau_y \widehat{S_\phi} \overset{1-1}{\longleftrightarrow} \mathcal{X}_1[z]$ of (6.11)(a) is given as follows. If $\tau = \{\delta_1, \dots, \delta_n\}$ then*

$$\tau_y \tau \to \alpha(\delta_1, -\delta_2, \delta_3, \dots, (-1)^{n-1}\delta_n). \tag{9.2}$$

In particular the case of pure strong real forms is given by $z = (-1)^{n-1} I$, so it is natural to take $\alpha = i^{n-1}$, so $y = x_b$ and $\tau_y = 1$. Then the bijection is

$$\widehat{S_\phi} \ni \tau = \{\delta_1, \dots, \delta_n\} \to i^{n-1}(\delta_1, -\delta_2, \dots, (-1)^{n-1}\delta_n) \in \mathcal{X}_1[z(\rho^\vee)]. \tag{9.3}$$

This is an easy exercise in the definitions.

Let $\tau_0 = \{1, -1, \dots, (-1)^n\} \in \widehat{S_\phi}$. If $\tau = \{\delta_1, \dots, \delta_n\}$ let $p(\tau), q(\tau)$ be the number of δ_i equal to $1, -1$ respectively.

LEMMA 9.2. *In the setting of the Lemma suppose* $\tau, \tau' \in \widehat{S_\phi}$ *and* $\tau_y\tau \to x, \tau_y\tau' \to x'$ *via the bijection* (9.2).

Then x, x' *define the same strong real form if and only if* $p(\tau\tau_0) = p(\tau'\tau_0)$. *They map to the same weak real form if and only if* $p(\tau\tau_0) = p(\tau'\tau_0)$ *or* $p(\tau\tau_0) = q(\tau'\tau_0)$.

As discussed in Section 8 the choice of $\alpha = \sqrt{\beta}$ amounts to a labelling of the strong real forms as $U(p,q)$ or $U(q,p)$. That is, we may *define* the strong real form $U(p,q)$ to be those $x \in \mathcal{X}_1[z]$ for which the dimension of the α-eigenspace is p. With this convention we have:

COROLLARY 9.3. $\tau_y\tau \in \tau_y\widehat{S_\phi}$ *goes to a discrete series representation of the strong real form* $U(p(\tau\tau_0), q(\tau\tau_0))$.

The most natural case is $z = z(\rho^\vee)$, i.e. $\beta = (-1)^{n-1}$, in which case it is natural to take $\alpha = i^{n-1}$. Then $U(p,q)$ is the strong real form corresponding to those x with i^{n-1} occuring as an eigenvalue of multiplicity p.

Note that each weak real form $U(p,q)$ occurs twice (as the strong real forms $U(p,q)$ and $U(q,p)$) unless $p = q$. The total number of discrete series representations is therefore $\sum_{p=0}^n \binom{n}{p} = 2^n = |\widehat{S_\phi}|$.

Note that $x \to -x$ is an automorphism of $\mathcal{X}_1[z]$, taking the strong real form $U(p,q)$ to $U(q,p)$. It is helpful to consider the effect of this automorphism on our basepoints, i.e. generic discrete series representations. We consider the pure case $z = z(\rho^\vee)$. Others are similar.

It is not hard to see that for $\tau \in \widehat{S_\phi}$, $\pi(\tau)$ is generic if and only if $x_\tau = \pm x_b$. If $n = 2m$, $\pm x_b$ are conjugate, so we are getting two generic discrete series representations of the same strong real form $U(m,m)$. If $n = 2m+1$ then $\pm x_b$ are not conjugate; these correspond to generic discrete series representations, one each of the two strong real forms $U(m+1,m)$ and $U(m,m+1)$.

It is helpful to express the bijection (5.3) in more familiar parameters. This is primarily an exercise in the definitions. We illustrate this with the examples of GL(4) and GL(3).

9.1. Example: GL(4). Consider pure strong real forms of $\mathrm{GL}(4,\mathbb{C})$, so take $z = -I$. In the setting of (9.3) we have $\alpha = -i$ and $y = x_b = -i\mathrm{diag}(1,-1,1,-1)$, We work at infinitesimal character $(4,3,2,1)$, which is a central shift of $\rho = (3/2, 1/2, -1/2, -3/2)$.

The quasisplit group is $U(2,2)$. Write a Harish-Chandra parameter for $U(2,2)$ as $(a,b;c,d)$. This indicates that, writing the positive roots as $\{e_i - e_j \,|\, 1 \le i < j \le 4\}$, the positive compact roots are $e_1 - e_2, e_3 - e_4$, and $e_2 - e_3$ is noncompact. Write $\pi(a,b;c,d)$ for the discrete series representation with this Harish-Chandra parameter. Then $\pi(4,2;3,1)$ is a generic discrete series representation: the simple roots are $e_1-e_3, e_3-e_2, e_2-e_4$, all of which are noncompact. The other generic discrete series representation with this infinitesimal character is $\pi(3,1;4,2)$.

We need to fix a generic discrete series representation of $U(2,2)$: choose $\pi(4,2;3,1)$. We then easily compute the following table, showing the bijections $\widehat{S_\phi}\to\mathcal{X}_1[z(\rho^\vee)]$ of (9.3) and $\widehat{S_\phi}\to\Pi_{z(\rho^\vee)}(\phi)$ of (6.11)(b).

Recall labelling the strong real forms as $U(p,q), U(q,p)$ requires a choice of $\sqrt{-1}$, which it was most natural to take to be $i^{n-1}=-i$. Therefore *define* the strong real form $U(p,q)$ to be the conjugacy class of x with eigenvalue $-i$ of multiplicity p.

$U(2,2)$

$\tau\in\widehat{S_\phi}$	$x\in\mathcal{X}_1[z(\rho^\vee)]$	HC parameter
{1,1,1,1}	-i(1,-1,1,-1)	(4,2;3,1)
{1,-1,-1,1}	-i(1,1,-1,-1)	(4,3;2,1)
{1,1,-1,-1}	-i(1,-1,-1,1)	(4,1;3,2)
{-1,-1,1,1}	-i(-1,1,1,-1)	(3,1;4,2)
{-1,-1,-1,-1}	-i(-1,1,-1,1)	(3,2;4,1)
{-1,1,1,-1}	-i(-1,-1,1,1)	(2,1;4,3)

The representations of the strong real form $U(3,1)$ are:

$U(3,1)$

$\tau\in\widehat{S_\phi}$	$x\in\mathcal{X}_1[z(\rho^\vee)]$	HC parameter
$\{1,-1,1,1\}$	-i(1,1,1,-1)	(4,3,2;1)
$\{1,-1,-1,-1\}$	-i(1,1,-1,1)	(4,3,1;2)
$\{1,1,1,-1\}$	-i(1,-1,1,1)	(4,2,1;3)
$\{-1,-1,1,-1\}$	-i(-1,1,1,1)	(3,2,1;4)

The negatives of these three parameters give representations of the strong real form $U(1,3)$:

$U(1,3)$

$\tau\in\widehat{S_\phi}$	$x\in\mathcal{X}_1[z(\rho^\vee)]$	HC parameter
$\{-1,1,-1,-1\}$	-i(-1,-1,-1,1)	(1;4,3,2)
$\{-1,1,1,1\}$	-i(-1,-1,1,-1)	(2;4,3,1)
$\{-1,-1,-1,1\}$	-i(-1,1,-1,-1)	(3;4,3,1)
$\{1,1,-1,1\}$	-i(1,-1,-1,-1)	(4;3,2,1)

Finally for the compact strong real forms we have:

$\tau\in\widehat{S_\phi}$	$x\in\mathcal{X}_1[z(\rho^\vee)]$	G	HC parameter
$\{1,-1,1,-1\}$	-i(1,1,1,1)	$U(4,0)$	(4,3,2,1)
$\{-1,1,-1,1\}$	-i(-1,-1,-1,-1)	$U(0,4)$	(4,3,2,1)

As discussed above this entire table has an automorphism given by multiplication by -1, which has the effect of switching $U(p,q)$ and $U(q,p)$, and interchanging the two generic discrete series representations of $U(2,2)$.

9.2. Example: GL(3). This is a little different from the preceding example since 3 is odd. In this case $x_b = -(1,-1,1)$. Fix infinitesimal character $(3,2,1)$. In this case there is only one generic discrete series representation of $U(2,1)$, so there is no choice to make.

The result is:

$\tau \in \widehat{S_\phi}$	$x \in \mathcal{X}_1[z(\rho^\vee)]$	G	HC parameter
{1,1,1}	-(1,-1,1)	$U(2,1)$	(3,1;2)
{1,-1,-1}	-(1,1,-1)	$U(2,1)$	(3,2;1)
{-1,-1,1}	-(-1,1,1)	$U(2,1)$	(2,1;3)
{1,-1,1}	-(1,1,1)	$U(3,0)$	(3,2,1)
{-1,-1,-1}	-(-1,1,-1)	$U(1,2)$	(2;3,1)
{-1,1,1}	-(-1,-1,1)	$U(1,2)$	(1;3,2)
{1,1,-1}	-(1,-1,-1)	$U(1,2)$	(3;2,1)
{-1,1,-1}	-(-1,-1,-1)	$U(0,3)$	(3,2,1)

In this case the two generic discrete representations are $-(1,-1,1)$ of $U(2,1)$ and $-(-1,1,-1)$ of $U(1,2)$.

10. Endoscopy for $U(p,q)$

We consider the question at the end of Section 7 in the case of $U(p,q)$. Set $n = p+q$ and $G = \mathrm{GL}(n,\mathbb{C})$.

Recall (7.8): for $x \in \mathcal{X}_1$, $s \in S_\phi$ and $\epsilon = \pm$ we set

$$\widehat{S_\phi}(x,s)_\epsilon = \{\tau \in \widehat{S_\phi} \mid x_{\tau\tau_x} \sim x,\ \tau(s) = \epsilon\}. \tag{10.1}$$

Let V be the standard module of $\mathrm{GL}(n,\mathbb{C})$, with basis $\{v_1,\dots,v_n\}$. Then (cf. 5.1) $S_\phi \subset H^\vee \subset \mathrm{GL}(V)$ acts on V and its exterior algebra, write this action $s : \gamma \to s\cdot\gamma$. Write $S_\phi \simeq (\mathbb{Z}/2\mathbb{Z})^n$ so that $(s_1,\dots,s_n)\cdot v_i = s_i v_i$.

PROPOSITION 10.1. *Given x, s and ϵ, choose α such that $x^2 = \alpha^2 I$. Let*

$$\delta = (\tau_x/\tau_{\alpha I})(s)\epsilon = \pm 1. \tag{10.2}$$

Let r be the dimension of the α-eigenspace of x. Then there is a natural bijection between $\widehat{S_\phi}(x,s)_\epsilon$ and a basis of the δ-eigenspace of s acting on $\Lambda^{n-r}(V)$.

REMARK. Both δ and r depend on α, so write δ_α, r_α. The δ_α-eigenspace of s acting on $\Lambda^{n-r_\alpha}(V)$ is in bijection with the $\delta_\alpha\det(s)$-eigenspace of s acting on $\Lambda^{n-(n-r_\alpha)}(V)$. It is easy to see $\delta_\alpha\det(s) = \delta_{-\alpha}$ and $r_\alpha = n - r_{-\alpha}$, confirming that the statement is independent of the choice of α.

For an explicit description of the sign $(\tau_x/\tau_{\alpha I})(s)$ see (10.8).

LEMMA 10.2. *Fix $0 \le r \le n$. Let $S = (\mathbb{Z}/2\mathbb{Z})^n$, embedded diagonally in $GL(V) = GL(n,\mathbb{C})$. Let $\tau = diag(\delta_1,\dots,\delta_n)$ ($\delta_i = \pm 1$) be a character of S, and let $q(\tau)$ be the number of δ_i equal to -1. Suppose $s \in S$ and $\epsilon = \pm 1$.*

There is a natural bijection between a basis of

$$\{\gamma \in \Lambda^r(V) \mid s\cdot\gamma = \epsilon\gamma\} \tag{10.3(a)}$$

and

$$\text{(10.3)(b)} \qquad \{\tau\in\widehat{S}\,|\,q(\tau)=r,\ \tau(s)=\epsilon\}$$

Proof. The proof is elementary. Suppose $1\le i_1<\cdots<i_r\le n$ are the indices for which $\delta_{i_j}=-1$. Then $\tau(s)=\prod_j s_{i_j}$. Map τ to $\gamma=v_{i_1}\wedge v_{i_2}\wedge\cdots\wedge v_{i_r}$. Then $s\cdot\gamma=\prod_j s_{i_j}\gamma$. It is easy to see this is a bijection. □

PROOF OF THE PROPOSITION. The fact that $x^2=\alpha^2 I$ implies $x=\alpha(x_1,\dots,x_n)$ with $x_i\in\pm1$. Two such elements are conjugate if and only if the dimension of their α-eigenspaces agree. Thus

$$\text{(10.4)}\quad \widehat{S_\phi}(x,s)_\epsilon=\{\tau\in\widehat{S_\phi}\,|\,\text{dim of the }\alpha\text{-eigenspace of }x_{\tau\tau_x}=r,\tau(s)=\epsilon\}.$$

With notation as in Section 9 write $\tau=\{\delta_1,\dots,\delta_n\}$. A short calculation shows that

$$\text{(10.5)} \qquad x_{\tau\tau_x}=\alpha(\delta_1x_1,\dots,\delta_nx_n).$$

Therefore if we let $\mu_x=\{x_1,\dots,x_n\}\in\widehat{S_\phi}$ then

$$\text{(10.6)} \qquad \widehat{S_\phi}(x,s)_\epsilon=\{\tau\in\widehat{S_\phi}\,|\,p(\tau\mu_x)=r,\tau(s)=\epsilon\}$$

where $p(*)$ is the number of times 1 occurs. Let $\tau'=\tau\mu_x$, so

$$\text{(10.7)} \qquad \widehat{S_\phi}(x,s)_\epsilon=\{\tau'\mu_x\,|\,\tau'\in\widehat{S_\phi},\ p(\tau')=r,\ \tau'(s)=\mu_x(s)\epsilon\}.$$

Replace the condition $p(\tau')=r$ with $q(\tau')=n-r$. It is straightforward to see that $\mu_x=\tau_x/\tau_{\alpha I}$. The result follows from the Lemma. □

From the proof we see that if we write $x=\alpha(x_1,\dots,x_n)$ with $x_i\in\pm1$, and $s=\mathrm{diag}(s_1,\dots,s_n)$ then

$$\text{(10.8)} \qquad (\tau_x/\tau_{\alpha I})(s)=\prod_{\{i\,|\,x_i=-1\}} s_i$$

References

[1] Jeffrey Adams, Lifting of characters. In *Harmonic analysis on reductive groups* (Brunswick, ME, 1989), volume 101 of *Progr. Math.*, pages 1–50. Birkhäuser Boston, Boston, MA, 1991.

[2] Jeffrey Adams, Guide to the atlas software: Computational representation theory of real reductive groups. In Representation Theory of Real Reductive Groups, Proceedings of Conference at Snowbird, July 2006, *Contemp. Math. Amer. Math. Soc.*, 2008.

[3] Jeffrey Adams, Dan Barbasch & David A. Vogan, Jr., The Langlands classification and irreducible characters for real reductive groups, volume 104 of *Progress in Mathematics*, Birkhäuser Boston Inc., Boston, MA, 1992.

[4] Jeffrey Adams & Fokko du Cloux, Algorithms for representation theory of real reductive groups. *J. Inst. Math. Jussieu*, 8(2):209–259, 2009.

[5] A. Borel, Automorphic *L*-Functions, volume 33 of *Proc. Symp. Pure Math.* American Math. Soc., Providence, Rhode Island, 1979.

[6] Robert E. Kottwitz, Sign changes in harmonic analysis on reductive groups. *Trans. Amer. Math. Soc.* 278(1):289–297, 1983.

[7] Robert E. Kottwitz, Shimura varieties and λ-adic representations. In *Automorphic forms, Shimura varieties, and L-functions,* Vol. I (Ann Arbor, MI, 1988), volume 10 of *Perspect. Math.,* pages 161–209. Academic Press, Boston, MA, 1990.

[8] R. Langlands. On the Classification of Irreducible Representations of Real Algebraic Groups. Number 31 in *Mathematical Surveys and Monographs,* American Mathematical Society, Providence, 1989.

[9] R.P. Langlands & D. Shelstad, On the definition of transfer factors. *Math. Ann.,* 278(1-4):219–271, 1987.

[10] D. Shelstad, L-indistinguishability for real groups. *Math. Ann.,* 259(3):385–430, 1982.

[11] D. Shelstad, Tempered endoscopy for real groups. III. Inversion of transfer and L-packet structure. *Represent. Theory,* 12:369–402, 2008.

[12] Diana Shelstad, Orbital integrals and a family of groups attached to a real reductive group. *Ann. Sci. École Norm. Sup. (4),* 12(1):1–31, 1979.

[13] David A. Vogan, Jr., Gel′fand-Kirillov dimension for Harish-Chandra modules. *Invent. Math.,* 48(1):75–98, 1978.

[14] David A. Vogan, Jr., Representations of reductive Lie groups. ICM Series. American Mathematical Society, Providence, RI, 1988. A plenary address presented at the International Congress of Mathematicians held in Berkeley, California, August 1986, Introduced by Wilfried Schmid.

[15] David A. Vogan, Jr., The local Langlands conjecture. In *Representation theory of groups and algebras,* volume 145 of *Contemp. Math.,* pages 305–379. Amer. Math. Soc., Providence, RI, 1993.

Unramified representations of unitary groups
[M.III.C]

Alberto Mínguez

Introduction

Let F be a non-Archimedean locally compact field, of residual characteristic p and let E/F be an unramified quadratic extension. Denote by $U(n)$ an unramified unitary group attached to the extension E/F.

The goal of this chapter is to make explicit the transfer of *unramified* representations in the following cases:

(1) Quadratic base change from $U(n)$ to $GL_n(E)$.
(2) Endoscopic transfer from $U(a) \times U(n-a)$ to $U(n)$.

These results are used in chapter [L.IV.A] of this volume to construct automorphic representations. We recall the classification of spherical representations in terms of Satake parameters for any reductive group and then make it explicit for linear and unitary groups (not necessarily unramified).

1. Notation

1.1. Let F be a non-Archimedean locally compact field, of residual characteristic p. For every finite extension L over F we will denote by $\mathcal{O}_L$ its ring of integers, $\mathfrak{p}_L$ its maximal ideal, ϖ_L a uniformizing parameter and k_L its residue field. We denote by q_L the cardinality of k_L.

Let W_F be the absolute Weil group of F and, if L is a finite Galois extension over F, let $W_{L/F}$ be the Weil group of L/F. If L/F is a Galois extension, let $\Gamma(L/F)$ be the Galois group. The symbol Γ will denote $\Gamma(\overline{F}/F)$,

Partially supported by EPSRC grant EP/G001480/1, MTM2007-66929 and FEDER. These notes came out of a workshop on unitary groups which took place at the University of East Anglia in the winter of 2007 and whose participants were S. Stevens, V. Sécherre and M. Miyauchi. This paper could have not been written without their help. I would like to thank M. Harris for inviting me to contribute to this book and the other authors for having made it possible. Thanks also to G. Chenevier, for his useful remarks on the text, and to H. Saito and G. Henniart for all helpful conversations and their support on numerous occasions.

where $\overline{\mathrm{F}}$ is an algebraic closure of F. The norm and the trace maps will be denoted by $\mathrm{N}_{\mathrm{L/F}}$ and $\mathrm{tr}_{\mathrm{L/F}}$ respectively.

1.2. Let G be a connected reductive group over F. As usual, if there is no confusion, we will not distinguish between G and the group G(F) of F-points of G. Fix a minimal F-parabolic subgroup P_0 of G and let M_0 be a Levi factor of P_0 defined over F. We denote by $\mathcal{W}_{\mathrm{G}}$ the *spherical* Weyl group, defined to be

$$\mathcal{W}_{\mathrm{G}} = \mathcal{N}_{\mathrm{G}}(\mathrm{M}_0)/\mathrm{M}_0,$$

where $\mathcal{N}_{\mathrm{G}}(\mathrm{M}_0)$ is the normalizer of M_0 in G. A parabolic subgroup P of G will be called standard if it contains P_0. In this case we will denote by M_{P} the unique Levi factor of P containing M_0, by N_{P} the unipotent radical of P, P^- the opposite of P (such that $\mathrm{P} \cap \mathrm{P}^- = \mathrm{M}_{\mathrm{P}}$) and $\mathrm{N}_{\mathrm{P}}^- = \mathrm{N}_{\mathrm{P}^-}$. Set $\mathrm{N}_0 = \mathrm{N}_{\mathrm{P}_0}$.

1.3. In this chapter, all representations are assumed to be smooth and admissible – that is, by a representation of G we understand a pair (π, V) where V is a vector space over $\mathbb{C}$ and π is a group homomorphism from G into GL(V) such that the stabilizer of every vector in V is an open subgroup of G and, for every compact open subgroup K of G, the space V^{K} of K-invariant vectors in V is finite dimensional.

We denote by Irr(G) the set of isomorphism classes of irreducible representations of G. Given π a representation of G we will denote by $\widetilde{\pi}$ the contragredient representation of π and let $\mathrm{JH}(\pi)$ be the multiset of irreducible constituents of π.

1.4. Let P be a standard parabolic subgroup of G and let (τ, V) be a representation of the Levi factor M_{P} of P, regarded as a representation of P on which N_{P} acts trivially. We denote by $i_{\mathrm{P}}^{\mathrm{G}}(\tau)$, the representation of G unitarily induced from τ; this is the representation by right translation in the space of functions $\phi : \mathrm{G} \to \mathrm{V}$ satisfying the following conditions:

(a) One has $\phi(pg) = \delta_{\mathrm{P}}^{1/2}(p)\tau(p)\phi(g)$, for $p \in \mathrm{P}$, $g \in \mathrm{G}$ and where δ_{P} is the modulus function.
(b) There exists a compact open subgroup K of G such that $\phi(gk) = \phi(g)$ for $g \in \mathrm{G}$ and $k \in \mathrm{K}$.

The factor $\delta_{\mathrm{P}}^{1/2}$ is there to ensure that $i_{\mathrm{P}}^{\mathrm{G}}(\tau)$ is unitary if τ is (hence the term unitary induction). The functor $i_{\mathrm{P}}^{\mathrm{G}}$ preserves finite length, *i.e.* if τ is a finite length representation of M_{P}, then $i_{\mathrm{P}}^{\mathrm{G}}(\tau)$ is also a representation of finite length of G.

1.5. A representation π of G is called cuspidal if it is not a composition factor of any representation of the form $i_{\mathrm{P}}^{\mathrm{G}}(\tau)$ with P a proper parabolic subgroup of G and τ a representation of M_{P}. This is equivalent to every coefficient of π being compactly supported modulo the center of G (see, for example, [**BZ1**, 3.21].

A cuspidal datum is a pair (M, ρ) where M is a Levi subgroup of G and ρ is a cuspidal representation of M. Two cuspidal data, (M, ρ), (M', ρ') are conjugate if there exists $g \in \mathrm{G}$ such that

$$\mathrm{Ad}\, g : \mathrm{M} \xrightarrow{\sim} \mathrm{M}',$$
$$\mathrm{Ad}\, g : \rho \xrightarrow{\sim} \rho'.$$

If π is an irreducible representation of G, there exists, up to conjugacy, a unique cuspidal datum (M, ρ) such that π is a composition factor of $i_{\mathrm{P}}^{\mathrm{G}}(\rho)$. We call it the cuspidal support of π and write it $\mathrm{supp}(\pi)$.

1.6. Let K be a compact open subgroup of G. The Hecke algebra of G relative to K, denoted by $\mathcal{H}(\mathrm{G}, \mathrm{K})$, is the space of compactly supported functions $f : \mathrm{G} \to \mathbb{C}$ which are left and right K-invariant. This is an algebra under the operation of convolution, relative to some choice of Haar measure μ_{G} on G. It has a unit element, denoted by e_{K}, which is $\mu_{\mathrm{G}}^{-1}(\mathrm{K})$ times the characteristic function of K.

The Hecke algebra $\mathcal{H}(\mathrm{G})$ is then defined as

$$\mathcal{H}(\mathrm{G}) := \bigcup_{\mathrm{K}} \mathcal{H}(\mathrm{G}, \mathrm{K}),$$

where K runs through a basis of neighborhoods of 1 consisting of compact open subgroups.

Representations of G on a complex vector space V correspond bijectively to the non-degenerate representations of $\mathcal{H}(\mathrm{G})$ on V, where the action of $\mathcal{H}(\mathrm{G})$ on V is defined as usual by

$$\pi(f)v = \int_{\mathrm{G}} f(g)\pi(g)v \, d\mu_{\mathrm{G}} g. \tag{1.1}$$

1.7. Let (B, T) be a Borel pair, *i.e.*, a pair consisting of a maximal split torus T of G and a Borel subgroup B containing T. It gives rise to a reduced based root datum

$$\psi(\mathrm{G}, \mathrm{B}) = (X^*(\mathrm{T}), \Delta^*, X_*(\mathrm{T}), \Delta_*)$$

where $X^*(\mathrm{T})$ (resp. $X_*(\mathrm{T})$) is the character (resp. co-character) group, $\Delta^* \subset X^*(\mathrm{T})$ is the subset of simple roots of T which are positive with respect to B and Δ_* is the set of co-roots associated to the roots in Δ^*. Up to canonical isomorphism, $\psi(\mathrm{G}, \mathrm{B})$ is independent of the choice of (B, T).

Let ${}^L\mathrm{G}$ be the *L-group* of G, that is ${}^L\mathrm{G}$ is the semi-direct product of $\widehat{\mathrm{G}}$ – the complex connected reductive group whose reduced based root datum is dual to that of G – and the Weil group W_{F}. See [**Bo1**] or the introduction of this volume [**Har**] for more information on the L-group.

Let H and G be two connected reductive groups over F. A homomorphism $\xi : {}^L\mathrm{H} \to {}^L\mathrm{G}$ is called an L-homomorphism if

(i) it is a homomorphism over W_{F}, that is, the following diagram commutes

$$\begin{array}{ccc} {}^{L}\mathrm{H} & \xrightarrow{\ \xi\ } & {}^{L}\mathrm{G} \\ \big\downarrow & & \big\downarrow \\ W_F & = & W_{\mathrm{F}} \end{array}$$

where the vertical arrows are the projection onto the factor W_{F} of the L-group,

(ii) ξ is continuous, and

(iii) the restriction of ξ to $\widehat{\mathrm{H}}$ is a complex analytic homomorphism $\xi : \widehat{\mathrm{H}} \to \widehat{\mathrm{G}}$.

The principle of functoriality predicts that, at least when G is quasi-split, to each such L-homomorphism there is associated a correspondence from irreducible representations of H to irreducible representations of G. The original idea comes from the fact studied in this chapter: we will see the lifting of unramified representations of H to unramified representations of G in the case where H and G are unramified groups.

2. Spherical representations

In this section, given a maximal compact subgroup K of G, we recall the classification of K-*spherical* representations of G – that is, irreducible representations of G having a non-zero K-fixed vector – in terms of Satake parameters. All the proofs can be found in the articles of Cartier [**Car**] and Borel [**Bo1**] at Corvallis. One can also consult the original paper of Satake [**Sat**].

2.1. Let K be a good, special, maximal compact subgroup of G. The original definition of such a compact subgroup appears in [**BT**] in terms of the apartment of G. There always exists such a group. The relevant properties of such a group that we need in the sequel are:

***Iwasawa decomposition*:** $\mathrm{G} = \mathrm{P}_0\mathrm{K}$ (a maximal compact subgroup satisfying to this property is called *good*, [**BT**, 4.4.1.]).

***Cartan decomposition*:** $\mathrm{G} = \mathrm{K}\Lambda^{+}\mathrm{K}$, where Λ^{+} is a subset of $\mathrm{M}_0/\mathrm{M}_0 \cap \mathrm{K}$ as defined in [**BT**, §4.4.3.].

***Iwahori decomposition*:** There exists an Iwahori subgroup $\mathrm{I} \subset \mathrm{K}$ such that $\mathrm{I} = (\mathrm{I} \cap \mathrm{N}_0^{-})(\mathrm{I} \cap \mathrm{M}_0)(\mathrm{I} \cap \mathrm{N}_0)$ (unique factorization).

The subgroup $\mathrm{M}_0 \cap \mathrm{K}$ is the unique maximal compact subgroup of M_0; it is normal in M_0 and we have $\mathrm{M}_0/\mathrm{M}_0 \cap \mathrm{K} \simeq \mathbb{Z}^d$ where d is the rank of G.

REMARK. In the case where M_0 is a F-split torus, then $\mathrm{M}_0(\mathrm{F})$ is isomorphic to $\mathrm{F}^{\times d}$ and the map taking $\alpha^\vee$ to $\alpha^\vee(\varpi_{\mathrm{F}})$ induces an isomorphism between the group of co-characters $X_*(\mathrm{M}_0)$ and $\mathrm{M}_0/\mathrm{M}_0 \cap \mathrm{K}$.

Denote by $X^{\mathrm{un}}(\mathrm{M}_0)$ the set of unramified characters of M_0, that is, the characters of M_0 which are trivial on $\mathrm{M}_0 \cap \mathrm{K}$. The Weyl group $\mathcal{W} = \mathcal{W}_{\mathrm{G}}$ acts on $X^{\mathrm{un}}(\mathrm{M}_0)$ by

$$(w\chi)(m) = \chi(w^{-1}mw),$$

for $\chi \in X^{\mathrm{un}}(\mathrm{M}_0)$, $m \in \mathrm{M}_0$ and $w \in \mathcal{W}$. If we fix $m_1, \ldots, m_d$ some basis of $\mathrm{M}_0/\mathrm{M}_0 \cap \mathrm{K}$, we get an isomorphism:

$$(2.1) \qquad \begin{array}{ccc} \mathbb{C}^{\times d} & \xrightarrow{\sim} & X^{\mathrm{un}}(\mathrm{M}_0) \\ z & \mapsto & \chi_z, \end{array}$$

given by $\chi_z(m_i) = z_i$, for $i = 1, \ldots, d$ and $z = (z_1, \ldots, z_d)$.

REMARK. The unramified characters of M_0 are hence in bijection with the $\mathbb{C}$-rational points of a torus defined over $\mathbb{C}$. This particular fact is not an unimportant accident, as we will see.

Let $\chi \in X^{\mathrm{un}}(\mathrm{M}_0)$. By the Iwasawa decomposition we easily check that the space of K-fixed vectors $i_{\mathrm{P}_0}^{\mathrm{G}}(\chi)^{\mathrm{K}}$ is one-dimensional and generated by the *canonical spherical vector*

$$\psi_{\chi,\mathrm{K}} : \mathrm{G} \to \mathbb{C}, \qquad mnk \mapsto \delta_{\mathrm{P}}^{1/2}(m)\chi(m),$$

for $m \in \mathrm{M}_0$, $n \in \mathrm{N}_0$ and $k \in \mathrm{K}$.

As the functor of taking K-invariants is exact, there is one unique irreducible composition factor of $i_{\mathrm{P}_0}^{\mathrm{G}}(\chi)$ which is K-spherical. Thus we get a map

$$(2.2) \qquad X^{\mathrm{un}}(\mathrm{M}_0) \to \{\text{K-spherical representations of G}\}.$$

THEOREM 2.1. *The map defined by equation* (2.2) *induces a bijection*

$$X^{un}(\mathrm{M}_0)/\mathcal{W} \xrightarrow{\sim} \{\text{K-}\textit{spherical representations of } \mathrm{G}\}.$$

One can prove injectivity with a light version of Bernstein-Zelevinski geometric lemma [**Ca1**]. Using the Iwahori decomposition and the Borel-Matsumoto theorem [**Bo2**] [**Mat**] (see also [**Ca3**]), one gets that the map is surjective. Remark that the inverse map is given by $\pi \mapsto \mathrm{supp}(\pi)$.

2.2. Another way of defining an inverse map is by using the Satake isomorphism [**Sat**]. Consider the spherical Hecke algebra $\mathcal{H}(\mathrm{M}_0, \mathrm{M}_0 \cap \mathrm{K})$. For elementary reasons it is a polynomial algebra

$$\mathcal{H}(\mathrm{M}_0, \mathrm{M}_0 \cap \mathrm{K}) \simeq \mathbb{C}\left[x_1^{\pm 1}, \ldots, x_d^{\pm 1}\right]$$

where x_i corresponds to the characteristic function of $m_i\,(\mathrm{M}_0 \cap \mathrm{K})$, and we recall that $(m_1, \ldots, m_d)$ is a fixed basis of $\mathrm{M}_0/\mathrm{M}_0 \cap \mathrm{K}$.

The Weyl group $\mathcal{W}$ acts on $\mathcal{H}(\mathrm{M}_0, \mathrm{M}_0 \cap \mathrm{K})$ in the obvious manner: for $f \in \mathcal{H}(\mathrm{M}_0, \mathrm{M}_0 \cap \mathrm{K})$, $w \in \mathcal{W}$, $m \in \mathrm{M}_0$,

$${}^{w}f(m) = f(w^{-1}mw).$$

Fix dn the Haar measure on N_0, such that the measure of $\mathrm{N}_0 \cap \mathrm{K}$ is 1.

THEOREM 2.2. *The $\mathbb{C}$-algebra homomorphism (the Satake transform):*

$$\mathcal{S} : \mathcal{H}(\mathrm{G}, \mathrm{K}) \longrightarrow \mathcal{H}(\mathrm{M}_0, \mathrm{M}_0 \cap \mathrm{K})$$

given by

$$(\mathcal{S}f)(m) = \delta_{\mathrm{P}_0}^{1/2}(m) \int_{\mathrm{N}_0} f(mn)\, dn = \delta_{\mathrm{P}_0}^{-1/2}(m) \int_{\mathrm{N}_0} f(nm)\, dn,$$

is injective with image $\mathcal{H}(\mathrm{M}_0, \mathrm{M}_0 \cap \mathrm{K})^{\mathcal{W}}$, the algebra of $\mathcal{W}$-invariant elements of $\mathcal{H}(\mathrm{M}_0, \mathrm{M}_0 \cap \mathrm{K})$.

Let χ be any unramified character of M_0 and fix a Haar measure on M_0 such that $\int_{\mathrm{M}_0 \cap \mathrm{K}} dm = 1$. The map (Fourier transform) $f \mapsto \int_{\mathrm{M}_0} f(m)\chi(m)\, dm$ is an algebra homomorphism from $\mathcal{H}(\mathrm{M}_0, \mathrm{M}_0 \cap \mathrm{K})$ to $\mathbb{C}$, and, by varying χ, we get in this way all such homomorphisms.

Define a linear map $\omega_\chi : \mathcal{H}(\mathrm{G}, \mathrm{K}) \to \mathbb{C}$ by

$$\omega_\chi(f) = \int_{\mathrm{M}_0} \mathcal{S}f(m)\chi(m)\, dm.$$

PROPOSITION 2.3. *Any algebra homomorphism from $\mathcal{H}(\mathrm{G}, \mathrm{K})$ into $\mathbb{C}$ is of the form ω_χ for some unramified character χ of M_0. Moreover, one has $\omega_\chi = \omega_{\chi'}$ if, and only if, there exists an element $w \in \mathcal{W}$ such that $\chi' = w\chi$.*

2.3. Now let π be a K-spherical representation of G. As we have supposed that the compact subgroup K is special, the Hecke algebra $\mathcal{H}(\mathrm{G}, \mathrm{K})$ is commutative. The subspace of all K-fixed vectors in π, being irreducible, is then one-dimensional.

Thus, we obtain an algebra homomorphism:

$$\lambda_\pi : \mathcal{H}(\mathrm{G}, \mathrm{K}) \to \mathbb{C},$$

defined by $\lambda_\pi(f)v_0 = \pi(f)v_0$, with v_0 any K-fixed vector, and where the action of $\mathcal{H}(\mathrm{G}, \mathrm{K})$ on the representation space of π is defined by (1.1). One can check, using for example [**Car**, §1.5], that we also have

$$\lambda_\pi(f) = \mathrm{tr}\,(\pi(f))\,. \tag{2.3}$$

So, by Proposition 2.3, there exists an unramified character χ_π, unique up to conjugacy by $\mathcal{W}$ such that

$$\lambda_\pi = \omega_{\chi_\pi}.$$

The next proposition is proved in [**Gar**].

PROPOSITION 2.4. *The map $\pi \mapsto \chi_\pi$ is the inverse of the homomorphism defined in Theorem 2.1.*

Proof. Let π be a K-spherical representation and let $\chi \in X^{\mathrm{un}}(\mathrm{M}_0)$ be such that π is a subrepresentation of $i_{\mathrm{P}_0}^{\mathrm{G}}(\chi)$. Let ψ be the canonical spherical vector. Then for any $f \in \mathcal{H}(\mathrm{G},\mathrm{K})$, and any $g \in \mathrm{G}$ we have

$$\begin{aligned}\lambda_\pi(f)\psi(g) &= i_{\mathrm{P}_0}^{\mathrm{G}}(\chi)(f)\psi(g)\\ &= \int_{\mathrm{G}} f(h)\psi(gh)\,dh.\end{aligned}$$

Normalizing the Haar measure on K to be 1, we deduce that:

$$\begin{aligned}\lambda_\pi(f) &= \lambda_\pi(f)\psi(1)\\ &= \int_{\mathrm{G}} f(h)\psi(h)\,dh\\ &= \int_{\mathrm{P}_0}\int_{\mathrm{K}} f(pk)\psi(pk)\delta_{\mathrm{P}_0}^{-1}(p)\,dp\,dk\\ &= \int_{\mathrm{P}_0} f(p)\psi(p)\delta_{\mathrm{P}_0}^{-1}(p)\,dp\\ &= \int_{\mathrm{P}_0} f(p)\chi(p)\delta_{\mathrm{P}_0}^{-1/2}(p)\,dp\\ &= \int_{\mathrm{N}_0}\int_{\mathrm{M}_0} f(nm)\chi(nm)\delta_{\mathrm{P}_0}^{-1/2}(nm)\,dn\,dm\\ &= \int_{\mathrm{M}_0} \mathcal{S}f(m)\chi(m)\,dm\\ &= \omega_\chi.\end{aligned}$$

□

Depending upon the choice of some basis $(m_1,\dots,m_d)$ for the quotient $\mathrm{M}_0/\mathrm{M}_0\cap\mathrm{K}$, the *Satake parameters* attached to π are the images

$$\chi_\pi(m_1),\dots,\chi_\pi(m_d).$$

2.4. A connected reductive group is said to be *unramified* over F if it is quasi-split and splits over an unramified extension of F. One can consult [**Ca2**] where these groups and their spherical representations are treated in detail. The special properties of these groups are:

(1) A connected reductive group G is unramified if, and only if, it has *hyperspecial* maximal compact subgroups [**Tit**, 1.10.2.]. The original definition of these compact subgroups appears in [**Tit**, 1.10.] in terms of hyperspecial points in the Bruhat-Tits building of G, generalizing the concept of special compact subgroups.
(2) There exists a group scheme X over $\mathcal{O}_\mathrm{F}$ such that $\mathrm{G} = X_\mathrm{F}$ and X_{k_F} is a connected reductive group. The compact subgroup $X(\mathcal{O}_\mathrm{F})$ is hyperspecial [**Tit**, 3.8.1.].
(3) The action of W_F on $\widehat{\mathrm{G}}$ factors through the projection of W_F onto $\Gamma(\mathrm{F}^{\mathrm{un}}/\mathrm{F})$, where F^{un} is the maximal unramified extension of F in $\overline{\mathrm{F}}$. It is hence determined by the action of the Frobenius element.

(4) Denote by $\mathrm{T} = \mathrm{M}_0$ a maximal torus contained in a Borel subgroup $\mathrm{B} = \mathrm{P}_0$ of G. Let A be a maximal split subtorus of T and K a good hyperspecial maximal compact subgroup of G. Then the embedding of A into T induces an isomorphism of the lattices

$$\mathrm{A}/\mathrm{A}\cap\mathrm{K} \xrightarrow{\sim} \mathrm{T}/\mathrm{T}\cap\mathrm{K}. \tag{2.4}$$

This last property gives us the idea for using the L-group for classifying spherical representations. The dual group of A is a complex torus $\widehat{\mathrm{A}}$. So we have a set of canonical isomorphisms:

$$\begin{aligned} \mathrm{Hom}\left(\mathrm{T}/\mathrm{T}\cap\mathrm{K}, \mathbb{C}^\times\right) &= \mathrm{Hom}\left(\mathrm{A}/\mathrm{A}\cap\mathrm{K}, \mathbb{C}^\times\right) \\ &= \mathrm{Hom}\left(X_*(\mathrm{A}), \mathbb{C}^\times\right) \\ &= \mathrm{Hom}\left(X^*(\widehat{\mathrm{A}}), \mathbb{C}^\times\right), \end{aligned} \tag{2.5}$$

which is, by definition, the group of points of $\widehat{\mathrm{A}}$. Here, the first equality comes from property (2.4), the second one from the fact that A is split and Remark 2.1, and the third equality by definition of the dual group.

The embedding of A into T gives rise to a surjection from $\widehat{\mathrm{T}}$ to $\widehat{\mathrm{A}}$, which by (2.5) associates to each element in $\widehat{\mathrm{T}}$ an unramified character of T. We have an action coming from the Weyl group on $X^{\mathrm{un}}(\mathrm{T})$ and action of the Frobenius element on $\widehat{\mathrm{T}}$. Let's see that these actions are compatible.

2.5. Suppose until the end of this section that G is an unramified group. We say that a representation of G is K-*unramified* if it is K-spherical where K is a hyperspecial compact subgroup of G.

REMARK. The notion of unramified representation depends on the choice of the hyperspecial maximal compact subgroup K. However, the hyperspecial maximal compact subgroups form a single orbit under the action of the adjoint group G_{ad} of G (*cf.* [**Tit**, 2.5]).

We will denote by $\mathrm{Irr}^{\mathrm{K\text{-}un}}(\mathrm{G})$ the set of isomorphism classes of K-unramified representations of G. If there is no confusion with the choice of K we will simply denote it by $\mathrm{Irr}^{\mathrm{un}}(\mathrm{G})$

Fix an element $\mathfrak{F} \in W_{\mathrm{F}}$ whose projection to $\Gamma(\mathrm{F}^{\mathrm{un}}/\mathrm{F})$ is the Frobenius element. We say that two elements g_1' and g_2' of $\widehat{\mathrm{G}}$ are $\mathfrak{F}$-*conjugate* if there exists $h \in \widehat{\mathrm{G}}$ such that $g_2' = h^{-1} g_1' h^{\mathfrak{F}}$.

The following theorem was first stated and proved in [**La1**]. See also [**La2**] and [**Bo1**].

THEOREM 2.5. (1) *Every semisimple* $\widehat{\mathrm{G}}$-*conjugacy class in* $\widehat{\mathrm{G}} \rtimes \mathfrak{F}$ *contains an element of the form* $t' \rtimes \mathfrak{F}$ *with* $t' \in \widehat{\mathrm{T}}$.

(2) *The surjection* $\widehat{\mathrm{T}} \to \widehat{\mathrm{A}}$ *which associates to each element* $t' \in \widehat{\mathrm{T}}$, *by* (2.5), *an unramified character* $\chi_{t'}$ *of* T *is such that two elements* t_1' *and* t_2' *of* $\widehat{\mathrm{T}}$ *are* $\mathfrak{F}$-*conjugate if, and only if, the unramified characters*

$\chi_{t'_1}$ and $\chi_{t'_2}$ of T *are conjugate under the action of the Weyl group* $\mathcal{W}$.

2.6. Combining Theorems 2.1 and 2.5, we deduce that K-unramified representations are in bijective correspondence with the $\widehat{\mathrm{G}}$-conjugacy classes of semisimple elements in $\widehat{\mathrm{G}} \rtimes \mathfrak{F}$. Furthermore, each such class can be represented by an element of the form $(t, \mathfrak{F})$, with $t \in \widehat{\mathrm{T}}$ fixed under $\mathfrak{F}$. To sum up, when G is an unramified group the set $\mathrm{Irr}^{\text{K-un}}(\mathrm{G})$ is in canonical bijection with:

(1) $X^{\mathrm{un}}(\mathrm{M}_0)/\mathcal{W}$.
(2) $\mathrm{Hom}\,(\mathcal{H}(\mathrm{G},\mathrm{K}), \mathbb{C})$.
(3) Semi-simple $\widehat{\mathrm{G}}$-conjugacy classes in the coset $\widehat{\mathrm{G}} \rtimes \mathfrak{F}$.
(4) Equivalence classes of unramified L-parameters of G, that is, commuting diagrams

$$\begin{array}{ccc} \Gamma(\mathrm{F}^{\mathrm{un}}/\mathrm{F}) & \xrightarrow{\phi} & {}^{L}\mathrm{G} \\ \downarrow & \swarrow & \\ \Gamma(\mathrm{F}^{\mathrm{un}}/\mathrm{F}) & & \end{array}$$

where the vertical arrow is the identity map from $\Gamma(\mathrm{F}^{\mathrm{un}}/\mathrm{F})$ to itself and $\phi(w)$ is semisimple, for all $w \in \Gamma(\mathrm{F}^{\mathrm{un}}/\mathrm{F})$.

For (4), remark that an unramified parameter ϕ is determined by the semisimple element $\phi(\mathfrak{F}) = g_\phi \rtimes \mathfrak{F}$.

2.7. Let H and G be two unramified connected reductive groups over F and let

$$\xi : {}^{L}\mathrm{H} \to {}^{L}\mathrm{G}$$

be an L-homomorphism. We deduce a map from semi-simple conjugacy classes in $\widehat{\mathrm{H}} \rtimes \mathfrak{F}$ to semi-simple conjugacy classes in $\widehat{\mathrm{G}} \rtimes \mathfrak{F}$. Thus if we fix some hyperspecial maximal compact subgroups $\mathrm{K_H}$ and $\mathrm{K_G}$ in H and G respectively, using the preceding results, we deduce a lift, called the *natural unramified lift*:

$$\widetilde{\xi} : \mathrm{Irr}^{\mathrm{K_H}\text{-un}}(\mathrm{H}) \to \mathrm{Irr}^{\mathrm{K_G}\text{-un}}(\mathrm{G}). \tag{2.6}$$

We will make this lift explicit in some special cases in section 4.

We get also a natural map from $\mathrm{Hom}(\mathcal{H}(\mathrm{H},\mathrm{K_H}),\mathbb{C})$ to $\mathrm{Hom}(\mathcal{H}(\mathrm{G},\mathrm{K_G}),$ $\mathbb{C})$ and hence we deduce a lift from the relative Hecke algebras:

$$b(\xi) : \mathcal{H}(\mathrm{G},\mathrm{K_G}) \to \mathcal{H}(\mathrm{H},\mathrm{K_H}). \tag{2.7}$$

By equation (2.3), this map is characterized by the property $\mathrm{tr}\,(\pi\,(b(f))) = \mathrm{tr}\left(\widetilde{\xi}(\pi)(f)\right)$, for $\pi \in \mathrm{Irr}^{\mathrm{K_H}\text{-un}}(\mathrm{H})$ and $f \in \mathcal{H}(\mathrm{G},\mathrm{K_G})$.

3. Basic structure of linear and unitary groups

3.1. Linear groups. $\mathrm{GL}_n(\mathrm{F})$ is by definition the multiplicative group of invertible matrices in $\mathrm{End}_{\mathrm{F}}(\mathrm{F}^n)$. It is endowed with the inherited topology. The identity has a countable basis of neighborhoods that are compact open subgroups; in particular, $\mathrm{GL}_n(\mathrm{F})$ is a locally compact topological group. As such, its Haar measure are left and right invariant (the group is unimodular).

3.1.1. Denote by $\mathrm{GL}_n(\mathcal{O}_{\mathrm{F}})$ the subgroup of $\mathrm{GL}_n(\mathrm{F})$ of elements $g \in \mathrm{End}_{\mathcal{O}_{\mathrm{F}}}(\mathcal{O}_{\mathrm{F}}^n)$ such that $\det g$ is a unit in $\mathrm{F}^\times$.

THEOREM 3.1. *$\mathrm{GL}_n(\mathrm{F})$ contains a unique conjugacy class of maximal compact subgroups, and each such subgroup is open. One element in this class is $\mathrm{GL}_n(\mathcal{O}_{\mathrm{F}})$.*

3.1.2. Let $\alpha = (n_1, \dots, n_r)$ be a partition of the integer n. We denote by M_α the subgroup of $\mathrm{GL}_n(\mathrm{F})$ of invertible matrices which are diagonal by blocks of size n_i and P_α the subgroup of upper triangular matrices by blocks of size n_i. A standard parabolic subgroup of $\mathrm{GL}_n(\mathrm{F})$ is a subgroup of the form P_α and its Levi factor is M_α.

Thus a minimal Levi factor M_0 is the subgroup of diagonal matrices $\mathrm{M}_{(1,\dots,1)}$, which is isomorphic to the product of n copies of $\mathrm{F}^\times$. For convenience, we will write elements in M_0 by $\mathrm{diag}(\lambda_1, \dots, \lambda_n)$, with $\lambda_i \in \mathrm{F}^\times$ for $1 \leq i \leq n$. The spherical Weyl group is isomorphic to the group of permutations $\mathcal{S}_n$. Here $\mathcal{S}_n$ acts on M_0 by permutations on the matrix entries λ_i, $1 \leq i \leq n$.

3.1.3. The dual group of $\mathrm{GL}_n(\mathrm{F})$ is $\mathrm{GL}_n(\mathbb{C})$. To verify this we identify $X^*(\mathrm{M}_0)$ and $X_*(\mathrm{M}_0)$ with $\mathbb{Z}^n$ under the standard pairing $\langle e_i, e_j \rangle = \delta_{ij}$, and let

$$\Delta^* = \Delta_* = \{e_i - e_{i+1} : 1 \leq i \leq n-1\}.$$

The Galois action on the dual group is trivial since $\mathrm{GL}_n(\mathrm{F})$ is a split group, thus ${}^L\mathrm{GL}_n(\mathrm{F}) = \mathrm{GL}_n(\mathbb{C}) \times W_{\mathrm{F}}$ (direct product).

3.1.4. The set of isomorphism classes of unramified representations is, by Theorem 2.1, in bijection with the n-tuples $(\chi_1, \dots, \chi_n)$ of unramified characters of $\mathrm{F}^\times$ up to permutation. To such an n-tuple we associate the conjugacy class in $\mathrm{GL}_n(\mathbb{C})$ of the diagonal element $\mathrm{diag}(\chi_1(\varpi), \dots, \chi_n(\varpi))$, where ϖ is a uniformizing parameter of F.

3.2. Unitary groups. Let E/F be a quadratic extension and let σ be the non trivial element of $\Gamma(\mathrm{E}/\mathrm{F})$. We also write $\sigma(x) = \overline{x}$. The character of order two of $\mathrm{F}^\times$ associated to E/F by local class field theory will be denoted by $\omega_{\mathrm{E}/\mathrm{F}}$, *i.e.* $\omega_{\mathrm{E}/\mathrm{F}}$ is trivial on $\mathrm{N}_{\mathrm{E}/\mathrm{F}}(\mathrm{E}^\times)$. If χ is a character of $\mathrm{E}^\times$, $\overline{\chi}$ will denote the character $\overline{\chi}(x) = \chi(\overline{x})$.

Let V be an n-dimensional vector space over E. A *hermitian form* on V is a pairing

$$h : \mathrm{V} \times \mathrm{V} \to \mathrm{E}$$

that is σ-linear in the first variable, linear in the second variable:

$$h(\alpha v, \beta w) = \overline{\alpha}\beta h(v, w)$$

and satisfies $h(w, v) = \sigma(h(v, w))$ for $v, w \in \mathrm{V}$ and $\alpha, \beta \in \mathrm{E}$. We always assume h to be non-degenerate, *i.e.* for $v \in \mathrm{V}$, $v \neq 0$, there exists $w \in V$ such that $h(v, w) \neq 0$. We say that two hermitian vector spaces are *isometric* if there is an E-linear isomorphism between them that identifies the hermitian forms. Such a map is called an *isometry*. The group of isometries of a hermitian space into itself is called a unitary group, *i.e.* $\mathrm{U}(V)$ is the subgroup of $g \in \mathrm{GL}(\mathrm{V})$ that preserve h:

$$h(g(v), g(w)) = h(v, w), \qquad \text{for } v, w \in \mathrm{V}.$$

This relation defines an algebraic group over F. We say that $n = \dim_{\mathrm{E}} \mathrm{V}$ is the degree of the unitary group $\mathrm{U}(\mathrm{V})$. A hermitian space is called anisotropic if, for all $v \in \mathrm{V}$, $v \neq 0$, we have $h(v, v) \neq 0$. A subspace W of V is called totally isotropic if $h(w, w) = 0$ for all $w \in \mathrm{W}$. If V and V′ are two hermitian vector spaces, one constructs the hermitian vector space (orthogonal sum) $\mathrm{V} \perp \mathrm{V}'$ in the obvious way.

EXAMPLE 3.2. (1) For $n = 1$, let $a \in \mathrm{F}$. We define the hermitian space $\mathrm{E}(a)$ as the E-vector space of dimension 1 where the hermitian form h is defined by:

$$h(e, e') = \overline{e} a e'.$$

(2) For $n = 2$, the hyperbolic plane H over E is the E-vector space of dimension 2 with the product:

$$h\left((e_1, e_2), (e'_1, e'_2)\right) = \overline{e}_1 e'_2 + \overline{e}_2 e'_1.$$

(3) For $n = 2$, the anisotropic hermitian space $\mathrm{W}_2(a_1, a_2)$ is the E-vector space of dimension 2, $\mathrm{W}_2(a_1, a_2) = \mathrm{E}(a_1) \perp \mathrm{E}(a_2)$ with a_1, a_2 not equal to zero and $-a_1/a_2 \notin \mathrm{N}_{\mathrm{E}/\mathrm{F}}(\mathrm{E}^{\times})$. All anisotropic hermitian spaces of dimension 2 are isometric.

In general, by a theorem of Landherr [**L**], for each n there are exactly two different classes of isomorphism of n-dimensional hermitian spaces over E:

(1) For $n = 2m + 1$ odd, let $\mathrm{V}^{\pm} \simeq mH \perp \mathrm{W}^{\pm}$, where $\mathrm{W}^{\pm} \simeq \mathrm{E}(a)$ (see example 3.2) depending on whether $a \in \mathrm{N}_{\mathrm{E}/\mathrm{F}}(\mathrm{E}^{\times})$ or not.
(2) For $n = 2m$ even, let $\mathrm{V}^{+} \simeq mH$ and $\mathrm{V}^{-} \simeq (m-1)H \perp \mathrm{W}_2^{-}$ where W_2^{-} is an anisotropic space of dimension 2.

This decomposition is called the Witt decomposition. The number of hyperbolic planes appearing in this decomposition is called the Witt index $w(\mathrm{V})$ of V.

3.2.1. If n is odd, $\mathrm{U}(\mathrm{V}^+)$ is isomorphic to $\mathrm{U}(\mathrm{V}^-)$ and it is always a quasi-split group. We will denote it by $\mathrm{U}(n)$. If n is even, then $\mathrm{U}(\mathrm{V}^+)$ is not isomorphic to $\mathrm{U}(\mathrm{V}^-)$. We usually write $\mathrm{U}(\mathrm{V}^+) = \mathrm{U}(m, m)$; it is a quasi-split group while $\mathrm{U}(\mathrm{V}^-)$ is not. A unitary group U is thus unramified if, and only if, E/F is an unramified extension and U is isomorphic to $\mathrm{U}(\mathrm{V}^+)$ (see Paragraph 2.4).

3.2.2. At least if $p \neq 2$ (*cf.* [**Hij**]), the number of conjugacy classes of maximal compact subgroups of a unitary group $\mathrm{U}(\mathrm{V})$ is equal to $w(\mathrm{V}) + 1$. By [**Tit**], two of them consist of special compact subgroups and, when $\mathrm{U}(\mathrm{V})$ is unramified, they are also hyperspecial if the degree of $\mathrm{U}(\mathrm{V})$ is even while, if the degree is odd, just one conjugacy class of maximal compact subgroups consists of hyperspecial compact subgroups.

3.2.3. Let (V, h) be a hermitian space. A self dual flag Φ is a decreasing sequence of spaces

$$\mathrm{V} = \mathrm{V}_{-d} \supsetneq \mathrm{V}_{1-d} \supsetneq \cdots \supsetneq \mathrm{V}_{-1} \supsetneq \mathrm{V}_0 \supseteq \mathrm{V}_0^{\perp} \supsetneq \mathrm{V}_1 \supsetneq \cdots \supsetneq \mathrm{V}_d = \{0\},$$

where, for $i = -d, \dots, d$, $i \neq 0$, we have

$$\mathrm{V}_i^{\perp} := \{v \in \mathrm{V} : h(v, v_i) = 0, \forall v_i \in \mathrm{V}_i\} = \mathrm{V}_{-i}.$$

For such a flag Φ and $i = 1, \dots, d$, we can always choose a totally isotropic hermitian subspace $\mathrm{W}_i \subset \mathrm{V}_{-i}$ such that $\mathrm{V}_{-i} = \mathrm{V}_{-i+1} \oplus \mathrm{W}_i$, and a hermitian space W_0 such that $\mathrm{V}_0 = \mathrm{V}_0^{\perp} \oplus \mathrm{W}_0$.

Parabolic subgroups of $\mathrm{U}(\mathrm{V})$ are stabilizers of self-dual flags. For a given flag Φ, denote by P_Φ its associated standard parabolic subgroup. Then, the Levi factor of P_Φ is isomorphic to

$$\mathrm{M}_\Phi \simeq \prod_{i=1}^{d} \mathrm{Aut}_{\mathrm{E}}(\mathrm{W}_i) \times \mathrm{U}(\mathrm{W}_0).$$

In particular, for any hermitian space V of Witt index $w(\mathrm{V}) = m$, the minimal Levi subgroup M_0 of the unitary group $\mathrm{U}(V)$ is isomorphic to m copies of $\mathrm{E}^\times$ times $\mathrm{U}(\mathrm{V}_{\mathrm{an}})$, where

$$\mathrm{U}(\mathrm{V}_{\mathrm{an}}) = \begin{cases} 1, & \text{if } \dim \mathrm{V} = 2m \ (\mathrm{V} \simeq \mathrm{V}^+), \\ \mathrm{U}(1), & \text{if } \dim \mathrm{V} = 2m+1 \\ \mathrm{U}(\mathrm{W}_2^-), & \text{if } \dim \mathrm{V} = 2m+2 \ (\mathrm{V} \simeq \mathrm{V}^-), \end{cases}$$

For convenience, we will write elements in M_0 by $\mathrm{diag}(\lambda_1, \dots, \lambda_m, u_0)$, with $\lambda_i \in \mathrm{E}^\times$ and $u_0 \in \mathrm{U}(\mathrm{V}_{\mathrm{an}})$. The spherical Weyl group $\mathcal{W}$ is isomorphic to $\mathcal{S}_m \rtimes \mathbb{Z}_2^m$. Here $\mathcal{S}_m$ acts on M_0 by permutations on the matrix entries λ_i, $1 \leq i \leq m$. If c_i is the non-trivial element of the i-th copy of $\mathbb{Z}_2$, then c_i changes λ_i into $\overline{\lambda_i}^{-1}$.

3.2.4. Let K be a good special maximal compact subgroup of $\mathrm{U}(\mathrm{V})$. The set of K-spherical representations of $\mathrm{U}(\mathrm{V})$ is, by Theorem 2.1, in canonical bijection with the m-tuples $(\chi_1, \dots, \chi_m)$ of unramified characters of $\mathrm{E}^\times$ where m is the Witt index $w(\mathrm{V})$ of V and two m-tuples $(\chi_1, \dots, \chi_m)$ and

$(\chi'_1, \ldots, \chi'_m)$ correspond to the same K-spherical representation if there exists a permutation $s \in \mathcal{S}_m$ such that, for $1 \le i \le m$, $\chi'_{s(i)}$ equals χ_i or $\overline{\chi_i}^{-1}$.

3.2.5. Let U(n) be a unitary group of degree n (quasi-split or not). Over E, U(n) is isomorphic to $\mathrm{GL}_n(\mathrm{E})$, and U$(n)$ is an outer form of $\mathrm{GL}_n(\mathrm{E})$. It follows that $\widehat{\mathrm{U}(n)} = \mathrm{GL}_n(\mathbb{C})$ and the Galois action factors through $\Gamma(\mathrm{E}/\mathrm{F})$.

Let Φ_n be the $n \times n$ matrix whose ij entry is $(-1)^{i+1}\delta_{i,n-j+1}$;

$$\Phi_n = \begin{pmatrix} 0 & 0 & \ldots & 0 & 1 \\ 0 & 0 & \ldots & -1 & 0 \\ \vdots & \vdots & \iddots & \vdots & \vdots \\ 0 & (-1)^n & \ldots & 0 & 0 \\ (-1)^{n+1} & 0 & \ldots & 0 & 0 \end{pmatrix} \tag{3.1}$$

Then $g \mapsto \Phi_n {}^t g^{-1} \Phi_n^{-1}$ is the unique outer automorphism of $\mathrm{GL}_n(\mathbb{C})$ which preserves the standard splitting (defined with respect to the upper-triangular Borel subgroup and the standard basis for the simple root spaces). Hence the non-trivial element σ of $\Gamma(\mathrm{E}/\mathrm{F})$ acts on $\widehat{\mathrm{U}(n)} = \mathrm{GL}_n(\mathbb{C})$ by this automorphism

$$\sigma(g) = \Phi_n {}^t g^{-1} \Phi_n^{-1}.$$

An action of W_F on $\widehat{\mathrm{U}(n)}$ is defined by projection onto $\Gamma(\mathrm{E}/\mathrm{F})$. The L-group of U(n) is the semi-direct product of $\widehat{\mathrm{U}(n)}$ with W_F with respect to this action.

3.2.6. Suppose U(n) is an unramified group of degree n – that is, *cf.* Paragraph 3.2.1, E/F is an unramified extension and there exists a positive integer m (the Witt index) such that U(n) is isomorphic to U$(2m+1)$ or to U(m,m). Fix a hyperspecial maximal compact subgroup K of U(n) and denote by w_σ a fixed element of $W_{\mathrm{E}/\mathrm{F}}$ whose projection to $\Gamma(\mathrm{E}/\mathrm{F})$ is σ. Then $W_\mathrm{F} = W_\mathrm{E} \cup w_\sigma W_\mathrm{E}$.

As we have seen in Paragraph 3.2.4, K-unramified representations are classified by m-tuples $(\chi_1, \ldots, \chi_m)$ of unramified characters of $\mathrm{E}^\times$, where two m-tuples $(\chi_1, \ldots, \chi_m)$ and $(\chi'_1, \ldots, \chi'_m)$ correspond to the same K-unramified representation if, and only if, there exists a permutation $s \in \mathcal{S}_m$ such that, for $1 \le i \le m$, $\chi'_{s(i)}$ equals χ_i or χ_i^{-1}.

Fix a maximal torus T in U(n) and a maximal split torus A in T. Then T is isomorphic to $\mathrm{Res}_{\mathrm{E}/\mathrm{F}}\mathrm{A}$ so that $\widehat{\mathrm{A}} \simeq \mathbb{C}^{\times m}$ and $\widehat{\mathrm{T}} \simeq \mathbb{C}^{\times 2m}$. The diagonal embedding A into T gives rise to a natural projection form $\widehat{\mathrm{T}}$ onto $\widehat{\mathrm{A}}$ given by $(t_1, t_2) \mapsto t_1 t_2^{-1}$ where $t_i \in \mathbb{C}^{\times m}$ for $i = 1, 2$.

So, by Paragraph 2.4, to an m-tuple $(\chi_1, \ldots, \chi_m)$ of unramified characters of $\mathrm{E}^\times$ we can associate the conjugacy class in ${}^L\mathrm{U}(n)$ of the diagonal element

$$\operatorname{diag}\left(\chi_1^{1/2}(\varpi),\ldots,\chi_m^{1/2}(\varpi),\chi_m^{-1/2}(\varpi),\ldots,\chi_1^{-1/2}(\varpi)\right)\rtimes w_\sigma,\quad \text{if } n=2m \text{ is even},$$

$$\operatorname{diag}\left(\chi_1^{1/2}(\varpi),\ldots,\chi_m^{1/2}(\varpi),1,\chi_m^{-1/2}(\varpi),\ldots,\chi_1^{-1/2}(\varpi)\right)\rtimes w_\sigma,\quad \text{if } n=2m+1 \text{ is odd}.$$

Of course the w_σ-conjugacy class does not depend on the choice of the square root of the $\chi_i(\varpi)$ for $1 \leq i \leq m$.

4. Some lifting problems

In this section, we make explicit, in terms of Satake parameters, *base change* lifting and *endoscopic transfer*. We will deal only with the unramified case, that is we will suppose that E/F is a quadratic unramified extension and, for n a positive integer, we will write $\mathrm{U} = \mathrm{U}(n)$ the quasi-split unitary group of degree n. So, if we denote by θ the automorphism $g \mapsto \Phi_n\, {}^t\overline{g}^{-1}\Phi_n^{-1}$ of $\mathrm{GL}_n(\mathrm{E})$ with Φ_n defined as in (3.1), then U is isomorphic to the subgroup of fixed points by θ in $\mathrm{GL}_n(\mathrm{E})$. We still fix w_σ an element of $W_{\mathrm{E}/\mathrm{F}}$ whose projection to $\Gamma(\mathrm{E}/\mathrm{F})$ is σ.

Denote by μ the unique unramified character of $\mathrm{E}^\times$ of order 2.

4.1. Quadratic base change.

4.1.1. Set $\mathrm{G} = \mathrm{Res}_{\mathrm{E}/\mathrm{F}}(\mathrm{U})$. The dual group of G is $\widehat{\mathrm{G}} = \mathrm{GL}_n(\mathbb{C}) \times \mathrm{GL}_n(\mathbb{C})$ [**Bo1**, §I.5], where the Weil group acts on $\widehat{\mathrm{G}}$ through its projection onto $\Gamma(\mathrm{E}/\mathrm{F})$ and σ acts by $\sigma(x,y) = \left(\Phi_n\, {}^t y^{-1}\Phi_n^{-1}, \Phi_n\, {}^t x^{-1}\Phi_n^{-1}\right)$ with Φ_n as in (3.1).

There is a natural bijection between w_σ-conjugacy classes of $\widehat{\mathrm{G}}$ and semi-simple conjugacy classes of $\mathrm{GL}_n(\mathbb{C})$, sending the conjugacy class of $(g_1, g_2) \rtimes w_\sigma$ in ${}^L\mathrm{G}$ to the conjugacy class of $(g_1\Phi_n\, {}^t g_2^{-1}\Phi_n^{-1})$ in $\mathrm{GL}_n(\mathbb{C})$. This bijection reflects the fact that G(F) is isomorphic to U(E).

4.1.2. There are two natural L-homomorphisms (the *base change lifts*):

$$BC: {}^L\mathrm{U} \to {}^L\mathrm{G} \qquad (4.1)$$
$$BC': {}^L\mathrm{U} \to {}^L\mathrm{G}$$

defined by $BC(g,w) = (g,g,w)$ and $BC'(g,w) = \alpha(w)BC(g,w)$ where $\alpha(w)$ is the 1-cocycle defined by

$$\alpha(w) = \begin{cases} (\mu(w), \mu(w)), & \text{if } w \in W_{\mathrm{E}}, \\ (\mu(w_0), -\mu(w_0)), & \text{if } w = w_0 w_\sigma, w_0 \in W_{\mathrm{E}}. \end{cases}$$

where we regard here μ as a character of W_{E} via local class field theory.

4.1.3. We now make base change explicit for unramified representations. We fix a hyperspecial maximal compact subgroup K in U . We deduce, by (2.6), two morphisms:

$$\begin{aligned} \widetilde{BC} &: \mathrm{Irr}^{\mathrm{K\text{-}un}}(\mathrm{U}) \to \mathrm{Irr}^{\mathrm{un}}(\mathrm{G}) \\ \widetilde{BC'} &: \mathrm{Irr}^{\mathrm{K\text{-}un}}(\mathrm{U}) \to \mathrm{Irr}^{\mathrm{un}}(\mathrm{G}). \end{aligned} \tag{4.2}$$

THEOREM 4.1. (1) *Let π be a* K*-unramified representation of* $\mathrm{U}(n)$ *and denote by* $(\chi_1(\varpi),\dots,\chi_m(\varpi))$ *its Satake parameters. Then the Satake parameters of* $\widetilde{BC}(\pi)$ *are*

$$(\chi_1(\varpi),\dots,\chi_m(\varpi),\chi_m^{-1}(\varpi),\dots,\chi_1^{-1}(\varpi)), \quad \text{if } n=2m \text{ is even},$$
$$(\chi_1(\varpi),\dots,\chi_m(\varpi),1,\chi_m^{-1}(\varpi),\dots,\chi_1^{-1}(\varpi)), \quad \text{if } n=2m+1 \text{ is odd}.$$

(2) $\widetilde{BC'}(\pi)=\mu(\det)\otimes\widetilde{BC}(\pi)$.

Proof. Let π be a K-unramified representation of $\mathrm{U}(n)$ and $(\chi_1,\dots,\chi_m)$ its Satake parameters. By 3.2.6, π is also parametrized by the conjugacy class in ${}^L\mathrm{U}$ of the diagonal element

$$(t,w_\sigma)=\begin{cases} \mathrm{diag}\left(\chi_1^{1/2}(\varpi),\dots,\chi_m^{1/2}(\varpi),\chi_m^{-1/2}(\varpi),\dots,\chi_1^{-1/2}(\varpi)\right)\rtimes w_\sigma, \\ \hfill \text{if } n=2m, \\ \mathrm{diag}\left(\chi_1^{1/2}(\varpi),\dots,\chi_m^{1/2}(\varpi),1,\chi_m^{-1/2}(\varpi),\dots,\chi_1^{-1/2}(\varpi)\right)\rtimes w_\sigma, \\ \hfill \text{if } n=2m+1. \end{cases}$$

The conjugacy class $BC(t,w_\sigma)=(t,t,w_\sigma)$ in ${}^L\mathrm{G}$ can be regarded, by 4.1.1, as the conjugacy class of $(t\Phi_n t^{-1}\Phi_n^{-1})$ in $\mathrm{GL}_n(\mathbb{C})$, that is, by the conjugacy class of the element

$$\mathrm{diag}\left(\chi_1(\varpi),\dots,\chi_m(\varpi),\chi_m^{-1}(\varpi),\dots,\chi_1^{-1}(\varpi)\right), \quad \text{if } n=2m \text{ is even},$$
$$\mathrm{diag}\left(\chi_1(\varpi),\dots,\chi_m(\varpi),1,\chi_m^{-1}(\varpi),\dots,\chi_1^{-1}(\varpi)\right), \quad \text{if } n=2m+1 \text{ is odd}.$$

We deduce the first part of the theorem with 3.2.6. To prove (2) recall that, by the Langlands correspondence, the cocycle α corresponds to the character of G defined by $g\mapsto\mu\circ\det g$. □

COROLLARY 4.2. $\widetilde{BC}$ *is an injective map.*

Proof. Let π and π' be two K-unramified representations of U and $(\chi_1(\varpi),\dots,\chi_m(\varpi))$ and $(\chi_1'(\varpi),\dots,\chi_m'(\varpi))$ respectively their Satake parameters. Suppose $\widetilde{BC}(\pi)\simeq\widetilde{BC}(\pi')$. Then, up to a permutation in $\mathcal{S}_{2m}$, we have that the sets $(\chi_1,\dots,\chi_m,\chi_m^{-1},\dots,\chi_1^{-1})$ and $(\chi_1',\dots,\chi_m',\chi_m'^{-1},\dots,\chi_1'^{-1})$ are equal. Hence there exists a permutation $s\in\mathcal{S}_m$ such that, for $1\le i\le m$, $\chi_{s(i)}'$ equals χ_i or χ_i^{-1}. Thus, $\pi\simeq\pi'$.

□

4.1.4. For any representation σ of G denote by σ^θ the representation $g \mapsto \sigma(\theta(g))$. We say that σ is θ-invariant if $\sigma \simeq \sigma^\theta$. One could naively think that the image of $\widetilde{BC}$ is the set of θ-invariant representations of G. But this latter set is bigger as it also contains some representations coming from endoscopic groups (see next chapter). For example, the unramified representation of $GL_2(E)$ with Satake parameters $(1, \mu)$ is θ-invariant and comes (see Paragraph 4.2.3 for more details) from the endoscopic group $U(1) \times U(1)$ where in the first factor we take the standard base change $\widetilde{BC}$ and in the second we use the twisted base change $\widetilde{BC'}$.

The following proposition characterizes the image of $\widetilde{BC}$:

Proposition 4.3. *Let σ be an unramified representation of* G. *There exists an unramified representation π of* U *such that $\sigma = \widetilde{BC}(\pi)$ if, and only if, $\sigma \simeq \sigma^\theta$ and σ has trivial central character.*

Proof. If $\sigma \simeq \sigma^\theta$, then the Satake parameter of σ is of the form $d = (x_1, \dots, x_n)$, avec $(x_1, \dots, x_n)$ equal to $(x_1^{-1}, \dots, x_n^{-1})$ up to rearrangement. Moreover, the fact that σ has trivial central character implies that $x_1 x_2 \dots x_n = 1$ so the number of -1 in the sequence $(x_1, \dots, x_n)$ is always even. Thus we can rearrange the x_i so that

$$d = \left(y_1, \dots, y_r, 1, \dots, 1, y_r^{-1}, \dots, y_1^{-1}\right),$$

which is clearly in the image of the standard base change. □

4.2. Endoscopic transfer.

4.2.1. Recall from [**Ro1**, 4.6.1.] that, for a unitary group U of degree n, the elliptic endoscopic groups are the quasi-split unitary groups $H = U(a) \times U(b)$ where a and b are positive integers with $a+b=n$. The embedding ${}^L H \to {}^L U$ depends on the choice of characters μ_a and μ_b of $E^\times$ extending respectively the characters $\omega_{E/F}^a$ and $\omega_{E/F}^b$. Se also the introduction to this book [**Har**, 5.5].

Then the embedding $\xi_{\mu_a,\mu_b} : {}^L H \to {}^L U$ is defined by (*cf.* [**Ro2**, 1.2.]):

$$\begin{aligned}
(g_1, g_2) \rtimes 1 &\mapsto \begin{pmatrix} g_1 & \\ & g_2 \end{pmatrix} \rtimes 1 \\
1 \rtimes w &\mapsto \begin{pmatrix} \mu_b(w) 1_a & \\ & \mu_a(w) 1_b \end{pmatrix} \rtimes w \qquad \text{for } w \in W_E \\
1 \rtimes w_\sigma &\mapsto \begin{pmatrix} \Phi_a & \\ & \Phi_b \end{pmatrix} \Phi_n^{-1} \rtimes w_\sigma,
\end{aligned}$$

where Φ_m, $m = a, b, n$ is defined as in (3.1) and we regard μ_a and μ_b as characters of W_E via local class field theory.

4.2.2. Fix some hyperspecial maximal compact subgroups K_a, K_b and K in $U(a)$, $U(b)$ and $U(n)$ respectively. We suppose here that μ_a and μ_b are unramified characters, that is, for $i = a, b$:

$$\mu_i = \begin{cases} \mu, & \text{if } i \equiv 1 \mod 2 \\ 1, & \text{otherwise.} \end{cases} \tag{4.3}$$

The following theorem is now straightforward:

THEOREM 4.4. *Let π be a K_a-unramified representation of $\mathrm{U}(a)$ and π' be a K_b-unramified representation of $\mathrm{U}(b)$. Let a', b' be the Witt indexes of $\mathrm{U}(a)$ and $\mathrm{U}(b)$ respectively. Denote by $(\chi_1(\varpi), \ldots, \chi_{a'}(\varpi))$ and $(\chi'_1(\varpi), \ldots, \chi'_{b'}(\varpi))$ respectively the Satake parameters of π and π'. Then the Satake parameters of $\widetilde{\xi}_{\mu_a,\mu_b}(\pi, \pi')$ (see (2.6)) are:*

$$(\mu_b\chi_1(\varpi), \ldots, \mu_b\chi_{a'}(\varpi), \mu_a\chi'_1(\varpi), \ldots, \mu_a\chi_{b'}(\varpi)),$$
if a and b are not both odd integers,
$$(\mu\chi_1(\varpi), \ldots, \mu\chi_{a'}(\varpi), \mu(\varpi), \mu\chi'_1(\varpi), \ldots, \mu\chi_{b'}(\varpi)),$$
if a and b are both odd integers.

REMARK. Notice that the rank of $H = \mathrm{U}(a) \times \mathrm{U}(b)$ is the same as that of U unless a, b are both odd. In the second (exceptional) case, we have set $\mu_a = \mu_b = \mu$ and the rank of H is one less than that of U.

4.2.3. For global purposes, we study now the split case. Using the local Langlands correspondence [**HT**], [**Hen**], this can be done in a much greater generality, but we shall restrict ourselves just to an easy example. Let $\mathrm{H} = \mathrm{GL}_a(\mathrm{E}) \times \mathrm{GL}_b(\mathrm{E})$ and set $\mathrm{G} = \mathrm{GL}_n(\mathrm{E})$ with $n = a + b$ and let μ_a and μ_b be unramified characters of $\mathrm{E}^\times$. We define an L-homomorphism $\xi_{\mu_a,\mu_b} : {}^L\mathrm{H} \to {}^L\mathrm{G}$ by

$$(g_1, g_2) \mapsto \begin{pmatrix} \mu_b(\varpi)g_1 & \\ & \mu_a(\varpi)g_2 \end{pmatrix}. \tag{4.4}$$

It is now clear that if π and π' are unramified representations of $\mathrm{GL}_a(\mathrm{E})$ and $\mathrm{GL}_b(\mathrm{E})$ respectively and $(\chi_1(\varpi), \ldots, \chi_a(\varpi))$ and $(\chi'_1(\varpi), \ldots, \chi'_b(\varpi))$ are their respective Satake parameters, then the Satake parameters of $\widetilde{\xi}_{\mu_a,\mu_b}(\pi, \pi')$ are:

$$(\mu_b\chi_1(\varpi), \ldots, \mu_b\chi_a(\varpi), \mu_a\chi'_1(\varpi), \ldots, \mu_a\chi_b(\varpi)). \tag{4.5}$$

Hence $\widetilde{\xi}_{\mu_a,\mu_b}(\pi, \pi')$ is a composition factor of the parabolically induced representation

$$i^{\mathrm{G}}_{\mathrm{P}_{(a,b)}}\big((\pi\, \mu_b \circ \det) \otimes (\pi'\, \mu_b \circ \det)\big) \tag{4.6}$$

where we see $(\pi\, \mu_b \circ \det) \otimes (\pi'\, \mu_b \circ \det)$ as a representation of the Levi subgroup $\mathrm{H} \simeq \mathrm{M}_{(a,b)}$ of $\mathrm{P}_{(a,b)}$. In particular, if π and π' are unitary representations, by Theorem A.2, the representation (4.6) is irreducible and hence isomorphic to $\widetilde{\xi}_{\mu_a,\mu_b}(\pi, \pi')$.

4.2.4. Fix as in Paragraph 4.2.2 some hyperspecial maximal compact subgroups K_a, K_b and K in $\mathrm{U}(a)$, $\mathrm{U}(b)$ and $\mathrm{U}(n)$ respectively. Suppose that μ_a and μ_b satisfy the condition:

$$\mu_i = \begin{cases} \mu, & \text{if } i \equiv 1 \mod 2 \\ 1, & \text{otherwise.} \end{cases}$$

For every positive integer i, denote until the end of this section by G_i the group $\mathrm{Res}_{\mathrm{E}/\mathrm{F}}(\mathrm{U}(i))$. Using now Theorems 4.1 and 4.4, and (4.5) we see that we have a commutative diagram:

$$\begin{array}{ccc} \mathrm{Irr}^{\mathrm{K}_a\times\mathrm{K}_b\text{-un}}(\mathrm{U}(a)\times\mathrm{U}(b)) & \xrightarrow{\widetilde{\xi}_{\mu_a,\mu_b}} & \mathrm{Irr}^{\mathrm{K}\text{-un}}(\mathrm{U}(n)) \\ \downarrow \widetilde{BC} & & \downarrow \widetilde{BC} \\ \mathrm{Irr}^{\mathrm{un}}(\mathrm{G}_a(\mathrm{F})\times\mathrm{G}_b(\mathrm{F})) & \xrightarrow{\widetilde{\xi}_{\mu_a,\mu_b}} & \mathrm{Irr}^{\mathrm{un}}(\mathrm{G}_n(\mathrm{F})) \end{array} \tag{4.7}$$

We have a similar diagram for the twisted base change $\widetilde{BC'}$.

4.2.5. In this paragraph we see the dual picture of Theorems 4.1, 4.4 and diagram (4.7).

We denote, as usual, $C_c^\infty(\mathrm{GL}_n(\mathrm{E}))$ the space of locally constant, compactly supported, complex functions on $\mathrm{GL}_n(\mathrm{E})$. Let P_α be a standard parabolic subgroup of $\mathrm{GL}_n(\mathrm{E})$. Denote M_α its standard Levi factor and N_α the unipotent radical of P_α. For each $f \in C_c^\infty(\mathrm{GL}_n(\mathrm{E}))$, define the *constant term* along P_α by

$$f^{\mathrm{P}_\alpha}(m) := \delta_{\mathrm{P}_\alpha}^{1/2}(m) \int_{\mathrm{N}_a} \int_{\mathrm{GL}_n(\mathcal{O}_\mathrm{E})} f(kmnk^{-1})\,dk\,dn, \tag{4.8}$$

for every $m \in \mathrm{M}_\alpha$. We have that $f^{\mathrm{P}_\alpha} \in C_c^\infty(\mathrm{M}_\alpha)$.

Let $\xi_\alpha = {}^L\mathrm{M}_\alpha \hookrightarrow {}^L\mathrm{GL}_n$ be the L-morphism that trivially extends the canonical identification of $\widehat{\mathrm{M}_\alpha}$ as Levi factor of $\widehat{\mathrm{GL}_n}$. The following proposition is well known (see for example [**AC**, p.32-33] or [**Shi**, 3.3]):

Proposition 4.5. (1) *For any $\pi \in \mathrm{Irr}(\mathrm{M}_\alpha)$, $\mathrm{tr}\,\pi(f^{\mathrm{P}_\alpha}) = \mathrm{tr}\, i_{\mathrm{P}_\alpha}^{\mathrm{GL}_n(\mathrm{E})}(\pi)(f)$.*

(2) *If $f \in \mathcal{H}(\mathrm{GL}_n(\mathrm{E}), \mathrm{GL}_n(\mathcal{O}_\mathrm{E}))$, then f^{P_α} is the image of f under the map*

$$\mathcal{H}(\mathrm{GL}_n(\mathrm{E}), \mathrm{GL}_n(\mathcal{O}_\mathrm{E})) \to \mathcal{H}(\mathrm{M}_\alpha, \mathrm{M}_\alpha \cap \mathrm{GL}_n(\mathcal{O}_\mathrm{E}))$$

which is dual to ξ_α (see Paragraph 2.7).

Corollary 4.6. (1) *Let a, b, n be some positive integers with $n = a+b$ and let μ_a and μ_b be the unramified characters of $\mathrm{E}^\times$ defined by (4.3). Write, for any positive integer i, K_i' some maximal compact subgroup of G_i. Set $b(\xi_{\mu_a,\mu_b}) : C_c^\infty(\mathrm{G}) \to C_c^\infty(\mathrm{H})$ the morphism defined by*

$$b(\xi_{\mu_a,\mu_b})(f)(m_1, m_2) = (\mu_b \circ \det(m_1))\,(\mu_a \circ \det(m_2))\, f^{\mathrm{P}_{(a,b)}}(m_1, m_2),$$

where $m_1 \in \mathrm{G}_a(\mathrm{F}), m_2 \in \mathrm{G}_b(\mathrm{F})$ *and* $f \in C_c^\infty(\mathrm{G}_n(\mathrm{F}))$.

Then the restriction of $b(\xi_{\mu_a,\mu_b})$ *to* $\mathcal{H}(\mathrm{G}_n(\mathrm{F}), \mathrm{K}'_n)$ *is the dual map of the L-morphism* $\xi_{\mu_a,\mu_b} : {}^L(\mathrm{G}_a \times \mathrm{G}_b) \to {}^L\mathrm{G}_n$ *defined by* (4.4).

(2) *With notations as in Paragraph 4.2.4, we have a commutative diagram, dual to* (4.7)*:*

$$\begin{array}{ccc} \mathcal{H}(\mathrm{U}(n), \mathrm{K}) & \xrightarrow{b(\xi_{\mu_a,\mu_b})} & \mathcal{H}(\mathrm{U}(a) \times \mathrm{U}(b), \mathrm{K}_a \times \mathrm{K}_b) \\ {\scriptstyle b(BC)}\uparrow & & \uparrow{\scriptstyle b(BC)} \\ \mathcal{H}(\mathrm{G}_n(\mathrm{F}), \mathrm{K}') & \xrightarrow{b(\xi_{\mu_a,\mu_b})} & \mathcal{H}(\mathrm{G}_a(\mathrm{F}) \times \mathrm{G}_b(\mathrm{F}), \mathrm{K}'_a \times \mathrm{K}'_b) \end{array} \tag{4.9}$$

Appendix A. On the classification of irreducible representations of linear groups

By the work of Silberger [**Sil**] and Borel-Wallach [**BW**], extending the results of Langlands to the p-adic case, the problem of classifying irreducible representations of a connected reductive group G over F is reduced to the study of *tempered* representations. These representations appear as composition factors of parabolically induced representations of discrete series representations of Levi subgroups of G. So identifying the tempered dual of G consists of two problems:

(1) Determine the discrete series representations of the Levi subgroups (in terms of cuspidal representations or by the theory of types).
(2) Decompose the resulting parabolically induced representations.

Neither problem is resolved in any generality. The general theory of irreducible representations of GL_n over a non-Archimedean local field, however, is well understood. In this appendix, we recall the construction of irreducible representations in terms of cuspidal data. For a more detailed exposition and historical notes one can consult [**Moe**], see also [**Rod**], [**BZ1**] and [**Ze1**].

A.1. Discrete series. For a general reductive group, one does not know how to classify the discrete series in terms of cuspidal representations. But for $\mathrm{GL}_n(\mathrm{F})$ this has been done by Bernstein and Zelevinsky [**Ze1**]. An understanding of cuspidal representations in terms of types is due to Bushnell and Kutzko [**BK**].

Let ρ be a cuspidal representation of $\mathrm{GL}_r(\mathrm{F})$ and let a be a positive integer. We denote by $\delta(a, \rho)$ the unique irreducible quotient of the parabolically induced representation:

$$i_{\mathrm{P}}^{\mathrm{GL}_n(\mathrm{F})} \left(\rho|\det|^{-\frac{a-1}{2}} \otimes \rho|\det|^{-\frac{a-3}{2}} \otimes \cdots \otimes \rho|\det|^{\frac{a-1}{2}}\right),$$

where P denotes the standard parabolic subgroup associated to the partition $(r, r, \ldots, r)$.

THEOREM A.1. (1) *All representations of the form $\delta(a, \rho)$ where a is a positive integer and ρ is a unitary cuspidal representation of* $\mathrm{GL}_r(\mathrm{F})$ *are irreducible discrete series representations.*

(2) *Conversely, let π be an irreducible discrete series representation of* $\mathrm{GL}_n(\mathrm{F})$. *There exist a unique divisor r of n and a unique irreducible unitary cuspidal representation ρ of* $\mathrm{GL}_r(\mathrm{F})$ *such that, if we set $a = \frac{n}{r}$, then π is isomorphic to $\delta(a, \rho)$.*

A.2. Tempered representations. The parabolically induced representation of a discrete series representation, in the case of $\mathrm{GL}_n(\mathrm{F})$, is irreducible (see, for example [**Jac**]). We have a deeper result, proved by Bernstein [**Be1**]:

THEOREM A.2. *Let* P *be a standard parabolic subgroup of* $\mathrm{GL}_n(\mathrm{F})$ *and let* $\mathrm{M_P}$ *be its Levi factor. Let ρ be an irreducible unitary representation of* $\mathrm{M_P}$. *Then* $i_{\mathrm{P}}^{\mathrm{GL}_n(\mathrm{F})}(\rho)$ *is irreducible.*

A similar theorem for the inner forms of $\mathrm{GL}_n(\mathrm{F})$ has been proved by V. Sécherre [**Sec**].

A.3. Irreducible representations. The understanding of the unitary dual of $\mathrm{GL}_n(\mathrm{F})$ is due to Tadić [**Tad**]. The Langlands correspondence in this case is due to Harris-Taylor [**HT**] and Henniart [**Hen**]. The Langlands quotient theorem, in this case, reads:

THEOREM A.3. *Let π be an irreducible representation of* $\mathrm{GL}_n(\mathrm{F})$. *There exist a partition $\alpha = (n_1, \dots, n_r)$ of n and, for $1 \leq i \leq r$, a unique (up to isomorphism) tempered representation τ_i of* $\mathrm{GL}_{n_i}(\mathrm{F})$ *and a unique real number t_i with*

$$t_1 > t_2 > \cdots > t_r,$$

such that π is the unique irreducible quotient of

$$i_{\mathrm{P}_\alpha}^{\mathrm{GL}_n(\mathrm{F})}\left(\tau_1|\det|^{t_1} \otimes \tau_2|\det|^{t_2} \otimes \cdots \otimes \tau_r|\det|^{t_r}\right).$$

We note $\pi = L\left(\tau_1|\det|^{t_1}, \dots, \tau_r|\det|^{t_r}\right)$ and, if $\pi' = L(\tau'_1|\det|^{t'_1}, \dots, \tau'_{r'}|\det|^{t'_{r'}})$ is another representation of $\mathrm{GL}_n(\mathrm{F})$, *then π is isomorphic to π' if, and only if, $r = r'$ and for all $1 \leq i \leq r$, $\tau_i|\det|^{t_i} \simeq \tau'_i|\det|^{t'_i}$.*

References

[AC] J. Arthur, L. Clozel, Simple Algebras, Base Change, and the Advanced Theory of the Trace Formula, *Annals of Math. Studies* **120** (1989).

[Be1] I.N. Bernstein, P-invariant distributions on $\mathrm{GL}(N)$ and the classification of unitary representations of $\mathrm{GL}(N)$ (non-Archimedean case), Lie group representations II (College Park, Md., 1982/1983), *Lecture Notes in Math.*, vol. 1041, Springer, Berlin, 1984, p. 50–102. **47**, 1986, 180–192.

[Be2] I.N. Bernstein, Representations of p-adic groups, Notes by K.E. Rumelhart, Harvard Univ. 1992.

[BZ1] I.N. Bernstein, A.V. Zelevinsky, Representations of the groups $GL(n, F)$ where F is a local non-archimedean Field, *Uspekhi Mat. Nauk.*, Vol 31, No. 3, 1976, 74–75.

[BZ2] I.N. Bernstein & A.V. Zelevinsky, Induced Representations of Reductive p-Adic Groups I, *Ann. Scient. Ec. Norm. Sup.*, 4e serie, t. 10, 1977, 441–472.

[BR1] D. Blasius & J.D. Rogawski, Fundamental lemmas for U(3), *The zeta functions of Picard modular surfaces,* p. 363–394, Univ. Montréal, (1992).

[BR2] D. Blasius & J.D. Rogawski, Zeta-functions of Shimura varieties. Motives (Seattle, WA, 1991), 525–571, *Proc. Sympos. Pure Math.* **55**, Part 2, Amer. Math. Soc., Providence, RI, (1994).

[Bo1] A. Borel, Automorphic L-functions, Automorphic forms, representations and L-functions (*Proc. Sympos. Pure Math.*, Oregon State Univ., Corvallis, Ore., 1977), Part 2. *Proc. Sympos. Pure Math.*, XXXIII. 27–61, Amer. Math. Soc. Providence, R.I. (1979).

[Bo2] A. Borel, Admissible representations of a semi-simple group over a local field with vector fixed under an Iwahori subgroup, *Inv. Math* **35** (1976), pp. 233–259.

[BW] A. Borel & N. Wallach, Continuous Cohomology, Discrete Subgroups, and representations of Reductive Groups, *Annals of Mat. Studies,* no. **94**, Princeton University Press,Princeton, NJ, (1980).

[BT] F. Bruhat & J. Tits, Groupes réductifs sur un corps local: I. Données radicielles valuées, *Publ. Math. I.H.E.S.* **41**, (1972) p. 5–251.

[BK] C.J. Bushnell & P.C. Kutzko, The admissible dual of GL(N) via compact open subgroups, Princeton University Press,Princeton, NJ, (1993).

[Car] P. Cartier, Representations of p-adic groups: a survey, Automorphic forms, representations and L-functions (*Proc. Sympos. Pure Math.*, Oregon State Univ., Corvallis, Ore., 1977), Part 1. *Proc. Sympos. Pure Math.*, XXXIII, p. 111–155. Amer. Math. Soc. Providence, R.I. (1979).

[Ca1] W. Casselman, Introduction to the theory of admissible representations of p-adic reductive groups, preprint.

[Ca2] B. Casselman, A companion to Macdonald's book on p-adic spherical functions, preprint.

[Ca3] B. Casselman, The unramified principal series of p-adic groupsI. The spherical function, *Compositio Math.* **40**, Fasc. 3, p. 387–406, (1980).

[Gar] P. Garrett, Satake parameters versus unramified principal series, Notes, 1999, Available at http://www.math.umn.edu/ garrett/m/v/satake_urps.pdf.

[Har] M. Harris, Introduction to "The stable trace formula, *Shimura Varieties, and arithmetic applications,* volume I," this volume.

[HT] M. Harris & R. Taylor, The geometry and cohomology of some simple Shimura varieties, *Annals of Math. Studies* **151** (2001).

[Hen] G. Henniart, Une preuve simple des conjectures de Langlands pour GL_n sur un corps p-adique, *Inv. Math.* **130** (2000), 439–455.

[Hij] H. Hijikata, Maximal compact subgroups of some p-adic classical groups, Yale University 1964.

[Jac] H. Jacquet, Generic representations , *Non-commutative harmonic analysis,* pp. 91–101. *Lect. Notes in Math.*, Vol. **587** (1977).

[JL] H. Jacquet & R.P. Langlands Automorphic forms on GL(2), *Lect. Notes in Math.* Vol. **114** (1970).

[L] W. Landherr, Äquivalenze Hermitscher formen über einem beliebigen algebraischen zahlkörper, *Abh. Math. Sem. Hamb.*, **11** (1936), 245–248.

[LR] R.P. Langlands & D. Ramakrishnan, The description of the theorem, *The zeta functions of Picard modular surfaces,* p. 255–301, Univ. Montréal, (1992).

[La1] R.P. Langlands, Letter to André Weil (1967).

[La2] R.P. Langlands, Problems in the theory of automorphic forms, *Lectures in modern analysis and applications,* LNM **170** Springer (1970).

[Mac] I.G. Macdonald, Spherical functions on a p-adic Chevalley group, *Bull. Amer. Math. Soc.* **74**, p. 520–525 (1968).

[Mat] I.G. Matsumoto, Analyse Harmonique dans les Systèmes de Titis Bornologiques de Type Affine, *Springer Lecture Notes* **590** Berlin (1977).

[Moe] C. Mœglin, Representations of p-adic groups: a survey, Representation Theory and Automorphic Forms (Proc. Sympos. Pure Math., Edinburgh, Scotland 1996). *Proc. Sympos. Pure Math.* **61**, p. 303–319 (1997).

[Rod] F. Rodier, Représentations de $GL(n, k)$ où k est un corps p-adique, *Séminaire Bourbaki no 587* (1982), Asterisque **92-93**, 201–218.

[Sat] I. Satake, Theory of spherical functions on reductive algebraic groups over p-adic fields, *Publ. Math. I.H.E.S.* **18** (1963) p. 5–69.

[Sec] V. Sécherre, Proof of the Tadiċ conjecture U0 on the unitary dual of $\mathrm{GL}_m(\mathrm{D})$, *J. Reine Angew. Math.* **626** (2009), p. 187–204.

[Shi] S.W. Shin, Galois representations arising from some compact Shimura varieties, prepublication 2009.

[Sil] A. Silberger, The Langlands quotient theorem for p-adic groups, *Math. Ann.* **236** (1978) p. 95–104.

[Ro1] J.D. Rogawski, Automorphic representations of unitary groups in three variables, *Annals of Mathematics Studies* **123**, Princeton University Press, Princeton, NJ, (1990).

[Ro2] J.D. Rogawski, Analytic expression for the number of points mod p, *The zeta functions of Picard modular surfaces,* p. 65–109, Univ. Montréal, (1992).

[Tad] M. Tadić, Unitary dual of p-adic $\mathrm{GL}(n)$. Proof of Bernstein conjectures, *Bull. Amer. Math. Soc. (N.S.)* **13**, no. 1, 39–42 (1985).

[Tit] J. Tits, Reductive groups over local fields, Automorphic forms, representations and L-functions (*Proc. Sympos. Pure Math.,* Oregon State Univ., Corvallis, Ore., 1977), Part 1. *Proc. Sympos. Pure Math.,* XXXIII. 29–69. Amer. Math. Soc. Providence, R.I. (1979).

[Zel] A.V. Zelevinsky, Induced Representations of Reductive p-Adic Groups II, *Ann. Scient. Ec. Norm. Sup.,* 4e serie, t. 13, 1980, 165–210.

Institut de Mathématiques de Jussieu Université Paris 6. 175, rue de Chevaleret, 75013 Paris, France

E-mail address: minguez@math.jussieu.fr

URL: http://www.institut.math.jussieu.fr/~minguez/

Les facteurs de transfert pour les groupes unitaires [L.III.D]

J.-P. Labesse

1. Introduction

Langlands et Shelstad ont proposé dans [LS] une construction des facteurs de transfert. Nous allons expliciter cette construction dans un cas très particulier pour les groupes unitaires.

Soit F un corps de nombres et soit G un groupe réductif défini sur F. Rappelons que les facteurs de transfert entre un groupe G et un groupe endoscopique H ne sont pas uniques. Deux choix de facteurs peuvent différer localement par des constantes, qui sont des nombres complexes de module 1, dont le produit sur toutes les places vaut 1 et par un caractère unitaire de $H(\mathbb{A}_F)$ trivial sur $H(F)$. La détermination de ce caractère est formulée dans [LS] en terme de plongement de L-groupes, lorsqu'un tel plongement existe. Observons qu'un tel plongement existe pour les groupes endoscopiques des groupes unitaires. Un plongement étant fixé le facteur de transfert adélique est alors unique. Réciproquement, la donnée d'un facteur de transfert détermine le plongement (à conjugaison près).

Nous expliciterons le facteur de transfert, pour les groupes endoscopiques d'un groupe unitaire sur les éléments semi-simples réguliers appartenant au tore diagonal, lorsque les groupes unitaires sont définis par rapport à une forme hermitienne diagonale. Dans ce cas la formule est très simple. Ceci suffit à déterminer le facteur de transfert sur tous les autres tores et le plongement de L-groupes correspondant.

2. Notations

Dans toute la suite F est un corps local ou global de caractéristique zéro. On note C_F le groupe multiplicatif $F^\times$ lorsque F est local et C_F désigne le groupe des classes d'idèles $\mathbb{A}_F^\times/F$ lorsque F est global. Sauf mention expresse du contraire (ce sera le cas au §7) on notera $|x|$ la valeur absolue normalisée d'un élément x du groupe multiplicatif d'un corps local ou d'un idèle. Soit

E une extension quadratique. On note $x \mapsto \overline{x}$ l'action du groupe de Galois de E/F.

Soit G un groupe unitaire relatif à E/F. On suppose que G est une forme intérieure de U_m^* le groupe unitaire quasi-déployé à m variables. Soit

$$H = U_a^* \times U_b^*$$

avec $m = a+b$, un groupe endoscopique. Pour simplifier légèrement l'exposé on supposera de plus que ab est pair (ce qui est automatique lorsque m est impair).

Dans ce qui suit il est commode de prendre les formes hermitiennes, définissant les divers groupes unitaires qui nous intéressent sous forme diagonale. On peut alors plonger H dans U_m^* sous forme diagonale par blocs. On peut aussi réaliser le tore

$$T = U(1)^m$$

comme sous-groupe des matrices diagonales T_G dans G et T_H dans H. On fixe désormais ces plongements de $T(F)$, $G(F)$ et $H(F)$ dans $GL_m(E)$ ce qui permet d'identifier T, T_G et T_H et on aura

$$T(F) \subset G(F) \cap H(F)$$

Dans toute la suite γ désigne un élément de $T(F)$, vu comme élément diagonal de $GL_m(E)$, de la forme :

$$\gamma = x \oplus y = \mathrm{diag}(\mathrm{x}_1, \cdots, \mathrm{x}_a, \mathrm{y}_1 \cdots, \mathrm{y}_b) \ .$$

Les x_i et les y_j sont des élément de $E^\times$ de norme 1 dans $F^\times$. On considérera, par abus de notation, γ comme un élément de $G(F)$ mais aussi de $H(F)$ via les isomorphismes ci-dessus.

Sur la clôture algébrique, H est un sous-groupe de Levi de G. On notera $\mathfrak{h}$ l'algèbre de Lie de H et $\mathfrak{g}$ l'algèbre de Lie de G. On choisit un ordre sur les racines de T dans $\mathfrak{g}$ et on note $\mathfrak{n}$ l'algèbre de Lie du radical unipotent du sous-groupe parabolique standard de sous-groupe de Levi H. Nous supposerons que les racines de T dans $\mathfrak{n}$ sont les

$$\gamma^{\alpha_{i,j}} = x_i/y_j \ .$$

3. Facteurs de transfert et diagonale

Soit F un corps local. Considérons $\gamma_G \in G(F)$, $\gamma_H \in H(F)$ et

$$\gamma \in T(F) \subset G(F) \cap H(F) \ ,$$

avec γ_G régulier stablement conjugué de γ dans $G(F)$ et γ_H stablement conjugué de γ dans $H(F)$. Soit Δ un facteur de transfert. Le choix d'un plongement de L-groupes détermine un facteur de transfert à une constante près. Un facteur de transfert Δ est défini dans [LS] comme un produit de facteurs Δ_* dont la définition suppose des choix auxiliaires, en particulier les a-data et les χ-data. Nous utiliserons la notation $\Delta_*(\gamma)$ plutôt que $\Delta_*(\gamma_H, \gamma_G)$ lorsque le facteur Δ_* ne dépend que de γ. Le facteur Δ_{III_1}, qui est le seul à ne pas dépendre seulement de γ, ne peut être défini lorsque

G n'est pas quasi-déployé que comme un facteur "relatif" (cf. [LS, p 246]) : sa définition fait intervenir une autre paire (γ'_H, γ'_G), que nous omettrons pour ne pas alourdir la notation. Avec ces conventions, on a pour notre situation

$$\Delta(\gamma_H, \gamma_G) = c\,\Delta_I(\gamma)\,\Delta_{II}(\gamma)\,\Delta_{III_1}(\gamma_H, \gamma_G)\,\Delta_{III_2}(\gamma)\,\Delta_{IV}(\gamma)$$

où c est une constante. Le facteur $\Delta_I(\gamma)$ est indépendant de $\gamma \in T(F)$ et est un signe. Un changement de plongement de L-groupes modifie Δ via le facteur Δ_{III_2} par un caractère du groupe $H(F)$. On a

$$\Delta_{IV}(\gamma) = |\det(1 - \mathrm{Ad}(\gamma)\,|\,\mathfrak{g}/\mathfrak{h})|^{1/2} = \prod_{i,j} \frac{|(x_i - y_j)|}{|x_i y_j|^{1/2}} .$$

Un facteur de transfert doit vérifier

$$\Delta(\gamma_H, \gamma_G) = \kappa(x)\Delta(\gamma, \gamma)$$

où $x \in G(\overline{F})$ est tel que

$$x^{-1}\gamma_G\, x = \gamma$$

et où κ est le caractère endoscopique définissant H. Pour spécifier un facteur de transfert $\Delta(\gamma_G, \gamma_H)$ entre G et H pour le tore T il suffit donc de donner sa valeur sur la diagonale $\Delta(\gamma, \gamma)$.

4. Facteurs de transfert de Kottwitz et séries discrètes

Pour les tores elliptiques dans un groupe unitaire, sur les réels, la définition des facteurs de transfert est presque évidente. En effet, pour déterminer le transfert des pseudo-coefficients de représentations des séries discrètes entre $G(\mathbb{R})$ et $H(\mathbb{R})$ on est amené à comparer leurs intégrales orbitales et donc à comparer les formules pour certaines combinaisons linéaires de caractères de séries discrètes sur $G(\mathbb{R})$ et sur $H(\mathbb{R})$ (cf. [Lab]). Or on sait que la formule du caractère pour une série discrète est, avec des notations familières, de la forme

$$\Theta_\pi(\gamma) = (-1)^{q(G)} \frac{\sum_{w \in W_{\mathbb{R}}^G} \varepsilon(w) e^{<w(\mu), X>}}{\sum_{w \in W_{\mathbb{C}}^G} \varepsilon(w) e^{<w(\rho_G), X>}}$$

lorsque $\gamma = \exp(X) \in T(\mathbb{R})$ si π est une série discrète de paramètre μ, et ρ_G est la demi-somme des racines positives de T dans G pour un ordre convenable. On voit alors qu'un candidat naturel pour le facteur de transfert, en un couple de points (γ_H, γ_G) représentés par la même matrice diagonale

$$\gamma = \exp(X)$$

est le quotient des dénominateurs de Weyl pour γ^{-1} :

$$(-1)^{q(G)-q(H)} \frac{\sum_{w \in W_{\mathbb{C}}^G} \varepsilon(w) e^{-<w(\rho_G), X>}}{\sum_{w \in W_{\mathbb{C}}^H} \varepsilon(w) e^{-<w(\rho_H), X>}}$$

ce qui peut s'écrire

$$(-1)^{q(G)-q(H)} e^{<(\rho_H-\rho_G),X>} D_{\mathfrak{n}}(\gamma)$$

avec

$$D_{\mathfrak{n}}(\gamma) = \det(1 - \mathrm{Ad}(\gamma) \mid \mathfrak{n}) = \prod_{\alpha}(1-\gamma^\alpha)$$

où le produit porte sur les racines positives α de T dans G qui ne sont pas dans H. Mais ce candidat naturel ne définit pas une fonction sur le groupe car

$$X \mapsto e^{<(\rho_H-\rho_G),X>}$$

définit un caractère d'un revêtement double du tore $T(\mathbb{R})$ qui ne descend pas à $T(\mathbb{R})$ en général. Cette obstruction peut aussi se voir comme suit : le paramètre μ d'une série discrète pour G est de la forme

$$\mu = \lambda_G + \rho_G$$

où λ_G est un poids de T. Mais il n'est pas vrai en général que l'on puisse écrire

$$\mu = \lambda_H + \rho_H$$

où λ_H est un poids de T. Le transfert des séries discrètes de H à G suppose donc une translation sur le paramètre μ.

L'impossibilité de prolonger au groupe le facteur de transfert naturel, se reflète du côté dual : l'action du groupe de Galois sur $\widehat{G}$ n'induit pas sur le sous-groupe $\widehat{H}$ l'action souhaitée. Toutefois, dans notre cas, le plongement naturel de $\widehat{H}$ dans $\widehat{G}$ peut se prolonger en un plongement de L-groupes après torsion par un 1-cocycle du groupe de Weil.

Un facteur de transfert est obtenu en tordant le "candidat naturel" par le caractère d'une représentation de dimension 1 d'un revêtement de $H(\mathbb{R})$ de sorte que l'expression descende sur $T(\mathbb{R})$. Il n'y a pas unicité : une telle torsion n'est définie que modulo un caractère arbitraire de $H(\mathbb{R})$. On obtient ainsi la formule pour les facteurs de transfert proposée par Kottwitz dans [K] :

$$\Delta^K(\gamma,\gamma) = (-1)^{q(G)-q(H)}\chi(\gamma)D_{\mathfrak{n}}(\gamma)$$

où χ est un caractère convenable de $T(\mathbb{R})$ (cf. [K, p 184]). Le caractère dépend du plongement de L-groupes et, réciproquement, un choix admissible de ce caractère détermine le plongement.

Supposons de plus $G = U^*$ quasi-déployé ; dans ce cas Langlands et Shelstad définissent un facteur qui est canonique à un signe près dépendant du choix d'un épinglage et, comme toujours, du choix d'un plongement de L-groupes. Notons le $\Delta_{\mathbb{R}}^{LS}$. Les "a-data" utilisés dans [LS] pour la construction des facteurs Δ_I et Δ_{II} doivent être ici des nombres complexes imaginaires purs a_α que l'on peut choisir égaux à $i = \sqrt{-1}$ pour $\alpha > 0$. En comparant, terme à terme le facteur défini dans [LS] et celui de Kottwitz (une telle comparaison est faite en détail plus loin) on a

$$\Delta_{\mathbb{R}}^{LS} = \pm(i)^{ab}\Delta^K$$

et comme nous supposons ab pair les facteurs coïncident au signe près.

5. Un facteur de transfert explicite

Nous allons généraliser l'expression donnée par Kottwitz à tous les corps locaux ainsi qu'au cas global. Soit $\gamma \in T(F) \subset H(F) \subset U^*(F)$ avec

$$\gamma = \mathrm{diag}(\mathrm{x}_1, \cdots, \mathrm{x}_\mathrm{a}, \mathrm{y}_1 \cdots, \mathrm{y}_\mathrm{b}) .$$

Posons

$$d(\gamma) = \prod_{i,j}(y_j - x_i) \qquad \text{et} \qquad \eta(\gamma) = \prod_{i,j} x_i y_j .$$

On remarque que η est la valeur en γ d'un caractère rationnel de H et que

$$d(\gamma) = (y_1 \cdots y_b)^a \prod_{i,j}(1 - \gamma^{\alpha_{i,j}}) = (y_1 \cdots y_b)^a D_{\mathfrak{n}}(\gamma)$$

où les $\alpha_{i,j}$ sont les racines de T dans $\mathfrak{n}$.

On note $\mu_{E/F}$ le caractère d'ordre 2 de C_F associé à l'extension E/F par le corps de classe.

Proposition 5.1. *On suppose ab pair. Soit ξ un caractère unitaire qui prolonge à C_E le caractère $\mu_{E/F}$ de C_F et soit ζ un caractère de $H(F)$ si F est local ou de $H(\mathbb{A}_F)$ trivial sur $H(F)$ si F est global. Si γ est un élément de $T(F)$ lorsque F est local ou un élément de $T(\mathbb{A}_F)$ dans le cas global, tel que $d(\gamma)$ soit inversible, alors*

$$\Delta_{\xi,\zeta}(\gamma, \gamma) = \zeta(\gamma)\xi(d(\gamma)) \,.\, \Delta_{IV}(\gamma)$$

est la restriction à la diagonale de T d'un facteur de transfert entre G et H. Réciproquement, la restriction à la diagonale de T de tout facteur de transfert est de cette forme (à une constante près dans le cas local).

Preuve: Le fait que nos facteurs soient construits au moyen de caractères des classes d'idèles montre que c'est un choix globalement compatible : le produit sur toutes les places vaut 1 sur les éléments rationnels. Montrons maintenant que notre facteur coïncide avec un facteur de Langlands-Shelstad localement partout. Il convient de montrer que le produit

$$c\Delta_I \Delta_{II} \Delta_{III_1} \Delta_{III_2}$$

évalué sur la diagonale en γ est égal à $\zeta(\gamma)\xi(d(\gamma))$ pour un choix convenable de la constante et du plongement de L-groupes. Étudions d'abord le facteur Δ_{II}. Pour le calculer on doit choisir des "a-data" et des "χ-data". On choisit $\tau \in E^\times$ tel que $\overline{\tau} = -\tau$ et on pose

$$a_\alpha = \tau \qquad \text{et} \qquad \chi_\alpha = \xi$$

pour les racines positives α de T dans G. Avec un tel choix on a

$$\Delta_{II}(\gamma) = \prod_\alpha \xi(\frac{\gamma^\alpha - 1}{\tau})$$

le produit étant sur les racines positives α de T dans G qui n'apparaissent pas dans H. Donc

$$\Delta_{II}(\gamma) = \xi(-1/\tau)^{ab}\xi(D_{\mathfrak{n}}(\gamma)) .$$

Le facteur $\Delta_I(\gamma)$ dépend, entre autre, du choix des "a-data". Nous ne chercherons pas à calculer ce facteur ; il nous suffit de savoir que $\Delta_I(\gamma)$ est un signe indépendant de γ. Sur la diagonale le facteur Δ_{III_1} est aussi une constante égale à ± 1. On choisit c de sorte que

$$c\Delta_I\Delta_{III_1} = \xi(-\tau)^{ab} .$$

Le facteur Δ_{III_2} est un caractère de $T(F)$ qui dépend du choix des χ-data et du plongement de L-groupes. Compte tenu de notre choix de Δ_{II} un tel caractère de $T(F)$ est la restriction d'un caractère de $H(F)$: en effet Δ et Δ_{II} sont invariants sous le groupe de Weyl (absolu) de H ; il en est alors de même pour Δ_{III_2}. Donc, pour un choix convenable du plongement de L-groupes dépendant du choix de ξ et de ζ, et qui sera explicité dans la proposition 8.1, on peut supposer que

$$\Delta_{III_2}(\gamma) = \zeta(\gamma)\xi(\prod_j y_j^a)$$

et comme

$$d(\gamma) = D_{\mathfrak{n}}(\gamma) \prod_j y_j^a$$

on obtient l'égalité cherchée. □

6. Une application de Hilbert 90

Les considérations de ce paragraphe sont un cas particulier de [LS, Lemma 3.3.D].

LEMME 6.1. *Supposons ab pair. Alors, sur l'ensemble des $\gamma \in T(F)$ tels que $d(\gamma) \neq 0$ la fonction d est congrue modulo $F^\times$ à un caractère de $H(F)$ à valeurs dans $E^\times/F^\times$. On a un analogue adélique : soit F un corps global, alors sur l'ensemble des éléments $\gamma \in T(\mathbb{A}_F)$ tels que $d(\gamma)$ soit un idèle, la fonction d est congrue modulo C_F à un caractère de $H(\mathbb{A}_F)$ à valeurs dans C_E/C_F.*

Preuve: Considérons $\gamma \in T(F)$ avec $d(\gamma) \neq 0$. Les x_i et les y_j sont des éléments de $E^\times$ de norme 1 ; il existe des α_i et des β_j dans $E^\times$ bien déterminés modulo $F^\times$ tels que

$$x_i = \frac{\alpha_i}{\overline{\alpha_i}} \qquad \text{et} \qquad y_j = \frac{\beta_j}{\overline{\beta_j}}$$

et donc

$$d(\gamma) = \prod_{i,j} \frac{(\beta_j\overline{\alpha_i} - \overline{\beta_j}\alpha_i)}{\overline{\alpha_i\beta_j}}$$

Comme on a supposé ab pair le numérateur, qui est non nul par hypothèse de régularité, est à valeurs dans $F^\times$. Le produit

$$\delta(\gamma) = \prod_{i,j} \alpha_i \beta_j$$

est un élément de $E^\times$ bien défini par γ à un facteur de $F^\times$ près. Donc δ est un caractère à valeurs dans $E^\times/F^\times$. On conclut en observant que

$$d(\gamma) \equiv \delta(\gamma) \qquad \text{modulo} \qquad F^\times .$$

Dans le cas adélique la preuve est similaire.

□

On pourra observer que

$$\eta(\gamma) = \frac{\overline{\delta(\gamma)}}{\delta(\gamma)} \qquad \text{et que} \qquad \Delta_{IV}(\gamma) = |d(\gamma)|\,|\eta(\gamma)|^{-1/2} .$$

Nous allons expliciter la dépendance de $\Delta_{\xi,\zeta}$ lorsque ξ varie. On dispose d'une suite exacte

$$1 \to C_F \to C_E \to U(1)_F \to 1$$

où $U(1)_F$ est le groupe de $U(1,F)$ des points sur F du groupe unitaire en une variable lorsque F est local et le quotient $U(1,\mathbb{A}_F)/U(1,F)$ lorsque F est global. Soit ε un caractère du groupe $U(1)_F$. Compte tenu de la suite exacte ci-dessus, ε définit un caractère $\widetilde{\varepsilon}$ de C_E/C_F en posant pour $\alpha \in C_E$:

$$\widetilde{\varepsilon}(\alpha) = \varepsilon(\alpha/\overline{\alpha}) .$$

Lemme 6.2. Un changement du choix pour ξ modifie le facteur de transfert $\Delta_{\xi,\zeta}$ par un caractère de $H(F)$ si F est local et par un caractère de $H(\mathbb{A}_F)$ trivial sur $H(F)$ si F est global. De façon précise, soit $\xi' = \widetilde{\varepsilon}\,\xi$ un autre prolongement à C_E du caractère $\mu_{E/F}$, alors

$$\Delta_{\xi',\zeta}(\gamma,\gamma) = \Delta_{\xi,\zeta'}(\gamma,\gamma)$$

avec

$$\zeta'(\gamma) = \varepsilon \circ \eta(\gamma) \, . \, \zeta(\gamma) .$$

Preuve: On a

$$\Delta_{\xi',\zeta}(\gamma,\gamma) = \widetilde{\varepsilon}(d(\gamma))\Delta_{\xi,\zeta}(\gamma,\gamma)$$

Il suffit alors d'observer que d'après le lemme 6.1 et les remarques qui le suivent on a

$$\widetilde{\varepsilon}(d(\gamma)) = \widetilde{\varepsilon} \circ \delta(\gamma) = \varepsilon \circ \eta(\gamma)$$

et on rappelle que $\eta(\gamma)$ est un caractère rationnel de H.

□

7. Le cas $E/F = \mathbb{C}/\mathbb{R}$

Considérons le cas $F = \mathbb{R}$ et $E = \mathbb{C}$ et supposons que $\xi(z) = z/|z|_0$ pour $z \in \mathbb{C}^\times$ où $|z|_0$ est le module du nombre complexe z (on prendra garde que $|z|_\mathbb{C} = |z|_0^2$). Pour un tel choix, on remarque qu'alors

$$|\eta(\gamma)|_0 = 1$$

et on a

$$\xi(d(\gamma)) \,.\, |d(\gamma)|_0 |\eta(\gamma)|_0^{-1/2} = d(\gamma) \,.$$

Les facteurs de transfert de Kottwitz peuvent donc s'écrire sous la forme suivante : si

$$\gamma = \gamma_H = \gamma_G$$

est un élément de $T(F)$ on a

$$\Delta^K(\gamma, \gamma) = (-1)^{q(G)-q(H)} \chi(\gamma) d(\gamma) \prod_j y_j^{-a}$$

où χ est un caractère de $H(\mathbb{R})$. Donc $\Delta_{\xi,\zeta}(\gamma, \gamma)$ est, au signe près, le facteur de transfert de Kottwitz pour le caractère

$$\chi(\gamma) = \zeta(\gamma) \xi(\prod_j y_j^a) \,.$$

Le facteur de transfert $\Delta_{\xi,\zeta}$ est compatible avec les choix utilisés par ailleurs dans le livre, notamment dans [C] et [CHL] : c'est au signe près un facteur de Kottwitz aux places réelles (voir la remarque après la Proposition 5.1 de [CHL]). Le produit sur toutes les places vaut 1 sur les éléments rationnels. Enfin, aux places finies non ramifiées, le transfert d'une représentation sphérique est sphérique si ξ et ζ sont non ramifiés en cette place.

8. Plongement de L-groupe associé

Il nous reste à préciser le plongement entre L-groupes associé à ce facteur de transfert.

Rappelons tout d'abord la forme des plongements (voir par exemple [Rog]). Soit G un groupe unitaire à m variables relativement à E/F. Son groupe dual $\widehat{G}$ est le groupe $GL(m, \mathbb{C})$. Son L-groupe est le produit semi-direct

$$\widehat{G} \rtimes W_F$$

où W_F agit via son image dans $\mathrm{Gal}(E/F)$. On notera ρ_G cette action. Soit σ l'élément non trivial de $\mathrm{Gal}(E/F)$; il agit sur $\widehat{G} = GL(m, \mathbb{C})$ par

$$\rho_G(\sigma)(x) = J_m \, {}^t x^{-1} J_m^{-1}$$

où J_m est la matrice

$$\begin{pmatrix} 0 & \cdots & 0 & -1 \\ 0 & & 1 & 0 \\ \vdots & & \vdots & \vdots \\ (-1)^m & \cdots & 0 & 0 \end{pmatrix}.$$

Le groupe dual de H

$$\widehat{H} \simeq GL(a,\mathbb{C}) \times GL(b,\mathbb{C})$$

peut être réalisé comme sous-groupe diagonal par blocs de

$$\widehat{G} = GL(m,\mathbb{C}) \ .$$

C'est le centralisateur de

$$s = \mathbf{1}_a \oplus -\mathbf{1}_b \ .$$

On posera $J_G = J_m$ et $J_H = J_a \oplus J_b$. Un plongement

$$\varphi : {}^L H \to {}^L G$$

définit un 1-cocycle c de W_F à valeurs dans $\widehat{G}$:

$$\varphi(1 \rtimes w) = c(w) \rtimes w$$

tel que pour tout $w \in W_F$ et $h \in \widehat{H}$ on ait

$$\rho_H(w)h = \mathrm{Ad}\, c(w) \circ \rho_G(w)h \ .$$

Ceci détermine $c(w)$ modulo $Z(\widehat{H})$, le centre de $\widehat{H}$, à gauche. Soit $w_\sigma \in W_F$ un élément qui se projette sur le générateur σ du groupe de Galois $\mathrm{Gal}(E/F)$. On a nécessairement

$$c(w_\sigma) = z_\sigma J_H J_G^{-1} \qquad \text{avec } z_\sigma \in Z(\widehat{H})$$

et

$$c(w) \in Z(\widehat{H}) \qquad \text{pour } w \in W_E$$

puisque W_E agit trivialement sur $\widehat{G}$. La restriction à W_E du 1-cocycle est donc un homomorphisme à valeurs dans $Z(\widehat{H})$, qui peut être vu comme un homomorphisme, que nous noterons μ, de $C_E \simeq W_E^{ab}$ dans $Z(\widehat{H})$:

$$\varphi(h \rtimes w) = h\mu(w) \rtimes w$$

pour $w \in W_E$. On observe que w_σ^2 est un élément de W_E dont l'image dans $C_E \simeq W_E^{ab}$ appartient à C_F. Un changement du représentant w_σ de σ dans W_F, modifie l'image de w_σ^2 par une norme de C_E dans C_F. Mais, on a

$$\mu(w_\sigma^2) = c(w_\sigma) J_G \, {}^t c(w_\sigma)^{-1} J_G^{-1}$$

On observe que ${}^t J_p = J_p^{-1}$ donc

$$\mu(w_\sigma^2) = J_H^2 \, J_G^{-2}$$

et comme $J_p^2 = (-1)^p \mathbf{1_p}$ et $m = a + b$ on a

$$\mu(w_\sigma^2) = (-1)^b \mathbf{1}_a \oplus (-1)^a \mathbf{1}_b$$

qui est indépendant du choix de w_σ. Il en résulte que μ doit être trivial sur l'image des normes de C_E dans C_F. Réciproquement, la donnée de

$$\mu = \mu_a \mathbf{1}_a \oplus \mu_b \mathbf{1}_b$$

trivial sur l'image des normes de C_E dans C_F et vérifiant

$$\mu(w_\sigma^2) = (-1)^b \mathbf{1}_a \oplus (-1)^a \mathbf{1}_b$$

détermine un plongement φ à conjugaison près.

Nous devons donc expliciter l'homomorphisme μ associé à notre facteur de transfert. On notera

$$\widetilde{\zeta}_1 \mathbf{1}_a \oplus \widetilde{\zeta}_2 \mathbf{1}_b$$

le composé avec $W_E \to C_E$ de l'homomorphisme de C_E/C_F dans $Z(\widehat{H})$ associé au paramètre de Langlands de ζ. C'est dire que pour $\gamma = x \oplus y$ diagonal dans $H(F)$ on a

$$\zeta(\gamma) = \zeta(x \oplus y) = \zeta_1(\det(x))\zeta_2(\det(y))$$

et on a posé

$$\widetilde{\zeta}_i(\alpha) = \zeta_i(\alpha/\overline{\alpha}) .$$

Proposition 8.1. *Pour le facteur de transfert*

$$\Delta_{\xi,\zeta}(\gamma,\gamma) = \zeta(\gamma)\xi(d(\gamma)) \,.\, \Delta_{IV}(\gamma)$$

l'homomorphisme μ associé est

$$\mu(w) = \widetilde{\zeta}_1(w)\xi(w)^b \mathbf{1}_a \oplus \widetilde{\zeta}_2(w)\xi(w)^a \mathbf{1}_b$$

pour $w \in W_E$ où, par abus de notation, on a encore noté ξ le composé du caractère ξ avec l'application naturelle $W_E \to C_E$.

Preuve: Supposons tout d'abord ζ trivial. D'après [LS, § 2.5, 2.6 et 3.5] le plongement associé à notre choix des "χ-data" et de Δ_{III_2} est défini par un homomorphisme μ de la forme

$$\mu(w) = \xi(w)^{n_1} \mathbf{1}_a \oplus \xi(w)^{n_2} \mathbf{1}_b .$$

On pourrait faire le calcul des n_i en suivant [LS] pas à pas mais il sera plus économique de procéder comme suit. Si ξ est remplacé par $\xi' = \widetilde{\varepsilon}\xi$ alors d'après le lemme 6.2 on a

$$\Delta_{\xi',\zeta}(\gamma,\gamma) = \varepsilon \circ \eta(\gamma) \Delta_{\xi,\zeta}(\gamma,\gamma)$$

avec

$$\eta(\gamma) = \prod_{i,j} x_i y_j$$

et

$$\varepsilon \circ \eta(\gamma) = \varepsilon(\prod_{i,j} x_i y_j) = \widetilde{\varepsilon}(\prod_{i,j} \alpha_i \beta_j) .$$

Le paramètre de Langlands pour ce caractère est l'homomorphisme de W_E dans $Z(\widehat{H})$

$$w \mapsto \widetilde{\varepsilon}(w)^b \mathbf{1}_a \oplus \widetilde{\varepsilon}(w)^a \mathbf{1}_b .$$

La covariance entre ξ et μ implique que nécessairement $n_1 = b$ et $n_2 = a$ soit

$$\mu(w) = \xi(w)^b \mathbf{1}_a \oplus \xi(w)^a \mathbf{1}_b \ .$$

Le cas général pour un ζ arbitraire en résulte immédiatement.

□

REMARQUE 8.2. Le choix du facteur de transfert correspond au choix du plongement endoscopique ainsi qu'à la forme précise du transfert de fonctions de Hecke, à l'œuvre dans le Theorem 9.3.3.5 de [H]. L'homomorphisme μ qui détermine le plongement endoscopique correspond au choix de caractères μ_a, μ_b dans [H, 5.5] (voir aussi [M]).

Dans les articles [M] et [CHL] où on considère le cas $a = 1, b = n$ avec n pair, nous choisissons $\mu_1 = 1$, $\mu_n = \xi^{-1}$ (voir la discussion dans [CHL, (3.2)]) et donc

$$\mu(w) = \mathrm{diag}(1, \xi(\mathrm{w})^{-1}, \cdots, \xi(\mathrm{w})^{-1}) \ .$$

Le facteur de transfert correspondant est

$$\Delta_{\xi,\zeta}(\gamma, \gamma) = \xi(x^{n/2} \det(y))^{-1} \xi(d(\gamma)) \, . \, \Delta_{IV}(\gamma)$$

pour

$$\gamma = x \oplus y = \mathrm{diag}(\mathrm{x}, \mathrm{y}_1 \cdots \mathrm{y}_\mathrm{n}) \ .$$

C'est également le plongement utilisé par Rogawski [Rog] et dans le volume de Montréal, pour le cas $a = 1, b = 2$.

Références

[C] L. Clozel, Identités de caractères en la place archimédienne, *Ce volume.*

[CHL] L. Clozel, M. Harris, & J.P. Labesse, Endoscopic Transfer, *Ce volume.*

[H] M. Harris, An introduction to the stable trace formula, *Ce volume.*

[K] R. Kottwitz, Shimura Varieties and ℓ-adic Representations, *Perspectives in Math.* **10** Academic Press (1988), pp. 161–210.

[Lab] J.-P. Labesse, Introduction to Endoscopy, *Ce volume.*

[LS] R.P. Langlands & D. Shelstad, On the Definition of Transfer Factors, *Math. Annalen* **278** (1987). p 219–271.

[M] A. Minguez, Unramified representations of unitary groups, *Ce volume.*

[Rog] J. Rogawski, Automorphic Representations of Unitary Groups in Three Variables, *Annals of Math. Studies* **123** (1990).

INSTITUT MATHÉMATIQUE DE LUMINY, UMR 6206 FRANCE

Section IV

Some examples of stable transfer

Introduction to section IV

Laurent Clozel, Michael Harris, and Jean-Pierre Labesse

The next three chapters illustrate the results and techniques developed in this book through several applications of the sort for which the stable trace formula was originally developed. We have not striven for maximum generality; on the contrary, we have simplified as much as possible, in order to focus on the underlying structure of the comparison of trace formulas while reducing combinatorial considerations to a minimum. The three chapters are interrelated and lead to a construction of compatible families of Galois representations attached to cuspidal cohomological automorphic representation of $GL(n)$ over a CM field, subject to a symmetry property.

Labesse's chapter [L.IV.A] generalizes earlier results of [L], [CL], and [HL] on the base change comparing cohomological representations of $GL(n)$ over a CM field, satisfying the symmetry property mentioned above, with discrete series L-packets of unitary groups. The proof of the fundamental lemma for endoscopy, already in the case of unitary groups treated by Laumon and Ngô, permits the extension of these earlier results with no local hypotheses except at archimedean places. The restriction to representations of cohomological type, and the hypothesis that the CM field contains a totally real field strictly containing $\mathbb{Q}$, permits both the application of Arthur's simple twisted trace formula for $GL(n)$, from which the most difficult terms have been eliminated, and the elimination of all but the simplest of the twisted endoscopic groups for $GL(n)$ in the stabilization of the twisted trace formula (cf. Théorème 4.12 of [L.IV.A]).

Arthur's general multiplicity conjectures for the discrete spectra of automorphic forms on a group G are expressed in terms of parameters defined as homomorphisms of the conjectural Langlands group with values in the L-group of G. When G is a classical group, Arthur has suggested in [A2] the replacement of these hypothetical parameters by *ad hoc* parameters constructed as collections of cuspidal automorphic representations of various $GL(n_i)$, subject to symmetry conditions adapted to G. This point of view, which was already implicit in Rogawski's treatment of the group $U(3)$ ([R]; cf. also [LR]) is developed for symplectic and special orthogonal groups in [A1]. Labesse's chapter allows us to do the same for the unitary groups considered here, provided we restrict our attention to cohomological representations.

The second chapter, due to the three authors, works out the details of endoscopic transfer from $H = U(n) \times U(1)$ to $G = U(n+1)$, applied to cohomological representations – so we can use collections of cuspidal representations of $GL(n_i)$ as parameters, thanks to the results of the first chapter. Endoscopic transfer is defined at the level of L-packets. We have imposed additional simplifying hypotheses at non-archimedean primes (cf. [CHL.IV.B, 1.2]) that guarantee that local L-packets are singletons except at infinity; this allows us to dispense with conjectures about the structure of local L-packets of unitary groups[1], to state our results unconditionally, and in particular to show that the representations of interest occur with multiplicity (at most) one in the automorphic spectrum, a property used in the final chapter.

It may be helpful to remind novices that the goal of the comparison of trace formulas is to show that the comparison of harmonic analysis on the groups H and G is compatible with the comparison of Langlands parameters with values in ${}^LH$ and ${}^LG$; in other words, that the comparison of trace formulas explicitly realizes Langlands functoriality in the special case of endoscopy. Functoriality is a local property. In our simplified situation, there are three kinds of places – archimedean, non-archimedean places that split in the CM extension of its maximal totally real field), and non-archimedean places where the automorphic representation is unramified; the last two classes have a substantial overlap. Functoriality for non-archimedean places is verified according to the predictions in §9.3 of the Introduction to this volume; at split places this is elementary, whereas at unramified places this is precisely the content of the Fundamental Lemma; cf. Mínguez' chapter [M.III.C], as well as the chapters of Subsection II.C. Finally, functoriality at archimedean places for endoscopic transfer is a special case of Shelstad's theory, recalled from several different perspectives in the chapters of Renard, Clozel, and Adams. The main technical steps in the second chapter are calculations of archimedean signs. There are two such signs: the first corresponds to the stable part of the trace formula for G and involves the calculation of normalized intertwining operators; the second involves rewriting Shelstad's formulas for endoscopic transfer of archimedean discrete series in explicit terms for the group H considered above, following a first rewriting due to Kottwitz.

The third chapter, due to the three authors again, applies the considerations of the previous chapters to the determination of the associated Galois representations that appear in the cohomology of Shimura varieties of unitary type. We have chosen to follow the general procedure developed by Kottwitz in his fundamental papers [K1,K2], appealing to theorems rather

[1]These have recently been proved by Moeglin, but the statements are far from elementary and would distract from the main purpose of these sections.

than conjectures and using cuspidal automorphic representations of $GL(n_i)$ as parameters. As in [K1], the result is stated at unramified places only. An alternative construction due to Shin follows the book of Harris and Taylor and determines local Galois representations at all places. Given the results of the first two chapters, the methods of [K1] should in principle yield the desired conclusions immediately. The length of the third chapter is due in the first place to the difference between the similitude group used to define the Shimura variety and the unitary group considered in the two preceding chapters. A second and more serious complication is due to a gap in the literature. The formulas for functorial transfer of automorphic representations are local, and as such depend on the choice of local transfer factors. The local factors for transfer of orbital integrals are defined up to constants (cf. the discussion in §7, §8 of the introduction to this volume), subject to a global reciprocity condition that determines their product uniquely. These factors determine the formulas for transfer of local representations, and in particular the statement of the fundamental lemma at unramified places, as well as Shelstad's formulas for transfer of discrete series. For this spectral transfer, we have used the most natural local choices at finite places and Kottwitz' version of Shelstad's signs at archimedean places. But we have been unable to find any statement in the literature that allows us to conclude that these choices are compatible with the global reciprocity condition mentioned above, even in the very simplified case we have chosen to treat, where there is in principle only one natural choice of sign at finite places. As a result, we only know that our choice is correct up to sign. This is the sign ε that appears in the proof of Corollary 3.3.8 of [CHL.IV.B] and reappears as the sign $*$ in the proof of Lemma 4.7 of the final chapter. Shin's treatment faced an analogous sign problem. Fortunately, because $n \neq 1$, the two possible outcomes can be easily distinguished, and this ambiguity has no effect on the final results.

References

[A1] J. Arthur, The Endoscopic Classification of Representations: Orthogonal and Symplectic Groups (to appear).

[A2] J. Arthur, An introduction to the trace formula, in Harmonic analysis, the trace formula, and Shimura varieties, Clay Math. Proc. **4** Amer. Math. Soc., Providence, RI (2005) 1–263.

[CL] L. Clozel & J.-P. Labesse, Changement de base pour les représenta-1tions cohomologiques de certains groupes unitaires, *Astérisque,* **257** (1999), 121–136.

[HL] M. Harris & J.-P. Labesse, Conditional base change for unitary groups, *Asian J. Math.* **8** (2004), 653–684.

[K1] R. Kottwitz, Shimura varieties and λ-adic representations, in *Automorphic Forms, Shimura Varieties, and L-functions*, New York: Academic Press (1990), Vol. 1, 161–210.

[K2] R. Kottwitz, Points on some Shimura varieties over finite fields, *JAMS* **5** (1992) 373–444.

[L] J.-P. Labesse, Cohomologie, stabilisation et changement de base, *Astérisque,* **257** (1999) 1–116.

[LR] R.P. Langlands & D. Ramakrishnan, eds.: The zeta functions of Picard modular surfaces, Université de Montréal, Centre de Recherches Mathématiques, Montreal (1992).

[R] J. Rogawski, Automorphic Representations of Unitary Groups in Three Variables, *Annals of Math. Studies*, **123** (1990).

Membre de l'Institut Universitaire de France, Université Paris-Sud

Membre de l'Institut Universitaire de France, Institut de Mathématiques de Jussieu, U.M.R. 7586 du CNRS

Professeur émérite, Institut de Mathématiques de Luminy, U.M.R. 6206 du CNRS

Changement de base CM et séries discrètes
[L.IV.A]

J.-P. Labesse

1. Présentation

Soit F un corps de nombres totalement réel et E une extension quadratique totalement imaginaire de F. On s'intéresse aux groupes unitaires $\mathbf{U}$ relatifs à E/F. L'objectif est d'étudier le changement de base quadratique E/F pour les représentations automorphes de $\mathbf{U}$ qui sont cohomologiques aux places archimédiennes.

Cet article généralise des résultats et des techniques déjà présents dans [C3],[CL1] et [HL]. Contrairement aux articles précédents, le théorème principal 5.1 ne fait appel à aucune hypothèse supplémentaire de ramification, ni sur le groupe $\mathbf{U}$, ni sur les représentations aux places finies ; toutefois nous supposerons que F est de degré au moins 2 sur $\mathbb{Q}$.[1] Pour les théorèmes de multiplicité un 5.4 et de stabilité 5.5, nous supposerons de plus que les représentation sont non ramifiées à toutes les places finies inertes.

L'outil essentiel est la stabilisation d'une formule des traces tordue. Des progrès spectaculaires ont été faits récemment. La preuve du lemme fondamental, due à Laumon et Ngô pour les groupes unitaires [LN], a été généralisée par Ngô à tous les groupes réductifs ; la variante pondérée est annoncée par Chaudouard et Laumon. Ces travaux, complétés par ceux de Waldspurger ([W1] et [W2]), font que la stabilisation dans le cas non tordu, obtenue par Arthur [A6], est devenue inconditionnelle pour tous les groupes réductifs. Toutefois la preuve de la stabilisation dans le cas tordu n'a pas encore été rédigée en toute généralité, quoique le principal obstacle – à savoir le lemme fondamental tordu [W3], ainsi que sa variante pondérée – soit levé.

Nous proposons une version ad hoc de la stabilisation sous des hypothèses simplificatrices : en nous limitant au cas des représentations automorphes ne faisant intervenir que des séries discrètes aux places réelles

[1]Les résultats principaux (section 5) devraient encore être vrais dans le cas $F = \mathbb{Q}$, mais ce cas n'est pas traité ici : il demanderait un travail supplémentaire (voir les remarques 3.12 et 5.2). Nous espérons y revenir dans un article ultérieur.

pour un groupe unitaire nous donnons, pour le changement de base, une démonstration de la stabilisation pour la trace dans le spectre discret indépendante de la preuve du lemme fondamental pondéré. On en déduit des propriétés du changement de base sous la forme nécessaire pour les applications qui sont au cœur de ce livre. L'hypothèse que F est de degré $d \geq 2$ sur $\mathbb{Q}$, permet l'utilisation de la formule des traces simple et allège ainsi beaucoup les preuves et les pré-requis.

On aurait pu, comme dans [BLS], se contenter d'utiliser la formule des traces sous sa forme primitive (non invariante) et éviter ainsi d'avoir recours à la forme invariante de la formule des traces d'Arthur dont l'établissement est plus technique. C'était le point de vue adopté dans des versions antérieures de ce texte, mais d'une part cela alourdit un peu les arguments donnés ici et d'autre part, puisque nous prenons la formule des traces comme donnée, cela n'aide en rien pour la compréhension des preuves ; nous y avons renoncé.

Comme déjà dit plus haut ce papier reprend des techniques familières. Nous avons cependant fait des rappels assez longs pour le confort du lecteur inexpérimenté, la littérature sur le sujet étant vaste et parfois difficilement pénétrable, mais aussi pour fixer les notations. Nous ferons ces rappels dans le cadre d'un espace tordu sous un groupe réductif général mais nous donnerons la forme explicite des divers objets pour les cas utiles dans nos applications.

Nous avons proposé des preuves aussi élémentaires que possible et, sauf dans le dernier paragraphe, nous n'avons fait appel qu'a des résultats assez anciens d'Arthur concernant la formule des traces. De même en ce qui concerne le transfert et le lemme fondamental nous avons indiqué, chaque fois que cela était possible, des arguments anciens et plus élémentaires que ceux récemment obtenus par Laumon, Ngô et Waldspurger.

C'est un plaisir de remercier le rapporteur qui, par une lecture soigneuse du manuscrit, a permis la correction de nombreuses bourdes.

2. Groupes et espaces en question

2.1. Groupes unitaires et espaces associés. Les exemples essentiels pour toute la suite sont les suivants. Soit F un corps de nombres totalement réel et E une extension quadratique totalement imaginaire de F. Soit α_n l'automorphisme non trivial de $GL(n)$ qui fixe l'épinglage canonique :

$$\alpha_n(x) = J_n \ {}^t x^{-1} J_n^{-1}$$

avec $x \mapsto {}^t x$ la transposition et

$$J_n = \begin{pmatrix} 0 & \cdots & 0 & -1 \\ 0 & \cdots & 1 & 0 \\ \vdots & \cdots & \vdots & \vdots \\ (-1)^n & \cdots & 0 & 0 \end{pmatrix}$$

Soit $x \mapsto \overline{x}$ l'automorphisme non trivial du groupe de Galois de E/F. Le groupe

$$\mathbf{G}_n = \mathrm{Res}_{E/F}\, GL(n)$$

admet un automorphisme θ_n d'ordre 2 :

$$\theta_n(x) = \alpha_n(\overline{x}) \ .$$

On notera $\widetilde{\mathbf{G}}_n$ l'ensemble défini par la classe de θ_n dans le groupe produit semi-direct $\mathbf{G}_n \rtimes < \theta_n >$:

$$\widetilde{\mathbf{G}}_n = \mathbf{G}_n \rtimes \theta_n \ .$$

C'est un espace tordu suivant la terminologie de [Lab5] rappelée ci-dessous. On note $\mathbf{U}_n^* = \mathbf{U}^*(n, E/F)$ le sous-groupe des points fixes de θ_n dans $\mathbf{G}_n$. C'est le groupe unitaire quasi-déployé attaché à E/F en dimension n. On notera $\mathbf{U}_n$ une forme intérieure de $\mathbf{U}_n^*$. Pour alléger les notations on omettra parfois les indices n si le contexte est clair.

2.2. Espaces tordus. Pour l'étude de la formule des traces tordue, qui est l'outil essentiel de cet article, il est commode d'utiliser le langage des espaces tordus introduit dans [Lab5]. Cela permet en particulier de traiter de manière uniforme le cas tordu et le cas ordinaire c'est-à-dire le cas d'un groupe. C'est une variante du point de vue utilisé par Arthur, dans [Ar2] par exemple, où il développe la formule des traces invariante pour une composante non neutre d'un groupe non connexe.

Rappelons qu'un espace tordu (dans la catégorie des ensembles) est la donnée d'un groupe G, d'un G-espace principal homogène $\widetilde{G}$ à gauche et d'une application

$$\widetilde{\mathrm{Ad}} : \widetilde{G} \to \mathrm{Aut}\,(G)$$

qui est G-équivariante c'est-à-dire que pour tout $x \in G$ et tout $\delta \in \widetilde{G}$ on a

$$\widetilde{\mathrm{Ad}}(x\delta) = \mathrm{Ad}(x) \circ \widetilde{\mathrm{Ad}}(\delta)$$

où $\mathrm{Ad}(x)$ est comme d'habitude l'automorphisme intérieur défini par x. Choisissons un élément δ_0 de $\widetilde{G}$ et posons

$$\theta = \widetilde{\mathrm{Ad}}(\delta_0) \in \mathrm{Aut}\,(G) \ .$$

On définit une action à droite de G sur $\widetilde{G}$ en posant

$$x\delta_0\, y = x\theta(y)\delta_0 \ ,$$

ce qui fournit une action par conjugaison de G sur $\widetilde{G}$ et la notion de classe de G-conjugaison dans $\widetilde{G}$. Ces actions sont indépendantes du choix de δ_0. On a un isomorphisme d'espace tordus

$$\widetilde{G} \simeq G \rtimes \theta \subset G \rtimes \mathrm{Aut}\,(G)$$

défini par $x\delta_0 \mapsto x \rtimes \theta$ pour $x \in G$, mais en général cet isomorphisme dépend du choix de δ_0. En identifiant, via cet isomorphisme, G et $\widetilde{G}$ à des sous-ensembles du groupe $G \rtimes \mathrm{Aut}\,(G)$ on a

$$\delta_0\, x\, \delta_0^{-1} = \theta(x)\ .$$

On appellera représentation d'un espace tordu dans un espace vectoriel V la donnée d'un couple $(\widetilde{\pi}, \pi)$ où $\widetilde{\pi}$ est une application

$$\widetilde{\pi} : \widetilde{G} \to GL(V)$$

et π une représentation de G dans V :

$$\pi : G \to GL(V)$$

vérifiant pour $x, y \in G$ et $\delta \in \widetilde{G}$

$$\widetilde{\pi}(x\delta y) = \pi(x)\widetilde{\pi}(\delta)\pi(y)\ .$$

La donnée de $\widetilde{\pi}$ détermine π : on dira que π est la restriction de $\widetilde{\pi}$ à G. Si V est un espace de Hilbert on dira que $\widetilde{\pi}$ est unitaire si $\widetilde{\pi}$ prend ses valeurs dans le groupe unitaire de V.

Les représentations π que nous rencontrerons seront soit unitaires soit de dimension finie ; dans tous les cas le lemme de Schur sera valable. On dira que $\widetilde{\pi}$ est G-irréductible si π est irréductible. Dans ce cas on a

$$\pi(\theta(x)) = \widetilde{\pi}(\delta_0)\pi(x)\widetilde{\pi}(\delta_0)^{-1}$$

et donc

$$\pi \circ \theta \simeq \pi$$

et on dit que π est θ-invariante (ou θ-stable). Réciproquement, dire que l'automorphisme θ préserve une représentation irréductible (π, V) signifie qu'il existe un opérateur $I_\theta \in GL(V)$ tel que

$$I_\theta\ \pi(x) = \pi(\theta(x))\, I_\theta\ .$$

Ceci permet de prolonger π en une représentation $\widetilde{\pi}$ de $\widetilde{G}$ en posant

$$\widetilde{\pi}(x\delta_0) = \pi(x) I_\theta\ ;$$

toutefois un tel prolongement n'est pas canonique car I_θ n'est défini par π qu'à un scalaire près.

2.3. Cas des groupes réductifs. On définit de manière analogue la notion d'espace tordu dans la catégorie des variétés algébriques. Par un abus de notation classique on notera par la même lettre une variété algébrique et son ensemble de points sur une clôture algébrique $\overline{F}$ choisie une fois pour toutes.

Dans toute la suite G sera un groupe réductif connexe défini sur un corps de nombres F et $\widetilde{G}$ un G-espace tordu. Nous supposerons donné un isomorphisme, défini sur F

$$\widetilde{G} \to G \rtimes \theta$$

où θ est un F-automorphism de G. On choisit θ de sorte qu'il préserve un sous-groupe parabolique P_0 et un sous-groupe de Levi $M_0 \subset P_0$ minimaux sur F. On notera le plus souvent δ_0 (mais aussi parfois w_0) l'élément de $\widetilde{G}(F)$ image réciproque de $1 \rtimes \theta$ par l'isomorphisme. On dispose de l'espace tordu $\widetilde{G}(\mathbb{A}_F)$ des points adèliques de $\widetilde{G}$. Tout élément $y \in \widetilde{G}(\mathbb{A}_F)$ est de la forme

$$y = x\delta_0$$

avec $x \in G(\mathbb{A}_F)$.

On note A_G le tore déployé maximal de la restriction des scalaires à $\mathbb{Q}$ du centre Z_G de G. On note $\mathfrak{A}_G$ la composante neutre de $A_G(\mathbb{R})$ et $\mathfrak{a}_G$ son algèbre de Lie. On suppose que l'automorphisme induit par θ sur $\mathfrak{A}_G$ est d'ordre fini. On notera $\mathfrak{A}_{\widetilde{G}}$ (resp. $\mathfrak{a}_{\widetilde{G}}$) le sous-groupe (resp. la sous-algèbre) des points fixes sous θ dans $\mathfrak{A}_G$ (resp. $\mathfrak{a}_G$).

On dit qu'un élément $\delta \in \widetilde{G}$ est semi-simple si $\widetilde{\mathrm{Ad}}(\delta)$ est un automorphisme semi-simple. On notera G^δ son centralisateur et I_δ le centralisateur stable [Lab5, Définition II.1.3] qui est un sous-groupe de G^δ contenant sa composante neutre. Dans les exemples ci-dessous on aura $I_\delta = G^\delta$ le centralisateur d'un élément semi-simple étant connexe. On dit qu'un élément $\delta \in \widetilde{G}(F)$ est elliptique (cf. [Lab5, section IV.1]) si il est semi-simple et si de plus $\mathfrak{A}_{I_\delta} = \mathfrak{A}_{\widetilde{G}}$.

Les groupes adèliques seront munis de la mesure de Tamagawa et on notera $\tau(G)$ le nombre de Tamagawa de G. Les centralisateurs stables des éléments semi-simples étant quasi-connexes (au sens de [Lab4]) mais non connexes en général il est nécessaire de généraliser la notion de mesure de Tamagawa pour de tels groupes ; c'est ce qui est fait dans [KS] pour les centralisateurs des éléments fortement réguliers (ce sont des groupes diagonalisables) et dans [Lab4] pour le cas général. Mais, dans les exemples qui nous occuperons ici, ce sera inutile les centralisateurs étant connexes.

Pour les applications que nous avons en vue ici on aura soit

(A) $\qquad \widetilde{G} = \widetilde{\mathbf{G}}_n = \mathbf{G}_n \rtimes \theta \qquad$ avec $\qquad G = \mathbf{G}_n \qquad$ et $\qquad \theta = \theta_n$

soit

(B) $\qquad \widetilde{G} = G = \mathbf{U} \qquad$ et $\qquad \theta = 1$.

On rencontrera aussi des produits d'espaces tordus et de groupes d'un des types ci-dessus. Dans tous les cas $\delta_0 = 1 \rtimes \theta$.

Dans ces exemples les centralisateurs des éléments semi-simples sont connexes et le groupe $\mathfrak{A}_{\widetilde{G}}$ est trivial. En particulier, un élément semi-simple est elliptique si son centralisateur possède un tore maximal anisotrope.

3. Formule des traces

3.1. Le noyau intégral de la formule des traces. On reprend les notations du cas général au paragraphe 2.3. L'espace homogène

$$X_G = \mathfrak{A}_G G(F)\backslash G(\mathbb{A}_F)$$

est de volume fini ; il est compact si G_{der}, le groupe dérivé, est anisotrope. La représentation régulière gauche ρ de $G(\mathbb{A}_F)$ dans $L^2(X_G)$ admet un prolongement naturel en une représentation $\widetilde{\rho}$ de $\widetilde{G}(\mathbb{A}_F)$; elle est définie par

$$\widetilde{\rho}(g\delta_0)\varphi(x) = \varphi(\theta^{-1}(x\,g))$$

pour $\varphi \in L^2(X_G)$, $g \in G(\mathbb{A}_F)$, $\delta_0 \in \widetilde{G}(F)$, $\theta = \widetilde{\mathrm{Ad}}(\delta_0)$ et $x \in X_G$ ce qui peut aussi s'écrire

$$\widetilde{\rho}(y)\varphi(x) = \varphi(\delta_0^{-1}\,x\,y)$$

si $y = g\delta_0$. La représentation $\widetilde{\rho}$ est indépendante du choix de δ_0. Une représentation automorphe irréductible π du spectre discret, c'est-à-dire réalisée dans un sous-espace $G(\mathbb{A}_F)$-invariant de $L^2(X_G)$, et qui est de plus θ-invariante, admet un prolongement canonique $\widetilde{\pi}$ à $\widetilde{G}(\mathbb{A}_F)$, dit prolongement automorphe, par restriction de $\widetilde{\rho}$ à l'espace de π.

On considère une fonction $\phi \in \mathcal{C}_c^\infty(\widetilde{G}(\mathbb{A}_F))$ et l'opérateur $\widetilde{\rho}(\phi)$ défini par la représentation régulière gauche de $\widetilde{G}(\mathbb{A}_F)$ dans $L^2(X_G)$. L'opérateur $\widetilde{\rho}(\phi)$ est donné par le noyau intégral

$$K(x,y) = \int_{z\in\mathfrak{A}_G} \sum_{\delta\in\widetilde{G}(F)} \phi(x^{-1}\delta\,z\,y)\;dz\;.$$

Lorsque X_G est compact la formule des traces est l'égalité entre l'intégrale sur la diagonale du noyau que l'on peut développer en une somme indexée par les classes de conjugaisons dans $\widetilde{G}(F)$, appelée développement géométrique, et d'autre part la trace de $\widetilde{\rho}(\phi)$ développée suivant la décompositions spectrale de $L^2(X_G)$.

Dans le cas général où G_{der} n'est plus nécessairement anisotrope, la formule des traces est l'égalité du développement géométrique et du développement spectral pour la "trace renormalisée" de cet opérateur. Du point de vue géométrique une renormalisation est nécessaire car l'intégrale du noyau sur la diagonale est divergente. Du point de vue spectral une renormalisation est nécessaire car l'opérateur $\widetilde{\rho}(\phi)$ n'est traçable que dans le spectre discret. Or la décomposition spectrale comporte en général un spectre continu.

La renormalisation se fait au moyen de troncatures qui dépendent de divers choix : outre le choix d'un sous-groupe de Levi minimal θ-invariant M_0, il convient de choisir un bon sous-groupe compact maximal K. Cette dépendance implique que la distribution obtenue n'est pas invariante par conjugaison. Pour pallier cet inconvénient un avatar invariant de cette distribution a été construit par Arthur (cf. [Ar2]). La preuve dans le cas tordu en est désormais inconditionnelle grâce au théorème de Paley-Wiener scalaire tordu de Delorme et Mezo [DM]. On notera $T^{\widetilde{G}}(\phi)$ cette trace renormalisée (elle est notée $I^{\widetilde{G}}(\phi)$ chez Arthur). On renvoie le lecteur à [Ar5] pour une introduction détaillée.

3.2. Termes géométriques. Le développement géométrique pour la trace renormalisée est donné par l'intégrale du noyau sur la diagonale mais dont on a retranché des contre-termes pour la rendre convergente puis invariante. Il est la somme de deux termes :

$$T^{\widetilde{G}}(\phi) = T_e^{\widetilde{G}}(\phi) + T_{ne}^{\widetilde{G}}(\phi)$$

Le premier terme $T_e^{\widetilde{G}}(\phi)$ est la contribution des éléments elliptiques, le second est la contribution des éléments non elliptiques.

Le terme $T_e^{\widetilde{G}}(\phi)$ est la distribution invariante définie par l'intégrale absolument convergente sur la diagonale de la partie du noyau donnée par la somme sur l'ensemble $\widetilde{G}(F)_e$ des éléments elliptiques dans $\widetilde{G}(F)$:

$$T_e^{\widetilde{G}}(\phi) = \int_{X_G} \left(\int_{\mathfrak{A}_G} \sum_{\delta \in \widetilde{G}(F)_e} \phi(x^{-1}\delta\, z\, x)\ dz \right) dx\ .$$

On pose

$$\phi^0(\delta) = \int_{\mathfrak{A}_{\widetilde{G}}} \phi(\delta\, z)\, dz$$

et

$$\mathcal{O}_\delta(\phi) = \int_{I_\delta(\mathbb{A}_F)\backslash G(\mathbb{A}_F)} \phi(x^{-1}\delta\, x)\ d\dot{x}$$

désigne l'intégrale orbitale. Un calcul élémentaire (voir par exemple [Lab3, 4.1]) montre que

Proposition 3.1.

$$T_e^{\widetilde{G}}(\phi) = J(\widetilde{G}) \sum_{\delta \in \widetilde{G}_e} a^{\widetilde{G}}(\delta)\mathcal{O}_\delta(\phi^0)$$

où $\widetilde{G}_e$ est un ensemble de représentants des classes de conjugaison dans $\widetilde{G}(F)_e$. Le nombre $a^{\widetilde{G}}(\delta)$ est défini par

$$a^{\widetilde{G}}(\delta) = \iota(\delta)^{-1}\ \mathrm{vol}\ (\mathfrak{A}_{\widetilde{G}} I_\delta(F)\backslash I_\delta(\mathbb{A}_F))$$

où $\iota(\delta)$ est l'ordre du quotient $G^\delta(F)/I_\delta(F)$; enfin

$$J(\widetilde{G}) = |\det\ (\theta - 1|\mathfrak{a}_G/\mathfrak{a}_{\widetilde{G}})|\ .$$

Dans les exemples qui interviendrons dans la suite, le groupe $\mathfrak{A}_{\widetilde{G}}$ est trivial et comme de plus les centralisateurs sont connexes, on aura donc simplement

$$T_e^{\widetilde{G}}(\phi) = J(\widetilde{G}) \sum_{\delta \in \widetilde{G}_e} \tau(I_\delta)\mathcal{O}_\delta(\phi)\ .$$

On prendra garde que le facteur $J(\widetilde{G})$ n'apparaît pas dans les formules d'Arthur par suite d'un choix légèrement différent de l'opérateur dont on calcule la trace renormalisée. Nous suivons ici les conventions de [KS], [Lab3] et [Lab5].

Le second terme $T_{ne}^{\widetilde{G}}(\phi)$ fait intervenir les avatars invariants des intégrales orbitales pondérées des éléments semi-simples non elliptiques et des variantes d'icelles pour les éléments ayant une composante unipotente non triviale. Nous n'en donnerons pas l'expression ici ; pour son étude nous renvoyons le lecteur à [Ar5] par exemple. Pour les applications que nous avons en vue on choisira des fonctions ϕ telles que ce second terme $T_{ne}^{\widetilde{G}}(\phi)$ soit nul.

3.3. Termes spectraux. L'expression spectrale pour la trace renormalisée comporte une partie discrète et une partie continue

$$T^{\widetilde{G}}(\phi) = T_{disc}^{\widetilde{G}}(\phi) + T_{cont}^{\widetilde{G}}(\phi) \ .$$

Dans les applications que nous avons en vue nous choisirons des fonctions ϕ telles que la partie continue, qui est l'avatar invariant d'une expression faisant intervenir des intégrales de caractères pondérés, sera nulle :

$$T_{cont}^{\widetilde{G}}(\phi) = 0 \ .$$

Ici encore on renvoie le lecteur à [Ar5] pour une étude de ces termes.

La partie discrète $T_{disc}^{\widetilde{G}}$ est une distribution qui contient, entre autre, la trace dans le spectre discret.

$$T_{G,disc}^{\widetilde{G}}(\phi) = \text{ trace } \left(\widetilde{\rho}(\phi)|L^2_{disc}(X_G)\right) = \sum_{\widetilde{\pi}\in\Pi_{disc}(\widetilde{G})} m(\widetilde{\pi}) \text{ trace } \widetilde{\pi}(\phi)$$

où $\Pi_{disc}(\widetilde{G})$ est l'ensemble des classes d'équivalence de représentation $\widetilde{\pi}$ de l'espace tordu $\widetilde{G}(\mathbb{A}_F)$ dont la restriction π à $G(\mathbb{A}_F)$ reste irréductible et qui interviennent dans le spectre discret

$$L^2_{disc}(X_G) \ .$$

En effet, seules les représentations $\widetilde{\pi}$ qui par restriction à $G(\mathbb{A}_F)$ restent irréductibles donnent une contribution non triviale. Enfin $m(\widetilde{\pi})$ est la multiplicité de $\widetilde{\pi}$ dans le spectre discret.

Nous avons omis la sommation partielle – utilisée chez Arthur – suivant les modules des caractères infinitésimaux à l'infini, désormais inutile puisque, grâce à Müller, nous savons que le spectre discret est traçable et donc la série est absolument convergente [Mü]. En fait grâce aux travaux récents de Finis, Lapid et Müller on sait maintenant que le développement spectral tout entier est absolument convergent [FLM].

Remarque 3.2. Soit $\widetilde{\pi}$ une représentation de $\widetilde{G}(\mathbb{A}_F)$ dont la restriction π à $G(\mathbb{A}_F)$ reste irréductible. Si on note $m(\pi)$ la multiplicité de π dans le spectre discret, on a

$$m(\widetilde{\pi}) \leq m(\pi) \ .$$

Comme n'y a pas unicité du prolongement de π à $\widetilde{G}(\mathbb{A}_F)$ il n'est pas clair à priori que l'on ait l'égalité $m(\widetilde{\pi}) = m(\pi)$ sauf bien entendu si $G = \widetilde{G}$ ou encore si $m(\pi) \leq 1$, ce qui est le cas lorsque $G = \mathbf{G}_n$. On renvoie le lecteur à [Lab3, section 4.4, pages 106-107] pour une discussion de cas plus généraux.

D'autres termes discrets, quoique provenant du spectre continu, peuvent apparaître dans la trace renormalisée ; nous allons les décrire. On note $\mathcal{L}^0$ l'ensemble des sous-groupes de Levi de G contenant le sous-groupe de Levi minimal θ-invariante fixé M_0. On note $W^{\widetilde{G}}$ l'ensemble de Weyl de $\widetilde{G}$ c'est-à-dire $W^G \rtimes \theta$. Soit $M \in \mathcal{L}^0$ un sous-groupe de Levi de G ; on notera $W^{\widetilde{G}}(M)$ le quotient de l'ensemble des $s \in W^{\widetilde{G}}$ tels que $s(M) = M$ par W^M, le groupe de Weyl de M. On observera que $W^G(M)$ est un groupe. On notera $W^{\widetilde{G}}(M)_{reg}$ le sous-ensemble des s tels que

$$\det\,(s - 1|\mathfrak{a}_M/\mathfrak{a}_G) \neq 0 \;.$$

On fixe un tel s, et soit Q un sous-groupe parabolique de G, admettant M comme sous-groupe de Levi. Nous allons rappeler la définition de l'opérateur d'entrelacement $\mathbf{M}_{Q|s(Q)}(0)$ et de l'opérateur $\rho_Q(s,0,\phi)$ utilisés par Arthur dans [Ar2, § 4]. Soit φ une fonction sur $G(\mathbb{A}_F)$ invariante à gauche par $Q(F)N(\mathbb{A}_F)$ et telle que pour tout $x \in G(\mathbb{A}_F)$ la fonction

$$m \mapsto \varphi(mx)$$

soit une forme automorphe du spectre discret de M. On pose, avec des notations standard,

$$\varphi_\lambda(x) = e^{<\lambda + r_Q|H_Q(x)>}\varphi(x)$$

où r_Q est la demi-somme des racines positives pour Q. On notera $V_Q(\lambda)$ l'espace des φ_λ. Soit w_s un représentant de s dans $\widetilde{G}(F)$. Par abus de notation on notera encore s l'automorphisme de M défini par w_s. Arthur introduit [Ar2, p 516] un opérateur $\rho_Q(s,\lambda,y)$ qui envoie $V_Q(\lambda)$ dans $V_{s(Q)}(s(\lambda))$:

$$\rho_Q(s,\lambda,y)\varphi_\lambda(x) = \varphi_\lambda(w_s^{-1}x\,y) \;.$$

On définit un opérateur entre $V_Q(\lambda)$ et $V_Q(s(\lambda))$ en le composant avec l'opérateur d'entrelacement usuel $\mathbf{M}_{Q|s(Q)}(s(\lambda))$ (cf. [Ar1, pp. 1292]) :

$$\mathbf{M}_{Q|s(Q)}(s(\lambda))\rho_Q(s,\lambda,y)\varphi_\lambda(x) = \int_{N_s(\mathbb{A}_F)} \varphi_\lambda(w_s^{-1}n_s\,x\,y)\,dn_s$$

où

$$N_s = N \cap s(N)\backslash N \;.$$

L'intégrale ne converge que pour λ dans le translaté d'un cône mais admet un prolongement méromorphe qui, d'après [Langlands2], est holomorphe en $\lambda = 0$. On obtient ainsi un endomorphisme de $V_Q(0)$ qui, par intégration contre ϕ fournit l'opérateur

$$\mathbf{M}_{Q|s(Q)}(0)\rho_Q(s,0,\phi) \;.$$

On pose

$$T^{\widetilde{G}}_{M,disc}(\phi) = \sum_{s \in W^{\widetilde{G}}(M)_{reg}} |\det\,(s-1|\mathfrak{a}_M/\mathfrak{a}_G)|^{-1}\;\mathrm{trace}\;(\mathbf{M}_{Q|s(Q)}(0)\rho_Q(s,0,\phi)) \;.$$

L'énoncé suivant est emprunté à [Ar2, § 4] (voir aussi [Ar5] pages 133 et 237).

PROPOSITION 3.3. *La partie discrète de la formule des traces d'Arthur peut s'écrire :*

$$T^{\widetilde{G}}_{disc}(\phi) = \sum_{M\in\mathcal{L}^0} \frac{|W^M|}{|W^G|}\, T^{\widetilde{G}}_{M,disc}(\phi)$$

ou si on préfère

$$T^{\widetilde{G}}_{disc}(\phi) = \sum_{M\in\mathcal{L}^0/W^G} \frac{1}{|W^G(M)|}\, T^{\widetilde{G}}_{M,disc}(\phi)$$

où la somme porte sur un ensemble de représentants des orbites de W^G dans $\mathcal{L}^0$.

Comme pour le développement géométrique, notre formule et celle d'Arthur, diffèrent par le facteur $J(\widetilde{G})$. En effet, le déterminant de $(s-1)$ est calculé chez Arthur sur le quotient $\mathfrak{a}_M/\mathfrak{a}_{\widetilde{G}}$ alors que nous le calculons sur le quotient $\mathfrak{a}_M/\mathfrak{a}_G$.

3.4. Normalisation d'Arthur. Soit π^M une représentation automorphe de M appartenant au spectre discret. Soit comme ci-dessus un sous-groupe parabolique Q de Levi M et de radical unipotent N. On notera $V_Q(\lambda,\pi^M)$ le sous-espace de $V_Q(\lambda)$ formé des fonctions φ_λ où φ est telle que pour tout $x\in G(\mathbb{A}_F)$ la fonction

$$m \mapsto \varphi(mx)$$

soit une forme automorphe de l'espace de π^M. Supposons que l'opérateur

$$\mathbf{M}_{Q|s(Q)}(0)\rho_Q(s,0,y)$$

laisse stable $V_Q(0,\pi^M)$. C'est le cas si et seulement si π^M est s-invariante, c'est-à-dire que l'automorphisme s préserve l'espace de π^M et

$$s(\pi^M)\simeq \pi^M\ .$$

En d'autres termes, π^M se prolonge, au moyen du prolongement automorphe, en une représentation $\widetilde{\pi^M}$ de $\widetilde{M}_s(\mathbb{A}_F)$ où $\widetilde{M}_s$ est l'espace tordu engendré par M et w_s. Seules de telles représentations peuvent fournir une contribution non triviale à la formule des traces. On notera alors

$$I_Q(\widetilde{\pi^M})(y)$$

la restriction de l'opérateur ci-dessus à $V_Q(0,\pi^M)$. Nous dirons que la représentation $I_Q(\widetilde{\pi^M})$ de $\widetilde{G}(\mathbb{A}_F)$ ainsi obtenue est définie au moyen de la normalisation d'Arthur.

Il convient parfois de remplacer l'opérateur d'entrelacement qui intervient dans la définition de l'opérateur $I_Q(\widetilde{\pi^M})(y)$ par un opérateur normalisé de sorte que

$$I_Q(\widetilde{\pi^M})(y) = n(\widetilde{\pi^M})R_Q(\widetilde{\pi^M})(y)$$

où $n(\widetilde{\pi^M})$ est un facteur de normalisation. (C'est nécessaire par exemple si on souhaite utiliser des opérateurs locaux satisfaisant de bonnes équations fonctionnelles ; voir par exemple [Ar4,§2]). Ce qui précède nous permet de reformuler la proposition 3.3 :

PROPOSITION 3.4. *La partie discrète du développement spectral de la formule des traces est donnée par*

$$T^{\widetilde{G}}_{disc}(\phi) = \sum_{M\in\mathcal{L}^0/W^G} \sum_{s\in W^L(M)_{reg}} \sum_{\widetilde{\pi^M}\in\Pi_{disc}(\widetilde{M_s})} a^{\widetilde{G}}_{disc,\widetilde{M_s}}(\widetilde{\pi^M}) \text{ trace } R_Q(\widetilde{\pi^M})(\phi)$$

avec

$$a^{\widetilde{G}}_{disc,\widetilde{M_s}}(\widetilde{\pi^M}) = \frac{m(\widetilde{\pi^M})\, n(\widetilde{\pi^M})}{|\det (s-1|\mathfrak{a}_M/\mathfrak{a}_G)|\ |W^G(M)|}$$

où $\Pi_{disc}(\widetilde{M_s})$ est l'ensemble des classes d'équivalence de représentations de l'espace tordu adèlique

$$\widetilde{M_s}(\mathbb{A}_F)$$

qui restent irréductibles par restriction à $M(\mathbb{A}_F)$ et $m(\widetilde{\pi^M})$ est la multiplicité de $\widetilde{\pi^M}$ dans le spectre discret de M.

On observera qu'en général les coefficients $a^{\widetilde{G}}_{disc,\widetilde{M_s}}$ ne sont ni entiers ni même positifs : par exemple pour

$$\widetilde{G} = G = GL(2)$$

et M le sous-groupe diagonal, un coefficient $-1/4$ apparait ainsi dans un terme spectral de la formule des traces classique.

Pour la formule des traces sur $\widetilde{G} = G$, de tels termes pour $M \neq G$ seront nuls dans nos applications puisque l'on ne considérera comme fonctions à l'infini que des pseudo-coefficients de séries discrètes (cf.[Ar3] p.268). Mais pour $\widetilde{G} = \widetilde{\mathbf{G}}_n$ ces termes donneront, même dans le cas "simple" (au sens du paragraphe 3.9), des contributions non triviales.

3.5. Normalisation automorphe et terme constant. La représentation de $G(\mathbb{A}_F)$ dans $V_Q(\lambda, \pi^M)$ peut aussi se réaliser comme représentation automorphe au moyen de séries d'Eisenstein : pour φ comme ci-dessus on pose

$$E(x,\varphi,\lambda) = \sum_{\gamma\in Q(F)\backslash G(F)} \varphi_\lambda(\gamma x)$$

la série ne convergeant que pour λ dans le translaté d'un cône et on définit

$$E(x,\varphi) = E(x,\varphi,0)$$

par prolongement analytique, ce qui a un sens car, d'après [Langlands2] les séries d'Eisenstein n'ont pas de singularités pour les valeurs imaginaires pures de λ. Comme dans la section 3.1, on prolonge la représentation régulière droite dans l'espace des séries d'Eisenstein en posant pour $y \in \widetilde{G}(\mathbb{A}_F)$

$$\widetilde{\rho}(y)E(x,\varphi,\lambda) = E(w^{-1}\,x\,y,\varphi,\lambda)$$

où w peut être choisi arbitrairement dans $\widetilde{G}(F)$ compte tenu de l'invariance à gauche par $G(F)$ de la série d'Eisenstein (w était noté δ_0 dans 3.1). C'est ce que nous appelons la normalisation automorphe. On a, pour λ dans le cône de convergence

$$\widetilde{\rho}(y)E(x,\varphi,\lambda) = \sum_{\gamma\in Q(F)\backslash G(F)} \varphi_\lambda(\gamma w^{-1}xy)$$

soit encore

$$\widetilde{\rho}(y)E(x,\varphi,\lambda) = \sum_{\xi\in \widetilde{G}(F)/Q(F)} \varphi_\lambda(\xi^{-1}xy)$$

Supposons maintenant que π^M est une représentation automorphe cuspidale pour M. Un calcul classique (voir par exemple [Langlands1, Lemma 3 pp. 237]) montre que le terme constant de E le long de Q

$$C_Q E(x,\varphi,\lambda) = \int_{N(F)\backslash N(\mathbb{A}_F)} E(nx,\varphi,\lambda)dn$$

est égal (pour λ dans le cône de convergence) à :

$$\sum_{w_i\in W^G(M)} \int_{N_i(\mathbb{A}_F)} \varphi_\lambda(w_i^{-1}n_ixy)\,dn_i$$

où

$$N_i = N\cap w_iNw_i^{-1}\backslash N\ .$$

Maintenant, par prolongement analytique cette expression a un sens et est encore vraie pour $\lambda = 0$. Le même calcul montre que le terme constant de $\widetilde{\rho}(y)E$:

$$C_Q\widetilde{\rho}(y)E(x,\varphi,\lambda) = \int_{N(F)\backslash N(\mathbb{A}_F)} \widetilde{\rho}(y)E(nx,\varphi,\lambda)dn$$

est égal à :

$$\sum_{w_j\in W^{\widetilde{G}}(M)} \int_{N_j(\mathbb{A}_F)} \varphi_\lambda(w_j^{-1}n_jxy)\,dn_s$$

où N_j est défini au moyen de w_j comme ci-dessus. Supposons maintenant de plus que M est tel que le groupe $W^G(M)$ soit trivial. Sous ces hypothèses l'expression pour le terme constant le long de Q de E se réduit à un seul terme :

$$C_QE(x,\varphi,\lambda) = \varphi_\lambda(x)\ .$$

Dans ce cas $W^{\widetilde{G}}(M)$ est aussi un singleton et l'expression du terme constant de $\widetilde{\rho}(y)E$ se réduit au terme relatif à w_s :

$$C_Q\widetilde{\rho}(y)E(x,\varphi,\lambda) = \int_{N_s(\mathbb{A}_F)} \varphi_\lambda(w_s^{-1}n_sxy)\,dn_s$$

qui est, par définition

$$I_Q(\widetilde{\pi^M})(y)\varphi_\lambda(x) \ .$$

En résumé on a la

Proposition 3.5. *Supposons M tel que le groupe $W^G(M)$ soit trivial alors, si π^M est cuspidale, on a dans le sous-espace des séries d'Eisenstein associées à π^M :*

$$C_Q \circ \widetilde{\rho}(y) = I_Q(\widetilde{\pi^M})(y) \circ C_Q \ .$$

En d'autres termes, la normalisation d'Arthur est induite, via le terme constant, par la normalisation automorphe.

3.6. Normalisation de Whittaker. Supposons maintenant G quasi-déployé sur F et muni d'un épinglage. Soit N_0 le radical unipotent du sous-groupe de Borel de l'épinglage. Considérons un caractère ψ de $N_0(F)\backslash N_0(\mathbb{A}_F)$, en position générale. On appelle représentations générique une représentation irréductible (π, V) admettant une fonctionnelle de Whittaker c'est-à-dire une forme linéaire non nulle $\Lambda : V \to \mathbb{C}$ telle que pour $v \in V$ et $n \in N_0(\mathbb{A}_F)$ on ait

$$\Lambda(\pi(n)v) = \psi(n)\Lambda(v) \ .$$

Une fonctionnelle globale est un produit de fonctionnelles locales. On sait qu'une telle fonctionnelle, si elle existe, est unique, localement partout, à un scalaire non nul près. Lorsque $G = GL(n)$, qui est le seul cas dont nous aurons besoin, ceci est dû, independemment, à Piateskii-Shapiro et Shalika [Shal]. Le cas général est dû à Rodier [Rod].

Supposons donnée une fonctionnelle de Whittaker. Si θ est un automorphisme de G qui préserve une représentation irréductible (π, V) du groupe $G(\mathbb{A}_F)$ il existe un opérateur $I_\theta \in GL(V)$ défini à un scalaire près tel que

$$I_\theta \ \pi(x) = \pi(\theta(x)) \, I_\theta \ .$$

Si de plus θ préserve le couple (N_0, ψ) et compte tenu de l'unicité il existe une constante c_θ telle que

$$\Lambda \circ I_\theta = c_\theta \Lambda$$

On appelle normalisation de Whittaker l'opérateur I_θ tel que $c_\theta = 1$.

L'intégrale de Whittaker (appelée aussi coefficient de Fourier)

$$W(x, E(\bullet, \varphi)) = \int_{N_0(F)\backslash N_0(\mathbb{A}_F)} E(nx, \varphi)\overline{\psi}(n)\, dn$$

fournit, si elle n'est pas identiquement nulle, une fonctionnelle de Whittaker sur l'espace de la représentation engendrée par les séries d'Eisenstein $E(\bullet, \varphi)$ lorsque φ varie en posant

$$\Lambda(E(\bullet, \varphi)) = W(e, E(\bullet, \varphi))$$

où e est l'élément neutre de $G(\mathbb{A}_F)$. Supposons de plus que l'automorphisme θ préserve l'épinglage. On notera w_0 (plutôt que δ_0) un élément de $\widetilde{G}(F)$

qui le représente. Supposons de plus que ψ est θ-invariant. On fait agir $y \in \widetilde{G}(\mathbb{A}_F)$ sur la fonction W par

$$\widetilde{R}(y)W(x, E(\bullet,\varphi)) = W(w_0^{-1}xy, E(\bullet,\varphi))$$

ce qui, compte tenu de la θ-invariance de N_0 et ψ, s'écrit encore

$$\int_{N_0(F)\backslash N_0(\mathbb{A}_F)} E(w_0^{-1}nxy,\varphi)\overline{\psi}(n)\,dn = W(x,\widetilde{\rho}(y)E(\bullet,\varphi))$$

c'est à dire que l'on a

$$\widetilde{R}(y)W(x, E(\bullet,\varphi)) = W(x,\widetilde{\rho}(y)E(\bullet,\varphi)) \ .$$

On observe que

$$\widetilde{R}(w_0)W(e, E(\bullet,\varphi)) = W(e, E(\bullet,\varphi))$$

et donc

$$\Lambda(\widetilde{\rho}(w_0)E(\bullet,\varphi)) = \Lambda(E(\bullet,\varphi))$$

Ceci montre que la normalisation automorphe induit la normalisation de Whittaker globale, si la fonctionnelle définie par l'intégrale de Whittaker est non nulle.

On définit de manière analogue, localement partout, la normalisation de Whittaker. Aux places non ramifiées la normalisation de Whittaker donne un opérateur qui laisse fixe le vecteur non ramifié ; on en déduit que la normalisation globale est le produit des normalisations locales. En résumé on a la proposition suivante :

Proposition 3.6. *Soit π^M une représentation automorphe cuspidale pour M, stable par s. Supposons que M est tel que $W^G(M)$ soit trivial. Supposons de plus que l'intégrale de Whittaker n'est pas identiquement nulle. Alors, la normalisation d'Arthur de l'action de $\widetilde{G}(\mathbb{A}_F)$ (définie au moyen de l'opérateur d'entrelacement non normalisé) dans $V_Q(\pi^M)$ est compatible avec la normalisation de Whittaker localement partout.*

3.7. Un cas particulier. La proposition suivante sera utilisée dans [CHL, section 4].

Proposition 3.7. *Lorsque $\widetilde{G} = \widetilde{\mathbf{G}}_{n+1}$ avec $n \geq 2$, la somme des contributions, à la partie discrète de la formule des traces, des sous-groupes de Levi M isomorphes à $\mathbf{G}_1 \times \mathbf{G}_n$ est*

$$\frac{1}{2}\sum_{\widetilde{\pi^M}\in\Pi_{disc}(\widetilde{M_s})} \text{trace } I_Q(\widetilde{\pi^M})(\phi)$$

où $s = \theta_M$ et la somme porte sur les $\widetilde{\pi^M}$ dans le spectre discret de $\widetilde{M_s}$, soit encore

$$\frac{1}{2}\sum_{\widetilde{\pi^M}\in\Pi_{disc}(\widetilde{M_s})} n(\widetilde{\pi^M}) \text{ trace } R_Q(\widetilde{\pi^M})(\phi) \ .$$

Si R_Q est défini au moyen de la normalisation de Langlands-Shahidi [Shah] pour les opérateurs d'entrelacement et si de plus π^M est cuspidale on a

$$n(\widetilde{\pi^M}) = 1 \ .$$

Enfin la normalisation d'Arthur coïncide avec la normalisation de Whittaker localement partout.

Preuve: Tout d'abord on utilise que pour les groupes du type $GL(n)$ on dispose du théorème de multiplicité un pour le spectre discret. C'est classique pour le spectre cuspidal [Shal] et cela s'étend à tout le spectre discret grâce à Mœglin et Waldspurger [MW]. Par ailleurs, dans notre cas

$$W^{\widetilde{G}}(M) = W^{\widetilde{G}}(M)_{reg}$$

est réduit à un élément s : la classe de θ^*_M, et on a

$$|\det\,(s-1|\mathfrak{a}_M/\mathfrak{a}_G)| = 2 \ .$$

De plus, si

$$\pi^M = \tau \boxtimes \chi$$

avec τ une représentation cuspidale de $\mathbf{G}_n$ et χ un Größencharacter pour E, le facteur de normalisation de Langlands-Shahidi $n(\widetilde{\pi^M})$ est la valeur en $t=0$ de

$$\frac{L(\tau\otimes\chi^{-1}, 1-t)}{L(\tau\otimes\chi^{-1}, 1+t)}$$

où L est la fonction L de Godement-Jacquet pour $GL(n)$. Comme τ est cuspidale et $n\geq 2$ la fonction L est holomorphe partout et on a

$$n(\widetilde{\pi^M}) = 1 \ .$$

La dernière assertion résulte de 3.5 combinée avec la proposition 3.6 : il convient pour l'appliquer de vérifier que l'intégrale de Whittaker est non identiquement nulle. Dans notre cas cette intégrale a été calculée par Shahidi [Shah, pp. 352] : elle s'exprime au moyen de l'inverse de la fonction L ci-dessus. Dire qu'elle est identiquement nulle équivaut à dire que la fonction L a un pôle en $t=0$; mais comme déjà observé τ est cuspidale et comme $n\geq 2$ la fonction L est une fonction entière.

□

3.8. Un lemme. Nous aurons également besoin du lemme ci-dessous. Supposons que

$$\widetilde{G} = \widetilde{\mathbf{G}}_n = \mathbf{G}_n \rtimes \theta_n$$

et considérons un sous-groupe de Levi de la forme

$$M = \mathbf{G}_{n_1} \times \cdots \times \mathbf{G}_{n_r}$$

On pose

$$\theta_M = \theta_{n_1} \times \cdots \times \theta_{n_r}$$

LEMME 3.8. Soit $s \in W^{\mathbf{G}}(M)_{reg}$ et soit

$$\pi^M = \pi_1 \boxtimes \cdots \boxtimes \pi_r$$

une représentation de M telle que $s(\pi^M) \simeq \pi^M$ alors

$$\theta_M(\pi^M) \simeq \pi^M$$

Preuve: L'action de s sur $\mathfrak{a}_M/\mathfrak{a}_G$ est le produit de l'action induite par un élément w du groupe de Weyl de $\mathbf{G}_n$ et de la symétrie $a \mapsto -a$ qui est induite par θ_M. Pour que s soit régulier il est nécessaire que -1 ne soit pas valeur propre de l'action induite par w. Mais w induit une permutation des facteurs de π. Pour que la valeur propre -1 n'apparaisse pas il est nécessaire que cette permutation ne comporte que des cycle de longueur impaire. Comme par ailleurs θ_M est d'ordre 2, et donc d'ordre premier à l'ordre des cycles, la condition s régulier impose que pour chaque facteur π_i on ait

$$\theta_{n_i}(\pi_i) \simeq \pi_i$$

□

3.9. Formule des traces simple. Soit v une place de F. On dit qu'une fonction $\phi_v \in \mathcal{C}_c^\infty(\widetilde{G}(F_v))$ est cuspidale si les intégrales orbitales $\mathcal{O}_\delta(\phi_v)$ des éléments $\delta \in \widetilde{G}(F_v)$ semi-simples réguliers et non-elliptiques sont nulles.

On dira que, pour un choix convenable de la fonction test ϕ, la formule des traces est simple si seuls apparaissent des termes elliptiques dans le développement géométrique et des termes discrets dans le développement spectral. Pour cela nous ferons appel à des fonctions cuspidales aux places réelles.

La première apparition du principe qui gouverne les propositions ci-dessous (empruntées pour l'essentiel à Arthur [Ar2, Theorem 7.1]) se trouve dans le chapitre 16 de Jacquet-Langlands. Donnons tout d'abord la proposition spectrale.

PROPOSITION 3.9. *Soit F est un corps totalement réel. Soit ϕ_∞ une fonction produit de fonctions cuspidales en chaque place réelle. On a alors*

$$T^{\widetilde{G}}(\phi) = T^{\widetilde{G}}_{disc}(\phi)\ .$$

Preuve: La décomposition spectrale de la formule des traces invariante d'Arthur ne comporte que des termes discrets dès que ϕ est cuspidale en une place [Ar2, Theorem 7.1.(a)].

□

Voici maintenant la proposition géométrique.

PROPOSITION 3.10. *Soit F est un corps totalement réel de degré ≥ 2. Supposons que ϕ_∞ est un produit de fonctions cuspidales en chaque place réelle. Si de plus en une place finie v la fonction ϕ_v est à support régulier on a alors*

$$T^{\widetilde{G}}(\phi) = T^{\widetilde{G}}_e(\phi)$$

Preuve: Les contributions non semi-simples sont éliminées par la condition de support régulier en une place finie. Dans ces conditions, l'annulation de $T_{ne}^{\widetilde{G}}(\phi)$ lorsque ϕ est cuspidale en deux places est un théorème d'Arthur [Ar2, Theorem 7.1.(b)]

□

COROLLAIRE 3.11. Si F est un corps totalement réel de degré ≥ 2 avec ϕ comme dans la proposition 3.10 ci-dessus on a

$$T^{\widetilde{G}}(\phi) = T_e^{\widetilde{G}}(\phi) = T_{disc}^{\widetilde{G}}(\phi) \ .$$

REMARQUE 3.12. Dans le cas $F = \mathbb{Q}$ avec ϕ_∞ cuspidale et ϕ_v à support régulier en une place finie on aurait

$$T^{\widetilde{G}}(\phi) = T_e^{\widetilde{G}}(\phi) + T_{ss}^{\widetilde{G}}(\phi) = T_{disc}^{\widetilde{G}}(\phi)$$

où $T_{ss}^{\widetilde{G}}(\phi)$ est la contribution des éléments semi-simples non elliptiques. Cette distribution est, dans ce cas, une combinaison linéaire de termes qui sont le produit des avatars invariants des intégrales orbitales pondérées en la place réelle et des intégrales orbitales ordinaires aux autres places.

4. Endoscopie et changement de base

Cette section contient des rappels sur l'endoscopie en général. Nous ferons également le calcul des divers objets endoscopiques qui interviendront dans nos application. L'endoscopie tordue permet d'étudier le changement de base cyclique qui nous préoccupe ici.

4.1. Préliminaires cohomologiques. Pour la cohomologie galoisienne nous utiliserons les notations classiques : si A est un groupe algébrique défini sur un corps F on pose

$$\mathbf{H}^*(F, A) := \mathbf{H}^*(\mathrm{Gal}(\overline{F}/F), A(\overline{F}))$$

et suivant les notations introduites dans [KS] et reprises dans [LBC] et [Lab5] on pose si F est un corps global et A un tore

$$\mathbf{H}^*(\mathbb{A}_F/F, A) := \mathbf{H}^*(\mathrm{Gal}(\overline{F}/F), A(\overline{\mathbb{A}_F})/A(\overline{F})) \ .$$

Pour le calcul des caractères et des groupes endoscopiques nous utiliserons les groupes de cohomologie abélianisée introduits dans le livre [Lab3], auquel nous renvoyons pour leur définition et leurs propriétés (voir aussi [Lab6]). Il nous suffira, pour les exemples traités ici, de rappeler que si un groupe réductif connexe A admet un groupe dérivé A_{der} simplement connexe alors sa cohomologie abélianisée n'est autre que la cohomologie de son cocentre C_A (le tore quotient de A par son groupe dérivé A_{der}) :

$$\mathbf{H}^i_{ab}(*, A) \simeq \mathbf{H}^i(*, C_A)$$

Plus généralement si $A \to B$ est un morphisme entre deux groupes réductifs connexes dont les groupes dérivés sont simplement connexes et dont les cocentres sont C_A et C_B, la cohomologie abélianisée du complexe $[A \to B]$

peut se calculer au moyen de l'hypercohomologie usuelle du complexe de tores $C_A \to C_B$:

$$\mathbf{H}^i_{ab}(*, A \to B) \simeq \mathbf{H}^i(*, C_A \to C_B) .$$

Soient F un corps totalement réel, E/F une extension quadratique totalement imaginaire et $\mathbf{U}_n$ un groupe unitaire forme intérieure de $\mathbf{U}^*_n$ le groupe unitaire quasi-déployé de dimension n relatif à E/F. On a en particulier

$$\mathbf{H}^i_{ab}(*, \mathbf{U}_n) = \mathbf{H}^i(*, \mathbf{U}_1) \qquad \text{et} \qquad \mathbf{H}^i_{ab}(*, \mathbf{G}_n) = \mathbf{H}^i(*, \mathbf{G}_1) .$$

Il sera utile d'observer que, par l'isomorphisme de Tate Nakayama, on a

$$\mathbf{H}^1(\mathbb{A}_F/F, \mathbf{U}_1) = \hat{\mathbf{H}}^{-1}(\mathrm{Gal}(E/F), \mathbb{Z})$$

(où $\hat{\mathbf{H}}^*$ désigne le groupe de cohomologie de Tate) et

$$\hat{\mathbf{H}}^{-1}(\mathrm{Gal}(E/F), \mathbb{Z}) \simeq \mathbb{Z}/2\mathbb{Z}$$

puisque l'action du groupe de Galois de E/F sur $\mathbb{Z}$ est ici l'action non triviale. On a donc

$$\mathbf{H}^1_{ab}(\mathbb{A}_F/F, \mathbf{U}_n) \simeq \mathbb{Z}/2\mathbb{Z}$$

De même, si F_v est un corps local tel que $E_v = E \otimes F_v$ soit un corps on a

$$\mathbf{H}^1_{ab}(F_v, \mathbf{U}_n) \simeq \mathbb{Z}/2\mathbb{Z}$$

On en déduit que

$$\ker^1_{ab}(F, \mathbf{U}_n) = \ker^1(F, \mathbf{U}_1) = 1$$

où par définition

$$\ker^1_*(F, G) = \ker[H^1_*(F, G) \to H^1_*(\mathbb{A}_F, G)] .$$

4.2. Caractères et groupes endoscopiques. Soit $\delta \in \widetilde{G}(F)$ un élément elliptique et soit I_δ son centralisateur stable ; les caractères endoscopiques relatifs à δ sont les caractères, notés κ, du groupe d'hypercohomologie abélianisé adèlique

$$\mathbf{H}^0_{ab}(\mathbb{A}_F/F, I_\delta \to G)$$

triviaux sur l'image de $\mathbf{H}^0_{ab}(\mathbb{A}_F, G)$. Un tel couple (δ, κ) est appelé paire endoscopique. Une paire endoscopique détermine un espace endoscopique elliptique (cf. [Lab5, IV.3]). Le choix d'un facteur de transfert complète la donnée endoscopique. On supposera, pour simplifier l'exposé, que les espaces endoscopiques sont simplement les groupes endoscopiques et qu'il n'est pas nécessaire de faire appel à des extensions centrales de ces groupes pour définir le transfert. Ce sera le cas dans nos applications.

On dira que la paire (δ, κ) est dominée par (δ_0, κ_0) si, à conjugaison stable près, I_δ est un sous groupe de I_{δ_0} et si κ est induit par κ_0 via l'homomorphisme

$$\mathbf{H}^0_{ab}(\mathbb{A}_F/F, I_\delta \to G) \to \mathbf{H}^0_{ab}(\mathbb{A}_F/F, I_{\delta_0} \to G)$$

On dira que (δ, κ) est (δ', κ') sont équivalentes si il existe une paire (δ_0, κ_0) qui les domine. Deux paires équivalentes définissent le même groupe endoscopique. Pour déterminer tous les groupes endoscopiques, il suffit de considérer les éléments réguliers ; ils définissent les paires minimales. On pourrait aussi se contenter de considérer les paires maximales. Le groupe endoscopique principal est celui associé aux paires (δ, κ) où κ est trivial.

Dans les exemples qui nous intéressent l'équivalence entre paires suffit à déterminer l'équivalence entre donnée endoscopiques : dans nos exemples toute paire endoscopique (δ_0, κ_0) maximale est telle que le centralisateur de δ_0 est une forme intérieure du groupe endoscopique (quasi-déployé par définition) associé à la classe de (δ_0, κ_0) ; cette description très simple n'est malheureusement pas valable en général. Nous allons maintenant expliciter ces groupes endoscopiques en invoquant, comme il est classique, le calcul de la cohomologie galoisienne au moyen de l'isomorphisme de Tate-Nakayama. Dans toute la suite de ce paragraphe, F est un corps totalement réel, E/F est une extension quadratique totalement imaginaire.

Considérons tout d'abord le cas du changement de base c'est-à-dire le cas où $\widetilde{G} = \widetilde{\mathbf{G}}_n$. Soit $\delta \in \widetilde{\mathbf{G}}_n(F)$ un éléments elliptique. Soit I_δ son centralisateur dans $\mathbf{G}_n$. Les caractères endoscopiques, relatifs à δ sont par définition les caractères de

$$\mathbf{H}^0_{ab}(\mathbb{A}_F/F, I_\delta \to \mathbf{G}_n)$$

triviaux sur l'image de $\mathbf{H}^0_{ab}(\mathbb{A}_F, \mathbf{G}_n)$. Mais on a une suite exacte

$$\mathbf{H}^0_{ab}(\mathbb{A}_F/F, \mathbf{G}_n) \to \mathbf{H}^0_{ab}(\mathbb{A}_F/F, I_\delta \to \mathbf{G}_n) \to \mathbf{H}^1_{ab}(\mathbb{A}_F/F, I_\delta) \to 1$$

et une application surjective

$$\mathbf{H}^0_{ab}(\mathbb{A}_F, \mathbf{G}_n) \to \mathbf{H}^0_{ab}(\mathbb{A}_F/F, \mathbf{G}_n) \to 1$$

Le groupe des caractères endoscopiques est donc le dual de Pontryagin du groupe $\mathbf{H}^1_{ab}(\mathbb{A}_F/F, I_\delta)$. Pour simplifier la discussion supposons que δ est régulier de sorte que I_δ est un tore anisotrope. Compte tenu de la suite spectrale de Hochschild-Serre, le théorème 90 de Hilbert montre que

$$\mathbf{H}^1(\mathbb{A}_F/F, I_\delta) = \mathbf{H}^1(\mathrm{Gal}(E/F), I_\delta(\mathbb{A}_E)/I_\delta(E))$$

et en particulier c'est un 2-groupe. Le groupe endoscopique associé à un caractère κ de ce groupe de cohomologie admet pour tore maximal un tore isomorphe à I_δ et un système de coracines défini au moyen des δ-orbites de coracines pour $\widetilde{G}$ sur lesquelles κ est trivial via l'isomorphisme de Tate-Nakayama (cf. [Lab5, II.4 et IV.3]). Il est plus classique de le formuler comme suit : par dualité de Tate Nakayama un caractère κ correspond à un élément

$$s \in GL(n, \mathbb{C}),$$

avec $s^2 = 1$. Soit p le nombre de $+1$ et q le nombre de -1 comme valeurs propres de s. Le groupe endoscopique H associé à κ a pour groupe dual $\hat{H}$ la composante neutre du centralisateur de s, et donc

$$\hat{H} \simeq GL(p, \mathbb{C}) \times GL(q, \mathbb{C})$$

avec $p+q=n$ et, compte tenu de l'action de Galois, on a

$$H \simeq \mathbf{U}_p^* \times \mathbf{U}_q^* \ .$$

Dans le cas particulier du tore $I_\delta \simeq (\mathbf{U}_1)^n$ tous les couples (p,q), et donc tous les groupes $\mathbf{U}_p^* \times \mathbf{U}_q^*$, apparaissent comme groupes endoscopiques.

On prendra garde que les éléments s et $-s$ correspondent à des caractères endoscopiques inéquivalents si $p \neq q$. Ils définissent des groupes endoscopiques isomorphes mais les facteurs de transfert seront distincts (voir par exemple [HL] pour la discussion de $s = \pm 1$ et [Rog] pour le cas général, du point de vue plongements de L-groupe). On en déduit que lorsque $\widetilde{G} = \widetilde{\mathbf{G}}_n$ les classes d'équivalence de données endoscopiques sont en bijection avec l'ensemble des couples (p,q) avec $p+q=n$.

Traitons maintenant le cas de l'endoscopie ordinaire pour les groupes unitaires. Soit $\delta \in \mathbf{U}_n(F)$ elliptique et soit I_δ son centralisateur. Les caractères endoscopiques relatifs à δ sont, par définition, les caractères du groupe

$$\mathbf{H}^0_{ab}(\mathbb{A}_F/F, I_\delta \to \mathbf{U}_n)$$

triviaux sur l'image de $\mathbf{H}^0_{ab}(\mathbb{A}_F, \mathbf{U}_n)$ mais on a une suite exacte

$$\begin{aligned}\mathbf{H}^0_{ab}(\mathbb{A}_F/F, \mathbf{U}_n) \to \mathbf{H}^0_{ab}(\mathbb{A}_F/F, I_\delta \to \mathbf{U}_n)& \\ \to \mathbf{H}^1_{ab}(\mathbb{A}_F/F, I_\delta) \to \mathbf{H}^1_{ab}(\mathbb{A}_F/F, \mathbf{U}_n)&\end{aligned}$$

et compte tenu de ce que

$$\ker^1_{ab}(F, \mathbf{U}_n) = \ker^1(F, \mathbf{U}_1) = 1$$

on a une application surjective

$$\mathbf{H}^0_{ab}(\mathbb{A}_F, \mathbf{U}_n) \to \mathbf{H}^0_{ab}(\mathbb{A}_F/F, \mathbf{U}_n) \to 1$$

Les caractères endoscopiques relatifs à I_δ sont donc les caractères du groupe

$$\ker[\mathbf{H}^1_{ab}(\mathbb{A}_F/F, I_\delta) \to \mathbf{H}^1_{ab}(\mathbb{A}_F/F, \mathbf{U}_n)] \ .$$

On a vu plus haut que

$$\mathbf{H}^1_{ab}(\mathbb{A}_F/F, \mathbf{U}_n) = \mathbf{H}^1(\mathbb{A}_F/F, \mathbf{U}_1) \simeq \mathbb{Z}/2\mathbb{Z} \ .$$

Donc, ici encore un caractère endoscopique κ correspond à un $s \in GL(n, \mathbb{C})$ d'ordre 2 et les groupes endoscopiques sont de la forme $H = \mathbf{U}_p^* \times \mathbf{U}_q^*$, mais cette fois les éléments s et $-s$ correspondent au même caractère endoscopique. On en déduit que lorsque $\widetilde{G} = \mathbf{U}_n$ les classes d'équivalence de données endoscopiques sont en bijection avec l'ensemble des couples (p,q) avec $p+q=n$ et $p \geq q$.

4.3. κ-intégrales orbitales. Dans les exemples qui nous préoccupent la conjugaison stable pour des éléments semi-simples est la conjugaison sur la clôture algébrique. Pour le cas général on renvoie le lecteur à [Lab5]. Si $\widetilde{G}$ est un espace tordu sous un groupe réductif connexe G et $\delta \in \widetilde{G}(F)$ un élément semi-simple rationnel son intégrale orbitale stable est définie par

$$\mathcal{SO}_\delta(\phi) = \int_{(I_\delta \backslash G)(\mathbb{A}_F)} e(x^{-1} I_\delta\, x)\phi(x^{-1}\delta\, x)\, d\dot{x}$$

où $e(x^{-1}I_\delta\, x)$ est le signe de Kottwitz. Ce signe vaut 1 pour les éléments réguliers. Plus généralement, soit κ un caractère endoscopique relatif à δ ; on définit une fonction sur l'ensemble

$$(I_\delta \backslash G)(\mathbb{A}_F) = \mathbf{H}^0(\mathbb{A}_F, I_\delta \to G)$$

en composant κ avec le morphisme naturel

$$\mathbf{H}^0(\mathbb{A}_F, I_\delta \to G) \to \mathbf{H}^0_{ab}(\mathbb{A}_F/F, I_\delta \to G)\ .$$

On note encore κ ce composé et on pose

$$\mathcal{O}^\kappa_\delta(\phi) = \int_{(I_\delta \backslash G)(\mathbb{A}_F)} \kappa(x) e(x^{-1} I_\delta\, x)\phi(x^{-1}\delta\, x)\, d\dot{x}\ .$$

4.4. Norme. Le transfert repose sur la notion de norme entre classes de conjugaisons semi-simples dans $\widetilde{G}$ et classes de conjugaisons stables d'un groupe endoscopique H. Nous allons en rappeler la définition pour les cas utilisés dans la suite. Dans les cas qui nous préoccupent ici, les définitions sont simples car, contrairement au cas général, on peut identifier H à un sous groupe de $\mathbf{U}^*_n$, qui est le groupe endoscopique principal, qui est lui même un sous groupe de $\mathbf{G}_n$ (voir §4.2), et de plus, les groupes dérivés étant simplement connexes, les centralisateurs des éléments semi-simples sont connexes. On renvoie le lecteur à [KS] et [Lab5] pour le cas général. Nous définirons en même temps la notion d'élément $\widetilde{G}$-H-régulier.

Considérons tout d'abord le cas $\widetilde{G} = G = \mathbf{U}$ où $\mathbf{U}$ est une forme intérieure d'un groupe unitaire quasi-déployé $\mathbf{U}^*$. Notons

$$\psi : \mathbf{U} \to \mathbf{U}^*$$

un isomorphisme sur la clôture algébrique. Soit H un groupe endoscopique de $\mathbf{U}$. On note φ un plongement

$$\varphi : H \to \mathbf{U}^*\ .$$

Soit $\gamma \in H$ semi-simple et $\delta^* = \varphi(\gamma)$; on dira que γ est G-H-régulier si I_γ, le centralisateur de γ dans H, et I_{δ^*}, le centralisateur de δ^* dans $\mathbf{G}$, sont isomorphes sur la clôture algébrique. On dira qu'un élément elliptique $\gamma \in H(F)$ est une norme de $\delta \in \mathbf{U}(F)$ si $\delta^* = \varphi(\gamma)$ est conjugué à $\psi(\delta)$ dans $\mathbf{U}^*$ sur la clôture algébrique.

Passons maintenant au cas $\widetilde{G} = \widetilde{\mathbf{G}}$. On remarquera d'abord que pour $\delta = x\delta_0 \in \widetilde{\mathbf{G}}$ on a $\delta^2 = x\theta(x) \in \mathbf{G}$. Soit H un groupe endoscopique pour $\widetilde{\mathbf{G}}$. On note φ un plongement

$$\varphi : H \to \mathbf{U}^* \subset \mathbf{G}\ .$$

On dira qu'un élément semi-simple $\gamma \in H$ est $\widetilde{G}$-H-régulier s'il existe un élément $\delta^* \in \widetilde{\mathbf{G}}$ tel que $\varphi(\gamma) = (\delta^*)^2$ et tel que I_γ et I_{δ^*} soient isomorphes sur la clôture algébrique. On dira qu'un élément elliptique $\gamma \in H(F)$ est une norme de $\delta \in \widetilde{\mathbf{G}}(F)$ si $\varphi(\gamma)$ est conjugué, sur la clôture algébrique, à δ^2.

Pour l'endoscopie ordinaire et pour le changement de base qui sont les cas qui nous préoccupent ici, l'existence d'une norme repose sur un résultat ancien de Steinberg généralisé par Kottwitz [K1, lemma 5.8].

4.5. Transfert. Soit $\mathcal{E}$ une donnée endoscopique et soit H le groupe endoscopique associé à $\mathcal{E}$. Dans nos exemples une donnée endoscopique est la donnée d'un groupe $H \simeq \mathbf{U}_p^* \times \mathbf{U}_q^*$, d'un plongement

$$\varphi : H \to \mathbf{U}_n^*$$

d'un caractère κ de

$$\mathbf{H}^1_{ab}(\mathbb{A}_F/F, H)$$

définissant H comme groupe endoscopique et d'un facteur de transfert $\Delta_v^{\mathcal{E}}$ pour toute place v de F.

Les facteurs de transfert $\Delta_v^{\mathcal{E}}$ sont des fonctions sur l'ensemble des couples (δ, γ) d'éléments semi-simples réguliers dans

$$\widetilde{G}(F_v) \times H(F_v)$$

où γ est une norme de δ. Le facteur de transfert est invariant par conjugaison stable de $\gamma \in H(F_v)$ et on a

$$\Delta_v^{\mathcal{E}}(x^{-1}\delta\, x, \gamma) = \kappa_v(x)\Delta_v^{\mathcal{E}}(\delta, \gamma)$$

pour tout $x \in (I_\delta \backslash G)(F_v)$. De plus le produit sur toutes les places vaut 1 sur les couples rationnels :

$$\prod_v \Delta_v^{\mathcal{E}}(\delta, \gamma) = 1$$

si $\delta \in \widetilde{G}(F)$ et $\gamma \in H(F)$ est une norme de δ.

Les facteurs de transfert ont été construits par Langlands et Shelstad ; ils supposent un choix qui peut se décrire au moyen d'un plongement entre L-groupes pour H et G ; c'est ce point de vue adopté par Langlands et Shelstad (cf.[KS]). Lorsque κ est trivial le facteur de transfert peut être choisi trivial. Les facteurs de transfert pour l'endoscopie ordinaire des groupes unitaires sont explicités sur le tore $\mathbf{U}_1^n$ dans [Lab7].

Considérons une fonction $\phi \in \mathcal{C}_c^\infty(\widetilde{G}(\mathbb{A}_F))$. On appelle transfert endoscopique de ϕ sur H une fonction $\phi^{\mathcal{E}} \in \mathcal{C}_c^\infty(H(\mathbb{A}_F))$ (parfois notée simplement ϕ^H), telle que l'on ait localement partout l'identité :

$$\mathcal{SO}_\gamma(\phi_v^{\mathcal{E}}) = \Delta_v^{\mathcal{E}}(\delta,\gamma)\mathcal{O}_\delta^\kappa(\phi_v)$$

si $\gamma \in H(F_v)$ est une norme de $\delta \in \widetilde{G}(F_v)$ semi-simple et régulier et

$$\mathcal{SO}_\gamma(\phi_v^{\mathcal{E}}) = 0$$

si γ est $\widetilde{G}$-H-régulier, mais n'est pas une norme. On en déduit que globalement

$$\mathcal{SO}_\gamma(\phi^{\mathcal{E}}) = \mathcal{O}_\delta^\kappa(\phi)$$

lorsque $\gamma \in H(F)$ est une norme de $\delta \in \widetilde{G}(F)$ et

$$\mathcal{SO}_\gamma(\phi^{\mathcal{E}}) = 0$$

si γ est $\widetilde{G}$-H-régulier, mais n'est pas localement partout une norme.

On appelle transfert stable, le transfert sur le groupe endoscopique principal (c'est-à-dire le groupe endoscopique correspondant au caractère endoscopique trivial) avec facteur de transfert trivial. Rappelons que $\mathbf{U}_n^*$ est le groupe endoscopique principal pour $\widetilde{\mathbf{G}}_n$. Il est usuel de dire que $\phi \in \mathcal{C}_c^\infty(\widetilde{\mathbf{G}}_n(\mathbb{A}_F))$ et $f \in \mathcal{C}_c^\infty(\mathbf{U}_n^*(\mathbb{A}_F))$ sont associées si f est un transfert stable de ϕ. Le lemme suivant est un cas particulier du théorème 4.3 dont la preuve est ancienne et beaucoup plus élémentaire.

Lemme 4.1. Pour toute place finie v et toute $\phi_v \in \mathcal{C}_c^\infty(\widetilde{\mathbf{G}}_n(F_v))$ il existe $f_v \in \mathcal{C}_c^\infty(\mathbf{U}_n^*(F_v))$ associée. Réciproquement toute f_v est associée à une ϕ_v.

Preuve: Le transfert sur le groupe endoscopique principal c'est-à-dire l'existence du transfert pour le changement de base stable est connu depuis longtemps [Lab3] : il se ramène par descente au cas des formes intérieures ce qui est acquis grâce à des travaux de Waldspurger. Observons de plus qu'il résulte de [Lab3 ; 2.5.3] que, dans notre cas, la norme pour le changement de base est surjective et que donc pour toute f_v sur $\mathbf{U}^*(F_v)$ il existe ϕ_v sur $\widetilde{\mathbf{G}}(F_v)$ telle que f_v et ϕ_v sont associées.

□

Aux places non ramifiées on a un énoncé plus précis ; le lemme fondamental pour toute l'algèbre de Hecke sphérique affirme que le transfert est compatible à la fonctorialité, au moins pour les représentations sphériques :

Théorème 4.2. Soit v une place finie non ramifié dans E/F. Soit h une fonction dans l'algèbre de Hecke sphérique $\mathcal{H}(\mathbf{G}_n(F_v))$. Considérons la fonction sur $\widetilde{G}(\mathbb{A}_F)$:

$$\phi_v(x \rtimes \theta_n) = h(x)$$

On dispose de l'homomorphisme de changement de base

$$b : \mathcal{H}(\mathbf{G}_n(F_v)) \to \mathcal{H}(\mathbf{U}_n^*(F_v))$$

Alors le transfert stable est compatible à l'homomorphisme de changement de base, c'est-à-dire que ϕ_v est associée à $f_v = b(h)$. De même pour tout groupe endoscopique H de $\mathbf{U}_n^*$ on dispose d'un homomorphisme

$$\psi_H : \mathcal{H}(\mathbf{U}_n^*(F_v)) \to \mathcal{H}(H(F_v))$$

compatible au transfert.

Preuve: Le lemme fondamental stable pour le changement de base, dû à Kottwitz [K3] pour l'élément neutre, s'étend à toute l'algèbre de Hecke sphérique par les techniques de Clozel [C2] ou Labesse [Lab1]. Le lecteur prendra garde qu'un argument clef dans la preuve de [C2] (et repris dans [Lab1]) est incorrect précisément pour les groupes unitaires en dimension impaire. Cet argument est corrigé dans l'erratum rédigé par Clozel et donné en appendice [C4]. Pour l'endoscopie ordinaire des groupes unitaires ceci est dû à Hales [Hal] compte tenu de la validité du lemme fondamental pour l'élément unité qui résulte de [LN] et [W2].

□

Nous aurons également besoin de tous les transferts endoscopiques pour les groupes unitaires. Nous donnons ci-dessous le théorème général.

THÉORÈME 4.3. Pour toute fonction $\phi \in \mathcal{C}_c^\infty(\widetilde{G}(\mathbb{A}_F))$ et toute donnée endoscopique $\mathcal{E}$ il existe une fonction $\phi^{\mathcal{E}} \in \mathcal{C}_c^\infty(H(\mathbb{A}_F))$, qui est un transfert endoscopique de ϕ.

Preuve: Pour les places réelles l'existence du transfert est dû à Shelstad dans le cas non tordu et à Renard en général. Toutefois, pour les cas étudiés ici on dispose d'une preuve élémentaire rappelée ci-dessous. Pour les places finies Waldspurger a montré, dans une longue série de travaux parmi lesquels nous citerons [W1], [W2] et [W3], que l'existence du transfert, ainsi que la validité du lemme fondamental aux places non ramifiées pour les unités de l'algèbre de Hecke en caractéristique zéro, résulte du lemme fondamental en caractéristique positive qui lui-même est désormais un théorème grâce aux travaux de Laumon et Ngô pour les groupes unitaires [LN] et généralisés par Ngô à tous les groupes réductifs.

□

4.6. Fonctions de Lefschetz et transfert. Pour un groupe de Lie $G_\infty = G(F \otimes \mathbb{R})$ on notera $\mathfrak{g}$ son algèbre de Lie et K_G un sous-groupe compact maximal. Soit V une représentation de dimension finie de $\mathbf{U}_\infty$. On appelle fonction d'Euler-Poincaré sur $\mathbf{U}_\infty$ relativement à V une fonction f_V lisse et à support compact sur $\mathbf{U}_\infty$ telle que

$$\text{trace}\ \sigma(f_V) = \sum (-1)^q \dim H^q(\mathfrak{u}, K_{\mathbf{U}}; \sigma \otimes \check{V})$$

pour toute représentation admissible σ de $\mathbf{U}_\infty$. Dans le cas du groupe quasi-déployé $\mathbf{U}^*$ la fonction d'Euler-Poincaré sera notée f_V^*.

Le transfert endoscopique pour les fonctions d'Euler-Poincaré, et plus généralement pour les pseudo-coefficients de séries discrètes, sur un groupe unitaire aux places réelles est construit de façon élémentaire dans [Lab6, § 7.2] (reproduit dans ce volume).

Considérons $W = V \otimes V^\theta$ et soit $\widetilde{W}$ un prolongement de W à $\widetilde{\mathbf{G}}_\infty$. On note $\phi_{\widetilde{W}}$ une fonction de Lefschetz pour $\widetilde{\mathbf{G}}_\infty$ attachée à $\widetilde{W}$. Par définition $\phi_{\widetilde{W}}$ est une fonction lisse et à support compact sur $\widetilde{\mathbf{G}}_\infty$ telle que pour toute représentation admissible $\widetilde{\pi}$ de $\widetilde{\mathbf{G}}_\infty$

$$\text{trace } \widetilde{\pi}(\phi_{\widetilde{W}}) = \sum (-1)^q \text{ trace } (\theta_c | H^q(\mathfrak{g}, K_{\mathbf{G}}; \widetilde{\pi} \otimes \check{\widetilde{W}}))$$

où θ_c est l'involution de Cartan relative à $K_{\mathbf{G}}$.

La construction de fonctions d'Euler-Poincaré est classique ; pour celles de Lefschetz on renvoie le lecteur à [Lab2] (voir aussi [C3]).

Lemme 4.4. La fonction $\frac{1}{a}\phi_{\widetilde{W}}$ admet comme transfert stable $\frac{1}{b^*}f_V^*$ où $a = 2^{nd}$ et b^* est l'ordre des L-paquets de séries discrètes pour $\mathbf{U}_\infty^*$.

Preuve: Le cas des coefficients triviaux traité dans [CL1] Corollaire A.1.2 a été complété (et certaines bévues corrigées) dans [CL2] qui traite le cas général. Rappelons en l'argument. On observe, en reprenant les calculs de [CL1] pages 120-121, que si γ est elliptique régulier de norme δ les intégrales orbitales de $\phi_{\widetilde{W}}$ et f_V sont données par

$$\mathcal{O}_\delta(\phi_{\widetilde{W}}) = \mathcal{O}_\gamma(f_V) = \text{ trace } (\gamma | V) .$$

Les éléments semi-simples non elliptiques ont des intégrales orbitales nulles. En passant aux intégrales orbitales stables il résulte de [CL2] que $\frac{1}{a}\phi_{\widetilde{W}}$ admet comme transfert stable $\frac{1}{b^*}f_V^*$ où $a = 2^{nd}$ est l'ordre du groupe de 1-cohomologie galoisienne d'un tore maximal compact (or un tel tore est produit de $n \times d$ copies de $\mathbf{U}(1)$) et b^* est l'ordre des L-paquets de séries discrètes pour $\mathbf{U}_\infty^*$.

□

Remarque 4.5. On a $b^* = (b_0^*)^d$ où b_0^* est le coefficient binomial $\binom{n}{p}$ avec $p = [\frac{n}{2}]$.

Soit $\mathbf{U}$ une forme intérieure de $\mathbf{U}^*$; on note f_V une fonction d'Euler-Poincaré pour $\mathbf{U}$ relativement à V. Une variante de la preuve ci-dessus fournit le

Lemme 4.6. La fonction $\frac{1}{b}f_V$ admet comme transfert stable $\frac{1}{b^*}f_V^*$ où b est l'ordre des L-paquets de séries discrètes pour $\mathbf{U}_\infty$.

Lemme 4.7. Il y a une seule représentation unitaire irréductible générique π_W de $\mathbf{G}_\infty \simeq GL(n, \mathbb{C})^d$ à cohomologie non triviale dans W et on a

$$\text{trace } \widetilde{\pi}_W(\phi_{\widetilde{W}}) = \pm a ,$$

le signe dépendant du choix du prolongement de π_W à l'espace tordu.

Preuve: On supposera que θ_c agit sur $\widetilde{W}$ par permutation des facteurs. Par définition des fonctions de Lefschetz on a

$$\text{trace } \widetilde{\pi}(\phi_{\widetilde{W}}) = \sum (-1)^q \text{ trace } (\theta_c | H^q(\mathfrak{g}, K; \widetilde{\pi} \otimes \check{\widetilde{W}})) \ .$$

Pour un groupe complexe les représentations unitaires irréductibles génériques sont des séries principales. Le calcul de leur $\mathfrak{g}$-K-cohomologie est classique ; il se fait au moyen de la suite spectrale de Hochschild-Serre (cf. [BW] theorem 3.3 page 93) et on voit qu'il existe une unique représentation π_W de la série principale, dont la cohomologie, en tant qu'espace vectoriel gradué, est non nulle. Elle est calculée dans [C1, lemme 3.14] ainsi que dans [C3, p.113] : à un décalage de degré $l(s)$ près (où $l(s)$ est la longueur d'un certain élément s du groupe de Weyl), la cohomologie est isomorphe à l'algèbre extérieure de $\mathfrak{a}$, l'algèbre de Lie de la partie déployée du tore maximal ; c'est une algèbre de Lie abélienne de dimension nd. L'automorphisme θ_c agit par $(-1)^p$ sur $\wedge^p\mathfrak{a}$. La somme alternée des traces est donc la dimension de l'algèbre extérieure soit $2^{nd} = a$; c'est donc, au signe près, le nombre cherché.

□

4.7. Transfert et terme constant. Soit Q un sous-groupe parabolique standard de $\mathbf{G}$ de sous-groupe de Levi M et de radical unipotent N. On note $\widetilde{M}$ l'espace tordu engendré par M et θ_M. Soit v une place non ramifiée et soit h une fonction dans l'algèbre de Hecke sphérique sur $\mathbf{G}(F_v)$. Le terme constant de h le long de Q est défini par

$$h_Q(m) = \delta_Q(m)^{1/2} \int_{N(F_v)} h(mn)\, dn$$

où δ_Q est la fonction module pour Q. Considérons la fonction sur $\widetilde{\mathbf{G}}(\mathbb{A}_F)$ définie par

$$\phi_v(x \rtimes \theta_n) = h(x) \ .$$

On appellera terme constant de ϕ_v le long de Q la fonction sur $\widetilde{M}(F_v)$ définie par

$$\phi_{v,Q}(m \rtimes \theta_M) = h_Q(m) \ .$$

Soit λ un caractère non ramifié de $M_0(F_v)$ le sous-groupe diagonal de $\mathbf{G}_n(F_v)$. Soit π_λ la représentation de la série principale de $\mathbf{G}_n(F_v)$ et π_λ^M la représentation de la série principale de $M(F_v)$ définies par induction parabolique normalisée par λ.

Supposons que $w(\lambda) = \lambda$ pour un $w \in W^{\widetilde{M}}$ alors $\theta_M(\pi_\lambda^M) \simeq \pi_\lambda^M$ et on peut prolonger π_λ^M en une représentation $\widetilde{\pi^M}_\lambda$ de $\widetilde{M}(F_v)$ en imposant que l'opérateur $\widetilde{\pi^M}(\theta_M)$ fixe les vecteurs invariants sous le sous-groupe compact hyperspécial (ce qui est la normalisation naturelle dans le cas non ramifié). De même on prolonge π_λ en une représentation $\widetilde{\pi}_\lambda$ de $\widetilde{\mathbf{G}}(F_v)$

LEMME 4.8. Dans ces conditions on a

$$\text{trace } \widetilde{\pi}_\lambda(\phi_v) = \text{ trace } \widetilde{\pi^M}_\lambda(\phi_{v,Q}) \ .$$

Preuve: L'induction par étages montre que

$$\text{trace } \pi_\lambda(h) = \text{ trace } \pi_\lambda^M(h_Q) \ .$$

L'opérateur associé à θ_M agit trivialement sur le vecteur invariant par le compact hyperspécial ; on aura donc

$$\text{trace } \widetilde{\pi^M}_\lambda(\phi_{v,Q}) = \text{ trace } \pi_\lambda^M(h_Q) \ .$$

De même on a

$$\text{trace } \widetilde{\pi}_\lambda(\phi_v) = \text{ trace } \pi_\lambda(h)$$

ce qui fournit l'égalité souhaitée.

□

Soit $\mathbf{U}$ un groupe unitaire et soit H un groupe endoscopique pour $\mathbf{U}$. On sait que H est un produit de groupes unitaires quasi-déployées. Le groupe H_E obtenu par extension des scalaires à E est un produit de groupes linéaires. On pose

$$M^H = \mathrm{Res}_{E/F} H_E \ .$$

C'est un sous-groupe de Levi de $\mathbf{G}$. Soit θ_H l'automorphisme de M^H défini par l'action de Galois. On note $\widetilde{M}^H$ l'espace tordu défini par θ_H.

On suppose choisi pour chaque groupe endoscopique H un facteur de transfert Δ^H. Soit $f \in \mathcal{C}_c^\infty(\mathbf{U}(\mathbb{A}_F))$; on note f^H un transfert de f sur H. Le transfert dépend du choix du facteur de transfert mais nous avons omis cette dépendance dans la notation. Ceci n'engendre aucune ambiguïté puisqu'un groupe endoscopique pour $\mathbf{U}$ ne figure que dans une seule classe d'équivalence de données endoscopiques.

PROPOSITION 4.9. *(i) Il existe pour chaque H une fonction $\widetilde{f}^H$ sur $\widetilde{M}^H(\mathbb{A}_F)$, admettant f^H comme transfert stable sur $H(\mathbb{A}_F)$. On notera ϕ la fonction $\widetilde{f}^H$ lorsque $H = \mathbf{U}^*$.*
(ii) Considérons une place non ramifiée telle que ϕ_v est la translatée d'une fonction dans l'algèbre de Hecke sphérique. Alors, à torsion près par un caractère du déterminant sur chaque facteur de M^H, dépendant du choix du facteur de transfert Δ^H, la fonction $\widetilde{f}_v^H$ a les mêmes intégrales orbitales stables que le terme constant ϕ_{v,Q^H} de ϕ_v le long du sous-groupe parabolique standard Q^H de sous-groupe de Levi M^H.

Preuve: L'existence de $\widetilde{f}^H$ résulte de la surjectivité de la norme pour le changement de base stable dans notre cas (Lemme 4.1). Lorsque $H = \mathbf{U}^*$ cela revient à dire qu'il existe $\phi \in \mathcal{C}_c^\infty(\widetilde{\mathbf{G}}(\mathbb{A}_F))$ de sorte que $f = \phi^{\mathbf{U}^*}$. Le lemme fondamental pour toute l'algèbre de Hecke (théorème 4.2) montre que le transfert est compatible à la fonctorialité au moins pour les représentations sphériques. On conclut en invoquant le lemme 4.8.

□

4.8. Stabilisation des termes elliptiques. La stabilisation des termes elliptiques de la formule des traces est due à Langlands et Kottwitz [K2] dans le cas standard. Elle a été étendue au cas tordu dans [KS] pour les éléments elliptiques fortement réguliers. Enfin [Lab5] traite, dans le cas tordu, tous les éléments elliptiques. La stabilisation des termes elliptiques suppose l'existence du transfert, ce qui était une conjecture lors de la rédaction des articles cités ci-dessus.

La stabilisation complète de la formule des traces dans le cas ordinaire (i.e. non tordu) est due à Arthur ; elle utilise le lemme fondamental pondéré dont de la preuve vient d'être annoncée par Chaudouard et Laumon. Nous n'y ferons pas appel. En effet dans notre situation, qui pourtant invoque le cas tordu, de nombreux arguments se simplifient. Comme dans [Lab3] nous utiliserons des stabilisations conditionnelles indépendantes de la stabilisation d'Arthur et du lemme fondamental pondéré.

Si H est un groupe réductif connexe on note ST_e^H la distribution stable

$$ST_e^H(f) = \tau(H) \sum_\gamma \tilde{\imath}(\gamma)^{-1} \mathcal{SO}_\gamma(f^0)$$

où la somme porte sur un ensemble de représentants des classes de conjugaisons stables d'éléments elliptiques et $\tilde{\imath}(\gamma)$ est l'ordre du groupe des points rationnels du groupe quotient H^γ/I_γ, c'est-à-dire du groupe des composantes connexes du centralisateur de γ. On rappelle que les groupes adèliques sont munis de la mesure de Tamagawa et que $\tau(H)$ désigne le nombre de Tamagawa de H. Dans nos applications on aura simplement :

$$ST_e^H(f) = \tau(H) \sum_\gamma \mathcal{SO}_\gamma(f)$$

puisque les centralisateurs des éléments elliptiques seront connexes et les espaces vectoriels $\mathfrak{a}_H$ triviaux.

Si H est un groupe endoscopique appartenant à une donnée endoscopique $\mathcal{E}$ pour un espace tordu $\widetilde{G}$ on notera $ST_e^{\mathcal{E}}$ la variante de la distribution ST_e^H où on ne somme que sur les γ elliptiques qui satisfont de plus la condition de $\widetilde{G}$-H-régularité. Dans les applications que nous donnons ici, les deux distributions coïncident.

Pour chaque donnée endoscopique $\mathcal{E}$ on définit des nombres rationnels

$$\iota(\mathcal{E}) = \iota(\widetilde{G}, H) = \frac{J(\widetilde{G})\tau(G)}{\lambda(\widetilde{G}, H)c(\widetilde{G}, H)\ d(\widetilde{G})\tau(H)}$$

où τ désigne le nombre de Tamagawa, $J(\widetilde{G})$ est le déterminant de $1-\theta$ dans le quotient $\mathfrak{a}_G/\mathfrak{a}_{\widetilde{G}}$, l'entier $d(*)$ désigne l'ordre d'une obstruction cohomologique [Lab5, IV.1.3] et $c(*,*)$ est le quotient de deux mesures de Haar canoniques sur les algèbre de Lie isomorphes $\mathfrak{a}_{\widetilde{G}}$ et $\mathfrak{a}_H$. Enfin $\lambda(\widetilde{G}, H)$ est l'ordre du groupe des automorphismes extérieurs du couple $(\widetilde{G}, H)$. L'énoncé ci-dessous est emprunté à [Lab5].

THÉORÈME 4.10. Pour $\phi \in \mathcal{C}_c^\infty(\widetilde{G}(\mathbb{A}_F))$, la partie elliptique de la formule des traces $T_e^{\widetilde{G}}(\phi)$ est donnée par

$$T_e^{\widetilde{G}}(\phi) = \sum_{\mathcal{E}} \iota(\mathcal{E}) ST_e^{\mathcal{E}}(\phi^{\mathcal{E}})$$

où la somme porte sur les classes d'équivalence de données endoscopiques et où $\phi^{\mathcal{E}}$ est le transfert de ϕ sur le groupe endoscopique H.

4.9. Calcul des constantes. Nous allons calculer les constantes $\iota(\mathcal{E})$ qui interviennent dans le théorème 4.10 pour les exemples utiles ici.

PROPOSITION 4.11. *Si* $H = \mathbf{U}_p^* \times \mathbf{U}_q^*$ *avec* $p + q = n$, $p \neq q$ *et* $pq \neq 0$ *on a*

$$\iota(\widetilde{\mathbf{G}}_n, H) = \iota(\mathbf{U}_n, H) = \frac{1}{2}$$

Si $pq = 0$ *c'est-à-dire* $H = \mathbf{U}_n^*$ *on a*

$$\iota(\widetilde{\mathbf{G}}_n, H) = \iota(\mathbf{U}_n, H) = 1$$

Enfin, si $H = \mathbf{U}_p^* \times \mathbf{U}_q^*$ *avec* $p = q$ *on a*

$$\iota(\widetilde{\mathbf{G}}_n, H) = \iota(\mathbf{U}_n, H) = \frac{1}{4}$$

Preuve: Rappelons que $\mathfrak{a}_G$ désigne l'algèbre de Lie de la partie déployée de la restriction à $\mathbb{Q}$ du centre de G et $\mathfrak{a}_{\widetilde{G}}$ la sous algèbre des points fixes sous θ. On rappelle que

$$\iota(\widetilde{G}, H) = \frac{J(\widetilde{G})\tau(G)}{\lambda(\widetilde{G}, H)c(\widetilde{G}, H)\ d(\widetilde{G})\tau(H)}$$

Mais ici $\mathfrak{a}_{\mathbf{G}_n}$ est de dimension 1 et θ agit par -1. On en déduit que $\mathfrak{a}_{\widetilde{\mathbf{G}}}$ est nulle de même que $\mathfrak{a}_{\mathbf{U}}$ et $\mathfrak{a}_H$; donc

$$J(\widetilde{\mathbf{G}}_n) = 2 \quad \text{et} \quad c(\widetilde{\mathbf{G}}, H) = 1\ .$$

Il est bien connu que

$$\tau(\mathbf{G}_n) = 1 \quad , \quad \tau(\mathbf{U}_n^*) = 2\ ,$$

donc $\tau(H) = 2$ si $pq = 0$ et 4 sinon ; on a toujours $d(G) = 1$, donc $d(\mathbf{U}_n) = 1$ et on voit facilement que $d(\widetilde{\mathbf{G}}_n) = 1$. Enfin $\lambda(\widetilde{\mathbf{G}}_n, H) = 1$ si $H = \mathbf{U}_p^* \times \mathbf{U}_q^*$ avec $p \neq q$ alors que $\lambda(\widetilde{\mathbf{G}}_n, H) = 2$ si $p = q$.

□

4.10. Notion de changement de base. Notons $\mathcal{L}_F$ le groupe de Langlands dont l'existence dans le cas global est conjecturale. Pour les corps locaux archimédiens $\mathcal{L}_F$ est le groupe de Weil W_F. Pour les corps locaux non archimédiens $\mathcal{L}_F$ est le groupe de Weil-Deligne $WD_F = W_F \times SL(2,\mathbb{C})$. Un paramètre de Langlands est la donnée d'un L-homomorphisme admissible

$$\psi : \mathcal{L}_F \to {}^L G$$

et sa classe de conjugaison définit un élément $[\psi]$ dans

$$\mathbf{H}^1(\mathcal{L}_F, \widehat{G})$$

Un tel élément est associé (conjecturalement) à un L-paquet de représentations admissibles (automorphes dans le cas global).

Le changement de base pour une extension E/F est la correspondance entre L-paquets de représentations de G sur F et ceux pour

$$G_E = \mathrm{Res}_{E/F} G$$

définie au moyen des paramètres de Langlands (lorsque ceux-ci sont définis) par la restriction

$$\mathbf{H}^1(\mathcal{L}_F, \widehat{G}) \to \mathbf{H}^1(\mathcal{L}_E, \widehat{G})$$

ou, si on préfère, par l'application

$$\mathbf{H}^1(\mathcal{L}_F, \widehat{G}) \to \mathbf{H}^1(\mathcal{L}_F,\ \mathrm{Ind}_{\mathcal{L}_E}^{\mathcal{L}_F} \widehat{G})$$

obtenue en composant la restriction avec l'isomorphisme de Shapiro

$$\mathbf{H}^1(\mathcal{L}_E, \widehat{G}) \to \mathbf{H}^1(\mathcal{L}_F,\ \mathrm{Ind}_{\mathcal{L}_E}^{\mathcal{L}_F} \widehat{G})\ .$$

Supposons que E/F est une extension galoisienne cyclique. Le changement de base peut alors être vu comme un transfert endoscopique lorsque G est quasi-déployé. En effet, dans ce cas, G est le groupe endoscopique principal (c'est-à-dire correspondant au caractère endoscopique trivial) pour l'espace tordu $\widetilde{G} = G_E \rtimes \theta$ où θ est l'automorphisme de G_E défini par un générateur du groupe cyclique $\mathrm{Gal}(E/F)$. Et donc, du point de vue de la formule des traces, le changement de base cyclique est associé à la stabilisation de la formule des traces pour l'espace tordu $\widetilde{G}$.

Le cas qui nous intéresse ici correspond à E/F quadratique $G = \mathbf{U}_n^*$ et $G_E = \mathbf{G}_n$. Lorsque $G = \mathbf{U}_1^*$ le changement de base pour les caractères n'est autre que l'application

$$\chi \mapsto \chi \circ N_\theta$$

où N_θ est la norme associée à θ :

$$N_\theta(x) = x\,\theta(x)$$

avec

$$\theta(x) = \overline{x}^{-1}\ .$$

Ceci permet de définir immédiatement le changement de base pour les représentations de la série principale des groupes locaux quasi-déployés. Comme déjà observé plus haut, pour les représentations sphériques des groupes locaux non ramifiés le changement de base peut aussi être défini grâce à l'isomorphisme de Satake. Aux places où l'extension quadratique E/F est déployée, et où donc

$$G_{E_v} \simeq G_v^2$$

le changement de base de σ_v est simplement

$$\sigma_v \otimes \sigma_v$$

(Pour une étude plus détaillée on renvoie le lecteur à l'article [Minguez] de ce volume).

Soient $\sigma = \otimes\sigma_v$ une représentation automorphe irréductible pour G et $\pi = \otimes\pi_v$ une représentation automorphe irréductible pour G_E ; on dit que π est un changement de base faible de σ si, pour presque toute place finie v non ramifiée (pour les groupes et les représentations) π_v est le changement de base de σ_v (en tant que représentations sphériques).

La notion de changement de base fort pour les représentations automorphes, c'est-à-dire de changement de base défini localement au moyen du morphisme de changement de base pour le L-groupe, suppose définie la paramétrisation de Langlands locale et la notion de L-paquet. Cela sera étudié pour nos exemples plus loin (paragraphe 5.3).

Le changement de base, stricto sensu, entre $\mathbf{U}_n^*$ et $\mathbf{G}_n$ décrit ci-dessus est parfois appelé "changement de base stable" pour le distinguer de la variante dite "instable" qui est attachée à la donnée endoscopique admettant encore $H = \mathbf{U}^*$ comme groupe endoscopique mais pour le caractère κ non trivial, correspondant à $s = -1$ dans le groupe dual.

Supposons tout d'abord F local et considérons un caractère de

$$\mathbf{G}_1(F) = E^\times$$

obtenu de la façon suivante : soit $\mu_{E/F}$ le caractère de $F^\times$ associé à E/F par le corps de classe c'est-à-dire le caractère trivial sur les normes (au sens usuel : $N_{E/F}(x) = x\,\overline{x}$) mais non trivial. Soit μ est un prolongement quelconque de $\mu_{E/F}$ à $E^\times$. On obtient ainsi tous les caractères qui sont θ_1-stable mais ne sont pas un changement de base : il ne sont pas de la forme $\chi \circ N_\theta$.

Le choix du prolongement μ correspond au choix d'un facteur de transfert (cf. [Lab7]) pour le transfert endoscopique correspondant à $s = -1$ dans le groupe dual. Par composition avec le déterminant, μ définit un caractère, encore noté μ, de $\widetilde{\mathbf{G}}_n$. Si $\phi \in \mathcal{C}_c^\infty(\widetilde{G}(F))$ admet f comme transfert stable sur $\mathbf{U}$ alors $\mu^{-1}\phi$ admet f comme transfert instable sur $\mathbf{U}$. Dualement si π est le changement de base (ou transfert stable) de σ alors $\pi \otimes \mu$ est le transfert spectral de σ correspondant à $s = -1$ dans le groupe dual. Ce transfert est parfois appelé "changement de base instable". Cette terminologie est un peu dangereuse et devrait être évitée ; au surplus, contrairement au cas stable, ce transfert n'est pas canonique : il dépend du choix de μ. On prendra garde que les images de ces deux transferts ne sont pas nécessairement disjointes (cf. [HL]).

Ces constructions ont un analogue global évident.

4.11. Termes elliptiques et changement de base. Nous devons maintenant rappeler un résultat technique essentiel, appelé identité de changement de base :

THÉORÈME 4.12. Si ϕ_∞ est une fonction de Lefschetz, et si f est associée à ϕ (c'est-à-dire $f = \phi^{\mathbf{U}^*}$) on a

$$T_e^{\widetilde{\mathbf{G}}_n}(\phi) = ST_e^{\mathbf{U}^*}(f) .$$

Preuve: Ce résultat est déjà utilisé dans [CL1] et dans [HL]. Rappelons-en la preuve. Soit ϕ_∞ une fonction de Lefschetz ; on sait que ϕ_∞ est stabilisante [CL2]. Ceci signifie que ϕ_∞ est cuspidale et que les κ-intégrales orbitales $\mathcal{O}_\delta^\kappa(\phi_\infty)$ sont nulles pour tout δ semi-simple $\mathbb{R}$-elliptique sauf, peut-être, si $\kappa = 1$. De plus le lemme A.2.1 de [CL1] affirme que $\{\infty\}$ l'ensemble des places archimédiennes est un ensemble $(\mathbf{G}_n, \mathbf{U}^*)$ essentiel au sens de [Lab3]. Le corollaire de ces observations est que seul le groupe endoscopique principal $H = \mathbf{U}^*$ contribue à la stabilisation de la partie elliptique de la formule des traces tordue pour $\widetilde{\mathbf{G}}$. On a donc

$$T_e^{\widetilde{\mathbf{G}}}(\phi) = \iota(\widetilde{\mathbf{G}}, \mathbf{U}^*) ST_e^{\mathbf{U}^*}(f)$$

L'assertion

$$\iota(\widetilde{\mathbf{G}}, \mathbf{U}^*) = 1$$

résulte de la proposition 4.11 et est reprise de [CL1, A.3.1]. □

REMARQUE 4.13. En reprenant les arguments de [CL1, Lemme A.2.1] on voit que ce théorème est encore essentiellement vrai plus généralement si on remplace $\mathbf{U}^*$ par un groupe réductif G quasi-déployé sur F et $\widetilde{\mathbf{G}}$ par l'espace tordu associé au groupe obtenu à partir de G par extension puis restriction des scalaires pour E/F sous les hypothèses suivantes :

– le groupe dérivé de G est simplement connexe,

– le cocentre de G est un produit d'un tore déployé et d'un produit de tores $\mathbf{U}(1)$ associés à E/F.

Toutefois la présence d'un tore déployé introduirait une constante dans l'égalité.

5. Applications

Dans toute la suite, F est un corps totalement réel de degré $d \geq 2$, E/F est une extension quadratique totalement imaginaire et $\mathbf{U}$ un groupe unitaire forme intérieure du groupe quasi-déployé $\mathbf{U}^*$ relatif à E/F.

5.1. Le résultat principal.

THÉORÈME 5.1. Soit $f \in \mathcal{C}_c^\infty(\mathbf{U}(\mathbb{A}_F))$. On suppose que f_∞ est un pseudo-coefficient de série discrète. Alors, avec les notations du paragraphe 4.7, il existe pour chaque groupe endoscopique H attaché à $\mathbf{U}$ (ou plus correctement, pour chaque donnée endoscopique) une fonction f^H sur $H(\mathbb{A}_F)$ et une fonction $\widetilde{f}^H$ sur $\widetilde{M}^H(\mathbb{A}_F)$, qui en chaque place réelle est une fonction de Lefschetz (pour un système convenable de coefficients), admettant f^H comme transfert (pour le changement de base stable) et on a l'identité suivante :

$$T_{disc}^{\mathbf{U}}(f) = \sum \iota(\mathbf{U}, H)\, T_{disc}^{\widetilde{M}^H}(\widetilde{f}^H)\ .$$

Preuve: Étant donné $f_\infty \in \mathcal{C}_c^\infty(\mathbf{U}(F \otimes \mathbb{R}))$, on note f^H un transfert sur H. On peut choisir pour f_∞^H une combinaison linéaire de fonctions d'Euler-Poincaré. L'existence de $\widetilde{f}^H$ est établie dans la proposition 4.9. Le fait qu'aux places archimédiennes on puisse choisir pour $\widetilde{f}_\infty^H$ des fonctions de Lefschetz résulte de ce qu'on peut choisir une fonction d'Euler-Poincaré comme transfert f_∞^H du pseudo-coefficient de série discrète f_∞ ; il convient ensuite d'invoquer le lemme 4.4. Sous nos hypothèses simplificatrices on dispose de la formule pour le changement de base stable 4.12

$$T_e^{\widetilde{\mathbf{G}}}(\phi) = ST_e^{\mathbf{U}^*}(f^{\mathbf{U}^*})$$

avec $\phi = \widetilde{f}$ et donc plus généralement

$$T_e^{\widetilde{M}^H}(\widetilde{f}^H) = ST_e^H(f^H)\ .$$

On a d'autre part la formule de stabilisation 4.10

$$T_e^{\mathbf{U}}(f) = \sum \iota(\mathbf{U}, H)\, ST_e^H(f^H)$$

d'où on déduit l'identité :

$$T_e^{\mathbf{U}}(f) = \sum \iota(\mathbf{U}, H)\, T_e^{\widetilde{M}^H}(\widetilde{f}^H)$$

Supposons de plus pour l'instant que la fonction f_v est à support régulier en une place finie v ; grâce au corollaire 3.11 on obtient l'identité

$$T_{disc}^{\mathbf{U}}(f) = \sum \iota(\mathbf{U}, H)\, T_{disc}^{\widetilde{M}^H}(\widetilde{f}^H)$$

pour la famille de fonctions considérées. On choisit comme place v une place finie, non ramifiée pour F et déployée dans E. En une telle place on a

$$\mathbf{U}_v^* = GL(n, F_v)\ .$$

Pour f_v on considère une fonction élémentaire au sens de [Lab1] fortement régulière (f_v est définie par un élément d'un tore déployé de $GL(n)$, dont les valeurs propres ont des modules deux à deux distincts). Les arguments de [Lab1] montrent que l'égalité

$$T_{disc}^{\mathbf{U}}(f) = \sum \iota(\mathbf{U}, H)\, T_{disc}^{\widetilde{M}^H}(\widetilde{f}^H)$$

qui est vraie si f_v est élémentaire fortement régulière fournit la même identité mais avec pour f_v une fonction dans l'algèbre de Hecke sphérique. Il convient essentiellement d'observer que si on varie f_v, et donc aussi $\widetilde{f}_v^H$, dans l'une ou l'autre de ces familles de fonctions alors, compte tenu de la proposition 4.9, les traces $T_{disc}^{\mathbf{U}}(f)$ ainsi que $T_{disc}^{\widetilde{M}^H}(\widetilde{f}^H)$ s'expriment au moyen de sommes finies de caractères du tore maximal déployé de $GL(n, F_v)$. En variant v on obtient l'identité cherchée en général.

□

Remarque 5.2. L'hypothèse sur le degré de F n'intervient ici que pour pouvoir invoquer le corollaire 3.11. Elle n'est sans doute pas nécessaire pour 5.1 mais nous n'avons pas rédigé la preuve lorsque $F = \mathbb{Q}$.

Soit σ une représentation automorphe cuspidale pour $\mathbf{U}$ et telle que σ_∞ est une série discrète. Nous dirons que σ vérifie hypothèse $(*)$ si la propriété suivante est vraie :
$(*)$ *Soit f_∞ un pseudo-coefficient pour σ_∞. Les représentations σ' du spectre discret de* $\mathbf{U}$ *avec $\sigma'_f \simeq \sigma_f$ sont telles que les entiers* trace $\sigma'_\infty(f_\infty)$ *sont tous de même signe que* trace $\sigma_\infty(f_\infty)$

On observe que $(*)$ est vérifiée si le paramètre de σ_∞ est assez régulier pour que les seules représentations unitaires avec le même caractère infinitésimal sont les séries discrètes du L-paquet de σ_∞. En particulier $(*)$ est toujours vérifiée si $\mathbf{U}_\infty$ est compact. La propriété $(*)$ a été établie pour toutes les représentations à cohomologie de certain groupes unitaires anisotropes par Kottwitz [K4].

Corollaire 5.3. Soit σ une représentation du spectre discret pour $\mathbf{U}$ qui vérifie la propriété $(*)$. On suppose que σ_∞ est une représentation de la série discrète. Alors il existe une partition

$$n = n_1 + \cdots + n_r$$

et une collection π_i de représentation automorphes discrètes et θ_{n_i}-invariantes telles que

$$\pi = \pi_1 \boxplus \cdots \boxplus \pi_r$$

définisse un changement de base faible pour σ[2]. Aux places non ramifiées ou décomposées la correspondance $\sigma_v \mapsto \pi_v$ est donnée par le changement de base local. Aux places archimédiennes les π_v sont à cohomologie (pour le système de coefficients décrit en 4.4) et la correspondance $\sigma_v \mapsto \pi_v$ est compatible au changement de base de leurs caractères infinitésimaux. Si nous supposons de plus les π_i cuspidales alors π_∞ est le changement de base de σ_∞.

Preuve: On se place dans les hypothèses de 5.1. Soient σ et σ' deux représentations automorphes. On dira que σ' est faiblement équivalente à σ si $\sigma'_v \simeq \sigma_v$ pour presque toute place v. Les classes d'équivalence pour cette relation

[2]La notation $\pi_1 \boxplus \cdots \boxplus \pi_r$ est empruntée à [Langlands3].

seront appelées paquets automorphes. En séparant les chaînes de valeurs propres de Hecke dans l'identité de formules des traces donnée par 5.1 et compte tenu de la proposition 4.9, on sépare les paquets automorphes et obtient une identité raffinée où seules interviennent à gauche les représentations appartenant à un même paquet automorphe. On a choisi comme f_∞ un pseudo-coefficient de série discrète, donc seul le spectre discret contribue à la partie discrète de la trace (cf.[Ar3] p.268). Les représentations σ' du paquet automorphe défini par σ contribuent à la partie discrète de la formule des traces avec le coefficient $m(\sigma')$ qui est la multiplicité de σ' dans le spectre discret. C'est un entier positif. Sous la propriété $(*)$ les nombres les trace $\sigma'_\infty(f_\infty)$ sont tous de même signe. Il résulte de l'indépendance linéaire des caractères aux places finies que la distribution

$$f^\infty \mapsto \sum m(\sigma') \text{ trace } \sigma'(f_\infty \otimes f^\infty)$$

où la somme porte sur les σ' du paquet automorphe défini par σ, est non nulle. C'est dire que le paquet automorphe défini par σ contribue non trivialement à l'identité raffinée. Le second membre de l'identité raffinée est donc aussi non nul. On peut maintenant invoquer la description par Mœglin et Waldspurger du spectre discret pour $GL(n)$ et d'autre part la classification de Jacquet-Shalika des représentation automorphes pour $GL(n)$ pour conclure que seules des représentations induites pour un seul sous-groupe de Levi M interviennent à droite (bien entendu, sauf si $M = \mathbf{G}_n$, un tel sous-groupe de Levi intervient lui dans plusieurs groupes M^H). Par induction parabolique, de M^H à $\mathbf{G}_n$, après torsion de l'induisante par un caractère dépendant du choix du facteur de transfert entre $\mathbf{U}$ et H, on obtient des représentations sur $\mathbf{G}_n$ qui, compte tenu de la séparation de caractères de Hecke déjà effectuée, sont presque partout équivalentes et donc en fait équivalentes au vu de la rigidité pour les représentations cuspidales de $GL(n)$ (c'est-à-dire qu'une représentation automorphe cuspidale de $GL(n)$ est déterminée par sa classe d'équivalence presque partout) et de la description par Mœglin et Waldspurger du spectre discret pour $GL(n)$. On notera π cette représentation. C'est le changement de base faible de σ. La première assertion résulte alors du lemme 3.8. Les assertions sur la correspondance locale aux places décomposées ou non ramifiées résultent immédiatement des propriétés du transfert. Comme f_∞ est un pseudo-coefficient de la série discrète σ_∞ on a déjà observé que les fonctions f_∞^H peuvent être choisies comme des fonction d'Euler-Poincaré et les fonctions $\widetilde{f}_\infty^H$ sont des fonctions de Lefschetz pour un système convenable de coefficient pour chaque $\widetilde{M}^H$. Donc seules des représentations à cohomologie peuvent contribuer. Pour la dernière assertion on remarque que le transfert est compatible aux multiplicateurs d'Arthur et donc préserve le caractère infinitésimal (voir [Lab3, section 3.5]). Mais ceci ne permet pas de conclure que π_∞ est le changement de base de σ_∞ sauf si le paramètre est assez régulier. Toutefois, si nous nous limitons à des représentations de $\mathbf{G}_n$ induites de cuspidales ces représentations sont (conjecturalement tempérées et à tout le moins) génériques. Il en résulte que π_∞ est une représentation cohomologique (pour le système de coefficients convenable) tempérée ; c'est donc le changement de base de la série discrète σ_∞ car, d'après 4.7, il y a pour $\mathbf{G}_n(F_\infty)$ une seule représentation unitaire générique à cohomologie (pour le système de coefficients convenable).

□

5.2. Représentations à changement de base cuspidal. Nous supposons de plus dans ce paragraphe que E/F est une extension partout non ramifiée aux places finies, et que $\mathbf{U}$ est quasi-déployé en toute place finie inerte.

THÉORÈME 5.4. Soit π une représentation automorphe cuspidale de $\mathbf{G}_n$ qui est θ-invariante et cohomologique à l'infini. Alors π est un changement de base faible pour une représentation automorphe du spectre discret de $\mathbf{U}$ cohomologique à l'infini. Soit σ une telle représentation. On suppose que σ est cuspidale, que σ_∞ est une représentation de la série discrète et que σ_v est sphérique en toute place v finie inerte. Alors, au moins si σ vérifie la propriété $(*)$, la représentation σ apparaît avec multiplicité un.

Preuve: D'après 5.1 on a

$$T_{disc}^{\mathbf{U}}(f) = \sum \iota(\mathbf{U}^*, H)\, T_{disc}^{\widetilde{M}^H}(\widetilde{f}^H)$$

pour des fonctions choisies comme il convient à l'infini. En séparant les chaînes de valeurs propres de Hecke pour $\mathbf{G}_n$ on obtient une identité raffinée et on peut alors, compte tenu de la proposition 4.9, invoquer la classification de Jacquet-Shalika pour s'assurer que les contributions endoscopiques n'interfèrent pas avec les paquets dont le changement de base est cuspidal. Compte tenu de la rigidité pour $GL(n)$ et puisque par hypothèse π est cuspidale, cohomologique à l'infini et θ-invariante, π contribue non trivialement à la formule des traces pour $\widetilde{G}$ (pour la famille de fonctions utilisée dans 5.1) et on obtient en particulier l'identité

$$\sum_{\sigma'} m(\sigma') \text{ trace } \sigma'(f) = \text{ trace } \widetilde{\pi}(\phi)$$

où la somme porte sur les σ' intervenant, avec multiplicité $m(\sigma')$, dans le spectre discret pour $\mathbf{U}$ dont le changement de base faible est π. Le second membre étant non identiquement nul l'existence de représentations σ' en résulte. Supposons maintenant que σ, l'une de ces représentations, est cuspidale, que σ_∞ est une représentation de la série discrète et que σ_v est sphérique en toute place v finie inerte. Comme on a séparé les caractères de Hecke et que par hypothèse toutes les places finies inertes sont non ramifiées

σ'_f est uniquement déterminée par π et l'identité raffinée peut encore s'écrire

$$\sum_{\sigma'_\infty} m(\sigma'_\infty \otimes \sigma_f) \text{ trace } \sigma'_\infty(f_\infty) = \varepsilon \text{ trace } \widetilde{\pi}_\infty(\phi_\infty)$$

où ε est un signe dépendant de choix locaux. On choisit pour f_∞, au signe $(-1)^{q(\mathbf{U})}$ près, un pseudo-coefficient pour la série discrète σ_∞ de $\mathbf{U}(F \otimes \mathbb{R})$ de même caractère infinitésimal qu'une représentation de dimension finie V de sorte que f_∞ a les mêmes intégrales orbitales stables que

$$\frac{1}{b} f_V$$

On considère sur $\widetilde{\mathbf{G}}_\infty$ une fonction égale à une constante près à une fonction de Lefschetz associée à $W = V \otimes V^\theta$

$$\phi_\infty = \frac{1}{a} \phi_{\widetilde{W}}$$

(autrement dit ϕ_∞ est au signe près un pseudo-coefficient tordu pour $\widetilde{\pi}_W$). Compte tenu de 4.4 et 4.6 l'identité ci-dessus s'applique à ce couple de fonctions. Rappelons que l'on a choisi f_∞ telle que

$$\text{trace } \sigma_\infty(f_\infty) = (-1)^{q(\mathbf{U})}$$

et par hypothèse

$$m(\sigma_\infty \otimes \sigma_f)$$

est un entier non nul. Donc, compte tenu de l'hypothèse $(*)$ le premier membre

$$\sum_{\sigma'_\infty} m(\sigma'_\infty \otimes \sigma_f) \text{ trace } \sigma'_\infty(f_\infty)$$

est un entier non nul. Le second membre est alors aussi non nul et compte tenu de 4.7, on a nécessairement $\pi_\infty = \pi_W$. Le second membre est donc égal à ± 1. Il en résulte que seule σ_∞ peut contribuer au premier membre et que

$$m(\sigma_\infty \otimes \sigma_f) = 1$$

□

THÉORÈME 5.5. Soit σ^* une représentation automorphe cuspidale pour $\mathbf{U}^*$ admettant un changement de base faible π qui est cuspidal. On suppose que σ^*_∞ est une représentation de la série discrète et que σ^*_v est sphérique en toute place v finie inerte. Alors, si l'hypothèse $(*)$ est vérifiée pour $\mathbf{U}^*$ il existe une représentation cuspidale σ pour $\mathbf{U}$ telle que $\sigma_f = \sigma^*_f$ et σ_∞ a le même caractère infinitésimal que σ^*_∞. En particulier elle est cohomologique. Lorsque le caractère infinitésimal de σ^*_∞ est assez régulier alors on sait de plus que σ_∞ est une série discrète et toutes les séries discrètes du même L-paquet apparaissent ainsi.

Preuve: En reprenant la preuve de 5.4 on voit que après séparation des caractères de Hecke on a

$$m(\sigma^*) \text{ trace } \sigma^*(f^*) = \text{ trace } \widetilde{\pi}(\phi)$$

et que

$$\sum_{\sigma} m(\sigma) \text{ trace } \sigma(f) = \text{ trace } \widetilde{\pi}(\phi)$$

où la somme porte sur les σ intervenant, avec multiplicité $m(\sigma)$, dans le spectre cuspidal pour $\mathbf{U}$ dont le changement de base faible est π, lorsque f et ϕ ont pour transfert stable f^*, les trois fonctions étant choisies cuspidales aux places archimédiennes. En particulier $\sigma_f = \sigma_f^*$ et l'hypothèse $(*)$ pour $\mathbf{U}^*$ étant supposée vérifiée l'identité raffinée peut encore s'écrire compte tenu de 5.4

$$\sum_{\sigma_\infty} m(\sigma_\infty \otimes \sigma_f) \text{ trace } \sigma_\infty(f_\infty) = \text{ trace } \sigma_\infty^*(f_\infty^*) = (-1)^{q(\mathbf{U}^*)} .$$

Le premier membre est donc non nul ce qui implique l'existence d'une représentation σ pour $\mathbf{U}$ avec $\sigma_f \simeq \sigma_f^*$ et σ_∞ de même caractère infinitésimal que σ_∞^*. La dernière assertion du théorème s'obtient en faisant varier le pseudo-coefficient.

□

5.3. Changement de base fort. Dans cette section nous abandonnons le parti pris du reste du texte qui était de donner des démonstrations aussi élémentaires que possible, quitte à restreindre la généralité des résultats. Nous ferons ici appel à l'ensemble des résultats d'Arthur, qui culmine dans [A6], sur la stabilisation de la formule des traces ainsi qu'aux résultats récents de Colette Mœglin [Mœ] sur la définition et les propriétés des L-paquets pour les groupes unitaires quasi-déployés.

On notera ST_{disc}^H la partie discrète du développement spectral de la formule des traces stable. On notera $n(\sigma)$ la multiplicité moyenne (ou multiplicité stable) de σ dans le développement spectral de la formule des traces stable de sorte que

$$ST_{disc}^H(f) = \sum_{\sigma} n(\sigma) \text{ trace } \sigma(f)$$

La stabilisation de la formule des traces [A6] fournit l'identité :

Proposition 5.6.

$$T_{disc}^{\mathbf{U}}(f) = \sum_{H} \iota(\mathbf{U}, H) ST_{disc}^H(f^H)$$

On dispose alors d'un analogue stable de 3.11 :

Proposition 5.7. *Si F est un corps totalement réel de degré ≥ 2 avec ϕ comme dans la proposition 3.10 ci-dessus on a*

$$ST_e^{\mathbf{U}^*}(f) = ST_{disc}^{\mathbf{U}^*}(f) .$$

Compte tenu de 4.12 on en déduit le

THÉORÈME 5.8. Soit F un corps totalement réel de degré ≥ 2. Soit

$$\phi \in \mathcal{C}_c^\infty(\widetilde{\mathbf{G}}_n(\mathbb{A}_F))$$

où ϕ_∞ est une fonction de Lefschetz, admettant f comme transfert stable sur $\mathbf{U}$. On a

$$T_{disc}^{\widetilde{G}}(\phi) = ST_{disc}^{\mathbf{U}^*}(f)\ .$$

On observera que la conjonction de 5.6 et 5.8 redonne 5.1. Ici encore l'hypothèse sur le degré de F n'est sans doute pas utile, mais nous n'avons pas rédigé la preuve pour le cas $F = \mathbb{Q}$.

On dira qu'une représentation σ de $\mathbf{U}^*(\mathbb{A}_F)$ est stablement discrètement automorphe si σ contribue non trivialement à la partie discrète de la formule des traces stable pour $\mathbf{U}^*$. On prendra garde que de telles représentations ne sont pas nécessairement automorphes. Par séparation des valeurs propres des opérateurs de Hecke dans 5.8 et compte tenu de la rigidité pour $GL(n)$ on obtient une identité raffinée

$$\text{trace } \widetilde{\pi}(\phi) = \sum_\sigma n(\sigma) \text{ trace } \sigma(f)$$

où $n(\sigma)$ est un nombre rationnel et où la somme porte sur les représentations de $\mathbf{U}^*$ qui sont stablement discrètement automorphes et dont le changement de base faible est π. On en déduit des identités analogues en chaque place :

$$\text{trace } \widetilde{\pi}_v(\phi_v) = \sum_\sigma n(\widetilde{\pi}_v, \sigma_v) \text{ trace } \sigma_v(f_v)$$

Comme la correspondance $\phi_v \mapsto f_v$ est surjective l'indépendance linéaire des caractères montre que les coefficients $n(\widetilde{\pi}_v, \sigma_v)$ ne dépendent que de $\widetilde{\pi}_v$.

On suppose désormais π générique ou, ce qui est équivalent, induite parabolique irréductible à partir de représentations cuspidales. Soit v une place finie. Suivant [Mœ] on définit le L-paquet de représentations de $\mathbf{U}_v^*$ dont le changement de base est π_v comme l'ensemble des σ_v telles que $n(\widetilde{\pi}_v, \sigma_v) \neq 0$. On renvoie le lecteur à l'article de Colette Mœglin pour une étude fine de ces L-paquets. Elle montre en particulier que les L-paquets sont disjoints et compatibles au transfert endoscopique dans le cas "θ-discret". On passe au cas général en invoquant la compatibilité à l'induction parabolique. Il est probable qu'en fait $n(\widetilde{\pi}_v, \sigma_v)$ est un entier égal à 1 ou 0 si $\widetilde{\pi}_v$ est générique et normalisé par le modèle de Whittaker mais ceci n'est pas montré dans l'article.

On a vu en 5.3 que si σ est une représentation du spectre discret pour $\mathbf{U}^*$ qui vérifie la propriété $(*)$ et telle que σ_∞ est une représentation de la série discrète, alors il existe une partition

$$n = n_1 + \cdots + n_r$$

et une collection π_i de représentation automorphes discrètes et θ_{n_i}-invariantes telles que

$$\pi_1 \boxplus \cdots \boxplus \pi_r$$

définisse un changement de base faible pour σ.

THÉORÈME 5.9. Soit σ et π comme dans 5.3. Supposons de plus π cuspidale (en particulier $r = 1$). Alors, en chaque place finie v où $\mathbf{U}_v$ est quasi-déployé, σ_v appartient au L-paquet dont le changement de base est π_v. Aux places archimédiennes π_∞ est le changement de base de σ_∞ et π_∞ est cohomologique pour le système de coefficients décrit en 4.4.

Preuve: On sépare les caractères de Hecke dans 5.6 et 5.8 et on procède comme dans 5.3. Comme nous supposons π cuspidale seul $M^H = \mathbf{G}_n$ intervient et on obtient une identité raffinée où seuls les assertions aux places finies sont alors une conséquence immédiates de la définition des L-paquets. Par contre, aux places archimédiennes les identités de traces ne sont obtenues que pour les fonctions de Lefschetz. On sait donc déjà que seules des représentations cohomologiques (pour le système de coefficients décrit en 4.4) interviennent. Comme nous nous limitons à des représentations induites de cuspidales de $GL(n)$ ces représentations sont (conjecturalement tempérées et à tout le moins) génériques. Il en résulte que π_∞ est une représentation cohomologique (pour le système de coefficients convenable) tempérée ; c'est donc le changement de base de la série discrète σ_∞ car, d'après 4.7, il y a pour $\mathbf{G}_n(F_\infty)$ une seule représentation unitaire générique à cohomologie (pour le système de coefficients convenable).

□

Références

[Ar1] J. Arthur, On a family of distributions obtained from Eisenstein series II : Explicit formulas, *Amer. J. Math.* **104** (1988), pp. 1289–1336

[Ar2] J. Arthur, The invariant trace formula II, *J. Am. Math. Soc.* **1** (1988), pp. 501–554

[Ar3] J. Arthur, The L^2 - Lefschetz numbers of Hecke operators, *Invent. Math.* **97** (1989), no. 2, pp. 257–290

[Ar4] J. Arthur, Intertwining Operators and Residues I. Weighted Characters, *J. Funct. Analysis.* **84** (1989), pp. 19–84

[Ar5] J. Arthur, An introduction to the trace formula, *Harmonic analysis, the trace formula, and Shimura varieties, Clay Math. Proc.* **4** Amer. Math. Soc., Providence, RI (2005), pp. 1–263

[A6] J. Arthur, A stable trace formula. III. Proof of the main theorems, *Ann. of Math.* **158** (2003), pp. 769–873

[BLS] A. Borel J.-P. Labesse J. Schwermer, On the cuspidal cohomology of S-arithmetic groups of reductive groups over number fields, *Compositio Math.* **102** (1996), pp. 1–40

[BW] A. Borel N. Wallach, *Continuous Cohomology, Discrete Subgroups, and Representations of Reductive Groups*, *Annals of Math. Studies* **94** Princeton Univ. Press (1980)

[C1] L. Clozel, Motifs et formes automorphes, Automorphic forms, Shimura varieties and L-functions, *Perspectives in Math.* **10** (1990), pp. 77–159

[C2] L. Clozel, The fundamental lemma for stable base change, *Duke Math. J.* **61** (1990), pp. 255–302.

[C3] L. Clozel, Représentations galoisiennes associées aux représentations automorphes autoduales de GL(n), *Inst. Hautes âtudes Sci. Publ. Math.* **73** (1991), pp. 97–145

[C4] L. Clozel , Appendice : An erratum, *Ce volume* (2009)

[CL1] L. Clozel J.-P. Labesse, Changement de base pour les représentations cohomologiques de certains groupes unitaires, *Astérisque* **257** (1999), pp. 121–136

[CL2] L. Clozel J.-P. Labesse, Orbital integrals and distributions, *Preprint* (2008)

[CHL] L. Clozel M. Harris J.-P. Labesse, Endoscopic transfer, *Ce volume* (2009)

[DM] P. Delorme P. Mezo, A twisted invariant Paley-Wiener theorem for real reductive groups, *Duke Math. J.* **144** (2008), pp. 341–380

[FLM] T. Finis E. Lapid W. Müller, The spectral side of Arthur's trace formula, *Proc. Natl. Acad. Sci. USA* **106** (2009), no. 37, pp. 15563–15566

[Hal] Thomas Hales, On the fundamental lemma for standard endoscopy : reduction to unit elements., *Canad. J. Math.* **47** (1995), pp. 974–994

[HL] M. Harris J.-P. Labesse, Conditional base change for unitary groups, *Asian J. Math.* **8** (2004), pp. 653–684

[K1] R. Kottwitz, Rational conjugacy classes in reductive groups, *Duke Math. J.* **49** (1982), pp. 785–806

[K2] R. Kottwitz, Stable trace formula : elliptic singular terms, *Math. Ann.* **275** (1986), pp. 365–399

[K3] R. Kottwitz, Base change for unit elements of Hecke algebras, *Compositio Math.* (1986), pp. 237–250

[K4] R. Kottwitz, On the λ-adic representations associated to some simple Shimura varieties, *Invent. Math.* **108** (1992), pp. 653–665

[KS] R. Kottwitz D. Shelstad, Foundations of twisted endoscopy, *Asterisque* **255** (1999)

[Lab1] J.-P. Labesse, Le lemme fondamental pour le changement de base stable, *Duke Math. J.* **61** (1990), pp. 519–530

[Lab2] J.-P. Labesse, Pseudo-coefficients très cuspidaux et K-théorie, *Math. Ann.* **291** (1991), pp. 607–616

[Lab3] J.-P. Labesse, Cohomologie, stabilisation et changement de base, *Asterisque* **257** (1999), pp. 1-116

[Lab4] J.-P. Labesse, Nombres de Tamagawa des groupes réductifs quasi-connexes, *Manuscripta Math.* **104** (2001) pp. 407–430

[Lab5] J.-P. Labesse, Stable Twisted Trace Formula : Elliptic Terms, *Jour. Inst. Math. Jussieu* **3** (2004), pp. 473–530

[Lab6] J.-P. Labesse, Introduction to Endoscopy, *Ce volume.*

[Lab7] J.-P. Labesse, Les facteurs de transfert pour les groupes unitaires, *Ce volume* (2009)

[Langlands1] R.P. Langlands, Eisenstein Series (AMS Conference, Boulder 1965), *Proc. Sympos. Pure Math.* **9** (1966), pp. 235–252

[Langlands2] R.P. Langlands, On the functional equation satisfied by Eisenstein Series, *Lecture Notes in Math.* **544** Springer (1976)

[Langlands3] R.P. Langlands, Automorphic representations, Shimura varieties and motives. Ein Märchen. (AMS Conference, Corvallis 1977), *Proc. Sympos. Pure Math.* **33** Part 2, (1979), pp. 205–246

[LN] G. Laumon B.C. Ngô, Le lemme fondamental pour les groupes unitaires, *Ann. of Math.* **168** (2008), pp. 477–573

[Minguez] E. Minguez, Unramified representations of unitary groups, *Ce volume* (2009)

[Mœ] C. Mœglin, Classification et changement de base pour les séries discrètes des groupes unitaires p-adiques, *Pacific J. Math.* **233** (2007), no. 1, pp. 159–204

[MW] C. Mœglin J.-L. Waldspurger, Le spectre résiduel de GL(n), *Ann. Sci. ENS* **22** (1989), pp. 605–674

[Mü] W. Müller, The trace class conjecture in the theory of automorphic forms. II, *Geom. Funct. Anal.* **8** (1998), pp. 315–355

[Rod] F. Rodier, Whittaker models for admissible representations of reductive p-adic groups (AMS Conference, Williamstown 1972), *Proc. Symp. Pure Math.* **26** (1973), pp. 425–430

[Rog] J. Rogawski, Automorphic Representations of Unitary Groups in Three Variables, *Annals of Math. Studies* **123** (1990).

[Shah] F. Shahidi, On certain L-functions, *Amer. J. Math.* **103** (1981), pp. 297–355

[Shal] J. Shalika, The multiplicity one theorem for GL_n, *Annals of Math.* **100** (1974), pp. 171–193

[W1] J.-L. Waldspurger, Le lemme fondamental implique le transfert, *Compositio Math.* **105** (1997), pp. 153–236

[W2] J.-L. Waldspurger, Endoscopie et changement de caractéristique, *J. Inst. Math. Jussieu* **5** (2006), pp. 423–525

[W3] J.-L. Waldspurger, L'endoscopie tordue n'est pas si tordue, *Mem. Amer. Math. Soc.* **194** (2008)

Institut Mathématique de Luminy, UMR 6206 France

Appendice : Un erratum

Laurent Clozel

Un argument-clé de la démonstration du Lemme fondamental pour le changement de base stable [**3**] , repris dans [**6**], repose sur une assertion erronnée. Nous donnons l'argument correct ; le reste de la démonstration de [**3**], ou de celle de [**6**] , reprises depuis par plusieurs auteurs dans d'autres contextes, est inchangé.

Soit G un groupe semi-simple non ramifié sur un corps p-adique F ; on suppose G de type adjoint et quasi-simple. Alors G provient par extension des scalaires d'un schéma en groupes semi-simple $\mathcal{G}$, lisse, sur $\mathcal{O}_F$, dont la fibre spéciale est semi-simple ([**8**], p.55). En particulier, la fibre spéciale est connexe. Le sous-groupe compact $K = \mathcal{G}(\mathcal{O}_F)$ est alors maximal; un tel sous-groupe maximal est dit hyperspécial. Un sous-groupe hyperspécial est le stabilisateur d'un point hyperspécial de l'immeuble de G, cf. [**8**] , p. 35; cette définition s'applique à tout groupe réductif et connexe sur F.

Si $P = MN$ est un sous-groupe parabolique de G, défini sur F, il est conjugué à un tel sous-groupe qui provient par extension des scalaires d'un parabolique $\mathcal{P} = \mathcal{M}\mathcal{N}$ défini sur $\mathcal{O}$. Alors $\mathcal{M}$ est réductif et $\mathcal{M}(\mathcal{O}) \subset M(F)$ est hyperspécial. Ceci justifie les arguments de récurrence ("descente parabolique") de [**3**] , par exemple au début du §3 (p. 268), ou p. 270 (b).

La démonstration du (crucial) Lemme 3.5 de [**3**], dont nous ne reproduisons pas l'énoncé, reposait sur le résultat suivant. Soit $B \subset G$ un sous-groupe de Borel défini sur F, $B = TN$, soit χ un caractère non ramifié de T et considérons l'induite normalisée

$$I(\chi) = \mathrm{ind}_{TN}^{G}(\chi \otimes 1)$$

où l'induction est unitaire. (On écrit G pour $G(F)$ quand cela ne cause pas de confusion).

THÉORÈME 1. Supposons G adjoint. Si χ est unitaire, $I(\chi)$ est irréductible.

Avant de donner la démonstration, notons que ce résultat est naturel d'après la description de Langlands pour les R-groupes. Cf. Arthur [**1**], où ceci est étendu aux données non tempérées. Soit en effet ${}^LG$ le groupe dual :

$$^LG = \widehat{G} \rtimes \Gamma$$

où $\widehat{G}$ est simplement connexe, muni des données usuelles $\widehat{B} \supset \widehat{T}$ et où $\Gamma = Gal(\overline{F}/F)$ opère en fait par un quotient fini C du groupe de Galois non ramifié, de générateur γ. La représentation $I(\chi)$ est associée à un homomorphisme $\phi : W_F \to {}^LG$ compatible avec $W_F \to C$ et non ramifié, donc à un élément semi-simple (s, γ) de $\widehat{G} \rtimes C$. D'après Langlands, le cardinal du R-groupe, i.e., le nombre de composantes de $I(\chi)$, est égal au cardinal de S/S^0, S étant le centralisateur de (s, γ) dans $\widehat{G}$ et S^0 sa composante neutre. La conjugaison par (s, γ) étant un automorphisme semi-simple, et $\widehat{G}$ étant simplement connexe, ce centralisateur est en fait connexe d'après un théorème de Steinberg ([**7**], Theorem 8.1) et la représentation doit être irréductible.

Nous allons vérifier que le Théorème 1 est conséquence directe de résultats de D. Keys [**4**] , [**5**].

Soit G_{sc} le revêtement simplement connexe de G et $\phi : G_{sc} \to G$ l'isogénie naturelle. De nouveau, G_{sc} a une structure entière naturelle, compatible avec celle de G ; soit K_{sc} le sous-groupe compact de G_{sc} associé. La décomposition de Bruhat montre que $G = \phi(G_{sc})B$; la restriction de $I(\chi)$ à G_{sc} est donc l'induite $I(\chi_{sc})$ où χ_{sc}, un caractère de l'image inverse T_{sc} de T, est obtenu par composition. La démonstration de Keys repose sur l'étude de la réductibilité des induites $I(\chi_{sc})$.

Avant de poursuivre, évitons une terminologie certes classique, mais qui peut prêter à confusion. Bruhat et Tits, dans [**2**] et les articles antérieurs, définissent les "bons sous-groupes compacts maximaux" de G (un groupe réductif sur F). Ce sont ceux pour lesquels la décomposition d'Iwasawa est vérifiée ([**2**], (4.4.1)). Nous ne connaîtrons que des sous-groupes (compacts maximaux) *spéciaux* et *hyperspéciaux*. On renvoie à [**8**], [**2**] pour la définition des sous-groupes spéciaux. Nous nous autorisons une dernière fois la terminologie antérieure :

PROPOSITION 2. (Bruhat-Tits). Dans G_{sc}, un sous-groupe compact maximal est bon si et seulement si il est spécial.

C'est la Proposition 4.4.6 de [**2**] , l'homomorphisme de revêtement pour G_{sc} étant l'identité. (Le fait que $id : G_{sc} \to G_{sc}$ est de type connexe au sens de [**2**], (4.1.3) résulte de [**8**], 1.13).

Cela étant, le résultat principal se déduit dans la plupart des cas du théorème suivant, effectivement démontré par Keys [**4**] :

THÉORÈME 3 (Keys). (G_{sc} non ramifié, simplement connexe, χ_{sc} non ramifié unitaire). Toute composante irréductible de $I(\chi_{sc})$ a un vecteur fixe par un sous-groupe compact maximal spécial K de G_{sc}. De plus ce vecteur (et donc la sous-représentation associée) est unique.

L'unicité, qui n'est pas explicite dans [**4**], résulte aussitôt du fait que $G_{sc} = KB_{sc}$ d'après la Proposition 2. L'erreur de [**3**] - voir la preuve du Lemme 3.2, partie (a) - consiste à attribuer à Keys le même énoncé, en remplaçant "spécial" par "hyperspécial".

Revenons à la démonstration du Théorème 1. Il résulte du Théorème 2 si le groupe adjoint G permute transitivement les sous-groupes spéciaux de G_{sc}. C'est vrai si G est déployé ([**8**], p.47) ; dans tous les cas, si G est non ramifié, les sous-groupes hyperspéciaux sont permutés transitivement (loc. cit.). De plus, il y a un critère combinatoire simple pour savoir si deux compacts spéciaux K et K' sont permutés par G : les sommets associés s, s' du diagramme de Dynkin relatif local Δ de G_{sc} ([**8**], 2.4) doivent être permutés par un automorphisme de Δ préservant les données de dimension relatives à s, s'. Cf. [**8**], p.47.

Considérons donc les groupes quasi-déployés, non déployés et non ramifiés de type A_n, D_n et E_6. Nous utilisons évidemment les tables de Tits ([**8**], p.60-66). Dans le cas de A_n (n impair) , donc des groupes unitaires de rang pair, les sommets spéciaux sont hyperspéciaux (${}^2A'_{2m-1}$ p.62). De même pour D_n (${}^2D_n$ p. 65) et E_6 (${}^2E_6$ p. 66) ainsi que D_4 trialitaire (${}^3D_4$ p. 65).

Il nous reste à considérer le cas de A_{2n}, donc des groupes spéciaux unitaires non ramifiés $SU(2n+1)$. Dans ce cas G_{sc} contient deux sous-groupes compacts maximaux, l'un spécial et l'autre hyperspécial, évidemment non conjugués (${}^2A'_{2n}$ p. 63). Pour ces seuls groupes - mais ils ont quelque intérêt pour les démonstrations de ce volume - l'argument de [**3**] ne conclut pas.

Dans ce cas, Keys a déterminé explicitement les R-groupes des séries principales unitaires. Ecrivons désormais G pour le groupe spécial unitaire, donc simplement connexe. Le tore T (dans G, évidemment) s'identifie à $(E^\times)^n$, E/F étant l'extension non ramifiée définissant le groupe unitaire. Une série principale est donnée par n caractères de $E^\times$. On les suppose unitaires et non ramifiés. On note χ cette donnée. Soit $I(\chi)$ la représentation induite. Le résultat suivant complète notre démonstration :

Théorème 4 (Keys). *Si les χ_i sont unitaires et non ramifiés, $I(\chi)$ est irréductible.*

C'est le théorème du §8 de [**5**] ; l'énoncé donné par Keys est malheureusement imprécis. Pour être complet nous esquissons donc sa démonstration dans le cas qui nous intéresse. (Nous renvoyons le lecteur à son article pour les détails et les définitions relatives aux R-groupes).

Démonstration : Le groupe de Weyl W de (G, T) s'identifie à $\mathfrak{S}_n \ltimes \pm 1^n$, $\mathfrak{S}_n$ opérant par permutations et le générateur de coordonnée i de $S := (\pm 1)^n$ envoyant z_i sur z_i^{-c}, où on note c l'automorphisme galoisien.

Le système de racines restreint de (G, T) s'identifie au système donné, dans $\mathbf{R}^n$, par

$$\Delta = \{\pm e_i \pm e_j (i \neq j), \pm e_i, \pm 2e_i\} \ ,$$

les racines positives étant les $e_i - e_j (i < j), e_i, 2e_i$. Soit W_χ le stabilisateur de χ dans W. Soit $\Delta' = \{\alpha \in \Delta : s_\alpha \in W_\chi \ et\ \mu_\alpha(\chi, 0) = 0\}$ (cf. [**5**], p.20). Le nombre de composantes de $I(\chi)$ est le cardinal du groupe

$$R = \{w \in W_\chi : w\alpha > 0 \text{ si } \alpha \in \Delta' \text{ et } \alpha > 0\}.$$

Les calculs de mesures de Plancherel en rang 1 ([**5**], p.127) donnent :

(A1) $\alpha = 2e_i \in \Delta' \Leftrightarrow \chi_i \chi_i^c = 1$
(A2) $\alpha = e_i - e_j \in \Delta' \Leftrightarrow \chi_i \chi_j^{-1} = 1$
(A3) $\alpha = e_i + e_j \in \Delta' \Leftrightarrow \chi_i \chi_j^c = 1$,

l'équivalence (A1) n'étant vraie que dans le cas non ramifié (loc. cit.). De plus ([**5**], Lemma 1 p.128), Keys montre que $R = R \cap S$. On peut supposer que $\chi_i = \chi_i^{-c}$ pour les indices i allant de 1 à r exactement. Noter que dans notre situation χ_i^c est en fait égal à χ_i. Alors $S_\chi := S \cap W_\chi$ est égal à $(\pm 1)^r$ opérant sur les r premières coordonnées. Puisque d'après (A1) $2e_i \in \Delta'$ pour $i \leq r$, on voit que $R \cap S = R \cap S_\chi = \{1\}$. □

Références

[1] J. Arthur, A note on L-packets, *Pure and Appl. Math. Quarterly,* **2** (2006), 199–217.

[2] F. Bruhat & J. Tits, Groupes réductifs sur un corps local I : données radicielles valuées, *Publ. Math. IHES* **41** (1971), 5–252.

[3] L. Clozel, The fundamental lemma for stable base change, *Duke Math. J.* **61** (1990), 255–302.

[4] D. Keys, Reducibility of unramified unitary principal series representations of p-adic groups and class-1 representations, *Math. Ann.* **260** (1982), 397–402.

[5] D. Keys, Principal series representations of special unitary groups over local fields, *Compositio Math.* **51** (1984), 115–130.

[6] J.-P. Labesse, Le lemme fondamental pour le changement de base stable, *Duke Math. J.* **61**, 1990, 519–530.

[7] R. Steinberg, Endomorphisms of linear algebraic groups, *Mem. Amer. Math. Soc.* **80**, 1986.

[8] J. Tits, Reductive groups over local fields, in Borel, Casselman eds., Automorphic forms, representations and L-functions, *Proc. Symp. Pure Math.* **33**, Part I, AMS, 1977, 29–70.

Endoscopic transfer
[CHL.IV.B]

Laurent Clozel, Michael Harris, and Jean-Pierre Labesse

Introduction

The goal of this chapter is to generalize a result of Blasius and Rogawski [BR] on the transfer of cohomological tempered automorphic representations of $H' = U(2)$ to $G = U(3)$, when H' is replaced by $U(n)$ for any even $n = 2m$ and G is a specific inner form of $U(2m+1)$. Endoscopic transfer properly requires an automorphic representation, or rather an L-packet, of an endoscopic group of the form $H = U(a) \times U(b)$, with $a+b = n+1$; so here H should be $U(n) \times U(1)$. However, the representations of the factors $U(n)$ and $U(1)$ play different roles here. In subsequent books we will want to attach compatible families of ℓ-adic representations to cohomological automorphic representations of $GL(n)$ that descend to H'. The additional character χ of $U(1)$ serves as a factor of adjustment to guarantee that the resulting L-packet of G corresponds to an n-dimensional piece of the ℓ-adic cohomology of the corresponding Shimura variety. This is not always possible, and when $n = 1$ [BR] determines exactly when the character χ can be chosen to obtain an L-packet of G of the desired form. This is the result we generalize. We only treat situations in which the ramification at finite primes is as simple as possible, which allows us to concentrate on the complications at archimedean primes. Our calculation is almost an immediate consequence of the results of [C.III.A] and [L.IV.A].

We thank G. Chenevier for a careful reading that has allowed us to correct a number of omissions and misprints and for his request that we include Theorem 4.7, whose proof he was thoughtful enough to provide. We also thank Paul-James White for several helpful remarks, and especially for pointing out that the earlier versions of the regularity hypothesis 1.3 was not strong enough to apply the results of [L.IV.A]. We are grateful to F. Shahidi for suggesting improvements in the presentation of the calculations in §4.4. Finally, we thank the anonymous referee for a comprehensive reading that has considerably improved the clarity of the text.

1. Conventions for unitary groups

Let F be a totally real field, $\mathcal{K}/F$ a totally imaginary quadratic extension, $\eta = \eta_{\mathcal{K}/F}$ the corresponding quadratic character of $\mathbf{A}_F^\times/F^\times$ (and equivalently of $Gal(\bar{F}/F)$, $d = [F : \mathbb{Q}]$, $c \in Gal(\mathcal{K}/F)$ the non-trivial Galois automorphism. Let n be a positive integer and $\mathcal{G} = \mathcal{G}_n$ be the algebraic group $R_{\mathcal{K}/\mathbb{Q}}GL(n)_{\mathcal{K}}$. Let $\mathfrak{g} = Lie(\mathcal{G}(\mathbb{R}))$, $K_\infty \subset \mathcal{G}(\mathbb{R})$ a maximal compact subgroup. We consider cuspidal automorphic representations Π of $\mathcal{G}$ satisfying the following two hypotheses:

Hypotheses 1.1. Writing $\Pi = \Pi_\infty \otimes \Pi_f$, where Π_∞ is an admissible $(\mathfrak{g}, K_\infty)$-module, we have

(i) (Regularity) There is a finite-dimensional complex algebraic irreducible representation $W(\Pi) = W_\infty$ of $\mathcal{G}(\mathbb{R})$ such that

$$H^*(\mathfrak{g}, K_\infty; \Pi_\infty \otimes W_\infty) \neq 0.$$

(ii) (Polarization) The contragredient $\Pi^\vee$ of Π satisfies

$$\Pi^\vee \xrightarrow{\sim} \Pi \circ c.$$

Remark. These are natural generalizations of Hecke characters of type A_0, inasmuch as present methods do not allow us to attach Galois representations to more general *algebraic* representations, in the sense defined by Clozel in [C90]. The polarization condition is unnecessarily restrictive; the more general condition would be

$$\Pi^\vee \xrightarrow{\sim} \Pi \circ c \otimes \xi$$

where ξ is a Hecke character that factors through the norm to the idèles of F. Two questions arise:

(1) Can the theory developed below be extended to this more general setting?
(2) Can the authors or their friends find a convincing name (rather than acronym such as RP for Regular Polarized) for this class of automorphic representations?

We are hopeful at least with regard to (1).

The following hypotheses will remain in force throughout this chapter.

Simplifying Hypotheses 1.2.

(1.2.0) The degree n is even.
(1.2.1) $\mathcal{K}/F$ is unramified at all finite places (in particular $d > 1$).
(1.2.2) Π_v is spherical (unramified) at all non-split non-archimedean places v of $\mathcal{K}$.
(1.2.3) The degree $d = [F : \mathbb{Q}]$ is even.

In Book 2, we will associate a compatible system of Galois representations, occurring in the cohomology of Shimura varieties, to representations

Π satisfying these hypotheses and (when n is even) those satisfying Hypotheses 1.3 below. If n is odd the Shimura varieties are attached to unitary groups of vector spaces of dimension n. When n is even, and only then, this is not always possible, and we must resort to endoscopy. Thus we assume (1.2.0). However, all the results of this chapter remain true for odd n, when appropriately modified, but the appropriate modifications require the introduction of additional notation. The remaining hypotheses allow us to simplify the analysis of the stable trace formula, and in particular to sidestep the problem of classification of L-packets for p-adic unitary groups, important for the classification of automorphic representations but not essential for the construction of Galois representations.

The irreducible representation $W(\Pi)$ factors over the set Σ_F of real embeddings of F

$$W(\Pi) = \otimes_{v\in\Sigma_F} W_v,$$

where W_v is an irreducible representation of

$$GL(n, \mathcal{K} \otimes_{F,v} \mathbb{C}) \xrightarrow{\sim} GL(n, \mathbb{C}) \times GL(n, \mathbb{C})$$

each factor associated to a prime (say $\tilde{v}, \tilde{v}^c$) of $\mathcal{K}$ extending v. Each factor is parametrized by its highest weight, denoted

$$\mu(\tilde{v}) = (\mu_1(\tilde{v}) \geq \mu_2(\tilde{v}) \geq \cdots \geq \mu_n(\tilde{v}))$$

and $\mu(\tilde{v}^c)$. Moreover, the polarization condition implies that the two factors are dual, or equivalently that

$$\mu_i(\tilde{v}^c) = -\mu_{n+1-i}(\tilde{v}). \tag{1.2.4}$$

STRONG REGULARITY HYPOTHESIS 1.3. For every $v \in \Sigma_F$, the highest weight $\mu(v)$ is sufficiently regular, i.e. $|\mu_i(\tilde{v}) - \mu_j(\tilde{v})| > 3$ if $i \neq j$.

REMARK. It should be enough to assume a weaker version of 1.3 – $|\mu_i(\tilde{v}) - \mu_j(\tilde{v})| > 0$—for a single $v \in \Sigma_F$, but this probably requires results of Book 2. The present formulation allows us to make use of the multiplicity one results in the form given in [L.IV.A], which do not make use of the full strength of the stable trace formula.

1.1. Existence of hermitian spaces with fixed local invariants. The following lemmas are special cases of the well known classification of hermitian spaces.

LEMMA 1.4.1. There exists a hermitian space $V_1/\mathcal{K}$ of dimension $n+1$ relative to the extension $\mathcal{K}/F$ such that, letting $G = U(V_1)$,

(1.4.1.1) For all finite places v of F, $G(F_v)$ is quasi-split and splits over an unramified extension of F_v; in particular, $G(F_v)$ contains a hyperspecial maximal compact subgroup.

(1.4.1.2) For all real places v of F , with the exception of one place v_0, $G(F_v)$ is compact; $G(F_{v_0}) \xrightarrow{\sim} U(1, n)$.

LEMMA 1.4.2. There exists an n-dimensional hermitian space $V_0/\mathcal{K}$ relative to the extension $\mathcal{K}/F$ such that the unitary group $G_0 = U(V_0)$ satisfies

(1.4.2.1) For all finite places v, $G_0(F_v)$ is quasi-split and splits over an unramified extension of F_v; in particular, $G_0(F_v)$ contains a hyperspecial maximal compact subgroup.

(1.4.2.2) For all real places v, $G_0(F_v)$ is compact.

Moreover, G_0 is unique up to isomorphism.

LEMMA 1.4.3. Under Hypotheses 1.2 there exists an n-dimensional hermitian space $V_2/\mathcal{K}$ relative to the extension $\mathcal{K}/F$ such that, letting $G_2 = U(V_2)$,

(1.4.3.1) For all finite places v of F, $G_2(F_v)$ is quasi-split and splits over an unramified extension of F_v; in particular, $G_2(F_v)$ contains a hyperspecial maximal compact subgroup.

(1.4.3.2) For all real places v, with the exception of one place v_0, $G_2(F_v)$ is compact; $G_2(F_{v_0}) \xrightarrow{\sim} U(2, n-2)$.

Sketch of proofs. In dimension $n+1$, which is odd, it is known that there is no Hasse obstruction to the existence of a unitary group with prescribed local factors.

In Lemmas 1.4.2 and 1.4.3, the hermitian spaces are of even dimension. Kottwitz' explicit description of the Hasse obstruction in the even-dimensional case (cf. [H.I.A], Prop. 4.A.3) implies the existence of the groups G_0 and G_2.

We let U^* denote the quasi-split inner form of G_0 (or G_2). In Book II G is denoted G_1.

1.5 Vanishing of the obstruction to stable descent. The central character ξ_Π of Π is a Hecke character of $\mathcal{K}$ satisfying the analogues of hypotheses 1.1(i) and (ii), and in particular

$$\xi_\Pi^{-1} = \xi_\Pi \circ c.$$

It follows that the restriction of ξ_Π to the idèles of F is a Hecke character trivial on norms from $\mathcal{K}$, in other words

$$\xi_\Pi \mid_{\mathbf{A}_F^\times} = \eta^u, u = 0, 1. \tag{1.5.1}$$

Moreover, ξ_Π is obtained by base change from a Hecke character of $U(1)$ if and only if $u = 0$. This global condition can be checked at any place of F that does not split in $\mathcal{K}$. In particular, it suffices to check the condition at an archimedean place, and it follows from (1.2.4) that $u = 0$. Thus ξ_Π is locally a base change everywhere, in other words:

LEMMA 1.5.2. Let v be an inert place of F, w the place of $\mathcal{K}$ dividing v. Then there is a character β_v of $U(1)_v = \ker N_{\mathcal{K}_w/F_v} : \mathcal{K}_w^\times \to F_v^\times$ such that, for all $z \in \mathcal{K}^\times$

$$\xi_{\Pi,w}(z) = \beta_v(z/c(z)).$$

2. Review of stable descent

For any finite prime v of F, let U_v^* denote a quasisplit unitary group over v, relative to the extension $\mathcal{K}/F$ (a split étale extension if v is a split prime). Write $\Pi_v := \bigotimes_{w|v} \Pi_w$, where w runs over primes of $\mathcal{K}$. We can construct a formal descent π_v of Π_v as follows:

2.1. v **split.** Then

$$GL(n, \mathcal{K}_v) \xrightarrow{\sim} GL(n, F_v) \times GL(n, F_v)$$

and the polarization condition implies for the local components of Π:

$$\Pi_v = \Pi_1 \otimes \Pi_2$$

with $\Pi_2 \xrightarrow{\sim} \Pi_1 \circ {}^t g^{-1}$. The group $U_v^*(\xrightarrow{\sim} GL(n, F_v))$ embeds (up to conjugacy) via

$$g \mapsto (g, {}^t g^{-1})$$

We set $\pi_v = \Pi_1$, which is well-defined.

2.2. v **inert.** Then Π_w is unramified by (1.2.2) and invariant under the outer automorphism corresponding to descent to U_v^*. In the notation of [M.III.C], this implies that Π_w is in the image of either the base change map $\widetilde{BC}$ or the twisted base change $\widetilde{BC}'$. Bearing in mind that n is *even*, one verifies that Π_w is in the image of $\widetilde{BC}$. Indeed, Theorem 4.1 of [M.III.C] shows that $\widetilde{BC}'$ and $\widetilde{BC}$ coincide in this case. We let π_v be the spherical representation of U_v^* such that $\widetilde{BC}(\pi_v) = \Pi_w$. The existence and properties of this formal base change are recalled in [M.III.C], where in particular it is proved that *this property characterizes* π_v *uniquely.*

REMARK. When n is odd, $\widetilde{BC}'$ and $\widetilde{BC}$ do not coincide, but can be distinguished by their central characters: Theorem 4.1 of [M.III.C] shows that a θ-stable spherical representation is in the image of $\widetilde{BC}$ if and only if its central character is a stable base change. It then follows from Lemma 1.5.2 that Π_w is in the image of $\widetilde{BC}$ for odd n as well.

In each of the above cases we choose an open compact subgroup $K_v \subset U_v^*$ such that $\pi_v^{K_v} \neq 0$. In case (2.2), and for all split v outside a finite set $S = S(\Pi)$ of split primes, π_v is spherical with respect to a hyperspecial maximal compact subgroup K_v, and we choose such a group K_v. For $v \in S(\Pi)$ choose $K_w \subset GL(n, \mathcal{K}_w)$ so that $\Pi_w^{K_w} \neq 0$, and let $K_v \subset U_v^*$ be the corresponding subgroup under the isomorphism $U_v^* \xrightarrow{\sim} GL(n, \mathcal{K}_w)$.

Let $K_f = K_{\Pi,f} = \prod_v K_v$. Letting π_f be the restricted tensor product of the π_v defined above, we then have $\pi_f^{K_f} \neq 0$. Using the theory of the conductor for $GL(n)$, we can even assume $\dim \pi_f^{K_f} = 1$, but this seems to be irrelevant. We recall the results of Labesse on stable descent [L.IV.A,Theorems 5.4]:

THEOREM 2.3. Let Π satisfy Hypotheses 1.1-1.3. Let $\mathcal{A}(G_?)$ be the space of (necessarily cuspidal) automorphic forms on (the anisotropic unitary group) $G_?$, where $G_? = G_0, G_2, U^*$.

(a) There is exactly one representation π_∞ of $G_{0,\infty}$ such that $\pi_\infty \otimes \pi_f$ occurs in $\mathcal{A}(G_0)$, and it occurs with multiplicity one.

(b) There are $\frac{n(n-1)}{2}$ distinct representations π_j of $G_{2,\infty}$ such that $\pi_j \otimes \pi_f$ occurs in $\mathcal{A}(G_2)$, each with multiplicity one.

(c) Finally, Π descends to some L-packet $\tau = \{\tau_i(\Pi)\}$ of U^* with $(\tau_i(\Pi))_f = \pi_f$ for all i, each with multiplicity one.

The representations at the real primes are described as follows. Let v be a real prime of F. The coefficient system W has a factor W_v, a representation of $GL(n, \mathcal{K} \otimes_{F,v} \mathbb{C})$, described above. We write $W = W' \otimes W''$ with respect to the product decomposition of $GL(n, \mathcal{K} \otimes_{F,v} \mathbb{C})$ of §1. The group $G_?(F \otimes_v \mathbb{C})$ embeds naturally in $GL(n, \mathcal{K} \otimes_v \mathbb{C})$ and, by the polarization condition, the restriction W_+ of W' or W'' to $G_?(F_v)$ is the same. Thus we have an irreducible complex representation W_+ of $G_?(F_v)$, or equivalently of its compact inner form.

In case (a), the factor at v of π_∞ is the *dual* of W_+. In cases (b) and (c), $\tilde{W}_+$ defines, as in [C.III.A, beginning of §2], an L-packet of discrete series of $G_2(F_v)$ or $U^*(F_v)$ at the place(s) v where this group is non-compact. The representations π_j of (b) are, with multiplicity one, the constituents of this L-packet. In case (c) we get, with multiplicity one, the tensor products over all real primes of these constituents.

REMARK. For this chapter, only (c) of 2.3 is really needed. However, we note here that (a) is verified in [L.IV.A] without hypothesis 1.3.

We write τ_f instead of $(\tau_i(\Pi))_f$; of course τ_f is just another name for the π_f constructed above.

3. Endoscopic parameters

3.1. Conventions. In what follows we let H denote the quasi-split F-group $U^* \times U(1)$, where U^* is the quasi-split form of $U(n)$ introduced at the end of §1. We view H as an elliptic endoscopic group for the group G of Lemma 1.4, cf. [H.I.A], 5.5. The transfer of parameters, and of automorphic representations, depends on the choice of an L-homomorphism

$$\xi : {}^L H \to {}^L G$$

which we normalize as in [H.I.A, (5.5)]. We recall the formulas here. Fix a Hecke character μ of $\mathcal{K}$ whose restriction to the idèles of F is the quadratic character $\eta_{\mathcal{K}/F}$ and which is unramified at all finite places of $\mathcal{K}$ that do not split over F. (This is possible by (1.2.1)). On $\hat{H} = {}^L H^0 = GL(n) \times GL(1)$, we have

$$(3.1.1) \qquad \xi(h_n, h_1) = \begin{pmatrix} h_n & 0 \\ 0 & h_1 \end{pmatrix} \in \hat{G} \subset {}^L G, \quad h_n \in GL(n), h_1 \in GL(1)$$

The Hecke character μ can be viewed as a character of $W_{\mathcal{K}}$ by class field theory, and for $w \in W_{\mathcal{K}}$ we let

$$(3.1.2) \qquad \xi(w) = \begin{pmatrix} \mu(w) I_n & 0 \\ 0 & 1 \end{pmatrix} \times w \in \hat{G} \rtimes W_{\mathcal{K}}$$

Finally, if $w_\sigma \in W_F$ is a representative of the non-trivial coset of $W_{\mathcal{K}}$, we set

$$(3.1.3) \qquad \xi(w_\sigma) = \begin{pmatrix} \Phi_n & 0 \\ 0 & 1 \end{pmatrix} \cdot \Phi_{n+1}^{-1} \times w_\sigma.$$

Here for any m, Φ_m is the anti-diagonal $m \times m$-dimensional matrix with entries $(\Phi_m)_{ij} = (-1)^{i-1}\delta_{i,m-j+1}$ [cf. M. III.C., 3.2.5].

Let Π be a representation satisfying Hypotheses 1.1–1.3, and let $\tau = \tau_\Pi$ be the L-packet of U^* of Theorem 2.3 (c). Let χ be a character of $U(1)$, and let

$$\tilde{\chi}(z) = \chi(z/c(z)) : \mathcal{K}^\times \backslash \mathbf{A}_{\mathcal{K}}^\times \to \mathbb{C}^\times$$

be the base change of χ to a Hecke character of $GL(1)_{\mathcal{K}}$. The goal of this chapter is to analyze the endoscopic transfer of the L-packet $\tau \times \chi$ for the endoscopic group H to an L-packet, denoted $\pi(\tau, \chi)$, of G.

3.2. Local endoscopic transfer at non-archimedean places. Expanding the set $S = S(\Pi)$ if necessary, we may assume that χ (and equivalently $\tilde{\chi}$) is unramified for $v \notin S$. We may also assume that all primes at which χ ramifies are split in $\mathcal{K}/F$; thus S consists only of split primes. We consider factorizable test functions $f = \otimes_v f_v$ on $G(\mathbf{A}_F)$, and write $f = f_\infty \otimes f_f$. Let $G_v = G(F_v)$ and let $K'_v \subset G_v$ be a hyperspecial maximal compact subgroup for each finite $v \notin S$, so that $K'_v = \prod_v K'_v$ is an open compact subgroup of $G((\mathbf{A}_F^f)^S)$. For $v \in S$ the endoscopic transfer $\pi_v(\tau_v, \chi_v)$ of $\tau_v \times \chi_v$ is defined as an explicit induced representation in [M.III.C, 4.2.3; cf. H.I.A, (9.3.2)]: identifying G_v with $GL(n+1, F_v)$, we have

$$(3.2.1) \qquad \pi_v(\tau_v, \chi_v) = I_{P_n}^{G_v}(\Pi_v \otimes (\mu_v \circ \det) \times \tilde{\chi}_v),$$

where P_n is the standard maximal parabolic subgroup with Levi factor $GL(n, F_v) \times GL(1, F_v)$, $I_{P_n}^{G_v}$ denotes normalized induction, and the induced representation is necessarily irreducible.[1] Note that the transfer depends on the homomorphism ξ, and particularly on the local component μ_v of μ at v. For $v \notin S$ we assume f_v belongs to the Hecke algebra of K'_v-bi-invariant functions. For $v \in S$ we choose a K'_v so that $\pi_v(\tau_v, \chi_v)^{K'_v} \neq 0$. One can make a more judicious choice using the theory of types, but this is not necessary.

[1]Because Π is cuspidal unitary, hence Π_v is generic and unitary, and $\tilde{\chi}_v$ is also unitary.

Combining the theorems of Laumon-Ngô, Hales, and Waldspurger recalled in part II (cf. a summary in [H.I.A, §6]), we know that there is a test function f^H on $H(\mathbf{A}_F^f)$ such that

$$\sum_{\gamma \sim \gamma_H} \Delta_f(\gamma_H, \gamma) O_\gamma(f) = SO_{\gamma_H}(f^H) \tag{3.2.2}$$

whenever $\gamma \in G(\mathbf{A}_F^f)$ and $\gamma_H \in H(\mathbf{A}_F^f)$ are associated and $\Delta_f(\bullet, \bullet)$ is the corresponding product of local transfer factors. Moreover, for $v \notin S$, f_v and f_v^H are related by

$$f_v^H = b_v(\xi)(f_v), \;\; b_v(\xi) : \mathcal{H}(G_v, K_v') \to \mathcal{H}(H_v, K_v \times K_{1,v}) \tag{3.2.3}$$

where, in the notation of [M.III.C, (2.7)], $b_v(\xi)$ is the homomorphism of Hecke algebras corresponding to the L-homomorphism ξ and $K_{1,v} \subset U(1)_v$ is the unique maximal compact subgroup (cf. [H.I.A, (9.3.3.4)] and [M.III.C]). For v split, the relation between f_v and f_v^H is explicitly given by Theorem 9.3.3.4 of [H.I.A] (in terms of distributions), or in terms of functions, by:

$$f_v^H = (\mu_v \circ \det)(h_n) f^{P_{n,1}}(h_n, h_1), \tag{3.2.4}$$

cf. [M.III.C, Corollary 4.7] for a more general formula.

3.3. Local endoscopic transfer at real places. Now let v be a real place. The analogue of transfer formula (3.2.2) is in this case due to Shelstad, and is recalled in [C.III.A]. We apply it when f_v is a pseudocoefficient for a member of the discrete series L-packet attached to the local representation $W(\Pi)_v$. Let $\phi_G : W_{\mathbb{R}} \to {}^L G$ be the Langlands parameter of this L-packet. As in [C.III.A], §3, let ϕ_H be a discrete series parameter for H_v with $\phi_G = \xi \circ \phi_H$. We assume ϕ_H dominant (formula (3.14) in [C.III.A]) and ϕ_G *regular*. Then

$$(\Theta_{\phi_H}, f_v^H) = \sum_{\pi \in \Pi_G(\phi_G)} \Delta_v(\phi_H, \pi)(\Theta_\pi, f_v) \tag{3.3.1}$$

where the transfer factors $\Delta_v(\phi_H, \pi)$ are signs calculated explicitly by Clozel, following Shelstad and Kottwitz, in [C.III.A., Theorem 3.4].

We only need to determine the local transfer factors explicitly in two situations. Notation and terminology is as in [C.III.A].

Compact places. Assume $v \neq v_0$, so that G_v is compact. Then $\Pi_{G_v}(\phi_G)$ is a singleton, which we denote simply by π_v. Let $\omega \in \Omega$ be the unique element such that $\omega^{-1} r$ is dominant, with r the parameter associated to ϕ_G in [C.III.A, before (3.14)]. Then

$$\Delta_v(\phi_H, \pi_v) = \det \omega. \tag{3.3.2}$$

Indeed, Theorem 3.4 in [C.III.A] yields

$$\Delta_v(\phi_H, \pi_v) = < a_{\omega_\sigma}, s > \det \omega$$

where $s = \begin{pmatrix} I_n & \\ & -1 \end{pmatrix} \in \hat{G}$. Here $\sigma \in \Omega/\Omega_{\mathbb{R}}$, the quotient of the complex Weyl group by the real Weyl group, so we can take $\sigma = 1$. At any rate, G_v

being compact, the cocycle a_ω is trivial [C.III.A, before (3.12)]. Thus

$$\Delta_v(\phi_H, \pi_v) = \det \omega. \tag{3.3.3}$$

Note that ω depends on v. We set $\delta_v := \det \omega$.

The non-compact place. We assume $v = v_0$ and choose ϕ_H – i.e., we choose χ, because the representation of U^* will be fixed – so that ϕ_G is in dominant position. Then with our notation, which is that of Theorem 3.4 in [C.III.A], $\omega = 1$. The transfer factor is determined as follows. Recall that $G_v = U(1,n)$, $\Omega = \mathfrak{S}_{n+1}$, $\Omega_{\mathbb{R}} = \mathfrak{S}_n$ (acting on the last n letters). The discrete series L-packet of G_v associated to ϕ comprises the representations

$$\pi(\phi, \sigma^{-1}B_0)$$

[C.III.A, after (3.11)], $\sigma \in \Omega/\Omega_{\mathbb{R}} \simeq \{1, 2, \ldots, n+1\}$, the bijection being given by $\sigma \mapsto \sigma \cdot 1$. We identify σ with $\sigma \cdot 1$ and write $\pi(\phi, i)$ for $\pi(\phi, \sigma^{-1}B_0)$. Then, for a suitable choice of holomorphic structure[2] on the symmetric space of G:

$$\pi(\phi, 1) \text{ is a holomorphic discrete series} \tag{3.3.4}$$

$$\pi(\phi, n+1) \text{ is an antiholomorphic discrete series.} \tag{3.3.5}$$

Since $s = \begin{pmatrix} I_n & \\ & -1 \end{pmatrix}$, formula (3.13) of [C.III.A] now yields

$$\Delta_v(\phi_H, \Pi(\phi, i)) =< a_\sigma, s >= \{ \begin{matrix} 1 & i \leq n \\ -1 & i = n+1 \end{matrix}. \tag{3.3.6}$$

Thus

LEMMA 3.3.7. The discrete series L-packet $\Pi_G(\phi_G)$ is partitioned into two subsets Π^+ and Π^- such that

(1) Π^+ has n elements, Π^- has 1 element;
(2) $\Delta_v(\phi_H, \pi) = 1$ $(\pi \in \Pi^+)$, -1 $(\pi \in \Pi^-)$.
(3) Π^- is the unique antiholomorphic discrete series.

In conclusion, we have

COROLLARY 3.3.8. Let $\delta \in \pm 1$. Fix an archimedean L-packet τ_∞ of the first factor U^*_∞ of H_∞. There is a choice of local characters χ_v for $v \in \Sigma_F$, or equivalently a character χ_∞ of the second factor $U(1)_\infty$ of H_∞, so that, letting $\phi_H(\tau_\infty, \chi_\infty)$ be the corresponding Langlands parameter, the product over the archimedean places

$$\Delta_{v_0}(\phi_H(\tau_\infty, \chi_\infty), \pi_{v_0}) \cdot \prod_{v \neq v_0} \Delta_v(\phi_H(\tau_\infty, \chi_\infty), \pi_v) = \delta \Leftrightarrow \pi_{v_0} \in \Pi^+.$$

Here for $v \neq v_0$ we write $\Pi_{G_v}(\phi_G) = \{\pi_v\}$.

---

[2]In the applications to Galois representations, this will be specified more precisely in Book 2.

Moreover, we can also assume that the transfer $\Pi_{G_v}(\phi_G)$ has the property that its infinitesimal character equals that of a finite-dimensional representation of G with **regular** highest weight.

Proof. As in the proof of Lemma 3.3.7, we choose χ_{v0} such that $\xi \circ \phi_H(\tau_{v0}, \chi_{v0})$ is a dominant discrete series parameter for G. Then

$$\delta_{v0} = \Delta_{v0}(\phi_H(\tau_{v0}, \chi_{v0}), \pi_{v0}) = 1$$

for $\pi_{v0} \in \Pi^+$ (cf. Lemma 3.3.7). We fix a $v_1 \neq v_0$ in Σ_F and for $v \notin \{v_0, v_1\}$, we choose χ_v arbitrarily again, and let $\delta_v = \Delta_v(\phi_H(\tau_v, \chi_v), \pi_v)$. Let $\varepsilon = \prod_{v \neq v_1} \delta_v$. Thus we need to show that, for an appropriate choice of χ_{v_1}, we have

$$\Delta_{v_1}(\phi_H(\tau_{v_1}, \chi_{v_1}), \pi_{v_1}) = \varepsilon \cdot \delta.$$

The left-hand side is the determinant of the permutation ω such that $\omega^{-1} r(\xi \circ \phi_H(\tau_{v_1}, \chi_{v_1}))$ is dominant. Recall that there is a Grössencharacter μ in the definition of ξ, cf. (3.1.2). Let $p_\tau = (p'_1, \ldots, p'_n)$ be the (integral) parameter of τ_{v_1}, m that of μ_{v_1}, p' that of χ_{v_1}. Then

$$r = (m + p'_1, \ldots, m + p'_n, p') := (p_1, \ldots, p_{n+1}).$$

Note that $m \equiv \frac{1}{2}$ (mod 1), cf. [C.III.A, §3.1]. For instance, we could take $m = \frac{1}{2}$. There are now two cases to consider.

Assume first that $\varepsilon\delta = 1$. It will suffice to choose $p' = p_{n+1}$ such that r is dominant and satisfies the strong regularity hypothesis. The parameter $(p_1, \ldots, p_n)$ being so by assumption, it will be enough to take p' very negative.

Assume $\varepsilon\delta = -1$. We seek ω with sign -1 such that

$$p_{\omega(1)} > \cdots > p_{\omega(n+1)}.$$

We know that $p_1 > \cdots > p_n$; moreover, since the *weight* satisfies the Strong Regularity Hypothesis 1.3, we have $p_i \geq p_{i+1} + 4$ for all i. We can thus choose $p' = p_{n+1}$ such that

$$p_i - 1 > p_{n+1} > p_{i+1} + 1 \tag{3.3.9}$$

for i such that the corresponding shuffle permutation has sign -1. This proves the first part of the Corollary, and the final assertion follows from (3.3.9) and Hypothesis 1.3. □

REMARK. The proof shows that the assumed regularity of the highest weight is much too strong a condition. For instance, the argument applies if $p_i \geq p_{i+1} + 2$ for two successive values $(i, i+1)$ with i *odd*. Nevertheless, the regularity is useful elsewhere, for example in the proof of multiplicity one in [L.IV.A, Theorem 5.4], a result needed in the next chapter.

4. Global calculation

4.1. We let f be a test function on $G(\mathbf{A}_F)$ for which [L.IV.A, Théorème 5.1] is valid. Hence we assume that $[F:\mathbb{Q}] \geq 2$ and that f is a pseudo-coefficient of discrete series at the real primes. Let H be an endoscopic group for G and let f^H be a transfer of f. By our choice of f_∞ the transfer f_∞^H can be taken to be a linear combination of pseudo-coefficients of discrete series (cf. Shelstad [Shel1], Clozel-Delorme [CD]). Since f_∞^H intervenes only through its stable orbital integrals, this linear combination may be chosen stable. Then f_∞^H is, up to a scalar, an Euler-Poincaré function ([CD, Théorème 3], [L, Proposition 9]). Fix H. Let φ^H be a function on $H(\mathbf{A}_\mathcal{K})$ corresponding to f^H by base change. This is legitimate in our situation because base change for Hecke algebras is surjective, cf. [M.III.C, Cor. 4.2]. At the archimedean places φ^H is a twisted Euler-Poincaré function, as in [CL] (see also [L.IV.A, §4.6]).

By [L.IV.A, Théorème 5.1] there is an identity

$$T_{disc}^G(f) = \sum \iota(G,H)\, T_{disc}^{\widetilde{M}_H}(\varphi^H \times \theta) \tag{4.1.1}$$

where H runs over the endoscopic groups for G.[3] The notation is essentially that of Labesse: $\widetilde{M}_H$ is the "espace tordu" associated to $(H(\mathbf{A}_\mathcal{K}),\theta)$; cf. [L.IV.A, §4.7], where θ is the Galois automorphism. We have written $\varphi^H \times \theta$ for the function on the non-identity component $\widetilde{M}_H$ obtained from φ^H by translation by θ. Recall that Labesse writes $\tilde{f}^H$ for our $\varphi^H \times \theta$.

Such an identity can also be obtained by composing the Proposition 5.6 of [L.IV.A] which recalls Arthur's stabilization of the discrete part of the trace formula for G:

$$T_{disc}^G(f) = \sum \iota(G,H)\, ST_{disc}^H(f^H) \tag{4.1.2}$$

and the Theorem 5.8 of [L.IV.A] which states the stabilization of the discrete part of the twisted trace formula for Lefschetz functions (also called twisted Euler-Poincaré functions) on M_H

$$T_{disc}^{\widetilde{M}_H}(\varphi^H \times \theta) = ST_{disc}^H(f^H)\ . \tag{4.1.3}$$

but the proof of both results uses more prerequisites.

Since G is anisotropic,

$$T_{disc}^G(f) = T^G(f) = \sum_\pi \text{trace}\ \pi(f) \tag{4.1.4}$$

where $\{\pi\}$ decomposes $L^2(G(F)\backslash G(\mathbf{A}_F))$. With this choice of f_∞, we now have:

LEMMA 4.2. *We can choose our test function f_f on $G(\mathbf{A}_F^f)$ such that*

(1) *For all finite v, f_v is bi-invariant under the subgroup K_v' of §3.2;*

[3] To simplify notation we use θ for the outer automorphism appropriate to the group, in this case H.

(2) For all finite v, $Tr(f_v; \pi_v(\tau_v, \chi_v)) \neq 0$;
(3) $T_{disc}^{\widetilde{M}_H}(\varphi^H \times \theta) = 0$ unless $H = U(n+1)^*$ or $H = U(n)^* \times U(1)$

Proof. Note that by the choice of f_∞, (4.1.4) and all terms in (4.1.1) now range over a finite set of representations once the level of f_f is fixed. Consider a term of (4.1.1) indexed by an endoscopic group $H = U^*(a) \times U^*(b)$, $a+b = n+1$. Again we adapt Labesse's notation from [L.IV.A, §4.7]. Thus $\widetilde{M}_{a,b}$ will denote the "espace tordu" $GL(a)_{\mathcal{K}} \times GL(b)_{\mathcal{K}} \rtimes \theta$ with the Galois conjugation θ acting on both factors. However, unlike Labesse, we denote by φ^H a function on the adelic group $H(\mathbf{A}_{\mathcal{K}}) = GL(a, \mathbf{A}_{\mathcal{K}}) \times GL(b, \mathbf{A}_{\mathcal{K}})$. Then we have further [L.IV.A, Prop. 3.4]

$$T_{disc}^{\widetilde{M}_{a,b}}(\varphi^H \times \theta) = \sum_{\Pi} c_\Pi \text{ trace } (\Pi(\varphi^H) I_\theta). \tag{4.2.1}$$

On the right-hand side, Π runs over a "discrete" set of representations of

$$GL(a, \mathbf{A}_{\mathcal{K}}) \times GL(b, \mathbf{A}_{\mathcal{K}})$$

such that Π and $\Pi \circ \theta$ are isomorphic, θ being "complex conjugation" with respect to H_F; I_θ is the naturally defined, global intertwining operator $\Pi \simeq \Pi \circ \theta$; and $c_\Pi \neq 0$. Let us recall what these representations are. We may as well consider only one factor $GL(a, \mathbf{A}_{\mathcal{K}})$; change notation and write Π for a representation of this group. The condition on Π is that

(1) (tempered case) Π should be of the form $\Pi_1 \boxplus \cdots \boxplus \Pi_r$, Π_i a cuspidal representation of $GL(a_i, \mathbf{A}_{\mathcal{K}})$, $a_1 + \dots a_r = a$, $\Pi_i \simeq \Pi_i^\theta$, and $\Pi_i \neq \Pi_j$. Here $\boxplus$ denotes block parabolic induction,
(2) (general case) More generally, Π has a similar decomposition, but with Π_i the Speh representation $Sp(P_i, b_i)$ where $a_i = c_i b_i$ and P_i is cuspidal for $GL(c_i, \mathbf{A}_{\mathcal{K}})$. (Cf. [A, Theorem 26.2] or [MW]).

For a justification, cf. [L.IV.A, Lemme 3.8]). The condition $\Pi_i \neq \Pi_j$ is not explicitly mentioned there, but is automatic for representations such that

$$\text{trace } \Pi_\infty(\varphi_\infty^H \times \theta) \neq 0$$

as follows from the regularity of discrete series parameters (cf. [C.III.A, (3.6)]). As we will see, the case (2) is irrelevant for us.

Now fix the level of f_f, for instance assume it is bi-invariant under $K' = \prod K'_v$. Consider our representation

$$\pi_f(\tau, \chi) = \bigotimes_{v \nmid \infty}{}' \pi_v(\tau_v, \chi_v) \ .$$

For $H = U^*(n) \times U(1)$, (4.2.1) contains a term such that, for $v \notin S$, in the notation of [H.A.I, §9.3.3], the associated character of (the unramified Hecke algebra) $\mathcal{H}(H_v)$ is, composed with the endoscopic morphism $\mathcal{H}(G_v) \to \mathcal{H}(H_v)$ of [H.A.1, (9.3.3.4)], the character of $\mathcal{H}(G_v)$ given by π_v. On the right-hand side, it is simply the twisted trace associated to $(\Pi \otimes (\mu \circ \det)) \times \chi$

(§1). For $H = U^*(n+1)$, there is a similar term of the form (1), which is now composite, associated to an induced representation $(\Pi \otimes (\mu \circ \det)) \boxplus \chi$.

We can certainly choose f verifying (1), (2) of Lemma 4.2. We may vary f at the primes $v \notin S$, preserving (2).

Now consider the terms (4.2.1) for $H = U^*(a) \times U^*(b)$, and a or $b > 1$. By our choice of f, thus of f^H, φ_∞^H is a twisted Euler-Poincaré function [CL,Theorem 5.1]. In particular, for fixed level, the sum in (4.2.1) is again finite. By a theorem of Jacquet-Shalika [JS], the characters of

$$\mathcal{H}_{\mathcal{K}}^S = \bigotimes_{w \notin S} {}'\mathcal{H}(GL(n, \mathcal{K}_w), GL(n, \mathcal{O}_w))$$

associated to *all* terms in (4.2.1) for *all* H with $(a,b) \neq (n,1)$ or $(n+1,0)$ are linearly independent. (For instance, the non-tempered terms of type (2) above have different Hecke eigenvalues by the Jacquet-Shalika estimate for the Hecke eigenvalues of cusp forms; if a or $b > 1$ terms of type (1) do not contain a cuspidal factor of dimension n and we apply [JS, II, Theorem 4.4]). This proves the Lemma. □

The same proof shows the stronger statement:

COROLLARY 4.3. Let $\mathcal{W}$ be a finite set of irreducible finite-dimensional representations of $G(\mathbb{R})$. We can choose f_f as in Lemma 4.2 to satisfy the additional conditions

(4) Suppose π is an automorphic representation of G of level $\prod_v K'_v$, the product taken over all finite places of F, such that π_∞ is cohomological relative to some $W \in \mathcal{W}$. Then $Tr(f_f, \pi_f) = 0$ unless $\pi_f = \otimes'_v \pi_v(\tau_v, \chi_v)$, in which case $Tr(f_f, \pi_f) = 1$.

The hypotheses on π imply that π belongs to a finite set of automorphic representations. The assertion now follows from the standard argument of separation by Hecke eigenvalues that is needed in every application of the trace formula.

Henceforward, we assume f_f to be chosen as in 4.2 and 4.3. We write

$$\pi_f(\tau_f, \chi_f) = \bigotimes_{v \nmid \infty}{}' \, \pi_v(\tau_v, \chi_v) \, .$$

The next two subsections determine the contributions $T_{disc}^{\widetilde{M}_H}(\varphi^H \times \theta)$ for

$$H = U(n+1)^* \qquad \text{and} \qquad H = U(n)^* \times U(1) \, ,$$

respectively, when f_∞ is chosen to be a pseudo-coefficient of a fixed discrete series representation. These contributions will be controlled by base-change, using Labesse's results in [L.IV.A].

4.4. Calculation of $T_{disc}^{\widetilde{M_H}}(\varphi^H \times \theta)$, $H = U(n+1)^*$. Let $H = U(n+1)^*$, so that $\iota(G, H) = 1$. By the Jacquet-Shalika classification, the representation $\pi_f(\tau_f, \chi_f)$ cannot base-change to the finite part of a cuspidal automorphic representation of $M_H = GL(n+1, \mathcal{K})$. It does, however, base-change to the (finite part of an) Eisenstein representation on $GL(n+1, \mathcal{K})$ induced from the standard maximal parabolic subgroup P_n already encountered in (3.2). In what follows we write P instead of P_n. The components τ and χ are respectively θ_n-stable and θ_1-stable, where θ_i is the outer automorphism of $GL(i, \mathcal{K})$ considered in [L.IV.A]. Thus the corresponding automorphic representation, occurring in the space of Eisenstein series on $GL(n+1, \mathcal{K})$, say $\Pi = \Pi(\tau, \chi)$, is θ_{n+1}-stable. Moreover, although it belongs to a one-dimensional complex continuous family of automorphic representations, $\Pi(\tau, \chi)$ is even θ_{n+1}-discrete in the sense that it is the only fixed point in the family under θ_{n+1}. It will, then, contribute a discrete term (4.4.2) to the left-hand side of (4.2.1).

Now let f_∞ be a pseudocoefficient of a discrete series representation π_∞ in the L-packet $\Pi_G(\phi_G)$ with parameter $\phi_G = \xi \circ \phi_H$, as in (3.3). We assume the strong regularity hypothesis 1.3 and we choose χ to satisfy the last condition of Corollary 3.3.8. It follows that f_∞ has non-zero trace only on discrete series representations of G_∞. In particular, f_∞ satisfies hypothesis (*) of [L.IV.A, §5], and we can apply Theorems 5.3 and 5.4 of (*loc. cit.*).

The function φ_∞^H, being associated to f_∞^H, must be such that

$$\text{trace } (\Pi_\infty(\varphi_\infty^H \times \theta)) = 1, \tag{4.4.1}$$

the 1 on the right-hand side being the trace of f_∞^H in the L-packet for $H(F_\infty)$ associated to Π_∞. Here Π_∞ has been extended to the twisted space using the Whittaker normalization: $\Pi_\infty(\theta)$ is the Whittaker-normalized intertwining operator for $\theta = \theta_{n+1}$, as in [C.III.A,§2] (see the remarks at the end of §2 of that chapter). *Up to a scalar*, φ_∞^H is a Lefschetz function for Π_∞.

Using the finiteness arguments in the proof of Lemma 4.2 and the fact that φ_∞^H traces only in cohomological representations, we can now refine our choice of f^H (and associated φ^H) so that only one class of terms survives in the expression of [L.IV.A, Proposition 3.3], namely

$$T_{disc}^{\widetilde{M_H}}(\varphi^H \times \theta) = |W(M)|^{-1} T_{M,disc}^{\widetilde{M_H}}(\varphi^H \times \theta)$$

where $M = GL(n) \times GL(1)$. Since $n \geq 2$ one has $|W(M)| = 1$ so that we get (see [L.IV.A., Proposition 3.7])

$$T_{disc}^{\widetilde{M_H}}(\varphi^H \times \theta) = \frac{1}{2} \sum_{\Pi} \text{trace } I_P(\Pi, s)(\varphi^H \times \theta)$$

where $I_P(\Pi, s)(\varphi^H \times \theta)$ is the operator that shows up in in [L.IV.A.§3.4] defined using Arthur's normalization. We have chosen f^H so that only the term $\Pi = \Pi(\tau, \chi)$ survives. Hence we get:

$$T_{disc}^{\widetilde{M}_H}(\varphi^H \times \theta) = \frac{1}{2}\text{trace } I_P(\Pi(\tau,\chi), s)(\varphi^H \times \theta) \tag{4.4.2}$$

PROPOSITION 4.4.3. Let f_∞ be a pseudocoefficient of a discrete series representation $\pi_\infty \in \Pi_G(\phi_G)$ as above, and let f_f satisfy the conditions of (4.2) and (4.3). Then, if $H = G^* = U(n+1)^*$,

$$T_{disc}^{\widetilde{M}_H}(\varphi^H \times \theta) = \frac{1}{2}.$$

Proof. In view of (4.4.2) we are left to show that

$$\text{trace } I_P(\Pi(\tau,\chi), s)(\varphi^H \times \theta) = 1\ .$$

Implicit in the definition is a globally defined intertwining operator

$$I_\theta = I_P(\Pi, s)(\theta)$$

using Arthur's normalization (cf. [L.IV.A, §3.4]). By our choice of φ^H we know by [C.III.A. §2] that the assertion would be true if I_θ were defined using Whittaker's normalization. The proposition is now equivalent to showing that, in this case, Arthur's normalization and Whittaker's normalization coincide. This is established in [L.IV.A, Proposition 3.7]. An alternative proof is given below.

Now

$$\Pi(\tau,\chi) = ind_{P(\mathbf{A}_\mathcal{K})}^{GL(n+1,\mathbf{A}_\mathcal{K})}(\tau_\mathcal{K} \times \chi_\mathcal{K}) := I_P^G(\tau_\mathcal{K} \times \chi_\mathcal{K})$$

where $\tau_\mathcal{K}$ is the representation Π of §1 – the base-change to $GL(n, \mathbf{A}_\mathcal{K})$ of the L-packet τ of (3.1) – and $\chi_\mathcal{K} = \chi \circ N_{\mathcal{K}^\times/U(1)}$ is the base-change of χ (denoted $\tilde{\chi}$ above), the notation $N_{\mathcal{K}^\times/U(1)}$ being self-explanatory.

The automorphism θ_{n+1} acts on $GL(n+1, \mathbf{A}_\mathcal{K})$, and $\theta_n \times \theta_1$ acts on the Levi subgroup $GL(n) \times GL(1)$. There is a tautological intertwining operator

$$T_\theta : I_P^G(\tau_\mathcal{K} \times \chi_\mathcal{K}) \to I_{{}^\theta P}^G({}^\theta\tau_\mathcal{K} \times {}^\theta\chi_\mathcal{K}) \tag{4.4.4}$$

where ${}^\theta P$ is now of type $(1, n)$. The groups P and ${}^\theta P$ contain the unipotent radical of the Borel subgroup, and T_θ sends the natural Whittaker functional W' (on the right-hand side of (4.4.2), with respect to a choice of additive character) to its analogue W on the left-hand side.

We now have to send $I_{{}^\theta P}^G({}^\theta\tau_\mathcal{K} \times {}^\theta\chi_\mathcal{K})$ back to $I_P^G(\tau_\mathcal{K} \times \chi_\mathcal{K})$, and this is done by the standard unnormalized intertwining operator (integration on a suitable unipotent subgroup), composed with the involution $\theta_M = \theta_n \times \theta_1$. Note that according to [L.IV.A, Lemme 3.8]

$$\theta_M(\tau_\mathcal{K} \times \chi_\mathcal{K}) \simeq \tau_\mathcal{K} \times \chi_\mathcal{K}.$$

Further, θ_M fixes a Whittaker functional on the space of the inducing representation and therefore also on the space of the induced representation.

Let us denote by

$$A : I^G_{\theta P}({}^\theta\tau_K \times {}^\theta\chi_K) \to I^G_P(\tau_K \times \chi_K)$$

the unnormalized intertwining operator. It is defined, *a priori,* for a complex parameter σ such that $Re(\sigma) > 0$. We take σ real and consider $A(\sigma)$. Following Langlands and Shahidi [Sh1], and using the results in [JS] and [Sh2], we normalize $A(\sigma)$ in the form

$$A(\sigma) = n(\Pi, \sigma)N(\sigma),$$

where

$$n(\Pi,\sigma) = \frac{L(\tau_K \otimes \chi_K^\vee, \sigma)}{\varepsilon(\tau_K \otimes \chi_K^\vee, \sigma)L(\tau_K \otimes \chi_K^\vee, \sigma+1)}, \tag{4.4.6}$$

the L and ε factors being the Rankin-Selberg L-function and ε-factor (in this case for $GL(n) \times GL(1)$) as in [JS], including the archimedean factors. Note that the equality of the local factors in [JS] and [Sh2] is proved by Shahidi, in [Sh3] in the p-adic case and in [Sh4] in the real case. The functional equation

$$L(\tau_K \otimes \chi_K, \sigma) = \varepsilon(\tau_K \otimes \chi_K, \sigma)L(\tau_K^\vee \otimes \chi_K^\vee, 1-\sigma)$$

here gives the following result:

$$n(\Pi,\sigma) = \frac{L(\tau_K^\vee \otimes \chi_K, 1-\sigma)}{L(\tau_K \otimes \chi_K^\vee, 1+\sigma)}.$$

But τ_K and χ_K are unitary representations, thus are isomorphic to the duals of their complex conjugates (complex conjugation acting on the *coefficients*), and σ is real. We thus obtain

$$n(\Pi,\sigma) = \frac{L(\tau_K \otimes \chi_K^\vee, 1-\sigma)}{L(\tau_K \otimes \chi_K^\vee, 1+\sigma)}.$$

Arthur's formula involves the value of this function at $\sigma = 0$, and thus

$$n(\Pi, 0) = 1,$$

since the global L-function is holomorphic at $s = 1$.

We still need to calculate the effect of $N(0)$ on the Whittaker functionals. The normalized operator is a product of normalized local operators. At the finite places where our data are unramified, we know that $N(0)$ acts trivially on the vector fixed by the hyperspecial maximal compact subgroup (Lemma 6.1 in [KS]; note that the special assumptions in that paper are removed in [Sh2]) and that it acts by 1 on the Whittaker functional. At split places, a simple calculation using [Sh2] shows that this is still true. (One could also use the argument that follows, completing [Sh1, *loc. cit.*], by the fact that it is now known that his local coefficients $c_\psi(s,\Pi)$ admit analogous formulas to those that we use below at real places.)

Now consider a real place v. If f is a function in the local induced representation

$$I^G_{\theta P}({}^\theta\tau_w \times {}^\theta\chi_w) = I^G_{\theta P}(\tau_w \times \chi_w)$$

for w dividing v, and if the natural Whittaker functions (for which the previous arguments, relative to induction, apply) on the two induced representations are denoted Wh, Shahidi [Sh1, p. 298] shows the formula

$$< Wh, f >= c(\sigma, \tau \otimes \chi) < Wh, A(\sigma) f >$$

where we have dropped the subscripts on τ and χ and the factor $c(\sigma, \tau \otimes \chi)$, which depends *locally* on an additive character ψ of $\mathcal{K}_w$, is given by

$$c(\sigma, \tau \otimes \chi) = c_\psi(\sigma, \tau \otimes \chi) = \varepsilon(\tau \otimes \chi^\vee, \psi, \sigma) \times \frac{L(\tau^\vee \otimes \chi, 1-\sigma)}{L(\tau \otimes \chi^\vee, \sigma)}.$$

We now return to the formula (4.4.6) for the normalizing factor. We have (*locally*) $N(\sigma) = n(\sigma) A(\sigma)$, thus

$$< Wh, N(\sigma) f >= n(\sigma) c(\sigma, \tau \otimes \chi) < Wh, f > .$$

The scalar product is thus

$$\varepsilon(\tau \otimes \chi^\vee, \psi, \sigma) \times \frac{L(\tau^\vee \otimes \chi, 1-\sigma)}{L(\tau \otimes \chi^\vee, \sigma)} \times \frac{L(\tau \otimes \chi^\vee, \sigma)}{\varepsilon(\tau \otimes \chi^\vee, \psi, \sigma) L(\tau \otimes \chi^\vee, \sigma + 1)}$$

At $\sigma = 0$, this equals

$$\frac{L(\tau^\vee \otimes \chi, 1)}{L(\tau \otimes \chi^\vee, 1)}$$

But τ and χ are (locally) unitary, thus $\tau^\vee = \bar{\tau}$, $\chi^\vee = \bar{\chi}$, and since $\sigma = 1$ is real we have

$$L(\tau^\vee \otimes \chi, 1) = L(\bar{\tau} \otimes \bar{\chi}^\vee, 1) = L(\tau \otimes \chi^\vee, 1).$$

Now, by the choice of our functions locally everywhere at finite places and thanks to the compatibility of the Whittaker normalization with the spectral transfer [C.III.A, §2] we have

$$\text{trace } \Pi_v(\varphi_v^H \times \theta) = \text{trace } \tau_v(f_v)$$

if π_v is the base-change of τ_v (known to exist for unramified representations and at split or archimedean places) extended to the twisted space using the Whittaker normalisation [C.III.A,§2]. In particular, with I_θ as defined by Arthur (cf. (4.2.2) above)

$$\text{trace } (I_P(\Pi, s)(\varphi^H \times \theta)) = 1$$

if $\Pi = \Pi(\tau, \chi)$. □

Note that this theorem does not depend on the choice of χ, nor does it depend on the choice of π in the L-packet $\Pi_G(\phi_G)$. This will not be the case in the following section.

4.5. Calculation of $T^{\widetilde{M_H}}_{disc}(\varphi^H \times \theta)$, $H = U(n)^* \times U(1)$. Now let $H = U(n)^* \times U(1)$. Then $\iota(U^*, H) = \frac{1}{2}$ [L.IV.A, Proposition 4.11]. In this case the proof of Lemma 4.2 shows that the analogue of (4.4.2) is simply

$$T^{\widetilde{M_H}}_{disc}(\varphi^H \times \theta) = T^{\widetilde{M_H}}_{M_H,disc}(\varphi^H \times \theta). \tag{4.5.1}$$

For the precise meaning of the notation in (4.5.1), see [L.IV.A,§3.3]. In words, the only $\theta_n \times \theta_1$-discrete representations that contribute are the θ-stable representations in the discrete spectrum for

$$M_H = GL(n, \mathcal{K}) \times GL(1, \mathcal{K})$$

Hypothesis 1.3 guarantees that these representations are in fact cuspidal, and we use this fact without comment in what follows.

To determine the contribution of

$$\iota(U^*, H)\, T^{\widetilde{M_H}}_{disc}(\varphi^H \times \theta) = \frac{1}{2} T^{\widetilde{M_H}}_{M_H,disc}(\varphi^H \times \theta)$$

to (4.1.1), it suffices to determine φ^H whose base-change is f^H. The computation of f^H has been done in (3.2) and (3.3), and we just collect the answers. The result depends this time on the choice of $\pi \in \Pi_G(\phi_G)$. Refer to Corollary 3.3.8, let $\delta = 1$ and choose χ_∞ so that

$$\Delta_{v_0}(\phi_H(\tau_\infty, \chi_\infty), \pi_{v_0}) \cdot \prod_{\sigma \neq v_0} \Delta_\sigma(\phi_H(\tau_\infty, \chi_\infty), \pi_\sigma) = \delta \Leftrightarrow \pi_{v_0} \in \Pi^+ \tag{4.5.2}$$

and consistent with the hypothesis already used in §4.4, namely

STRONG REGULARITY HYPOTHESIS 4.5.3. The infinitesimal character of the discrete series L-packet $\Pi_G(\phi_G)$ equals that of a finite-dimensional representation of G with regular highest weight.

Then

PROPOSITION 4.5.4. *With this choice of χ_∞, let f_∞ be a (very cuspidal) pseudocoefficient of a discrete series representation $\pi_\infty = \otimes_{\sigma \in \Sigma_F} \pi_\sigma$. Let f_f be as above. Then*

$$\iota(U^*, H)\, T^{\widetilde{M_H}}_{M_H,disc}(\varphi^H \times \theta) = \begin{cases} \frac{1}{2} & \text{if} \quad \pi_{v_0} \in \Pi^+ \\ -\frac{1}{2} & \text{if} \quad \pi_{v_0} \in \Pi^- \end{cases}$$

Proof. By the choice of f_f we see that there is only one term in the left-hand side of (4.5.1):

$$T^{\widetilde{M_H}}_{M_H,disc}(\varphi^H \times \theta) = \text{trace}\,(\Pi'(\varphi^H \times \theta)) \tag{4.5.5}$$

where $\Pi' = \Pi \otimes \chi$, and where $I_\theta = \Pi'(\theta)$ (acting on the space of automorphic forms on $GL(n, \mathbf{A}_\mathcal{K}) \times \mathbf{A}_\mathcal{K}^\times$) stabilizes a Whittaker vector [Cl,§2].

Assume $\pi_{v0} \in \Pi^+$ and $f_\infty = \otimes_v f_v$ with f_{v0} a pseudocoefficient of π_{v0}. According to the results in [C.III.A,§3] recalled in this chapter, the stable trace

$$\text{trace } \Pi_{H,\infty}(f_\infty^H) = \sum_{\pi'_\infty \in \Pi_{H,\infty}} \text{trace } \pi'_\infty(f_\infty^H)$$

is equal to 1 when $\Pi_{H,\infty}$ is the L-packet of discrete series $\Pi_H(\phi_H)$ defined by the Langlands parameter ϕ_H. It is 0 for other discrete series L-packets (this was not stated in [C.III.A] but follows easily from Shelstad's construction [Shel1] of f_∞^H, or from a simple argument with infinitesimal characters). Assume moreover the Strong Regularity Hypothesis 1.3. Then f_∞^H has non-zero trace only in discrete series representations. Now by [C.III.A, §2], since I_θ is Whittaker normalized, we have

$$\text{(4.5.6)} \qquad \text{trace } (\Pi'(\varphi^H \times \theta)) = \sum_{\pi'_\infty \in \Pi_H(\phi_H)} \text{trace } \pi'_\infty(f_\infty^H)$$

using that with our choice of f_f the trace of f_f^H is 1 on all representations of $H(\mathbf{A})$ involved. (We extend the results, trivially, to take care of the one-dimensional factor.)

In view of (4.5.5) and (4.5.6) we obtain the first part of the Proposition. If π_{v0} belongs to (the singleton) Π^-, the same argument applies, now with

$$\text{trace } \Pi_{H,\infty}(f_\infty^H) = -1$$

yielding the requisite equality in that case. □

Combining the last two propositions, we conclude:

THEOREM 4.6. Let Π be a cuspidal automorphic representation of $\mathcal{G} = \mathcal{G}_n$ satisfying Hypotheses 1.1, 1.2, and (strong regularity) 1.3. Define $G = U(V_1)$ as in Lemma 1.4.1. Then there exists a Hecke character χ of $U(1)$ such that the representation $\pi_\infty \otimes \pi_f(\tau_f, \chi_f)$ occurs with multiplicity one (resp. multiplicity zero) in the automorphic spectrum of G for any $\pi_\infty \in \Pi^+ \subset \Pi_G(\phi_G)$ (resp. $\pi_\infty \in \Pi^-$). Here the partition $\Pi_G(\phi_G) = \Pi^+ \coprod \Pi^-$ is as in Lemma 3.3.7, and depends on ϕ_G, hence on χ_∞.

The arguments given above apply without change if the group G is replaced by any anisotropic unitary group in $n+1$ variables which is quasi-split at all finite places and either $U(n,1)$ or compact at each archimedean place. The article of Bellaïche and Chenevier[4] about the determination of the sign of the Galois representations constructed in these books requires the following additional case.

Assume now that G is the unitary group associated to a totally definite hermitian space of odd dimension $n+1$ over $\mathcal{K}/F$. The letter χ still denotes a character of $U(1)$ such that χ_v is unramified if v is finite and inert. We have $\chi_v(z) = z^{-a_v}$ for some $a_v \in \mathbb{Z}$ if v is archimedean. Let Π be a cuspidal

[4]The sign of Galois representations attached to automorphic forms for unitary groups, *Compositio Math.*, in press.

automorphic representation of $\mathcal{G}_n$ satisfying Hypotheses 1.1, 1.2, and 1.3. The unitary group G being compact at each real place, the Langlands parameter ϕ_G defined in 3.3 defines a unique (finite dimensional) irreducible representation π_∞ of $\prod_{v \text{ real}} G(F_v)$.

THEOREM 4.7. If a_v is big enough for each real place v, the representation $\pi_\infty \otimes \pi_f(\tau_f, \chi_f)$ occurs in the automorphic spectrum of G (and with multiplicity one).

Indeed, the analysis of §4 applies verbatim to this case, the only difference occuring in §4.5. There is no choice now for the element $\pi \in \Pi_G(\phi_G) = \{\pi_\infty\}$ and Proposition 4.5.4 has to be replaced by the following observation. Assume that for each real place v of F, the integer a_v is big enough so that the endoscopic L-parameter ϕ_G at v is dominant "in the natural order" (the permutation ω_v of 3.3 is then the identity), then $\iota(U^*, H)\ T_{disc}^{\widetilde{M}_H}(\varphi^H \times \theta) = \frac{1}{2}$. Indeed, by the proof of Prop. 4.5.4 and the character identity (3.3.1), whose right-hand side reduces to the term π_∞ now, it is enough to show that $\Delta_v(\phi_H, (\pi_\infty)_v) = 1$ for each v. But this follows from (3.3.2) as ω_v is the identity, and concludes the proof. Note that the conclusion of the statement would hold as well under the "opposite" assumption "$-a_v$ is big enough at each real place v", as then ω_v would be an $n+1$-cycle, whose signature is 1 again as n is even.

5. A remark on transfer factors

Identities such as (4.1.1) are true on the nose, not up to a constant, and therefore rely on a precise definition of the correspondence $f \mapsto f^H$. This has been defined by Langlands and Shelstad [LS]. Recall that we have fixed the L-group embedding. Then, locally the transfer factors are defined up to a constant; a precise form depends on some auxiliary choices. Globally, they must satisfy a product formula proved in §6 of [LS]. The global (or adelic) transfer factor, denoted Δ, is uniquely determined.

In our situation we have at hand a "canonical" spectral transfer at all places as will be recalled below. Such a local spectral transfer is associated with a uniquely defined local geometric transfer factor that will be denoted Δ_v^{can}. We need to show that

$$\Delta = \prod_v \Delta_v^{can} .$$

Labesse has given in [L.III.D] an explicit form of the adelic factor on the torus which is a product of $U(1)$. This factor is given as a product of local factors we shall denote Δ_v^L. There are constants λ_v such that, on that torus

$$\Delta_v^L = \lambda_v \Delta_v^{can} .$$

We have to compute these constants and prove that their product is 1.

When dealing with a real place v, the spectral transfer we want to use is described in [C.III.A] and it is shown to be given by the geometric Kottwitz' transfer factors Δ_v^K. This means that:

$$\Delta_v^{can} = \Delta_v^K \ .$$

On the other hand it is shown in [L.III.D] that

$$\Delta_v^L = (-1)^{q(G_v)-q(H_v)}\Delta_v^K$$

with

$$q(G_v) = \frac{1}{2}(\dim G_v - \dim K_v^G)$$

and where K_v^G is a maximal compact subgroup of G_v. In our case G_v is either $U(n,1)$ (at one place) or $U(n+1)$ (at all other places). In all cases we have

$$(-1)^{q(G_v)-q(H_v)} = (-1)^p$$

when $n = 2p$ which yields

$$\Delta_\infty^L = (-1)^{pd}\Delta_\infty^K$$

where d is the degree of F. Since d is assumed to be even we have

$$\Delta_\infty^L = \Delta_\infty^{can} \ .$$

At finite split places endoscopic groups are Levi subgroups and we want to use the obvious spectral transfer: parabolic induction up to a twist by the character imposed by the choice of the L-group embedding. Hence the geometric transfer amounts to taking constant terms along Levi subgroups times a twist by a character compatible with the chosen L-group embedding. The geometric transfer factor is nothing but this character. It is immediate that we have again

$$\Delta_v^L = \Delta_v^{can} \ .$$

At finite inert primes, assumed to be unramified, we want to use the spectral transfer which sends unramified representations to unramified representations in a way compatible with the L-group embedding. Such a transfer is induced by the Langlands-Shelstad transfer factor Δ_v^{LS} computed using unramified data; indeed, this is precisely the content of the Fundamental Lemma (cf. the discussion of Theorem 9.3.3.5 of [H.I.A]). One can either compute directly from their definition or, more simply, use Waldspurger's explicit computation of these factors [W] to see that again

$$\Delta_v^L = \Delta_v^{LS} = \Delta_v^{can} \ .$$

As a conclusion we see that

$$\lambda = \prod_v \lambda_v = 1$$

and hence the automorphic transfer is compatible with our local "canonical" spectral transfer at all places.

References

[A] J. Arthur, An introduction to the trace formula, in Harmonic Analysis, the Trace Formula, and Shimura Varieties, *Clay Mathematics Proceedings*, **4** (2005).

[BR] D. Blasius & J. Rogawski, Motives for Hilbert modular forms, *Inventiones Math.* **114** (1993) 55–87.

[C.III.A] L. Clozel, Identités de caractères en la place archimédienne, this volume, III.A.

[CD] L. Clozel & P. Delorme. Le théorème de Paley-Wiener invariant pour les groupes de Lie réductifs II, *Ann. Scient. E.N.S.* **23** (1990) 193–228.

[CL] L. Clozel & J.-P. Labesse. Orbital integrals and distributions, to appear in Shahidi volume.

[H.I.A] M. Harris, An introduction to the stable trace formula, this volume, I.A.

[HL] M. Harris & J.-P. Labesse. Conditional base change for unitary groups, *Asian J. Math.* **8**, (2004) 653–683.

[JS] H. Jacquet & J. Shalika. On Euler products and the classification of automorphic representations. I. *Amer. J. Math.* **103** (1981) 499–558; II. *Amer. J. Math.* **103** (1981) 777–815.

[KS] C.D. Keys & F. Shahidi. Artin L-functions and normalization of intertwining operators, *Ann. Scient. E.N.S.* **21** (1988) 67–89.

[L] J.-P. Labesse, Pseudo-coefficients très cuspidaux et K-théorie, *Math. Ann.* **291** (1991) 607–616.

[L.IV.A] J.-P. Labesse, Changement de base CM et séries discrètes, this volume, IV.A.

[L.III.D] J.-P. Labesse, Un exemple simple de facteurs de transfert, this volume, III.D.

[LS] R.P. Langlands Jr. & D. Shelstad. On the definition of transfer factors, *Math. Ann.* **278** (1987) 219–271.

[M.III.C] E. Minguez A. Unramified representations of unitary groups, this volume, III.C.

[MW] C. Moeglin & J.-L. Waldspurger. Le spectre résiduel de $GL(n)$, *Ann. Scient. E.N.S.* **22** (1989) 605–674.

[R.II.A] D. Renard, Endoscopy for real reductive groups, this volume, R.II.A.

[Sh1] F. Shahidi, On certain L-functions, *Am. J. Math.* **103** (1981) 297–355.

[Sh2] F. Shahidi, A proof of Langlands' conjecture on Plancherel measures: Complementary series for p-adic groups, *Ann. Math.* **132** (1990) 270–330.

[Sh3] F. Shahidi, Fourier transforms of intertwining operators and Plan-1cherel mesures for GL(n), *Am. J. Math.* **100** (1984) 67–111.

[Sh4] F. Shahidi, Local coefficients as Artin factors for real groups, *Duke Math. J.* **32** (1985) 973–1007.

[Shel1] D. Shelstad, L-indistinguishability for real groups, *Math. Annalen* **259** (1982) 385–430.

[Shel2] D. Shelstad, Tempered endoscopy for real groups I. Geometric transfer with canonical factors, in Representation theory of real reductive Lie groups, *Contemp. Math.* **472**, 215–246, Amer. Math. Soc., Providence, RI (2008).

[W] J.-L. Waldspurger, Les facteurs de transfert pour les groupes classiques : un formulaire (manuscript 2010).

Université Paris-Sud, Membre, Institut Universitaire de France

Institut de Mathématiques de Jussieu, U.M.R. 7586 du CNRS, Membre, Institut Universitaire de France

Institut de Mathématiques de Luminy, U.M.R. 6206 du CNRS

Construction of automorphic Galois representations, I
[CHL.IV.C]

Laurent Clozel, Michael Harris, and Jean-Pierre Labesse

Introduction

Our goal in this concluding chapter is to apply the results of [L.IV.A] and [CHL.IV.B], together with Kottwitz' results in [K1,K2], to obtain a rapid construction of the compatible system of ℓ-adic Galois representations attached to a cohomological cuspidal automorphic representation Π of $GL(n)$ over a CM field, satisfying the symmetry condition corresponding to base change from a unitary group. The Galois representations are shown to be *weakly associated* to Π, in the sense that the characteristic polynomials of Frobenius at almost all places at which Π is unramified bear the expected relation to Π. We thus extend the results of [K3] to Shimura varieties attached to untwisted unitary groups, where endoscopy is an issue; the results of [C3] are generalized to Π not assumed to satisfy a local ramification hypothesis.

We have restricted our attention to Π satisfying an excessively strong regularity hypothesis at archimedean places, in order to guarantee the property (*) of [L.IV.A], which implies that the corresponding representations occur in the spectrum of unitary groups with multiplicity one, leading to simplifications in statements as well as proofs. The construction can be extended to cohomological representations not satisfying this hypothesis by a p-adic continuity argument sketched several years ago by one of the authors and now completed in [CH].

Kottwitz' approach has been reconsidered and updated in the book [M] of S. Morel. The emphasis of [M], however, is on the new phenomena arising in non-compact Shimura varieties, and only the case of imaginary quadratic fields is treated there.

The construction of even-dimensional Galois representations follows the strategy used in [BR] when $n = 2$, and necessarily breaks down when Π fails to satisfy a weak regularity condition. This condition arises as well, and for the same reason, in the article [S] of Shin, which gives a somewhat different construction of the Galois representations attached to Π, under precisely this weak version of regularity. Shin's results are more complete than ours: his Galois representations are shown to satisfy a strong version of compatibility with the local Langlands correspondence at finite places. His results thus generalize the combined results of [HT] and [TY]. An expository version of [S] should appear in Book 2.

1. Cohomology of unitary Shimura varieties, following Kottwitz

Notation is as in [CHL.IV.B]. In particular, F is a totally real field, $\mathcal{K}/F$ is a totally imaginary quadratic extension, $\eta = \eta_{\mathcal{K}/F}$ the corresponding quadratic character of $\mathbb{A}_F^\times$, $d = [F : \mathbb{Q}]$, $c \in Gal(\mathcal{K}/F)$ the non-trivial Galois automorphism. We assume $\mathcal{K} = F{\cdot}E$ for some imaginary quadratic extension E, chosen compatibly with the simplifying hypotheses of [CHL.IV.B, (1.2)]; we make this more explicit in (1.2) below. The quadratic character for $E/\mathbb{Q}$, a character of $\mathbb{A}^\times/\mathbb{Q}^\times$, is denoted $\eta_{E/\mathbb{Q}}$. Except where otherwise indicated, n denotes a positive even integer.

1.1. Automorphic representations of similitude and unitary groups. The results of [L.IV.A] and [CHL.IV.B] relate automorphic representations of unitary groups and θ-stable representations of linear groups, under the simplifying hypotheses. We will be constructing Galois representations in the cohomology of Shimura varieties attached to unitary similitude groups, and the formalism developed by Kottwitz in [K1] and [K2] is adapted to similitude groups rather than unitary groups. Automorphic representations of similitude and unitary groups can be related without redoing the comparisons of the two previous chapters, as in [C3] and [HT].

In this section we let G be the unitary group denoted $U(V_1)$ in [CHL.IV.B, 1.4.1], the stabilizer of the hermitian form $<\bullet, \bullet>$ on the $n+1$-dimensional space V_1. Let $GU = GU(V_1)$ be the algebraic group over $\mathbb{Q}$ defined by

$$GU(R) = \{g \in GL(V_1 \otimes_{\mathbb{Q}} R) \mid \; < gv, gw >= \nu(g) < v, w >, \; v, w \in V_1 \otimes_{\mathbb{Q}} R\}$$

for any $\mathbb{Q}$-algebra R, where $\nu(g) \in R^\times$. There is a short exact sequence

$$1 \to G \to GU \xrightarrow{\nu} \mathbb{G}_m \to 1 \qquad (1.1.1)$$

of algebraic groups over $\mathbb{Q}$, but G is also the restriction of scalars of a group over F that we also denote G.

Let $T = R_{E/\mathbb{Q}}\mathbb{G}_{m,E}$ and let $T^1 = \ker N : T \to \mathbb{G}_{m,\mathbb{Q}}$ where $N = N_{E/\mathbb{Q}}$ is the norm homomorphism. Thus T^1 is the group $U(1)$ for the quadratic extension $E/\mathbb{Q}$ and in particular can be identified with a subgroup of the center of G. There is an injective homomorphism

$$T^1 \to T \times G; \; t \mapsto (t, t^{-1})$$

that gives rise to a short exact sequence of algebraic groups

$$1 \to T^1 \to T \times G \xrightarrow{\phi} GU \to 1 \qquad (1.1.2)$$

where $T \to GU$ is the obvious inclusion in the center. Let Z denote the center of GU. Then

$$Z(\mathbb{Q}) = \{x \in \mathcal{K}^\times \mid N_{\mathcal{K}/F}(x) \in \mathbb{Q}^\times\}$$

and $Z(R)$ is determined similarly for any $\mathbb{Q}$-algebra R. Let $A = \mathbb{R}_+^\times$ be the obvious diagonal subgroup of $Z(\mathbb{A})$, associated to $\mathbb{Q}^\times \subset \mathcal{K}^\times$. This is also the identity component of the real points of the split center of GU. The group $Z(\mathbb{Q})A$ is closed in $Z(\mathbb{A})$ and $Z(\mathbb{A})/Z(\mathbb{Q})A$ is compact. Write $\Gamma = GU(\mathbb{Q})\cdot Z(\mathbb{A})$. Then $\Gamma = GU(\mathbb{Q})A\omega$, for some compact subset $\omega \subset Z(\mathbb{A})$, and it follows that Γ is closed. Let

$$\mathbf{S} = G(\mathbb{A})GU(\mathbb{Q})Z(\mathbb{A}) = \Gamma.G(\mathbb{A});$$

this is an open (so closed) subgroup of $GU(\mathbb{A})$, because the map $Z \times G \to GU$ is open locally, and surjective on the groups of $\mathbb{Z}_p$-points at unramified primes by Lang's theorem.

LEMMA 1.1.3. When $n+1$ is odd

$$GU(\mathbb{A}) = \mathbf{S}\ .$$

Proof. By the above discussion the quotient

$$C = GU(\mathbb{A})/\mathbf{S}$$

is the quotient of a subgroup of $\mathbb{G}_m(\mathbb{A})$, the image of ν, by a subgroup containing the image of rational elements and norms of idèles of E. Since $\dim V_1$ is odd and at least 3 the map ϕ is surjective locally at all but a finite set S of finite primes, namely the primes of $\mathbb{Q}$ that are not split in E and ramify in $\mathcal{K}$ (cf. [HT], p. 201; a proof is given in the appendix.) Hence C is a finite group of exponent 2. A non trivial element in C can be represented by an element of $GU(\mathbb{Q}_S)$. But by weak approximation it is in the image of $GU(\mathbb{Q})$. In other words C is trivial, and $GU(\mathbb{A}) = \mathbf{S}$. □

In particular, if we temporarily let $\Sigma = G(\mathbb{A}) \cap \Gamma$ then, for $n+1$ odd

$$\Gamma\backslash GU(\mathbb{A}) \simeq \Sigma\backslash G(\mathbb{A})$$

PROPOSITION 1.1.4. Assume $n+1$ is odd. Let π be an irreducible automorphic representation of GU whose restriction to G contains an irreducible representation σ. Suppose σ occurs with multiplicity one in the discrete spectrum for G. Then π occurs with multiplicity one in the discrete spectrum for GU.

Moreover, if χ is the central character of π, then π is the only automorphic representation of GU containing σ with central character χ.

Proof. For any Hecke character χ of $Z(\mathbb{A})$, let

$$\mathcal{A}(GU,\chi) = L^2(\Gamma\backslash GU(\mathbb{A}),\chi)$$

be the space of automorphic forms on $GU(\mathbb{A})$, square-integrable modulo $Z(\mathbb{A})$, on which $Z(\mathbb{A})$ acts by χ, or equivalently Γ acts by χ extended trivially to $GU(\mathbb{Q})$. We have

$$\mathcal{A}(GU,\chi) = I_\chi := Ind_\Gamma^{\mathbf{S}}\chi,$$

since $\mathbf{S} = GU(\mathbb{A})$. We may assume that π has central character χ, thus the multiplicity we need to determine is

$$m_{GU}(\pi) = \dim Hom_{GU(\mathbb{A})}(\pi, \mathcal{A}(GU,\chi)) = \dim Hom_{\mathbf{S}}(\pi, Ind_\Gamma^{\mathbf{S}}\chi).$$

Similarly, letting $\Sigma = G(\mathbb{A}) \cap \Gamma$, we have, $\mathcal{A}(G)$ being $L^2(G(\mathbb{Q})\backslash G(\mathbb{A}))$:

$$\mathcal{A}(G) = \oplus_\xi \mathcal{A}(G,\xi)$$

where for any character ξ of Σ trivial on $G(F)$, $\mathcal{A}(G,\xi) = L^2(\Sigma\backslash G(\mathbb{A}),\xi)$ is defined by analogy with $\mathcal{A}(GU,\chi)$. As above, we can identify

$$\mathcal{A}(G,\xi) = Ind_\Sigma^{G(\mathbb{A})}\xi.$$

Now suppose ξ is the restriction to Σ of χ and that σ occurs in $\mathcal{A}(G,\xi)$. Observe that since

$$\Gamma\backslash GU(\mathbb{A}) \simeq \Sigma\backslash G(\mathbb{A})$$

then

$$\mathcal{A}(G,\xi) = Res_{G(\mathbb{A})} I_\chi.$$

and

$$m_G(\sigma) := \dim Hom_{G(\mathbb{A})}(\sigma, \mathcal{A}(G,\xi)) = \dim Hom_{G(\mathbb{A})}(\sigma, Res_{G(\mathbb{A})} I_\chi).$$

But

$$\dim Hom_{G(\mathbb{A})}(\sigma, Res_{G(\mathbb{A})} I_\chi) = \sum_\pi \dim Hom_{G(\mathbb{A})}(\sigma, Res_{G(\mathbb{A})}\pi)$$

where π runs over representations of $GU(\mathbb{A})$ in I_χ, counted with multiplicity. Since, by hypothesis

$$\dim Hom_{G(\mathbb{A})}(\sigma, \mathcal{A}(G,\xi)) = 1$$

there is exactly one such representation π and

$$\dim Hom_{GU(\mathbb{A})}(\pi, I_\chi) = 1$$

and we are done. □

1.1.5. Remark. If $n+1$ is even, then the group C defined as above is still a group of exponent 2 but is not in general trivial, cf. the discussion at the end of the appendix. The argument above breaks down if $Res_{\mathbf{S}}\pi$ is reducible. If this happens, then it is necessarily the case that $\pi \xrightarrow{\sim} \pi \otimes \varepsilon$ for some non-trivial character of C. One expects then that π arises from some endoscopic group of GU, as in [LL].

1.2. Base change and descent for similitude groups. Let σ be an automorphic representation of G, with central character ξ_σ, and let ξ be a Hecke character of T (called χ in §1). By the results of the previous section, if

$$\xi(t) = \xi_\sigma(t), \forall t \in T^1(\mathbb{A}), \tag{1.2.1}$$

then the pair (σ, ξ) defines a representation $\sigma * \xi$ of the open subgroup $\mathbf{V} = im(\phi) \subset GU(\mathbb{A})$, occurring in $L^2(\phi(G(\mathbb{Q}) \times T(\mathbb{Q}))\backslash\mathbf{V})$. Condition (1.2.1) corresponds to the condition that the representation $\sigma \otimes \xi$ of $G(\mathbb{A}) \times T(\mathbb{A})$ factors via the map ϕ.

In what follows, n is even and π will be any automorphic representation of $GU(\mathbb{A})$ whose restriction to $im(\phi)$ contains $\sigma * \xi$ (and hence its restriction to $G(\mathbb{A})$ contains σ). We recall the special hypotheses in force in the two previous chapters.

SPECIAL HYPOTHESES 1.2.2.

(1.2.2.1) $\mathcal{K}/F$ is unramified at all finite places (in particular $d > 1$).
(1.2.2.2) Π_v is spherical (unramified) at all non-split non-archimedean places v of $\mathcal{K}$.
(1.2.2.3) The degree $d = [F : \mathbb{Q}]$ is even.

SPECIAL HYPOTHESIS 1.2.3. The highest weight $\mu(v)$ is sufficiently far from the walls. This comes in three strengths:

(i) (Regularity) For at least one $v \in \Sigma_F$, $\mu(v)$ is regular, i.e. $\mu_i(\tilde{v}) \neq \mu_j(\tilde{v})$ if $i \neq j$.
(ii) (Strong regularity) $\mu(v)$ satisfies Hypothesis 1.3 of [CHL.IV.B] for all $v \in \Sigma_F$; i.e. $|\mu_i(\tilde{v}) - \mu_j(\tilde{v})| > 3$ if $i \neq j$.
(iii) (Weak regularity) For at least one $v \in \Sigma_F$, there is at least one **odd** i such that $\mu_i(\tilde{v}) > \mu_{i+1}(\tilde{v})$.

In order to apply the results of [CHL.IV.B], the main theorem 2.2.3 will be stated for strongly regular representations.

PROPOSITION 1.2.4. Let Π be a cuspidal automorphic representation of $GL(n+1)_{\mathcal{K}}$ that is θ_{n+1}-stable, cohomological at archimedean places, and satisfies the property (*) of [L.IV.A, §5]. Moreover, assume Π is spherical at all places inert in $\mathcal{K}/F$, and let σ be an automorphic representation of G, whose existence is guaranteed by [L.IV.A, 5.3], whose weak base change is isomorphic to Π. Assume the infinitesimal character of Π_∞ is strongly regular. Then

(1) σ_∞ is in the discrete series.
(2) For any σ'_∞ in the same L-packet, $\sigma'_\infty \otimes \sigma_f$ occurs in the automorphic spectrum of G with multiplicity one. Moreover, any automorphic representation σ' of G with weak base change isomorphic to Π is of this kind.

(3) There is a Hecke character ξ of T such that the central character ξ_Π of Π satisfies

$$\xi_\Pi(t) = \xi(t^c/t), \ \forall t \in T(\mathbb{A}) \subset \mathcal{K}^\times(\mathbb{A}) = Z_{GL(n+1)_\mathcal{K}}(\mathbb{A});$$

(4) Let σ' be any automorphic representation of G with weak base change isomorphic to Π. There exists an automorphic representation π of $GU(\mathbb{A})$ whose pullback to $G(\mathbb{A}) \times T(\mathbb{A})$ via the map ϕ of (1.1.2) contains $\sigma' \otimes \xi$. Any such π occurs in the discrete automorphic spectrum of GU with multiplicity 1.

Proof. We know that σ_∞ is cohomological, so the first assertion follows from the strong regularity hypothesis. Assertion (2) is a consequence of [L.IV.A,5.3]. The claim (3) is proved in [HT, VI.2.10]. Finally, (4) follows from (2) and Proposition 1.1.4. □

COROLLARY 1.2.5. Notation is as in Proposition 1.2.4. The set of automorphic representations π' of $GU(\mathbb{A})$ with $\pi'_f \xrightarrow{\sim} \pi_f$ has $n+1$ members. The restrictions of these π'_∞ to $G(\mathbb{R})$ are precisely the $n+1$ elements of the discrete series L-packet of σ_∞.

The analogue for the endoscopic representations considered in [CHL.IV.B] is the following. Notation is as in [CHL.IV.B].

PROPOSITION 1.2.6. Let τ be a cuspidal automorphic representation of $GL(n)_\mathcal{K}$ that is θ_n-stable, cohomological at archimedean places, and satisfies the property (*) of [L.IV.A, §5]. Moreover, assume τ is spherical at all places inert in $\mathcal{K}/F$. Assume the infinitesimal character of τ_∞ is strongly regular. Let χ be a θ_1-stable cohomological Hecke character of $GL(1)_\mathcal{K}$ chosen as in Theorem 4.6 of [CHL.IV.B]. Let σ be an automorphic representation of G, whose existence is guaranteed by [CHL.IV.B, 4.6], whose weak base change to $\mathcal{G}_{n+1} = GL(n+1)_\mathcal{K}$ is isomorphic to $\Pi = \tau \times \chi$.

Then

(1) σ_∞ is in the discrete series. Let $\Pi_G(\phi_G)$ denote the corresponding discrete series L-packet.

(2) For any σ'_∞ in $\Pi^+ \subset \Pi_G(\phi_G)$, $\sigma'_\infty \otimes \sigma_f$ occurs in the automorphic spectrum of G with multiplicity one. Moreover, any automorphic representation σ' of G with weak base change isomorphic to Π is of this kind. In particular, if $\sigma'_\infty \in \Pi^-$, then $\sigma'_\infty \otimes \sigma_f$ does not occur in the automorphic spectrum of G.

(3) There is a Hecke character ξ of T such that the central character ξ_Π of Π satisfies

$$\xi_\Pi(t) = \xi(t^c/t), \ \forall t \in T(\mathbb{A}) \subset \mathcal{K}^\times(\mathbb{A}) = Z_{GL(n+1)_\mathcal{K}}(\mathbb{A});$$

(4) Let σ' be any automorphic representation of G with weak base change isomorphic to Π. There exists an automorphic representation π of $GU(\mathbb{A})$ whose pullback to $G(\mathbb{A}) \times T(\mathbb{A})$ via the map ϕ of

(1.1.2) contains $\sigma' \otimes \xi$. Any such π occurs in the discrete automorphic spectrum of GU with multiplicity 1.

COROLLARY 1.2.7. Notation is as in Proposition 1.2.6. The set of automorphic representations π' of $GU(\mathbb{A})$ with $\pi'_f \xrightarrow{\sim} \pi_f$ has n members. The restrictions of these π'_∞ to $G(\mathbb{R})$ are precisely the n elements of the subset Π^+ of the discrete series L-packet of σ_∞.

1.3. Endoscopic groups of similitude groups. The L-group of GU in the Galois version is a semidirect product of $GL(n+1,\mathbb{C})^{[F:\mathbb{Q}]} \times GL(1,\mathbb{C})$ with $Gal(\overline{\mathbb{Q}}/\mathbb{Q})$. The factors of $GL(n+1,\mathbb{C})^{[F:\mathbb{Q}]}$ are indexed by embeddings $\alpha : \mathcal{K} \to \overline{\mathbb{Q}}$ above a fixed embedding of E, and the elements of $Gal(\overline{\mathbb{Q}}/E)$ act transitively on the factors of $GL(n+1,\mathbb{C})$ by permuting the indices in the tautological way, while leaving the factor $GL(1,\mathbb{C})$ untouched. Any complex conjugation $c \in Gal(\overline{\mathbb{Q}}/\mathbb{Q})$ acts by

$$(g_\alpha, x) \mapsto (x w_0 {}^t g_\alpha^{-1} w_0^{-1}, x) \tag{1.3.1}$$

where w_0 is the antidiagonal matrix with entries $(1,-1,\ldots,1)$.

The homomorphism $T \times G \to GU$ identifies by duality ${}^LGU$ with a subgroup of ${}^LT \times {}^LG$ that maps surjectively to ${}^LG$; let ϕ^* denote this map. Let (H', s', ξ') be an elliptic endoscopic datum for ${}^LGU$. Then $\xi = \phi^* \circ \xi' : H' \to {}^LG$ defines an elliptic endoscopic datum (H, s, ξ) with $s = \phi^*(s')$ and $H = \xi(H')$. This establishes a bijection between equivalence classes of elliptic endoscopic triples for GU and for G. [1]

Specifically, for any partition $n+1 = a+b$, there is an elliptic endoscopic group $GH = GH_{a,b}$ of GU which can be identified with the subgroup

$$\{\begin{pmatrix} g_1 & 0 \\ 0 & g_2 \end{pmatrix} \in (GL(a) \times GL(b)) \cap GU^*\} \tag{1.3.2}$$

where GU^* is the quasi-split inner form of GU and the homomorphism of L-groups from that of GH to that of GU depends on the choice of Hecke characters μ_a and μ_b, as in [H.I.A, (5.5)]. More precisely, the identification of the group (1.3.2) with the centralizer in ${}^LGU$ of the element

$$s'_{a,b} = ((diag(I_a, -I_b)_\alpha, 1),$$

in the above coordinates of $\widehat{GU}$, is given by a map ξ_{μ_a,μ_b} defined as in [H.I.A, (5.5)].

Letting H be the corresponding elliptic endoscopic group for G, there is a short exact sequence

$$1 \to T^1 \to T \times H \overset{\phi_H}{\to} GH \to 1 \tag{1.3.3}$$

just as in (1.1.2).

---

[1]This fails for $GU(n)$ for n even.

Lemma 1.3.4. The image of $\phi_H : T(\mathbb{A}) \times H(\mathbb{A}) \to GH(\mathbb{A})$ is of finite index in $GH(\mathbb{A})$. In particular, $GH(\mathbb{Q}) \cdot im(\phi_H) = GH(\mathbb{A})$.

Proof. Let $\nu : GH \to \mathbb{G}_m$ be the similitude homomorphism. Since $n+1$ is odd, one of (a, b), say b, is also odd. Thus

$$\nu(GH(\mathbb{A})) \supset \nu_b(GU(b)^*(\mathbb{A}))$$

where $GU(b)^*$ is the quasi-split group of rational unitary similitudes of size b and ν_b is the corresponding similitude factor. But this group contains $N(T(\mathbb{A}))$ as a subgroup of finite index, as we have already seen in §1.1 (cf. appendix for details). This implies the first statement, and the second statement follows by weak approximation, as in §1.1. □

The proof of Proposition 1.1.4 then applies without change to GH, and we obtain

Proposition 1.3.5. Let π be an automorphic representation of GH whose restriction to H contains a representation σ. Suppose σ occurs with multiplicity one in the discrete spectrum for H. Then π occurs with multiplicity one in the discrete spectrum for GH. Moreover, π is uniquely determined by its central character among automorphic representations containing σ.

In their study of Galois representations on the cohomology of Shimura varieties, Morel and Shin [M,S] do not verify the multiplicity one property. The multiplicity is left undetermined in Morel's formulas, whereas Shin makes use of Taylor's argument, published as [HT, VII.1.8], to obtain a Galois representation of the right dimension.

2. Statement of the main theorem

2.1. Global parameters. The following discussion has been strongly influenced by Arthur's ongoing work on functoriality for classical groups and Moeglin's analysis of local L-packets in the same setting. In this section, n is not assumed even, $G = U(V)$ and $GU = GU(V)$ (rational similitudes) for some n-dimensional hermitian space (*not* $n+1$-dimensional).

As in §4 below, Kottwitz' conjectural formula for the intersection cohomology of Shimura varieties is expressed in terms of Arthur's conjectural parametrization of discrete automorphic representations of the group GU by homomorphisms

$$\psi : \mathcal{L}_{\mathbb{Q}} \times SL(2, \mathbb{C}) \to {}^L GU,$$

where $\mathcal{L}_{\mathbb{Q}}$ is the hypothetical Langlands group. (Such a map is generally called an *Arthur parameter.*) As Arthur and Moeglin have emphasized in their respective contexts, for a classical group H whose L-group naturally embeds in ${}^L GL(n)$, representations of $\mathcal{L}$ can be replaced by collections of cuspidal automorphic representations Π_i of $GL(n_i)$ for some partition

$n = \sum_i n_i$, where the collection satisfies a group-theoretic symmetry condition replacing the condition that the (hypothetical) corresponding homomorphism $\psi = \oplus_i \psi_{\Pi_i}$ take values in the image of the L-group of H in ${}^L GL(n)$. This description, valid for (conjecturally) tempered parameters, can be extended to parameters that are non-trivial on the factor $SL(2, \mathbb{C})$, using the Moeglin-Waldspurger determination of the discrete automorphic spectrum of $GL(n)$.

The work of Arthur and Moeglin (for example [A1, Mo]) shows that conjectures of Arthur that seemed to depend on the existence and properties of $\mathcal{L}_{\mathbb{Q}}$ can be phrased unconditionally (in terms of packets of cuspidal representations Π_i) and then *proved*, at least for special orthogonal or symplectic groups (Arthur [A1], see also the final chapter of [A2]). Our job, in §4, will therefore consist in replacing, in Kottwitz' formulation, statements apparently depending on $\mathcal{L}_{\mathbb{Q}}$ by propositions that, in our simple cases, can be stated unconditionally and proved.

For example, let U temporarily be any inner form of $U(n)$ over F – for example $U = G$ in the notation of §1 – and let σ be an automorphic representation of U occurring in the discrete spectrum, with σ_∞ in the discrete series and satisfying the regularity property (*) of [L.IV.A].[2] Corollaire 5.3 of [L.IV.A] states that σ then admits a base change to $GL(n)$ of the form $\Pi_1 \boxplus \Pi_2 \boxplus \cdots \boxplus \Pi_r$ for some partition $n = \sum_i n_i$ as above, with each Π_i in the automorphic discrete spectrum of $GL(n_i)$. The symmetry condition is that each Π_i is θ_{n_i}-stable.

Moreover, since $\boxplus_i \Pi_i$ is cohomological (cf. [L.IV.A, §5]), its infinitesimal character is regular, which implies in particular that

$$\Pi_i \not\simeq \Pi_j \text{ for } i \neq j.$$

Since σ_∞ is tempered, it is presumably the case that each Π_i is in fact cuspidal, but Labesse does not verify this; however, his proof actually does provide enough information to draw this conclusion. The unordered collection

$$\{[\Pi_1], [\Pi_2], \ldots, [\Pi_r]\} \tag{2.1.1}$$

can be taken to be the parameter of σ; the brackets have been added to make the notation more parameter-like. The θ_{n_i}-stability of Π_i corresponds – formally, since one has no $\mathcal{L}$ – to the possibility of extending the parameter to a parameter with values in ${}^L U$. This begs the question of the uniqueness of such an extension, but the requirement that σ_∞ be in the discrete series determines the extension of the archimedean parameter in the Langlands classification (cf. the discussion in (3.2) of [C.III.A]), and this in turn (formally) determines the extension of the parameter of Π uniquely.

[2]Condition (*) is satisfied if σ_∞ is sufficiently regular, for example if the associated rational representation of U satisfies special hypothesis (1.2.3)(i).

The parameter for an automorphic representation of GU has in addition a Hecke character ξ (not dignified with brackets) of the imaginary quadratic field E—of the torus T of §1—subject to the compatibility condition (1.2.1). It follows from the last part of Proposition 1.1.4 that the pair (σ, ξ) determines a unique automorphic extension π of the representation $\sigma * \xi$ to $GU(\mathbb{A})$. Thus if (2.1.1) is the parameter of σ, we can take

$$\psi = \psi_\pi = \{[\Pi_1], [\Pi_2], \ldots, [\Pi_r], \xi\} \tag{2.1.2}$$

to be the parameter of the automorphic representation π of GU. The discussion in §1 shows that (2.1.2) is sufficient to determine ψ uniquely when n is odd. When n is even we leave the possible ambiguity open.

In order to apply Kottwitz's formulas one needs to be able to calculate with the group $\mathfrak{S}_\psi$, where ψ is a global Arthur parameter with values in the L-group of GU, where

$$S_\psi = Cent(\psi, {}^L GU)Z(\widehat{GU}) \subset {}^L(GU)$$

where $Cent(\psi, {}^L GU)$ is the centralizer of ψ in the L-group, and

$$\mathfrak{S}_\psi = S_\psi / S_\psi^0 Z(\widehat{GU}),$$

the quotient of the component group of S_ψ by the image of $Z(\widehat{GU})$.

What is this centralizer for our parameters? Consider first the *unitary* group $G = U(V_1)$ seen, as is GU, as a $\mathbb{Q}$-group, and a parameter $\psi = \{[\Pi_1], [\Pi_2], \ldots, [\Pi_r]\}$ for $GL(n)$ as in (2.1.1). The group $G/\mathbb{Q}$ is obtained by restriction of scalars from a unitary group G_1 over F. Then ${}^L(G_1/F) = GL(n, \mathbb{C})$. It is natural – and this is consistent with Arthur's principles for classical groups [A2, Chapter 30] – to assume that the centralizer of ψ is $\prod_{i=1}^r Z({}^L GL(n_i, \mathbb{C})) = \prod_i \mathbb{C}^\times$. (Indeed, this is appropriate because, as we have seen, the different Π_i are pairwise non-isomorphic.) Here we view ψ as a parameter $\mathcal{L}_\mathcal{K} \to GL(n, \mathbb{C}) = {}^L(G_1/\mathcal{K})$. Now if we view ψ (or more correctly, its extension to $\mathcal{L}_F$) as a parameter for G_1, the computation of its centralizer S_ψ translates into the condition that the elements of S_ψ commute with complex conjugation as defined in a number of earlier chapters, for example in (3.1) of [C.III.A]. As this acts in each factor $\mathbb{C}^\times$ as $z \mapsto z^{-1}$, we see that the centralizer C is $(\pm 1)^r$, and

$$\mathfrak{S}_\psi = C \cdot Z(\hat{G}_1)/Z(\hat{G}_1) = C/C \cap Z(\hat{G}_1) = (\pm 1)^r/(\pm I_n) \simeq (\pm 1)^{r-1}, \tag{2.1.3}$$

with $(\pm I_n) \subset Z(GL(n, \mathbb{C}))$ embedded diagonally. One views the elements of $\mathfrak{S}_\psi$ as diagonal matrices in $GL(n, \mathbb{C})$, up to multiplication by $\pm I_n$.

The computation of $\mathfrak{S}_\psi$, however, is invariant under restriction of scalars, by a variant of Shapiro's lemma which we do not have to justify, as this section is purely motivational. Therefore, we have computed $\mathfrak{S}_\psi$, when ψ is seen as a parameter for the $\mathbb{Q}$-group G. Finally, if we replace G by the group GU—with L-group described by (1.3.1), and with parameter $\psi = \{[\Pi_1], [\Pi_2], \ldots, [\Pi_r], \xi\}$—the computation of $\mathfrak{S}_\psi$ gives the same result; in other words, the presence of ξ makes no difference.

2.1.4. In the applications, we only consider $r = 1$ (the *stable case*), in which case $\mathfrak{S}_\psi$ is trivial, or $r = 2$ (the *endoscopic case*), with $\mathfrak{S}_\psi = \mathbb{Z}/2\mathbb{Z}$. In keeping with the conventions of previous chapters, we write $[\Pi]$ instead of $[\Pi_1]$ in the stable case and $([\tau], [\chi])$ instead of $\{[\Pi_1], [\Pi_2]\}$ in the endoscopic case, with τ cuspidal and θ_n-stable on $GL(n)$ and χ a θ_1-stable Hecke character of $GL(1)$. We always assume χ chosen to satisfy the hypotheses of [CHL.IV.B, Theorem 4.6].

Our special hypotheses are in force here and prescribe that $[\Pi]$, $[\tau]$, and $[\chi]$ all be spherical at places inert in $\mathcal{K}/F$. Then we can apply the results of §1.2. If $\psi = \{[\Pi_1], [\Pi_2], \ldots, [\Pi_r], \xi\}$, $r \leq 2$, then we fix π as in Propositions 1.2.4 and 1.2.6 and let $\Pi(\psi)$ denote the set of representations π' of $GU(\mathbb{A})$ with $\pi'_f \xrightarrow{\sim} \pi_f$. We have

$$(2.1.4.1) \qquad |\Pi(\psi)| = \begin{cases} n+1 & r = 1 \\ n & r = 2. \end{cases}$$

Remark 2.1.4.2. The choice of π fixes an element of the *near equivalence class* of π_f, that is to say the class of representations of $GU(\mathbf{A}^f)$ that are locally isomorphic to π_f almost everywhere. We believe this near equivalence class is a singleton under our hypotheses. For even dimensional unitary groups this need not be the case, but the choice of a hyperspecial maximal compact subgroup at all finite places should still single out an element of the near equivalence class. This question may have been treated explicitly in the literature, but we have been unable to find a complete reference.

2.2. Conjectures on Galois representations. For the purposes of this section we let n be any positive integer and let $\mathcal{G} = \mathcal{G}_n$ be the algebraic group $R_{\mathcal{K}/\mathbb{Q}}GL(n)_{\mathcal{K}}$. Let $\mathfrak{g} = Lie(\mathcal{G}(\mathbb{R}))$, $K_\infty \subset \mathcal{G}(\mathbb{R})$ the product of a maximal compact subgroup with the center $Z_{\mathcal{G}}(\mathbb{R})$. We consider cuspidal automorphic representations Π of $\mathcal{G}$ satisfying the following two hypotheses:

General Hypotheses 2.2.1. Writing $\Pi = \Pi_\infty \otimes \Pi_f$, where Π_∞ is an admissible $(\mathfrak{g}, K_\infty)$-module, we have

(i) (Regularity) There is a finite-dimensional irreducible representation $W(\Pi) = W_\infty$ of $\mathcal{G}(\mathbb{R})$ such that

$$H^*(\mathfrak{g}, K_\infty; \Pi_\infty \otimes W_\infty) \neq 0.$$

(ii) (Polarization) The contragredient $\Pi^\vee$ of Π satisfies

$$\Pi^\vee \xrightarrow{\sim} \Pi \circ c.$$

A special case of the conjectural global Langlands correspondence between automorphic representations of algebraic type [C2] and compatible systems of Galois representations is given by the following conjecture:

Conjecture 2.2.2. Let Π be a cuspidal automorphic representation of $\mathcal{G}$ satisfying Hypotheses 2.2.1. There is a compatible family of continuous representations

$$\{\rho_{\Pi,\lambda}\} : Gal(\overline{\mathbb{Q}}/\mathcal{K}) \to GL(n, E(\Pi)_\lambda),$$

as λ runs through non-archimedean completions of a certain number field $E(\Pi)$, satisfying

(a) $\rho_{\Pi,\lambda}$ is **geometric** in the sense of Fontaine and Mazur [FM]. More precisely, $\rho_{\Pi,\lambda}$ is unramified outside a finite set of places of $\mathcal{K}$, namely the set of places of ramification of Π and the set of primes of $\mathcal{K}$ dividing the residue characteristic ℓ of λ. Moreover, at places dividing ℓ, $\rho_{\Pi,\lambda}$ is de Rham, in Fontaine's sense, and in particular is Hodge-Tate.

(b) $\rho_{\Pi,\lambda}$ is **HT regular**: the Hodge-Tate weights of $\rho_{\Pi,\lambda}$ have multiplicity one.

(c) There is a non-degenerate bilinear pairing

$$\rho \otimes \rho \circ c \ \to E(\Pi)_\lambda \otimes \mathbb{Q}_\ell(1-n).$$

This correspondence has the following properties:

(i) For any finite place v prime to the residue characteristic ℓ of λ,

$$(\rho_{\Pi,\lambda} \mid_{\Gamma_v})^{Frob-ss} = \mathcal{L}(\Pi_v).$$

Here Γ_v is a decomposition group at v and $\mathcal{L}$ is the **normalized** local Langlands correspondence (denoted $r_\ell(\Pi_v)$ in [HT]). The superscript on the left denotes Frobenius semisimplification.

(ii) The representation $\rho_{\Pi,\lambda} \mid_{\Gamma_v}$ is de Rham for any v dividing ℓ and the Hodge-Tate numbers at v are explicitly determined by the highest weight of W_∞, by the formula given in [HT,Theorem VII.1.9], cf. §4.5, below.

(iii) If Π_v is unramified then $\rho_{\Pi,\lambda} \mid_{\Gamma_v}$ is crystalline.

Note that the correspondence $\mathcal{L}$ in (i) must include a "half-integral twist" for reasons of purity. We refer to the introduction of [HT] for a complete discussion.

Through the efforts of Shin, Bellaïche, Chenevier, and Sorensen, as well as the authors of the present article, this conjecture is now nearly completely established. The statement and an account of the proof of a slighly weaker version – Frobenius semisimplification in (i) is replaced by semisimplification when Π is not weakly regular, in the sense of 1.2.3 – should appear in Book 2. The main theorem of the present article is

THEOREM 2.2.3. Let Π be as in Conjecture 2.2.2. Assume Π is strongly regular, in the sense of 1.2.3. Then there is a number field $E(\Pi)$ and a compatible family of continuous representations

$$\{\rho_{\Pi,\lambda}\} : Gal(\overline{\mathbb{Q}}/\mathcal{K}) \to GL(n, E(\Pi)_\lambda)$$

satisfying (a), (b), (c), and (ii) of Conjecture 2.2.2. Moreover, for almost all finite places v of $\mathcal{K}$ prime to the residue characteristic ℓ of λ at which Π is unramified,

$$(2.2.4) \qquad (\rho_{\Pi,\lambda}\mid_{\Gamma_v})^{Frob-ss} = \mathcal{L}(\Pi_v).$$

Finally, if Π_v and $\mathcal{K}$ are unramified at all primes v dividing the residue characteristic ℓ, then $\rho_{\Pi,\lambda}\mid_{\Gamma_v}$ is crystalline at v.

This is a direct consequence of the approach developed by Kottwitz in [K1] and [K2], together with the results of [L.IV.A] and [CHL.IV.B]. The bulk of §4 is devoted to reformulating the latter results as a substitute for the appeal to the (still conjectural) Arthur parameters in [K1].

The following section constructs Galois representations of the appropriate dimension.

2.3. Local systems and Galois representations. In this section $\mathcal{G} = R_{\mathcal{K}/\mathbb{Q}}GL(n+1)_{\mathcal{K}}$ with n even, as in §1. We fix a complex embedding ι of the imaginary quadratic field E and let Σ be the CM type of $\mathcal{K}$ consisting of all extensions of ι to $\mathcal{K}$ (so the notation of (1.1) is no longer in force). The irreducible representation $W(\Pi)_{\mathbb{R}}$ of $R_{\mathcal{K}/\mathbb{Q}}GL(n+1)_{\mathcal{K}}(\mathbb{R})$ factors over the set Σ_F of real embeddings of F:

$$W(\Pi) = \bigotimes_{v\in\Sigma_F} W_v,$$

where W_v is an irreducible representation of $\mathcal{G}(\mathcal{K}\otimes_{F,v}\mathbb{R})\xrightarrow{\sim}GL(n+1,\mathbb{C})\times GL(n+1,\mathbb{C})$. The highest weight of W_v is denoted $\mu(v)$; it can be identified in the usual way with a pair of non-increasing n-tuples of non-negative integers $(\mu(\tilde{v}),\mu(\tilde{v}^c))$, where $\tilde{v}$ is the extension of $v\in\Sigma_F$ to an element of Σ and we write

$$(2.3.1) \qquad \mu(\tilde{v}) = (\mu_1(\tilde{v}) \geq \mu_2(\tilde{v}) \geq \cdots \geq \mu_n(\tilde{v})).$$

We assume $W(\Pi)$ is strongly regular, as in (1.2.3.ii), so that all the inequalities in (2.3.1) are in fact strict.

Let W_+ be the corresponding irreducible representation of $G(\mathbb{R})$, as in the discussion following Theorem 2.3 of [CHL.IV.B]. We use the same notation to denote an extension to $GU(\mathbb{R})$. This representation has a rational model over a finite extension $E(W)$ of E. For any rational prime ℓ and any ℓ-adic place $\lambda*$ of $E(W)$, we obtain a representation $GU(E(W)_{\lambda*})\rightarrow GL(W_{+,\lambda*})$ that defines a local system

$$\tilde{W}_{\lambda*} = \tilde{W}(\Pi)_{\lambda*}$$

in $E(W)_{\lambda*}$-vector spaces over the locally symmetric variety

$$(2.3.2) \qquad Sh(GU) = \varprojlim_{K_f} Sh(GU)_{K_f} = \varprojlim_{K_f} GU(\mathbb{Q})\backslash GU(\mathbb{A})/K_{\infty,GU}\cdot K_f.$$

This is a complex projective pro-variety of dimension n which can be endowed with a canonical model over the field $\mathcal{K}$, as we recall briefly in §3, below; this will be discussed at greater length in Book 2.

Now let $\psi = \{[\Pi_1], [\Pi_2], \ldots, [\Pi_r], \xi\}$ be a parameter for GU, as in §2.1. We assume $r \leq 2$; if $r = 2$ we assume $\Pi_1 = \tau$ is a θ_n-stable representation of $GL(n)$ and $\Pi_2 = \chi$ a θ_1-stable Hecke character of $GL(1)$. We assume the special hypotheses (1.2.2); in particular each Π_i satisfies (1.2.2.2). Define $\Pi(\psi)$ as in (2.1.4); by construction all the π_∞ are cohomological for the representation W_+. Moreover, all the automorphic representations π of GU in $\Pi(\psi)$ share a common finite part π_f. This representation has a model over a finite extension $E(\pi_f)$ of $E(W)$.[3]

PROPOSITION 2.3.3. Assume the coefficient representation $W(\Pi)$ is strongly regular, as in (1.2.3.ii).

Suppose $r = 1$, so that $\psi = \{[\Pi], \xi\}$. Then the space

$$H(Sh(GU), \tilde{W}_{\lambda*})[\pi_f] = Hom_{GU(\mathbf{A}^f)}(\pi_f, H^n(Sh(GU), \tilde{W}_{\lambda*}) \otimes_{E(W)} E(\pi))$$

is a free $E(W)_{\lambda*} \otimes_{E(W)} E(\pi)$-module of rank $n+1$. For each finite place λ of $E(\pi)$ dividing $\lambda*$, we thus have an $n+1$-dimensional representation

$$\rho_{\Pi,\xi,\lambda} : Gal(\overline{\mathbb{Q}}/\mathcal{K}) \to GL(n+1, E(\pi)_\lambda)$$

realized on the λ-component of $H(Sh(GU), \tilde{W}_{\lambda*})[\pi_f]$.

If $r = 2$, so that $\psi = \{[\tau], [\chi], \xi\}$, the space $H(Sh(GU), \tilde{W}_{\lambda*})[\pi_f]$ is a free $E(W)_{\lambda*} \otimes_{E(W)} E(\pi_f)$-module of rank n. For each finite place λ of $E(\pi_f)$ dividing $\lambda*$, we thus have an n-dimensional representation

$$\rho_{\tau,\xi,\lambda} : Gal(\overline{\mathbb{Q}}/\mathcal{K}) \to GL(n, E(\pi_f)_\lambda)$$

realized on the λ-component of $H(Sh(GU), \tilde{W}_{\lambda*})[\pi_f]$.

Proof. The representation $W(\Pi)$ defines a local system $\tilde{W}$ in $E(W)$ vector spaces over $Sh(GU)$, and for any complex embedding $\beta : E(W) \hookrightarrow \mathbb{C}$, we can consider the complex vector space

$$H(Sh(GU), \tilde{W}(\mathbb{C}))[\pi_f] = Hom_{GU(\mathbf{A}^f)}(\pi_f, H^n(Sh(GU), \tilde{W}) \otimes_{E(W)} \mathbb{C}).$$

We note that here, as in the similar formula of the Proposition, only the cohomology in degree n occurs since the local system $W(\Pi)$ is assumed strongly regular.

This is isomorphic by Matsushima's formula to

$$Hom_{GU(\mathbf{A}^f)}(\pi_f, H^n(\mathfrak{g}, K_\infty; \mathcal{A}(GU) \otimes W_+)) = \bigoplus_{\pi';\pi'_f \simeq \pi_f} H^n(\mathfrak{g}, K_\infty; \pi'_\infty \otimes W_+).$$

[3]In fact, because all the ramification is concentrated at places split in $\mathcal{K}/F$, it follows from the existence of the conductor for representations of $GL(n+1)$ that we can take $E(\pi_f)$ to be the field of definition of the isomorphism class of π_f. The ramified representations come, in fact, from $GL(n, F_v)$-factors, by hypothesis (1.2.2.2), and we apply Proposition 3.2 of [C2]. At inert primes the unitary group is unramified and the local unramified representation is defined over its field of rationality because the space of invariants of a hyperspecial subgroup has multiplicity one by an argument of Waldspurger, cf. [C2, p. 103].

Here $\mathcal{A}(GU)$ is the space of automorphic forms on GU, which we break up as a sum over π' with fixed finite part. Recall here that GU is anisotropic modulo center, so all automorphic representations of $GU(\mathbb{A})$ are essentially unitary. Since $W(\Pi)$ is assumed strongly regular, so is W_+; the representation-theoretic translation of the vanishing of cohomology in degrees different from n is that the only automorphic representations that contribute non-trivially are those π' for which π'_∞ is in the discrete series L-packet of GU associated to W_+. The result now follows from Corollaries 1.2.5 and 1.2.7 and the comparison theorem between Betti and étale cohomology.

□

3. Review of Shimura varieties attached to anisotropic unitary groups

We follow Kottwitz' conventions [K2, §5] in attaching a Shimura variety $Sh(V_1)$ to the PEL type defined by the group $GU(V_1)$ and an appropriate Shimura datum. For the sake of completeness, we briefly review the definition of $Sh(V_1)$ as a moduli space of abelian varieties with PEL structures. A more leisurely account of these Shimura varieties is contained in Book 2.

The definition of V_1 includes the choice of a unique real prime v_0 of F such that $U(V_{1,v_0})$ is not compact. We choose an extension of v_0 to a complex prime w_0 of $\mathcal{K}$ or, what is the same thing, a complex embedding of the quadratic imaginary field E. With our choices, the reflex field $E(GU(V_1), X(V_1))$ of $Sh(V_1)$ can be identified with $\mathcal{K}$, or rather, since it is by definition a subfield of $\mathbb{C}$, with the image of $\mathcal{K}$ under the complex embedding w_0. We assume the CM type Σ is chosen to contain w_0 and so that

(1) $V_1 \otimes_{\mathcal{K},w_0} \mathbb{C}$ is of signature $(1, n)$, whereas
(2) $V_1 \otimes_{\mathcal{K},w} \mathbb{C}$ is of signature $(0, 1+n)$ for $w \in \Sigma$, $w \neq w_0$.

We write GU instead of $GU(V_1)$. Fix an open compact subgroup $K = \prod_p K_p \subset GU(\mathbf{A}^f)$, and let $Sh_K = Sh(V_1)_K$ denote the corresponding Shimura variety at finite level. By this we mean the collection of complex points

$$GU(\mathbb{Q})\backslash X(V_1) \times GU(\mathbf{A}^f)/K.$$

Here K_∞ is the product of a maximal compact subgroup of $GU(\mathbb{R})$ with its center and $X(V_1)$ can be identified with the hermitian symmetric space $GU(\mathbb{R})/K_\infty$, and Sh_K is endowed with its canonical model. The constructions in [K2] allow us to identify Sh_K with (a part of) the set of complex points of a moduli space of abelian varieties of PEL type. This is determined as follows. Let C_1 be the matrix algebra $End_{\mathcal{K}}(V_1)$, viewed as a simple $\mathbb{Q}$-algebra, and let $*$ be the anti-involution on C_1 defined by the hermitian form on V_1. Thus GU is the algebraic group over $\mathbb{Q}$ whose points in any $\mathbb{Q}$-algebra R are given by

$$\{x \in C_1 \otimes_{\mathbb{Q}} R \mid xx^* \in R^\times\}.$$

Let $h_1 : \mathbb{C} \to GU(\mathbb{R})$ be a homomorphism such that $h_1(\bar{z}) = *(h_1(z))$ and such that the symmetric real-valued bilinear form $(v, h_1(i)w)$ on $V_1(\mathbb{R})$ is positive definite. Writing

$$V_1(\mathbb{R}) \xrightarrow{\sim} \bigotimes_{w\in\Sigma} V_{1,\sigma} = \otimes_{w\in\Sigma} V_1 \otimes_{\mathcal{K},w} \mathbb{C}, \tag{3.1}$$

it follows from our choice of Σ above that we can take

$$h_1(z) = (diag(z, \bar{z}I_n), \bar{z}I_{n+1}, \ldots, \bar{z}I_{n+1}). \tag{3.2}$$

We follow [K2] in defining the Shimura variety $Sh(V_1)$ to be that attached to the Shimura datum $(GU, X(V_1))$ where $X(V_1)$ is the conjugacy class of h_1^{-1}. In particular, $X(V_1)$ is not the conjugacy class of h_1, which was used mistakenly in [K1]. However, it is important to point out that the formula (19.6) of the number of fixed points in [K2] does yield the formula (mistakenly applied to $Sh(G, h_1)$) that is used as the basis of the stabilization in [K1], which is the basis of the calculations in the following section. Thus (3.2) is still the basis for the calculations in our §4.3.

As hermitian symmetric space $X(V_1)$ can be identified with the unit ball in $\mathbb{C}^{n-1}$. The data C_1 and h_1 are used in [K2] to define a moduli space of abelian varieties of PEL type, denoted simply S_K, which is smooth and projective over $Spec(\mathcal{O}_{\mathcal{K},p})$ provided p is unramified in $\mathcal{K}$ and K_p is a hyperspecial maximal compact subgroup. Then $S_K(\mathbb{C})$ can be identified with a disjoint union of $|ker^1(\mathbb{Q}, GU(V_1))|$ copies of $Sh_K(\mathbb{C})$.

We also introduce the similitude group $GU(V_2)$, defined as in (1.1) with V_1 replaced by the hermitian space V_2 of [CHL.IV.B, Lemma 1.4.3]. We can define a Shimura datum $(GU(V_2), X_2)$ and obtain a Shimura variety $Sh(V_2)$ whose reflex field is again identified with the image of $\mathcal{K}$ under w_0. Now $\dim V_2 = n$, the signatures are determined by

(1) $V_2 \otimes_{\mathcal{K},w_0} \mathbb{C}$ is of signature $(2, n-2)$, whereas
(2) $V_2 \otimes_{\mathcal{K},w} \mathbb{C}$ is of signature $(0, n)$ for $w \in \Sigma$, $w \neq w_0$.

and the homomorphism h_1 is replaced by a homomorphism h_2 which, in appropriate coordinates and with respect to the decomposition analogous to (3.1), is given by

$$h_2(z) = (diag(zI_2, \bar{z}I_{n-2}), \bar{z}I_n, \ldots, \bar{z}I_n). \tag{3.3}$$

In the next section we follow Kottwitz in determining the unramified factors of the zeta function of $Sh(V_1)$ associated to certain parameters for $GU(V_1)$, with the goal of determining the corresponding ℓ-adic Galois representations at almost all unramified places. The corresponding analysis for $Sh(V_2)$ will be carried out in the second book of this series, where it will be used to study p-adic Hodge-theoretic properties of the automorphic Galois representations constructed by p-adic analytic continuation of the representations constructed here.

4. Kottwitz's formula for Shimura varieties attached to anisotropic unitary groups

4.1. Preliminaries. Let $W_\infty = W(\Pi)$ and W_+ be as in §2.3. We use the same notation to denote an extension to $GU(\mathbb{R})$. This representation has a rational model over some number field $E(W)$. Let λ denote an ℓ-adic place of $E(W)$ and let W_λ denote the corresponding local system over $Sh(V_1)$, as in [K1] (where it is called $\mathcal{F}_\lambda$).

Let p be a rational prime such that K_p is a hyperspecial maximal compact subgroup of $GU(\mathbb{Q}_p)$. Choose a prime v of $\mathcal{K}$ – viewed as the reflex field of $Sh(V_1)$ – dividing p. In §10 of [K1] Kottwitz obtains what he calls a "conjectural description" of the cohomology

$$H^\bullet(Sh_K, W_\lambda) = \bigoplus_i (-1)^i H^i(Sh_K, W_\lambda)$$

where the cohomology in question is étale (ℓ-adic) cohomology and the description is as a representation of the group $Gal(\bar{\mathcal{K}}_v/\mathcal{K}_v) \times \mathcal{H}^p_K$, where $\mathcal{H}^p_K$ is the Hecke algebra of K-biinvariant functions on $GU(\mathbb{A}^p_f)$ (adeles trivial at p). Several remarks are in order.

4.2. Remarks.

(4.2.1) Kottwitz' conjectural description applies to the intersection cohomology $IH^\bullet$ of a compactification of Sh_K rather than the cohomology, but since Sh_K is (smooth[4] and) proper there is no difference. One of his conjectures concerns the validity of an extension of his calculations to the boundary of the Baily-Borel compactification of the Shimura variety under consideration and the relation to the non-elliptic terms in the trace formula for GU. This conjecture, which has been verified by S. Morel in a number of cases, is vacuous in the present situation. In particular, where Kottwitz refers to "the contribution of $P = G$" to the cohomology, the reader should bear in mind that in our situation there are no proper parabolic subgroups and $P = G$ represents the only contribution.

(4.2.2) The article [K1] is written for general Shimura varieties. In §3 of [K1] Kottwitz proposes a conjectural expression for the trace of powers of Frobenius, twisted by Hecke operators, based on an axiomatic description of the points on the Shimura variety over finite fields. This expression is proved for Shimura varieties of PEL types A and C in [K2]. In the case at hand, this should be sketched in Book 2.

(4.2.3) In sections 4-7 of [K1] Kottwitz stabilizes this expression, assuming the validity of the fundamental lemma. Actually, Kottwitz assumes two fundamental lemmas: the fundamental lemma for unstable

[4] This is true for K sufficiently small, which we can assume by taking K_p torsion free for some p split in in E; this is compatible with the standing hypotheses of [CHL.IV.B, 2.2].

cyclic base change, invoked in the discussion preceding [K1,(7.2)] and proved by Kottwitz in Appendix A to [M], and the fundamental lemma for endoscopy, invoked in the discussion preceding [K1, (7.1)]. These are now proved ([C1,Lab] for the first in the stable case, and [W, M] in the unstable case; [LN, N] for the second). This portion of Kottwitz' discussion should be described in more detail, taking these more recent developments into account, in Book 2.

(4.2.4) For our purposes, the final conjectural ingredient in Kottwitz' formula is Arthur's formula for the multiplicities of representations of $GU(\mathbb{A})$ in the discrete spectrum of automorphic forms on GU. This is still conjectural at the time of writing, though it may not be by the time this book appears in print. However, for the test functions we have in mind, a substitute for Arthur's multiplicity conjectures is provided by the explicit computation of the multiplicities for the parameters introduced in §2.1, specifically (2.1.1) and (2.1.2).

(4.2.5) In [K1, (9.6)], Kottwitz invokes a conjectural expression in terms of local (spectral) transfer factors for the (conjectured) pairing between elements of a global L-packet with elements in the corresponding component group $\mathfrak{S}_\psi$ (defined below). In our setting the global L-packets are defined in terms of local L-packets at the real prime, and our pairing is given by Shelstad's spectral transfer factors, so this conjecture is superfluous for us.

4.2.6 Notation for global parameters. As indicated in (4.2.4) above, we work with the parameters $\psi = \{[\Pi_1], \ldots, [\Pi_r], \xi\}$ of §2.1. In the applications, we only consider $r = 1$ (the *stable case*), in which case $\mathfrak{S}_\psi$ is trivial, or $r = 2$ (the *endoscopic case*), with $\mathfrak{S}_\psi = \mathbb{Z}/2\mathbb{Z}$. In order to be consistent with the previous chapters, we write $[\Pi]$ instead of $[\Pi_1]$ in the stable case and $([\tau], [\chi])$ instead of $\{[\Pi_1], [\Pi_2]\}$ in the endoscopic case, with τ cuspidal and θ_n-stable on $GL(n)$ and χ a θ_1-stable Hecke character of $GL(1)$. Our special hypotheses are in force here and prescribe that $[\Pi]$, $[\tau]$, and $[\chi]$ all be spherical at places inert in $\mathcal{K}/F$.

4.3. Translation of Kottwitz' formulas. We return to the notation of the earlier sections; thus $G = U(V_1)$ with $\dim V_1 = n + 1$. The results of the two previous chapters, together with the discussion of §1, provide unconditional spectral expressions for the traces of our test functions. The remainder of this section reviews the calculations in §10 of [K1] using the spectral parameters introduced in (4.2) in place of Arthur's conjectural multiplicity formulas. The reader should not be surprised to learn that, with this interpretation of Arthur's ψ, our calculations are consistent with his multiplicity conjectures.

Let f be a test function on GU. Then [K1, (8.3)] writes Arthur's conjectural spectral expression for the trace of f in the discrete spectrum as follows

$$(4.3.1) \qquad T_{disc}(f) = \sum_{[\psi]} |\mathfrak{S}_\psi|^{-1} \sum_{\pi \in [\psi]} \sum_{x \in \mathfrak{S}_\psi} \epsilon_\psi(x s_\psi) < x s_\psi, \pi > tr\ \pi(f)$$

We write $[\psi]$ instead of $\Pi(\psi)$ for the packet defined by ψ, the letter Π being otherwise employed. Instead of ψ our parameters are pairs $([\Pi], \xi)$ or $([\tau], [\chi], \xi))$ as described in (4.2.6). For uniformity we write Π in both cases when this is convenient; in the endoscopic case we have $\Pi = \Pi(\tau, \chi)$. We fix a sufficiently large set S of rational primes, outside which all our data are unramified, and we choose a function f_S that has trace zero on any representation ρ_S such that $\rho_\infty \otimes \rho_S \otimes \rho^S$ is automorphic for $GU(\mathbb{A})$ for any ρ^S and $\rho_S \neq \Pi_S$; this is possible by the arguments of [CHL.IV.B] and the multiplicity one property. We also assume *trace* $\Pi_S(f_S) = 1$. We then consider functions on $GU(\mathbf{A}^f)$ of the form $f_f = f_S \otimes f^S$ where f^S is a tensor product of functions unramified for the hyperspecial compact subgroup (for all $v \notin S$). For all but at most finitely many p, it is possible to choose f_f in such a way with the local component f_p equal to the characteristic function of the chosen hyperspecial maximal compact at p. This last condition is required for the validity of Kottwitz' formulas, and so we assume p is not in the (finite) set of primes for which it may be impossible to realize this condition.

Moreover, we choose f_∞ to eliminate cohomology outside the middle dimension, or more precisely we only consider automorphic representations with cohomology in regular coefficient systems; this is regularity in the sense of Special Hypotheses 2.3 (i). In particular, our ψ's are all parameters that should be tempered, and in particular are trivial on the $SL(2)$-factor, the terms s_ψ and ϵ_ψ of Arthur's formula are irrelevant, and (4.3.1) becomes

$$(4.3.2) \qquad T_{disc}(f) = \sum_{\psi = ([\Pi], \xi)} |\mathfrak{S}_{[\Pi],\xi}|^{-1} \sum_{\pi \in [\psi]} \sum_{x \in \mathfrak{S}_{[\Pi],\xi}} < x, \pi > tr\ \pi(f)$$

As explained in (4.2.3), we set $\mathfrak{S}_{[\Pi],\xi} = 1$ in the stable case and $\mathfrak{S}_{[\tau],[\chi],\xi} = \{\pm 1\}$ in the endoscopic case. In our setting, if ψ is a parameter factoring through $\hat{H}$ for an endoscopic group H then S_ψ is the group of θ-invariants in $Z(\hat{H})$ modulo the center of $\widehat{GU}$, as already seen. Then the pairing $< x, \pi >$ for $BC(\pi) = (\tau \times \chi, \xi)$ should in fact be $< x, \pi_\infty >$, again by our choice of f_f. We can define $< x, \pi_\infty >$ in terms of the transfer factors.

There is a similar expression to (4.3.2) for the discrete trace of test functions on an endoscopic group H, cf. [K1]. Bear in mind that $\Pi = \Pi(\tau, \chi)$ occurs both in the term for GU and in the term for H where H is isogenous to $U(n)^* \times U(1)$. In the expression for the stable trace of H, $([\tau], [\chi], \xi)$ is a stable parameter ψ_H, say, which we denote $([\tau], [\chi], \xi)_H$ so

$$(4.3.3) \qquad \mathfrak{S}_{\psi_H} = \mathfrak{S}_{([\tau],[\chi],\xi)_H} = 1.$$

Finally, Kottwitz' (conjectural) expression for the stable trace is a simplified version of (4.3.1) [K1, (8.5)]. Applying the additional simplifications of (4.3.2) and (4.3.3) we find simply

$$ST_{disc}(f) = \sum_{([\Pi],\xi)} \sum_{[\pi]\in([\Pi],\xi)} tr\ \pi(f). \tag{4.3.4}$$

Kottwitz' expression for the Lefschetz trace [K1, (10.1)] is the following. Let v be a place of $\mathcal{K}$ dividing a rational prime p where data are unramified. Then if $K_f = \prod K_p$ is hyperspecial at p, $Sh(GU)_{K_f} := S_{K_f}$ has good reduction at v [K2], and

$$trace(Frob_v^\alpha \times f_S \mid H^\bullet_{et}(S_{K_f}, \tilde{W}_\lambda))$$

is given by

$$\sum_{\mathcal{E}} \iota(GU,H) ST^H_{disc}(h) \tag{4.3.5}$$

where $h = h^p h^\alpha_p h_\infty$ is a certain test function on H with $h^p = (f^H_f)^p$ in our previous notation (where $f^p_f = f_S \times f_0^{S,p}$, f_S as above and f_0^S the unit of the Hecke algebra), h_∞ determined by Shelstad's character formulas, and h^α_p is the key to Kottwitz' approach to the zeta function. We have added a superscript ^{H} in (4.3.5) for the sake of clarity.

We note that in [K1] the stable trace $ST^H_{disc}(h)$ is instead a partial sum indexed only by the (G,H)-regular elements in $H(\mathbb{Q})$ (cf. [H.I.A], discussion following remark 5.6, for the pertinent definitions). However, given the particular nature of the functions h_∞ (cf. [C.III.A], [K1]), the other terms of the expression for ST^H_{disc} vanish. This has been checked by Morel ([M] Theorem 6.2.1, Prop. 3.3.4); see [M, Remark 6.2.2].

Kottwitz then proceeds to rewrite this sum in terms of Arthur's ψ-packets, using Arthur's conjectures, finally deriving (4.3.1). In our case this is easy and has in essence been done in [CHL.IV.B], so we will simply bypass formula (4.3.1) and exhibit its analogue directly. What is important is that the expression (4.3.5) for the trace of Frobenius is *unconditional.*

First, the *geometric* expression for the trace [K1, 3.1] was proved by Kottwitz in [K2]. The fact that it implies (4.3.5) is proved in [K1,§7, Theorem 7.2]. One only has to check the assumptions of local harmonic analysis used in this section, which have now been verified (cf. (4.2.3)). What we now have to do is to check (4.3.4) for the H relevant to us, without relying on the arguments in [K1].

Thus, we first verify Kottwitz' formula (4.3.4) for the stable contribution.

LEMMA 4.3.6. *Let Π be a cuspidal automorphic representation of $GL(n+1)_{\mathcal{K}}$ satisfying Special Hypotheses 1.2.2 and the strong regularity hypothesis 1.2.3(ii). Then the conjectural formula (4.3.4) is valid provided f_∞ is a very cuspidal Euler-Poincaré function or pseudocoefficient for the* discrete series L-packet $([\Pi],\xi)_\infty$ of GU attached to $([\Pi_\infty],\xi_\infty)$ and f_f is biinvariant under the level subgroup K.

Proof. This is just a reformulation of Labesse's results in §5 of [L.IV.A] and their extension in §1.2 above. Indeed, under our hypotheses, the set of $[\pi] \in ([\Pi],\xi)$ with non-zero trace on f is of the form $\pi_\infty \otimes \pi_f$ where π_∞ belongs to the discrete series L-packet $([\Pi],\xi)_\infty$; the strong regularity hypothesis guarantees that there are no non-tempered cohomological representations with non-zero trace under f_∞. Then (4.3.4) is a restatement of Labesse's multiplicity one theorem, or more precisely our extension 1.2.4 of this multiplicity one theorem, in the present setting. □

The final step in Kottwitz' remarkable paper consists in rewriting (4.3.5) for the Frobenius trace as an expression (which, obviously, will not be stable) in terms of automorphic representations of $GU(\mathbb{A})$. This is the formula (10.3) in [K1]. It is deduced there from Arthur's conjectural expression (8.1)[5] for multiplicities of representations π of $GU(\mathbb{A})$ in $L^2(A_{GU}GU(\mathbb{Q})\backslash GU(\mathbb{A}))$, as well as the similar expression (8.5) for endoscopic groups. We can use Kottwitz' arguments as soon as (8.1) is verified in our situation, and in our terms which do not depend on conjectures.

Consider first the stable odd-dimensional case, so ψ is a pair $([\Pi],\xi)$ as in (4.2.6). In this case Kottwitz' (8.1) simply says the contribution of π_f, in the notation of Corollary 1.2.5, is the sum of the $\pi'_\infty \otimes \pi_f$ with multiplicity 1. This is true by [L.IV.A], cf. Proposition 1.2.4 and Corollary 1.2.5.

We choose $f = f_S \otimes f^S$ on $GU(\mathbb{A})$ as explained after (4.3.1). By the arguments in [L.IV.A] and [CHL.IV.B], the proper endoscopic contributions to (4.3.5) are zero.

Since p is not a prime of ramification, we have an unramified parameter $\phi_p : W_{\mathbb{Q}_p} \to {}^LGU$ for the representation π_p.[6] With the preceding remark, and letting $v \mid p$ as above, Kottwitz' expression of the Frobenius trace is simply

$$(4.3.7)\qquad trace(Frob_v^\alpha \times f_S \mid H^\bullet_{et}(S_{K_f},\tilde{W}_\lambda)) = C(Sh)tr(r_\mu(\phi_p)(Frob_v^\alpha))).$$

Here

$$C(Sh) = (-1)^{\dim Sh} Nv^{\frac{\dim Sh}{2}},$$

and the representation r_μ of ${}^LGU$ is associated to the datum μ obtained from Shimura's $h_1 : \mathbb{C}^\times \to GU(\mathbb{R})$ defining the Shimura variety (cf. §3). We recall this construction in our case. The group $GU(\mathbb{R})$ is isomorphic to the subgroup of $GU(1,n) \times GU(n+1)^{d-1}$ of d-tuples with equal similitude

---

[5]Otherwise unexplained formula numbers and notation refer to [K1].

[6]If we restrict our attention to p split in the imaginary quadratic field E, we can separate the contributions of the various places v of $\mathcal{K}$ dividing p, as in [HT]. In particular, we can treat v for which Π_v is unramified even if $\Pi_{v'}$ is ramified for some other v' dividing p.

factor, where $d = [F : \mathbb{Q}]$. We have taken $GU(1,n)$ consistently with the computations in [CHL.IV.B,§3], relying on [C.III.A]. As in §3, we take

$$h_1(z) = (diag(z, \bar{z}I_n), \bar{z}I_{n+1}, \dots, \bar{z}I_{n+1}; z\bar{z}),$$

where we have added an extra $\mathbb{C}^\times$ term at the end for consistency with the computation in the L-group. (Of course, the last term is determined, for $h_1 : \mathbb{C}^\times \to G(\mathbb{R})$, by the similitude factor.) Now, as was implicit in the dual calculation of §1.3,

$$GU(\mathbb{C}) \simeq GL(n+1, \mathbb{C})^d \times \mathbb{C}^\times$$

(with the similitude factor included) and the parameter μ deduced from h_1 is the cocharacter $\mathbb{G}_m \to GU$, defined over $\mathbb{C}$ or $\overline{\mathbb{Q}}$, given by

$$x \mapsto (diag(x, I_n), I_{n+1}, \dots, I_{n+1}; x).$$

If we recall that the reflex field of Sh is $\mathcal{K}$, we see that this defines a representation of ${}^LGU$ over $\mathcal{K}$, simply given, in the description of §1.3, by

$$(g_\alpha, x) \mapsto g_{\alpha_0} \otimes x \in GL(n+1, \mathbb{C}).$$

(Recall that $\mathcal{K}$ is embedded in $\mathbb{C}$ as a reflex field. We take $\overline{\mathbb{Q}}$ equal to the algebraic closure of $\mathbb{Q}$ in $\mathbb{C}$, and then g_{α_0} is defined in terms of this identification.)

We are now able to prove Theorem 2.2.3 in the odd-dimensional case.

PROPOSITION 4.3.8. There is a compatible system $r_\lambda(\xi)$ of λ-adic characters of $\mathcal{K}$ such that the representations

$$\rho_{\Pi,\lambda} = (\rho_{\Pi,\xi,\lambda})^\vee \otimes r_\lambda(\xi),$$

with $\rho_{\Pi,\xi,\lambda}$ as in Proposition 2.3.3, satisfy the conclusions of Theorem 2.2.3.

Proof. Lemma 4.3.6 is the verification that formula (4.3.7) is correct in the stable case. In view of formula (4.3.7), the proof is identical to that of [C3], with the additional simplification that the multiplicity $a = 1$ in [C3, (5.12)]. The character $r_\lambda(\xi)$ is computed explicitly in [HT] in the proofs of Theorems VII.1.3 and VII.1.9, where it is shown that $r_\lambda(\xi) = rec(\xi^c)$ (the character ψ of [HT] is here denoted ξ) with rec the notation of [HT] for the global reciprocity map associating ℓ-adic characters to Hecke characters. □

4.4. Proof of Theorem 2.2.3, assuming Special Hypotheses 1.2.2. We now describe Kottwitz' formula applied to a parameter $\psi = ([\tau], [\chi], \xi)$ in the endoscopic case (cf. (4.2.6)). Here τ is our given θ_n-stable representation of $GL(n, \mathbb{A}_\mathcal{K})$, and n is even. The choice of f cuts out the contribution in the stable discrete trace formula. This has been described in [CHL.IV.B,§4]. Arthur's multiplicity formula, stated by Kottwitz as [K1,(8.1)], has been proved as Theorem 4.6 of [CHL.IV.B], and extended to the similitude group in Proposition 1.2.6, for our standing choice of f_S. The formula is simply

$$
\begin{aligned}
m(\pi_\infty \otimes \pi_f) &= \frac{1}{|\mathfrak{S}_\psi|}[< 1, \pi_\infty > + < x, \pi_\infty >] \\
&= 0 \text{ or } 1.
\end{aligned}
\tag{4.4.1}
$$

The integer $m(\pi_\infty \otimes \pi_f)$ is the multiplicity in $L^2(A_{GU}GU(\mathbb{Q})\backslash GU(\mathbb{A}))$. As described above, $\mathfrak{S}_\psi \simeq \{\pm 1\}$, and it can be seen as the group generated by

$$s'_{n,1} = ((diag(I_n, -1)), 1),$$

cf. §1.3. In (4.4.1), π_∞ runs over the L-packet $\Pi^+_\infty \cup \Pi^-_\infty$ of Proposition 1.2.6. Formula (4.4.1) means that

$$
\begin{aligned}
&< 1, \pi_\infty > = 1 \\
&< x, \pi_\infty > = \pm 1 \ (\pi_\infty \in \Pi^\pm_\infty).
\end{aligned}
$$

For the compatibility with Kottwitz' computations, we note that this coincides with his formula (9.6):

$$< x, \pi >= \Delta_f(\psi_H, \pi_f) \cdot \Delta_\infty(\psi_H, \pi_\infty). \tag{4.4.2}$$

Here the finite factor $\Delta_f(\psi_H, \pi_f)$ is equal to 1, because all data are unramified or $GL(n)$, whereas $\Delta_\infty(\psi_H, \pi_\infty)$ is computed in [C.III.A] and [CHL.IV.B,§3.3].

Using only Kottwitz' geometric formula (4.3.5), Arthur's formula (4.4.1), combined with (4.4.2), and the fact that the only contributions of (4.3.5) for our function f are given by GU and H, we obtain

(4.4.3)

$$
\begin{aligned}
trace(Frob^\alpha_v \mid H^\bullet) = \frac{1}{2}C(Sh)[&trace\ r_\mu(\phi_p(Frob^\alpha_v)) \\
&+ < x, \pi >< x, \lambda_{\pi_\infty} > trace\ r_\mu(\phi_p(x \cdot Frob^\alpha_v))],
\end{aligned}
$$

cf. [K1,(10.3)]. The representation r_μ has already been defined. The coefficient

$$c_x =< x, \pi >< x, \lambda_{\pi_\infty} > \tag{4.4.4}$$

is shown by Kottwitz to be independent of the representation $\pi = \pi_\infty \otimes \pi_f$ where π_f is fixed. We now have to compute c_x.

LEMMA 4.4.5. The coefficient $c_x = 1$.

4.4.6. Proof of Lemma 4.4.5. Note that in (4.4.4), $< x, \pi >$ depends only on (real) representation theory, whereas $< x, \lambda_{\pi_\infty} >$ depends on the choice of data for the Shimura variety. We have $< x, \pi >=< x, \pi_\infty >$, and its value has been described above.

The pairing $< x, \lambda_{\pi_\infty} >$ is defined in §7 of [K1]. We have to return to the calculations of [C.III.A] and the description of the L-packet $\{\pi_\infty\}$. We work at the place $v_0 \mid \infty$ where G is of signature $(1, n)$. In [C.III.A,§3] and [CHL.IV.B], we have taken $G_{v_0} = U(1, n)$. For the purposes of this paragraph, we let GU denote the full similitude group, so that

$$GU(\mathbb{R}) = GU_{v_0} \times \prod_{v \neq v_0} GU(n+1); \; GU_{v_0} = GU(1, n).$$

The additional split torus factors do not affect the calculation. We have chosen in [C.III.A] an elliptic torus $T \subset G_{v_0}$, the diagonal torus, which defines $\tilde{T} \subset GU$. The representations $\pi_\infty \in \Pi_\infty$ were given by

$$\pi = \pi(\phi, \sigma^{-1} B_0),$$

where B_0 is the standard Borel subgroup in $G(\mathbb{C}) = GL(n, \mathbb{C})$ and $\sigma \in \Omega/\Omega_{\mathbb{R}} = \mathfrak{S}_{n+1}/\mathfrak{S}_1 \times \mathfrak{S}_n$. In [CHL.IV.B,§3], we had used the natural bijection

$$\Omega/\Omega_{\mathbb{R}} \leftrightarrow \{1, \ldots, n+1\}; \; \sigma \mapsto \sigma \cdot 1.$$

On the other hand, we have (relative to the place v_0) $\hat{G} = GL(n+1, \mathbb{C})$; we denote by $(\hat{B}, \hat{S})$ the standard Borel subgroup and maximal torus. Given $B = \sigma^{-1} B_0$, there is a unique isomorphism $j_B : \hat{T} \xrightarrow{\sim} \hat{S}$ sending the positive coroots for (B, T) to the positive roots for $(\hat{B}, \hat{S})$. Identifying both $\hat{T}$ and $\hat{S}$ with $(\mathbb{C}^\times)^{n+1}$, we see that for $B = \sigma^{-1} B_0$,

$$j_B : t \mapsto \sigma(t) \; (t \in (\mathbb{C}^\times)^{n+1}).$$

Now Shimura's parameter for GU – hence for GU_{v_0} – is given by

$$z \mapsto diag(z, \bar{z} I_n) \in \tilde{T}, z \in \mathbb{C}^\times.$$

This defines in the usual manner a 1-parameter group

$$\begin{aligned} \mu : \mathbb{G}_m &\to \tilde{T}/\mathbb{C}, \; \tilde{T}(\mathbb{C}) = (\mathbb{C}^\times)^{n+1} \times \mathbb{C}^\times \\ x &\mapsto (x, I_n, x). \end{aligned}$$

Kottwitz defines $< x, \lambda_{\pi_\infty} >$ as equal to $< x, j_B \circ \mu >$ for the natural identification between $X_*(\tilde{T}/\mathbb{C})$ and $X^*(\hat{\tilde{T}})$. Here $\mu = (1, 0, \ldots, 0, 1) \in X_*(\tilde{T}) \simeq \mathbb{Z}^{n+2}$ and $j_B \circ \mu = \sigma\mu = (\varepsilon_i, 1) \in \mathbb{Z}^{n+2}$, where

$$\begin{aligned} \varepsilon_i &= 1, \; i = \sigma \cdot 1 \\ &= 0, \; i \neq \sigma \cdot 1. \end{aligned}$$

Finally, since $x = (1, \ldots, 1, -1, 1)$, we obtain

$$\begin{aligned} < x, \lambda_{\pi_\infty} > &= 1, \; i \leq n \\ &= -1, \; i = n+1. \end{aligned}$$

Comparing this to our previous expression for $< x, \pi > = < x, \pi_\infty >$, this completes the proof of Lemma 4.4.5. We also see that the similitude factor plays no role in the calculation, justifying our decision to work with the full similitude group.

We finally obtain, always under Special Hypotheses 1.2.2,

LEMMA 4.4.7. There is a compatible system $r_\lambda(\xi,\mu_1)$ of λ-adic characters of $\mathcal{K}$ such that, for all p outside a finite set, the representations

$$\rho_{\mathcal{T},\lambda} = (\rho_{\mathcal{T},\xi,\lambda} \otimes r_\lambda(\xi,\mu_1)^{-1})^\vee,$$

with $\rho_{\mathcal{T},\xi,\lambda}$ as in Proposition 2.3.3, satisfy the conclusions of Theorem 2.2.3.

Proof. Let $r_\mu^\pm$ be the representation of the centralizer ${}^LH$ of $x = x_{n,1}$ in ${}^LGU$ on the $\pm$-eigenspace of x. It follows immediately from Lemma 4.4.5 and the expression (4.4.3) for $trace(Frob_v^\alpha \mid H^\bullet)$ that there is a sign $* \in \pm$ such that, for all p outside a finite set, the expression (4.4.3) is equal to $tr(r_\mu^*(\psi_{\pi_p}(Frob_v^\alpha)))$.

Recall that π_p is an extension to $GU(\mathbb{Q}_p)$ of an irreducible representation of $G(\mathbb{Q}_p) = \prod_{v|p} G(F_v)$ whose v component is the representation $\pi_v(\tau_v,\chi_v)$ defined in [CHL.IV.B]. We can identify an unramified local Langlands parameter ψ_p for $R_{F/\mathbb{Q}}G$ at p with the collection of unramified local Langlands parameters $(\psi_v, v \mid p)$ of G. Then the local Langlands parameter ψ_{π_p} defines by the map $\widehat{GU} \to \hat{G}$ the collection $(\psi_v = \psi_{\xi_{\mu_n,\mu_1}} \circ \psi_{\tau_v,\chi_v})$ of parameters of $\pi_v(\tau_v,\chi_v)$ (cf. [H.I.A, (5.5.4)]; this is completed by [H.I.A, (5.5.5)] if p does not split in E). The element $r_\mu(x_{n,1})$ commutes with these parameters, by construction, and its eigenvalue decomposition is the decomposition of the representation space of r_μ as the sum

$$r_\mu^+ \circ \psi_{\tau_v} \otimes \mu_1 \oplus r_\mu^- \circ \psi_{\chi_v} \otimes \mu_n.$$

Now if $* = +$, we have

$$tr(\rho_{\mathcal{T},\xi,\lambda}(Frob_v^\alpha)) = tr(r_\mu^+ \circ \psi_{\tau_v} \otimes \mu_1)(Frob_v^\alpha),$$

and the remainder of the argument follows as in the proof of Proposition 4.4.1. If $* = -$, on the other hand, we have

$$tr(\rho_{\mathcal{T},\xi,\lambda}(Frob_v^\alpha)) = tr(r_\mu^- \circ \psi_{\chi_v} \otimes \mu_n)(Frob_v^\alpha),$$

which is the Galois representation attached to a twist, depending on ξ (cf. Proposition 4.3.8), of the Hecke character $(\chi \otimes \mu_n)^{-1}$. But $\rho_{\mathcal{T},\xi,\lambda}$ is an n-dimensional representation, hence its trace on a dense subset of $Gal(\overline{\mathbb{Q}}/\mathcal{K})$ cannot equal the value of an abelian twist of a Hecke character. This completes the proof. □

4.5. Hodge-Tate numbers of automorphic Galois representations. In this section we complete the proof of Theorem 2.2.3 by verifying condition (ii) of Conjecture 2.2.2, always assuming strong regularity of Π_∞ as well as Special Hypotheses 1.2.2. We first need to state condition (ii) explicitly.

Fix a prime λ of the coefficient field $E(\pi)$, say of residue characteristic ℓ. Up to twisting by the ℓ-adic realization of an algebraic Hecke character – namely $r_\lambda(\xi)^{-1}$ if $r=1$, $r_\lambda(\xi,\mu_1)^{-1}$ if $r=2$ – the automorphic Galois representation $\rho_{\Pi,\lambda}$ constructed above is realized in the cohomology of the geometric λ-adic local system $\tilde{W}(\Pi)_\lambda$ on a Shimura variety, obtained in a standard way from the finite-dimensional representation $W(\Pi)$ introduced above. It is therefore of geometric type, in the sense of Fontaine and Mazur: each $\rho_{\Pi,\lambda}$ is unramified outside a finite set of places of $\mathcal{K}$, and at every place dividing the residue characteristic ℓ of λ, $\rho_{\Pi,\lambda}$ is de Rham. The latter fact is a consequence of the comparison theorems of p-adic Hodge theory. In particular the Hodge-Tate numbers can be read off from the Hodge numbers of the de Rham cohomology of the flat vector bundle $\tilde{W}(\Pi)$ associated to $W(\Pi)$. The comparison of $\tilde{W}(\Pi)_\lambda$ and $\tilde{W}(\Pi)$, and therefore the determination of the Hodge-Tate numbers from the highest weights $\mu(\sigma)$ of W_σ, presupposes a dictionary relating complex and ℓ-adic places of $\mathcal{K}$. In [HT] this is given by an isomorphism $\alpha : \bar{\mathbb{Q}}_\ell \xrightarrow{\sim} \mathbb{C}$ (in [HT] the isomorphism is denoted ι, but that letter is already used for a complex embedding of E). For what follows it suffices to identify the algebraic closure of $\mathbb{Q}$ in $\bar{\mathbb{Q}}_\ell$ with the field of algebraic numbers in $\mathbb{C}$. Then the ℓ-adic embeddings of $\overline{\mathbb{Q}}$, and in particular of $\mathcal{K}$, are identified with the complex embeddings; if s is an embedding of $\mathcal{K}$ in $\bar{\mathbb{Q}}_\ell$, we write α_s for the corresponding complex embedding.

Let s be an embedding of $\mathcal{K}$ in $\bar{\mathbb{Q}}_\ell$, $\mathcal{K}_s$ the (topological) closure of the image of s in $\bar{\mathbb{Q}}_\ell$, and let $D_{dR,s}$ denote Fontaine's functor from representations of $\Gamma_s = Gal(\bar{\mathbb{Q}}_\ell/\mathcal{K}_s)$ to filtered $\bar{\mathbb{Q}}_\ell \otimes_{\mathbb{Q}_\ell} E(\Pi)_\lambda$-modules:

$$D_{dR,s}(R) = (R \otimes_{\mathbb{Q}_p} B_{dR})^{\Gamma_s}.$$

The Hodge-Tate numbers of R (with respect to s) are the j such that $gr^j D_{dR,s}(R) \neq (0)$. Then in the situation of Theorem 2.2.3, the Hodge-Tate numbers of $\rho_{\Pi,\lambda}$ with respect to s are the j of the form

$$j = i - \mu_{n-i}(\alpha_s),\ i = 0, \dots, n-1. \tag{4.5.1}$$

This is to be compared to part 4 of Theorem VII.1.9 of [HT]; we have replaced the index p there (an unfortunate misprint gives p two incompatible meanings) by i.

Scholium 4.5.2. The set $HT(\Pi, \lambda)$ of Hodge-Tate numbers of $\rho_{\Pi,\lambda}$ satisfies the following properties:

(1) The set $HT(\Pi, \lambda)$ depends only on the identification α of complex and ℓ-adic places and on the archimedean components Π_∞, ξ_∞, $\mu_{1,\infty}$.
(2) The choice of $\mu_{1,\infty}$ can be made independently of Π. The choice of ξ_∞ depends only on the central character of the representation $W(\Pi)$.
(3) Let ρ_1 and ρ_2 be two continuous λ-adic representations of $Gal(\overline{\mathbb{Q}}/\mathcal{K})$ such that, for almost all primes v prime to ℓ,

$$Tr(\rho_1(Frob_v)) = Tr(\rho_2(Frob_v)).$$

Then the Hodge-Tate numbers of ρ_1 and ρ_2 coincide at all primes dividing ℓ.

Proof. Point (2) is easy to check, and point (3) is a simple consequence of Chebotarev density, since the Hodge-Tate numbers of a continuous representation R depend only on its semisimplification.

Point (1) is a bit more subtle. First consider the stable case, and let $\{\pi\}$ be the L-packet for GU indexed by $([\Pi], \xi)$. The calculation of the Hodge-Tate numbers for $Sh(V_1)$ is identical to that for the simple Shimura varieties considered in [HT]. The formula in Proposition III.2.1.6 of [HT]:

$$\dim gr^j D_{DR,\sigma}(R^i_\xi(\pi)) = \#ker^1(\mathbb{Q}, G_{\iota\sigma}) \sum_{\pi_\infty} m_{\iota(\pi)\otimes\pi_\infty} d(\pi_\infty), \tag{4.5.3}$$

where

$$d(\pi_\infty) = \dim H^{i-p}(LieQ_{\iota\sigma}, U_{\iota\sigma}, \pi_\infty \otimes \mu^p(\iota\xi))$$

and the other notation (different from the notation used here) is explained in [HT], shows that $HT(\Pi, \lambda)$ depends only on α and on the set of representations π_∞, which in turn depends only on Π_∞.

In the endoscopic case, we let $\tau = \Pi$, choose χ as above, and let $\{\pi\}$ be the corresponding L-packet for GU. The formula (4.5.3) remains valid and shows that $HT(\Pi, \lambda)$ depends on α (denoted ι in (4.5.3)) and on the set of representations π_∞ with non-trivial multiplicity $m_{\iota(\pi)\otimes\pi_\infty}$ in the L-packet $([\tau], [\chi], \xi)$. Our construction above shows that this depends in turn exclusively on $\tau_\infty = \Pi_\infty$ and χ_∞. But our choice of χ_∞ also depends only on Π_∞. So (1) is true in the endoscopic case as well. □

It is now easy to verify condition (ii) of Conjecture 2.2.2. It follows from (1) and (2) of (4.5.2) that, in order to determine $HT(\Pi, \lambda)$, it suffices to determine $HT(\Pi', \lambda)$ for any Π' satisfying the hypotheses of Theorem 2.2.3 with $\Pi'_\infty = \Pi_\infty$. In particular, we may assume Π' to satisfy the condition that, at some finite prime v_0, Π'_{v_0} is in the discrete series. Then [HT], following [C3] and [K3], constructs a Galois representation $\rho'_{\Pi',\lambda}$ satisfying the local-global compatibility (2.2.4) at almost all primes. By (3) of (4.5.2), this implies $\rho'_{\Pi',\lambda}$ and $\rho_{\Pi',\lambda}$ have the same Hodge-Tate weights. Condition (ii) thus follows from Theorem VII.1.9 of [HT].

4.5.4. The crystalline property. Finally, if Π_v is unramified at all primes dividing ℓ, and if ℓ is unramified in $\mathcal{K}$, it is claimed in Theorem 2.2.3 that $\rho_{\Pi,\lambda}\,|_{\Gamma_v}$ is crystalline. Indeed, in this case we can assume the compact open subgroup $K_\ell \subset GU(\mathbb{Q}_\ell)$ is hyperspecial maximal, which implies that the Shimura variety $Sh(V_1)$ has good reduction at all primes dividing ℓ (cf. [K2], §5, where Kottwitz relies on previous results of Langlands-Rapoport and Zink). It follows by the crystalline conjecture in p-adic Hodge theory (with $p = \ell$, see [Fa], [Ts]) that any representation in the cohomology $H^*(Sh(V_1), \tilde{W}(\Pi))$ is crystalline; here we also use the fact that $\tilde{W}(\Pi)$ is obtained by tensor operations from the cohomology of an abelian scheme over $Sh(V_1)$, cf. [HT, Lemma III.4.2]. Moreover, we are free to choose our Hecke characters ξ and μ_1 unramified at primes dividing ℓ. Thus the λ-adic characters $r_\lambda(\xi)^{-1}$ and $r_\lambda(\xi,\mu_1)^{-1}$ are also crystalline. This completes the verification of the crystalline property as stated in Theorem 2.2.3.

The stronger property conjectured in Conjecture 2.2.2 (iii) requires a more careful analysis of the integral models of $Sh(V_1)$, and is deduced as in [HT] by Shin for the representations constructed above [S].

5. Removal of Special Hypotheses 1.2.2

Let now $\mathcal{K}/F$ be any CM quadratic extension of a totally real field, and let Π be an automorphic representation of $GL(n,\mathcal{K})$ satisfying Hypothesis 2.2.1. Let $\mathcal{I}$ be a set of cyclic Galois extensions K/F of prime degree q_K. We say $\mathcal{I}$ is S-*general* if, for any $v \notin S$ and any finite extension M/F, there is $K \in \mathcal{I}$ in which v splits completely which is linearly disjoint from M. See [So] for more details.

PROPOSITION 5.1. *There is a finite set S of places of F and an S-general collection $\mathcal{I}$ of totally real quadratic extensions F_i/F such that, for each $F_i \in \mathcal{I}$, letting $\mathcal{K}_i = F_i \cdot \mathcal{K}$, Π_i the base change of Π to $\mathcal{K}_i$, the triple $(F_i, \mathcal{K}_i, \Pi_i)$ satisfies Special Hypotheses 1.2.2. Moreover, we can assume that, for every $v \in S$ and every $F_i \in \mathcal{I}$, either v splits in $\mathcal{K}/F$ or the unique extension of v to $\mathcal{K}$, denoted $v_\mathcal{K}$, splits in $\mathcal{K}_i$.*

Proof. Let S be the set of primes of v at which (1.2.2.1) or (1.2.2.2) fails: either v ramifies in $\mathcal{K}/F$ or v stays prime in $\mathcal{K}$ and the corresponding component $\Pi_{v_\mathcal{K}}$ is ramified. We take $\mathcal{I}$ to be the set of totally real quadratic extensions F_i/F with the property that, for all $v \in S$, $F_{i,v} \xrightarrow{\sim} \mathcal{K}_v$. It is obvious that this set has the properties claimed. □

THEOREM 5.2. *Let Π be an automorphic representation of $GL(n,\mathcal{K})$ satisfying the hypotheses of Conjecture 2.2.2 and the strong regularity hypothesis of 1.2.3 (ii). Then Theorem 2.2.3 is valid for Π.*

Proof. The theorem is deduced from the proof of Theorem 2.2.3 in §4, in which Special Hypotheses 1.2.2, using Proposition 5.1. exactly as in [HT], pp. 229-232. We omit the details, since the more complicated case of a general solvable extension is treated in Sorensen's notes [So]. □

Remark 5.3. We can also extend Theorem 5.2 to CM fields $\mathcal{K}$ that do not contain imaginary quadratic fields. As in [HT], it suffices to replace $\mathcal{K}$ by $\mathcal{K} \cdot E_i$ for a sufficiently general family of imaginary quadratic fields E_i, then to use the patching argument of [HT], pp. 229-232. We omit the details, since the argument is identical to the one developed at length in [HT].

Appendix

We verify the claims about local surjectivity of the map $\phi : T \times G \to GU$ made in the course of the proof of Lemma 1.1.3, and the analogous claims for endoscopic groups in §1.3. Let p be a rational prime, $\Gamma_p = Gal(\bar{\mathbb{Q}}_p/\mathbb{Q}_p)$. Consider the long exact cohomology sequence attached to the short exact sequence (1.1.1):

$$GU(\mathbb{Q}_p) \xrightarrow{\nu} \mathbb{Q}_p^\times \to H^1(\Gamma_p, G) \to H^1(\Gamma_p, GU) \to \cdots . \tag{A.1}$$

If p splits in E then $G \xrightarrow{\sim} GL(n, \mathbb{Q}_p)$ and $H^1(\Gamma_p, G) = 1$ by Hilbert's Theorem 90, so ν is surjective; but one can also show easily that $\phi_p : T(\mathbb{Q}_p) \times G(\mathbb{Q}_p) \to GU(\mathbb{Q}_p)$ is surjective in that case. We therefore assume p is inert in E. Denote by $\mathcal{K}_p$, F_p, E_p the tensor products of the corresponding fields with $\mathbb{Q}_p$. We have in all degrees n the exact sequences of *groups*

$$GU(\mathbb{Q}_p) \xrightarrow{\nu} \mathbb{Q}_p^\times \xrightarrow{g} H^1(\mathbb{Q}_p, G) \to H^1(\mathbb{Q}_p, GU) \to \cdots . \tag{A.2}$$

Now $H^1(\mathbb{Q}_p, G)$ classifies hermitian forms on free rank n $\mathcal{K}_p$-modules (hermitian relative to F_p), so is isomorphic to $F_p^\times/N_{\mathcal{K}/F}\mathcal{K}_p^\times$. An easy check shows that the map $\mathbb{Q}_p^\times \to F_p^\times/N_{\mathcal{K}/F}\mathcal{K}_p^\times$ is the obvious one. Also, $H^1(\mathbb{Q}_p, GU)$ classifies hermitian forms up to $\mathbb{Q}_p$-similitude. If $\alpha \in F_p^\times/N_{\mathcal{K}/F}\mathcal{K}_p^\times$ is the invariant of a form h, the invariant of the similar form λh, for $\lambda \in \mathbb{Q}_p^\times$, is $\lambda^n \alpha$. If n is odd and p is unramified in $\mathcal{K}$, we see that the map $g : \mathbb{Q}_p^\times \to H^1(\mathbb{Q}_p, G)$ factors through an injection $\mathbb{Q}_p^\times/N_{E_p/\mathbb{Q}_p}E_p^\times \to H^1(\mathbb{Q}_p, G)$. Thus

$$\nu(GU(\mathbb{Q}_p)) \subset N_{E_p/\mathbb{Q}_p}E_p^\times = \nu \circ \phi(T(\mathbb{Q}_p) \times G(\mathbb{Q}_p))$$

which implies that

$$\phi(T(\mathbb{Q}_p) \times G(\mathbb{Q}_p)) = GU(\mathbb{Q}_p)$$

provided p is unramified in $\mathcal{K}$. In particular, ϕ is locally surjective at all but finitely many places.

It is well-known that the real group $GU(r, s)$ is connected if $r \neq s$, which easily implies that ϕ is locally surjective at the real place.

This completes the verification of the claims made in the proof of 1.1.3. The proof for the group $GH_{a,b}$ is similar. Here the right half of (A.2) is replaced by

$$H^1(\mathbb{Q}_p, H_{a,b}) \to H^1(\mathbb{Q}_p, GH_{a,b})$$

where the first group classifies pairs of hermitian forms (h_a, h_b) on spaces of dimension (a, b), respectively, and the second classifies pairs (h_a, h_b) as before up to simultaneous $\mathbb{Q}_p$-similitude $(h_a, h_b) \sim (\lambda h_a, \lambda h_b)$. Since either a or b is odd, one again sees that the image of ν is contained in $N_{E_p/\mathbb{Q}_p}E_p^\times$ and one concludes as before.

For completeness, we treat the case of even $n = 2m$. Assume V is a quasi-split hermitian space of dimension n over F_p, so that $GU(\mathbb{Q}_p)$ is the rational similitude group of the form with matrix

$$J_{2m} = \begin{pmatrix} 0 & I_m \\ I_m & 0 \end{pmatrix}.$$

The map

$$\mathbb{Q}_p^\times \to GU(\mathbb{Q}_p);\ t \mapsto diag(I_m, t \cdot I_m)$$

splits the similitude map, hence $\nu : GU(\mathbb{Q}_p) \to \mathbb{Q}_p^\times$ is surjective in the even-dimensional quasi-split case. On the other hand, the image of $\nu \circ \phi$ is always contained in $\{a \in \mathbb{Q}_p^\times \mid val_p(a) \in 2\mathbb{Z}\}$ when p is unramified in $\mathcal{K}$. It follows that the quotient $C = GU(\mathbb{A})/\mathbf{S}$, with $\mathbf{S}$ defined as in the proof of Proposition 1.1.4, is in general highly non-trivial in the even case.

References

[A1] J. Arthur, The Endoscopic Classification of Representations: Orthogonal and Symplectic Groups (book to appear).

[A2] J. Arthur, An introduction to the trace formula, in Harmonic analysis, the trace formula, and Shimura varieties, Clay Math. Proc. **4** Amer. Math. Soc., Providence, RI (2005) 1–263.

[BC] J. Bellaïche, & G. Chenevier. p-adic families of Galois representations and higher rank Selmer groups, *Astérisque* **324** (2009).

[BR] D. Blasius & J. Rogawski. Motives for Hilbert modular forms, *Inventiones Math.* **114** (1993) 55–87.

[CH] G. Chenevier & M. Harris. Construction of automorphic Galois representations, II, manuscript at http://fa.institut.math.jussieu.fr/node/45.

[C1] L. Clozel, The fundamental lemma for stable base change, *Duke Math. J.* **61** (1990) 255–302.

[C2] L. Clozel, Motifs et formes automorphes, in Automorphic forms, Shimura varieties and L-functions, *Perspectives in Math.* **10** (1990) 77–159.

[C3] L. Clozel, Représentations galoisiennes associées aux représentations automorphes autoduales de GL(n), *Publ. Math. I.H.E.S.* **73** (1991) 97–145.

[C.III.A] L. Clozel, Identités de caractères en la place archimédienne, this volume, C.III.A.

[CHL.IV.B] L. Clozel, M. Harris & J.-P. Labesse, Endoscopic transfer, this volume, CHL.IV.B.

[CHT] L. Clozel, Harris, M. & R. Taylor. Automorphy for some ℓ-adic lifts of automorphic mod ℓ Galois representations, *Publ. Math. IHES* **108** (2008) 1–181.
(1979) 247–290.

[Fa] G. Faltings, p-adic Hodge theory, *J.A.M.S* **1** (1988) 255–299.

[FM] J.-M. Fontaine & B. Mazur. Geometric Galois representations, in J. Coates, et al., ed. Elliptic curves, modular forms, and Fermat's last theorem, Hong Kong (1993) 41–78.

[H.I.A] M. Harris, Introduction to the stable trace formula, this volume, H.I.A.

[HT] M. Harris & R. Taylor. The geometry and cohomology of some simple Shimura varieties, *Annals of Mathematics Studies* **151** (2001).

[K1] R. Kottwitz & R. Shimura varieties and λ-adic representations, in Automorphic Forms, Shimura Varieties, and L-functions, New York: Academic Press (1990), Vol. 1, 161–210.

[K2] R. Kottwitz, Points on some Shimura varieties over finite fields, *J.A.M.S* **5** (1992) 373–444.

[K3] R. Kottwitz, On the λ-adic representations associated to some simple Shimura varieties, *Inv. Math.* **108** (1992) 653–665.

[K4] R. Kottwitz, Stable trace formula: cuspidal tempered terms, *Duke Math. J.* **51** (1984) 611–650.

[Lab] J.-P. Labesse, Le lemme fondamental pour le changement de base stable, *Duke Math. J.* **61** (1990) 519–530.

[L.IV.A] J.-P. Labesse, Changement de base CM et séries discrètes, this volume, L.IV.A.

[LL] J.-P. Labesse & R.P. Langlands, L-indistinguishability for $SL(2)$. *Canad. J. Math.* **31** (1979) 726–785.

[L] R.P. Langlands, Les débuts d'une formule des traces stable, *Publications de l'Université Paris 7* **13** (1983).

[LN] G. Laumon & B. C. Ngô. Le lemme fondamental pour les groupes unitaires, *Annals of Math.* **168** (2008) 477–573.

[Mo] C. Moeglin, Classification des séries discrètes pour certains groupes classiques p-adiques, *Harmonic Analysis, Group Representations, Automorphic Forms and Invariant Theory*, Vol. 12, Lecture Notes Series, Institute of Mathematical Sciences, National University of Singapore (2007) 89–150.

[M] S. Morel, On the Cohomology of Certain Non-Compact Shimura Varieties, *Annals of Math. Studies* **173** (2010).

[N] B.C. Ngô, Le lemme fondamental pour les algèbres de Lie, *Publ. Math. IHES* **111** (2010) 1–169.

[R] J. Rogawski, Automorphic Representations of Unitary Groups in Three Variables, *Annals of Math. Studies* **123** (1990).

[S] S.W. Shin, Galois representations arising from some compact Shimura varieties, *Ann. of Math* (in press).

[So] C.M. Sorensen, Patching, to appear in Book 2.

[TY] R. Taylor & T. Yoshida. Compatibility of local and global Langlands correspondences, *J. Am. Math. Soc.* **20** (2007) 467–193.

[Ts] T. Tsuji, p-adic étale cohomology and crystalline cohomology in the semistable reduction case, *Invent. Math.* **137** (1999) 233–411.

[W] J.-L. Waldspurger, L'endoscopie tordue n'est pas si tordue, *Mem. AMS*, **194** (2008).

Université Paris-Sud, Membre, Institut Universitaire de France

Institut de Mathématiques de Jussieu, U.M.R. 7586 du CNRS, Membre, Institut Universitaire de France

Institut de Mathématiques de Luminy, U.M.R. 6206 du CNRS

Made in United States
Orlando, FL
04 November 2026

90207621R00297